Reinforced Concrete Fundamentals

Reinforced Concrete Fundamentals

THIRD EDITION

PHIL M. FERGUSON
Professor of Civil Engineering
The University of Texas at Austin

John Wiley & Sons, Inc. New York • London • Sydney • Toronto

Library of Congress Cataloging in Publication Data:
Ferguson, Phil Moss, 1899–
Reinforced concrete fundamentals

Includes bibliographies
1. Reinforced concrete construction. I. Title

TA683.2.F4 1973 624'.1834 72–8871
ISBN 0–471-25741-9

Printed in the United States of America

10 9 8 7

Preface

Research progress in reinforced concrete has been substantial since the publication of the 1963 ACI Building Code and the second edition of this book. Much of the research reported by 1969 is reflected in the 1971 Code. Except for slabs—and possibly columns— this new information has often been local and specific, increasing the designer's ability for the particular case without unifying his basic understanding of the behavior of reinforced concrete.

The behavior of reinforced concrete is given a prominent place in this textbook, with many pictures of members at failure. Because reinforced concrete is largely semiempirical, the design engineer who understands how reinforced concrete behaves in approaching its ultimate resistance has an advantage in assessing the many new situations that he faces day to day.

Well over half the material of this edition is new or has been completely rewritten; the remainder has been reedited in detail. The strip method for slab design is presented in Chapter 12 as a formalization of approximate methods that have been long used in designing irregular slabs. Detailing for seismic design as outlined in Code Appendix A is covered in Chapter 18. The desirability of modifying allowable soil pressures and foundation design to match the strength design of the structure is emphasized in Chapter 7 on retaining wall design and in Chapter 15 on footings.

With the greatly increased coverage of the 1971 ACI Building Code, some limits on the scope of this book have been necessary. Only a simple introduction to prestressed concrete is included, composite member coverage is kept brief, shearheads and brackets are outlined rather than covered in depth, and Vierendeel trusses and shear walls are omitted.

Working stress design no longer has equal status with strength design (the ultimate strength design of past years). The transformed-area idea is initially given in Chapter 6 on serviceability, and Appendix C covers elementary analysis and the design of flexural members. Columns cannot be designed by basic working stress methods.

I do not always agree in detail with the 1971 Code but try to indicate my disagreement clearly while also presenting the Code requirements. The new ACI notation is used.

Although this book is designed for the beginning student in reinforced concrete, it covers material adequate for a second semester of work. Many detailed examples appear in the text, and there are numerous student problems at the end of most chapters. Since this is primarily a book on basic philosophy, behavior, and theory, design is included chiefly as a teaching tool.

In the required three-semester-hour course* at the University of Texas at Austin, emphasis is placed on general theory, discussed in the first five chapters, the retaining wall of Chapter 7 as a simple example that knits theory together, the continuous beams of Chapter 8 as a more complex application of theory, columns in Chapter 14, and footings in Chapter 15. Some of the more complex portions of these chapters are bypassed, and the remaining chapters are either completely omitted from this short course or are given superficial attention. The text contains ample material for a six-semester-hour course or part of a graduate course.

I am indebted to so many of my colleagues and to earlier writers that each cannot be individually acknowledged. However, the many helpful suggestions from Professors Richard W. Furlong and John E. Breen must be noted.

Readers over the years have been most cooperative in reporting any errors. I appreciate this and hope that the practice will continue. Errors can thus be corrected at each reprinting.

Phil M. Ferguson
Austin, Texas
June 1972

*The second half of the junior year.

Contents

Reinforced Concrete Fundamentals

1
Materials and Specifications

1.1 CONCRETE MATERIALS AND PRODUCTION

Concrete for reinforced concrete consists of aggregate bonded together in a paste made from portland cement and water. The paste fills most of the voids in the aggregate, and after the concrete is placed it hardens to form a solid structural material. A typical cross section of concrete is shown in Fig. 1.1*a*.

Although there are five standard portland cements, most concrete for buildings is made from Class I ordinary or standard cement (for concrete where the critical strength is needed in something like 28 days) or from Class III high-early strength cement (for concrete where strength is required in a few days). The heat generated by the different types of cements during the setting and hardening process varies widely, as indicated in Fig. 1.2. Where shrinkage and temperature stresses are important in the design, the volumetric change associated with these heat differences becomes significant. Air-entraining cement or admixtures for entraining air in the concrete are frequently used for greater workability or durability. Expansive cement to limit shrinkage is also on the market.

Aggregate consists of both fine and coarse aggregate, usually sand for the fine and gravel or crushed stone for the coarse aggregate. Lightweight aggregate made from expanded shale, slate, or clay has become increasingly important. Other aggregates, such as expanded slag, are also used. The size (and also the grading) of aggregate has an important influence on the amount of cement and water required to make 1 cu yd of concrete of a given consistency (Fig. 1.1*b*). It also exerts a major influence on bleeding, ease of finishing, shrinkage and permeability.

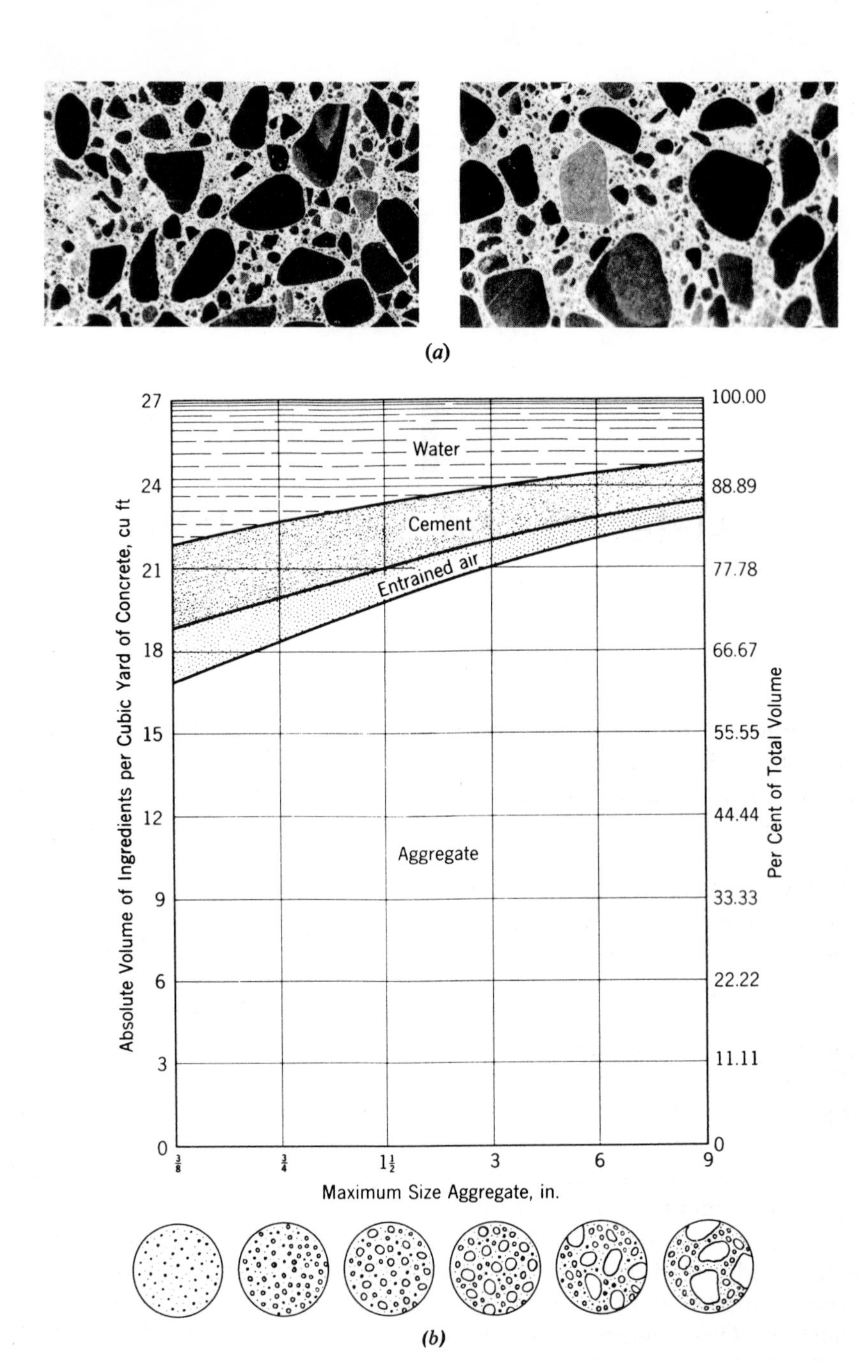

Fig. 1.1. Components of a concrete mix. (*a*) Cross sections of concrete showing coarse and fine aggregate separated by cement paste. (Courtesy Bureau of Reclamation.) (*b*) Quantities of each material in 1 cu yd of concrete. (From Ref. 1, Bureau of Reclamation.)

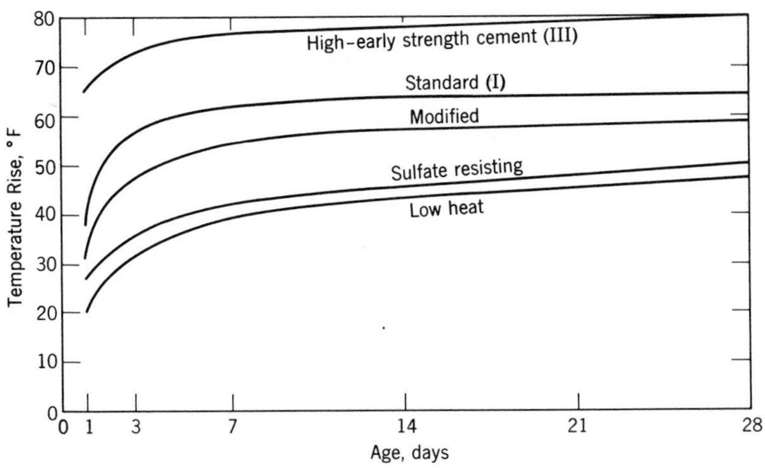

Fig. 1.2. Temperature rise in concrete for various types of cement, when no heat is lost. One barrel of cement per cubic yard. (From Refs. 2 and 1, ACI and Bureau of Reclamation.)

The quantity of water relative to that of the cement is the most important item in determining concrete strength. The effect of the water-cement ratio on strength is indicated in Fig. 1.3. The water is sometimes controlled indirectly and approximately by specifying the cement content in terms of sacks per cubic yard of concrete.

It is important that concrete have a workability adequate to assure its consolidation in the forms without excessive voids. This property is usually measured in the field by the slump test (Fig. 1.4*a*) or the Kelly ball test (Fig. 1.4*b*). The necessary slump may be small when vibrators are used to consolidate the concrete. For methods of designing concrete mixes the student is referred to the American Concrete Institute's "Recommended Practice for Selecting Proportions for Normal Weight Concrete" (ACI 211.1-70),[4] "Recommended Practice for Selecting Proportions for Structural Lightweight Concrete (ACI 211.2-69),"[5] or the Portland Cement Association booklet, "Design and Control of Concrete Mixes."[3]

Proper curing of concrete requires that the water in the mix not be allowed to evaporate from the concrete until the concrete has gained the desired strength. Figure 1.5 is representative of the variations in strength which can result from differences in curing. Temperature is also an important element in the rate at which concrete gains strength, low temperatures slowing up the process but raising the potential strength if normal temperature is restored, as indicated in Fig. 1.6. Early high temperatures lead to rapid setting and some permanent loss of strength potential.

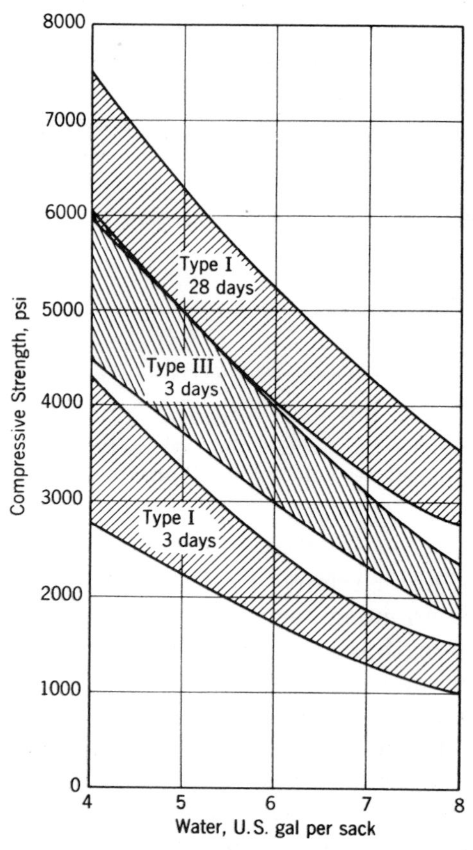

Fig. 1.3. Effect of water-cement ratio on strength of non-air-entrained concrete at different ages. All specimens moist cured at 70°F. (Modified from Ref. 3, Portland Cement Assn., earlier edition.)

1.2 COMPRESSIVE STRENGTH

Depending upon the mix (especially the water-cement ratio) and the time and quality of the curing, compressive strengths of concrete can be obtained up to 10,000 psi or more. Commercial production of concrete with ordinary aggregates is usually in the 3000 to 7000-psi range with the most common near 3000 to 4000 psi. On the other hand, highway departments often expect strengths of 4000 to 7000 psi. Because of the difference in aggregates, and to a lesser degree in cements, the same mix proportions result in substantially lower strengths in some sections of the country than in others. In these sections a lower water-cement ratio must be used.

(a)

(b)

Fig. 1.4. Testing for workability of concrete. (*a*) Slump test. (Courtesy Portland Cement Assn.) (*b*) Kelly ball test. The "ball" penetration is read on the graduated shaft by Professor Kelly of the University of California. This mix is quite stiff.

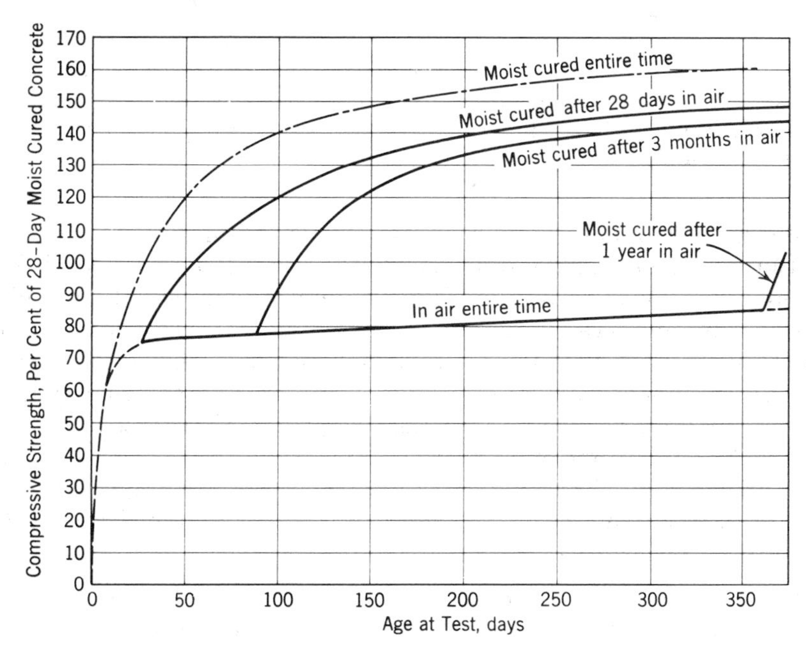

Fig. 1.5. Effect of curing conditions on strength of concrete. (From earlier edition of Ref. 3, Portland Cement Assn.)

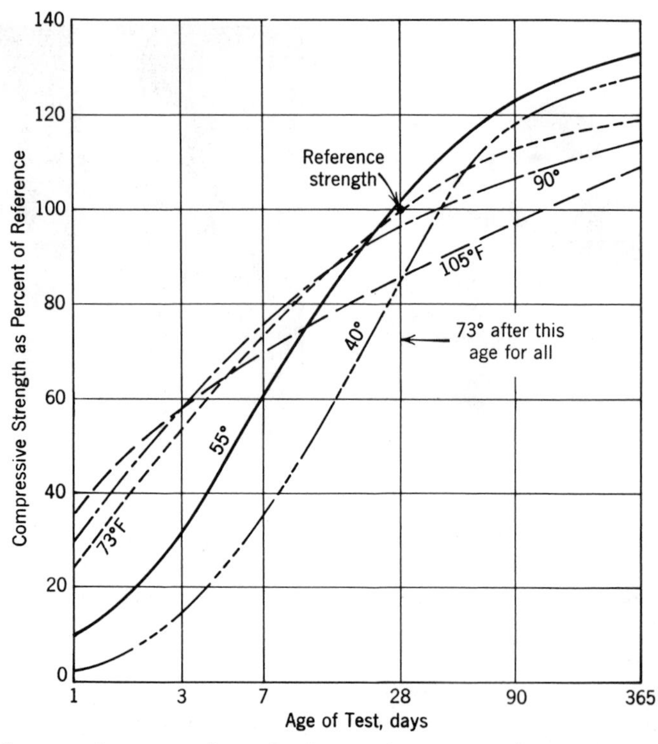

Fig. 1.6. Compression strength attained at various ages and temperatures as percent of 28-day strength under curing at 73°F. Type I cement. (Modified from Ref. 3, Portland Cement Assn.)

Compressive strength f_c' is based on standard 6-in. by 12-in. cylinders cured under standard laboratory conditions and tested at a specified rate of loading at 28 days of age. The designer should note that building concrete cured in place on the job will rarely develop as much strength as these standard cured cylinders. Separate cylinders should be made to check on the quality of curing if this is desired. These same cylinders, however, are not suitable for checking the quality of the mix.

The ACI Code, discussed in Sec. 1.10 following, specifies the average of two cylinders from the same sample tested at the same age (usually 28 days) for a strength test. It specifies the frequency at which such tests shall be made and sets up this criterion:

> 4.3.3. The strength level of the concrete will be considered satisfactory if the averages of all sets of three consecutive strength test results equal or exceed the required f_c' and no individual strength test result falls below the required f_c' by more than 500 psi.

Such a simple criterion can not always be perfect. The Code Commentary states that, even with the strength level and uniformity satisfactory, an occasional low strength may be indicated (at least once in 100 tests) and makes suggestions about allowance for these.

The *average* concrete strength for which a concrete mix must be designed must exceed f_c' by an amount which depends on the uniformity of plant production, that is, on how well the variations in operation are minimized. Where production records are available, the Code in 4.2.2.1 specifies the mix must be designed for 400 psi above f_c' if the standard deviation has been held to less than 300 psi (extremely good) with a larger increment for higher values. The requirement is 1200 psi above f_c' if the standard deviation is over 600 psi or if suitable records are not available.

It must be emphasized that f_c' for design is *not* to be considered as the average strength of job cylinder tests. The design f_c' is nearer a minimum than an average, but it is not an absolute minimum. An individual test may be nearly 500 psi low and still be acceptable if all averages of three consecutive tests are satisfactory.

With lightweight aggregates a mix design should definitely be made on the basis of trial batches. Many lightweight aggregates produce 3000 psi concrete and some easily give 6000 psi concrete under proper control.

A special treatment of concrete after it has set, which can lead to extremely high potential strengths, is currently in an interesting stage of development. Polymer concrete is made by impregnating ordinary concrete with a liquid organic monomer* which is subsequently polymerized by radiation or thermal catalytic methods. The result is greatly increased tension and compression strengths (up several hundred percent) accompanied by much improved impermeability, hardness, and durability.

1.3 TENSILE STRENGTH

The tensile strength of concrete is relatively low, about 10 to 15% of the compression strength, occasionally 20%. This strength is more difficult to measure and the results vary more specimen to specimen than those from compression cylinders. The modulus of rupture as measured from standard 6-in. square beams somewhat exceeds the real tensile strength. The value of $7.5\sqrt{f_c'}$ is often used for the modulus of rupture. The cylinder splitting test, often called the Brazilian test, is the most respected test for tensile strength. A compression force applied full length to opposite elements splits the cylinder open along the connecting diameter.

*Such as styrene or methyl methacrylate.

Lightweight concrete in many, but not all cases, has a lower tensile strength than ordinary weight concrete. The ACI Code accepts its tensile capacity at the ordinary value in computing shear and development of reinforcement if the splitting tensile strength f_{ct} is at least $6.7\sqrt{f_c'}$. For lower splitting strengths $f_{ct}/6.7$ must be substituted for $\sqrt{f_c'}$. Alternatively, in the absence of such tests, assumed shear resistance may simply be reduced by a factor of 0.75 for "all-lightweight" concrete and 0.85 for "sand-lightweight" concrete. The latter refers to lightweight concrete containing natural sand for the fine aggregate. Linear interpolation may be used for mixtures of natural sand and lightweight fine aggregate. Development length (Chapter 6) is modified in a similar way, the ACI Code factors being $6.7\sqrt{f_c'}/f_{ct}$, 1.33, and 1.18, since splitting strength enters here as a reciprocal.

1.4 SIGNIFICANCE OF LOW TENSILE STRENGTH

In a homogeneous elastic beam subjected to bending moment, one can calculate the bending stresses from $f = Mc/I$. The extreme fiber on one face carries compression, on the opposite face tension. If the beam is rectangular (or of any shape symmetrical about the centroidal axis), the maximum tensile stress equals the maximum compressive stress. In concrete construction, except in massive structures such as gravity dams and sometimes heavy footings, it is not economical to accept the low tensile strength of plain concrete as a limit on beam strength. It is generally more economical to make up a beam with compressive bending stresses carried by concrete and tensile bending stress carried entirely by steel reinforcing bars. It is not possible for concrete to cooperate with steel in carrying these tensile stresses except at very low and uneconomical values of steel stress. Except in prestressed concrete, when the steel stress reaches about 6000 psi the tensile concrete starts to crack and the steel soon thereafter must pick up essentially all the tension necessary to provide for the applied moment. Hence in ordinary reinforced concrete beams the tensile concrete is not assumed to assist in resisting the moment. Still, tension in the concrete does reduce the beam deflection considerably.

1.5 SHEAR STRENGTH

The shear strength of concrete is large, variously reported as from 35 to 80% of the compression strength. It is difficult to separate shear from other stresses in testing and this accounts for some of the variation reported. The lower values represent attempts to separate friction effects from true shears. The shear value is significant only in rare cases, since shear must ordinarily be limited to much lower values in order to protect the concrete against diagonal tension stresses (Chapter 4).

Diagonal tension stresses are often referred to as shear stresses, but this is actually a misnomer. If the student will keep in mind that true shear strength is rarely in question, it will not matter that the term "shear" is often loosely used for diagonal tension.

1.6 STRESS-STRAIN CURVE

Typical stress-strain curves for concrete cylinders on initial loadings are shown in Fig. 1.7. The first part of each curve is nearly a straight line, but there is some curvature at f_c equal to half the maximum value. The maximum stress is designated as f_c' (but note the statistical nature of the f_c' value to be used in design, as given in Sec. 1.2). The curve for low strength concrete has a long and relatively flat top. For high strength concrete the peak is sharper. Special techniques are necessary to establish these curves on the steep downward sections beyond the peak stress. Otherwise the testing machine characteristics result in sudden instead of gradual failure. Figure 1.8*b* illustrates this difficulty with high strength concretes.

At strains beyond the peak value of stress, considerable strength still exists. It will be noted that the cylinder strain occurring near maximum stress is nearly the same for all strengths of concrete, being roughly 0.002 in./in.* In a cylinder a maximum strain of something like 0.0025 measures the useful limit for all concretes except those of low strength or those made with lightweight aggregate. In beams of ordinary concrete, the shape of the cross section is a factor and the steeper the strain gradient the greater is the usable edge strain for a given f_c'. Observations show that unit strains of 0.0030 to 0.0045 normally occur before a beam fails; with f_c' over 6000 psi, the maximum observed strains are from 0.0025 to 0.0040. Tests[7] have proved conclusively that the stress-strain curve for the compression face of a beam is essentially identical with that for a standard test cylinder when stress is applied at the same rate, as shown in Fig. 1.8. However, confinement of the concrete by a spiral or by heavy ties can greatly increase the extreme strain value in compression.

Strictly speaking, concrete on initial loading has no fixed ratio of f_c/ϵ which truly justifies the term "modulus of elasticity." The initial slope of the stress-strain curve defines the initial or tangent modulus used with the

*Lightweight concrete with its lower initial modulus will give higher values.

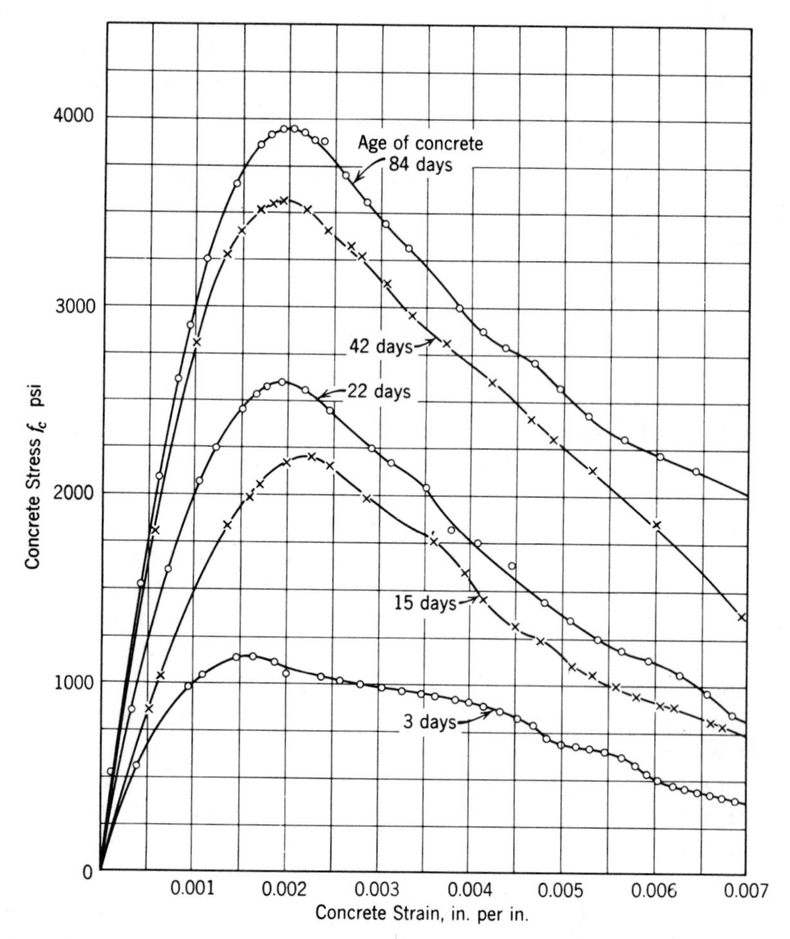

Fig. 1.7. Concrete stress-strain curves from compression cylinders. (From Ref. 6, Bureau of Reclamation.)

parabolic stress method and occasionally elsewhere. The slope of the chord (up to about $0.5f_c'$) determines the secant modulus of elasticity which is generally used in straight-line stress calculations (Fig. 1.9). When E or the term "modulus of elasticity" is used without further designation, it is usually the secant modulus which is intended.

The secant modulus in psi is taken in the Code as

$$E_c = w^{1.5}33\sqrt{f_c'}$$

where w is the weight of concrete in pounds per cubic foot for values

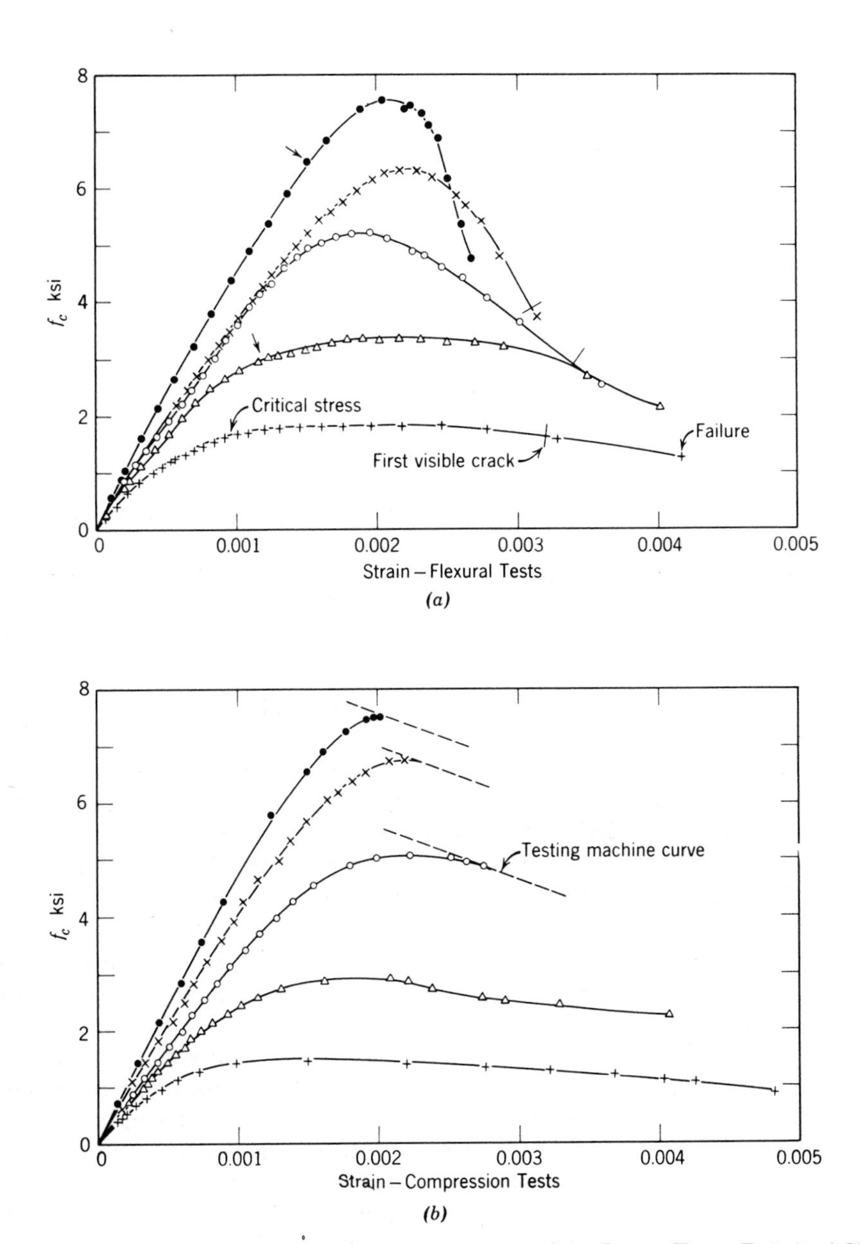

Fig. 1.8. Compression stress-strain curves at age of 28 days. (From Ref. 7, ACI.) (*a*) From flexural tests on 5 by 8 by 16-in prisms. (*b*) From direct compression tests on 6- by 12-in. cylinders for the concretes shown in (*a*).

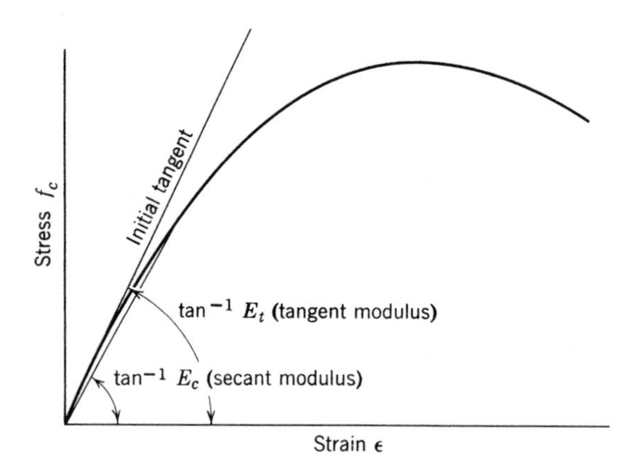

Fig. 1.9. Tangent and secant modulus of elasticity determinations.

between 90 and 155 pcf. For normal weight concrete, w may be taken as 145 pcf, which leads to $E_c = 57{,}400\sqrt{f_c'}$.

The use of a reduced modulus is one way to account for the time effects discussed in Sec. 1.7 and Sec. 1.8.

1.7 CREEP OF CONCRETE

The initial strain in concrete on first loading at low unit stresses is nearly elastic, but this strain increases with time even under constant load (Fig. 1.10*a*). This increased deformation with time is called creep and under ordinary conditions it may amount to more than the elastic deformation. Factors tending to increase creep include loading at an early age (while the concrete is still "green"), using concrete with a high water-cement ratio, and exposing the concrete to drying conditions. Concrete completely wet or completely dried out creeps only a little, and in general creep decreases with the age of the concrete.

At stresses up to the usual working stresses, creep is directly proportional to the unit stress; hence elastic and creep deformations are essentially proportional in plain concrete members. Under overload conditions this proportionality no longer holds; and in reinforced concrete the constant modulus of the steel causes strain readjustments with time. While these are minor for tension steel, compression steel stresses may be more than doubled by creep.

The creep rate is more rapid when the load is first applied and decreases somewhat exponentially with time, as shown in Fig. 1.10*b*.

Creep is also much reduced by delaying the loading until the concrete is more mature, as indicated by Fig. 1.10*c*. This figure shows that both the initial elastic strain and the creep are reduced when the loading is delayed.

Creep is one common cause of deflections which increase with time. In reinforced concrete, without compressive steel, the final deflection will usually be from 2.5 to 3.0 times the initial deflection. It has been suggested that long-time deflections be calculated on the basis of a reduced modulus having a value one-third that of the instantaneous modulus, but see Chapter 6 for better procedures. The Code requires (9.5.2.3)* that the additional long-time deflection for the sustained load be obtained by multiplying the immediate deflection by

$$(2 - 1.2A_s'/A_s) \ngtr 0.6$$

where A_s'/A_s is the ratio of compression reinforcement to tension reinforcement areas.

Creep deformation is not always harmful. Creep relaxes the stress effect of early deformation loadings; such stresses are so much reduced that creep in this situation is often given the special name of relaxation. For example, concrete stresses set up by differential settlement may be nearly eliminated where settlement occurs early in the life of the concrete, while it is still "green."

1.8 SHRINKAGE OF CONCRETE

As concrete loses moisture by evaporation, it shrinks. Since moisture is never uniformly withdrawn throughout the concrete, the differential moisture changes cause differential shrinkage tendencies and internal stresses. Stresses due to differential shrinkage can be quite large and this is one of the reasons for insisting on moist curing conditions. The larger the ratio of surface area to member cross section the larger will be the resulting shrinkage. Therefore large specimens shrink much less than small ones.

In plain concrete completely unrestrained against contraction, a uniform shrinkage would cause no stress; but complete lack of restraint and uniform shrinkage are both theoretical terms, not ordinary conditions. With reinforced concrete even uniform shrinkage causes stresses, compression in the steel, tension in the concrete.

*The section number in the 1971 ACI Code will often be listed in this book in this manner.

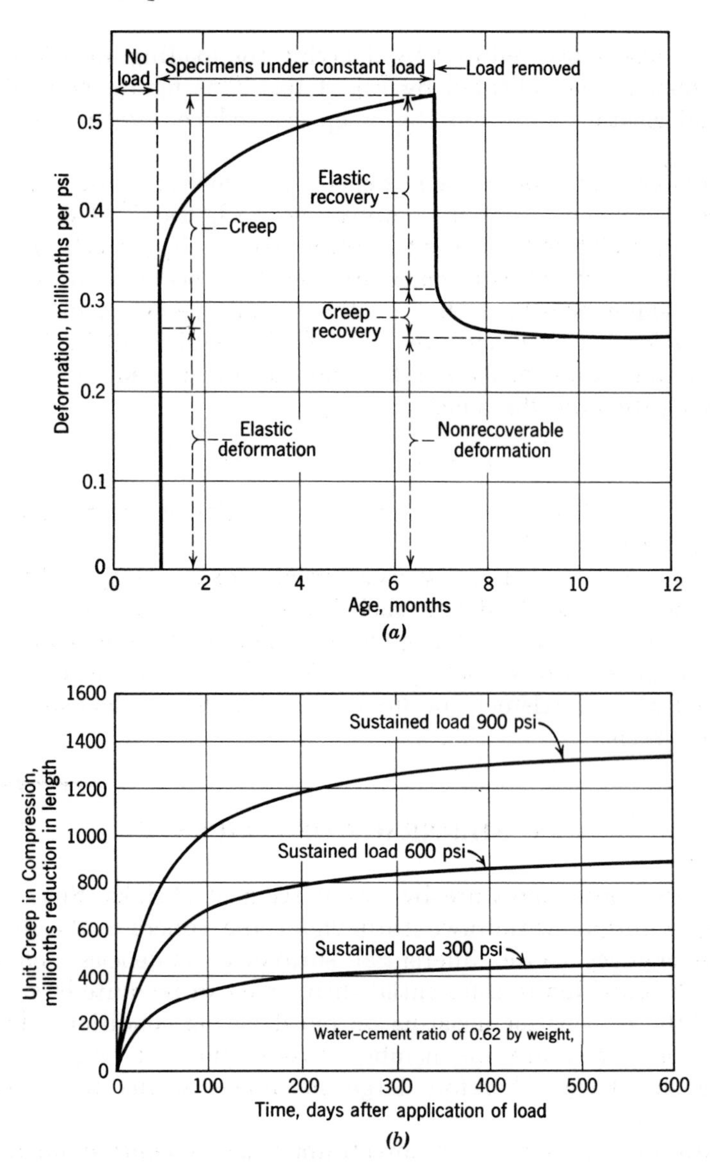

Fig. 1.10. Creep of concrete. (From Ref. 1, Bureau of Reclamation.) (*a*) Creep and elastic deformation. (*b*) Effect of unit stress on unit creep for identical concretes. (*c*) Effect of age at loading. (From Ref. 13, ACI).

Expansive cement utilizes this relationship. Because of expansive material in the cement, this concrete first expands a little. If it is partially

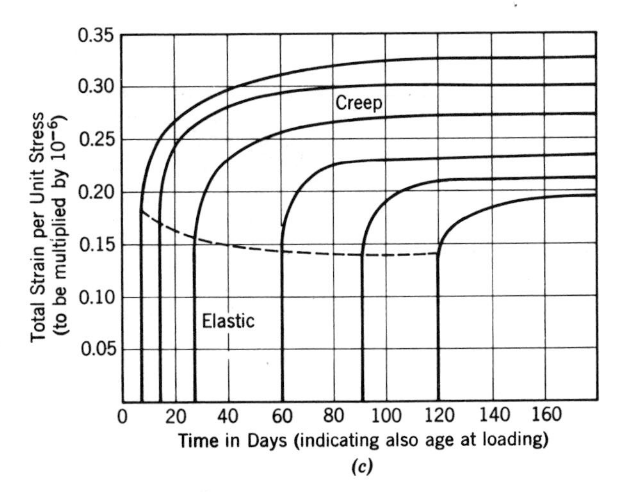

(*c*)

restrained by embedded reinforcement, tension builds up in this steel and compression in the concrete. As the concrete later shrinks and cools it comes to equilibrium with less change from its initial length.

In ordinary concrete the amount of shrinkage will depend on the exposure and the concrete. Exposure to wind greatly increases the shrinkage rate. A humid atmosphere will reduce shrinkage; a low humidity will increase shrinkage. Shrinkage is usually expressed in terms of the shrinkage coefficient s, which is the shortening per unit length. This coefficient varies greatly, with values commonly 0.0002 to 0.0006 and sometimes as much as 0.0010. An indication of how shrinkage varies with the water and cement content is given in Fig. 1.11*a* with shrinkage there expressed in terms of inches per 100 ft. This figure can only show trends since the amount of shrinkage differs with materials and drying conditions. In lightweight concrete early shrinkage is noticeably reduced by the water contained in the pores of the lightweight aggregate.

Shrinkage is, to a considerable extent, a reversible phenomenon. If concrete is soaked after it has shrunk, it will expand to nearly its original size, as indicated in Fig. 1.11*b*. The recovery is now known not to be *total*.[12]

Shrinkage is one common cause of deflections which increase with time. Only symmetrical reinforcement can prevent curvature and deflection from shrinkage (Chapter 6).

1.9 REINFORCING STEEL

Reinforcing bars are made from billet steel in many grades. Rail steel bars rerolled from old rails are also available. These principal steels have

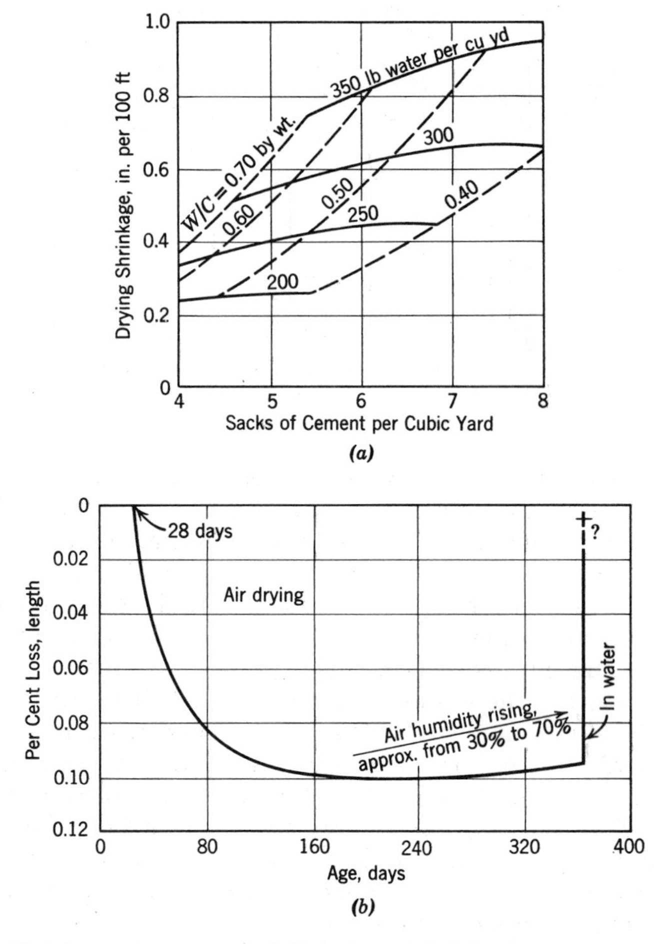

Fig. 1.11. Shrinkage of concrete. (*a*) Relation of shrinkage to water content. (From Ref. 2, ACI.) (*b*) Typical shrinkage-time curve starting with water-cured specimen (3 by 3 by 40-in.) 28 days old. Note recovery when again placed in water; constant volume reached in about 24 hours of soaking. (Replotted and modified from Ref. 8, The Engineering Foundation.)

the following yield points and ultimate strengths (Table 1.2). The modulus of elasticity E_s is usually taken as 29×10^6 psi.

To increase the bond between concrete and steel, projections called deformations are rolled on the bar surface as shown in Fig. 1.12, the pattern of these varying with the manufacturer. The deformations shown must

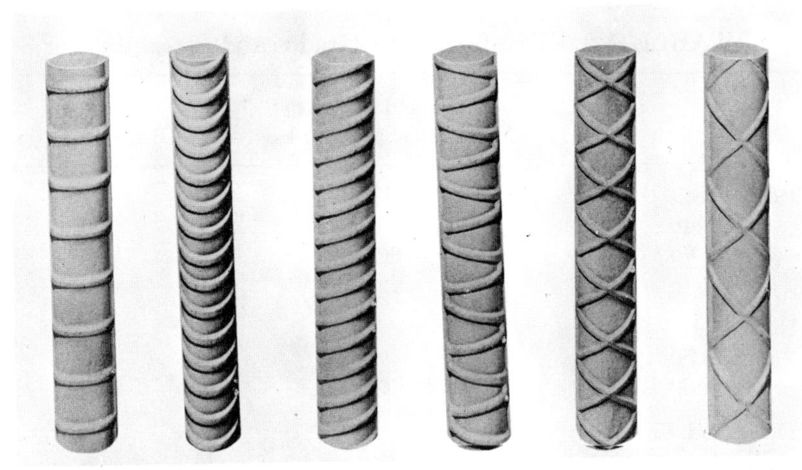

Fig. 1.12. Various types of deformed bars. (Courtesy Concrete Reinforcing Steel Inst.)

TABLE 1.1 Weight, Area, and Perimeter of Individual Bars

Current Bar Designation No.	Unit Weight per Foot, lb	Nominal Dimensions—Round Sections: Diameter, d_b in.	Cross-Sectional Area, A_b sq in.	Perimeter, in.
2[a]	0.167	0.250	0.05	0.786
3	0.376	0.375	0.11	1.178
4	0.668	0.500	0.20	1.571
5	1.043	0.625	0.31	1.963
6	1.502	0.750	0.44	2.356
7	2.044	0.875	0.60	2.749
8	2.670	1.000	0.79	3.142
9	3.400	1.128	1.00	3.544
10	4.303	1.270	1.27	3.990
11	5.313	1.410	1.56	4.430
14	7.65	1.693	2.25	5.32
18	13.60	2.257	4.00	7.09

[a] #2 is a plain bar and was not covered by ASTM A615–68 for deformed billet steel bars for reinforcement. However, it is reported that the 1971 revision covers both plain and deformed bars.

TABLE 1.2. Reinforcement-Grades and Strengths

	Min Yield Point or Yield Strength, f_y, ksi	Ultimate f_u, ksi
Billet steel		
Grade 40	40	70
Grade 60	60	90
Grade 75	75	100
Rail steel		
Grade 50	50	80
Grade 60	60	90
Deformed wire		
Reinf.	75	85
Fabric	70	80
Cold drawn wire		
Reinf.	70	80
Fabric	65, 56	75, 70

satisfy ASTM Spec. A615-68 to be accepted as *deformed* bars. The deformed wire has indentations pressed *into* the wire or bar to serve as deformations.

Except for spirals, only deformed bars, deformed wire, and fabric made from cold drawn or deformed wire may be counted for strength in reinforced concrete under the ACI Code.*

All standard bars are round bars, designated by size as #3 to #18, this number corresponding roughly to the bar diameter in eighths of an inch. The bar weights and nominal areas, diameters, and circumferences or perimeters are tabulated in Table 1.1. Since the area is calculated from the weight, including that of the deformations, it is a nominal area as far as minimum cross section is concerned. The perimeter is the circumference of this nominal circular area.

Welded wire fabric is increasingly important for slab and pavement reinforcing. Fabric is made of cold-drawn wires running in two directions and welded together at intersections. A more recent development is deformed wire fabric and the use of larger rod sizes made up into fabric mats. Plain wire fabric depends largely on the welded crosswires for development. Deformed wire fabric usually has fewer cross wires with lighter welds and depends more, but not wholly, on the wire deformations for development. Deformed wire up to about 0.6-in. in diameter is available.

*Plain bars are still used in many countries.

1.10 DESIGN CODES

A specification for reinforced concrete design may take the form of a code or a recommended practice. A code is written in the form of a law for enactment by public bodies such as city councils. It represents, usually, the minimum requirements necessary to protect the public from danger. It makes no attempt to specify the best practice, although it usually attempts to eliminate the most common mistakes, especially those involving safety. A recommended practice, on the other hand, attempts to define the best practice or at least to state satisfactory design assumptions and procedures. It may state reasons as well as methods.

The most significant reinforced concrete code in the United States is the "Building Code Requirements for Reinforced Concrete" (ACI 318–71). The Code is available from the American Concrete Institute, Box 4754, Redford Station, Detroit, Mich., 48219, at $5 per copy to ACI members ($6.50 to others). This code will be quoted frequently in this book, usually as the Code, or the ACI Code, or by a section reference number, as 4.3.3 was used in Sec. 1.2 of this chapter.

A Commentary on the 1971 Code is also available from the same address as the Code at $5 per copy to ACI members ($6.50 to others). Both Code and Commentary are available on a single order at $8 to ACI members ($11 to others). The Commentary is not a legal part of the Code but it attempts to indicate some of the reasoning behind the various sections of the Code.

It might be noted that this Code uses generally a new standard on notation that has been incorporated in this text. All Code symbols are defined in each Code chapter and then summarized in Code Appendix B.

Two general codes have considerable usage: the Pacific Coast Building Officials Conference's "Uniform Building Code" and the Building Officials Conference of America "Basic Building Code," the latter using the ACI Code. In addition, many cities write their own codes, making some modifications in these more standard codes. The designer must at the beginning of any project determine the code under which he is required to operate.

Highway bridges are normally designed under specifications prepared by the American Association of State Highway Officials (AASHO).

In Europe, in addition to the various national codes, the CEB-FIP* "International recommendations for the design and construction of concrete structures," dated June 1970, is highly regarded as a set of "Principles and Recommendations" valuable to code writing commissions.

*Comité Européen du Béton—Fédération Internationale de la Précontrainte.

1.11 STRESS VERSUS STRENGTH CONCEPTS

The curved stress-strain diagram for concrete, with all strains increasing with time because of creep, complicates theoretical analyses. At low stress levels elastic analysis is reasonable, that is, stress varies almost with the distance from the neutral axis. However, the reader should not confuse this with a feeling that such calculated stresses are therefore either elastic or exact. Shrinkage (usually nonuniform), creep, and cracking all complicate the stresses. Shrinkage tends to reduce compressive stress or increase tensile stress in typical cases. Creep increases the strains, whether tensile or compressive, in effect reducing the modulus of the concrete. Stresses computed elastically can only be index numbers, not real stresses.

Early investigators (before 1910) sensed this behavior and favored calculations based on ultimate strength, usually using the parabolic stress distribution of Fig. 1.13*a*. However, the apparent simplicity of the triangle of stress shown in Fig. 1.13*b* was strong enough to lead to the adoption of the so-called *working-stress design* (sometimes designated WSD).

In the 1930s field studies of strains in columns showed that the elastic concept of adjacent stresses in steel and concrete in proportion to the moduli values was not tenable; creep shifted much more of the initial load from concrete to steel. Tests on beams showed extreme face strains in compression, much in excess of the strain at peak stress, shown in Fig. 1.7 as approximately 0.002. It became clear that the descending part of the stress-strain curve for $\epsilon > 0.002$ was also important. In 1942 C. S. Whitney* presented a paper emphasizing this fact and showing how a probable stress-strain curve like those of Fig. 1.7 could, with reasonable accuracy,† be replaced with an artificial rectangular stress block simplification, as in Fig. 1.13*c*. With the rectangular stress block simplification, the 1956 ACI Code added an Appendix permitting *ultimate strength design* (USD) as an alternate to WSD. The 1963 Code gave the two methods equal standing, both in the body of the Code.

Because of the nonlinear response of columns and prestressed concrete, engineers have been pushed toward ultimate strength design. The 1971 Code is almost totally a *strength* code with strength meaning ultimate. However, it also permits an alternate method which for beams and slabs (not columns) is effectively a working stress method.

*Whitney was not the first to suggest the rectangular stress block, but he was the first in the USA to achieve some general acceptance.

†Accuracy with beams is excellent. With columns the results are more sensitive to the exact stress-strain curve assumed.

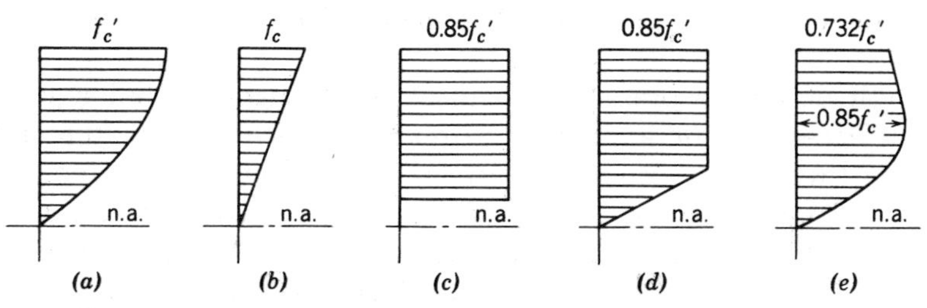

Fig. 1.13. Stress distributions assumed on compression side of beam. (*a*) Simple parabola (ultimate). (*b*) Straight line (working stress). (*c*) Rectangular stress block (ultimate). (*d*) Trapezoidal stress block (ultimate). (*e*) Parabolic, with straight line (ultimate).

For strength design the Code permits any reasonable stress distribution to be assumed, if it agrees with tests, such as those in Figs. 1.13*c* to 1.13*e* or others. Practice has moved to the rectangular stress block because of its simplicity. Research often uses the stress distribution of Fig. 1.13*e*. The use of the computer has also encouraged the development of a continuous function rather than this half parabola plus a straight line.

This book emphasizes the strength method and its advantages. Nevertheless, for two reasons it also includes working stress analysis briefly (Chapters 6, 16, and Appendix C). First, its assumption of linear stress distribution is appropriate for calculations of deflections under service load conditions and for crack investigations at service loads. Second, in a time of transition at least a minimum understanding of working stress design methods will improve communications between those trained in today's best methods and those more versed in older procedures.

1.12 SAFETY PROVISIONS

(a) Factor of Safety

A factor of safety is a concept of long standing, some reserve strength to care for unusual loads beyond those in the design. It has often been thought of as the ratio of yield stress to service load stress, although this concept is invalid where nonlinear responses exist. Correctly defined, the factor of safety is the ratio of the load which would cause collapse to the service or working load. Factor of safety is now such a misused term it almost requires a definition each time it is used.

(b) Factors Determining Safety

Strength design as presented in the Code recognizes that a factor of safety is necessary for a number of reasons which are individually important but which have no real interrelation. (1) Some are matters beyond the control of the engineer, such as the possibility of unforeseen future loads. (2) Others are in part under the engineer's control, such as: the quality of materials used, which the engineer can at least partially control by his specifications and inspection; the usual dimensional tolerances in the sizing of beams and areas of bars, and the placement of reinforcing, which can be closely inspected but which have practical limitations inherent in manufacture and field construction. (3) Others relate to the importance of the member in maintaining the integrity of the structure; a local slab failure is not as serious as the collapse of a column. (4) Still other factors vary with the degree of safety factor which is justified, such as the hazards to life and limb from the collapse of a school building as compared to that of a simple shed for storage of equipment or material; these relative hazards are considered as already reasonably recognized in the size of the live load required by all general building codes.

(c) Partial Factors of Safety

Safety has always been clearer when thought of in terms of partial safety factors. In design these might well be in terms of loads, construction practices, the quality of the materials used, and the importance of the member or structure.

The loads specified in codes are usually not the average or mean loads, but more nearly the maximum loads. The European concept of characteristic loads is excellent. A characteristic load is one that has only a 5% probability of ever being exceeded. If one used such a load the strength above this load could be smaller than if one started from the mean or average load concept. The difficulty lies in identifying and specifying this characteristic load.

Construction practices introduce an element of uncertainty. If columns can only be built to a $\frac{3}{8}$-in. tolerance (Code 7.3.2.1), it is well to note that a 11 $\frac{5}{8}$-in. square column is some 7% short of the area of a 12-in. square column.

Materials are also variables. The designer will not always find actually present at a critical spot the particular concrete strength he assumed in design, in spite of good controls.

All of the various overloads or deficiencies are less likely to occur together. Probability theory can predict the reduced probability of all bad aspects occurring together if the probability of each occurring separately is known. But here is always a basic problem; the imput data cannot be clearly established in many cases.

(d) Code Safety Philosophy

The ACI Code separates safety provisions into two parts in the hope that ultimately (not in the present Code) different factors may be established for different qualities of specifications and inspection, etc. The two factors now prescribed are load factors and ϕ factors.

Load factors attempt to assess the possibility that prescribed service loads may be exceeded. Obviously, a specified live load is more apt to be exceeded than a dead load which is largely fixed by the weight of the construction. The ultimate strength of the members must care for the total of all service loads each multiplied by its respective load factor; and the load factors are different in magnitude for dead load, live load, and wind or earthquake loading.

The ϕ factors are provided to allow for variations in materials, construction dimensions, and calculation approximations, that is, matters at least partially under the engineer's control. At present* ϕ varies only with the type of stress or member considered, that is, whether flexure or shear stress exists or whether a compression member is involved.

The general building code prescribes live loads which are assumed to cover adequately the relative hazards resulting from failure. For example, the prescribed live load for a school auditorium is properly oversafe compared to that for an office building. The general code also usually considers the probability of all areas being fully loaded at the same time.

(e) Load Factors

For dead and live loads the Code specifies that *design* loads, *design* shears, and *design* moments be obtained from service loads by using the relation:

$$U = 1.4D + 1.7L \qquad \text{(Code Eq. 9.1)}$$

where U represents any of these design strengths, D represents that for service dead load and L that for service live load.

*At some future time the ϕ factor might have a range of values for different type specifications or inspection provisions.

When wind W is included the total loading may be worse, but the chance of maximum wind occurring when an overload in both dead and live load exists is less than the chance of the overloads existing alone. Hence the Code uses an 0.75 coefficient on the sum of all three:

$$U = 0.75(1.4D + 1.7L + 1.7W) \qquad \text{(Code Eq. 9.2)}$$

where L must be considered *both* as zero and as its maximum value. If L does act to reduce the total, this also suggests that D and W might act in an opposite sense. If so, an overlarge D is on the unsafe side and the third Code condition becomes necessary:

$$U = 0.9D + 1.3W \qquad \text{(Code Eq. 9.3)}$$

This equation underestimates D when it opposes W. (Note that $1.3W$ is approximately $0.75 \times 1.7W$, as in Eq. 9.2 with only two significant digits.)

The Code (9.3) also gives load factors for earthquake, lateral earth pressure, lateral liquid pressure, vertical liquid load, impact, and special effects such as settlement, creep, shrinkage or temperature change, but these special cases are not essential to the discussion here.

(f) Code ϕ Factors

The ϕ factor is multiplied into what might be called the basic strength equation to obtain the reasonably dependable strength. The basic strength equation might be said to give the "ideal" strength assuming materials are as strong as specified, sizes are as shown on the drawings, bars are of full weight, that calculations of load, shear, and moment are exact, and that the strength equation itself is scientifically correct. The practical dependable strength of a particular member will often be something less, probably sometimes very much less, since all these factors vary over some range. How low the strength may go can only be determined by probability theory; and the variables are not yet clearly enough defined to place this approach above criticism. Furthermore, the adequacy of theory varies, being lower for shear than for flexure. The Code provides for these variables by using these simple ϕ factors for the several cases listed.

Flexure calculations	$\phi = 0.90$
Shear and torsion	$\phi = 0.85$
Spirally reinforced compression members	$\phi = 0.75$
Tied compression members	$\phi = 0.70$
Bearing on concrete	$\phi = 0.70$
Bending in plain concrete	$\phi = 0.65$

The ϕ is largest for flexure because the variability of steel is less than that of concrete; and all flexural members are specified to be designed for failure in tension. The ϕ values for columns are lowest (favoring the toughness of spiral columns a little over tied columns). Because columns fail in compression where concrete strength is critical, there is some small danger that the analysis may miss the worst combination of axial load and moment; and the column is critical in the building. For shear and torsion ϕ is intermediate since these depend on concrete strength, but on $\sqrt{f_c'}$ rather than f_c' itself; the validity of shear and torsion theory is also still questionable to some degree.

(g) Special Notation, $\overline{P}$, $\overline{M}$, $\overline{V}$.

The author here introduces the notations $M_u = \phi\overline{M}, V_u = \phi\overline{V}, P_u = \phi\overline{P}$, designating the "ideal" strength in moment, shear, or load capacity by $\overline{M}$, $\overline{V}$, or $\overline{P}$, symbols not used in the Code. Then $\overline{M} = M_u/\phi$, $\overline{V} = V_u/\phi$, $\overline{P} = P_u/\phi$, which permits designs to be made on the basis of $\overline{M}$, $\overline{V}$, and $\overline{P}$ as though one dealt with ideal materials, ideal dimensions, and ideal design relations.

(h) Author's Evaluation of Code Safety Provisions

The Code safety provisions now leave the way open to study overload conditions in buildings to see whether different load factors might be appropriate, as well as to study design, manufacturing, and construction procedures and specifications to see whether ϕ factors might not in the future be more varied to reflect different qualities of design and control. Progress might well be possible for both types of factors considered separately. Lumped together into a single factor of safety the variables are so interwoven as to confuse a reappraisal.

The overall factor of safety given by the Code load factors and ϕ factors is considered quite realistic. The detailed values used for each factor are much less satisfactory to the author. He considers both the load factors and the ϕ factors unreasonably high and hence unrealistic. Since the overall factor of safety is in effect given by the *ratio* of load factor to ϕ factor, the end design is not changed if both are equally too high; but high factors make rational modifications more difficult.

As an example of the present difficult situation with inflated factors, consider the lowering* in the 1971 Code of the load factor for D from 1.5 to

*From 1963 Code values.

1.4 and for L from 1.8 to 1.7, based on better specified materials control and added tightening up on methods of design and analysis. Although the author thoroughly approves the changes, the earlier inflated values of both load factors and ϕ factors forced the correction into the wrong factor. The stated reasons logically call for a change in ϕ rather than in load factors. But a ϕ factor which is already 0.9, meaning 90% of ideal strength, would become obviously utopian if increased another 6 to 7%. In fact, one might almost describe the present 0.9 ϕ factor as already uptopian. Since individual bars may legally be underweight, since d may legally be 0.25 in. in error (Code 7.3.2.1), since overall slab thickness will not always be exactly as specified, since "perfect" contract performance is more the exception than the rule, the author would consider 0.8 or 0.85 more realistic for ϕ in the case of flexure. While the 1971 combination of ϕ and load factor is better than the 1963 one, both factors are still too high in the author's opinion.

Lest these comments cast doubt on the overall safety provisions, the author repeats: The Code ϕ values with the Code load factors give good practical results; only the logic of the operation suffers from inflated values of both.

1.13 DUCTILITY IN REINFORCED CONCRETE CONSTRUCTION

Provision for ductility in reinforced concrete could, with considerable logic, be considered as part of the safety provisions, but probably it deserves the emphasis of separate consideration. Ductility in a structure or a member means the maintenance of strength while sizeable deformation or deflection occurs. The engineer seeks to build ductility into his structures for several reasons. In typical indeterminate structures, ductility permits a heavily stressed portion to continue to carry its capacity load while deforming enough to bring lesser stressed neighboring portions more definitely into the resistance pattern. In slabs and beams ductility means a warning of overloads will be present in the form of excessive cracking and deflection. Where energy must be absorbed, as in blast and earthquake situations,* ductility is particularly important.

Although concrete in compression is far from being elastic at higher stresses, it is basically a brittle material that crushes in compression. In beams and slabs ductility is obtained by using reinforcing steel, which is not brittle, to carry tension and then limiting the amount of tensile steel to

*A recent official study in the United States recommends that some earthquake provisions be included in design much more broadly than in the past, exempting only a portion of seven states, with none totally exempt.

that which will yield before the concrete crushes in compression. This method of control is discussed in Chapter 2.

Reinforced concrete columns are basically brittle under vertical load, although spiral columns (Chapter 14) do introduce limited ductility. Under lateral loads (sidesway) considerable ductility can be built into a column by proper detailing and the use of spirals. The detailing of reinforcement at the joints between beams and columns is critical in the development of ductility and the capacity to survive shock loadings.

Where ductility cannot be achieved in a detail or a design, added strength is necessary to insure that any potential failure be initiated at a more ductile section or in a more ductile manner. In the Code this concern shows, for example, in more conservative ϕ values for columns and in longer bar laps for tension splices. Chapter 18 based on Code Appendix A gives special provisions for seismic design details and is largely directed toward building ductility into the structural frame.

1.14 LIMIT DESIGN

Present strength design is based on moments and shears calculated from elastic analysis of members and frames. Strength evaluations of the cross sections, nevertheless, involve inelastic action at the critical design sections. The elastic analysis gives moments and shears generally on the safe side, but their use involves at least a philosophical inconsistency. Although it is not often computed, it should also be noted that the influence of the cracking of some of the members and the influence of the joints where beam and column have a common section can modify the usual* computed moments more than most designers realize.

In view of these problems, limit design concepts look very attractive. A limited arbitrary reassignment of moments is now permitted, more than under the 1963 Code, but limit design as such is still considered in the development stage in the United States. A number of countries in Europe have moved further in this direction. Limit design is discussed further in Chapter 9.

1.15 CALCULATION ACCURACY

Reinforced concrete is still not a precise theory and, hence, judgement is always an important factor in the application of this theory.

*Based on members with a constant EI, usually based on gross concrete sections.

A distinguished engineer and the chairman of the ACI committee responsible for the 1963 ACI Building Code, Raymond C. Reese, introduced his *CRSI Design Handbook*[11] with these comments on the accuracy of calculations:

> If the somewhat involved mathematical methods used in rigid frame analysis lead one to believe that the design of reinforced concrete structures requires a high degree of precision, the reverse is the case. Concrete is a job-made material, and control cylinders that do not vary more than 10% are remarkably good. Reinforcing bars are shop-made; yet variations in strength characteristics run 3% to 5%; rolled weights can vary $3\frac{1}{2}$%. Formwork is field-built; frequently a 2 × 8 or 2 × 10 (measuring, respectively, $7\frac{5}{8}$ in. and $9\frac{1}{2}$ in.) is used to form the soffit of an 8 in. or 10 in. beam. Bars that are held in place to an accuracy of between $\frac{1}{8}$ in. and $\frac{1}{4}$ in. are extremely well placed. Two-figure accuracy is sufficient for almost all problems in reinforced concrete design.
>
> Concrete is weak in tension; reinforcing steel is supplied to make up that deficiency; the time and effort of the designer is best spent in recognizing and providing for such tensions wherever they may exist, not in striving for a high degree of precision by carrying figures to an unmeaning number of significant places.
>
> On the other hand, the bulk of present computing is done on a 10-in. slide rule, reading easily to three significant figures. When numbers are subtracted, significant figures are often lost. It is, therefore, recommended, more for control of the computations, for ready checking, and to keep the computer alert, rather than for any effect on the completed structure, that figures be carried to three significant places or to the extent of a 10-in. slide rule. There is no point in computing loads to a fine determination only to lose the results in a moment computation, nor is it logical to carry moments to the suggested three significant figures when the loads were guessed to one-figure precision. For that reason, the following table is suggested as a rough guide, not as any hard and fast rule, but only to give some indication of a satisfactory procedure.
>
> RECORD VALUES TO THE FOLLOWING PRECISION
>
> Loads to nearest 1 psf; 10 plf; 100 lb concentration
> Span lengths to about 0.01 ft ($\frac{1}{8}$ in. = 0.01 ft)
> Total loads and reactions to 0.1 kip
> Moments to nearest 0.1 kip-in., if readable
> Individual bar areas to 0.01 in.^{2}
> Concrete sizes to $\frac{1}{2}$ in.
> Bar spacings to $\frac{1}{2}$ in. (supports are crimped at 1-in. intervals)
> Effective beam depth to 0.1 in.

As a practical limit the designer should note the construction tolerances on effective depth d and clear cover specified under Code 7.3.2:

d of 8 in. or less	$\frac{1}{4}$ in.
8 in. $< d <$ 24 in.	$\frac{3}{8}$ in.
d of 24 in. or more	$\frac{1}{2}$ in.

"but the cover shall not be reduced by more than one-third of the specified cover."

1.16 HANDBOOKS

Since this book is written primarily as a text book, it does not concern itself greatly with office practice and the use of design aids. Nevertheless, the student should be aware that curves and tables can speed up design considerably.

The *Ultimate Strength Design Handbook*, Vol. 1[15], published by ACI in 1967 as SP-17 has many useful tables and charts. It is currently under revision and will be available soon in revised form to agree with the 1971 Code.

The *CRSI Handbook Ultimate Strength Design*,[14], 2nd Ed., 1970, published by the Concrete Reinforcing Steel Institute has many tables of allowable design loads on various types of members, some with details of the reinforcing steel. This book will undoubtedly be revised to match the 1971 Code.

In the detailing of structures the *Manual of Standard Practice for Detailing Reinforced Concrete Structures* (ACI R315-65)[16] is most helpful as a guide to the clear presentation of the necessary details. It is a drafting rather than a design manual, although it calls attention to some important considerations in designing reinforced concrete.

SELECTED REFERENCES

1. *Concrete Manual*, U.S. Bureau of Reclamation, Denver, Colo, 7th ed., 1963.
2. *ACI Manual of Concrete Inspection*, ACI, Detroit, 5th ed., 1967.
3. "Design and Control of Concrete Mixes," Portland Cement Association, Chicago, 11th ed., 1968.
4. ACI Committee 211, "Recommended Practice for Selecting Proportions for Normal Weight Concrete (ACI 211.1-70)" *Jour. ACI, 66,* No. 8, Aug. 1969, p. 612.

5. ACI Committee 211, "Recommended Practice for Selecting Proportions for Structural Lightweight Concrete (ACI 211.2-69)," *Jour. ACI*, *66*, No. 5, May 1969, p. 365.
6. David Ramaley and Douglas McHenry, "Stress-Strain Curves for Concrete Strained Beyond Ultimate Load," *Lab. Rep. No. Sp-12*, U.S. Bureau of Reclamation, Denver, Colo., 1947.
7. Eivind Hognestad, N. W. Hanson, and Douglas McHenry, "Concrete Stress Distribution in Ultimate Strength Design," *Jour*, *ACI*, *27*, Dec. 1955; *Proc.*, *52*, p. 455.
8. J. L. Savage, Ivan E. Houk, H. J. Gilkey, and Fredrik Vogt, *Arch Dam Investigation*, Vol. II, *Tests of Models of Arch Dams and Auxiliary Concrete Tests*, Engineering Foundation, New York, 1934.
9. "Specifications for Minimum Requirements for the Deformations of Deformed Steel Bars for Concrete Reinforcement," *ASTM Spec. A* 305–56*T*, ASTM, Philadelphia, 1956. Replaced 1968 by Ref. 10.
10. "Standard Specification for Deformed Billet-Steel Bars for Concrete Reinforcement," *ASTM Spec. A615-68*, ASTM, Philadelphia, 1970.
11. Raymond C. Reese, *CRSI Design Handbook*, Concrete Reinforcing Steel Institute, Chicago, 2nd ed., 1957.
12. Richard A. Helmuth and Danica H. Turk, "The Reversible and Irreversible Drying Shrinkage of Hardened Portland Cement and Tricalcium Silicate Pastes," *Journal*, *PCA* Research and Development Laboratories, *9*, No. 2, May 1967.
13. Lev Zetlin, Chas. H. Thornton, and I. Paul Lew, "Internal Straining of Concrete," *Designing for Effect of Creep, Shrinkage, and Temperature in Concrete Structures*, SP-27, Amer. Concrete Inst., Detroit, 1970, p. 323.
14. *CRSI Handbook Ultimate Strength Design*, 2nd ed., Concrete Reinforcing Steel Institute, Chicago, 1970.
15. *Ultimate Strength Design Handbook*, Vol. I, ACI, SP-17, Detroit, 1967.
16. Proposed 1970 Revision of *Manual of Standard Practice for Detailing Reinforced Concrete Structures* (*ACI R315–65*), *ACI*, Detroit, 1970.

2

Flexural Analysis of Beams

2.1 THE RESISTING COUPLE IN A BEAM

Statics shows that an external bending moment on any beam must be resisted by internal stresses which can be indicated as a resultant tension N_t and a resultant compression N_c. Unless there is axial load, summation of horizontal forces (Fig. 2.1) indicates that N_t must equal N_c and that together they form a couple. In a reinforced concrete beam the reinforcing steel is assumed to carry all the tension N_t (Sec. 1.4) and thus N_t is located at the level of the steel. The compressive force N_c is the resultant of compression stresses over some depth of beam and thus its location is not established from statics alone. Let the arm or distance between these N_t and N_c forces be designated as z, as shown in Fig. 2.1. The further depth below the steel is useful chiefly to fireproof the steel and protect it from moisture and to assist in bonding the reinforcing steel to the concrete; it does not influence these calculations. Further information is required to evaluate z.

The resisting couple idea can be applied to homogeneous beams. It is simpler for rectangular beams with straight-line distribution of stresses than is the use of $f = Mc/I$. For steel I-beams it is more awkward, but is useful for rough estimates based on neglect of the web area. For reinforced concrete beams it has the definite advantage of using the basic resistance pattern.

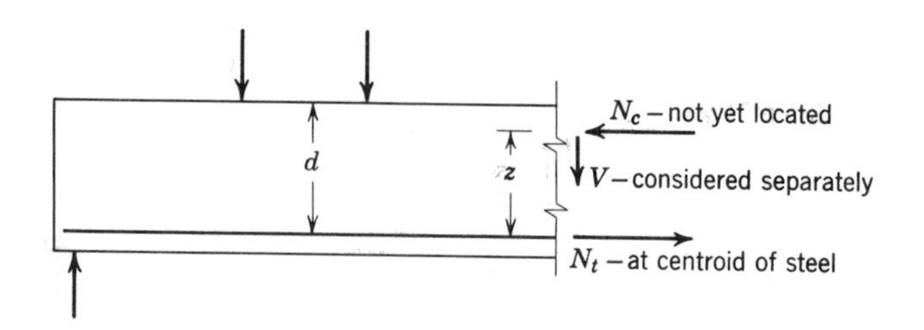

Fig. 2.1. Internal resisting couple in a beam.

2.2 DISTRIBUTION OF COMPRESSIVE STRESS

In reinforced concrete beams the compressive stress varies from zero at the neutral axis to a maximum at or near the extreme fiber. How it actually varies and where the neutral axis actually lies depend both on the amount of the load and on the history of past loadings. This variation is the result of several factors: (1) The spacing and depth of tension cracks depend upon whether the beam has been loaded before, and how heavily. (2) Shrinkage stresses and creep of concrete are important factors in relation to stress distribution, factors very difficult to include in an analysis. (3) Most important, the stress-strain curve for concrete as indicated in Figs. 2.2 and 1.9 is not a straight line.

Experiments confirm that, even for reinforced concrete cracked on the tension side of the beam, unit strains vary as the distance from the neutral axis,* the maximum strain increasing with the moment from ϵ_1 to ϵ_2, to $\epsilon_3, \ldots$ on to failure. The curve of Fig. 2.2 thus indicates that the compressive stress distribution on the first application of loading (without any shrinkage) would go through several stages. The neutral axis would shift location with the stress pattern to keep $N_t = N_c$. For example, the small shaded triangle $O\epsilon_1$ would represent the early stresses, as shown more conveniently in Fig. 2.3*a*. Likewise ϵ_2 would limit a larger nontriangular stress distribution as in Fig. 2.3*b*. The maximum unit stress would be f_c' in Fig. 2.3*c* with strain at ϵ_3 in Fig. 2.2. The maximum total compression in Fig. 2.3*d* would occur at some strain value ϵ_4 which is defined later. A detailed analysis would show that as the higher strains are reached, the stiffness of the concrete decreases. With the tension all carried in the elastic range by the steel, the neutral axis would have to drop as indicated in Fig. 2.3 to keep N_c increasing as rapidly as N_t.

*Exactly at a crack this is not quite true, but over an average gage length it is statistically true.

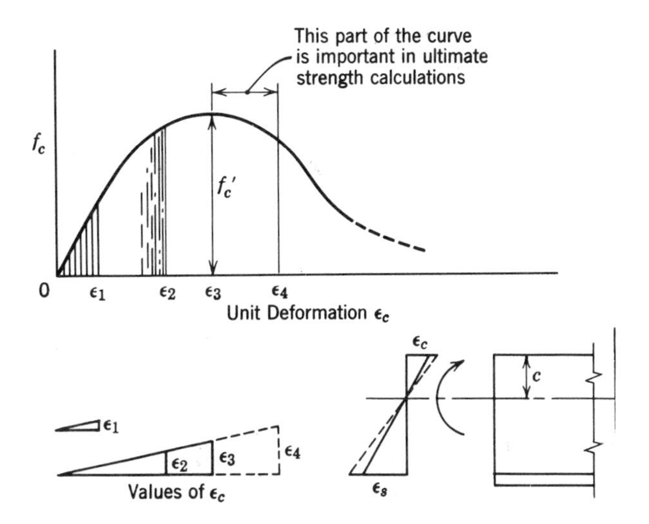

Fig. 2.2. Stress-strain curve for concrete of medium strength and possible strain-stress combinations.

Under service loads it probably is not feasible, and it may not even be possible, to do more than estimate rather crudely the approximate magnitude of real stresses. Certainly these are complicated by shrinkage and creep and depend upon how much cracking has been induced by previous loadings. Specifications formerly emphasized stresses at working loads, but laboratory tests of reinforced concrete beams show that actual deformations and stresses at working loads only faintly resemble values conventionally calculated on the basis of straight-line stresses (Fig. 2.3*a*). Hence design procedures have gradually shifted to ultimate strength methods, with checks at service load for deflection and cracking.

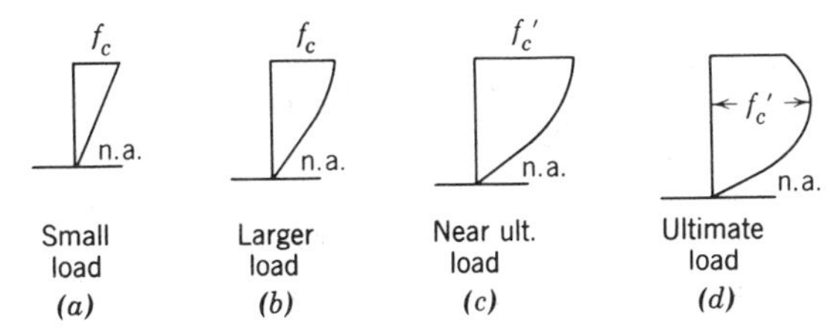

Fig. 2.3. Shifting stress pattern with increasing moment up to failure in compression. (Tension in concrete below neutral axis has been neglected.)

Fortunately the *ultimate* strength of reinforced concrete beams can be predicted or calculated with quite satisfactory accuracy. Design on the basis of (ultimate) strength is now standard in the 1971 ACI Code.

2.3 TENSION AND COMPRESSION FAILURES DUE TO BENDING MOMENT

Beams may fail from moment because of weakness in the tension steel or weakness in the compression concrete. Failure from stresses primarily related to shear will be considered later and separately, as in the case of steel construction, since it is only in very deep and relatively short beams that these influence bending strength.

Most beams are weaker in their reinforcing steel than in their compression concrete. Such beams fail under a load slightly larger than that which makes $N_t = f_y A_s$, where f_y is the yield-point strength of the steel and A_s is the area of the steel. Since $f_y A_s$ normally represents the entire usable strength of steel, the further increase in moment resistance needs explanation. When the steel first reaches its yield point, the compressive stress distribution may be like that previously illustrated in Fig. 2.3*b* or *c*. A slight additional load causes the steel to stretch a considerable amount. The increasing steel deformation in turn causes the neutral axis to rise (when the tension is on the bottom) and the center of compression N_c therefore moves upward, Fig. 2.4*a*. This increase in the arm z between N_c and N_t gives an increased resisting moment $N_t z$ even though N_t is essentially unchanged.* The rising of the neutral axis also reduces the area under compression and thereby increases the unit compressive stress required to develop the nearly constant value of N_c. This process continues until the reduced area fails in compression, as a secondary effect. These changes (neglecting any tension in the concrete) are summarized in Fig. 2.4*a*. This type of failure (inverted because in a negative moment region) is shown in Fig. 2.4*b*.

Such an *underreinforced* beam shows greatly increased deflection after the steel reaches the yield point, giving adequate warning of approaching beam failure; the steel, being ductile, will not actually pull apart even at failure of the beam. The increase in z will rarely be more than a few percent, say 3 to 5%. If the concrete stress is very high before the steel stress f_s reaches the yield-point value of f_y, the increase in z may be very small.

*When the steel has no sharp yield point, N_t increases above the yield strength enough to make calculations based on a yielding steel conservative. The same problem arises when a short yield plateau introduces a strain hardening region.

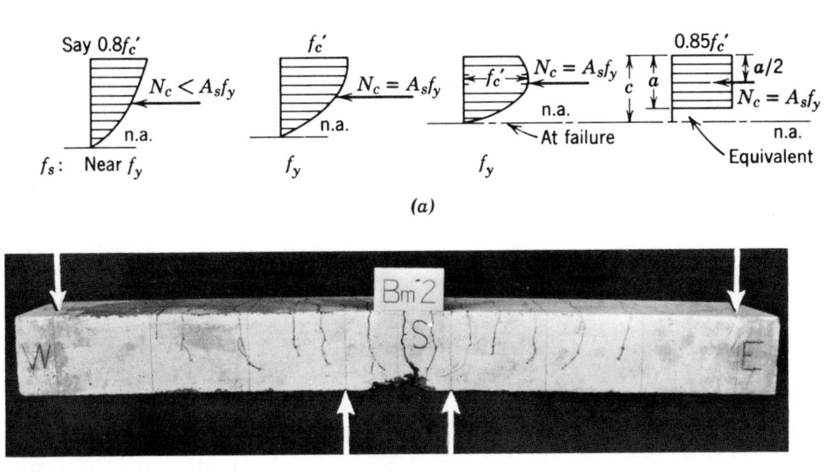

Fig. 2.4. Flexural failure in tension. (*a*) Variation in compressive stresses as beam approaches failure in tension. (*b*) Beam after failure under negative moment (tension on top).

If the concrete reaches its full compressive strength just as the steel reaches its yield-point stress, the beam is said to be a *balanced* beam (at failure). Such a beam requires very heavy steel and is rarely economical.

Before a beam fails from weakness in compression, the top elements of the beam shorten considerably under the final increments of load, causing the neutral axis to move lower down in the beam. Such movement of the neutral axis increases the area of concrete carrying compression and importantly increases the total N_c that can be carried (Fig. 2.5). The greatly increased N_c is offset slightly by a reduced arm z. The concrete finally fails suddenly and often explosively in compression, and the steel stress remains below the yield point unless the beam is exactly balanced. These stress changes, again ignoring any tension in the concrete, are summarized in Fig. 2.5. Such a beam is called an *overreinforced* beam.

For design purposes this real final stress distribution may be replaced adequately by an equivalent rectangle of stress (pioneered in this country by Whitney) of intensity $0.85f_c'$ and depth a as shown in the final sketches in Figs. 2.4a and 2.5. For rectangular beams, the shaded area of the rectangular stress block should equal that of the real stress block and their centroids should be at the same level. The value of a given in the Code is intended to give this result and is recommended:

For $f_c' \lesseqgtr 4000$ psi $\quad a = 0.85c$

For $f_c' > 4000$ psi, reduce the 0.85 factor* linearly at the rate of 0.05 per 1000 psi excess over 4000 psi.

*This factor is designated β_1.

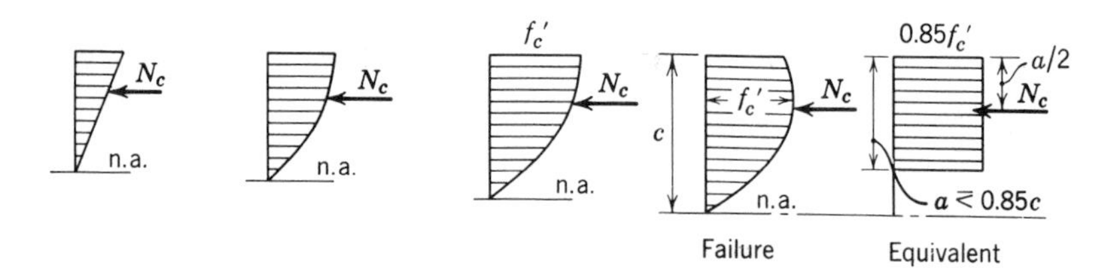

Fig. 2.5. Variation in compressive stresses as beam approaches failure in compression. See text for values of a.

2.4 ANALYSIS VERSUS DESIGN

In analysis, whether for actual stresses or for permissible moments, the engineer deals with given beams, known for both dimensions and steel. He has no control over the location of the neutral axis, which lies at a definite but initially undefined depth.

In design, loads and ultimate stresses are known and some or all the dimensions remain to be fixed. In this case the designer has some control over the location of the neutral axis. He can shift it where he wants it, to the extent that his change in dimensions changes the depth of rectangular stress block or magnitude of N_t.

The student should understand clearly this fundamental difference between analysis and design problems. Analysis for various types of beams will be discussed first, then design.

2.5 THE BALANCED RECTANGULAR BEAM

The balanced beam in ultimate strength design is not a practical beam, but the concept is fundamental to the philosophy of the Code.

The Code (10.3.2) specifies:

> For flexural members and . . . , the reinforcement ratio, ρ, shall not exceed 0.75 of that ratio which would produce balanced conditions for the section under flexure without axial load.

The ratio $\rho = A_s/bd$, where A_s is the area of steel reinforcement and b and d are shown in Fig. 2.6a. The Code (10.2.3) requires that the concrete strain be taken at 0.003, as also shown in Fig. 2.6a for the balanced condition.

These regulations represent an attempt to assure the ductile failure produced by yielding of steel as compared to the brittle type of failure occurring when the failure is in compression.

The analysis of the balanced beam starts from the strain triangles at failure as shown in Fig. 2.6a. The steel strain will be f_y/E_s and the maximum

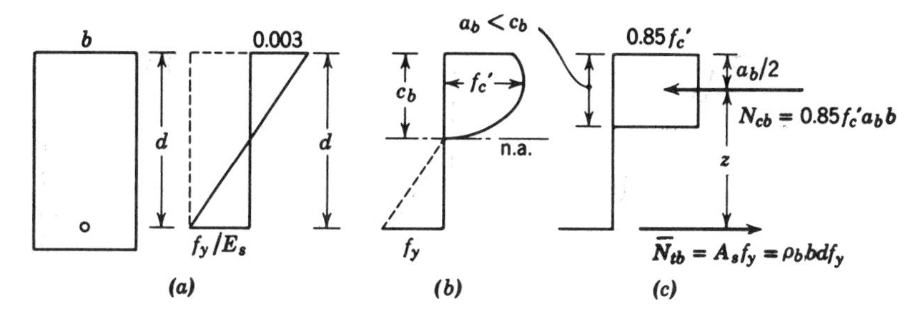

Fig. 2.6. Balanced beam. (*a*) Strains. (*b*) Stresses. (*c*) Equivalent stresses and forces.

concrete strain 0.003, which is conservatively in agreement with test observations. The neutral axis can be located from the strain triangles, most simply by considering the compression strain triangle and the large dotted triangle. For $E_s = 29 \times 10^6$ psi:

$$c_b = \frac{0.003}{0.003 + f_y/E_s} d = \frac{87{,}000}{87{,}000 + f_y} d \tag{2.1}$$

Experimental work has established the necessary depth a of stress block to go with a given neutral axis, or conversely the neutral axis distance c to go with a given stress block depth a. For $f_c' \lessgtr 4000$ psi, $a = 0.85c$, with or without balanced conditions. For $f_c' > 4000$ psi see Sec. 2.3 (last paragraph).

The balanced reinforcement ratio $\rho_b = A_{sb}/bd$ can now be established from $\overline{N}_{tb} = \overline{N}_{cb}$* but the following numerical approach is recommended to the student as presenting a clearer picture. Figure 2.6*c* uses $\overline{N}_{tb}$ and $\overline{N}_{cb}$ which can be regarded as a kind of ideal force which disregards the ϕ factor. Then $N_{tb} = \phi\overline{N}_{tb}$, $N_{cb} = \phi\overline{N}_{cb}$, and $M_{ub} = \phi\overline{M}_b$. For example, for $f_c' = 3000$ psi, $f_y = 40{,}000$ psi, $a_b = 0.85c_b$. Then for a rectangular beam of depth d and width b:

$$c_b = \frac{87{,}000}{87{,}000 + 40{,}000} d = 0.685d$$

$$a_b = 0.85c_b = 0.85 \times 0.685d = 0.582d$$

$$\overline{N}_{cb} = 0.85f_c'(0.582d)b = 0.495f_c'bd.$$

Since $\overline{N}_{tb} = \overline{N}_{tc}$, $0.495f_c'bd = A_{sb}f_y = (\rho_bbd)f_y$

$$0.495 \times 3000 = \rho_b \times 40{,}000$$

$$\rho_b = 0.0371$$

*See Sec. 1.12f for explanation of $\overline{N}_t$, $\overline{N}_c$, $\overline{M}$, and similar values. The strength reduction factor is *not* included.

Algebraically, this is Code Eq. 8.1, with β_1 representing a/c:*

$$\rho_b = 0.85\beta_1 \frac{f_c'}{f_y} \frac{87{,}000}{87{,}000 + f_y}.$$

Although there is little practical use for the balanced moment (except in columns), it is simple to establish from the resisting couple of the beam.

$$z = d - a_b/2 = d - 0.291d = 0.709d$$

$$\overline{M}_b = \overline{N}_{tb}z = (0.0371bd)f_y(0.709d)$$

$$= 0.0263f_ybd^2 = 1050bd^2$$

or

$$\overline{M}_b = \overline{N}_{cb}z = 0.495f_c'db(0.709d)$$

$$= 0.352f_c'bd^2 = 1050bd^2$$

$$M_{ub} = \phi\overline{M}_b = 0.9 \times 1050bd^2 = 945bd^2$$

This example has been followed through in detail because the methods also apply to nonbalanced beams when either c, $\overline{N}_t$, or a is known.

Whitney, in pioneering this method, assumed a balanced beam would carry $\overline{M} = 0.33f_c'bd^2$, independent of f_y, provided f_c' was at least 3000 psi. The present Code approach reduces this balanced moment for high strength steels and for f_c' above 4000 psi, but in the particular example given the Code value is larger than Whitney's.

Other shapes of beams such as T-beams, double-reinforced beams, and irregular shapes, can likewise be evaluated as balanced beams. The neutral axis is at the same relative depth in every case, being established from strains rather than stresses. The procedures with varying widths and with compression reinforcement are discussed starting in Sec. 2.11.

2.6 TENSILE STEEL LIMITATION—FOR DUCTILITY

Since a balanced beam will ultimately fail suddenly in compression, the Code, as quoted in Sec. 2.5, limits the tensile reinforcement to a maximum of $0.75\rho_b$ both for beams and for some lightly loaded columns (Code 10.3.2). This assures an underreinforced beam that will fail in tension, as in Fig. 2.4, with reasonable ductility and some warning of approaching capacity.

In stating this restriction as a single simple rule, the Code penalizes and complicates the analysis of some double-reinforced beams which carry no axial load. The details of this problem, which is strictly legal and not structural, will be discussed in Sec. 2.14. Fortunately, it is rarely economi-

*In earlier editions the ratio a/c was called k.

cal to make beams so small that this rule becomes an operative limit. Both deflections and costs normally point toward larger beams with smaller steel ratios.

2.7 CODE MAXIMUM MOMENT FOR RECTANGULAR BEAMS

The limitation to $0.75\rho_b$ as a Code maximum is equally well described by $\overline{N}_t = 0.75\overline{N}_{tb}$ or $\overline{N}_c = 0.75\overline{N}_{cb}$ and for the single case of a rectangular beam also by $c = 0.75c_b$ or $a = 0.75a_b$. For the rectangular beam the last is the most direct form. Then the Code maximum moment and reinforcement ratio for $f_c' = 3000$ psi and $f_y = 40{,}000$ psi can be found as follows:

$$c_b = 0.685d, \; a_b = 0.582d, \text{ from Sec. 2.5}$$
$$\text{Max } a = 0.75 \times 0.582d = 0.436d$$
$$\text{Max } \overline{N}_c = 0.85f_c'(0.436d)b = 0.371f_c'bd$$
$$z = d - a/2 = d - 0.218d = 0.782d$$
$$\text{Max } \overline{M} = \overline{N}_c z = 0.371f_c'bd(0.782d) = 0.290f_c'bd^2$$
$$\text{Max } M_u = \phi\overline{M} = 0.9(0.290f_c'bd^2)$$
$$= 0.261f_c'bd^2 = 783bd^2 = k_m bd^2$$

Since $\overline{N}_t = \overline{N}_c$,

$$\rho bdf_y = \overline{N}_c = 0.371f_c'bd$$
$$\text{Max } \rho = 0.371f_c'/f_y = 0.371 \times 3/40 = 0.0278$$
$$\text{Max } \overline{M} = \overline{N}_t z = (0.0278bdf_y)(0.782d) = 0.0217f_y bd^2$$
$$\text{Max } M_u = \phi\overline{M} = 0.9 \times 0.0217f_y bd^2 = 0.0196f_y bd^2 = 784bd^2$$

This maximum M_u would, with more decimal places, be the same whether based on $\overline{N}_c$ or $\overline{N}_t$ since M_u represents a couple. For convenience, designate the constant 783* as k_m* and say maximum $M_u = k_m bd^2$. Table 2.1 lists a number of values of maximum k_m, maximum $\rho = 0.75\rho_b$, and maximum a/d for various combinations of f_c' and f_y; also for $\rho = 0.18f_c'/f_y$ with $f_y = 40{,}000$ psi as in Fig. 2.8. Values of $\bar{k}$ are usable with $\overline{M}$.

The legal Code limit for ρ is definitely $0.75\rho_b$. However, a designer can be as logical in permitting a greater ρ (with moment capacity of the member limited to that for $0.75\rho_b$) as in permitting the use of an actual f_y of 46 ksi where Grade 40 steel is specified. Accordingly, the author limits the k_m

*This constant carries a unit of psi. The symbol k_m is in accord with the new ACI notation, but is not yet standardized. In earlier editions it was called R_u; the notation K or K_u has also been used. The symbol $\bar{k}$ in Sec. 2.17 is identical with $\bar{k}$ in Table 2.1.

value quite strictly and is less concerned (in nonsiesmic work) about extra steel which can be ignored for strength.

TABLE 2.1. Limiting Constants for Rectangular Beams

Steel:	$f_y = 40{,}000$				$f_y = 50{,}000$				$f_y = 60{,}000$			
Concrete	k_m	$\bar{k}$	100ρ	a/d	k_m	$\bar{k}$	100ρ	a/d	k_m	$\bar{k}$	100ρ	a/d
$f_c' = 3000$	783	870	2.78	0.436	740	822	2.06	0.405	705	783	1.61	0.378
$f_c' = 4000$	1047	1164	3.72	0.436	987	1097	2.75	0.405	937	1041	2.14	0.378
$f_c' = 5000$	1244	1383	4.36	0.411	1181	1312	3.24	0.381	1115	1238	2.52	0.355
$f_c' = 6000$	1424	1582	4.90	0.386	1346	1496	3.64	0.357	1281	1423	2.83	0.333
$\rho = 0.18f_c'/f_y$	$0.145f_c'$	$0.161f_c'$	$\dfrac{0.18f_c'}{40{,}000}$	0.212								

2.8 ANALYSIS OF RECTANGULAR BEAMS

Only underreinforced beams, as discussed in Sec. 2.6, are permitted by the Code and their analysis is simple. First establish whether the given ρ, which is simply A_s/bd, is less than (or equal to) $0.75\rho_b$, the maximum ρ, tabulated in Table 2.1. (Of course, the equivalent check can be made later, at some risk to earlier calculations, in terms of k_m or of maximum a/d.) If $\rho < 0.75\rho_b$, the beam obviously fails primarily in tension. The resulting stretching of the steel will raise the neutral axis until the final secondary compression failure occurs at the compression strain ϵ_4 in Fig. 2.2, taken in the Code as 0.003. The usual analysis, however, avoids the necessity of using this strain directly for simple cases.

Since $\overline{N}_t = \overline{N}_c$, the ultimate moment is reached when the compressive area can just support a $\overline{N}_c$ equal to $A_s f_y$. By using the Code's (and Whitney's) simplification as shown in Fig. 2.7, the compression $\overline{N}_c$ can be evaluated as an equivalent block of uniform stress, of intensity $0.85f_c'$ and height a (Fig. 2.7*b*). Then $\overline{N}_c = \overline{N}_t$ becomes:

$$0.85f_c'ab = A_s f_y$$

$$a = \frac{A_s f_y}{0.85f_c'b} = \frac{\rho f_y d}{0.85f_c'}$$

where $\rho = A_s/bd$. The arm between $\overline{N}_c$ and $\overline{N}_t$ is $d - a/2$. Hence

$$\overline{M} = \overline{N}_t z = A_s f_y \left(d - \frac{a}{2}\right)$$

where $\overline{M}$ is the ultimate moment under ideal conditions and the value of a is found from the above relation. The dependable $M_u = \phi\overline{M} = 0.9\overline{M}$.

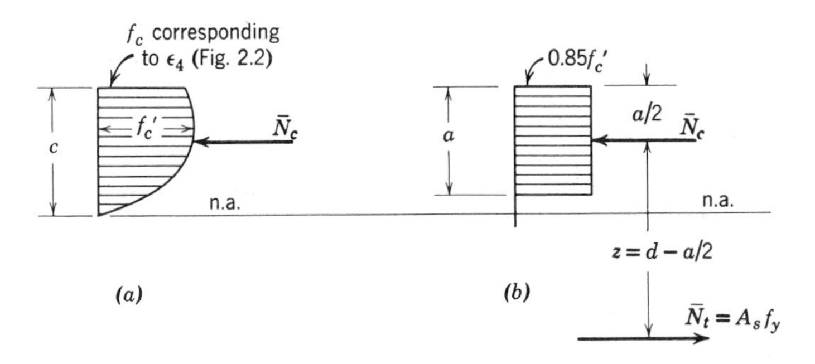

Fig. 2.7. The Code (or Whitney) compressive stress block. (*a*) Probable stress distribution. (*b*) Simplified model for calculations.

An alternate form for the formula for M_u in terms of $\omega = \rho f_y/f_c'$ is:

$$M_u = \phi f_c' bd^2 \omega \, (1 - 0.59\omega)$$

This is sometimes convenient and involves only an algebraic transformation of the above process. In terms of the steel, M_u could also be written:

$$M_u = \phi A_s f_y d \left(1 - \frac{0.59 \rho f_y}{f_c'}\right)$$

The author recommends for students the forms first developed above.

The Code places certain minimum steel requirements on beams.

To resist calculated moments the minimum reinforcement (Code 10.5.1) is $\rho_{\min} = 200/f_y$ unless *all* the flexural reinforcement for the member is at least 4/3 that required by analysis. This means $\rho_{\min}$ of 0.005 for Grade 40 and 0.0033 for Grade 60 bars. For joists or T-beams with webs in tension $\rho_{\min}$ is based on web width rather than flange width. For slabs of uniform thickness the minimum drops back to temperature and shrinkage requirements (Code 7.13) of 0.0020 of *gross* slab area for Grade 40 bars and 0.0018 for Grade 60 bars, with maximum bar spacing of five times the slab thickness but not more than 18 in.

The steel yield strength in calculations is limited to 80 ksi (Code 9.4). The Code specifies (3.5.1) that if the yield strength exceeds 60,000 psi the f_y used shall be that corresponding to a strain of 0.0035 determined by tests on full size bars.

The ratio of a/c used is to be decreased 0.05 for each 1000 psi of f_c' above 4000 psi (Code 10.2.7), as already noted at the close of Sec. 2.3, to reflect the more triangular or less curved nature of the stress-strain curve for these higher strength concretes (Fig. 1.8*a*).

TABLE 2.2. (Code Table 9.5*a*). Minimum Thickness of Beams or One-way Slabs Unless Deflections Are Computed[a]

Member	Minimum Thickness, h			
	Simply Supported	One End Continuous	Both Ends Continuous	Cantilever
	Members not supporting or attached to partitions or other construction likely to be damaged by large deflections.			
Solid one-way slabs	$l/20$	$l/24$	$l/28$	$l/10$
Beams or ribbed one-way slabs	$l/16$	$l/18.5$	$l/21$	$l/8$

[a] The span length l is in inches.

The values given in this table shall be used directly for nonprestressed reinforced concrete members made with normal weight concrete (w = 145 pcf) and Grade 60 reinforcement. For other conditions, the values shall be modified as follows:

(a) For structural lightweight concrete having unit weights in the range 90–120 lb per cu ft, the values in the table shall be multiplied by $1.65 - 0.005w$ but not less than 1.09 where w is the unit weight in lb per cu ft.

(b) For nonprestressed reinforcement having yield strengths other than 60,000 psi, the values in the table shall be multiplied by $0.4 + f_y/100{,}000$.

Under service loads, deflection of members must be considered and deflection calculations are discussed in Chapter 6. When the length-depth ratios of Table 2.2 (a copy of ACI Code Table 9.5*a**) are not satisfied, or when the member supports partitions that might be cracked, deflections must be calculated. This criterion does not necessarily separate members without deflection troubles from those with excessive deflections. It simply reflects, in a somewhat oversimplified fashion, a boundary where the designer must certainly become aware of deflection.

In the 1963 Code reference was made to checking deflections where ρ exceeded $0.18f_c'/f_y$. There is a greater design experience with members designed with approximately this steel ratio because this was a usual ρ under working stress methods. Larger reinforcement ratios imply smaller beam sizes than have been customary and hence larger deflections. Short-

*The Code is available from ACI as noted in Sec. 1.10.

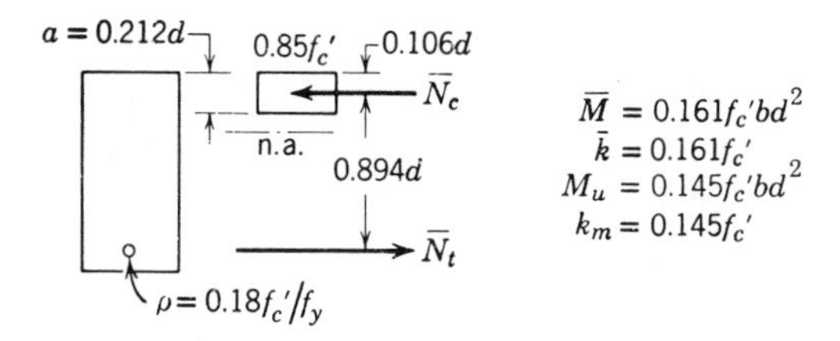

Fig. 2.8. Design constants for $\rho = 0.18f_c'/f_y$.

span beams can use larger ratios without deflection problems; long-span beams can present deflection problems even when a smaller ratio is used. Long-span beams, cantilever construction, and shallow members always require that special attention be given to deflections. As a guidepost from past experience, the criterion has value.

As a sample of the determination of beam constants for a given ρ, consider Fig. 2.8 and the following for $\rho = 0.18f_c'/f_y$:

$$\bar{N}_t = A_s f_y = \rho b d f_y = 0.18 f_c' b d$$
$$a = \bar{N}_t/(0.85 f_c' b) = (0.18 f_c' b d) \div (0.85 f_c' b) = 0.212d$$
$$z = d - 0.5 \times 0.212d = 0.894d$$
$$\bar{M} = \bar{N}_t z = 0.18 f_c' b d \times 0.894d = 0.161 f_c' b d^2$$
$$M_u = \phi \bar{M} = 0.9 \times 0.161 f_c' b d^2 = 0.145 f_c' b d^2$$
$$k_m = 0.145 f_c'$$

2.9 RECTANGULAR BEAM EXAMPLES

(*a*) A rectangular beam has $b = 11$ in., $d = 20$ in., $A_s = 3$- #8, $f_c' = 3000$ psi, Grade 40 steel. Calculate the ultimate moment capacity.

Solution

$\rho = 3 \times 0.79/(11 \times 20) = 0.0108 > \rho_{min} = 0.005$ O.K.

Ultimate $\bar{N}_t = A_s f_y = 3 \times 0.79 \times 40{,}000 = 95{,}000$ lb $= \bar{N}_c$

$95{,}000 = 0.85 f_c' b a = 0.85 \times 3000 \times 11a$

$a = 3.38$ in. $<$ maximum of $0.436d$ (Table 2.1) $= 8.72$ in. O.K.

$z = d - a/2 = 20 - 3.38/2 = 18.31$ in.

$\bar{M} = \bar{N}_t z = 95{,}000 \times 18.31/12{,}000 = 145$ k-ft

$M_u = \phi \bar{M} = 0.9 \times 145 = 131$ k-ft

The fact that tension controls this ultimate could have been established by two other methods: (1) by showing that $\rho < 0.0278$ (the maximum in

Table 2.1); (2) by showing that $M_u < 783bd^2$ (Table 2.1) or $\overline{M} < 870bd^2$. Since $a < 0.212d = 4.24$ in. (Fig. 2.8) deflections are probably no problem. The length-depth ratio is not available for checking deflection.

(*b*) If in (*a*) the steel is changed to 5- #11, calculate the ultimate moment capacity and corresponding f_s.

Solution

Ultimate $\overline{N}_t = A_s f_y = 5 \times 1.56 \times 40{,}000 = 312{,}000 \text{ lb} = \overline{N}_c$

$312{,}000 = 0.85 \times 3000 \times 11a$

$a = 11.1 \text{ in.} > 0.436 \times 20 = 8.72 \text{ in.}$ N.G.

This beam is not acceptable under the Code because a and A_s exceed the maximums shown in Table 2.1.

The author would tend to accept this beam based on evaluation on the basis of the maximum in Table 2.1, that is, on $a = 8.72$ in., in effect neglecting the excess A_s as discussed at the close of Sec. 2.7. The maximum k_m is a simpler equivalent approach.

$$M_u = k_m bd^2 = 783 \times 11 \times 20^2/12{,}000 = 287 \text{ k-ft}$$

$$z = \text{d} - a/2 = 20 - 8.72/2 = 15.64 \text{ in.}$$

$$\text{At } M_u,\ f_s = M_u/(A_s z) \doteq 287{,}000 \times 12/(5 \times 1.56 \times 15.64)$$

$$\doteq 28{,}200 \text{ psi}$$

Deflections may be large at this moment ($\rho \gg 0.18f_c'/f_y$), but the lowered f_s may help on this.

(*c*) How much steel can be effectively used in the beam of (*a*)? What is M_u for this case?

Solution

Table 2.1 shows the maximum percentage of steel as 2.78 and the corresponding k_m as 783. Hence

$A_s = \rho bd = 0.0278 \times 11 \times 20 = 6.12 \text{ in.}^2$

$M_u = k_m bd^2 = 783 \times 11 \times 20^2/12{,}000 = 287 \text{ k-ft}$

As an alternate, maximum $a = 0.436 \times 20 = 8.72$ in.

$\overline{N}_c = 0.85 \times 3000 \times 11 \times 8.72 = 245{,}000 \text{ lb}$

$A_s = \overline{N}_t/f_y = \overline{N}_c/f_y = 245{,}000/40{,}000 = 6.12 \text{ in.}^2$

$M_u = \phi\overline{M} = \phi N_c z = 0.9 \times 245{,}000(20 - 8.72/2)/12{,}000 = 287 \text{ k-ft}$

Deflection should be investigated.

2.10 SLABS

A one-way slab is simply a wide, shallow, rectangular beam insofar as analysis is concerned. The reinforcing steel is usually spaced uniformly over its width. For convenience a 1-ft width is generally taken for analysis or design, since loads are frequently specified in terms of load per square foot; on a 1-ft strip, this unit load becomes the load per linear foot. Since the slab can average the effect of steel over some width, the effective A_s may well correspond to a fractional number of bars in the 1-ft width. For example, #5 bars at 8-in. spacing, as in Fig. 2.9*a*, give $A_s/ft = 0.31 \times 12/8 = 0.465$ in.^{2}/ft for the 1-ft design strip, Fig. 2.9*b*.

2.11 ANALYSIS OF T-BEAMS

Because slabs and beams are ordinarily cast together as shown in Fig. 2.10, the beams are automatically provided with an extra width at the top which is called a flange. Such beams are known as T-beams. The portion below the slab is called the web.

Flange bending stress is not uniform from beam to beam, being largest over the web and tending to drop off with distance from the web. In part this results from the necessity of transferring all the flange flexural stress by longitudinal shearing stresses on vertical sections parallel to the web.

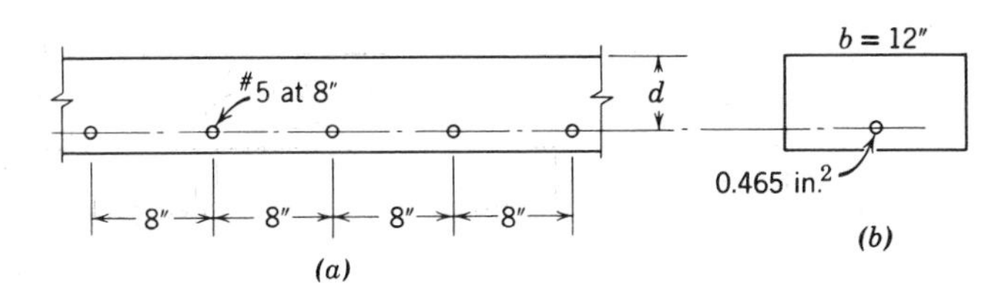

Fig. 2.9. Design strip and effective steel area in slab analysis. (*a*) Slab cross section. (*b*) Design strip.

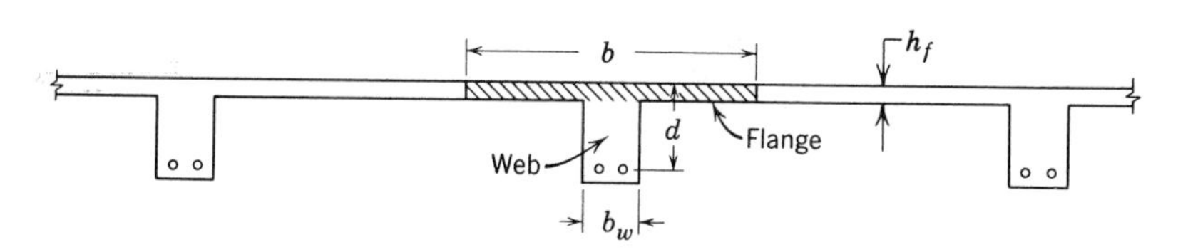

Fig. 2.10. T-beam as part of a floor system.

Codes use a reduced effective flange width at an assumed uniform stress distribution in lieu of a wider actual slab with nonuniform stresses. The Code (8.7.2) for symmetrical monolithically cast T-beams limits the flange width to be used in calculations to a maximum projection of eight times the slab thickness on each side of the web, or half way to the next beam, or a total width of one-fourth of the span, whichever is smallest.

T-beams are analyzed in much the same way as rectangular beams. The Code intends the same limitation (used in Sec. 2.6) to be placed on tension steel; that is, the tension steel should be limited to 0.75 that of the balanced section for the T-beam. This is usually no problem since the large flange area normally keeps compression on the concrete quite low and economy prevents the engineer from using excessive steel. Hence it is usually safe to assume (and check later) that $\overline{N}_t = A_s f_y$ and calculate the depth of stress block a which will provide an equal $\overline{N}_c$. If a, the resulting depth of stress block, is less than the slab (flange) thickness h_f, as shown in Fig. 2.11*a*, the entire analysis is identical with that of a very wide rectangular beam of width b.

If the area within the flange depth does not provide enough compression, the form of the calculation must be modified to take account of the narrower web width b_w below the flange (Fig. 2.11*b*). The total tension can be equated to the total compression to establish the depth of the stress block.

$$A_s f_y = 0.85 f_c'[a b_w + h_f(b - b_w)]$$

$$a = \frac{A_s f_y - 0.85 f_c' h_f (b - b_w)}{0.85 f_c' b_w}$$

The resultant compression $\overline{N}_c$ acts at the centroid of the shaded area in Fig. 2.11*b*.

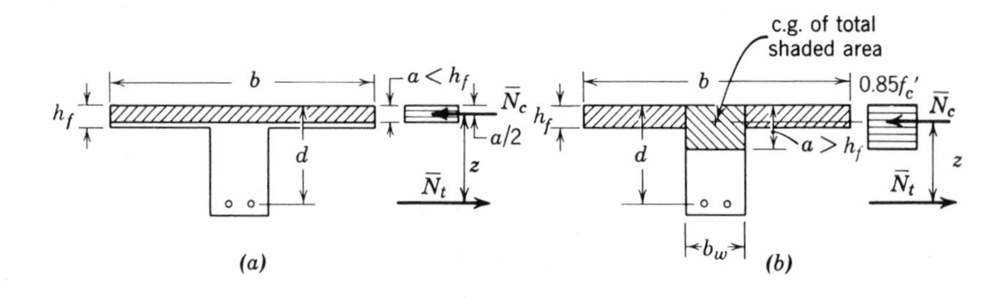

Fig. 2.11. Design of T-beams. (*a*) Acts as rectangular beam. (*b*) Direct analysis as T-beam.

Similarly, in checking A_s against the balanced reinforcement, the only change from the rectangular beam procedure is in noting that the area which establishes the balanced $\overline{N}_{cb}$ is not a single rectangle. The neutral axis, being determined from strains, is located at the same depth c_b as for the rectangular beam. The depth of stress block a_b is 0.85 c_b or less,* since tests indicate that the ratio a/c used for rectangular beams is accurate enough. The total compression $\overline{N}_{cb}$ is still $0.85f_c'$ times the compression area and this $\overline{N}_{cb}$ can be equated to $\overline{N}_{tb} = A_s f_y$ to establish the balanced A_s; or, more simply, the usable $\overline{N}_t = 0.75\overline{N}_{cb} = 0.75\overline{N}_{tb}$. The rectangular beam idea of $0.75a_b$ is not usable here because of the varying width.

2.12 T-BEAM EXAMPLES

For the T-beam shown in Fig. 2.12*a*, find the ultimate moment if $A_s = 3.00$ in.², steel is Grade 60, and $f_c' = 3000$ psi.

Solution

Since neither span nor beam spacing is given, the only available check on flange width is in terms of flange thickness.

$3.00/(8 \times 12) > \rho_{min} = 0.0033$ O.K.
Max flange overhang $= 8h_f = 16$ in. $>$ actual O.K.
Ultimate $\overline{N}_t = 3.00 \times 60{,}000 = 180{,}000$ lb
Max $\overline{N}_c$ within depth $h_f = 0.85 \times 3000 \times 32 \times 2 = 163{,}000$ lb

Since 180,000 > 163,000, the stress block extends below the flange far enough to pick up the remaining compression of 180,000 − 163,000 = 17,000 lb.

$$17{,}000 = 0.85 \times 3000 \times 8(a - 2)$$
$$a - 2 = 0.83 \text{ in.}, \qquad a = 2.83 \text{ in.}$$

The resultant $\overline{N}_c$ acts at the centroid of the compression area, which establishes z and leads to $M_u = \overline{N}_t z$ or $\overline{N}_c z$, as indicated in Figs. 2.11*b* and 2.12*b*.

$$\bar{y} = \frac{32 \times 2 \times 1 + 8 \times 0.83 \times 2.41}{64 + 6.64} = \frac{64 + 16}{70.6} = 1.13 \text{ in.}$$

$$z = 12 - 1.13 = 10.87 \text{ in.}$$
$$\overline{M} = 180{,}000 \times 10.87/12{,}000 = 163.3 \text{ k-ft.}$$
$$M_u = 0.9 \times 163.3 = 147 \text{ k-ft.}$$

*Less than 0.85 for $f_c' > 4000$ psi.

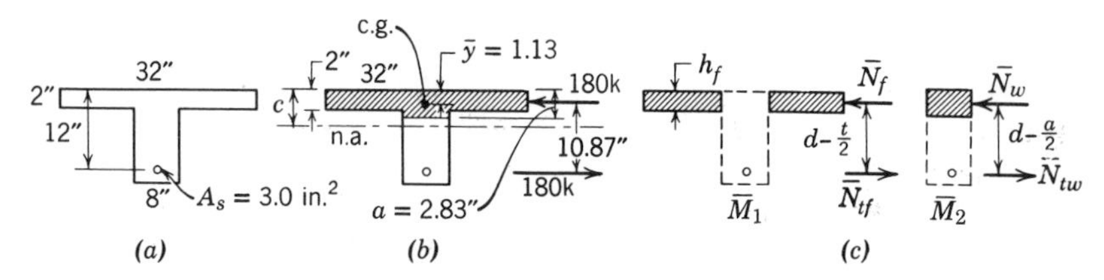

Fig. 2.12. T-beam of Sec. 2.12.

Check against balanced condition:

$$c_b = \frac{87{,}000}{87{,}000 + 60{,}000} \times 12 = 7.10 \text{ in.} \qquad a = 0.85 \times 7.10 = 6.04 \text{ in.}$$

$$\bar{N}_{cb} = 0.85 \times 3000(32 \times 2 + 8 \times 4.04) = 245{,}000 \text{ lb}$$

$$\text{Permissible } \bar{N}_c = 0.75 \times 245{,}000 = 184{,}000 \text{ lb} > 180{,}000 \qquad \text{O.K.}$$

As a convenient procedure the resisting moment is also calculated as the sum of two couples. The compression $\bar{N}_f$* on the outstanding flanges, as shown by the shaded areas in Fig. 2.12*c*, pairs with a corresponding tension to form couple $\bar{M}_1$. The remainder of the tension pairs with the compression $\bar{N}_w$ in the 8-in. web width to form the second couple $\bar{M}_2$.

$$\bar{N}_f = (32 - 8)2 \times 0.85 \times 3000/1000 = 122.7 \text{ k}$$
$$\bar{M}_1 = 122.7(12 - 2/2)/12 = 112.6 \text{ k-ft}$$
$$\bar{N}_w = 8 \times 2.83 \times 0.85 \times 3000/1000 = 57.6 \text{ k}$$
$$\bar{M}_2 = 57.6(12 - 2.83/2)/12 = 50.8 \text{ k-ft}$$
$$\bar{M} = \bar{M}_1 + \bar{M}_2 = 112.6 + 50.8 = 163.4 \text{ k-ft}$$
$$M_u = 0.9 \times 163.4 = 147 \text{ k-ft}$$

The minimum thickness (overall depth) should be checked against Table 2.2 to determine whether deflection is a probable concern. The old standard of $\rho = A_s/bd \lessapprox 0.18f_c'/f_y$ (now discarded from the Code) also gives a rough idea. The corresponding ρ in terms of the M_2 couple alone (Fig.2.12*c*) is often convenient for design tables.

2.13 ANALYSIS OF BEAMS WITH COMPRESSION STEEL

Beams are occasionally restricted in size to such an extent that steel is needed to help carry the compression. Compression reinforcement

*The complete symbol $\bar{N}_{cf}$ will be shortened here to $\bar{N}_f$ because there is no likely confusion. Since in flexure $\bar{N}_c = \bar{N}_t$ it is probable that subscripts *c* and *t* can often be dropped where a particular emphasis is not demanded.

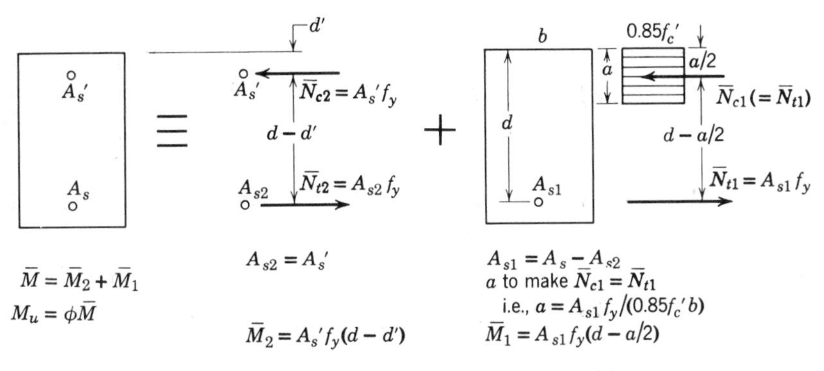

Fig. 2.13. Analysis of double-reinforced beam if all steel yields.

is also excellent for reducing the longtime deflection. For analysis it is convenient to consider two couples M_1 and M_2 similar to the approach used for the T-beam in Fig. 2.12*c*.

If the member is relatively deep the M_2* steel couple based on A_s' and A_{s2} should be evaluated first. Strictly, the extra compression available because of A_s' must also account for the loss of compression concrete displaced by A_s'. Then

$$\overline{M}_2 = A_s'(f_y - 0.85f_c')(d - d')$$
$$A_{s2} = A_s'(f_y - 0.85f_c')/f_y$$

If A_s' is small and f_c' is low, it is often close enough to neglect the displaced concrete, as follows (in Fig. 2.13):

$$\overline{M}_2 = A_s'f_y(d - d') \qquad \text{and} \qquad A_{s2} = A_s'$$

The remaining tension steel is the basis for the M_1 couple.

$$A_{s1} = A_s - A_{s2} \qquad \overline{M}_1 = A_{s1}f_y(d - a/2)$$
$$M_u = \phi(\overline{M}_2 + \overline{M}_1) = 0.9\ (\overline{M}_2 + \overline{M}_1)$$

The depth of the stress block a is calculated from $\overline{N}_{c1} = \overline{N}_{t1}$ or $0.85f_c'ba = A_{s1}f_y$.

If the member is shallow, especially a slab, there is a reasonable chance that the strain at the A_s' level will not develop f_y in compression and this must be verified early in the analysis.

1. Assume $A_{s2} \doteq A_s'$. Then $A_{s1} = A_s - A_s'$.
2. Calculate $a = A_{s1}f_y/(0.85f_c'b)$.

*The subscript 2 is used because the *usual* problem is in *design* where the M_2 couple is the second operation. It is well not to vary the numbering between the two cases.

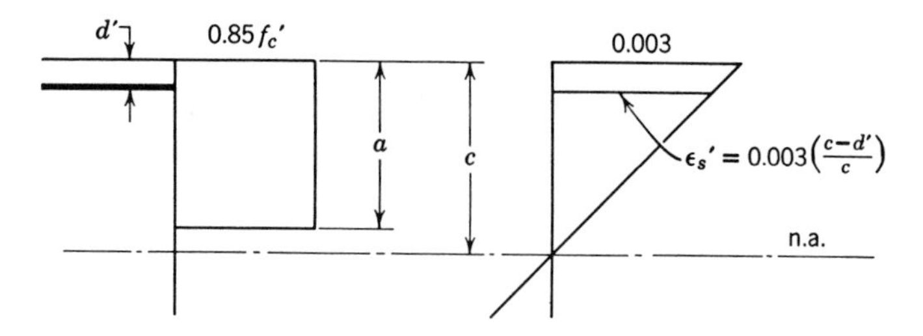

Fig. 2.14. Check on compression steel stress.

3. Calculate c, using the ratios between c and a at the close of Sec. 2.3.
4. Use the strain triangle of Fig. 2.14 to calculate

$$\epsilon_s' = 0.003(c - d')/c$$

5. If $E_s\epsilon_s' > f_y$, proceed to evaluate M_1 and M_2, using f_y with A_s'.

If $f_s' < f_y$ it becomes more important to evaluate $\overline{N}_{c2}$ and A_{s2} considering the displaced concrete, which in turn modifies the original assumed A_{s1}—an iteration problem in theory. However, the second example below sets up a possible quadratic equation that can be used.

Deflections are determined by top and bottom strains on the beam and hence can be measured by the $\overline{M}_1$ couple developed by A_{s1} and the concrete, ignoring the $\overline{M}_2$ couple developed by A_s' and A_{s2}.

2.14 DOUBLE-REINFORCED BEAM EXAMPLES

(*a*) A rectangular beam has $b = 11$ in., $d = 20$ in., $A_s = 5\text{-}\#11 = 7.80$ in.^{2}, $f_c' = 3000$ psi, Grade 40 steel, $A_s' = 2\text{-}\#11 = 3.12$ in.^{2}, with d' (cover to center line of A_s') of 2 in. Calculate the ultimate moment.

Solution

Neglecting concrete displaced by A_s', consider A_{s2} equal to A_s', namely, 3.12 in.^{2}

$$\begin{aligned}\overline{M}_2 &= A_s'f_y(d - d') \\ &= 3.12 \times 40{,}000(20 - 2)/12{,}000 = 187 \text{ k-ft}\end{aligned}$$

The remainder of the tension steel, $A_{s1} = A_s - A_{s2} = 7.80 - 3.12 = 4.68$ in.^{2}, works with the concrete, as in any rectangular beam, to develop an ultimate moment $\overline{M}_1$. Since $\overline{N}_t = \overline{N}_c$,

$$4.68 \times 40{,}000 = 0.85 \times 3000 \times 11a$$

$$a = 6.70 \text{ in.} < 0.436d = 8.72 \text{ in. (See Table 2.1)}$$

This check verifies that the beam has A_{s1} less than 0.75 of the balanced ratio for a simple rectangular beam. Extra A_{s2} and A_s' do not themselves lead to a brittle failure since this steel forms an auxiliary "beam" exhibiting great toughness, provided A_s' develops its f_y. Hence the author prefers this simple check on A_{s1} over the stricter and more complex Code requirement that the total ρ be limited to $0.75\rho_b$ with ρ_b computed for the entire double-reinforced beam. The legal Code check, however, considers that A_s' increases also the balanced moment capacity, the balanced M_2 steel couple then adding its tension steel to A_{sb} of the simple rectangular beam. The Code limit could then be written as $0.75\ (A_{s2b} + A_{sb}) = 0.75(A_s' + A_{sb})$, the last form being true only if f_s' reaches f_y. The last form is equivalent to $0.75A_s' + \rho bd$ where ρ is the maximum from Table 2.1. Applied to this example, this maximum permissible total A_s is $0.75 \times 3.12 + 0.0278 \times 11 \times 20 = 2.34 + 6.12 = 8.46$ in.², which is still satisfactory since only 7.80 in.² is actually used.

The stress in A_s' will now be checked.

Specified $a/c = 0.85$

$$c = 6.70/0.85 = 7.88 \text{ in.}$$
$$\epsilon_s' = 0.003 \times 5.88/7.88 = 0.00225$$
$$\epsilon_y = 40{,}000 \div 29{,}000{,}000 = 0.00138 < 0.00225$$

Therefore $f_s' = f_y$ and method is O.K.

$$z_1 = 20 - 6.70/2 = 16.65 \text{ in.}$$
$$\overline{M}_1 = A_{s1} f_y z_1 = 4.68 \times 40{,}000 \times 16.65/12{,}000 = 260 \text{ k-ft}$$
$$\overline{M} = \overline{M}_1 + \overline{M}_2 = 260 + 187 = 447 \text{ k-ft}$$
$$M_u = \phi\overline{M} = 0.9 \times 447 = 402 \text{ k-ft}$$

(*b*) A rectangular beam has $b = 11$ in., $d = 20$ in., $A_s = 3\text{-}\#11 = 4.68$ in.², $f_c' = 5000$ psi, Grade 60 steel, $A_s' = 1\text{-}\#11 = 1.56$ in.², with $d' = 2$ in. Calculate M_u.

Solution

Again neglecting concrete displaced by A_s', $A_{s2} = A_s' = 1.56$ in.²

$$A_{s1} = A_s - A_{s2} = 4.68 - 1.56 = 3.12 \text{ in.}^2$$
$$a = A_{s1} f_y/(0.85 f_c' b) = 3.12 \times 60/(0.85 \times 5 \times 11) = 4.00 \text{ in.}$$

From Sec. 2.3 $a/c = 0.85 - 0.05 = 0.80$

$$c = 4.00/0.80 = 5.00 \text{ in.}$$
$$\epsilon_s' = 0.003 \times 3.00/5.00 = 0.00180$$
$$\epsilon_y = 60/29{,}000 = 0.00207 > 0.00180,\ f_s' < f_y$$

Since $f_s' < f_y$, it is easy also to recognize the concrete displaced by A_s'. The compression concrete is valued at $0.85 \times 5 = 4.25$ ksi. An algebraic solution will be set up later but a trial-and-error technique seems equally efficient.

Try $f_s' = 52$ ksi (about f_y times the ratio ϵ_s'/ϵ_y)

$$\text{Effective } f_s'' = 52 - 4.25 = 47.8 \text{ ksi}$$
$$A_s'f_s'' = A_{s2}f_y, \qquad 1.56 \times 47.8 = A_{s2} \times 60$$
$$A_{s2} = 1.24 \text{ in.}^2$$

$$\text{Code max. } A_s = 0.75 \times 1.24 + 0.0252 \times 11 \times 20$$
$$= 0.93 + 5.54 = 6.47 \text{ in.}^2 \gg 4.68 \qquad \text{O.K.}$$

The author prefers the check on A_{s1}:

$$A_{s1} = 4.68 - 1.24 = 3.44 \text{ in.}^2$$
$$a = 3.44 \times 60/(0.85 \times 5 \times 11)$$
$$= 4.32 \text{ in.} < 0.355d = 7.10 \text{ in.} \qquad \text{O.K.}$$
$$c = 4.32/0.80 = 5.40 \text{ in.}$$
$$\epsilon_s' = 0.003 \times 3.40/5.40 = 0.00189$$
$$f_s' = 0.00189 \times 29{,}000 = 54.8 \text{ ksi vs. 52 ksi assumed}$$

This is probably close enough. Take $f_s' = 55$ ksi as accurate. This leads to $\overline{M}_2 = 118$ k-ft, $A_{s2} = 1.32$ in.^{2}, $A_{s1} = 3.36$ in.^{2}, $a = 4.32$ in., $\overline{M}_1 = 300$ k-ft, $\overline{M} = 418$ k-ft., and $M_u = 376$ k-ft.

The solution can be set up algebraically after it is discovered that $\epsilon_y > \epsilon_s'$, as shown in Fig. 2.15.

$$\overline{N}_{c1} + \overline{N}_{c2} = \overline{N}_t$$
$$0.85 \times 5 \times 11(0.80c) + 1.56[0.003 \times 29{,}000(c - 2)/c - 4.25]$$
$$= 4.68 \times 60$$
$$c = 5.41 \text{ in.}, a = 0.8 \times 5.41 = 4.32 \text{ in.}$$
$$\overline{M}_1 = 0.85 \times 5 \times 11 \times 4.32(20 - 0.5 \times 4.32)/12 = 299 \text{ k-ft}$$
$$\overline{M}_2 = 1.56[0.003 \times 29{,}000(5.41 - 2)/5.41 - 0.85 \times 5](20 - 2)/12$$
$$= 118.6 \text{ k-ft}$$
$$M_u = \phi(\overline{M}_1 + \overline{M}_2) = 0.9(299 + 118.6) = 376 \text{ k-ft}$$

2.15 TREATMENT FOR T-BEAMS IN 1963 CODE

It is possible and sometimes convenient to consider the overhanging part of the flange, shown dotted in Fig. 2.16, as removed and replaced by

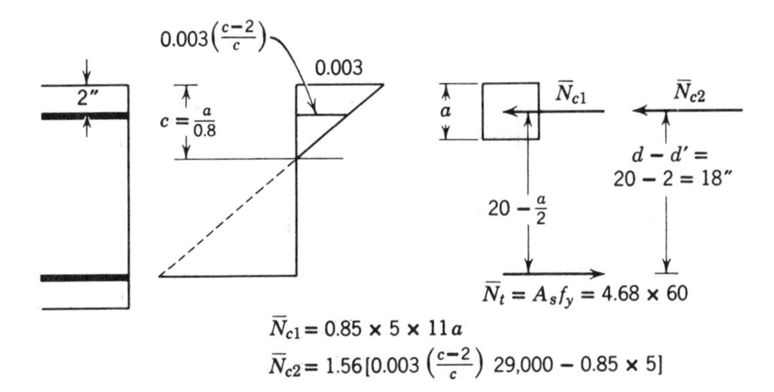

Fig. 2.15. Basic relations in algebraic solution for Sec. 2.14*b*.

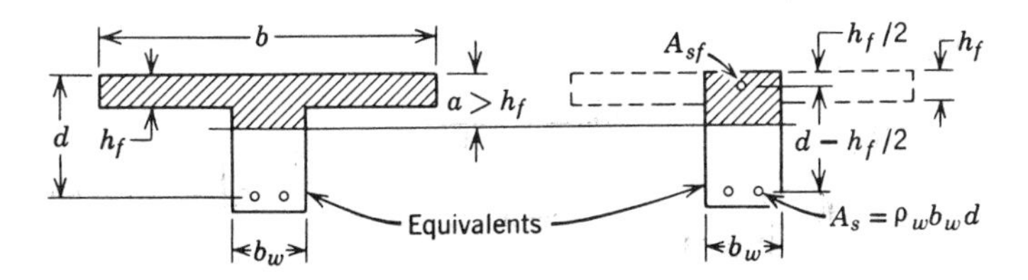

Fig. 2.16. T-beam treated as a "double reinforced" beam.

an imaginary area of compression steel A_{sf} having the same compression strength and located opposite the middle of the overhanging flange. Thus

$$A_{sf}f_y = 0.85f_c'h_f(b - b_w)$$
$$A_{sf} = 0.85f_c'h_f(b - b_w)/f_y{}^*$$

The resulting "double-reinforced rectangular beam" shown in Fig. 2.16*b* is analyzed exactly as indicated in Sec. 2.13 and Fig. 2.13 on the basis of two couples $\overline{M}_f$ (similar to the A_s' couple $\overline{M}_2$) and $\overline{M}_w$(similar to the rectangular beam couple $\overline{M}_1$).

This form of solution points out the fact that deflections are fully determined by the $\overline{M}_w$ couple, just as already discussed at the end of Sec. 2.13 for the $\overline{M}_1$ couple with the double-reinforced beam.

*Since this is an imaginary area of compression steel, no concrete is actually displaced and no deduction from f_y should be made for displaced concrete; nor is it necessary to check whether the strain at the level of A_{sf} is adequate to develop the f_y used.

2.16 ANALYSIS OF SPECIAL BEAM SHAPES

The general analysis presented in Secs. 2.5 and 2.7 and adapted to the T-beam in Secs. 2.11 and 2.12 is adaptable to any shape of cross section. The strain triangles can always be used to establish the balanced section neutral axis.* The stress block depth may be fixed from the a/c ratios of Sec. 2.3, and $\overline{N}_{cb}$ evaluated even for very irregular member outlines. The usable $\overline{N}_c$ or $\overline{N}_t$ should be restricted to 0.75 $\overline{N}_b$ as an upper limit. For this maximum $\overline{N}_t$, or such smaller $A_s f_y$ as may exist, the required depth of stress block a can be established, the resultant compression located, and the moment couple calculated. Since a also establishes c, values of $f_s < f_y$ can be taken into account when specific bar locations will not develop strains as large as ϵ_y.

2.17 TABLES AND CHARTS

Sample helpful tables and charts are included in Appendix D for rectangular beams and slabs, without compression steel. Tables D.1 and D.2 augment Table 2.1, Table D.1 giving k_m, c/d, a/d, and z/d for several common materials combinations. These are in terms of $\omega(= \rho f_y/f_c')$ and also in terms of ρ. The use of ω permits the grouping of data into a very compact form, with k_m tabulations here in psi units, over the entire usable range. Table D.2 shows values of coefficients which can be multiplied by f_c' to give $\bar{k}$ values, or by $0.9f_c'$ to give k_m values, each quite accurate in terms of ω. For example, the coefficient for $\omega = 0.015$ is on the second line in the sixth column of numbers, leading to $\bar{k} = 0.0149f_c'$ or $k_m = 0.0149 \times 0.9f_c'$. In using Table D.2 one must check ρ to be sure it is less than $0.75\rho_b$, since the limiting k_m value varies for different strengths of materials; Table D.1 automatically stops at that point.

Values of $\bar{k}$ are given graphically in Fig. D.1, along with the limiting $\omega_{\max}$ for various steel grades indicated. For a limited number of cases Fig. D.2 gives k_m values in terms of percentage of steel, 100ρ, along with values of z/d and a/d, and maximum values. These curves show the general influence of the materials and steel ratios.

In analysis the above tables and charts can be used by computing the actual ρ or ω, reading from the table or chart the corresponding value of k_m or $\bar{k}$, and then calculating $M_u = k_m bd^2$ or $\overline{M} = \bar{k}bd^2$. For the example in Sec. 2.9a:

$$\rho = 3 \times 0.79/(11 \times 20) = 0.0108$$

*Some uncertainty can arise when there are several layers or depths of tension reinforcing. Rather arbitrarily the author uses the yield strain on the steel most distant from the compression face to establish his balanced section.

Figure D.2, for $f_c' = 3000$ psi, $f_y = 40$ ksi, and $100\rho = 1.08$, shows $k_m =$ 350 psi, or

$$M_u = k_m bd^2 = 350 \times 11 \times 20^2/12{,}000 = 128.5 \text{ k-ft}$$

The one to two percent difference from the 131 k-ft found by basic calculation is a measure of the accuracy of curve reading. Alternatively, $\omega = \rho f_y/f_c' = 0.0108 \times 40{,}000/3000 = 0.144$, and Fig. D.1 shows $\bar{k} = 400$, $k_m = 0.9 \times 400 = 360$. Likewise, Table D.1 by interpolation for ρ gives $k_m = 347 + 22(0.0003/0.0008) = 355$. Finally, a third alternate, Table D.2 gives a coefficient of 0.1318, leading to $\bar{k} = 0.1318 \times 3000 = 395$, $k_m = 0.9 \times 395 = 356$. Any of the four values are reasonably close; only one need be found.

In the example of Sec. 2.9b, the 5- #11 provide more than $\rho = 0.75\rho_b$. Table D.2 will not automatically note this, but the other table and figures will.

The ACI Handbook given in the references at the end of Chapter 3, although primarily directed toward design, has many charts and tables that are also useful in analysis.

SELECTED REFERENCES

1. A. H. Mattock and L. B. Kriz, "Ultimate Strength of Nonrectangular Structural Concrete Members," *Jour. ACI*, Jan. 1961; *Proc.*, *57*, p. 737.

PROBLEMS

General Note: Tables 1.1 and 1.2 in Sec. 1.9 list bar sizes, areas, and strengths. Necessary load factors are given in Sec. 1.12*e* (usually Code Eq. 9.1 unless otherwise specified) and ϕ factors in Sec. 1.12*f*. Beam constants for many cases are given in Table 2.1.

Prob. 2.1. If $f_c' = 3000$ psi and steel is Grade 60, find M_u, the moment capacity of the beam of Fig. 2.17 if the A_s is:

(*a*) 2- #8. (*c*) 4- #9.
(*b*) 4- #8. (*d*) 4- #10.

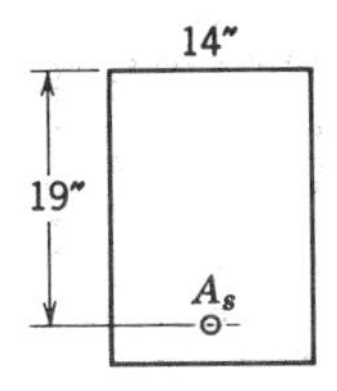

Fig. 2.17. Beam for Prob. 2.1.

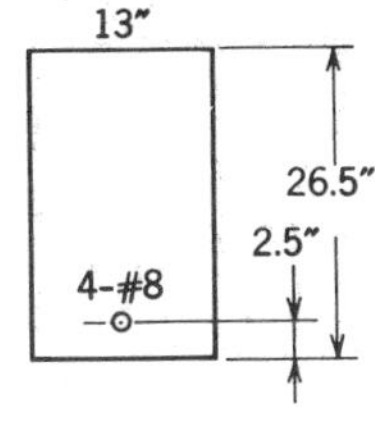

Fig. 2.18. Beam of Prob. 2.2.

Prob. 2.2. For $f_c' = 4000$ psi find moment capacity of the beam of Fig. 2.18, assuming:

(*a*) Grade 40 steel
(*b*) Grade 60 steel

Prob. 2.3. If $f_c' = 4000$ psi and steel is Grade 50, calculate the moment capacity of the beam of Fig. 2.19, assuming the steel as follows:

(*a*) 3- #8. (*c*) 4- #11.
(*b*) 6- #8. (*d*) 6- #11.

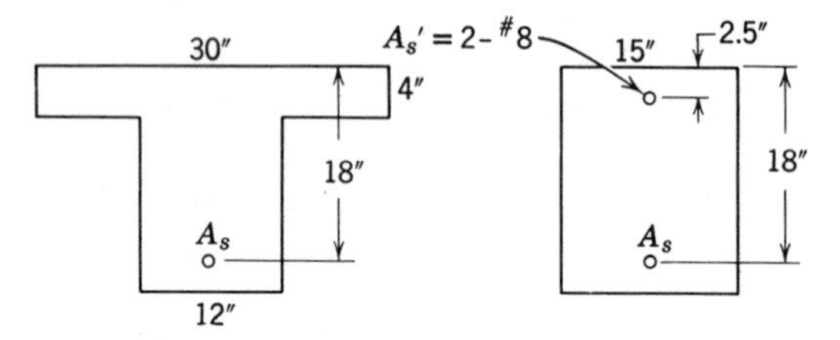

Fig. 2.19. Beam of Prob. 2.3.

Fig. 2.20. Beam of Prob. 2.4.

Prob. 2.4. If $f_c' = 4000$ psi and steel is Grade 40, calculate M_u on the beam of Fig. 2.20, assuming A_s is:

(*a*) 3- #8. (*c*) 4- #11.
(*b*) 6- #8. (*d*) 6- #11.

Prob. 2.5. Check the adequacy of the beam of Fig. 2.21 in flexure assuming $f_c' = 5000$ psi, Grade 40 bars, service live load of 1000 plf and dead load (including beam weight) of 500 plf. (*Suggestion:* Compare calculated M_u with required M_u.)

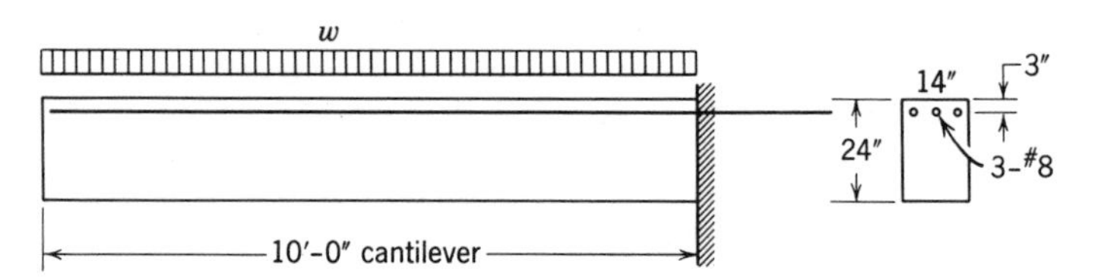

Fig. 2.21. Beam for Prob. 2.5.

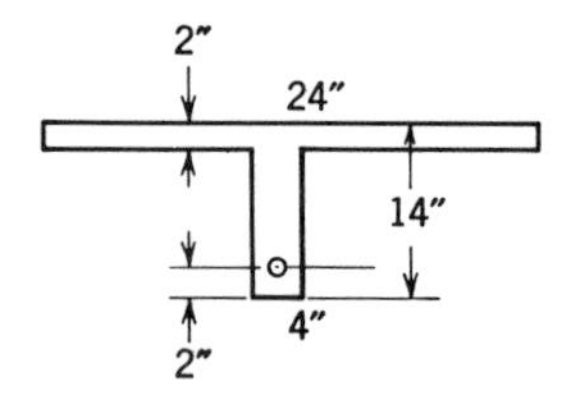

Fig. 2.22. Joist of Prob. 2.6.

Prob. 2.6. Find the moment capacity of the joist of Fig. 2.22 if $f_c' =$ 4000 psi and A_s is 2- #6 bars of Grade 60.

Prob. 2.7. Find the allowable moment on a 1-ft strip of the slab of Fig. 2.9 if the bars are made #7 at 7 in. on centers and $d = 5$ in., $f_c' = 3000$ psi, Grade 60 steel.

Prob. 2.8. Find allowable moment for the beam of Fig. 2.23 if $f_c' = 5000$ psi and Grade 60 steel is used. Ignore the lack of symmetry.

(*a*) Omitting the 2- #8 bars.
(*b*) Including the 2- #8 bars.

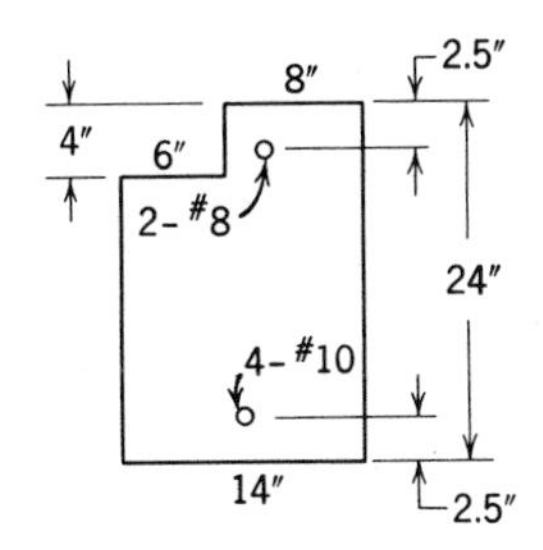

Fig. 2.23. Irregular beam for Prob. 2.8.

Prob. 2.9. Prob. 2.8 except consider the beam of Fig. 2.24.

(*a*) Omitting the 2- #9.
(*b*) Including the 2- #9.

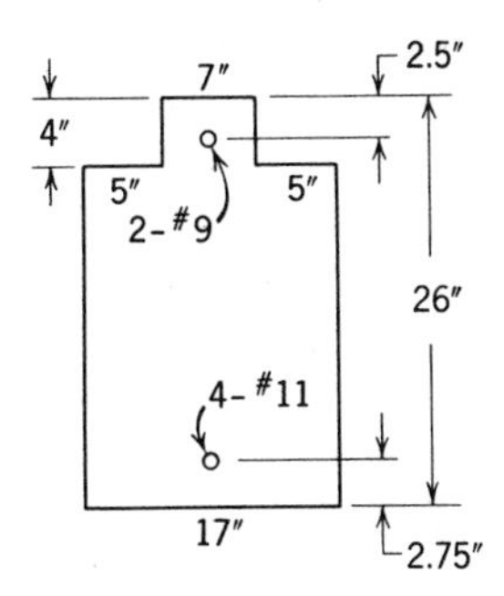

Fig. 2.24. Irregular beam for Prob. 2.9.

Prob. 2.10. If $f_c' = 5000$ psi and steel is Grade 60, what is the ultimate moment capacity of the beam of Fig. 2.25 when each leg is reinforced with: (*a*) 1- #7. (*b*) 1- #10.

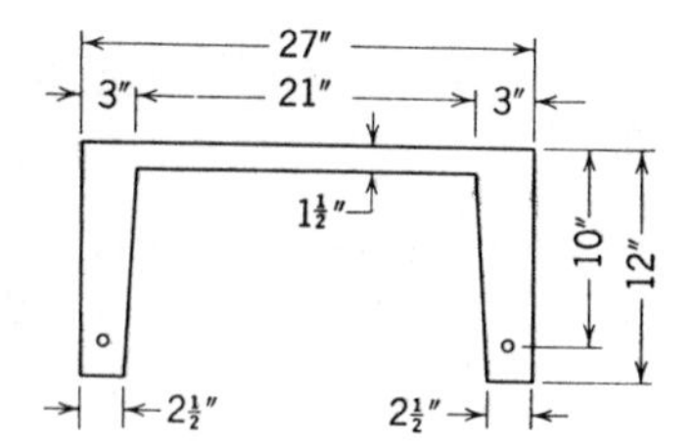

Fig. 2.25. Beam for Prob. 2.10.

Prob. 2.11. If the precast section of Fig. 2.26 has $f_c' = 5000$ psi, Grade 60 steel, 1- #8 bar in each leg, calculate the ultimate moment capacity. Flange $h_f = 1.5$ in.

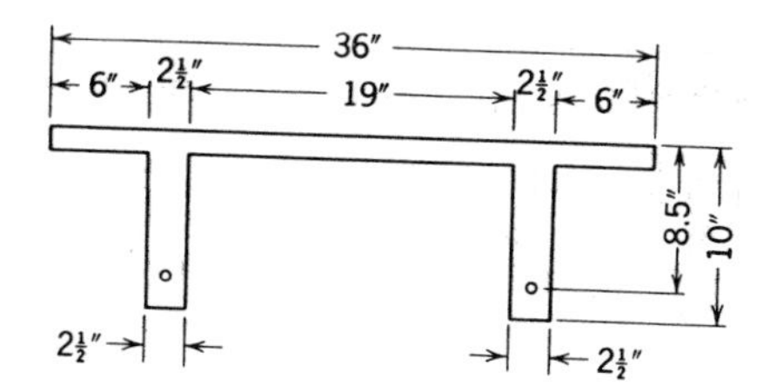

Fig. 2.26. Precast section for Prob. 2.11.

Prob. 2.12.

(a) Calculate the ultimate moment capacity (Fig. 2.27) of a strip of slab 12 in. wide if $f_c' = 4000$ psi and the steel is Grade 60.

(b) What would be the service or working live load per square foot if the slab were used on a 10-ft simple span? Use the Code load factors. Concrete weight may be taken at 150 pcf which includes the weight of reinforcement.

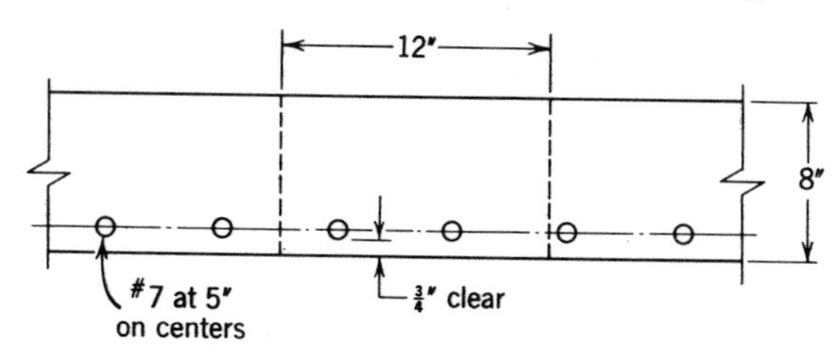

Fig. 2.27. Slab for Prob. 2.12.

Prob. 2.13. A rectangular beam has $b = 15$ in., $d = 18$ in., $f_c' = 3000$ psi, Grade 40 steel, $A_s = 6$- #11.

(a) Calculate the moment capacity.

(b) Calculate the ultimate moment if one adds $A_s' = 3$- #11 with cover d' to center of steel $= 2.5$ in.

3
Design for Flexure

3.1 GENERAL PROCEDURES

Analysis and design are related problems that nevertheless involve fundamental differences and require different techniques. It is essential that the student clearly understands these differences.

In analysis for allowable loads or moments the engineer deals with given beams, known both as to dimensions and steel. Each such case has a neutral axis unique to those particular dimensions and not dependent on the engineer's desires.

In design, loads and material strengths are known and some or all the dimensions remain to be fixed. Here the designer has some control over the location of the neutral axis. He can shift it where he wants it, to the extent that changes in dimensions can alter the balance between tension and compression areas.

After loads are known and the layout of the structure or structural element has been established,* the maximum bending moment can be determined. The member design for moment involves three separate steps:

1. Choice of beam cross section.
2. Choice of reinforcing steel at point or points of maximum moment.
3. Determination of points where bars are no longer needed for moment, that is, points for bending or stopping bars.

*The choice of proper design loads and the layout of a structure to support these loads efficiently is a major part of the design. The choice of loads and the general problem of framing layout are not covered in this text. Chapter 8 has a brief discussion of beam-and-girder framing and the proper arrangement of loading for maximum moment on continuous one-way slabs and beams. Various types of slab construction are discussed in Chapter 10.

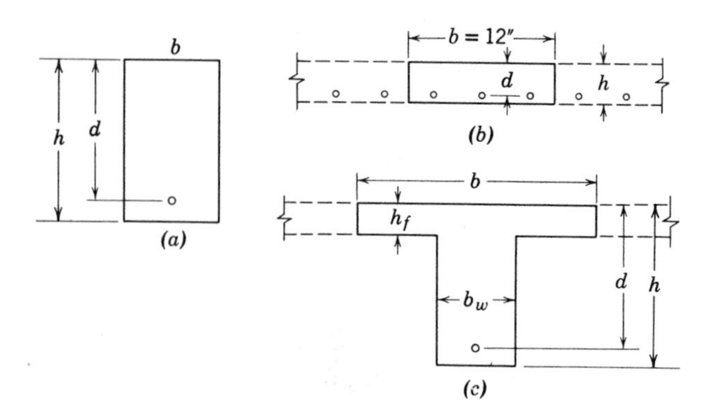

Fig. 3.1. Typical reinforced concrete beams. (*a*) Rectangular. (*b*) Slab. (*c*) T-beam.

Often other factors than moment determine the choice of the beam cross section, such as shear (Chapter 4). Such interrelationships will be discussed later.

In rectangular beams (Fig. 3.1*a*), both width b and depth d to steel are to be chosen. In slabs (Fig. 3.1*b*), the strip width b is fixed and only the depth d or the over-all thickness h can be selected. In T-beams (Fig. 3.1*c*), the flange thickness h_f and flange width b are determined, usually by a slab already designed, thus leaving the stem width b_w and the depth d to the steel to be selected. The choice of b_w and d is not usually determined by the moment alone.

In double-reinforced beams the compression steel is required because b and d have been chosen smaller than the normal requirements or because it is desirable to reduce creep deflections. However, when compression steel in definite amounts is desired for protection against deflection or otherwise, the size of double-reinforced beams can be based on moment requirements.

For any beam shape the design of steel, once the moment and beam size are known, is dependent on a proper choice of the arm z. In rectangular beams (and slabs) and in double-reinforced beams, the analytical solution for z is feasible, but the resulting quadratic equation is usually not conducive to an accurate slide-rule solution. Hence the author suggests that the cut-and-try procedure which is necessary for T-beams and irregular shapes be used for all shapes. The value of z is first estimated, the corresponding approximate A_s is calculated, and from this A_s a better value of z is established; in turn, this better z leads to a better and usually satisfactory value of A_s. Convergence is quite rapid.

The bending or stopping of bars is determined by the maximum moments to be resisted at various points along the beam. Bending of bars will be discussed separately from choice of beam and steel (Sec. 3.10).

3.2 RECTANGULAR BEAMS—MOMENT DESIGN

The economical dimensions for a rectangular beam are not sharply defined. A shallow beam is expensive because of the weight of steel required and on long spans it may run into deflection problems. A deep beam is more economical of steel but costs more for side forms and uses up headroom, which means greater story heights. The ratio of over-all depths to span listed in Table 2.2 in Sec. 2.8 (copied from Code Table 9.5*a*) can serve as a guide.

This chapter will deal with beam sizes established in various manners. It is almost wholly concerned with strength, slightly concerned with deflection, and not concerned with over-all economy. However, it will emphasize methods which conserve steel for the particular beam size chosen.

The maximum ρ ratios developed in Sec. 2.7 (Table 2.1) are absolute limits on the steel which may be considered effective because of danger from brittle-type failures.

The minimum reinforcement of $\rho = 200/f_y$ (Code 10.5.1) is necessary to insure that the member does not lose some strength when it first cracks. This requirement does not apply if the A_s used is 4/3 that required by analysis, because this implies a lowered probability that the cracking load will ever be reached, in other words, adequate safety. Slabs of uniform thickness are also exempt, but the author questions this exemption where shears are near the limiting value. A small ρ and a maximum shear make a bad combination.

3.3 RECTANGULAR BEAM EXAMPLES

(*a*) Design a minimum practical beam for service moments of 50 k-ft for D.L. and 100 k-ft for L.L. using $f_c' = 3000$ psi and Grade 60 steel.

Solution

The load factors of Code Eq. 9.1 (see Sec. 1.12*e*) are assumed to govern:

$$M_u = 1.4M_D + 1.7M_L = 1.4 \times 50 + 1.7 \times 100 = 240 \text{ k-ft}$$
$$\overline{M} = M_u/\phi = 240/0.9 = 267 \text{ k-ft}$$

From Table 2.1, for these materials, maximum $\bar{k} = 783$. The size of the beam can be found by equating $\bar{k}bd^2$ to the external $\overline{M}$ (or $k_m bd^2$ to M_u).

$$\overline{M} = \bar{k}bd^2 = 783\ bd^2 = 267 \times 1000 \times 12$$
$$\text{Reqd. } bd^2 = 4100 \text{ in.}^3$$

This requirement can be met by a wide range of sizes, for example:

If $b = 7.5$ in., $d = \sqrt{548} = 23.4$ in. $b = 10,\quad d = \sqrt{410} = 20.2$
$b = 8,\quad d = \sqrt{514} = 22.7$ $b = 11,\quad d = \sqrt{373} = 19.3$
$b = 9,\quad d = \sqrt{457} = 21.4$ $b = 11.5,\quad d = \sqrt{357} = 18.9$
$b = 9.5,\quad d = \sqrt{432} = 20.8$ $b = 12,\quad d = \sqrt{342} = 18.5$
$b = 13,\quad d = \sqrt{316} = 17.8$

The 7.5-in., 9.5-in., and 11.5-in. widths are listed because these are convenient widths for form lumber; forms are one of the most costly items in reinforced concrete construction. The narrowest beam shown may be too narrow to accommodate the reinforcing steel, it uses up the most headroom, and requires the greatest area of forms; but it uses the smallest volume of concrete and steel. A d/b ratio from 1.0 or 1.5 to 2.5 or 3.0 is usually considered in the desirable range, with the larger ratios for the bigger beams. Since deep beams often require increased story heights, more wall and partition height, and thus other costs than the cost of the beam, there is no simple textbook answer to the *best* depth. There are a number of satisfactory choices possible. For the purpose of this problem, $b = 9.5$ in. will be chosen, although a wider beam might be needed for a lower strength steel, with more bars then necessary.

If not exposed to the weather or the ground such a beam requires 1.5 in. of clear cover (Code 7.14.1.1) over the steel. The steel typically includes some stirrups as reinforcement as shown in Fig. 3.2, and #3 or #4 bars would be a reasonable size for this beam, say #3 bar stirrups. Then the cross section of the beam would appear as shown in Fig. 3.2*a*. The minimum over-all depth would be 20.8 + 0.5 (assuming bar diam. $d_b = 1$ in.) + 0.38 (stirrup) + 1.5 (cover) = 23.2 in. This is an impractical dimension to give a carpenter for forms and the only change permitted is to greater dimensions since maximum $\bar{k}$ was used originally.

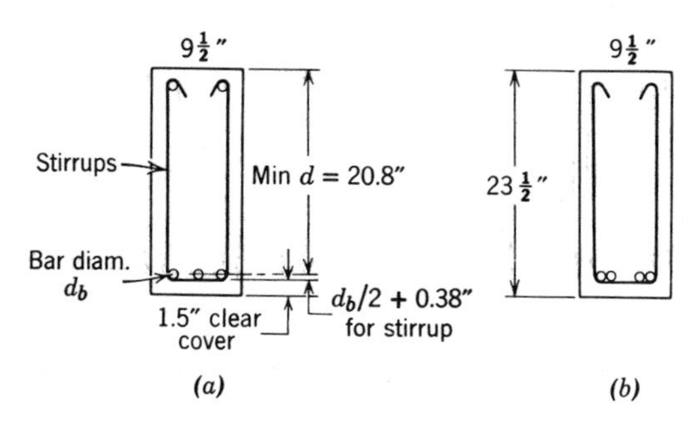

Fig. 3.2. Dimensions for beam of Sec. 3.3. (*a*) Minimums. (*b*) Final.

USE 9.5 in. by 23.5 in. (overall)

$d = 23.5 - 1.5 - 0.38 - 0.5$ (assumed $d_b/2$) $= 21.1$ in.

With $d = 21.1$ in., the beam is a little deeper than the theoretical minimum and thus has a greater z than the minimum, which means N_t and the balancing compression block depth a will be slightly reduced. Note that Table 2.1 shows maximum $a/d = 0.378$, or $a = 0.378 \times 20.8 = 7.84$ in.

Assume $a = 7.75$ in., $z = 21.1 - 0.5 \times 7.75 = 17.2$ in.

$$\overline{M} = A_s f_y z,\ A_s = 267 \times 12/(60 \times 17.2) = 3.10 \text{ in.}^2$$

$$a = A_s f_y/(0.85 f_c' b) = 3.10 \times 60/(0.85 \times 3 \times 9.5) = 7.68 \text{ in.}$$

This indicates the original guess is adequate, but this can be proved by completing the cycle:

$$z = 21.1 - 0.5 \times 7.68 = 17.3 \text{ in.}$$

$$A_s = 267 \times 12/(60 \times 17.3) = 3.09 \text{ in.}^2$$

The choice of bars cannot be final without checks on bar development (Chapter 5), but a selection can be made tentatively:

2- #11 $= 3.12$ in.^{2}	Development apt to be troublesome. d_b is 1.41 in. and this larger d_b would reduce available d by 0.2 in. or 1+%.
3- #10 $= 3.81$ in.^{2}	Wasteful
4- #9 $= 4.00$ in.^{2}	Wasteful, but 3 bars would be 3% short
4- #8 $= 3.16$ in.^{2}	O.K.
6- #7 $= 3.60$ in.^{2}	Wasteful
7- #6 $= 3.08$ in.^{2}	O.K. except probably requires two layers of bars, which requires increase in over-all beam depth

The clear spacing of bars in a layer must not be less than the nominal bar diameter, 1 in., or 4/3 of the nominal maximum aggregate size (Code 7.4.1 and 3.3.2).

Try 4- #8 for spacing. Clear width inside stirrups is $9.5 - 2 \times 1.5$ cover $- 2 \times 0.38$ stirrups $= 5.75$ in. The four bars cannot be spaced individually, since this would require 4×1 (for bars) $+ 3 \times 1$ (for spaces) $= 7$ in., assuming $\frac{3}{4}$ in. aggregate is used. However, if development is not too troublesome, two bars might be bundled together in each corner of the stirrups as shown in Fig. 3.2*b*.

Bars up to four in number can be bundled together under special limitations stated in Code 7.4.2, which include limitation to #11 maximum size bars, provision of ties or stirrups enclosing the bundled bars, etc. For spacing purposes the equivalent diameter of two bundled bars can be found from the diameter of an imaginary bar having the area of the two bundled bars, that is, $2 \times 0.79 = 1.58$ in.^{2}

$$d_b = \sqrt{1.58/0.786} = 1.42 \text{ in.}$$

The clear spacing between the two pairs of bars is $5.75 - 4 \times 1 = 1.75$ in., which exceeds the 1.42 in. minimum. Subject to development the arrangement is acceptable and economical. Deflections must be investigated (Chapter 6).

The student should note that the entire design process could be very compactly presented. This presentation has been greatly expanded by explanations.

(*b*) Redesign the beam in (*a*) to limit the reinforcement ratio to $\rho = 0.18f_c'/f_y$, for which constants were developed in Fig. 2.8.

Solution

Table 2.1 and Fig. 2.8 show $\bar{k} = 0.161f_c' = 0.161 \times 3000 = 483$

$\overline{M}$ as before $= 267$ k-ft
$\overline{M} = 267 \times 12{,}000 = 483\, bd^2$
$bd^2 = 267 \times 12{,}000/483 = 6640$
If $b = 9.5$ in., $d = \sqrt{700} = 26.5$ in.

Overall depth $h = 26.5 + 0.5$ in. (for $d_b/2$) $+ 0.38$ stirrup $+ 1.5$ in. cover $= 28.9$ in.

If $0.18f_c'/f_y$ is a judgment factor, d can be *either* decreased or increased at the engineer's option. Unless it infringes on headroom,

USE $b = 9.5$ in., $h = 29$ in., $d = 26.6$ in.

For the assumed $\rho = 0.18f_c'/f_y$, $a = 0.212d = 0.212 \times 26.6 = 5.64$ in.
Try $a = 5.6$ in., $z = 26.6 - 0.5 \times 5.6 = 23.8$ in.

$A_s \doteq 267 \times 12/(60 \times 23.8) = 2.25$ in.^{2}
$a = 2.25 \times 60/(0.85 \times 3 \times 9.5) = 5.56$ in. vs. 5.6 in. O.K.

Subject to proper bar development (Chap. 5),

USE 3- #8 $= 2.37$ in.^{2}

Minimum inside stirrup width required for bars singly spaced is 3 bars at 1 in. plus two clear spaces at 1 in. or a total of 5 in. vs. 5.75 in. available. O.K.

(*c*) For architectural reasons the beam in (*a*) and (*b*) is to be made 12 in. wide by 25 in. deep over-all. Find the necessary steel, assuming D.L. M unchanged.

Solution

Available $d = 25 - 1.5$ cover $- 0.38$ stirrup $- 0.5$ for $d_b/2 = 22.6$ in. If the preceding examples were not available, it would be necessary first to establish that the beam was adequate without compression steel.

Actual $\bar{k} = \bar{M}/bd^2 = 267 \times 12{,}000/(12 \times 22.6^2) = 521$. Table 2.1 shows the maximum $\bar{k}$ as 783. Therefore the beam can be much underreinforced, but one has no accurate value of a to work from, although it will usually fall in the range from $0.1d$ to $0.3d$ (still more for minimum size beams).

$$\begin{aligned} \text{Try } a &= 0.2 \times 22.6 = 4.52 \text{ in.} \\ z &= 22.6 - 2.26 = 20.3 \text{ in.} \\ A_s &\doteq 267 \times 12/(60 \times 20.3) = 2.63 \text{ in.}^2 \\ a &= 2.63 \times 60/(0.85 \times 3 \times 12) = 5.17 \text{ in.} \end{aligned}$$

The larger a will lower z and tend to increase A_s and a still further.

$$\begin{aligned} \text{Try } a &= 5.3 \text{ in.}, z = 22.6 - 2.65 = 19.9 \text{ in.} \\ A_s &\doteq 267 \times 12/(60 \times 19.9) = 2.68 \text{ in.}^2 \\ a &= 2.68 \times 60/(0.85 \times 3 \times 12) = 5.26 \text{ in.} \qquad \text{Say O.K.} \end{aligned}$$

This steel is awkward to fit nicely. Although some legitimately object to mixed bar sizes when the difference is only one bar size (because it is so easy to confuse the bars on the job),

USE 2- #7 plus 2- #8 = 2.78 in.^{2}

These bars need not be bundled in the 12-in. width available.

3.4 BEAMS WITH COMPRESSION STEEL—MOMENT DESIGN

Compression steel is used wherever a given beam requires more tension steel than the engineer thinks desirable in a rectangular beam, either from consideration of approaching brittle failure conditions or from consideration of excessive deflections. In either case the designer knows how much moment he must carry or desires to carry without compression steel and this moment is designated as the M_1 couple with steel A_{s1}. Any necessary additional moment M_2 can then be resisted by added tension steel A_{s2} and compression steel A_s' as sketched in Fig. 3.3. Whether A_s' works at

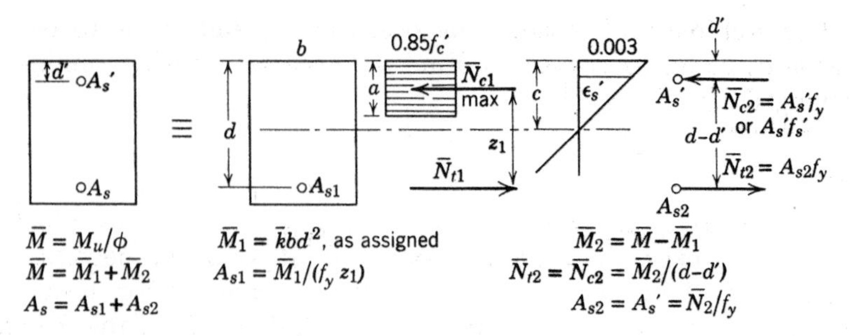

Fig. 3.3. Design of double-reinforced beam. Note that this calculation of A_s' requires modification as noted in the last paragraph of Sec. 3.5a to satisfy the legal wording of the Code.

yield stress or a lower stress depends upon the ϵ_s' strain, established by the neutral axis required with the M_1 couple.

When compression capacity requires the compression steel, the $\overline{M}_1$ couple is based on the $\bar{k}$ value of Table 2.1 and A_{s1} can be calculated from the corresponding ρ value (or from the a value, which establishes z for use in $A_{s1} = \overline{M}_1/f_y z$). It should be noted that the permissible $\bar{k}bd^2$ is normally so large that compression steel is rarely required for strength. When compression steel is used to reduce long time deflection, $\overline{M}_1$ and $\bar{k}_1$ can be based on the $\rho = 0.18f_c'/f_y$ values tabulated in Table 2.1 and Fig. 2.8, or the engineer can set up his own criterion following the method of Sec. 3.5*b*.

Whereas *analysis* of beams containing A_s' sometimes requires a cut-and-try procedure (Sec. 2.14*b*), the design procedure is always direct and relatively simple.

3.5 DOUBLE-REINFORCED BEAM EXAMPLES

(*a*) A double-reinforced beam with $b = 11$ in., d $= 20$ in., $f_c' = 3000$ psi, Grade 40 steel must carry a dead load moment of 136 k-ft and a live load moment of 150 k-ft. Calculate the required steel using the Code load factors.

Solution

Reqd. $M_u = 136 \times 1.4 + 150 \times 1.7 = 190 + 255 = 445$ k-ft
Reqd. $\overline{M} = M_u/0.9 = 494$ k-ft

As a rectangular beam without A_s', the maximum steel for A_{s1} is limited by Table 2.1 to 2.78% which develops

$\overline{M}_1 = \bar{k}bd^2 = 870 \times 11 \times 20^2/12{,}000 = 320$ k-ft
$A_{s1} = \rho_1 bd = 0.0278 \times 11 \times 20 = 6.14$ in.^{2}

As an alternate, $z_1 = d - a/2 = 20 - 0.436 \times 20/2 = 15.64$ in.

$$A_{s1} = \overline{M}_1/(f_y z_1) = 320 \times 12/(40 \times 15.64) = 6.14 \text{ in.}^2$$
$$\overline{M}_2 = \overline{M} - \overline{M}_1 = 494 - 320 = 174 \text{ k-ft}$$
$$\overline{N}_{t2} = \overline{N}_{c2} = \overline{M}_2/(d - d') = 174 \times 12/17.5 = 119 \text{ k}$$

on the assumption that d' is 2.5 in., that is, that the center line of A_s' is 2.5 in. from the compression face of the beam.

$$A_{s2} = 119/40 = 2.98 \text{ in.}^2$$
$$A_s = A_{s1} + A_{s2} = 6.14 + 2.98 = 9.12 \text{ in.}^2$$

Check to see if A_s' develops the yield strain of $40/29{,}000 = 0.00138$ by using strain triangles. Since $a = 0.436d = 8.72$ in., $c = 8.72/0.85 = 10.25$ in., and $\epsilon_s' = 0.003(10.25 - 2.5)/10.25 = 0.00224$, which is ample.

$$A_s' = \overline{N}_{t2}/f_y = 119/40 = 2.98 \text{ in.}^2$$

If the concrete displaced by A_s' is considered, the effective stress f_s'', that is, the increased value of the steel over the loss of the same area of concrete, becomes $f_y - 0.85f_c' = 40 - 0.85 \times 3 = 37.5$ ksi and

$$A_s' = \overline{N}_2/f_s'' = 119/37.5 = 3.18 \text{ in.}^2$$

Deflection may be large since A_{s1} is large. Check the 20 in. height of beam against Table 2.2 or calculate deflection.

As designed thus far this beam does not satisfy the Code steel limitation of $0.75\rho_b$, as already defined in Sec. 2.14*. In meeting this requirement, the $\bar{k}$ value from Table 2.1 leads to the proper A_{s1}; the $\overline{M}_2$ couple is, however, simply $\overline{M}_{2b}$ for the balanced beam. To hold A_{s2} not more than $0.75A_s'$, the designer can simply increase A_s' by 1/3, making $A_s' = 3.18/0.75 = 4.24$ in.^{2}, with the total A_s remaining unchanged. The author considers this additional reinforcement unnecessary unless the designer is legally under the Code and without freedom of discretion in the matter.

(*b*) Recalculate the reinforcing steel for the above case if it is desired to reduce deflections somewhat by limiting A_{s1} to $\rho_1 = 0.18f_c'/f_y$.†

*The author feels that the correction which follows is divorced from real beam behavior. Tests of double reinforced beams have always shown excellent ductility, even with high strength steels. High strength steel picks up added compression as ϵ_{cu} approaches 0.003 and delays failure until still higher strains are reached.

†Note that the method of Sec. 2.8 and Fig. 2.8 is available for any desired ρ_1, to give $\bar{k}_1$. Likewise any desired ρ' can lead to k_2. The sum $\bar{k}_1 + \bar{k}_2$ = total $\bar{k}$ resulting.

Solution

The total $\overline{M}$ is 494 k-ft, as calculated before. Table 2.1 and Fig. 2.8 show that this steel limitation gives $\overline{M}_1 = 0.161 f_c' bd^2$, $a = 0.212d$, and $z_1 = 0.894d$. Thus $\bar{k}_1 = 0.161 \times 3000 = 483$.

$$\overline{M}_1 = 483 \times 11 \times 20^2/12{,}000 = 177 \text{ k-ft}$$
$$A_{s1} = (0.18 \times 3/40)11 \times 20 = 2.97 \text{ in.}^2$$
$$\overline{M}_2 = 494 - 177 = 317 \text{ k-ft}$$
$$\overline{N}_{t2} = \overline{N}_{c2} = 317 \times 12/(20 - 2.5) = 217 \text{ k}$$
$$A_{s2} = 217/40 = 5.42 \text{ in.}^2$$
$$A_s = 2.97 + 5.42 = 8.39 \text{ in.}^2$$

For strain triangles, $a = 0.212 \times 20 = 4.24$ in. and $c = 4.24/0.85 = 4.98$ in. Then

$$\epsilon_s' = 0.003(4.98 - 2.5)/4.98 = 0.00150$$

Since $\epsilon_y = 40/29{,}000 = 0.00138$, the yield stress can be developed. Considering displaced concrete,

$$A_s' = 217/(40 - 0.85 \times 3) = 5.78 \text{ in.}^2$$

With shallow beams, ϵ_s' may be less than ϵ_y. In this case $f_s' = E_s \epsilon_s'$ is simply substituted for f_y in the calculation of A_s'. There is no cut-and-try in design of these beams.

The extra A_s' to satisfy the $0.75\rho_b$ Code requirement is not needed here since A_{s1} is far below the Table 2.1 allowable on the rectangular section.

Permissible A_{s1} (from example above)	6.14 in.^{2}
Allowable increase in A_s with A_{s2} of 5.42 = 0.75 × 5.42 =	4.07
Total permissible A_s =	10.21 in.^{2}

This compares with 8.39 in.^{2} actually used.

3.6 T-BEAMS—MOMENT DESIGN

The T-beam flange is usually determined by the slab design, and the web size (b_w and d) is often determined by shear or requirements other than moment. To some extent b_w, the stem width, is influenced by the number of reinforcing bars used for A_s, but the student probably lacks the experience at this stage to estimate this need in advance. In this section it is assumed that design for moment includes only the choice of the area of steel A_s, a check against brittle failure, and some consideration of deflection.

A flange is ordinarily assumed to provide enough compression area to eliminate the possibility of brittle failure in compression. This is frequently confirmed by a neutral axis which falls high in the flange and turns a T-beam design into the design of a wide rectangular beam. Actually, a T-beam must be quite deep relative to the slab and be quite heavily reinforced to make it really act as a T-beam in carrying compression. Even when the balanced condition would lead to a neutral axis deep in the beam (and a rather large value of $a = k_1c$), the brittleness concept leads to restricting the compression stress block to 0.75 of that at the balanced condition. Since it is rare that more than 0.25 of the total stress block area lies below the bottom of the flange, the usable stress block tends to be restricted to less than the total depth of the flange. Thus most T-beams are only rectangular beams in flexural behavior.

The cut-and-try determination of z, as for the rectangular beam, is recommended, as illustrated in Sec. 3.7. The Code provision of transverse steel in the flange (8.7.5) must not be overlooked:

> Where the principal reinforcement in a slab which is considered as the flange of a T-beam (not a joist in concrete joist floors) is parallel to the beam, transverse reinforcement shall be provided in the top of the slab. This reinforcement shall be designed to carry the design load on the portion of the slab required for the flange of the T-beam. The flange shall be assumed to act as a cantilever. The spacing of the bars shall not exceed five times the thickness of the flange nor in any case 18 in.

3.7 T-BEAM EXAMPLES

(*a*) The floor of Fig. 3.4 consists of a 4-in. slab supported by beams of 22-ft span cast monolithically with the slab at a spacing of 8 ft center to center. These beams have webs which are 12 in. wide and a depth of 19 in. to the center of steel plus the necessary cover over the steel. Dead load moment is 62 k-ft and live load moment is 120 k-ft. Using $f_c' = 3000$ psi,

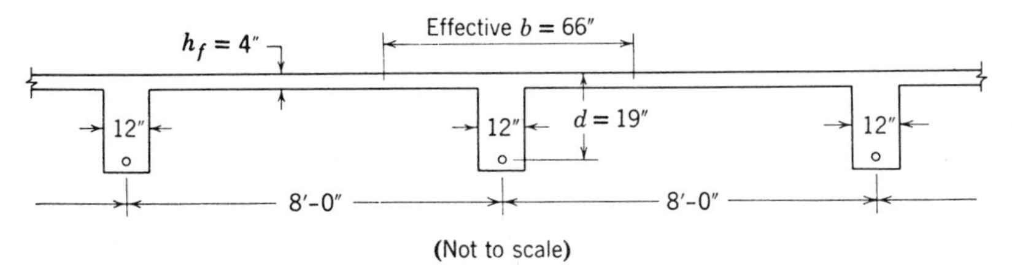

Fig. 3.4. Floor of Sec. 3.6.

Grade 60 steel, and the Code load factors of Code Eq. 9.1 calculate the required steel.

Solution

The effective flange width is defined by Code 8.7, or see Sec. 2.11 for this case.

Span/4 $= 22 \times 12/4 = 66$ in. ← Governs b
$8h_f$ overhang gives $b = 2 \times 8 \times 4 + 12 = 76$ in.
Beam spacing $= 8$ ft $= 96$ in.
$M_u = 1.4D + 1.7L = 1.4 \times 62 + 1.7 \times 120 = 291$ k-ft
$\overline{M} = M_u/\phi = 291/0.9 = 323$ k-ft

The lever arm z will usually be large and a good trial value will usually be at least $0.9d$ or $d - h_f/2$, whichever is the larger.

$0.9d = 0.9 \times 19 = 17.1$ in. $\qquad d - h_f/2 = 19 - 2 = 17$ in.
Trial $A_s = \overline{M}/f_y z = 323 \times 12/(60 \times 17.1) = 3.78$ in.^{2}

It must be determined whether the stress block is as deep as the flange. If $a = h_f$, available $\overline{N}_c = 0.85 \times 3 \times 66 \times 4 = 673$ k compared to the necessary $\overline{N}_c = \overline{N}_t = 3.78 \times 60 = 227$ k. Therefore the neutral axis is obviously high in the flange and the design becomes that for a rectangular beam 66 in. wide.

$$a = \frac{227}{0.85 \times 3 \times 66} = 1.35 \text{ in.,} \qquad z = 19 - 0.67 = 18.33 \text{ in.}$$

$A_s = 323 \times 12/(60 \times 18.33) = 3.52$ in.^{2}
$a = 3.52 \times 60/(0.85 \times 3 \times 66) = 1.26$ in.
$z = 19 - 0.63 = 18.37$ in. $\qquad$ Close enough
USE $A_s = 3.52$ in.^{2}

There can be no possibility of brittle compression failure with this small a. Deflection should be checked.

(b) Calculate A_s if the loads in Example a are increased to give $M_u =$ 1000 k-ft.

Solution

$\overline{M} = \overline{M}/\phi = 1000/0.9 = 1111$ k-ft

If $a = h_f$, available $\overline{N}_c = 673$ k, as in (a)
Approx. reqd. $\overline{N}_c = 1111 \times 12/17.1 = 778$ k
There $a > h_f$. Design as a true T-beam.

Approx. $A_s = 1111 \times 12/(60 \times 17.1) = 13.0$ in.^{2}
Compression block area $= 13.0 \times 60/(0.85 \times 3) = 306$ in.^{2}
Area of flange $= 66 \times 4 = 264$ in.^{2}
Area below flange $= 306 - 264 = 42$ in.^{2} or 3.50-in. depth

This area is shown in Fig. 3.5 and gives the centroid distance $\bar{y}$ as

$$\frac{66 \times 4 \times 2 + 42 \times 5.75}{306} = 2.51 \text{ in.}$$

$z = 19 - 2.51 = 16.49$ in.
$A_s = 1111 \times 12/(60 \times 16.49) = 13.4$ in.^{2}
Compression block area $= 13.4 \times 60/(0.85 \times 3) = 315$ in.^{2}
Area below flange $= 315 - 264 = 51$ in.^{2} or 4.25-in. depth

The change in $\bar{y}$ and z is reasonably small (about 0.15 in. more for z) and will be neglected here. The design would be satisfied with $A_s = 13.4$ in.^{2} except for the problem covered in the next pargraph.

The neutral axis is quite low and should be checked against the balanced condition.

$$\text{Balanced } c_b = \frac{87}{87 + 60} 19 = 11.24 \text{ in.}$$

$$a_b = 0.85 \times 11.24 = 9.55 \text{ in.}$$

Balanced compression block area $= 66 \times 4 + 12 \times 5.55 = 331$ in.^{2}

Limit compression block to $0.75 \times 331 = 248$ in.^{2} This compares to 315 in.^{2} needed above. It is definitely not desirable to use this T-beam for the indicated moment. However, if it is necessary, it can be accomplished by using A_{s1} as required to balance the 248 in.^{2} stress block, in which case this leads to $\overline{M}_1$ as for a rectangular beam, since 248 in.^{2} is less than bh_f. The additional moment $\overline{M}_2 = 1111 - \overline{M}_1$ should then be provided by A_s' and A_{s2} as for any double-reinforced beam.

Such a case with a T-beam is far from typical since most are not so heavily loaded with moment. The given moment might well call for a

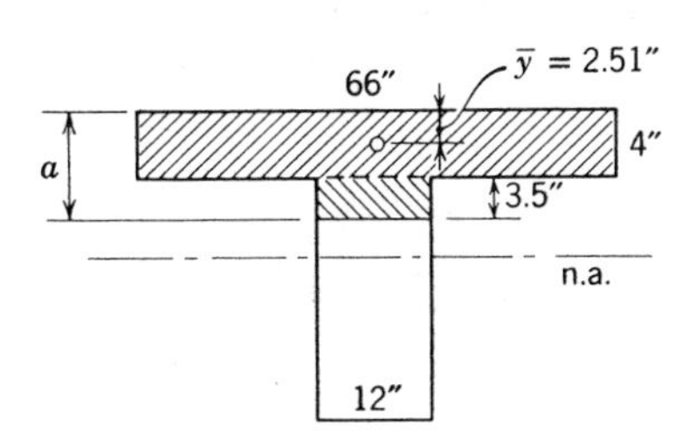

Fig. 3.5. Trial compression block.

heavier slab design which would automatically increase the moment capacity, or a deeper beam might be appropriate.

(*c*) Calculate A_s for the beam of (*b*) if d is increased to 23 in. and the span is reduced to 20 ft 8 in., making $b = 62$ in.

Solution

$\overline{M} = 1000/0.9 = 1111$ k-ft
Approx. reqd. $\overline{N}_c = 1111 \times 12/(23 - 2) = 634$ k
Reqd. area of stress block $= 634/(0.85 \times 3) = 249$ in.^{2}

Check against the limit to avoid brittle failure.

$$c_b = \left(\frac{87}{87 + 60}\right) 23 = 13.60 \text{ in.}$$

$a = 0.85 \times 13.60 = 11.55$ in.
Balanced stress block $= 62 \times 4 + 12\ (11.55 - 4) = 339$ in.^{2}

Maximum available area for design $= 0.75 \times 339 = 254$ in.^{2} The approximate required area of about 249 in.^{2} is barely available. Since the total flange area is 248 in.^{2}, the stress block extends below the flange by less than 1 in.^{2} The centroid of the stress block is essentially at mid-flange.

$z = 23 - 2.0 = 21$ in.
$A_s = 1111 \times 12/(60 \times 21) = 10.60$ in.^{2}

Stress block area $= 10.60 \times 60/(0.85 \times 3) = 250$ in.^{2}, which extends $2/12 = 0.17$ in. below flange.

$$\bar{y} = \frac{248 \times 2 + 2 \times 4.08}{250} = 2.01 \text{ in.,} \qquad z = 23 - 2.01 = 20.99 \text{ in.}$$

$A_s = 1111 \times 12/(60 \times 20.99) = 10.60$ in.^{2}

The example will be reworked to illustrate the equivalent double-reinforced beam idea presented in the 1963 Code. The overhanging flanges will be replaced with compression steel A_{sf} as sketched in Fig. 3.6.

$60A_{sf} = 0.85 \times 3(62 - 12)4 = 511$
$A_{sf} = 8.52$ in.^{2} at $d' = h_f/2 = 2$ in.

This steel will develop a moment $\overline{M}_f$:

$$\overline{M}_f = 8.52 \times 60(23 - 2)/12 = 894 \text{ k-ft}$$

This leaves $\overline{M}_w = 1111 - 894 = 217$ k-ft on the web section

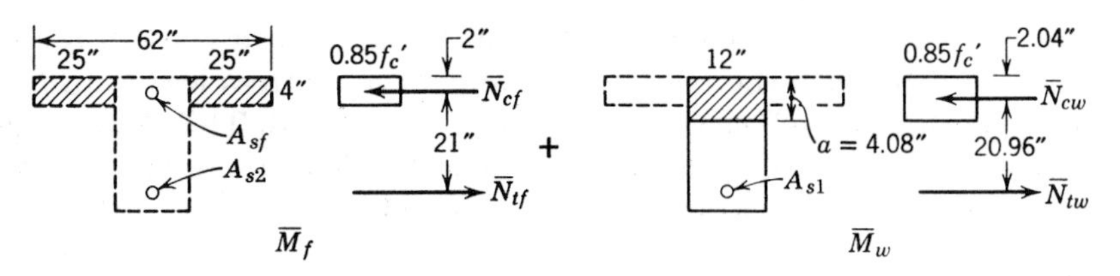

Fig. 3.6. T-beam as a "double-reinforced" beam. $M = \overline{M}_f + M_w$.

Assume $a = 4.3$ in., $z = 23 - 0.5 \times 4.3 = 20.85$ in.

$$\text{Approx. } A_{s1} = 217 \times 12/(60 \times 20.85) = 2.08 \text{ in.}^2$$
$$a = 2.08 \times 60/(0.85 \times 3 \times 12) = 4.08 \text{ in.}$$
$$z = 23 - 0.5 \times 4.08 = 20.96 \text{ in.}$$
$$A_{s1} = 217 \times 12/(60 \times 20.96) = 2.07 \text{ in.}^2$$
$$A_{s2} = A_{sf} = 8.52 \text{ in.}^2$$
$$A_s = 2.07 + 8.52 = 10.59 \text{ in.}^2\text{, essentially as before.}$$

Since the steel is essentially the maximum, deflection will require a check.

3.8 BEAMS OF SPECIAL SHAPES

As presented in the preceding sections, the criterion against brittle failure is perfectly general. The compression must be limited to 0.75 of $\overline{N}_{cb}$, or ρ must be limited to $0.75\rho_b$. When compression steel is used at full f_y stress, this portion of $\overline{N}_c$ does not contribute to a brittle failure but the Code does include it in $\overline{N}_c$ anyway.

For special-shaped beams, z can be estimated in order to calculate an approximate A_s; the compression area of the stress block can be determined to make $\overline{N}_c = \overline{N}_t$. The centroid of this area establishes a better value of z, which leads to a better A_s value.

Where webs are deep, flexural cracks will be conspicuous unless face steel is added. Code 10.6.5 requires for web depths greater than 3 ft an area of at least 10% of the tension steel be placed near the web faces at a spacing on centers of not more than 12 in., or the thickness of the web. Such reinforcement may be counted for flexure if strain compatibility is used to compute the effective stresses in each layer, as discussed in Sec. 3.9.

3.9 BEAMS WITH REINFORCING AT SEVERAL LEVELS

When reinforcing is placed in two layers, it is usually satisfactory to consider the f_y stress at the centroid of the steel. However, if the bars are in many layers or are distributed over all faces, as in a column section, some bars will be near the neutral axis and have a stress less than f_y. In such cases strains as well as stresses must be considered. This is not overly complex when strain triangles are used after the fashion already used for compression steel, as illustrated by Fig. 3.3 and the examples of Secs. 3.5 and 2.14.

3.10 STOPPING OR BENDING OF BARS—FOR MOMENT

(a) General Considerations

The maximum required A_s for a beam is needed only where the moment is maximum. Insofar as moment is concerned this steel may be reduced at points along the beam where only smaller moments exist. Two other stress considerations are also important in fixing the length of bars. First the individual bars require a specific length in which to develop their yield strength. Development length, the subject of Chapter 5, is as important as moment length; the designer must put the two aspects together. Second, when a bar is cut off where the neighboring bars are still carrying tension, a discontinuity in the tensile resistance occurs, flexural cracks open at earlier loads and become wider than normal, and shear strength is lowered, often with some loss of the beam ductility. Shear strength loss is discussed in Sec. 4.14. A bent bar anchored in the compression zone causes no loss in shear strength.

In this chapter only the moment aspect of stopping or bending bars is discussed, and this only for statically determinate members.

(b) Maximum Moment Curves

A study of where bars can be stopped, or bent away from the tension zone, must start with a study of the maximum moments which are possible at all points along the beam.

In some simple cases, the maximum moment diagram is simply the moment diagram for full load. In other cases, as for wheel loads (Fig. 3.7*a*), the determination of the maximum moment diagram requires the calculation of maximum moments at many points; the maximum moment diagram is then the envelope of the maximum moment values from the several loads, as in Fig. 3.7*b* for a single wheel load W. Each dashed triangle corresponds to a particular position of the load, the maximum moment at any point being given with the wheel at that point. The en-

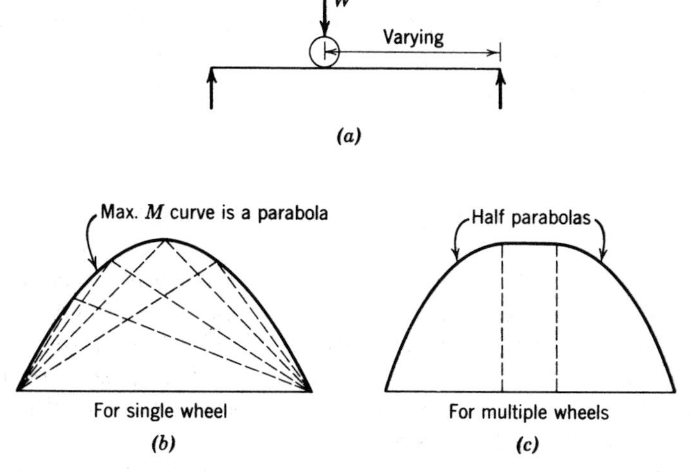

Fig. 3.7. Maximum moment curves.

velope in this case happens to be a parabola. With several wheels the curve approximates two half parabolas separated by a section of constant moment, as in Fig. 3.7*c*.

Only fixed loadings are illustrated in the remainder of this chapter. Continuous beams will be discussed in Chapter 8.

(c) Theoretical Bend Point or Cutoff Point

When a variable depth member is considered, the required A_s curve must be established by calculating the required A_s at representative points from the basic relation

$$A_s = \max \overline{M} \div (f_y z)$$

where z varies somewhat like d. This procedure is illustrated in Sec. 7.6.

With constant depth beams, the denominator becomes essentially a constant, the usual small variations in the ratio z/d being of little significance in this connection. Hence required A_s varies directly with the maximum moment, and the shape of the required A_s curve is identical with that of the maximum moment curve. The maximum moment curve may then be used as the required A_s curve, simply by changing the scale. The maximum ordinate corresponds to the maximum required A_s and this establishes the necessary correlation.

Bars can theoretically be stopped or bent wherever they are no longer needed for moment. The solution can be either graphical (for complex maximum moment diagrams), semigraphical for typical cases, or analytical where desired. The semigraphical process is illustrated in Sec. 3.10*f* and Sec. 3.10*g*.

(d) Arbitrary Requirements for Extending Bars

The ACI Code (12.1.4) requires that each bar be extended a distance d or 12 bar diameters beyond the point where it is "no longer required to resist flexure." This prohibits the cutting off of a bar at the theoretical minimum point, but can be interpreted as permitting bars to be *bent* at the minimum point. Three sound reasons can be advanced for this arbitrary extension, which the author designates as a.

First, when a bar is simply cut off (not bent away from the main steel), there is a large transfer of stress from this bar to those remaining. This stress concentration will cause a moment crack in the concrete at the end of the bar if the beam is carrying its working load. Bending the bar spreads out this concentration; extending the bar, without bending, removes the concentration from a point of maximum steel stress to a point of lower steel stress. (It should be noted that if a bar is cut off at the theoretical minimum point, the remaining bars necessarily work at maximum stress.)

The author gives greater weight to a second reason, which was well stated by an introductory statement in the long obsolete 1940 Joint Committee Specification:

> To provide for contingencies arising from unanticipated loads, yielding of supports, shifting of points of inflection, or other lack of agreement with assumed conditions governing the design of elastic structures, it is recommended that the reinforcement be extended at supports and at other points between supports as indicated. . . .

Thus interpreted, the arbitrary extension of the bar is the result of an envisioned possible extension of the maximum moment diagram. It would follow that bars must not even be bent until most of this extra length has been provided. Tests show, however, that bending bars is less dangerous than cutting them off in tension zones. The author would be satisfied to see *bends*, not cutoffs, at $d/2$ beyond the theoretical cutoff point. Even this is a more severe requirement than many engineers observe.

The third reason has developed out of diagonal tension tests. In Sec. 4.4f, from consideration of the free body of Fig. 4.2*b* it is pointed out that a diagonal crack leads to larger values of bar tension than would be predicted from moment alone. The larger tension is as serious as an excess of moment at the section. Such a condition argues against cutting off bars on the basis of the moment alone and points to the need for some extension beyond the theoretical bar cutoff point, especially where there is little or no web reinforcement against diagonal tension.

The ACI Code requires that at least one-third of the positive moment steel extend into the support in the case of simple spans and one-fourth in the case of continuous beams.* Custom usually increases these minimums. Also the Code requires that one-third of the negative moment reinforcement be extended past the point of inflection by the larger of: beam depth d, $12d_b$, or $l_n/16$ where l_n is the clear span. For span length the new ACI standard notation now permits only the lowercase letter, dropping the optional capital L used for many years. In script and on the typewriter, it may be desirable to use a script ℓ to distinguish it clearly from the number one.

A matter of design philosophy must now be introduced. The Code phrase quoted in the first sentence of this subsection is of long standing with only minor editorial changes. The author believes it has now become desirable to start thinking more of the arbitrary extension as an *actual* potential shift in the moment diagram. As moment diagrams from frame analysis tend to replace the approximate moment coefficients, these more clearly show the omission of the factors the Joint Committee listed; and these factors do cause real moments. The *idea* of a shifted moment diagram becomes essential in some form if we move toward redistribution of moments (Code 8.6) and limit design.

The practical effect of the substitution of a shifted moment diagram for an arbitrary extension is nil as far as lengths of bars *for moment* are concerned; but it makes a large difference in the magnitude of the stress considered to exist next to the cutoff point in the bars which continue further. It thus substantially influences the development length demands discussed in Chapter 5. Further discussion is continued in that chapter (Sec. 5.11 and Sec. 5.12).

(e) Nomenclature

It is common to refer to the first bar bent, second bar bent, and so forth; the first bar bent means the one bent nearest the point of maximum

*Note that Code 12.2.2, for a beam used as part of the primary lateral load system, requires this one-fourth be anchored for full f_y to give some ductility.

moment, typically, closest to the support for top bars and closest to mid-span for bottom bars.

Bars are often bent by pairs instead of singly. The first pair can be bent where the second bar is no longer required.

Bent bars are sometimes bent at points that make them available to resist diagonal tension. They might then be called shear or web reinforcement, but this is not a frequent practice in the United States.

(f) Examples Involving No Excess Steel, or Neglect of the Excess

(1) A uniformly loaded beam of 20-ft simple span requires 5- #9 bars, with $d = 15$ in. Considering moment alone, where can the first bar be cut off? The second bar?

Solution

The maximum moment diagram is simply the full load moment diagram and this parabola can be used as the required A_s diagram as in Fig. 3.8. The maximum ordinate becomes five bars. Steel areas could be used as ordinates, but the number of bars provides a more convenient unit when all bars are of the same size.

The first bar can theoretically be cut off at x_1 where only four bars are required. Since offsets to the center tangent vary as the square of the

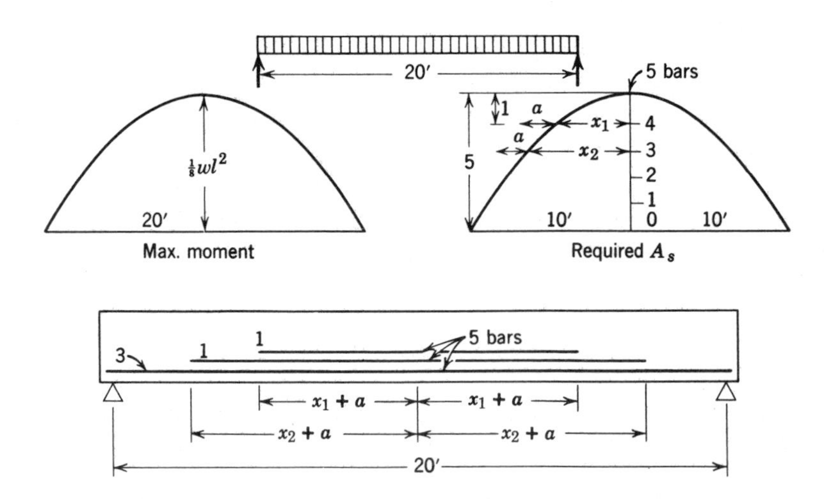

Fig. 3.8. Cutting off bars in uniformly loaded beam according to the moment diagram.

distances from the center,

$$x_1^2/10^2 = 1/5 \qquad x_1 = 10\sqrt{1/5} = 4.47 \text{ ft}$$

Since d of 1.25 ft exceeds $12d_b = 12 \times 1.13/12 = 1.13$ ft, use $a = 1.25$ ft for cutoff (or 0.62 ft for bend). The minimum distance from the center line of the span to the first bar cutoff is $4.47 + a = 4.47 + 1.25 = 5.72$ ft (or 5.09 ft to a bend point).

For the second bar stopped,

$$x_2^2/10^2 = 2/5 \qquad x_2 = 10\sqrt{2/5} = 6.33 \text{ ft}$$

The minimum distance from the center to the second bar cutoff is 6.33 + 1.25 ft = 7.58 ft, or 6.33 + 0.62 to a bend point.

Cutting off bars in a tension zone, as above, requires special shear calculations (Sec. 4.14). Also development lengths must be checked as in Sec. 5.12*c* and 8.15.

(2) A uniformly loaded cantilever beam 8 ft long requires 7- #7 bars when d is 18 in. Where can the first pair of bars be bent down for moment?

Solution

The maximum moment diagram is a half parabola. the maximum ordinate corresponding to seven bars of required A_s (Fig. 3.9). The first pair of bars can be bent where only five bars are required.

$$x^2/8^2 = 5/7 \qquad x = 8\sqrt{5/7} = 6.74 \text{ ft}$$

$d = 1.50$ ft $\quad 12d_b = 0.875$ ft $\quad a = 1.50$ ft for cutoff, or 0.75 ft for bend

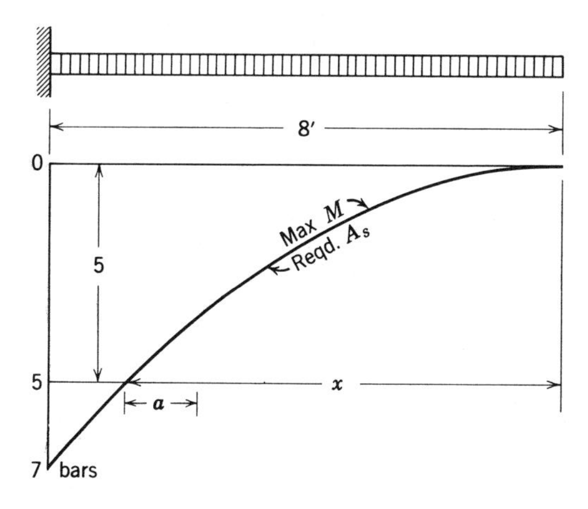

Fig. 3.9. Bending bars for cantilever beam.

The minimum distance from the support to the bend point for the first pair is $8 - 6.74 + a = 1.26 + 0.75 = 2.01$ ft. With the bar bent instead of cut off there is no loss of shear strength.

(g) Examples Permitting Reduced Lengths because of Excess Steel

(1) An 18-ft simple span beam with a fixed concentrated load at mid-span (negligible uniform load) and $d = 22$ in. requires 4.50 in.^{2} of steel but uses 5- #9 bars $= 5.00$ in.^{2} Where can the first pair of bars be stopped?

Solution

Figure 3.10 shows the maximum moment diagram used as a required A_s curve with the maximum ordinate marked as 4.50 in.^{2} $= 4.50$ bars required. The first pair of bars can be stopped where only three bars are required.

$x/9 = 1.5/4.5 \qquad x = 3.0$ ft

$d = 1.83$ ft $\qquad 12d_b = 1.13$ ft $\qquad a = 1.83$ ft for cutoff, or 0.92 ft for bend

The minimum distance from mid-span to the stop point for the first pair is $3.0 + a = 3.00 + 1.83 = 4.83$ ft.

(2) A uniformly loaded cantilever beam 8 ft long with $d = 18$ in. requires 3.85 in.^{2} and uses 7- #7 bars. Where can the first pair of bars be bent down?

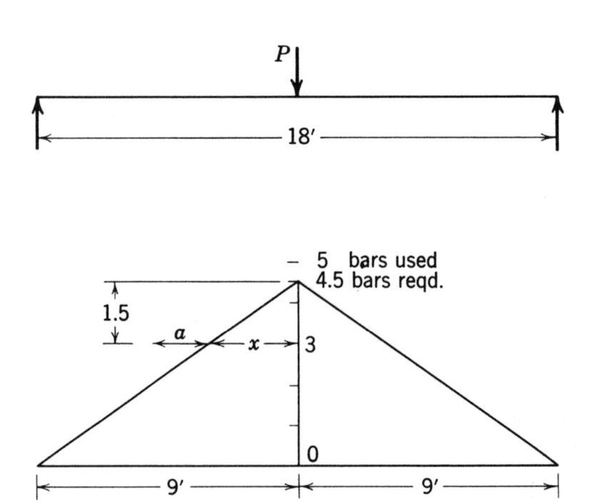

Fig. 3.10. Stopping bars, excess steel.

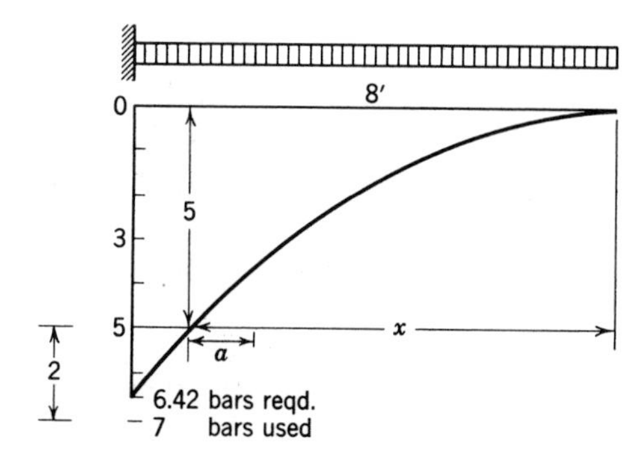

Fig. 3.11. Bending bars, excess steel, cantilever beam.

Solution

The maximum ordinate of the maximum moment diagram is designated as 3.85 in.^{2} = 3.85/0.60 = 6.42 bars required (Fig. 3.11). The desired point is where five bars are needed.

$x^2/8^2 = 5/6.42 \qquad x = 8\sqrt{5/6.42} = 7.05$ ft

$d = 1.50$ ft $\qquad 12\,d_b = 0.88$ ft $\qquad a = 1.50$ to cutoff, 0.75 ft to bend

The minimum distance from the support to the bend point for the first pair is $8 - 7.05 + a = 0.95 + 0.75 = 1.70$ ft.

3.11 DESIGN AIDS

Four available design aids for rectangular beams have already been discussed in Sec. 2.17, and Fig. D.3 and D.4 (see Appendix D) are additional curves particularly interesting for design. The latter are only one pair out of nine in Ref. 1 and cover only the combination of $f_y = 60$ ksi and $f_c' = 3000$ psi, one for slab elements 12 in. wide (the standard width for slab design) and one for an element 10 in. wide. The latter facilitates the use of proportional values for other desired beam widths.

The beam design of Sec. 3.3*a* will be used as a sample. The design M_u, including assumed dead load, is 240 k-ft. (The available $\bar{k}$ or k_m is available several places, for example, in Fig. D.2 for $f_y = 60$ ksi and $f_c' = 3000$ psi $\bar{k}$ shows as 783.) Figure D.4 shows for this moment and the maximum reinforcement ratio a needed effective depth d of about 20 in. with an A_s of 3.25 in.^{2} if a width of 10 in. were used. If one prefers 9.5 in., the width used in Sec. 3.3*a*, this would mean an equivalent moment on the 10 in. width of (10/9.5)240 = 253 k-ft and the use of 9.5/10 = 0.95 times the A_s shown for the 10 in. width. The 253 k-ft value of M_u on this figure shows minimum d of just under 21 in., compared to 20.8 in. in the example. For $d = 21$ in., the required A_s is 3.25 (from the chart) times 0.95 = 3.08 in.^{2}. This compares to a final $d = 21.1$ in. and a final required A_s of 3.09 in.^{2} in the example.

The use of Fig. D.3 for slabs is similar except that $b_w = 12$ in. for a slab would not be modified.

In either beams or slabs, the curves also give the required A_s for any depth chosen (greater than the minimum). The ρ of $0.18f_c'/f_y$ used in Sec. 3.3*b* is not shown on the chart of Fig. D.4, but if one uses $d = 26.6$ in. with the same 9.5 in. width, as in the example, the equivalent moment of 253 k-ft remains and one reads A_s (for $b = 10$ in.) as 2.30 in.^{2}. This leads to $0.95 \times 2.30 = 2.19$ in.^{2} required compared to 2.25 in.^{2} found in Sec. 3.3*b* and 2.37 in.^{2} actually used to fit bar sizes. If d is changed to 22.6 in. as in Sec. 3.3*c*, with $b = 12$ in., the equivalent moment is $240(10/12) = 200$ k-ft and the chart reads required $A_s = 2.20$ in.^{2} for a 10 in. width or a total of $2.20 \times 12/10 = 2.64$ in.^{2}, compared to 2.68 in.^{2} by calculation.

Design tables are in preparation for a revision of Ref. 1 which will probably be available by the time this text revision is issued. The 1967 tables are still usable for A_s' calculations except that the maximum possible for use is slightly altered by the 1971 Code, by a differing definition of the balanced ρ_b for this case. The design used in Sec. 3.5*a* has the same problem and calculation is nearly as fast as the use of tables. The revised tables may considerably simplify the solution of Sec. 3.5*b*.

For office use the ACI Handbook[1] has many such design aids. It also includes tables to assist in the choice of bars, with such tabulations as minimum web width to satisfy the minimum bar spacing rules.

The CRSI Handbook[2] is a very comprehensive set of tables giving directly the design live load capacity of many types of members, including beams, one-way slabs, joists, flat slabs, waffle slabs, columns, and footings, along with reinforcement details in many cases. Although now based on the 1963 Code it will undoubtedly be revised to be current. Beams, joists, one-way slabs, and footings are not generally changed in their basic design by the 1971 Code except for the change in the load factors from the 1963 values of $1.5D + 1.8L$ to the 1971 values of $1.4D + 1.7L$.

SELECTED REFERENCES

1. *Ultimate Strength Design Handbook*, Vol. 1, Special Publication No. 17, American Concrete Institute, Detroit, 1967.
2. *CRSI Handbook Ultimate Strength Design*, 2nd ed., Concrete Reinforcing Steel Institute, Chicago, 1970.

PROBLEMS

Prob. 3.1. In Prob. 2.5 calculate the exact steel that is needed for the loads given.

Prob. 3.2. In Prob. 2.5 calculate the exact steel that is needed if Grade 60 steel is used.

Prob. 3.3. Design the steel for the channel section of Prob. 2.10 and Fig. 2.25 if $M_u = 105$ k-ft and the steel is changed to Grade 40.

Prob. 3.4. A simple span rectangular beam 20 ft long is to carry a service load of 3400 plf made up of 1600 plf dead load (including beam weight) and 1800 plf live load. Use $f_c' = 4000$ psi, Grade 40 steel, and design a beam, subject to later check on shear and bond, for ρ approximately $0.18f_c'/f_y$. Make $b = 13$ in. and keep d in full inches. Choose bars.

Prob. 3.5. Assuming deflections are not considered serious for the usage contemplated, redesign the beam of Prob. 3.4 with $b = 13$ in. and the minimum effective depth (full inches) permitted under the Code.

Prob. 3.6. A continuous rectangular beam with $b = 14$ in., $d = 24$ in., $f_c' = 5000$ psi, Grade 60 steel, must care for $M_u = -1100$ k-ft. Design the necessary steel, using $d' = 3$ in. if A_s' is necessary.

Prob. 3.7. Redesign the depth in Prob. 3.6 to avoid use of A_s'. Design the steel for moment assumed unchanged.

Prob. 3.8. A rectangular beam 13 in. wide by 20 in. deep to center of steel must carry a total ultimate moment of 250 k-ft with $f_c' = 3000$ psi and Grade 40 steel. Find the required steel.

Prob. 3.9. A slab having $d = 4$ in. must carry a design moment of 3.9 k-ft per ft width for D.L. and 5.0 k-ft per foot for L. L. Find the required A_s per foot if $f_c' = 3000$ psi and steel is Grade 60. Specify bar size and spacing.

Prob. 3.10. A slab 7.5 in. thick with 1.5-in. cover to center of steel must carry its own weight and a service live load of 100 psf over a 15-ft simple span. Find the required A_s per foot if $f_c' = 3000$ psi and steel is Grade 40. Choose bars.

Prob. 3.11. From basic principles establish the value of the maximum steel ratio ρ for a rectangular beam with $f_c' = 3500$ psi and $f_y = 50{,}000$ psi.

Prob. 3.12. A rectangular beam 14 in. wide by 25 in. deep overall with 3-in. cover to center of steel must carry a design (factored) moment of 500 k-ft. Find the necessary steel if $f_c' = 4000$ psi and steel is Grade 50.

Prob. 3.13. Design a double-reinforced beam 13 in. wide for a total M_u = 500 k-ft using f_c' = 3000 psi, Grade 40 steel, cover 3 in. to center of steel, and basing the beam size on an approximate $\bar{k}$ of 800. Keep ρ-ρ' not more than $0.18 f_c'/f_y$, as in Sec. 3.5*b*.

Prob. 3.14. You wish to design a double reinforced rectangular beam for a known moment (as in Sec. 3.5*b*) with f_c' = 4000 psi and Grade 60 steel. First you want to choose such a beam size that $\rho_1 = 0.18 f_c'/f_y$ will be consistent with the use of ρ' = 0.01 for A_s'. Assuming d'/d = 0.12 and that A_s' bars yield, find the numerical $\bar{k}$ to use in establishing the beam size. (*Suggestion:* Base $\bar{k}$ on an equation for $M_1 + M_2$.)

Prob. 3.15. If the joist of Fig 2.22 must care for a design moment of 25 k-ft, calculate the required A_s. Take f_c' = 5000 psi and f_y = 60 ksi.

Prob. 3.16. A simple span beam with d = 20 in. and carrying uniform load over a 22-ft span requires 3.10 in.^{2} of steel and uses 6- #7 bars.
(*a*) Where can the first pair be bent up? The second pair?
(*b*) If the excess steel were neglected, where would these bends be permitted?

Prob. 3.17. A 21-ft simple span beam carries a large fixed concentrated load at 9 ft from the left end and requires A_s = 4.40 in.^{2} The steel used is 5-#9, d = 20 in. If the beam weight is disregarded, where can the first bar be bent up (each side of the load)? The third bar?

Prob. 3.18. A 12-ft cantilever beam carries a large concentrated load at its end, uniform load negligible. Required A_s = 3.60 in.^{2}, 4- #9 used, d = 19 in. Where can one pair of bars be stopped or bent down?

Prob. 3.19. A 9-ft cantilever beam with d = 16 in. carries only uniform load and requires A_s = 3.50 in.^{2} If 4-#9 are used, where can half of the bars be bent down?

Prob. 3.20. The beam of Fig. 3.12 with d = 18 in. is subject to a fixed dead load of 1000 plf and a movable live load of 2000 plf. The required positive moment steel is 4.50 in.^{2} and the required negative moment steel is 2.50 in.^{2} If the positive moment steel used is 5-#9 where can the first pair of positive steel bars be stopped? If the negative moment steel is 4-#8 bars, where can this be reduced to 2 bars? (Note that distances on each side of the support or reaction are needed. See Sec. 1.12*d* for load factors.)

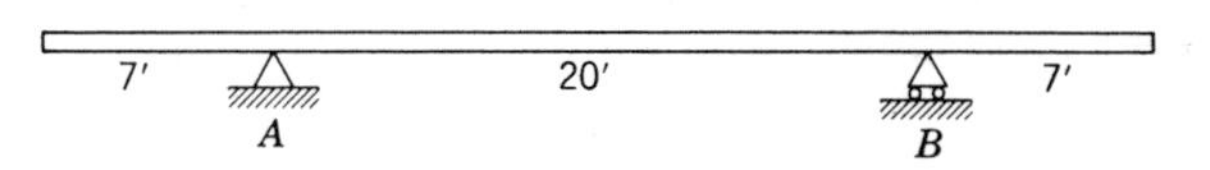

Fig. 3.12. Simple beam with overhanging ends for Prob. 3.20.

4
Shear and Torsion

SHEAR

4.1 SHEAR STRESS AND DIAGONAL TENSION

Concrete is relatively much weaker in tension than in compression, with real shear strength intermediate between the two. Most failures that would be termed shear failures are really diagonal tension failures, occasionally diagonal compression failures. While long usage has established shear stress as standard nomenclature, the phenomena involved will often be more clearly understood if diagonal tension is kept in mind.

Shear stresses as normally computed are, except sometimes in prestressed concrete, actually stress coefficients that are only nominally related to the actual stresses that are critical. The shear-friction theory is almost the only place shear itself as a critical stress is computed. Generally the nominal shear stress is considered simply as a measure of diagonal tension, or diagonal compression as a limiting state.

4.2 DIAGONAL TENSION BEFORE CRACKS FORM

In homogeneous beams diagonal stress can be analyzed by well-established relationships. Reinforced concrete beams, prior to the formation of cracks, probably have stresses quite similar to those of a homogeneous beam.* The diagonal tension stresses will be emphasized here.

*The notation of this section on basic strength of materials does not attempt to follow the new standard ACI notation. C and T are total forces; c, t, v, f_c, and f_t are unit stresses.

In the beam of Fig. 4.1*a* a small element at the neutral axis at *A* would be subject to a shearing stress v, but no bending stress. Figure 4.1*b* shows that such an element will develop unit diagonal tensile and compressive stresses of magnitude v. An element at *B* in Fig. 4.1*a* will have a compressive stress f_c in addition to shear. Such a stress produces diagonal compression, as shown in Fig. 4.1*c*. When Fig. 4.1*c* is combined with a shear effect similar to that in Fig. 4.1*b*, to obtain the total stress, the

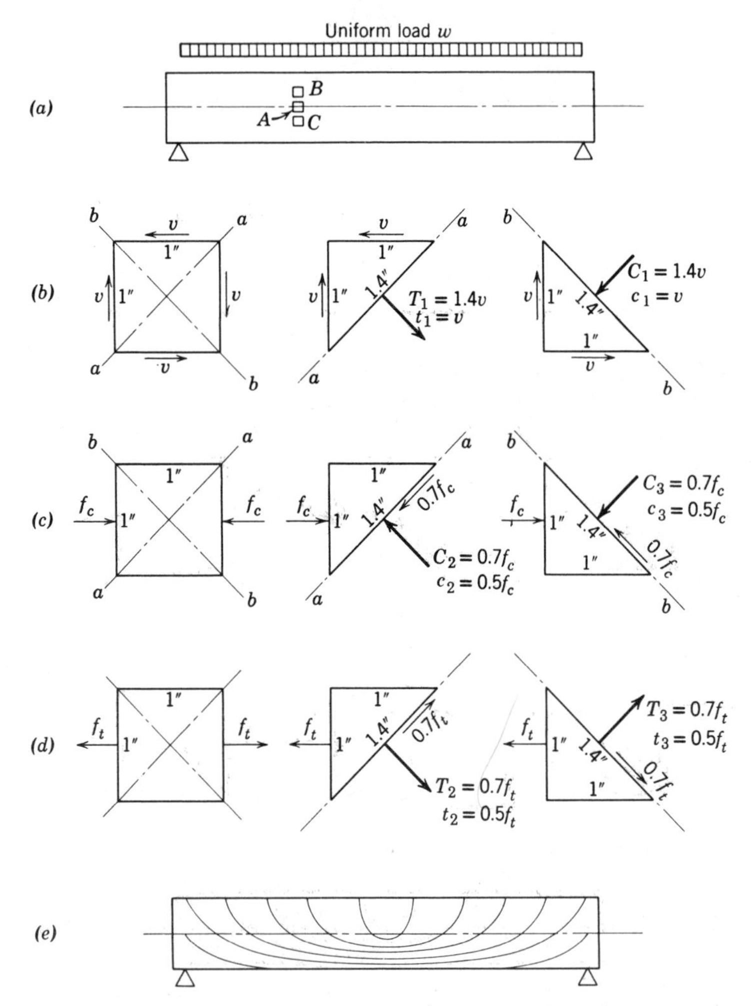

Fig. 4.1. Diagonal stresses in a homogeneous beam. (*a*) Typical beam under uniform load. (*b*) Analysis of stresses at *A*. (*c*) Analysis of added stresses at *B*. (*d*) Analysis of added stresses at *C*. (*e*) Tension stress trajectories.

diagonal tension on section *a-a* is reduced and the diagonal compression on section *b-b* is increased. Any vertical loading on top of the beam, such as the uniform load, leads to a vertical compression which combines similarly although rotated 90 degrees to that sketched in Fig. 4.1*c*.

Similarly, an element at *C* in Fig. 4.1*a* will carry a tension stress as well as a shear, leading to the added stresses shown in Fig. 4.1*d*. These combine to increase diagonal tension on section *a-a* and reduce diagonal compression on section *b-b*. The combined diagonal stresses are not maximum on sections *a-a* and *b-b*, tension being maximum on a steeper plane than *a-a* when the (horizontal) direct stress is tension, or on a flatter plane when the (horizontal) direct stress is compression. Likewise, a load hung from the bottom of the beam adds to the diagonal tension and tends to shift its direction to one a little steeper.

The relation developed in mechanics for maximum diagonal tension, adjusted to the notation used here is:

$$t = f/2 + \sqrt{(f/2)^2 + v^2}$$

In this relation t is the unit diagonal tension, f is the unit direct stress, taken as positive when it is tension, and v is the unit shear. The direction of the maximum diagonal tension is given by the relation:

$$\tan 2\theta = 2v/f$$

where θ is the angle t makes with the stress f, in this case with the horizontal.

Figure 4.1*e* illustrates the approximate trajectory of maximum tensile stresses in a homogeneous rectangular beam under uniform loading. In a reinforced concrete beam the pattern will be very similar until cracks open, either vertical cracks due to moment or inclined cracks due to diagonal tension. Diagonal tension cracks would be roughly perpendicular to the trajectories shown in Fig. 4.1*e* for a beam uniformly loaded. Diagonal tension cracks usually open at approximately 45° with the axis of the beam, starting typically from the top of a moment crack, but in short shear spans or deep beams starting independently near the neutral axis.

4.3 NOMINAL SHEAR STRESS

Since the diagonal tensile stress at the neutral axis is equal to the unit shear, this unit shear stress was formerly used as a measure of the diagonal tension. In a beam cracked by flexure, such a unit stress was only a nominal one; in effect the equation assumed vertical shear carried by the beam as though it were uncracked but without any flexural tension stresses existing on the concrete.

In 1962, the Joint ASCE-ACI Committee on Shear and Diagonal Tension reported[1] that it could not yet define clearly the failure mechanism in shear. It recommended that under these circumstances earlier methods be abandoned in favor of a simple average shear stress produced by the load causing the first diagonal cracking. The Code since 1963 has followed this report in its major outlines. Hence the nominal ultimate shear stress, as a measure of diagonal tension, is now to be computed as:

$$v_u = V_u/bd \tag{4.1}$$

with b_w substituted for I-beams or T-beams. For beam design the critical V_u is to be taken at a distance d from the face of support, with minor exceptions.

4.4 BEAM BEHAVIOR UNDER SHEAR LOADING—WITHOUT STIRRUPS*

(a) Failure Types, by Appearance

A shear failure appears to be least complicated when it occurs away from loads and reactions.[2] It is convenient to classify shear failures in terms of the distance between test load and reaction, a distance designated as shear span or a, Fig. 4.3a.

The simple *diagonal tension failure* of Fig. 4.2 occurs when the shear span is more than $3d$ or $4d$, under which conditions there is ample room for the full crack to develop to failure.

As the shear span is reduced, the load itself begins to influence the diagonal crack more and more, in a manner that increases shear capacity. The failure sketched in Fig. 4.3a is called a *shear-compression failure* and occurs when the shear span is from d to $2.5d$, with a rapidly reducing influence in the range of $2.5d$ to $4d$, as indicated in Fig. 4.9a.

When the shear span is less than d, the pattern changes to a *splitting failure* or compression failure at the reaction at a still higher load, and a tendency toward sudden failure, as sketched in Fig. 4.3b.

Since no overall basic theory can be presented, it is essential for general understanding to note details which may some day become parts in a more logical "theory."

*Stirrups are vertical U-shaped reinforcement that anchor in the compression face of the beam and enclose the flexural steel bars.

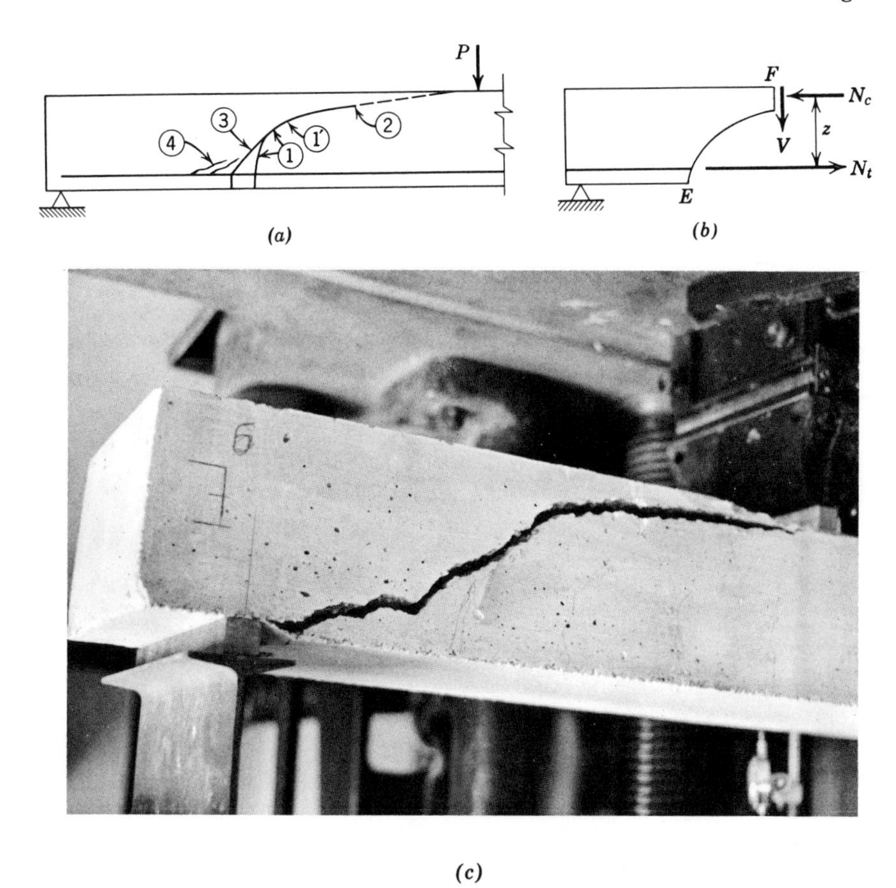

(c)

Fig. 4.2. Development of a diagonal tension crack when loads and reactions are far apart. (*a*) Diagram showing sequence in crack fromation. (*b*) Equilibrium sketch for portion of beam. (*c*) Failure of beam. The failure crack developed from the flexural crack faintly seen about one beam depth from the end. This crack turned gradually into the diagonal crack, as at 1 in the sketch. The final wide crack is comparable to 2–1–3–4 in (*a*) (The failure picture has been inverted to make this comparison easier).

(b) Diagonal Tension Failure

Always in the range of a/d above 2, and sometimes at lower a/d values, the diagonal crack starts from the last flexural crack and turns gradually into a crack more and more inclined under the shear loading, as noted in Fig. 4.2*a*. Such a crack does not proceed immediately to failure, although in some of the longer shear spans this either seems almost to be the case or an entirely new and flatter diagonal crack suddenly causes failure. More

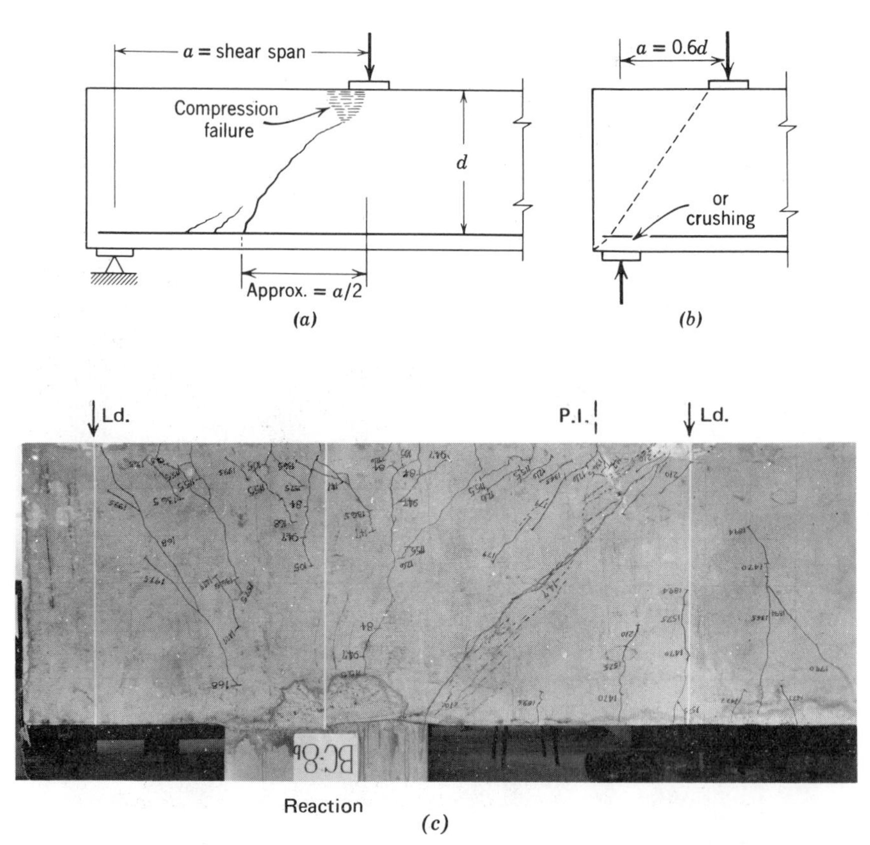

Fig. 4.3. (*a*) Shear-compression failure when shear span is small. Shear strength is increased. (*b*) Shear span less than *d*. (*c*) Failure in short shear span of semicontinuous beam, from load to reaction.

typically, the diagonal crack encounters resistance as it moves up into the zone of compression, becomes flatter, and stops at some point such as that marked 1′ in Fig. 4.2*a*. With further load, the tension crack extends gradually at a very flat slope until finally sudden failure occurs, possibly from point 2. Shortly before reaching the critical failure point at 2 the more inclined lower crack 3 will open back, at least, to the steel level and usually cracks marked 4 will develop. Cracks 3 and 4 are further discussed under (*f*) below. Figure 4.2*c* illustrates a failure with the start of the crack nearer the end than the usual $a/2$, this location resulting because two tension bars were cut off at the crack point. This crack is called a flexure-shear crack.

(c) Shear-Compression Failure

Very often the development of the diagonal crack described above is stopped by the presence of a nearby load, as indicated in Fig. 4.3a. Then the vertical compressive stresses under the load reduce the possibility of further tension cracking, and the vertical compressive stresses over the reaction likewise limit the bond splitting and diagonal cracking along the steel. Alternatively, a large shear in short shear spans (especially in I-beams) may initiate approximately a 45-degree crack across the neutral axis before a flexural crack appears. Such a crack crowds the shear resistance into a smaller depth and, by thus increasing the stresses, tends to be self-propagating until stopped by the load or reaction. With either start* a compression failure finally occurs adjacent to the load.

This type of failure has been designated as a shear-compression failure because the shaded area also carries most of the shear and the failure is caused by the combination. Such a failure can be expected to occur when the shear span a, as indicated in Fig. 4.3a, is less than four times the beam depth, or possibly a little less for lightweight concrete or very high strength concrete. When the shear span is small, the increased shear strength may be significant, the ultimate shear being over twice as much for $a = 1.5d$ as for $a = 3.0d$ (Fig. 4.9). The width of the critical crack, if there is no crack control steel, becomes large as the load increases, sometimes over one-eighth inch.

Occasionally with inadequate anchorage of the flexural steel beyond the crack, the small diagonal cracks (raveling cracks) will split out the bars before the compression failure occurs, this being called a shear-tension failure. The beam here acts somewhat as a tied arch with a thrust line from the load toward the reaction area, which demands full anchorage of the bars beyond the crack.

(d) Splitting or True Shear Failure

Where the shear span is less than the effective depth d, the shear is carried as an inclined thrust between load and reaction which almost eliminates ordinary diagonal tension concepts. Shear strength is much higher in such cases.[3] The final failure, as in Fig. 4.3b, becomes a splitting failure, almost like the vertical splitting of a compression test cylinder (which occurs when end friction is reduced) or it may fail in compression

*The second case is not always classified as shear-compression; the term web-shear crack is often used.

at the reaction. The analysis of such an end section is closely related to the analysis of a deep beam having a span of $2a$. Reference 3 has shown that it is the clear shear span between bearing and loading plates that is critical.

(e) Shear Transfer

The internal stress transfer in an uncracked beam (Fig. 4.1) is greatly complicated by the flexural cracking; and the flexural cracking itself is complicated by the presence of shear. In a constant moment region, flexural cracks are essentially perpendicular to the tensile reinforcement. In a shear zone, say at a/d of 4, the flexural cracks form first under the load and then, as moment increases, progressively farther from the load. The last-formed flexural cracks do not initially run as high into the web as the earlier ones and they tend to slope a little toward the load, this slope being more from the vertical when shear is important; then the critical one is like that in Fig. 4.2*a*.

Kani has spoken of these cracks as forming small cantilevers downward from the uncracked compression part of the beam, like prongs or teeth on a comb. In a shear zone each cantilever must pick up horizontal shear equal to the change in bar tension over that increment. Kani thought of the cantilever moment thus produced as eventually leading to progressive failure of these plain concrete cantilevers and thus shear failure of the beam.

In the absence of stirrups, vertical shear can be carried by (1) the uncracked compression concrete of the beam, (2) dowel action of the bars carrying shear across the cracks, and (3) shear between the cantilever teeth across the curved crack interface. Paulay and Fenwick in New Zealand established[4] that the aggregate interlock across these tension cracks accounted for substantial shear transfer, most effectively when cracks were small, but always significantly since the crack surface is far from smooth. They found at ultimate from 33% to 60% of the vertical shear was transferred by this interlock. Dowel action can be large but near ultimate appears to account for not more than 10% to 15% of the external shear. These investigators also showed that the agregate interlock shears changed Kani's simple cantilever into a very complex one with much lower moment at the critical section. The relation of the raveling cracks in Fig. 4.2*a* to the aggregate interlock is mentioned in the next subsection.

Beyond the last flexural (and shear) crack the shear is carried as an inclined thrust, actually a mixture of thrust and some flexural action and shear. As a/d becomes smaller, the shear failure begins to relate more and more to the strength of an inclined compression member carrying some bending and shear. The limiting case is the splitting or pure shear failure when a is less than d.

(f) Internal Stress Complications

A problem exists at the bottom of the crack in Fig. 4.2*a* (point E in Fig. 4.2*b*) which involves an increased bar stress, bond stress, and local shear stress. If the dowel action of the bar and the aggregate interlock along the crack are considered minor (and omitted), moments about N_c at F show that this moment M_F measures N_t, which is the tension at E*. Thus the steel at E suddenly picks up all the tension that to the left was carried by the concrete (a normal situation at the first flexural crack) and *also* all the normal tension increment that should develop from E to F. This creates a severe local bond stress (between steel and concrete) just to the left of E.

This large bond stress leads to a large localized unit shear stress just above the steel, and to the localized diagonal cracking (or raveling) indicated at point 4 in Fig. 4.2*a*. This diagonal cracking is also promoted by vertical tensile stress put into the concrete by whatever vertical load is carried across the crack by the dowel action of the bars and (Paulay and Fenwick show) chiefly by vertical shear transferred by interlock. This may explain why this local shear failure, always observed, indicates the approaching general failure, caused by the shift of this overload to a more critical spot.

Such localized raveling can extend and create the effect of a bond splitting failure. This raveling appears to have been prominent in some diagonal tension failures which occurred near the point of inflection of continuous beams in actual structures.

Other more obvious external loading influences on internal stresses are discussed in Sec. 4.7*c*.

(g) Interaction

Bond, shear, and moment resistance must not be regarded as independent responses to given loads. It is convenient for calculation purposes to treat each as a separate calculation. It is possible to design sections which in the laboratory will fail in any one of these three designated manners,

*The 1971 Code requirement of stirrups in beams where diagonal cracks might form will add vertical forces to this equilibrium sketch that reduce N_t at E and also restrict the raveling.

that is, specimens which appear to be free from any significant influence from the other two responses. However, when other combinations of dimensions and loading are used, unexpected interactions between these responses show up. The following three cases have already been noted.

1. In Sec. 4.4*b* it was noted that flexural cracking always precedes diagonal tension weakness, except for small a/d values. It must follow that the diagonal tension capacity is influenced by the height of these flexural cracks and the beam length which does crack. The smaller steel areas appropriate with the use of high strength steel result in deeper cracks and cracks over more of the length. On the other hand, the uncracked zone at a point of inflection is somewhat protected and stronger.
2. A study of the bond failure shown in Fig. 5.3*c* and the special stresses resulting from a diagonal crack (Fig. 4.2*b*) indicates that the diagonal crack has almost surely affected the bond resistance adversely.
3. The severe local bond conditions indicated in Fig. 4.3 contribute to the local diagonal cracking (raveling out) which lowers both bond resistance and shear resistance. The failure can be definitely bond, definitely shear, or a mixture.

Such interrelations, some actual and some possible, point up the need for further research. The change of bar stress must develop a bond stress over the total bar perimeter that transfers into the beam as an equal total horizontal shear over the total beam width (in pounds per linear inch of beam). Because of this geometric relationship between bond and shear stresses it is probable that neither can be well understood until such interactions are thoroughly explored.

Design provisions for beams are discussed beginning in Sec. 4.7.

4.5 ONE-WAY SLAB BEHAVIOR UNDER SHEAR LOADING

Since a one-way slab is really a wide shallow beam, no shear behavior different from that in beams should occur, but three comments are appropriate.

1. Shear stresses are usually low in one-way slabs, except where heavily loaded as in one-way footings under heavy walls.
2. Stirrups are rarely feasible to use in thin members like most slabs.
3. Concentrated loads require transverse reinforcement and develop shear demands similar to those discussed in the next section.

4.6 BEHAVIOR OF TWO-WAY SLABS AND FOOTINGS UNDER SHEAR LOADING

(a) Punching Shear

When a two-way slab is heavily loaded with a concentrated load or a column rests on a two-way footing, diagonal tension cracks form which encircle the load or column.[4] These are not visible on the surface, except as flexural cracks. Such cracks extend into the compression area of the slab and encounter resistance similar to the shear-compression condition shown for the beam in Fig. 4.3*a*. The slab or footing continues to take load and finally fails around and against the load or column, punching out a pyramid of concrete as indicated in Fig. 4.4. Diagonal cracks do not form further out from the load or column because of the rapid increase in the failure perimeter, which always totals eight times the distance from the center of load. The initial diagonal cracks thus proceed to failure in shear compression (or punching shear) directly around the load.

In compromising between the initial cracking location and the final punching shear condition at failure for different ratios between column (or load) diameter and footing (or slab) thickness, the Joint Committee[1] recommended a single-strength calculation at a pseudo-critical distance $d/2$ from the column face or edge of the load as shown in Fig. 4.5. Because there is no danger aside from the shear compression type of failure (with

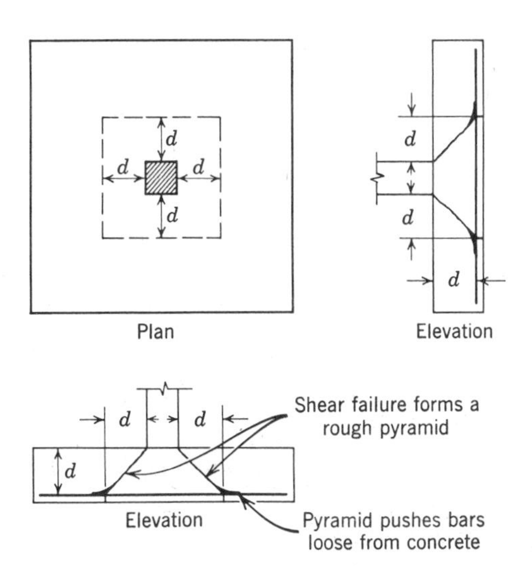

Fig. 4.4. A square column tends to shear out a pyramid from a footing.

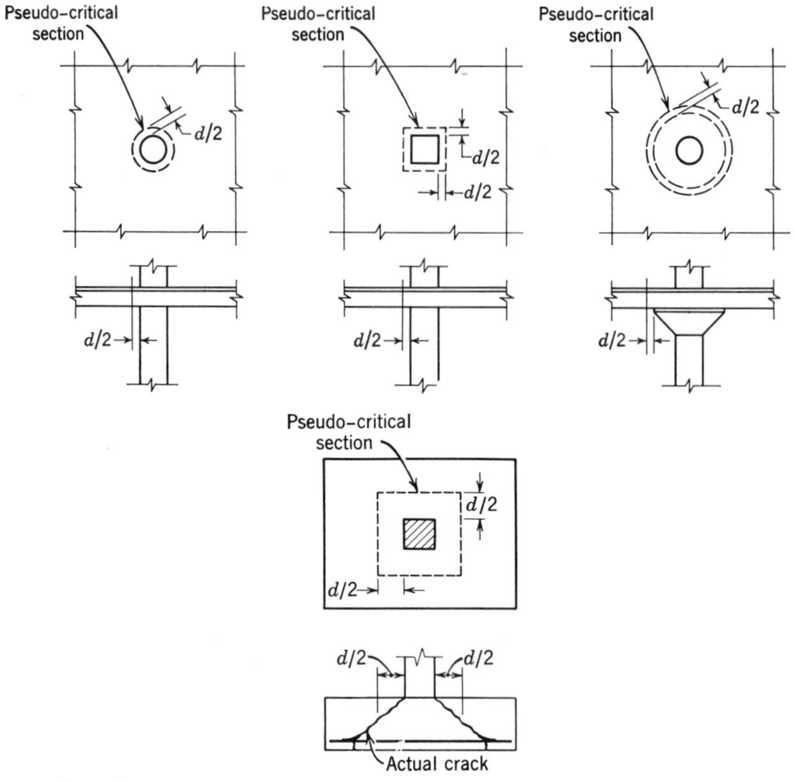

Fig. 4.5. Pseudocritical diagonal tension sections in slabs supported directly on columns and in footings.

the diagonal crack not visible from outside the concrete), these slab unit shear strengths are based on ultimate strengths (rather than cracking loads) and substantially exceed those of one-way slabs and beams. This increased strength is available *only when two-way bending* occurs at the load considered.

Shearhead reinforcement where slab thickness is inadequate is discussed in Sec. 10.23.

(b) Openings in Slabs which Influence Shear Strength

The Code suggests a simple way of handling openings near columns (Code 11.12). Radial lines are drawn from the edges of the openings to the center of the column as shown in Fig. 4.6. These radial lines intersect the pseudo-critical sections around the column, and all the resisting perimeter which is caught between such radial lines is considered as ineffective in resisting shears. If too much perimeter is lost, the designer must be sure adequate two-way bending is really present; otherwise the lower shears permitted in beams become the limiting values for the slab. The Joint Committee Report[1] on shear discusses openings in detail.

Shearhead reinforcement permits a different treatment for openings (Code 11.12*b*) because the support is moved farther from the column (Sec. 10.23).

(c) Transfer of Moment from Column into Slab

Another condition which adds locally to the shear around the column in flat plate construction arises when the column carries a moment (for example, the moment arising from lateral loads such as wind or earthquake) which must be resisted by the slab. The Code requires an analysis of the

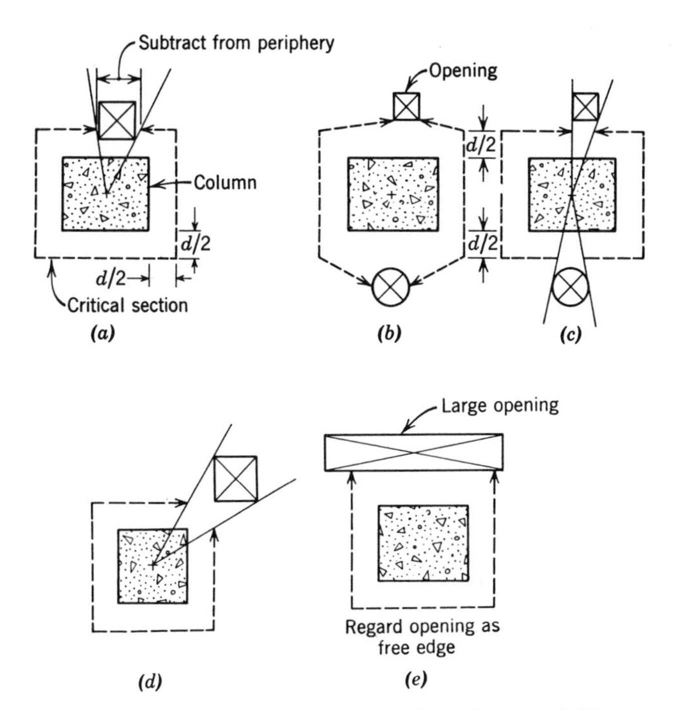

Fig. 4.6. Effect of openings on shear perimeter. (From Ref. 1, ACI.)

resulting shears. It recognizes that much of the moment is transferred directly into flexure in the slab, but allows for the extra punching type shear by specifying the following portion of the moment M_u transferred be used in the shear calculation:

$$\left(1 - \frac{1}{1 + \frac{2}{3}\sqrt{\frac{c_1 + d}{c_2 + d}}}\right) M_u$$

where c_1 is the column width parallel to span of the slab strip that will resist the moment by flexure and c_2 is in the perpendicular direction, as in Fig. 4.7. For a square column this moment becomes $0.4M_u$ applied as a torsion on the slab.

The ACI-ASCE Committee 326 in 1962 recommended essentially that the slab shear be analyzed as follows.*

1. Consider the usual critical shear perimeter at $d/2$ from the column to resist this shear as a torsion.
2. The resisting area becomes the four vertical sides of the slab cut by the perimeter. The axis of torsion is a horizontal line, through the column center for the usual interior column.
3. Calculate the polar moment of inertia of the resisting slab areas. From Fig. 4.7*a*

$$J = 2(b_1 d^3/12 + d b_1^{\,3}/12) + 2\, b_2 d (b_1/2)^2$$

4. Obtain the unit shear from Tx/J where T is the portion of M_u computed above and x is the distance from the torsion axis, with the maximum at $x = b_1/2$.
5. This torsional shear is to be added to the usual vertical shear as in Fig. 4.7*b*, quite similar to the P/A and Mc/I combination in elastic theory.

If one considers an edge column, the following adjustments are to be made.

1. The torsion axis must be the centroid of the three resisting slab faces.
2. The polar moment of inertia and shear area are correspondingly reduced.

*The proportion of the column moment assigned to torsion has been increased in the 1971 Code, as noted in the preceding paragraph.

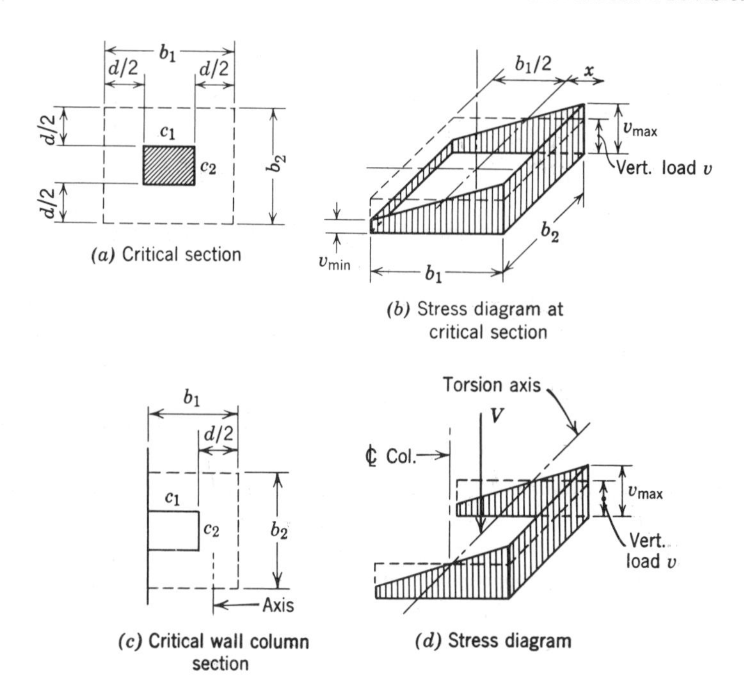

Fig. 4.7. Shears around a column which must introduce moment (and torsion) into a slab or accept moment from the slab.

4.7 GENERAL SHEAR PROVISIONS

(a) Safety Provisions for Shear

For shear the coefficient ϕ is to be taken as 0.85.

(b) Stirrups Required in Beams

The 1971 Code moved a little in the direction of long standing European practice in requiring stirrups in major beams, that is, in all flexural members (prestressed and nonprestressed) except:

1. Slabs and footings.
2. Joists, either one-way or two-way, as defined in Code 8.8.
3. Beams with:
   a. Total depth not exceeding 10 in.

b. Total depth not exceeding 2.5 times flange thickness.
c. Total depth not exceeding 0.5 times web width.

4. Lengths where $v_u < v_c/2$.

(c) Design Considerations Relating to Shear Span

As a result of the shear span effect, the worst position (on the basis of diagonal tension failure) for a concentrated load on the usual simple span test beam is not adjacent to the reaction but at some distance out on the span.

In test beams most of the extra resistance beyond initial diagonal cracking is created by the vertical compressive stress under the load and over the reaction. Loads or reaction applied as shears, as in Fig. 4.8, create very little vertical compression and add very little shear resistance.[2] If loads are applied below mid-depth of a girder, say, from precast beams bearing on a lower flange or bracket attached near the bottom of the girder, diagonal cracking occurs at lower loads and ultimate strength is reduced. Stirrups as hangers are essential in such cases.

Tests[2] have also shown that if the loads at a small a/d are applied as shear loads on the side of the beam, as in Fig. 4.8*a*, or if the reaction is picked up in shear (as when a beam frames into the side of a girder, as in Fig. 4.8*b*), not much increase in shear resistance is obtained as a result of the small shear spans.

(d) The Influence of Flexural Cracking on Shear Strength

Section 4.4*b* has pointed out, for an a/d of at least 2, that diagonal cracking always developed from an existing flexural crack. Thinking of the

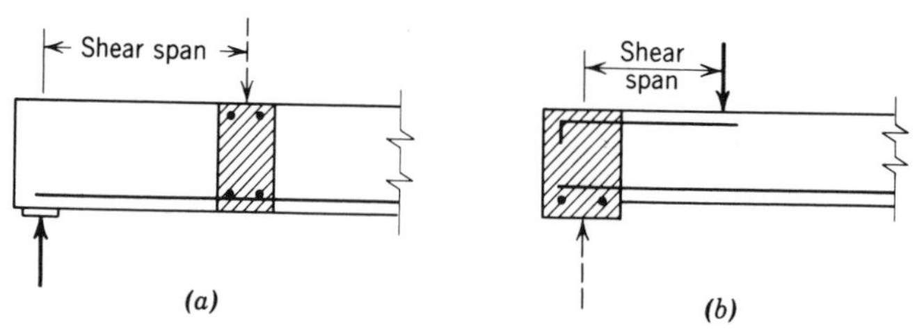

Fig. 4.8. Cases where small shear span does not significantly increase shear strength. (*a*) Girder loaded by shear from beam. (*b*) Beam supported by girder.

flexural compression zone as the region that tends to stop diagonal cracks, the author once wrote that the area around a point of inflection might be weak against diagonal tension. However, tests on simulated continuous beams have since shown that diagonal tension strength is a minimum where moment cracks have opened rather than nearer the point of inflection where moment cracks do not form. Shear weakness, on other than short shear spans and I-shaped prestressed concrete beams, develops from the *sequence* of flexural cracking followed by diagonal tension cracking.

4.8 DESIGN SECTIONS AND DESIGN STRESSES

The Joint Committee[1] was unable to incorporate the basic shear concepts into a mechanism of failure or an ultimate strength equation and was forced to proceed empirically in its report. Because (1) in large shear spans very little if any reserve strength existed after a diagonal tension crack formed, and because (2) the diagonal crack widths which were developed with short shear spans before the ultimate was reached were quite excessive in some cases, the Committee recommended that design be based on first diagonal crack strength. This recommendation eliminates many of the ultimate strength complications, but possibly more so than is desirable.

(a) Design Section for Beams and One-Way Slabs

Next to supports the Code (11.2.2) designates a length of member equal to the depth d for special treatment, provided the reaction introduces compression in the end of the member in the direction of the shear. In effect one then may calculate the unit shear at this distance d from the support and assume that the usual increase in shear over this distance may be ignored because shear strength in this zone is also somewhat increased. Do not ignore a heavy load that happens to fall within this distance d.

(b) Allowable Shear Stresses on Concrete

In general, design for torsion becomes a part of design for shear except that a torsional shear v_t not exceeding $1.5\sqrt{f_c'}$ may be ignored. In this book torsion combined with shear will be treated separately, starting with Sec. 4.19.

With torque negligible two slightly different permissible shear stresses are given. The first of these, $2\sqrt{f_c'}$, is an approximate value which will usually be adequate for design. The second is the more complete strength formula reported by the Joint Committee in 1962.

$$v_c = 1.9\sqrt{f_c'} + 2500\rho_w V_u d/M_u \ngtr 3.5\sqrt{f_c'} \qquad (4.2)$$

In this equation the value of the second term varies with the Vd/M ratio along the member, which is constant only for very simple loadings. It is provided that $M \nless Vd$ and thus Vd/M can never exceed unity. The term ρ_w is $A_s/b_w d$ for a T-beam and would simply be ρ in a rectangular beam. In general, the second term is small, but when Vd/M is unity and ρ_w is large, it can be appreciable. Thus its use in special cases may be economical. The total v_c is limited to $3.5\sqrt{f_c'}$. Although this equation was really empirically derived, it does incorporate the two variables ρ and M/Vd which, after f_c', seem to be the most significant in reported research. The equation has a good correlation with the observed strengths at the (diagonal) cracking load.

(c) Allowable Shear Stress on Joists

The value of v_c may be taken as 10 percent higher than the values listed in Sec. 4.8b.

(d) Allowable Shear Stresses when Axial Forces Are Present

Axial tension or compression can be even more influential on diagonal tension resistance than the vertical forces mentioned in Sec. 4.2. Overlooked axial tension from shrinkage or temperature variation plays a prominent part in most beam failures that occur in structures. The Code provides separate relations for axial tension and axial compression.

1. *Axial tension.* Where axial tension is large, as in the tension chord of a Vierendeel truss (no diagonals), the concrete may be cracked through in tension and unable to help the stirrups carry the tension aspects of shear. Stirrups must then carry the full shear. The permissible v_c can be evaluated from (Code 11.4.4)

$$v_c = 2(1 + 0.002N_u/A_g)\sqrt{f_c'}$$

   where N_u is *negative* for tension, with units of psi for N_u/A_g. For 500 psi axial tension this v_c drops to zero.

2. *Axial compression.* The axial compression forces actually available in a compression strut or column or compression chord of a Vierendeel

truss greatly improve its capacity to resist shear. Yet a recent earthquake in California showed serious shear problems in the columns. In spite of uncertainties as to acting shear and magnitude of uplift accompanying, it is evident that column shear is one of the earthquake hazards that must be investigated.

The Code gives two methods for evaluating v_c, the first being to use the normal shear equation [Eq. 4.2 in subsection (b) above] with a modified moment

$$M' = M_u - N_u(4h - d)/8 \gtreqless 0$$

where N_u is positive (for compression). The second term is the approximate moment of N_u about the center of resisting compression. The lowered moment increases Vd/M and in this case the Code does permit values of this ratio greater than unity. An upper limit for the resulting v_c is also specified as

$$v_c = 3.5\sqrt{f_c'}\sqrt{1 + 0.002N_u/A_g}$$

with N_u/A_g again in psi.

The alternate equation of Code 11.4.3 is simpler but normally leads to a lower allowable v_c:

$$v_c = 2(1 + 0.0005N_u/A_g)\sqrt{f_c'}$$

For 500 psi compression this gives $2.5\sqrt{f_c'}$ compared to $2\sqrt{f_c'}$ without compression. This equation is not required if the other is used; it is an alternate.

(e) Allowable Shear Stress on Two-way Slabs and Footings

These special provisions apply only where two-way bending occurs, as from a concentrated load on a slab or from a column load on a two-way footing. They do not apply to the slab shear adjacent to supporting beams or to one-way footings such as wall footings, where bending is primarily unidirectional.

In slabs with two-way bending the strength in shear is increased, as discussed in Sec. 4.6*a*. The Code recognizes this by fixing the permissble shear at $4\sqrt{f_c'}$ when measured around a periphery at a distance $d/2$ from the load or column (Fig. 4.5).

The Code also requires a check on a section all across a footing, as for a wide one-way slab, using the beam permissible stress and critical section as outlined in Sec. 4.8*b*. Apparently, this provision will govern only in rather narrow long footings where two-way action is at a minimum.

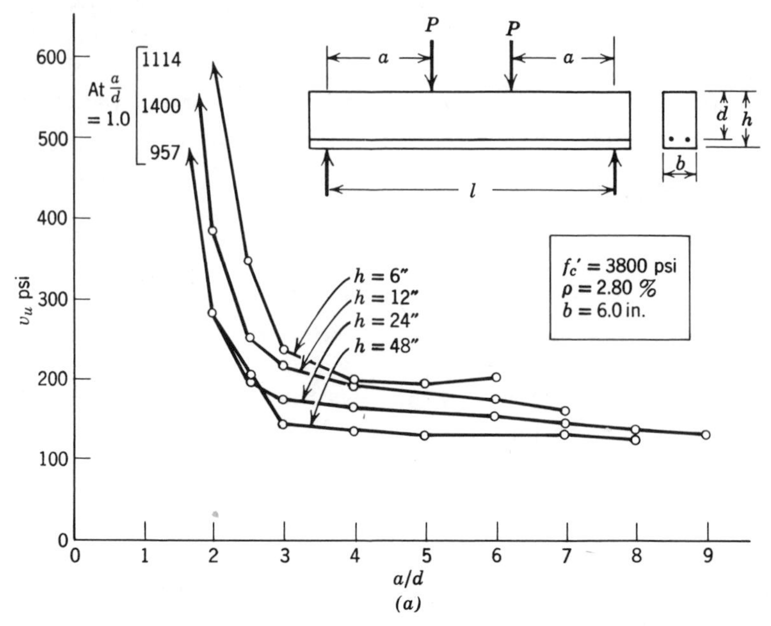

Fig. 4.9. Influence of several variables on shear resistance, without stirrups. (Adapted from Ref. 4 and 5, ACI.) (*a*) Influence of a/d and beam depth. (*b*) Influence of ρ_w of longitudinal steel, with $a/d \gtrdot 2.5$. (See opposite page.)

(f) Lightweight Aggregate Concretes

Some lightweight aggregate concretes have a lower shear resistance than ordinary aggregate concrete of the same compressive strength. The Code (11.3) provides alternate procedures, (1) allowable stress multiplied by an arbitrary 0.75 factor for "all-lightweight" and 0.83 for "sand-lightweight" concrete,* or (2) when f_{ct} is suitably specified the use of $f_{ct}/6.7 \lessdot \sqrt{f_c'}$ instead of $\sqrt{f_c'}$ in stating the allowable. The term f_{ct} is the average splitting tensile strength.

(g) Need for Crack Control Steel in Deeper Beams

A few years ago Kani raised the question, how safe are large beams?[5] He presented the curves of Fig. 4.9*a* and emphasized that with an overall height of 48 in. the shear at any a/d was less than with a 6-in. height. The

*"Sand-lightweight" is concrete using natural sand for fine aggregate.

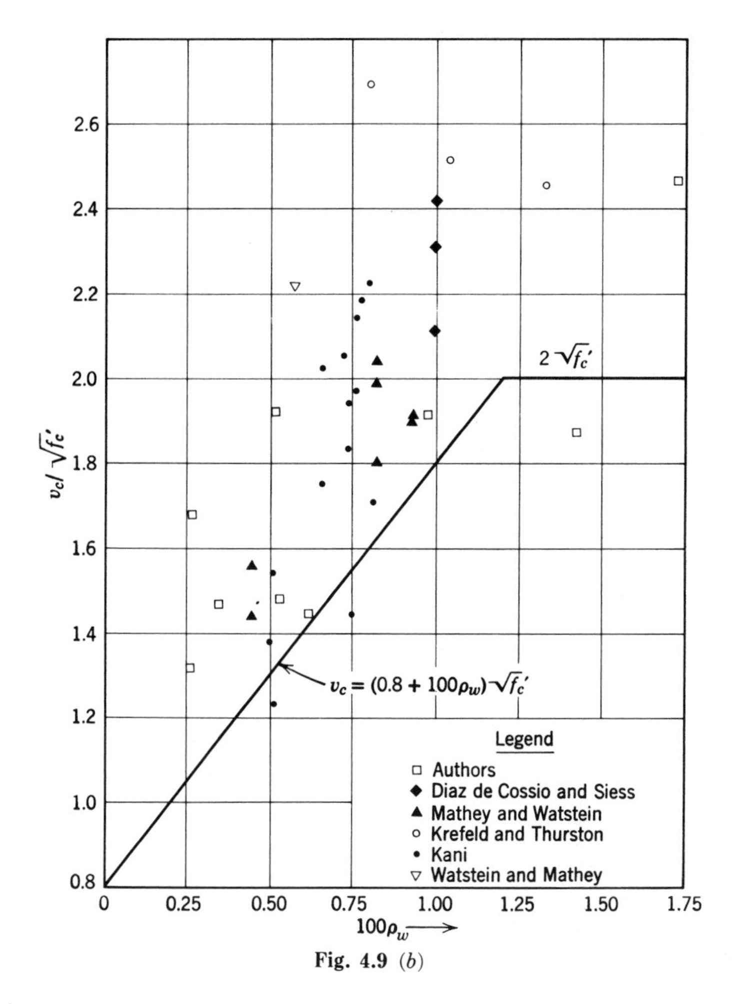

Fig. 4.9 (*b*)

author doubts that this is really a scale effect and feels that three specific conditions account for the differences; he also notes that the weakest beam did develop a shear stress of $2\sqrt{f_c'}$.

First, all beams were 6-in. wide, with the 48-in. beam crowded with 6 large bars (#9 and #10) to keep the high ρ chosen. Pictures show that this and the 24-in. beams failed in shear-tension, really a shear induced splitting problem; the 12-in. beam had no such problem.

Second, the different areas of steel and numbers of bars necessarily changed the crack spacing significantly, which probably influenced the failure.

Third, and probably the most significant, is the width to which even flexural cracks open in a large beam depth, away from the tension steel.

In a small beam the limited distance between the tension steel and the neutral axis controls the crack width. In contrast, the author was almost startled by the flexural crack widths in the first 36-in. beam he tested, cracks substantially wider toward mid-depth than at the steel level. There could be little aggregate interlock across such crack widths. This was probably one of the major weaknesses in Kani's deep beams.

Crack control steel is needed in all beams over 24-in. deep if appearance is of any importance. Horizontal bars or wire fabric are needed on each side for this purpose. Even a very small steel area is helpful. Probably #3 bars at 18-in. centers each face would have made quite a noticeable difference in the 48-in. beam. For a heavier beam, say one 24-in. thick, maybe #5 bars at 12 in. should be used, something like the temperature and shrinkage steel. Code 10.6.5 requires all beams more than 3-ft deep to have side face steel in the tension zone equal to 10% of A_s spaced not over 12 in. (or width of web if less).

(h) Special Shear Problems

Special problems, such as the ill effects of cutting off bars, members of varying depth, the shear-friction theory, and special members such as brackets and deep beams, are presented starting with Sec. 4.14, following the sections on stirrups.

4.9 NEEDED CODE CHANGES FOR SHEAR

(a) Reduced Shear Stress Where ρ Is Less Than 0.01

Tests have shown[6] that the Code overestimates shear resistance where flexural steel ρ is less than 0.01 and the a/d ratio is as large as 2.75. These lower strengths are probably the result of higher steel stresses and wider cracks that accompany low steel percentages, with resultant loss in the aggregate interlock (Sec. 4.4e). On the basis of Fig. 4.9*b* the author recommends that for all beams (except cantilevers uniformly loaded) the usual allowable stress value of $2\sqrt{f_c'}$ be reduced where $\rho < 0.012$ to

$$v_c = (0.8 + 100\rho_w)\sqrt{f_c'} \tag{4.3}$$

Cantilevers such as in wall footings have been excepted because they have lower flexural steel stresses at their critical shear section than is the case in the usual test specimen; test verification would be desirable.

(b) Reconsideration of Vd/M Term for Shear in Continuous Beams

Equation 6.2 of Sec. 4.8*b* was first proposed[1] in 1962 as an evaluation of the shear cracking stress, with the term Vd/M representing the influence of the shear span. Evaluated largely on the basis of simple span tests with concentrated loads, where Vd/M and d/a are interchangeable, its applicability to continuous beams rests on only scattering tests.

In a continuous span Vd/M and d/a are quite different and only d/a lends itself to a plausible physical interpretation. For the simple span with concentrated loads one can visualize a reducing shear span as making feasible an inclined strut to carry much of the shear;* the smaller a/d (larger d/a) means a steeper strut and one leaving less of the shear to cause diagonal tension problems. However, applied to a continuous beam, the equation seems to imply that the point of inflection acts as a reaction for the positive moment length and as a load point for the negative moment length; but the P.I. furnishes neither the reaction below nor the load above to produce the compression forces which make the simple span concept valid. Instead, as in Fig. 4.3*c*, the failure line goes from real load to real reaction and the P.I. seems to have shown no influence.

To think in terms of the slope of the end strut for a continuous beam, the equation must be in terms of d/a and not Vd/M. If there is a P.I. midway between load and reaction, the strut can be no steeper than if it were a simple span, as indicated in Fig. 4.10; but the use of Vd/M would mean the influence of this term was doubled by the change from simple span to continuous. This appears on the *unsafe* side, even for the diagonal cracking load.

Accordingly, for *continuous beams* the author now *recommends:*

1. Further study and tests to look again at the Joint Committee report equation.
2. That, in the interim, d/a be used in the equation instead of Vd/M.
3. That, for uniform load, d/a be used as 0.25 or the overall $2\sqrt{f_c'}$ be used, the particular d/a being a judgment decision.

4.10 WEB REINFORCEMENT

Web reinforcement is required in all major beams (Sec. 4.7*b*), over the lengths where the shear stress v_u exceeds half the allowable v_c as given in Sec. 4.8b. Stirrup size and spacing is based on $v_u - v_c$, with a lower limit of 50 psi.

*Rüsch and Leonhardt of Germany have separately made this point for years; the author now joins them.

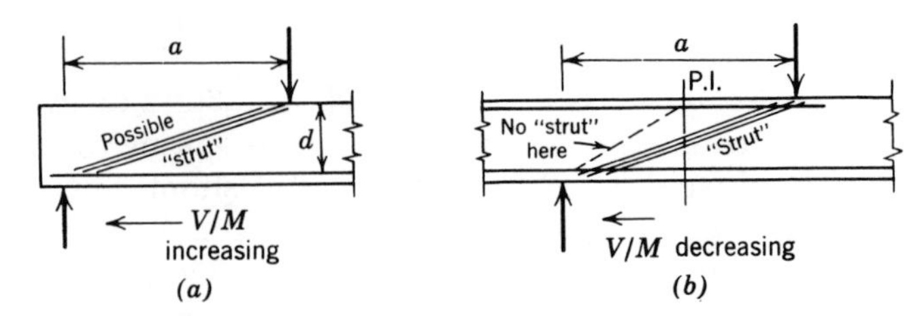

Fig. 4.10. Use of shear equation. (*a*) Simple span has $Vd/M = d/a$. (*b*) Continuous span $Vd/M = d/(\text{distance to P.I.}) \neq d/a$.

The most common web reinforcement consists of vertical stirrups, usually in U-shape, but occasionally in W-shape or in closed form, as shown in Fig. 4.11. Unfortunately, the vertical stirrup carries no significant stress until after a diagonal crack forms. It has little, if any, effect on the shear at which such cracks form. Once a diagonal crack opens, vertical stirrups act in tension to carry load from one side of the crack to the other. A common analogy considers the stirrups acting as tension verticals in a truss, with the concrete acting as compression diagonals, as shown in Fig. 4.11*c*. Tests show that the beam cannot fail by further opening of the diagonal crack until the stirrup stress passes the yield-point value. Even up to failure a portion of the shear is still carried by the concrete within the compression area above the crack.

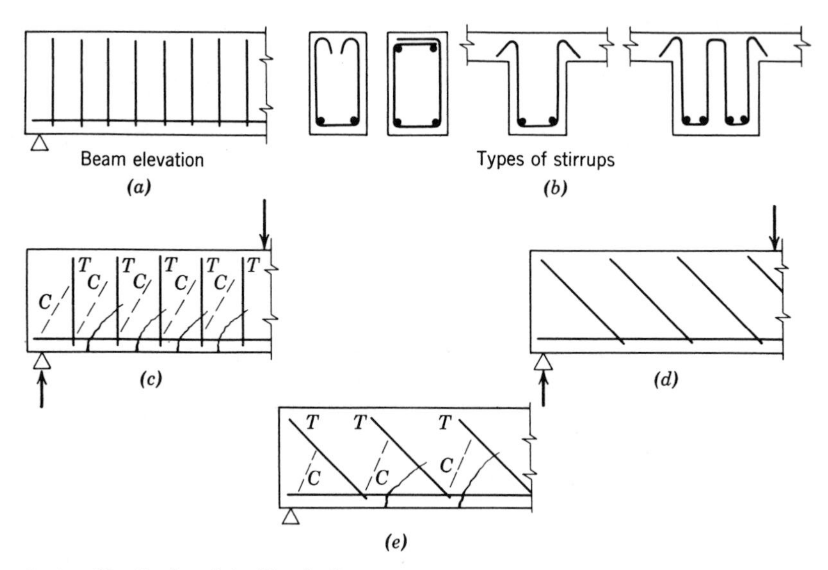

Fig. 4.11. Vertical and inclined stirrups.

European practice, which formerly assigned all shear strength to the stirrups and none to the concrete, is starting to move toward fewer stirrups as a result of low measured stirrup stresses in test beams.[12] They reason that (1) the compression members (concrete) act at flatter angles than 45° and thus deliver a smaller component to the stirrup verticals, and (2) the upper chord of the "truss" forms as an inclined thrust line which also leads to lower stress in the verticals. The second reason is similar to ACI's reasoning but is more specific as to *how* the concrete carries part of the shear.

Diagonal or inclined stirrups (Fig. 4.11*d*) are aligned more nearly with the principal tension stresses in the beam. They share in carrying this tension and slightly delay the formation of diagonal tension cracks. Inclined stirrups would be a preferred type where preassembled cages of steel are used, as in some present construction, but actually they are little used in this country. The truss analogy is still applicable, with stirrups acting as tension diagonals which alternate with concrete compression diagonals, as in a Warren type of truss (Fig. 4.11*e*).

Longitudinal bars are often bent up where no longer needed for moment. Such bent bars act also as inclined stirrups. Although many designs use bent-up longitudinal bars, only a few designers in the United States take the trouble to calculate their value as stirrups. One reason is that usually only a few bars are bent and these may not be conveniently spaced for use as web reinforcement.

Bent-up bars are normally bent at a 45° angle, but in some members, usually light joists, a flatter angle is used to secure some effect as diagonal tension reinforcement over a greater length.

One vital function of stirrups for which vertical stirrups are ideal is that of picking up any shear carried across a diagonal tension crack by dowel action of the longitudinal steel. They prevent the raveling out just above the steel at the end of a diagonal crack by transferring the dowel action load into a stirrup stress instead of into a vertical tension stress on the concrete.

In ordinary one-way slabs shear stresses are low. Possible shear reinforcement for flat plate construction is covered in Chapter 10.

4.11 SPACING RELATIONS FOR STIRRUPS

(a) Vertical Stirrups

Although a general formula for spacing of stirrups inclined at any angle is derived in Sec. 4.11*c*, the special case of vertical stirrups will be presented

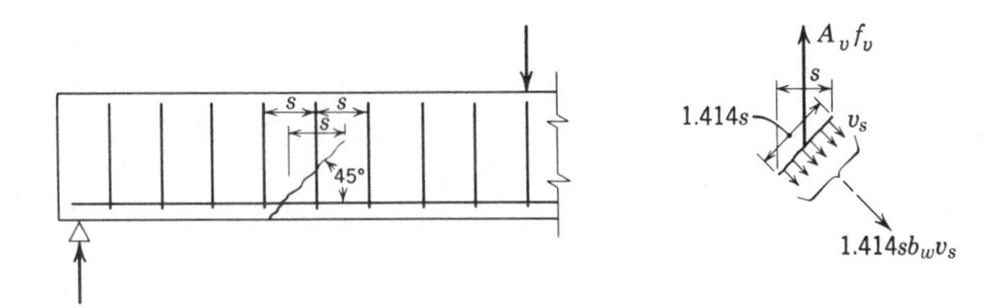

Fig. 4.12. Basis for calculation of vertical stirrups.

first because these are almost universally used in the U.S.A. The beam of Fig. 4.12 shows a stirrup intercepting a typical 45° diagonal tension crack. The stirrup is assumed to carry the vertical component of the diagonal tensile stresses originally acting across the crack over the horizontal length s. The horizontal component was automatically included in the design of the longitudinal A_s. The unit tension on the crack is taken as v_s and the area contributing stress to the stirrup is $1.414sb_w$, where b_w is the web thickness. The vertical component of concrete stress is $0.707(1.414sb_wv_s) = sb_wv_s$, which is equal to the vertical components of stirrup stress $= A_vf_y$.

$$sb_wv_s = A_vf_y \qquad s = \frac{A_vf_y}{b_wv_s} \tag{4.4}$$

The area of the stirrup includes two bar areas for a U-type stirrup. The width b_w for a T-beam becomes b when a rectangular beam is considered. The shear v_s is not the total shear on the beam since tests have shown some shear is carried by the concrete. No stirrups are required even in major beams until v_u exceeds $0.5\,v_c$ as given in Sec. 4.8, although some consider minimum stirrups through all beams good practice. In major beams minimum stirrups are also required where $v_c > v_u > 0.5v_c$. In any member stirrups are necessary for $v_s = v_u - v_c$.

The stirrup spacing formula can be written in terms of $\overline{V}_s$ where $\overline{V}_s = v_sb_wd$. The substitution of $\overline{V}_s/b_wd$ for v_s in Eq. 4.4 for s gives

$$s = \frac{A_vf_yd}{\overline{V}_s} \tag{4.5}$$

Stirrup spacing for stress varies inversely with v_s or $\overline{V}_s$, with the stirrup located in the middle of the length s it serves. The first stirrup should thus be placed at $s/2$ from the support, as in Fig. 4.13.

Stirrups must be spaced such that every 45° line representing a potential crack will be crossed in the tension side of the beam by at least one stirrup, as shown in Fig. 4.13a. This requirement limits the maximum s to $d/2$.

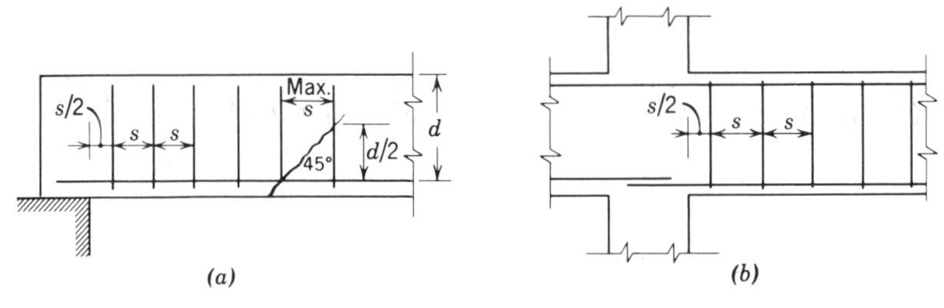

Fig. 4.13. Arrangement of vertical stirrups.

The Code (11.6.3) goes further to require that every such potential crack must intercept two stirrups whenever $v_u - v_c(=v_s)$ exceeds $4\sqrt{f_c'}$, which is equivalent to a maximum spacing of $d/4$. The Code permits the use of $v_u - v_c(=v_s)$ as large as $8\sqrt{f_c'}$.

Maximum stress on a stirrup is assumed to exist at mid-depth of the beam and the anchorage requirements* of Code 12.13 may limit either the size of stirrup bar that may be used or the stress f_y which may be utilized on still larger bars:

> 12.13.1—Web reinforcement shall be carried as close to the compression and tension surfaces of the member as cover requirements and the proximity of other steel will permit, and in any case the ends of single leg, simple U-, or multiple U-stirrup shall be anchored by one of the following means:
>
> 12.13.1.1 A standard hook plus an effective embedment of 0.5 l_d. The effective embedment of a stirrup leg shall be taken as the distance between the middepth of the member $d/2$ and the start of the hook (point of tangency).
>
> 12.13.1.2 Embedment above or below the middepth, $d/2$, of the beam on the compression side for a full development length l_d but not less than 24 bar diameters.
>
> 12.13.1.3 Bending around the longitudinal reinforcement through at least 180 deg. Hooking or bending stirrups around the longitudinal reinforcement shall be considered effective anchorage only when the stirrups make an angle of at least 45 deg with deformed longitudinal bars.

Development lengths are covered in Sec. 5.10, but in this connection the following l_d values are minimum values without a hook:

Bar	Grade 40	Grade 60	Lightweight Concrete
#3	12 in.	12 in.	12 in.
#4	12 in.	12 in.	More
#5	15 in.	15 in.	More

*Chapter 5 covers development lengths (l_d) which are also used for anchorage.

Anywhere stirrups are required, the mimimum stirrup area (two legs) is $A_v = 50b_w s/f_y$, slightly less than under the 1963 Code. However, plain $\frac{1}{4}$-in. diameter bars are no longer considered standard reinforcement. In many beams the use of #3 stirrups spaced at $d/2$ will automatically make the requirement more than under the earlier Code. Where deformed wire is available this may prove economical in some cases.

The author has found it much more convenient to measure the value of stirrups in terms of their contribution v_s to the unit shear capacity, that is, $v_s = \rho_n f_y$, where $\rho_n = A_v/(b_w s)$. The Code minimum is equivalent to a $\rho_n f_y$ of 50 psi or the capacity to handle v_s of 50 psi.

(b) Inclined Stirrups at 45°

For the 45° bent bars most frequently used, the spacing can be derived much as for vertical stirrups, that is, for the vertical component of the bent bar stress to resist the vertical component of the diagonal tensile stresses which existed where the crack forms. The forces are shown in Fig. 4.14*a*.

$$\text{Vertical component of concrete stress} = 0.707(v_s b_w s \times 0.707) = 0.5 v_s b_w s$$
$$\text{Vertical component of bar stress} = 0.707 A_v f_y$$
$$0.5 v_s b_w s = 0.707 A_v f_y$$
$$s = \frac{1.414 A_v f_y}{v_s b_w} = \frac{1.414 A_v f_y d}{V_s}$$

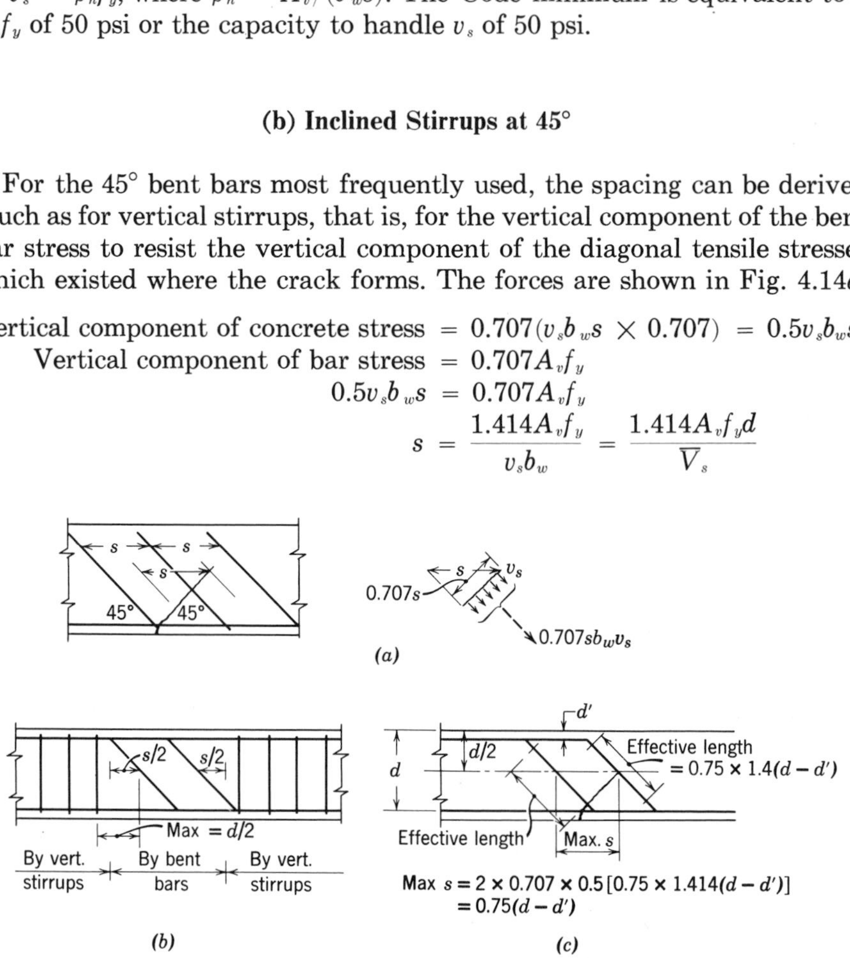

Fig. 4.14. Inclined stirrups at 45°. (*a*) Basis for calculation. (*b*) Arrangement in combination with vertical stirrups. (*c*) Maximum spacing to intercept cracks.

The spacing can thus be larger by some 41% more than for vertical stirrups, but it must also be noted that the inclined legs are longer in about the same proportion.

In considering the length of beam reinforced by a bent bar, the distance s is usually measured as $s/2$ each way from the bar at mid-depth of the beam. If bent bars are used over part of the length of the beam and vertical stirrups elsewhere, the spacing should be laid out together at mid-depth as in Fig. 4.14*b*. Care must be taken to intercept all potential 45° cracks at or below mid-depth of the beam.

Only the center three-fourths of the inclined bar is to be considered effective for web reinforcement (Code 11.6.2.4). When considering the maximum spacing rule, every inclined crack must intersect a bar within this effective length, as shown in Fig. 4.14*c*. This makes the maximum spacing 0.75 of the offset distance of bent bars, that is, $0.75(d - d')$ when v_s is not over $4\sqrt{f_c'}$, or half this when v_s exceeds $4\sqrt{f_c'}$.

(c) Inclined Stirrups at More or Less Than 45 Degrees

The general spacing formula is derived on a similar basis for the more general geometry of Fig. 4.15. The tension resisted acts over the distance y.

$$y = \frac{s \sin \alpha}{\sin (135 - \alpha)} = \frac{s \sin \alpha}{\sin (45^\circ + \alpha)} = \frac{s \sin \alpha}{\sin 45 \cos \alpha + \cos 45 \sin \alpha}$$

$$= \frac{s \sin \alpha}{0.707(\cos \alpha + \sin \alpha)}$$

$$\text{Vertical component of stress} = 0.707 v_s b_w y = \frac{v_s b_w s \sin \alpha}{\cos \alpha + \sin \alpha}$$

$$\text{Vertical component of bar stress} = A_v f_y \sin \alpha$$

$$\frac{v_s b_w s \sin \alpha}{\cos \alpha + \sin \alpha} = A_v f_y \sin \alpha$$

$$s = \frac{A_v f_y (\cos \alpha + \sin \alpha)}{v_s b_w} = \frac{A_v f_y d(\cos \alpha + \sin \alpha)}{V_s}$$

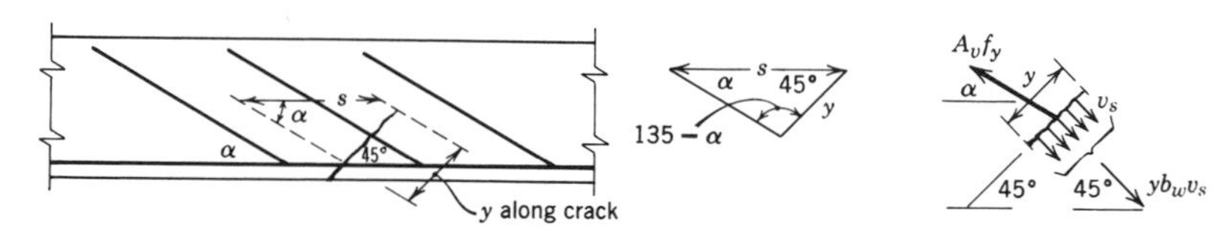

Fig. 4.15. General case of inclined stirrups.

This formula yields the special relations above for vertical and 45° stirrups if $\alpha = 90°$ and $\alpha = 45°$, respectively. The Code limits the use of this relation to a series of bars bent up at different points.

When only a single bar is bent up or when all bars are bent at the same point, the value for diagonal tension is discounted to

$$\overline{V}_s = A_v f_y \sin \alpha$$

This is equivalent to saying that the vertical component of the bar stress carries the shear $\overline{V}_s$. Such a bar is considered effective over the center three-fourths of its inclined length.

4.12 DESIGN OF VERTICAL STIRRUPS

The practical problem of designing stirrups requires:

1. Determination of maximum shears and the length over which stirrups are needed.
2. Choice of desirable size of stirrup bar.
3. Selection of a series of practical spacings.

The maximum shear diagram may involve partial as well as full span loads.

Stirrup bar sizes are almost never mixed in a given beam. Hence, with uniform loads, close spacings are required at points of maximum shear while at points of lesser shear the spacing may be limited by the maximum spacing, that is, the interception of all potential cracks. The stirrup size must be large enough to give a minimum spacing adequate to pass the aggregate readily and, for practical reasons, rarely as small as 2 in., usually 3 in. or more. Larger spacings are more economical if they do not involve too many spaces fixed by crack interception instead of by stress capacity.

For a member with a constant shear the stirrup spacing would be constant and preferably near the maximum permissible. When the shear varies, as is more usual, a sound general method is to calculate the required stirrup spacing at enough points to establish a stirrup spacing curve, including thereon the specification limits on maximum spacing. From this curve the practical spacings can be worked out. A detailed design of stirrups for a continuous T-beam span is given in Secs. 4.22c and 8.17.

The theoretical number of stirrups required for stress is often useful. For any v_s diagram (not total v_u diagram), such as Fig. 4.16*a*, the area under the diagram is a direct measure of the number of stirrups theoretically required. Since $s = A_v f_y \div b_w v_s$

$$A_v f_y = s b_w v_s = b_w \text{ (area of } v_s \text{ diagram for length } s)$$

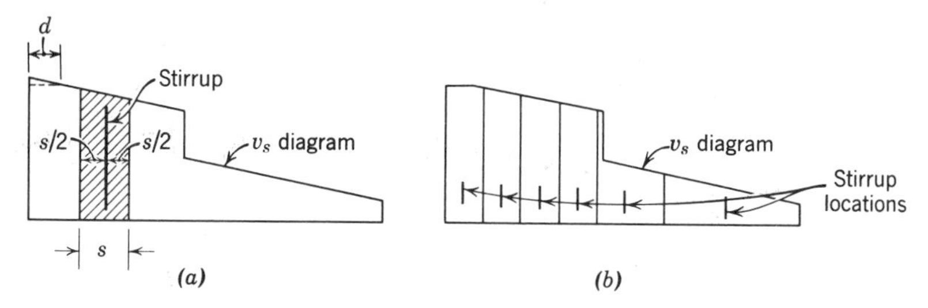

Fig. 4.16. Stirrup spacing related to area of v_s diagram.

Each stirrup thus cares for an equal area under the v_s diagram and the total number of stirrups needed is

$$n = \frac{b_w(\text{total area of } v_s \text{ diagram})}{A_v f_y}$$

This is a theoretical number. Spacings used will be stated in practical units, usually full inches (except for small dimensions, say, under 5 in.), and will thus average less than the theoretical spacings. There also will usually be some spaces kept smaller than the theoretical in order to intercept potential cracks. Extra stirrups will always be required beyond the theoretical in the length where $v_c > v_u > 0.5\, v_c$. The practical number of stirrups in a carefully designed section will usually be from three to six more than the theoretical number. Several designs are covered in detail in Sec. 4.22.

4.13 SPACING STIRRUPS FROM THE AREA OF THE SHEAR DIAGRAM

The above discussion indicates another possible method of spacing stirrups especially adaptable for irregular v_s diagrams. The area under the v_s diagram cared for by one stirrup is calculated first. The total area can then be subdivided into areas of this same size, as in Fig. 4.16*b*, with a stirrup placed at the centroid of each, except where maximum spacing controls.

If there were no length d with (specified) constant v_s next to the support, the v_s diagram would frequently be a triangle, an area easy to subdivide into equal pieces. If within the triangle there are n stirrups to be located at the centroid of n areas, the placement of the stirrups would be such as to subdivide each area into two equal parts, thereby creating $2n$ half areas, as in Fig. 4.17*a*. The simplest procedure then is based on subdividing the

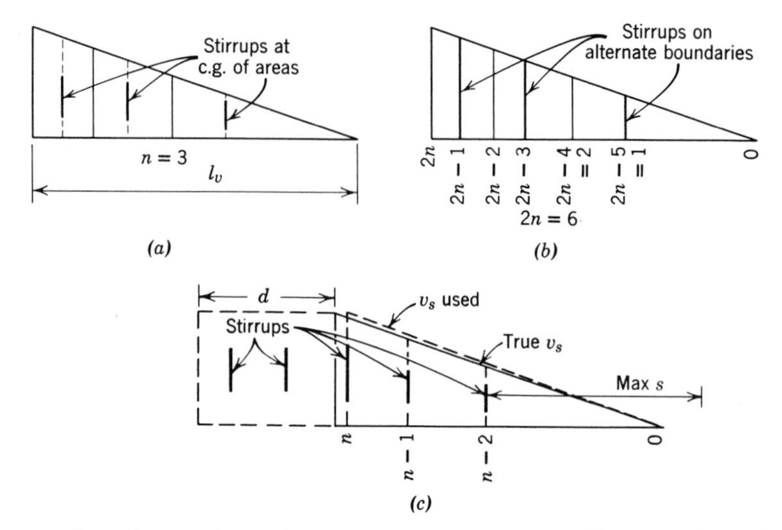

Fig. 4.17. Location of stirrups for triangular v_s diagram. (*a*) Theoretical. (*b*) Practical, using $2n$ areas. (*c*) With constant initial v_s zone, n areas become convenient.

triangle directly into $2n$ parts, as in Fig. 4.17*b*, placing the stirrups on alternate subdivisions.

Consider $2n$ equal areas from the triangle of Fig. 4.18, with boundaries at $2n, 2n - 1, 2n - 2, \ldots, 3, 2, 1$, located at distances from 0 equal to z_{2n}, $z_{2n-1}, z_{2n-2}, \ldots, z_3, z_2, z_1$. Since the total area of a triangle of base z, from the origin at 0, varies as z^2, the following ratios exist:

$$\frac{z_{2n-1}^2}{z_{2n}^2} = \frac{(2n-1)\text{ areas}}{(2n)\text{ areas}} = \frac{2n-1}{2n}$$

$$\frac{z_{2n-3}^2}{z_{2n}^2} = \frac{2n-3}{2n}$$

$$\vdots \qquad \vdots$$

$$\frac{z_1^2}{z_{2n}^2} = \frac{1}{2n}$$

In the denominators, $z_{2n} = l_v'$ (the triangle length) and the only unknowns in the equations are $z_{2n-1}, z_{2n-3}, z_{2n-5}, \ldots, z_1$. The above ratios can be rewritten in the form:

$$\frac{\sqrt{2n}}{l_v'} = \frac{\sqrt{2n}}{z_{2n}} = \frac{\sqrt{2n-1}}{z_{2n-1}} = \frac{\sqrt{2n-3}}{z_{2n-3}} = \cdots \frac{\sqrt{3}}{z_3} = \frac{1}{z_1}$$

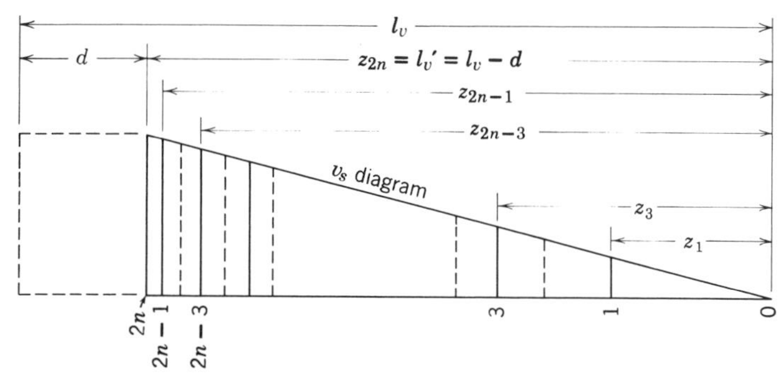

Fig. 4.18. Notation for slide-rule method of spacing stirrups.

which can very easily be set as a constant ratio on the slide rule, as shown in Fig. 4.19.

l_v' is set on the C scale opposite $2n$ on the A scale. The rider is moved to $2n - 1$ on the A scale to indicate z_{2n-1} on the C scale, and so forth. Obviously, the B and D scales could be used just as readily.

On slide rules having the expanded scale for square roots, as in Fig. 4.19*b*, $2n$ is set on the C scale opposite l_v' on the R_2 (or R_1) scale, the rider is moved to $2n - 1$ on the C scale to indicate z_{2n-1} on the R scale, and so forth.

Since stirrups are to be located only at alternate boundaries, only the z distances to these boundaries need be calculated.

With the uniform v_s for a distance d from the support the above can be simplified a little. The stirrups can first be located within the d distance with the last stirrup probably falling slightly within the triangle. Since this stirrup cares for an added $s/2$ distance, the first subdivision needed within the triangle is at the distance s from this stirrup (not $s/2$ from the end, as in the isolated triangle). This permits a solution based on n subdivisions, as indicated in Fig. 4.17*c*, with an example worked out in Sec. 4.22*d*.

4.14 SHEAR LOSS FROM CUTTING OFF BARS IN A TENSION ZONE

When flexural tension bars are cut off in the tension zone of a beam a sharp discontinuity in the steel is created; the sharpness is dependent on the percent of bars cut off. The terminated bar opens an early flexural crack at the cutoff point and then appears to act as an eccentric pull on the surrounding concrete which in some way changes the flexural crack into a

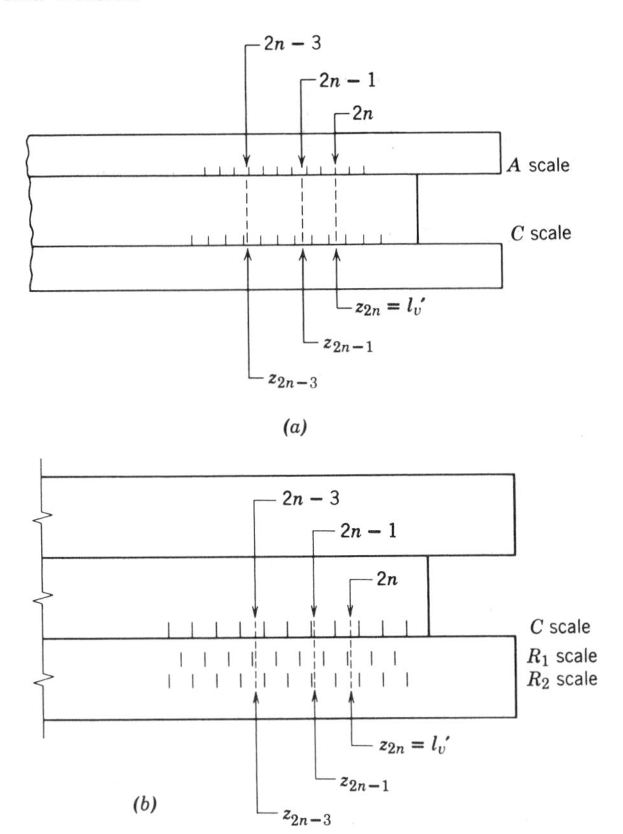

Fig. 4.19. Slide-rule setting for stirrup locations. (*a*) Simple slide rule. (*b*) Expanded square root scale.

diagonal crack, again prematurely. The mechanism is not quite clear, but the end result, a reduced shear strength, is well documented,[7] as in Fig. 4.20. The beams represented were designed to reach f_y and the Code shear strength simultaneously; hence the steel stress ratio (ordinate) is also the shear ratio.

The Code accordingly requires extra reinforcement where bars are cut off unless shear stresses are low at the cutoff point:

> 12.1.6—Flexural reinforcement shall not be terminated in a tension zone unless one of the following conditions is satisfied:
>
> 12.1.6.1 The shear at the cutoff point does not exceed two-thirds that permitted, including the shear strength of furnished web reinforcement.
>
> 12.1.6.2 Stirrup area in excess of that required for shear and torsion is provided along each terminated bar over a distance from the termination point equal to three-fourths the effective depth of the member. The excess

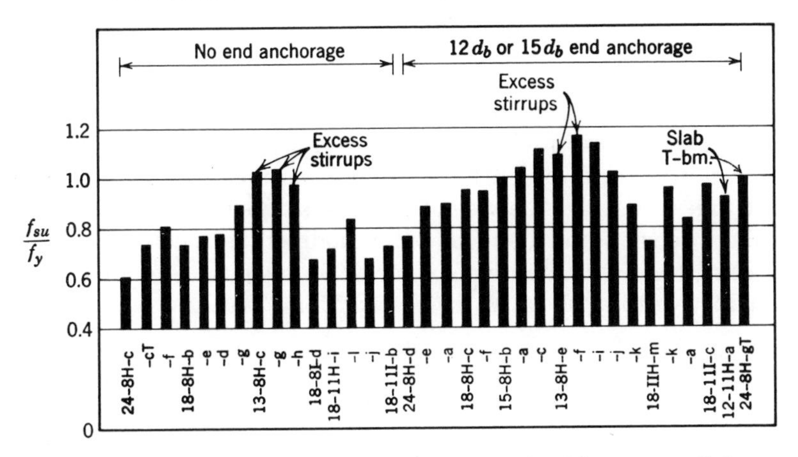

Fig. 4.20. Summary of beams with bars cut off and without remedial measures. (The five beams marked with arrows had excess $\rho_n f_y$ of 75 psi to satisfy minimum spacing rule.)

stirrups shall be proportioned such that their $(A_v/b_w s)f_y$ is not less than 60 psi. The resulting spacing s shall not exceed $d/8\beta_b$ where β_b is the ratio of the area of bars cut off to the total area of bars at the section.

12.1.6.3 For #11 and smaller bars, the continuing bars provide double the area required for flexure at the cutoff point and the shear does not exceed three-fourths that permitted.

The study mentioned above, of 64 beams,[7] shows that the usual extension of bars by $12d_b$ or $15d_b$ beyond the moment cutoff point kept the ill effects from being more severe. Without remedial steps only 2 out of 33 beams developed the design ultimate strength, with losses generally in the order of 15% to 25%, a few higher. In contrast, bars bent up caused no losses. Extra stirrups could erase losses, but seemingly at only 50 percent efficiency. The study design recommendations, necessarily a little conservative because of a phenomenon not quite understood, were:

> Designs should be based on a typical 30 percent loss of shear strength where bars are cut off in a beam, a 20 percent loss where heavy stirrups are already provided ($\rho_n f_y \gtrsim 130$ psi), and a 10 percent loss for slabs 12 in. or less in thickness. If remedial stirrups are used, they should provide an $\rho_n f_y$ adequate to care for twice the indicated shear deficiency and in no case less than 100 psi.

The extra stirrups are needed over the development length of the bars that are cut off.

Further investigation of the Code-required quantity of stirrups and their spacing is needed, since these provisions have never been experimentally verified. They showed first in the 1963 Code as a judgement decision based on quite limited data.

4.15 MEMBERS OF VARYING DEPTH

The relations for shear, bond, and even moment resistance must be modified for members in which the depth is varying, that is, members in which the bottom and top surfaces are not parallel. The moment effect is not large unless the angle between the faces is at least 10° or 15°, but shear for diagonal tension and bond may be modified as much as 30% by 10° slopes. The 1940 Joint Committee Specification gave the following formula for the effective total shear $\bar{V}_1$ to be used for $\bar{V}$ in the usual relations for v and u:

$$\bar{V}_1 = \bar{V} \pm \frac{\bar{M}}{d} (\tan c + \tan t)$$

where $\bar{V}$ and $\bar{M}$ are the external shear and moment to be resisted, d is the depth to tension steel, and t and c are the slope angles of the top and bottom of the beam as shown in Fig. 4.21. The plus sign before the parenthesis is used when the beam depth decreases as the moment increases, as in Fig. 4.21*a*, and the minus sign is used for the more usual case of increasing depth with increasing moment, as in Fig. 4.21*b*. When the sum of the angles becomes as much as 30°, such a formula becomes very inexact.

4.16 BRACKETS AND SHORT CANTILEVERS

The Code for the first time covers brackets and corbels (limited to a/d of unity or less) and the author includes other short cantilevers as an extension of the same general behavior. Their normal resistance near ultimate consists of a tension tie across the top with an inclined compression strut forming a triangle, with normal bending making only slight variations.

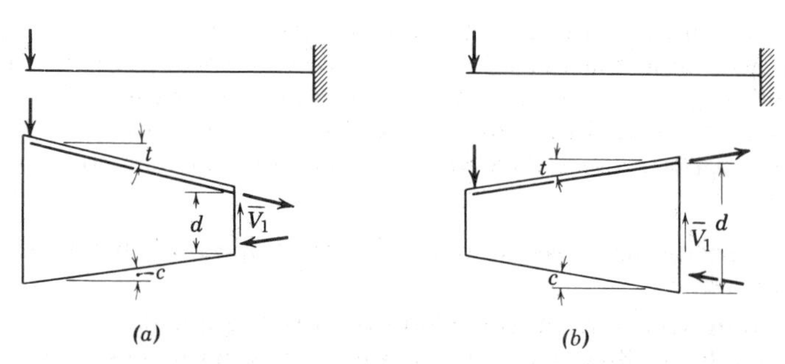

Fig. 4.21. Shear in beams of varying depth.

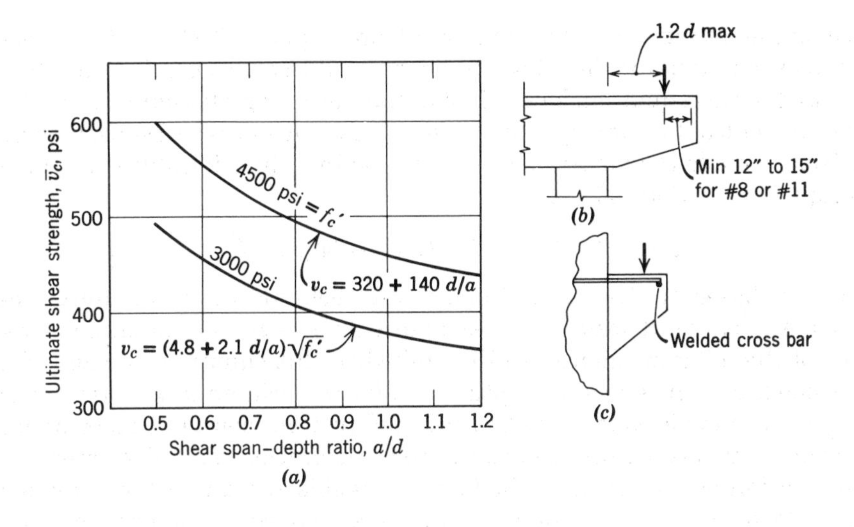

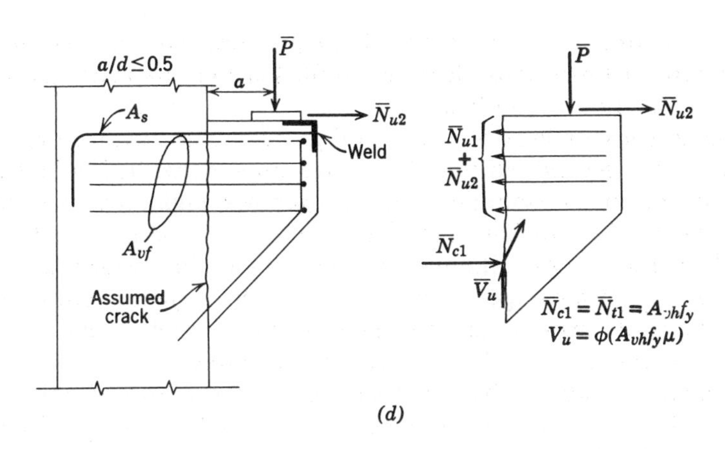

Fig. 4.22. Brackets and short cantilevers. (*a*) Author's unit shear strength, v_c. (*b*) Minimum end anchorage for values in (*a*). (*c*) Preferred end anchorage in short bracket. (*d*) Shear-friction applied to bracket design.

The inclination of this strut would determine the tension in the tie if it were simply a truss; the flexural calculation at the face of the column gives essentially the same tension, but the triangular truss idea emphasizes the anchorage problem; and this anchorage problem is the most critical one unless the bar extensions of Fig. 4.22*b* are possible.

The PCA bracket tests[8] took account of the high shrinkage and expansion stresses frequently occurring from beams supported on brackets. The Code, based on these tests, requires a horizontal force equal to $0.2V_u$ and provides a complex equation for the allowable shear, at the same time limiting the ρ to be used in this equation. The Commentary also suggests that the bracket plate be welded to the top bars where such tension is possible. Where provision is made to avoid the horizontal loading, the allowable shear may be simplified to (Code 11.14.3)

$$v_u = 6.5(1 - 0.5a/d)(1 + 64\ \rho_v)\sqrt{f_c'}$$

where ρ_v is based on the sum of the flexural steel plus lower horizontal ties of area A_{vh} (equal at least to $0.5A_s$ and distributed over the upper two-thirds of the effective depth) and is limited in this equation to $0.20f_y/f_c'$. The Code limits the effective depth d used at the column face to twice the depth at the outside edge of the bearing and the Commentary suggests that this bearing not start closer than 2 in. from the bracket face. The extension of the tension bars as close to the face as possible and an anchor provided by a welded cross bar of the same diameter is also recommended (Fig. 4.22*c*).

The problem area in a long bracket is the anchorage of bars because the variable depth (inclined strut idea) makes the bar tension nearly as critical at the load as at the column face.

If the anchorage can be extended as in Fig. 4.22*b*, the shear recommendations made in the second edition of this text and shown in Fig. 4.22*a* are still valid[9] in the author's opinion; they are simpler and a little less restrictive. For the longer brackets the author would normally calculate the required A_s at the face of column, keep the bars of a diameter that can be anchored (or try for a deeper bracket if necessary to make them smaller), and if a horizontal tension pull were expected add enough bars to resist it. Finally the shear should be checked and should rarely control. For a short bracket the shear may be the chief control on the bracket depth. For a heavy bracket, say more than 18 in. deep, additional horizontal U-bars or ties in the upper half will avoid wide cracks, add to the bracket strength, and even help with the moment. Vertical stirrups are essentially useless. The Code requirement is horizontal steel A_h equal to half the top A_s.

4.17 SHEAR-FRICTION FOR BRACKETS

The shear-friction theory (Code 11.15) is particularly useful for precast assemblies and composite construction.[10,11] The Code suggests this approach may be used for brackets cast-in-place (with the same reinforcement limitations of Code 11.14 provided $a/d \lesseqgtr 0.5$.

The shear-friction approach is to assume a shear plane already cracked as in Fig. 4.22*d*, with a coefficient of friction μ given in Code 11.15.4 (1.4

if cast monolithically). The depth must then be such that the shear stress on the area *bd* will not exceed $0.2f_c'$ or 800 psi. The necessary A_{vf} must be large enough to resist the normal force necessary to support the shear by friction:

$$A_{vf}f_y = \mu(V_u/\phi)$$

Finally A_{vf} is checked to see that it lies between $0.5A_s$ (Code 11.14.4) and $1.0A_s$ (Code 11.14.3), where A_s is the normal reinforcement for flexure plus that used for any horizontal component of load. Closed stirrups or ties are appropriate for A_{vf} and these should be distributed uniformly within two-thirds of d adjacent to A_s.

4.18 DEEP BEAMS

(a) General Case

Deep beams in Code 11.9 are totally new in the Code. Although the Code deals only with shear, a few general ideas first might be helpful. A deep beam is simply a member short enough to make shear deformations important in comparison to pure flexure; the Code considers a clear span l_n of up to $5d$. Plane sections in these beams do not remain essentially plane under loading; but this is also true at the end of every simple span beam. If one has a clear picture of how any beam fails in shear at $a/d = 1$, he can get a rough idea of how a deep beam of $l_n = 2d$ would fail in shear by imagining two such end sections of beam joined together with the load at the junction. At $a/d = 1$ in a long beam, flexure will not be of much interest; in the deep beam of $l_n = 2d$, flexure at the load is the critical flexure. If the load is applied to the top of the deep beam the lever arm z of the internal couple will be smaller than usual, say $0.8d$; if the load is applied to the bottom of the beam in tension, z may drop as low as $0.4d$ or $0.5d$. The Code provisions are limited to loading on top of the beam. Anchorage of tension steel becomes critical. Leonhardt[12,15] suggests *horizontal* hooks on tension bars (vertical compression helping to avoid splitting) or U-shaped bars lap spliced at midspan.

If such a deep beam is provided with stirrups that are able to deliver the bottom load to the upper part of the beam, the beam will behave nearly like a top loaded beam. It is always desirable in any kind of beam to provide hangers (stirrups) to pick up bottom loads in this fashion. Experimental evidence indicates that stirrups that must act as hangers *and* as web reinforcement need *not* be designed for the sum of the two requirements, but simply for the larger of the two.

Higher shear strengths are available when a/d is small, as already pointed out in Secs. 4.4*c,d*. For the deep beams a multiplier is given for application to the usual allowable v_c:

$$v_c = [3.5 - 2.5M_u/(V_u d)] \times [\text{usual formula values for } v_c]$$

where $[3.5 - 2.5M_u/(V_u d)]$ has here been designated as the magnifier, to be limited to a maximum of 2.5 and the magnified v_c also to be limited to $6\sqrt{f_c'}$. The values of this multiplier are plotted in Fig. 4.23*a* and should be applied to the usual allowable of $2\sqrt{f_c'}$ or to $1.9\sqrt{f_c'} + 2500\ \rho Vd/M$.

As pointed out in Sec. 4.9*b*, in a continuous span there is some question whether the variable that governs v_c is M/Vd as used above or a/d as the author prefers. The shorter the span the more serious is the possible difference. While not very significant in the usual formula for v_c, because the second term there is small compared to the first one, it appears very important in the multiplier term above. To be safe the author recommends that a/d be used in this multiplier for concentrated loads, and for uniform loads that the M_u/V_u used be that for a *simple* span under the loading used. The critical section is specified below.

The Code also limits v_u to $8\sqrt{f_c'}$ when $l_n/d \lesseqgtr 2$, and for larger l_n/d to

$$v_u = (2/3)\ (10 + l_n/d)\sqrt{f_c'}$$

which is plotted in Fig. 4.23*b*. So long as v_u on ordinary beams may be used up to $8\sqrt{f_c'} + v_c$ at an a/d of 1, the author regards this deep beam limit as probably too strict.

The critical beam section for shear is to be taken at $0.5a$ for concentrated loads and $0.15l_n$ for uniform load. For simple spans these appear quite reasonable values, closely kin to what is done on ordinary beams.

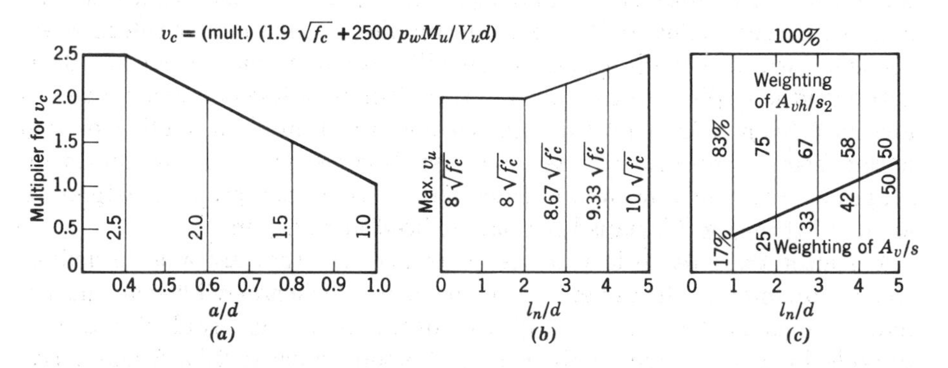

Fig. 4.23. Deep beam shear. (*a*) Multiplier for usual v_c. (*b*) Maximum total v_u. (*c*) Weighting of vertical and horizontal shear reinforcement.

Vertical stirrups are less effective in these beams because the shear cracks are steeper;[14] *horizontal* bars become more effective as a/d or l_n becomes smaller. The Code (11.9.5) accordingly expresses the shear resistance as the weighted sum of the two types of reinforcement in a modification of the basic equation:

$$\frac{A_v}{s}\left(\frac{1 + l_n/d}{12}\right) + \frac{A_{vh}}{s_2}\left(\frac{11 - l_n/d}{12}\right)$$
$$= [(A_v/s)(1 + l_n/d) + (A_{vh}/s_2)(11 - l_n/d)]/12 = (v_u - v_c)b_w/f_y$$

The horizontal reinforcement A_{vh} is weighted much the heavier, as shown in Fig. 4.23*c*. The Commentary suggests the precise weighting is not critical but that the weighted sum should be maintained. In addition the spacing may not be more than 18 in. nor $d/5$ for s or $d/3$ for s_h.

Note also the general flexural steel requirements for any beam deeper than 36 in. as discussed in Sec. 3.8 (Code 10.6.5), that is, $0.10A_s$ as face steel.

(b) Shear Walls

Shear walls are considered as cantilevers of deep beam type but carrying axial load, with horizontal wall length l_w as beam depth and the wall height h_w as the beam length. The nominal shear stress is computed as $v_u = V_u/(\phi\, td)$, with t the thickness and $d = 0.8\, l_w$, unless based on a strain compatibility analysis which considers all the tension reinforcement.

The critical design section for shear is at a height of $l_w/2$ or $h_w/2$, whichever is smaller, and the lower wall is reinforced as this section requires. The Code should be consulted for the allowable shear stresses for design (Code 11.16.2) and the reinforcement required if $v_u > v_c/2$. Typical wall reinforcement is considered adequate only if $v_u < v_c/2$.

TORSION

4.19 PURE TORSION

(a) General

Torsion may be important in some spandrel beams, in flat plates at exterior columns, in curved stair slabs or curved beams, and wherever large loads must be carried off the axis of the member.[19] In box girder bridges and in any isolated beam that must resist unbalanced loading, torsion is an important problem. When torque is small, producing no larger torsional shear stress than $v_{tu} = 1.5\sqrt{f_c'}$, the Code suggests that it be ignored. Where it is possible to devise framing that can take the load without torsion, this is a preferred solution (by the author). Because torque design is a new area in reinforced concrete, it is given more space here than the seriousness of the problem warrants.

(b) Elastic Theory of Pure Torsion

Elastic analysis of a round shaft under torsion is simple and results in tangential shearing stresses all around the member as shown on one diameter in Fig. 4.24*a*. The equation is $v_t = Tr/(\int r^2 dA)$. Any other shape is increasingly complex because the cross section warps and the member lengthens as it twists. For analysis and visualization the soap film, or membrane analogy is useful. A soap film is stretched across a hole cut in the top of a box to the shape of the member cross section. A small inside air pressure then raises a bubble over the hole. The volume between the bubble and the original plane, by analogy of the governing differential equations, is proportional to the total torque resistance and the slope of the membrane (which is under uniform tension) measures the unit shear stress, a steep slope representing a high stress.

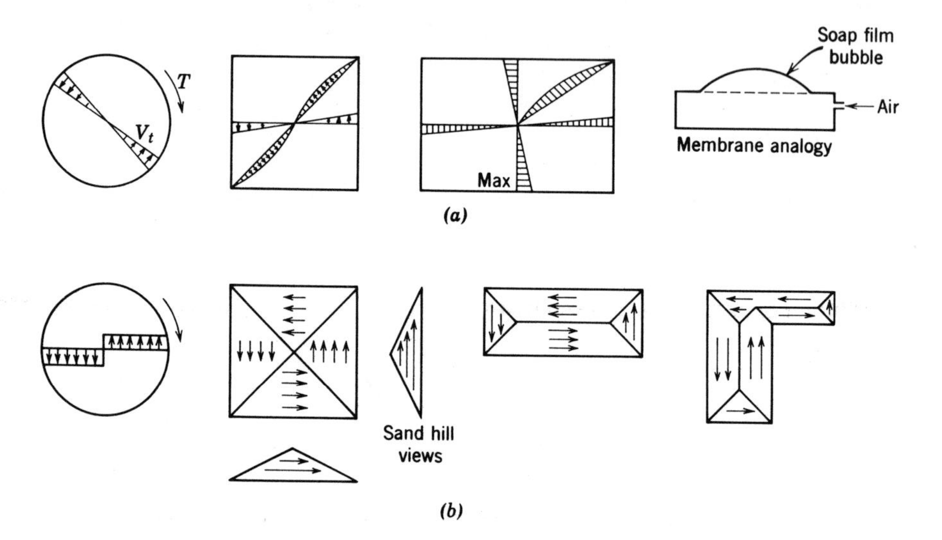

Fig. 4.24. Elastic and plastic analysis for pure torsion shears. (*a*) Elastic analysis. (*b*) Plastic analysis, v_t equal over entire cross section.

(c) Plastic Analysis for Pure Torsion

A plastic analysis assumes uniform shear intensity all around the surface and all over the cross section, as in Fig. 4.24*b*. The analysis can be envisioned in terms of the sand heap analogy. This gives uniform side slopes with the stress at 90 degrees to the slope, and again a volume proportional to the torque. For a slope equal to the maximum in the elastic analysis, the indicated volume and torque for plastic analysis is considerably larger. The equation can be written $v_t = T/J_p$ where J_p is twice the volume of the sand heap divided by the slope.

The final failure is a diagonal crack following around the surface to form a helix with about a 45° slope.

(d) Code Analysis

Neither of the above analyses is realistic, partially because of the properties of concrete but chiefly because concrete at design loading will be cracked. The best, that is, the most consistent calculation method has not yet been fully established. The author in early investigations found the two methods about equally consistent, but prefers the lower stress numbers the plastic method gives. The analysis is not fully adequate because it shows small, but consistent, differences in the ultimate stress for the *T*-beam, L-beam, and the rectangular beam, in a systematic decreasing manner which indicates the analysis does not reflect the shape properly. The value is also a little higher for uniform distributed torque load than for a concentrated torque load.

The Code uses an approximate relationship for the torque T_u that lies between the elastic and plastic analysis:

$$v_{tu} = T_u/[\phi\Sigma(x^2y)/3] = 3T_u/(\phi\Sigma x^2y)$$

where x is the short dimension of the rectangle and y the long one and Σ indicates the sum of all the component rectangles making up the beam cross section, Fig. 4.25. The beam shape may be subdivided to give the largest total.

4.20 TORSION COMBINED WITH SHEAR AND MOMENT

(a) Member Behavior Under Test

Because the crack patterns of torque, shear, and moment are quite different, it is necessary to test under all three in combination. To the

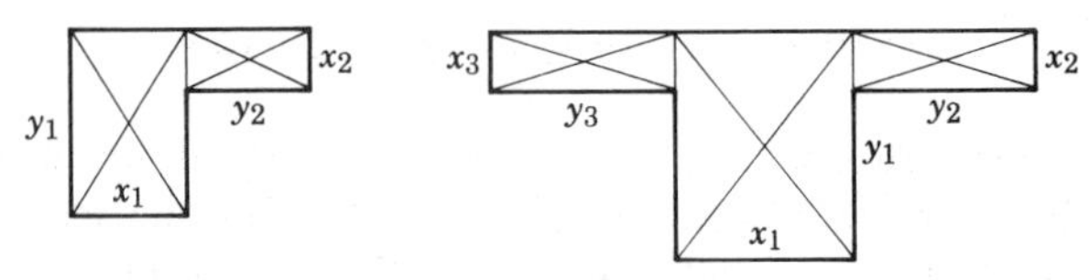

Fig. 4.25. Rectangles for Code v_{tu} calculation.

author test behavior seems best described physically and mathematically from the torque and shear interaction. On one web face the torque shear adds, and on the other subtracts, relative to the vertical shear stresses. A little torque causes a diagonal crack to open on one side of a beam earlier than on the other, but the shear failure pattern is not much altered, possibly a little steeper diagonal crack on the opposite face. The torque tends to make diagonal cracks out of the flexural cracks on the tension face. As the ratio of torque to shear is increased the failure pattern is more sensitive to the shape of the beam, and the tension face cracks become full diagonals.

With the rectangular beam without stirrups the load difference between initial diagonal crack and ultimate may not be large and the diagonal crack on the opposite face may be replaced by an inclined compression crack (evidence of a skew bending effect); or a compression crack on a skew plane may finally develop on the compression face of the beam and a steeper diagonal crack (nearly vertical or of reversed slope) may show on the opposite web face.

With a T-beam the failure mechanism is slower in developing (more load after first diagonal cracking) and is not yet well understood. The diagonal crack usually eventually opens on the far side of the web, always steeper, sometimes of reversed slope, dependent upon the ratio of torque to shear.

To the author torsion appears primarily as a special shear problem, one which is troublesome because many variables influence the available resistance, including beam shape, the stirrups present, and even the longitudinal beam steel. For rectangular beams progress has been made in defining ultimate strength in terms of skew bending, but this approach is not yet available for L- or T-beams.

When closed stirrups are added, which in practice must be everywhere that torsion is important, the spread between first diagonal crack load and ultimate load increases. The few diagonal cracks present in the test members without stirrups (Fig. 4.26*a*) become amazingly numerous with stirrups (Fig. 4.26*b*), member stiffness against rotation drops to from 3 to 7 percent of that before diagonal cracking,* and longitudinal steel stresses then

*Measured as the change in rotation angle ϕ relative to the torque increment applied, i.e., $d\phi/dT$, similar to the tangent modulus idea.

increase rapidly, in proportion to the angle. Most stirruped beams rotate so much before reaching ultimate strength as to appear worthless as structural members well before they fail. This feature of serviceability and what portion of this ultimate may be logical for design is under study at several universities.

(b) Strength Interaction, Without Stirrups

When torsion is calculated by plastic analysis of the gross (uncracked) concrete section, a pure torsion, without flexural shear, leads to a calculated v_{tc} of from $5\sqrt{f_c'}$ to $6\sqrt{f_c'}$, compared to about $2\sqrt{f_c'}$ for flexural shear cracking strength v_c, without any torsion.* The combinations of torque and shear lead to the ellipse of Fig. 4.27*a* which could be replaced with a circle (Fig. 4.27*b*) in terms of v_{tu}/v_{tc} and v_u/v_c. The circle is accepted generally in the United States and Canada, but not so widely in Europe. (The author also finds the ellipse appropriate for strengths with stirrups, both v_{tu} and v_u being then increased to v_{to} and v_o for any quantity of stirrups, as in Fig. 4.27*c*. This view is not generally accepted as yet.)

(c) Strength Interaction, with Stirrups, Code Design

When pure torsion tests are run in which closed *stirrups and longitudinal reinforcement are kept equal in volume*, which is the ideal ratio for pure torsion, the strength curve of Fig. 4.28*a* is found. The plotted points trace a straight line until an overreinforced condition is reached. When these strengths are projected back to zero stirrups, the straight line portion intersects at about v_{tc} of $2.4\sqrt{f_c'}$. The Torsion Committee concluded† that not all of $v_{tc} = 6\sqrt{f_c'}$ could be used with stirrups on the straight-line basis and the Code rules are based on this finding. The concrete stress usable for resisting torsion is limited to $2.4\sqrt{f_c'}$ and for general design the ellipse of Fig. 4.28*b* is used.

*That the ultimate v_{tc} might be three times the ultimate v_c long concerned the author, since each led to diagonal tension cracks. Finally he realized that neither calculation had a rational approach and that both numbers were simply crude stress coefficients, not stresses at all. Now he calls them "imaginary stresses" whenever someone raises the question. Both normally relate to flexurally cracked members, but both use properties of the uncracked section. A calculation based on imaginary areas (with sloppy theory as well) necessarily leads to imaginary stresses.

†Based on work by Dr. Hsu at the PCA Skokie Laboratory.

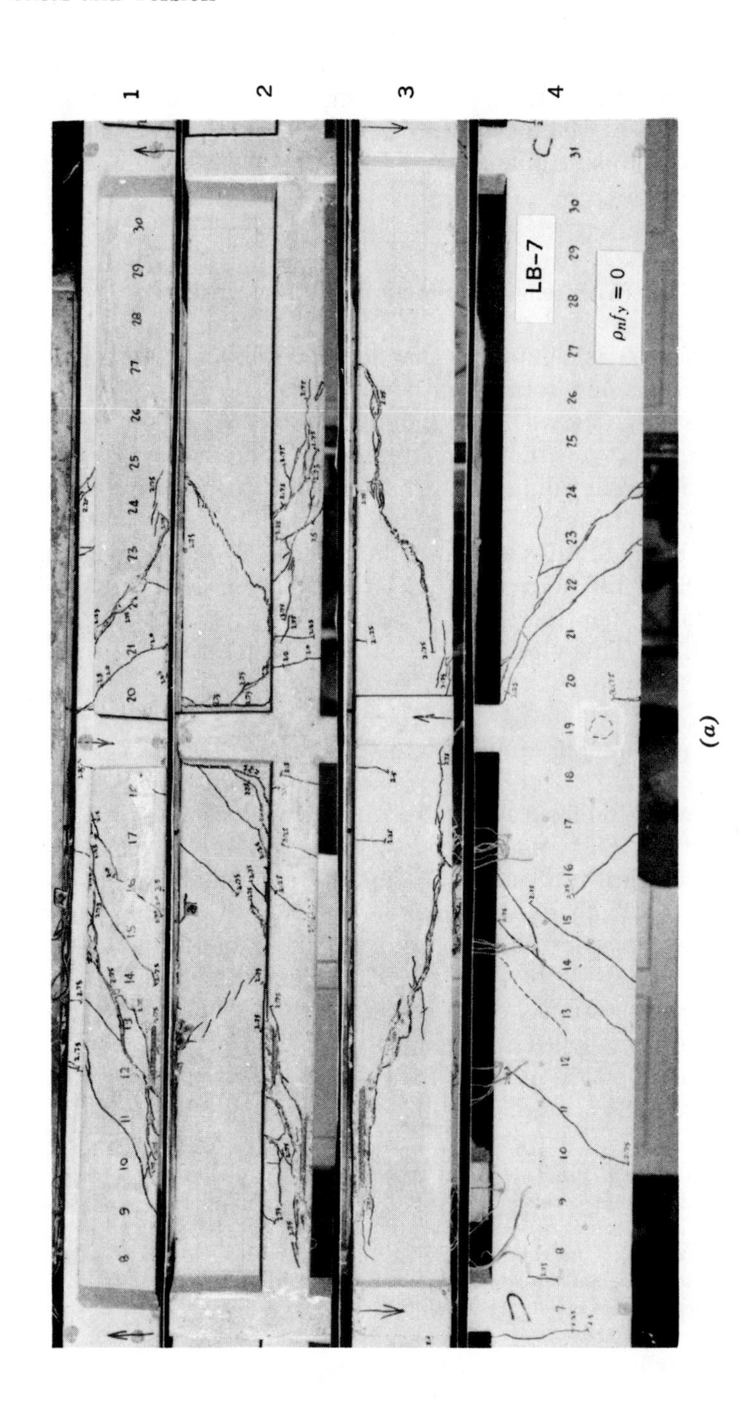

(a)

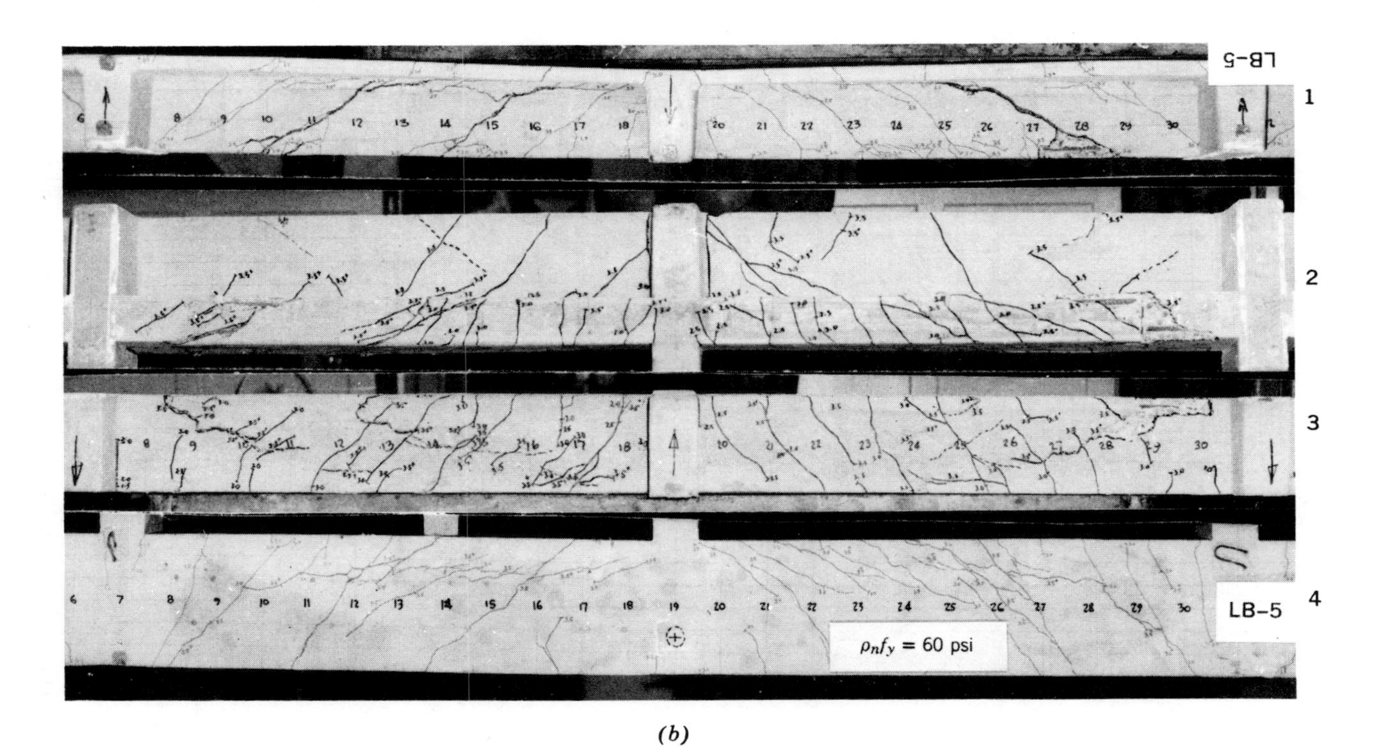

Fig. 4.26. L-beams with $b_w = 3$ in. loaded as semicontinuous beams with load eccentricity of 5 in. (*a*) No stirrups. (*b*) Light stirrups. Corresponding torque-rotation curves are shown for LB-7 and LB-5 in Fig. 4.29. The four views of each beam are: 1, web on flange side; 2, bottom of beam and flange; 3, side of web opposite flange; 4, top.

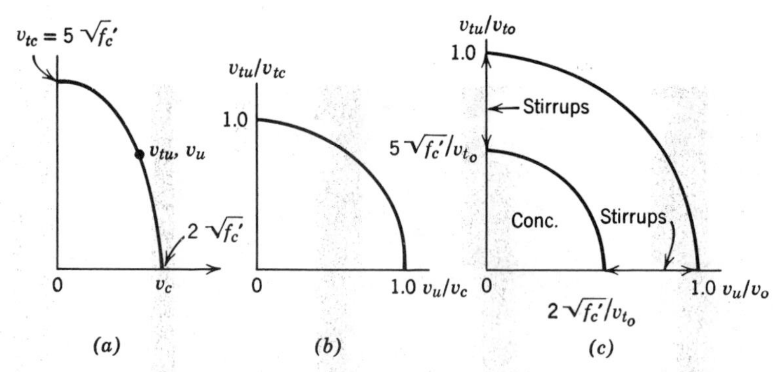

Fig. 4.27. Shear-torsion interaction diagram.

(d) Stirrup Design for Torsion

Although the Code allows $1.5\sqrt{f_c'}$ in torsion without any special provisions, above this value any combination of v_{tu} and v_u falling outside the ellipse requires the design of stirrups both for shear and for torsion. These limiting values without stirrups are given by these equations:

$$v_{tc} = \frac{2.4\sqrt{f_c'}}{\sqrt{1 + (1.2v_u/v_{tu})^2}}$$

$$v_c = \frac{2\sqrt{f_c'}}{\sqrt{1 + (v_{tu}/1.2v_u)^2}}$$

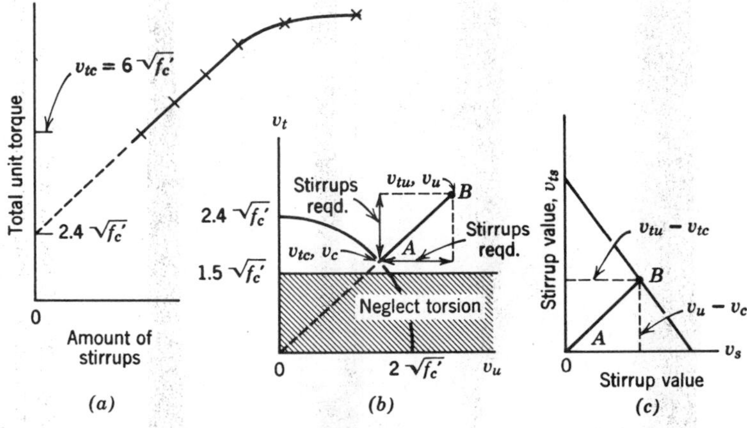

Fig. 4.28. Stirrup strength and design.

The critical section for torsion, as for shear, is a distance d from the face of support. If the design combination of stresses falls at point B in Fig. 4.28b, stirrups for torsion are to be based on the vertical projection of the overrun AB and for shear on the horizontal projection of AB. This stirrup requirement is equivalent to using the triangular interaction of Fig. 4.28c for the stirrup portion. The line AB from Fig. 4.28b can be superimposed on the triangular arrangement if the scales are each in terms of v_{tu} and v_u.

The minimum area of stirrups for torsion outside the ellipse in Fig. 4.28b must satisfy the equation

$$A_v + 2A_t = 50b_w s/f_y$$

where A_v is the (two-leg) area of stirrups for shear and A_t is the area of one leg of the stirrups for torsion.

For a given $v_{tu} - v_{tc}$ the required area for A_t is

$$A_t = \frac{(v_{tu} - v_{tc})\, s\Sigma x^2 y}{3\alpha_t x_1 y_1 f_y}$$

where

x_1 and y_1 = the narrow and long sides, respectively, of the closed stirrups

s = the stirrup spacing

f_y = the stirrup yield point

$\alpha_t = 0.66 + 0.33\, y_1/x_1 \lessapprox 1.50$

Torsion reinforcement must extend at least a distance $d + b$ beyond the point theoretically required.

(d) Longitudinal Steel for Torsion

When beams fail in torsion extra stress moves into the longitudinal bars. Torsion stirrups cannot act effectively unless a tension truss "chord" is available in each corner of a rectangular beam, since the tension in pure torsion may be needed in any (or all) corners as the diagonal crack wraps around the member.

The Code sets two requirements for extra longitudinal steel for torsion, the first being a volume of longitudinal steel for torsion equal to the volume of stirrups provided for torsion:

$$A_l = 2A_t(x_1 + y_1)/s$$

If the stirrup volume is very low, the following requirement defines a larger required A_l:

$$A_l = \left[\frac{400xs}{f_y}\left(\frac{v_{tu}}{v_{tu} + v_u}\right) - 2A_t\right]\frac{x_1 + y_1}{s}$$

where "$2A_t$ . . . need not be taken less than $50b_w s/f_y$."

The second equation is better understood in terms of its development. The Torsion Committee originally recommended that the torsion stirrups used not be less than

$$A_t = \frac{200\, b_w s}{f_y} \cdot \frac{v_{tu}}{v_{tu} + v_u}$$

and that longitudinal reinforcement A_l for torsion be used with a volume equal to that of the stirrups used for torsion. The Committee later agreed to the Code version which uses 50 instead of $200v_{tu}/(v_{tu} + v_u)$ on condition that A_l be increased enough to make up the deficiency in $2A_t$ brought about by the change. This requires something *more* for A_l than would have been required from the simpler equation. The result is a rather awkward equation. Its use is covered in the design of Sec. 4.23.

4.21 AUTHOR'S COMMENTS ON CODE PROVISIONS AND STATE-OF-ART

The European Committee for Concrete (CEB) strongly endorses a *linear* interaction between shear and torsion instead of the ACI circular interaction (for the concrete stress portion). The CEB members sustaining this position are quite experienced in the testing of large box type sections. The author suspects that the big box section may not show the efficient redistribution of stresses after diagonal cracking which compact L-beams and T-beams do show.

Some small boxes have been tested in this country, but the author believes these were in torsion rather than a loading that combined shear and moment. The author questions the resulting conclusion that the hollow rectangle and solid rectangle have the same strength in torsion *in the presence of flexural shear* that is always a part of practical cases. Having had no personal experience with box beam tests, he accepts the European (German and Swiss) views about box sections and their straight line interaction, while defending strongly the circular interaction for solid sections, especially L- and T-shapes.

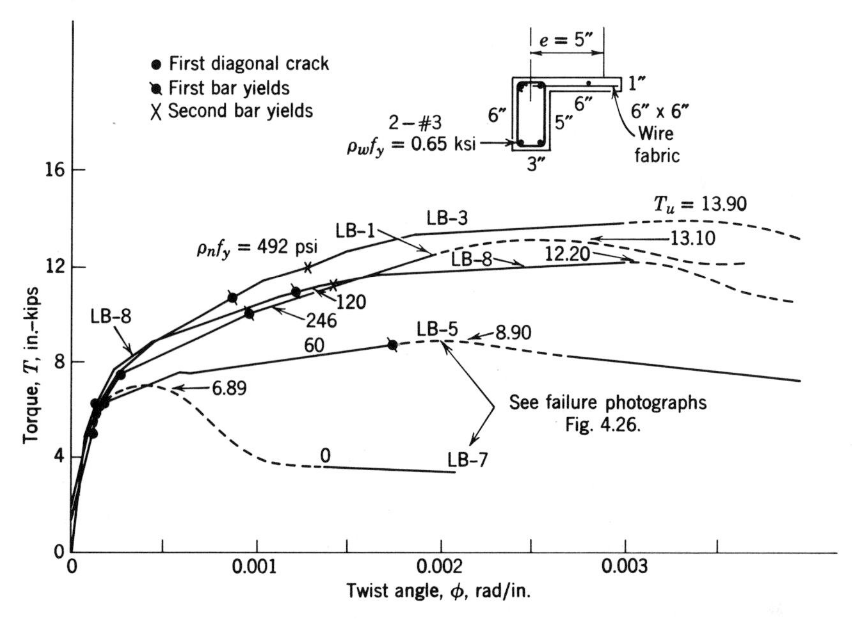

Fig. 4.29. Effect of amount of stirrups on T-ϕ curves. (f_c' from 4060 to 4400 psi) $\rho_n f_y = A_v f_y/(b_w s)$.

It is possible that, as the ratio of h/b for a rectangle increases to where it represents more of a wall-like section, the circular (Code) interaction may prove less tenable; an L-shape with h/b_w of 3 for the web, for example, shows at the diagonal *cracking load* (but not at ultimate) less of the circular interaction than does one with h/b_w of 2. It is also imaginable that the redistribution beyond initial diagonal cracking might well be much smaller for a big box than for the smaller solid section.

None of the author's tests with stirruped L-beams has given results similar to those of Fig. 4.28*a*, probably because he has never tested with the steel limitations there imposed, i.e., volume of longitudinal steel equal to volume of stirrup steel. When adequate longitudinal reinforcing for moment is used, the total longitudinal reinforcing is always much more than equal in volume to the transverse reinforcement.* A number of the author's tests have shown that stirrups valued at $\rho_n f_y$ = 60 psi (such as beam LB-5 in Fig. 4.29) did proportionately increase strength under combined torsion, shear, and flexure loading beyond the nominal v_{tc} value

*Note that the author compares *total* $A_s + A_s'$ with total stirrups, *not at all* the same as the Code A_l compared with A_t steel. The author's only limitation is to keep $A_s + A_s' \ngtr 1.25A_s$, which appears to be a safe rule.

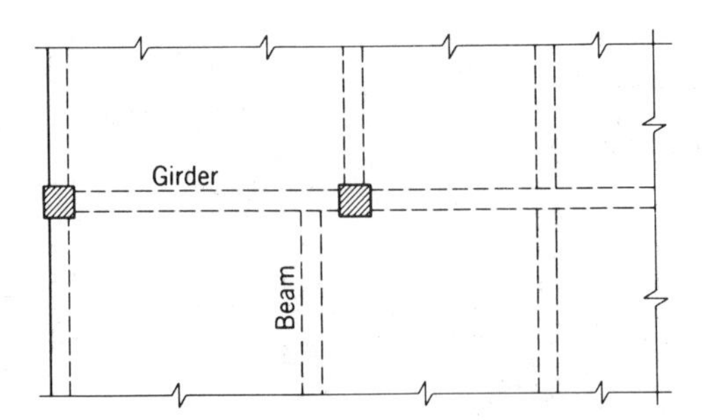

Fig. 4.30. This beam framing produces severe torsional stresses in girder between beam and adjacent column.

of $5\sqrt{f_c'}$.* He considers that the conditions which led to the straight line from $2.4\sqrt{f_c'}$ were the result of test beams designs that were not practical in real design. The design of longitudinal steel primarily for the moment automatically eliminates any need for the $2.4\sqrt{f_c'}$ limitation. A limit for v_{tc} of approximately $5\sqrt{f_c'}$ would be more appropriate and could also reduce the stirrup requirements. The use of an elliptical interaction for the stirrups would give a further saving, although at the present time, the author would not recommend this for box beams.

Continuous L-beams with stirrups[16] and continuous T-beams without stirrups[17] both seem to show better under distributed torque and shear loading than under constant torque and shear. Tested under T, V, and M, with critical T and V stresses assumed to be at a distance d from the support, the strengths showed at least 25% better than with the constant T and V. This increased strength probably results from the fact that critical stress combinations then exist over much shortened lengths of the beam, this possibly damping the spread of cracks after they initially form.

Since the Torsion Committee made its report around 1968, there has been time to originate and accumulate more information on torsion. Hence it would be a serious mistake to take the author's comments here as criticism of the Committee action at that time. The Committee then was making a bold new recommendation and was properly conservative in doing so.

A different problem, but still an important and a relatively unsolved one, is the proper calculation of the design torsion to be resisted. The European answer seems to be to neglect torsion when it is a secondary stress, but to design conservatively when static loads must be carried. (Care must still

*The increase in Fig. 4.29 was from T_u of 6.89 in.-k for LB-7 to 8.90 in.-k for LB-5, or 29 percent. These two beams are shown at failure in Fig. 4.26 where numerous fine cracks show with the light stirrups, which is desirable.

be used for the secondary stress condition of Fig. 4.30.) The simple use of elastic concepts for frame analysis from which the design torsion might be obtained tends to overemphasize torsional demands. Because torsional stiffness drops so much more upon the development of torsional cracking than it does with flexural cracking, possibly from 100 percent to 7 percent compared to a drop to possibly 50 percent for flexure, the author feels that elastic design alone is not adequate. After diagonal cracking very little additional torsion will be absorbed (in an indeterminate structure) from additional loading.

Divided opinion exists as to whether a spandrel beam should be designed for torsion as a rectangular or an L-beam. One investigator[18] noted large yield cracks in the slab at the face of spandrel web and no diagonal cracks in the slab, which caused him to consider the beam as a rectangular beam for torque resistance. However, he found on that basis a substantial *excess* torque strength for his "rectangular" beam.

The author notes that a slab has two functions in this case, one to deliver load to the beam and the other to help carry the torque that results. The slab probably works imperfectly to resist the torque at midspan where the torque is small, but near the column it should be more effective. Whether this is true or not, the slab forces the spandrel to twist about an axis at the level of the slab and not about the natural center of rotation as a rectangular beam. This alone increases the torque resistance through transverse bending. Also, after diagonal cracking, when the spandrel starts to lengthen under torsion, the monolithic slab tends to resist this lengthening and adds a counteracting axial force. In summary, the slab in this case probably works differently from the flange in a freestanding L-beam, but the flange does act significantly and effectively in the matter of strength. In fact, the Code limitation on flange overhang to $3h_f$ is about half of what might safely be used in the author's opinion.

From a behavior point of view a member size designed on the use of minimum stirrups would be good. Although the near elastic torque at service load (and even the member itself) might be uneconomically large, the diagonal crack would not show until much overload occurred. After cracking, the torque delivered by an indeterminate structure would not increase much with load; a low load factor on the torque calculated by elastic theory would then be appropriate.

A deformation rather than a torque load would govern in the case of Fig. 4.30. A relatively flexible girder (in torque) with medium stirrups might crack diagonally but the stirrups should then distribute the cracks and keep them from becoming noticeably large.

The author doubts that the full torque value of stirrups can actually be developed in a structure, because of the excessive rotations required to bring them fully into effect (Fig. 4.29). Possibly one needs to discount the last 40% of rotation capacity and the gain of resistance associated with it. This might mean discounting, say, the last 20% of stirrup strength, although what percentages are proper certainly needs to be further explored.

Longitudinal steel stress from torsion appears to be wholly a function of rotation angle (under combined loading) rather than a linear function of the torque carried. Before diagonal cracking this longitudinal stress is negligible, in the order of 2 or 3 ksi. It seems to be roughly the same for a given angle of twist regardless of whether the minimum or the maximum stirrups are used. Any necessary limitation the structure imposes on rotation at ultimate should thus relieve also the present demands for longitudinal steel added in proportion to the stirrups used for torsion.

Finally, design for torsion is totally new in the Code and rather new everywhere. It should not surprise anyone that such a complex field may take some time to stabilize.

EXAMPLES

4.22 EXAMPLES—SHEAR

(*a*) The beam of Fig. 4.31 will be checked for shear assuming $f_c' = 3000$ psi, and Grade 40 steel.

Solution

Beam weight $= (10 \times 13/144)\ 150 = 136$ plf
$d = 13 - 2 = 11$ in. $= 0.92$ ft

At d from support:

$w_u = 1.4w_d + 1.7w_l = 1.4 \times 136 + 1.7 \times 1060 = 1990$ plf
$V_u = 1990\ (6 - 0.92) = 10{,}100$ lb

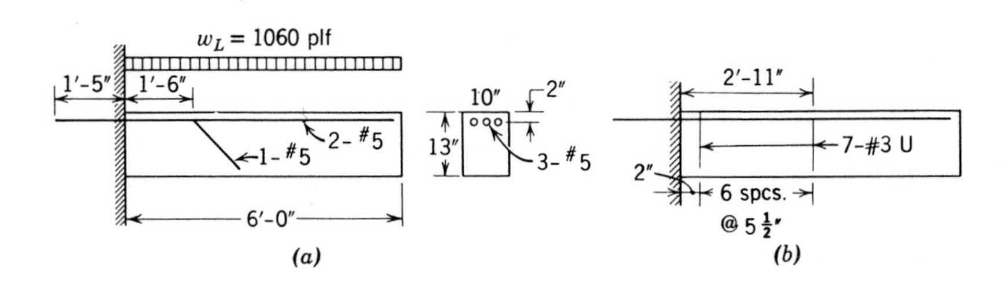

Fig. 4.31. Beam of Sec. 4.22*a*.

$\overline{V} = V_u/\phi = 10{,}100/0.85 = 11{,}880$ lb
$v_u = \overline{V}/bd = 11{,}880/(10 \times 11) = 108$ psi
Permissible $v_c = 2\sqrt{f_c'} = 2\sqrt{3000} = 109.6 \text{ psi} > v_u$

The concrete could carry this v_u without stirrups but Code 11.1.1d requires minimum stirrups (able to carry $v_s = 50$ psi) to a distance x from the support to where v_u drops to $v_c/2 = 109.6/2 = 55$ psi.

$x = 6(108 - 55)/108 = 2.94$ ft. $= 35.2$ in.

Try #3 U at $d/2 = 11/2 = 5.5$ in. (the maximum permissible spacing)

$A_v f_y / b_w s = 0.22 \times 40{,}000/(10 \times 5.5) = 160 \text{ psi} > 50$ psi. min. O.K.

USE 7- #3 at 2 in., 6 at 5.5 in. as sketched in Fig. 4.31*b*. The length of stirrup above mid-depth ($d/2$) must be checked against the development length l_d tabulated in Sec. 4.11a, which is 12 in. The standard hook (as in Fig. 4.32*c*) may be used with $l_d/2$ between mid-depth and the start of the hook.

$l_d/2 \ngtr 0.5d - \text{cover} - 3d_b = 11/2 - 1.5 - 1.12 = 2.88 \text{ in.} < 6$ in.*

This shows the stirrup can only be developed to roughly 74 percent (hook at nearly 50% as in Sec. 5.14 or exactly 50% in Code 12.13.1.1 plus some 24% straight); but say O.K. because only 50 psi of the 160 psi potential capacity is indicated as needed.

If v_c had proved critical and $2\sqrt{f_c'}$ a little deficient, the more exact equation for allowable v_c could have been used as follows:

Permissible $v_c = 1.9\sqrt{f_c'} + 2500\,\rho Vd/M$
$\rho = 3 \times 0.31/(10 \times 11) = 0.0084$, $M_u = 1990 \times 5.08^2/2 = 25.7$ k-ft
$v_c = 1.9\sqrt{3000} + 2500 \times 0.0084 \times 10.10 \times 11/(25.7 \times 12)$
$= 104.1 + 7.6 = 111.7$ psi

(*b*) Determine the stirrups for the beam of Fig. 4.32*a*, using Grade 40 steel and $f_c' = 3000$ psi.

*A projection of these numbers would indicate that d must be increased $2(6 - 2.88) = 6.24$ in., to 17.24 in., before a #3 stirrup would be developed, that is, have the equivalent of the 12 in. minimum which is demanded in either Grade 40 or Grade 60. However, Code 12.13.4 is more liberal for lapped U-stirrups carrying up to 9000 lb in each leg. These are permitted in a beam 18 in. deep (overall). The author considers this as a near waiver of the full 12-in. anchorage demand for #3 stirrups in such a beam depth. Steps are currently under way which hopefully will remove the minimum insofar as it relates to stirrups with hooks.

Solution

Critical stirrup spacing is to be calculated at $d = 19$ in. from support, 4.42 ft from mid-span.

Beam wt. $= (9 \times 22/144) \times 150 = 207$ plf
D.L. $V_u = 1.4(6500 + 207 \times 4.42) = 10{,}400$ lb
$V_u = 10{,}400 + 1.7 \times 15{,}000 = 35{,}900$ lb, $\overline{V} = 35{,}900/0.85 = 42{,}200$ lb
$v_u = 42{,}200/(9 \times 19) = 247$ psi $> v_c = 2\sqrt{3000} = 109.6$ psi $\therefore$ Stirrups
Allowable max. $v_u = 109.6 + 8\sqrt{3000} = 548$ psi O.K. (Code 11.6.4)

Just to the left of the load:

$V_{u3} = 1.4(6500 + 207 \times 3) + 1.7 \times 15{,}000 = 35{,}200$ lb
$\overline{V}_3 = 35{,}200/0.85 = 41{,}300$ lb, $v_3 = 41{,}300/(9 \times 19) = 247$ psi

The v_u and v_s diagrams are shown superimposed on Fig. 4.32*b*. Stirrups must be designed for $v_s = v_u - 2\sqrt{f_c'}$ over the 3-ft length and minimum stirrups must be used between the loads unless $v_u < v_c/2$. The variation in v_s is so small that all stirrups might well be designed for $v_s = 247 - 109.6 = 137$ psi. Code minimum is 50 psi stirrup capacity. Minimum stirrup size is #3. $v_s < 4\sqrt{f_c'} = 219$ psi (Code 11.6.3). Maximum $s = d/2 = 9.5$ in.
Try #3 U-stirrups, $A_v = 2 \times 0.11 = 0.22$ in.^{2}

$s_o = 0.22 \times 40{,}000/(137 \times 9) = 7.15$ in., say 7 in. < 9.5 in. max.

It is obvious that #4 U-stirrups would give a calculated spacing greater than the 9.5 in. permissible maximum. (The student might note the value of s_o varies directly with the bar area and hence #4 would give nearly double the #3 spacing.)

USE #3 U-stirrups.
Sketch (Fig. 4.32*c*) shows hook plus $l_d/2 = 12/2 = 6$ in. available in upper $d/2$. Stirrup development O.K. Next check beyond the load, the live load to the left being omitted to get maximum shear.

$$V_u = 1.4(6500 + 207 \times 3) + 1.7 \times 15{,}000 \times 3/12 = 16{,}310 \text{ lb}$$
$$\overline{V} = 16{,}310/0.85 = 19{,}100 \text{ lb}, \; v_{u3+} = 19{,}100/(9 \times 19) = 112 \text{ psi.}$$
$$v_{s3+} = 112 - 109.6 = 2 \text{ psi}$$

For this small v_s and also because $v_u > v_c/2$, use minimum stirrups across the middle length. By inspection #3 at 9.5 in. will care for more than the minimum 50 psi required by Code 11.1.2.

A formal stirrup spacing curve could be prepared as in Fig. 4.34, but it is totally unnecessary for this short length and this simple pattern of v_s

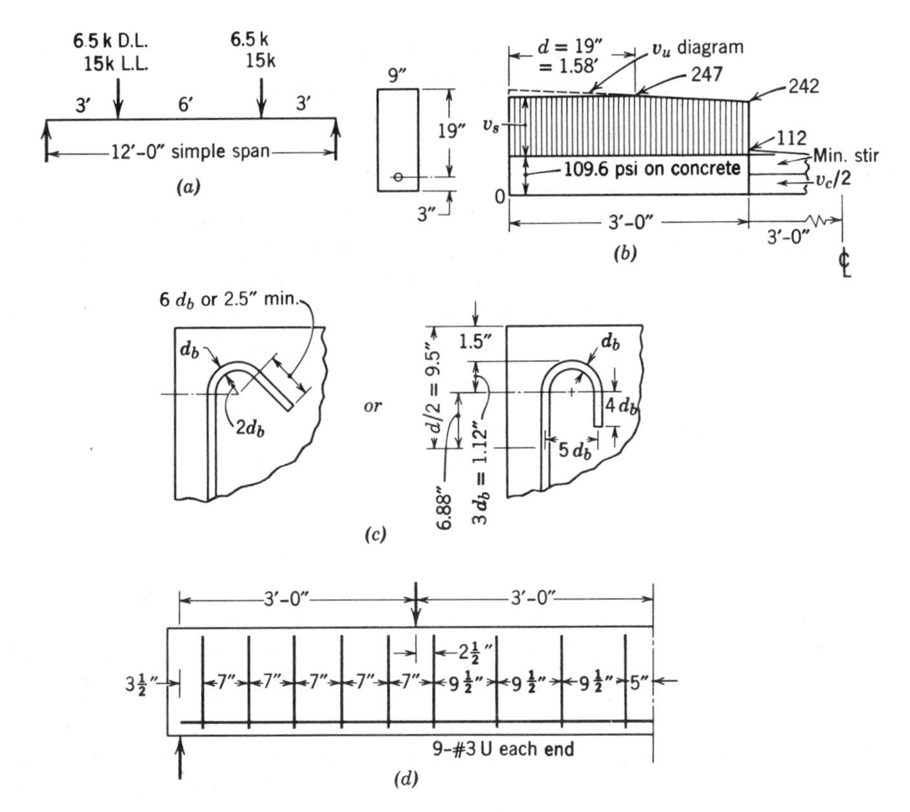

Fig. 4.32. Beam of Sec. 4.22*b*.

values. The s_o value calculated at the critical section must be used back to the support and the small change in v_s to $x = 3$ ft makes further calculations of s totally unnecessary.

USE 9- #3 U at 3½ in., 5 at 7 in., 3 at 9½ in. at each end. This arrangement is sketched in Fig. 4.32*d* and shows the last stirrup at the right 5 in. from the centerline instead of $s/2$, but say O.K.

(*c*) A continuous T-beam with $b_w = 9.5$ in., $d = 16.6$ in., $f_c' = 3000$ psi, Grade 60 steel must provide for the unit shears shown in Fig. 4.33 (These shears are calculated as in Sec. 8.17.) Design and space the necessary stirrups.

Solution

The critical v_s is at a distance d from the face of support where v_s is $438 - 109.6 = 328$ psi. Stirrups are needed for a distance l_v, out to the point where

$v_c = 2\sqrt{f_c'} = 109.6$ psi, the allowable without stirrups for computed stress. Minimum stirrups are also required where $v_u > v_c/2$.

$$\frac{l_v}{120} = \frac{495 - 109.6}{495 - 81} = \frac{385}{414}, \qquad l_v = 112 \text{ in.}$$

Stirrups must be designed for $v_s = v_u - 109.6$, as indicated by the shaded area in Fig. 4.33 and then be extended at maximum spacing to midspan.

For vertical stirrups, the maximum spacing permitted by Code 11.1.4*b* is $d/2 = 8.3$ in. for zones where v_s is $4\sqrt{f_c'} = 219$ psi or less, which is true only at 48 in. or more from the support. For this 48 in. Code 11.6.3 limits spacing to $d/4 = 4.1$ in. (Figure 4.34 plots 60 in., an error.)

For #3 U-stirrups,
$s_o = 2 \times 0.11 \times 60{,}000/(328 \times 9.5) = 4.23$ in. (revised below)

For #4 U-stirrups,
$s_o = 2 \times 0.20 \times 60{,}000/(328 \times 9.5) = 7.68$ in.

The #4 stirrups are ruled out by the need to satisfy a maximum 4.1 in. spacing. Also since the theoretical spacing increases to infinity at $l_v = 112$ in., the #4 stirrups would have an excessive number of spaces determined by the maximum spacing of 8.3 in. and this would be wasteful. The #3 stirrups must be investigated for available development length in the upper half depth $16.6/2 = 8.3$ in. The required $0.5\, l_d$ between mid-depth and start of hook (Sec. 4.11*a*) is 6 in. and a sketch similar to Fig. 4.32*c* shows the distance available is $d/2 - 1.5$ in. cover $- 3d_b = 8.3 - 1.50 - 3 \times 0.375 = 5.68 < 6$ in. Stirrup efficiency is thus reduced to 0.5 for the hook + $(5.68/6)0.5 = 0.973$ and the revised s_o becomes

$s_o = 0.973 \times 4.23 = 4.10$ in.

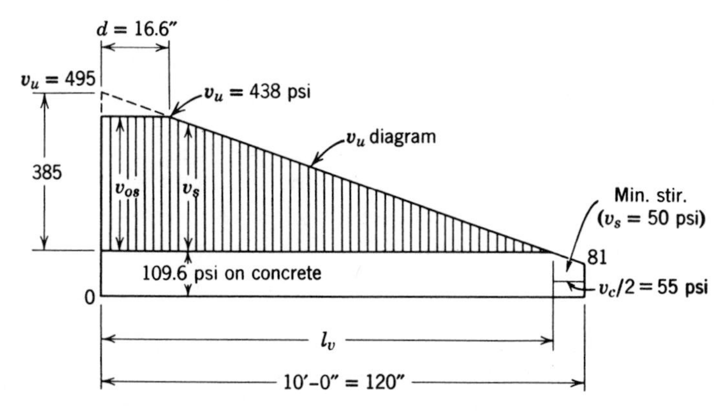

Fig. 4.33. v_s diagram for beam similar to Sec. 8.17.

The change is small but will be complied with.* (If Grade 40 stirrups were used the same l_d would be required simply as a minimun, not because of stress. In such a case, with the possible misplacement vertically of a stirrup very limited, the author would probably overlook this deficiency.)

The #3 stirrups will next be investigated at the practical maximum spacing of 8 in. with respect to whether they meet the 50 psi v_s requirement for capacity specified in Code 11.1.2.

$$v_s = A_s f_y / b_w s = 0.973 \times 0.22 \times 60{,}000/(9.5 \times 8) = 169 \text{ psi} \qquad \text{O.K.}$$

The detailed stirrup spacing curve will be developed for establishing the final spacings graphically. Theoretical spacings will be calculated at several points by noting that $s \propto 1/v_s$, this being simpler than numerical calculation of v_s values.

The critical spacing also applies for the 16.6 in. nearest the support. The remaining v_s triangle is $112 - 16.6 = 95.4$ in. long, which can be subdivided into four equal lengths of 23.9 in each. At these subdivision lines the shear v_s starting at the right, is successively $v_{so}/4$, $v_{so}/2$, $0.75v_{so}$, leading to stirrup spacings:

z from midspan	v_s	s
$8 + 23.9 = 31.9$ in.	$0.25v_{so}$	$4s_0 = 4 \times 4.10 = 16.4$ in. > 8.3 in.
$+ 47.8 = 55.8$	$0.50v_{so}$	$2s_0 = 2 \times 4.10 = 8.20$ in.
$+ 71.7 = 79.7$	$0.75v_{so}$	$1.33_{so} = 1.33 \times 4.10 = 5.47$ in.

These values are plotted in Fig. 4.34 and a theoretical curve sketched through the points. The maximum spacings drawn in at 8.3 and 4.1 in. show that the spacing curve governs only for a very short distance. Use the spacings marked at the bottom of the figure and indicated by the shaded practical spacing curve. It is obvious that these data have scarcely justified the formality of drawing the spacing curve.

To indicate the more general use of such a curve, consider now that the same spacing curve is required for a hypothetical case which does not demand the maximum spacing of the 4.1 in., as in Fig. 4.35.

The practical spacings are established graphically in the following fashion. Spacings equal to or just less than the theoretical are laid out horizontally, each stirrup indicated by a vertical line. A stair-step curve of the spacings actually used is also plotted at the same time, step by step. The vertical lines keep track of the actual stirrup locations and the stair-step curve indicates how these compare with the allowable curve values.

*Assuming beam height is at least 18 in., one might follow the footnote under Example (a) and use 100% efficiency here.

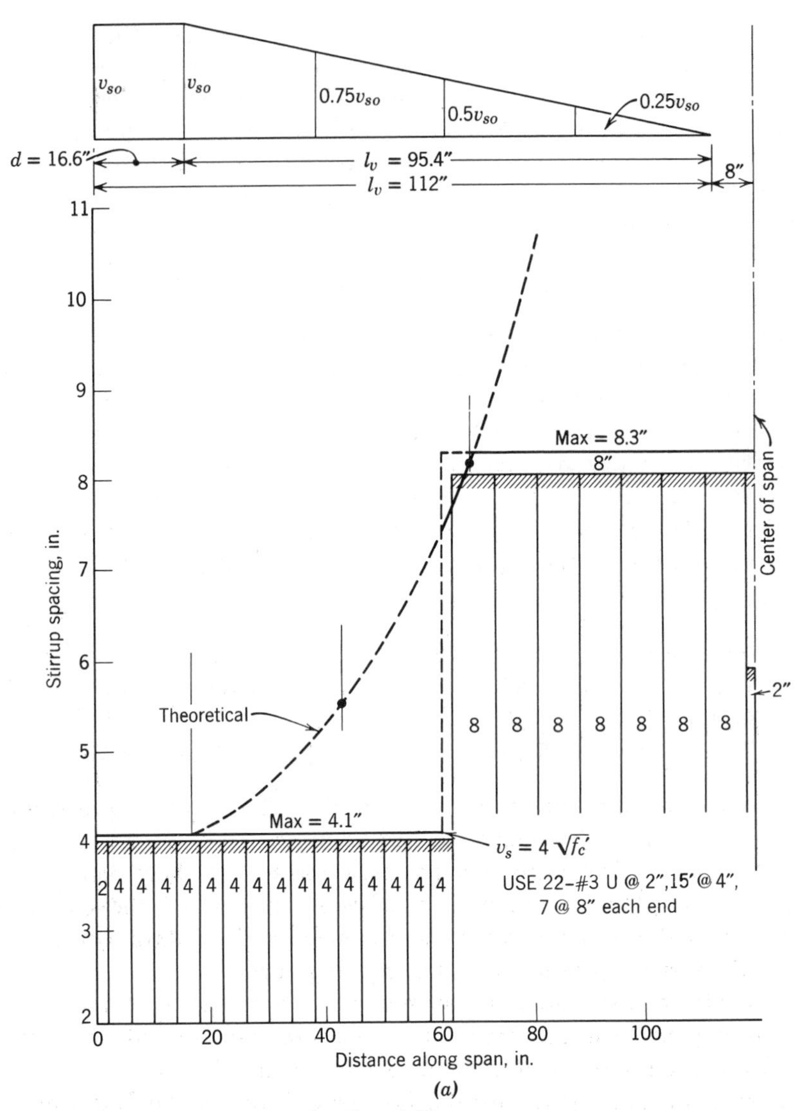

Fig. 4.34. Stirrup spacing curve for Sec. 4.22*c*.

The first space is necessarily a half (theoretical) space to put the stirrup in the center of the horizontal area it serves. Since quarter-inch dimensions are generally objectionable, the 4-in. spacing must be continued until 4.5 in. may be used, and some would continue until 5 in. could be used. It may appear that the 6 in. spacing is introduced prematurely. Careful analysis will show, however, that the criterion is not that the spacings used stay under the theoretical values at all points but that the area under the practical curve served by a single stirrup (that is, $s/2$ each side of the sitrrup) must not exceed the area under the theoretical curve. On this basis all the spacings used are very conservative and the 5-in. and 6-in. spacings could be stretched to 5.5 in. and 6.5 in., as shown dotted.

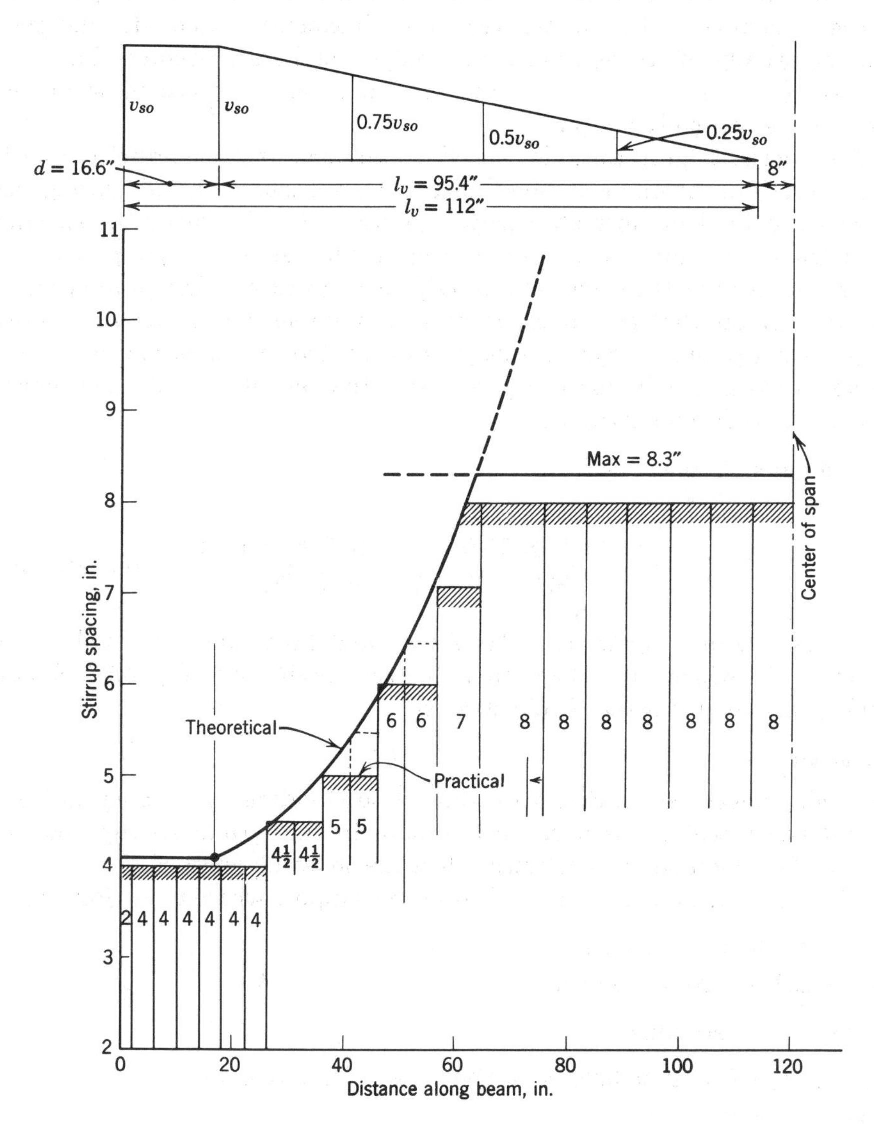

Fig. 4.35. Stirrup spacing curve, hypothetical alternate case. Record the results thus: USE 20-#3U at 2, 6 at 4, 2 at $4\frac{1}{2}$, 2 at 5, 2 at 6, 7, 6 at 8 each end plus 1-#3U at mid-span.

The 7-in. spacing puts a stirrup at 64 in. from the support and leaves 56 in. to be filled in with 8-in. spaces. Seven spaces exactly fit, which is an unusual result. The designer must usually stretch a few spaces (such as the 5- and 6-in. spaces just mentioned) or put in an extra stirrup at midspan. In the latter case he could use some of the extra coverage to simplify the earlier spacings. Stirrups are relatively inexpensive and the designer cannot usually afford as refined an analysis as here presented, but it is presented to give the student the reasoning he must use in whatever procedures he finally adopts.

An alternate procedure for spacing stirrups, less accurate but still satisfactory in the hands of one skilled at it, consists of (1) calculating the theoretical total number of stirrups required, (2) calculating the spacing at a few points, and (3) simply writing down a series of spacings which satisfy (2) and at the same time provide enough extra stirrups to account for the lengths that are governed by maximum spacing instead of stress. Extra stirrups are always necessary as illustrated by the design just completed with 20.5 stirrups compared to a theoretical, at 0.965 efficiency because of reduced stirrup l_d,

$$n = \frac{b_w(\text{area } v_s \text{ diag.})}{A_v f_y}$$

$$= \frac{9.5(328 \times 16.6 + 0.5 \times 328 \times 95.4)}{0.965 \times 2 \times 0.11 \times 60{,}000} = 15.9 \text{ stirrups}$$

(*d*) Given a simple span beam 24-in. wide and 18-in. depth to steel, $f_c' =$ 4000 psi, for which the v_s diagram is the triangle shown in Fig. 4.36, choose and space stirrups using Grade 40 steel.

Solution

The idea developed in Sec. 4.13, that of subdividing the v_s diagram into equal areas, will be used as an alternate to the formal stirrup spacing curve. This method is sometimes called the slide-rule method.

Shear is critical at a distance *d* from the support where $v_u = 262$ psi.

$$v_c = 2\sqrt{4000} = 126 \text{ psi}$$
$$v_s = 262 - 126 = 136 \text{ psi}$$

Try #4 U-stirrups.

$$s_o = A_v f_y / v_s b = 2 \times 0.20 \times 40{,}000/(136 \times 24) = 4.90 \text{ in.}$$
$$\text{Max } s = d/2 = 9 \text{ in.}$$

This maximum spacing would govern for nearly half of the length, since a

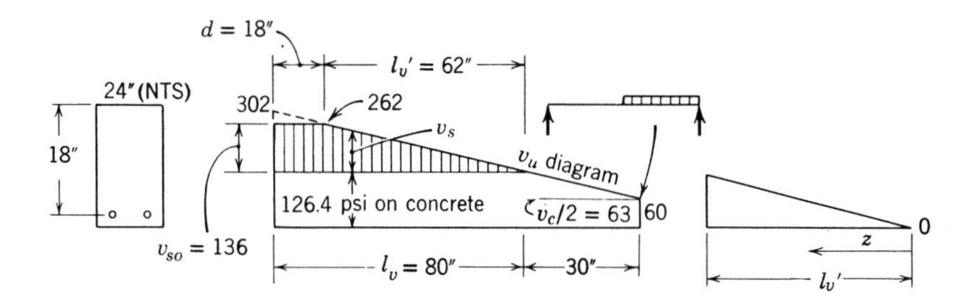

Fig. 4.36. v_s diagram for Sec. 4. 22*d*.

calculated *s* of 9 in. results from v_s of about 74 psi. A simple layout would result, but it would be wasteful of steel.

Try #3 U-stirrups.

$s_o = 2 \times 0.11 \times 40{,}000/(136 \times 24) = 2.69$ in.

This spacing is very close, nearly the minimum generally used. It appears preferable to use the #3 U-stirrups, although there is not much to choose between the two sizes. The small size probably calls for a smaller maximum spacing limit to give v_s at least 50 psi (Code 11.1.2).

$$\begin{aligned}\text{Max. } s &= A_v f_y / b v_s = 0.22 \times 40{,}000/(24 \times 50) \\ &= 7.3 \text{ in., say 7 in.} < d/2 \qquad \text{O.K.}\end{aligned}$$

USE #3 U-stirrups. Development length l_d checks O.K. as in Fig. 4.32*c*.

The slide rule solution of Sec. 4.13 is applicable only to a triangular area. One might calculate *n* for the triangular area and fill in the 18-in. constant v_s length (a little inefficiently) with stirrups at constant spacing as required. Instead, the stirrups in the 18-in. length will first be calculated and the triangle then adjusted to fit. Above, s_o was found to be 2.69 in., say $2\frac{1}{2}$ in.

To start, try 8-#3 at an initial space of 1 in., followed by 7 spaces at $2\frac{1}{2}$ in. This places the last of these stirrups at 18.5 in. and cares for an additional length of $s/2$ beyond. It will be simplest to visualize this $s/2$ as within the triangle, that is, to imagine the v_s triangle as starting at this last stirrup, and use the same 136 psi value of maximum v_s with an imaginary triangle length $l_v' = 62 - 0.5 = 61.5$ in. It will then be possible to calculate the theoretical number of stirrups *n*, subdivide the triangle into *n* parts, and put a stirrup at each subdivision rather than dealing with 2*n* parts.

$$n = \frac{24(0.5 \times 136 \times 61.5)}{2 \times 0.11 \times 40{,}000} = 11.5$$

TABLE 4.1. Stirrup Spacing from Slide-Rule Calculation (Sec. 4.22d)

For initial distance d: $s = 1$ in., 7 at 2.5 in.

Further Subdivision Points	z	s
$n = 11.5$	61.5 in.	
		2.5 in.
$n - 1 = 10.5$	59.0	
		3
$n - 2 = 9.5$	56.0	
		3
$n - 3 = 8.5$	53.0	
		3.5
$n - 4 = 7.5$	49.5	
		3.5
$n - 5 = 6.5$	46.0	
		3.5
$n - 6 = 5.5$	42.5	
		4
$n - 7 = 4.5$	38.5	
		4.5
$n - 8 = 3.5$	34.0	
		5.5
$n - 9 = 2.5$	28.5	
		6
$n - 10 = 1.5$	22.5	max $s = 7$ in.

Add 30 in. beyond triangle. Need 7-in. spacing for 22.5 + 30 in. = 52.5 in.
52.5/7 = 7.5 spaces

The v_s diagram will be subdivided into 11.5* equal areas. The slide-rule ratio is set up by lining up $z_n = 61.5$ in. on the C scale opposite $n = 11.5$ on the A scale, as in Fig. 4.19. The rider is moved to $n - 1 = 10.5$ on the A scale to read $z_{n-1} = 59.0$ in. on the C scale; then to $n - 2 = 9.5$ on the A scale, and so forth. It is suggested that z be read only as accurately as the stirrup spacing is to be stated, that is, to the nearest full inch if the designer wishes to avoid half-inch spacings. When tabulated as in Table 4.1 the calculations are completed by recording s as the difference between successive z values. The theoretical s given by $z_{2.5} - z_{1.5}$ equals the maximum

*11.5 equal areas sounds highly theoretical compared to simply using 12 areas. However, with the triangular v_s diagram, the theoretical spacing approaches infinity as v_s approaches zero, causing the maximum spacing rule to control. Hence this last area might as well be 0.5 of a full area; it is an unused area either way one proceeds.

spacing of 7 in. Hence this last 22.5 in. ($z_{1.5}$) plus the 30 in. to midspan (since $v_u > v_c/2$), total of 52.5 in., must be completed at 7 in. spacing, requiring $52.5/7 = 7.5$ more stirrups, say 7 spaces to last stirrup this side plus $s/2$ to midspan.

USE 24- #3 U-stirrups at 1 in., 8 at 2.5 in., 2 at 3 in., 3 at 3.5 in., 4 in., 4.5 in., 5.5 in., 6 in., 7 at 7 in., from each end.

4.23 DESIGN EXAMPLE—TORSION AND SHEAR

An 18-in. wide by 27-in. deep rectangular cantilever beam 9-ft long has been designed to pick up at its end a 20-ton hoisting load, which includes a proper allowance for impact. It is found operating conditions may place the load 10 in. off the axis of the beam to either side. Design the torsion and shear reinforcement, assuming Grade 60 reinforcement and f_c' of 3000 psi.

Solution

The critical section for both torsion and shear is at a distance d from the support.

$d = 27 - \text{say } 2.75 = 24.25$, say 24.3 in.

Beam weight $= (18 \times 27/144)\ 150 = 506$ plf

$V_u = 1.7\ (20 \times 2000) + 1.4 \times 506(9 - 2.02) = 72{,}900$ lb

$\overline{V} = 72{,}900/0.85 = 85{,}700$ lb

$v_u = 85{,}700/(18 \times 24.3)$

$= 196$ psi $<$ allowable

$T_u = 1.7(20 \times 2000)10 = 680{,}000$ lb-in.

$\overline{T} = 680{,}000/0.85 = 800{,}000$ lb-in.

$v_{tu} = 3\overline{T}/\Sigma x^2y$ or $\overline{T}/(\Sigma x^2y/3)$ from Code 11.7.2.

$(1/3)\Sigma x^2y = (1/3)18^2 \times 27 = 2920$ in.^{3}

$v_{tu} = 800{,}000/2920 = 274$ psi

Max. allow. $v_{tu} = 12\sqrt{f_c'}/\sqrt{1 + (1.2v_u/v_{tu})^2}$ from Code 11.7.7

$= 12 \times 54.8/1^+ > 274$ psi O.K.

Allow. $v_{tc} = 2.4\sqrt{f_c'}/\sqrt{1 + (1.2v_u/v_{tu})^2}$ from Code 11.7.5

$= 2.4 \times 54.8/\sqrt{1 + (1.2 \times 196/274)^2} = 99$ psi

$v_{ts} = v_{tu} - v_{tc} = 274 - 99 = 175$ psi

Assume #4 stirrups with 1.5 in. clear cover.

$\alpha_t = 0.66 + 0.33y_1/x_1$ from Code 11.8.2

$= 0.66 + 0.33(27 - 2 \times 1.75)/(18 - 2 \times 1.75) = 1.195$

$$A_t = \frac{(v_{tu} - v_{tc})s\Sigma x^2y/3}{\alpha_t x_1 y_1 f_y} \qquad \text{from Code 11.8.2}$$

$$2A_t = \frac{2 \times 175s \times 2920}{1.195 \times 14.5 \times 23.5 \times 60{,}000} = 0.0418s$$

For a given A_t, the spacing could be calculated directly, but it is better to combine with the stirrups for shear.

$$\begin{aligned}
\text{Allow. } v_c &= 2\sqrt{f_c'}/\sqrt{1 + (v_{tu}/1.2v_u)^2} \qquad \text{from Code 11.4.5} \\
&= 2 \times 54.8/\sqrt{1 + [274/(1.2 \times 196)]^2} = 71 \text{ psi} \\
v_s &= 196 - 71 = 125 \text{ psi} < 4\sqrt{f_c'}, \text{ max. } s = d/2 \\
A_v &= (v_u - v_c)b_w s/f_y = 125 \times 18s/60{,}000 = 0.0375s
\end{aligned}$$

For torsion plus shear the sum of $0.0418s$ and $0.0375s$ will be provided by one size of stirrup at a regular spacing. For #4 stirrups:

$$2 \times 0.20 = 0.0418s + 0.0375s = 0.0793s, \qquad s = 5.05 \text{ in.}$$

The Code requires a torsion spacing not greater than $(x_1 + y_1)/4 = (23.5 + 14.5)/4 = 9.5$ in., nor greater than 12 in.
Development of closed transverse torsion reinforcement is not specifically covered in the Code; the same provisions applied to stirrups for shear will be used (Sec. 4.11*a*).

$$d/2 = 24.3/2 = 12.3 \qquad l_d/2 = 6 \text{ in. for } \#4$$

Deduct cover and $3d_b$ from $d/2 = 12.1 - 1.5 - 3 \times 0.5 = 9.1$ in. > 6 in. This available depth plus the anchorage around the corner is O.K.

USE #4 closed stirrups at 5 in.

Calculate extra longitudinal reinforcement for torque.

$$A_l = 2A_t(x_1 + y_1)/s = 0.0418s(23.5 + 14.5)/s = 1.59 \text{ in.}^2$$

$$A_l = \left[\frac{400\ x\ s}{f_y}\left(\frac{v_{tu}}{v_{tu} + v_u}\right) - 2A_t\right]\frac{x_1 + y_1}{s} \qquad \text{from Code 11.8.4}$$

$$= \left[\frac{400 \times 18s}{60{,}000}\left(\frac{274}{274 + 196}\right) - 0.0418s\right]38/s = 1.11 \text{ in.}^2$$

The 1.59 in.² governs. The Code requires at least #3 longitudinal bars, spaced not farther apart than 12 in., and at least one bar in each corner of the stirrups. Assume approximately $1.59/3 = 0.53$ in.² at top, at bottom, and at midheight on the sides.

For A_l USE 1 − #5 = 0.31 in.² at midheight *each* side

ADD $(1.59 - 0.62)/2 = 0.48$ in.² *each* to top and to bottom bars. Because of beam width, there must be at least three bars on top and bottom face, that is, one at each stirrup corner and one midway between, since the spacing is limited to 12 in. The parts of A_l can be added to the longitudinal steel required for moment and the bars selected for the totals.

SELECTED REFERENCES

1. ACI-ASCE Committee 326, "Shear and Diagonal Tension," *Jour. ACI, Proc.*, 59, Jan, 1962, p. 1; Feb. 1962, p. 277; Mar. 1962, p. 352.
2. Phil M. Ferguson, "Some Implications of Recent Diagonal Tension Tests," *Jour ACI*, 28, Aug. 1956, June 1957; *Proc.*, 53, pp. 157, 1190.
3. H. A. R. dePaiva and C. P. Siess, "Strength and Behavior of Deep Beams in Shear," *Proc. ACSE*, Vol. 91, No. ST5, Part 1, Oct. 1965, p. 19.
4. R. C. Fenwick and T. Pauley, "Mechanisms of Shear Resistance of Concrete Beams," *Proc. ASCE*, Vol. 94, No. ST10, Oct. 1969, p. 2325.
5. G. N. J. Kani, "How Safe Are Our Large Reinforced Concrete Beams," *Jour. ACI, Proc.*, Vol. 64, No. 3, Mar. 1967, p. 128.
6. K. S. Rajagopalan and Phil M. Ferguson, "Exploratory Shear Tests Emphasizing Percentage of Longitudinal Steel," *Jour. ACI, Proc.*, Vol. 65, No. 8, Aug. 1968, p. 634.
7. Phil M. Ferguson and Syed I. Husain, "Strength Effect of Cutting Off Tension Bars in Concrete Beams," Research Report 80-1F, Center for Highway Research, University of Texas at Austin, June 1967, 37pp.
8. L. B. Kriz and C. H. Raths, "Connections in Precast Concrete Structures—Strength of Corbels," *Jour. Prestressed Concrete Institute*, Vol. 10, No. 1, Feb. 1965, p. 16.
9. Phil M. Ferguson, "Design Criteria for Overhanging Ends of Bent Caps—Bond and Shear," Research Report No. 52-1F, Center for Highway Research, The University of Texas at Austin, Aug. 1964.
10. J. A. Hofbeck, I. O. Ibrahim, and Alan H. Mattock, "Shear Transfer in Reinforced Concrete," *Jour. ACI, Proc.*, Vol. 66, No. 2, Feb. 1969, p. 119.
11. R. F. Mast, "Auxiliary Reinforcement in Precast Concrete Construction," *Proc. ASCE*, Vol. 94, No. ST6, June 1968, p. 1485.
12. F. Leonhardt, "Reducing the Shear Reinforcement in Reinforced Concrete Beams and Slabs," *Magazine of Concrete Research*, Vol. 17, No. 53, Dec. 1965, p. 187.
13. ACI Committee 438, "Tentative Recommendations for the Design of Reinforced Concrete Members to Resist Torsion," *Jour. ACI, Proc.*, Vol. 66, No. 1, Jan. 1969, p. 12.

14. Fung-Kew Kong, Peter J. Robins, and David F. Cole, "Web Reinforcement Effects on Deep Beams," *Jour. ACI, Proc.*, Vol. 67, No. 12, Dec. 1970, p. 1010.
15. F. Leonhardt and R. Walther, "Deep Beams," *Bulletin* 178, Deutscher Ausschuss fur Stahlbeton, Berlin, 1966, 159 pp (in German).
16. K. S. Rajagopalan and Phil M. Ferguson, "Distributed Loads Creating Combined Torsion, Bending, and Shear on L-Beams with Stirrups," *Jour. ACI, Proc.*, Vol. 69, No. 1, Jan. 1972, p. 46.
17. David J. Victor and Phil M. Ferguson, "Beams Under Distributed Load Creating Moment, Shear, and Torsion," *Jour. ACI, Proc.*, Vol. 65, No. 4, April 1968, p. 295.
18. John Minor and James O. Jirsa, "A Study of Bent Bar Anchorages," *Structural Research at Rice*, No. 9, Department of Civil Engineering, Rice University, Mar. 1971.
19. *Torsion of Structural Concrete*, SP-18, American Concrete Institute, Detroit, 1968, 505pp. (A collection of 19 papers.)

PROBLEMS

Note: The problems in this group relate only to shear and torsion stress; flexural considerations are not included, unless specifically stated. Whenever actual stresses are called for, these should be compared with the specified allowables of the Code.

Prob. 4.1. (a) For shear only and $v_c = 2\sqrt{f_c'}$, what is the permissible uniform live load in the beam of Fig. 4.37 if minimum stirrups are used? Assume $f_c' = 4000$ psi, Grade 60 steel, and $w_d = 1300$ plf (including beam weight). Where can the stirrups then be discontinued?
(b) Same as (a) except use Eq. 4.2 of Sec. 4.8*b* for v_c allowable.

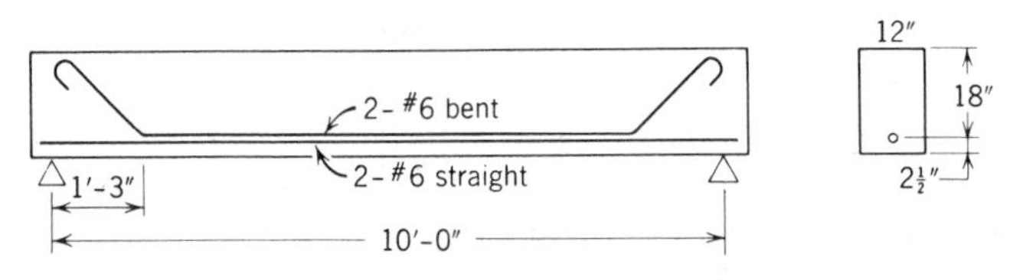

Fig. 4.37. Simple span beam for Prob. 4.1.

Prob. 4.2. In the beam of Fig. 4.38*a*, if the 20k is all live load, $f_c' = 4000$ psi, and Grade 60 bars are used, design the stirrups and on an elevation (as in Fig. 4.38) show their arrangement and spacing.

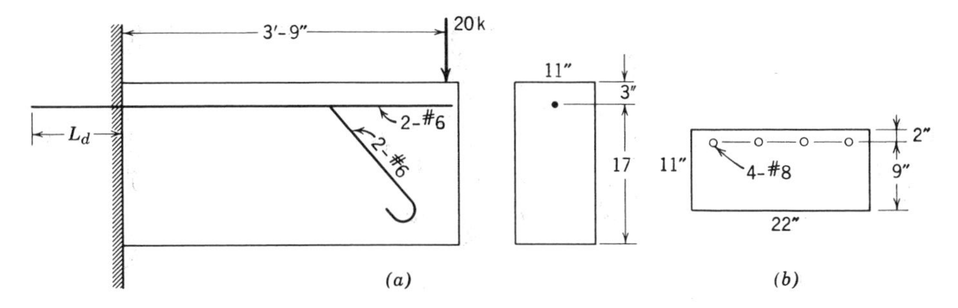

Fig. 4.38. Cantilever beam. (*a*) For Probs. 4.2. and 4.3. (*b*) For Prob. 4.4.

Prob. 4.3. If the load on the beam of Fig. 4.38*a* and Prob. 4.2 is changed to w_d = 2000 plf (including beam weight) and w_l = 4000 plf, and l becomes 4ft 4in., design the stirrups and sketch their arrangement.

Prob. 4.4. If the cross section of Fig. 4.38*b* is used on a 4ft 4in. cantilever, f_c' = 4000 psi, Grade 60 steel, under a dead load w_d = 2000 plf (including beam weight) and w_l = 4000 plf, check for shear and, if needed, choose and position stirrups.

Prob. 4.5. Evaluate V_u for the beam of Fig. 2.19.
(a) For no stirrups.
(b) For minimum stirrups; design these stirrups.
(c) For maximum stirrups; design these stirrups.

Prob. 4.6. Evaluate maximum V_u for the joist of Fig. 2.22 if f_c' = 5000 psi.

Prob. 4.7. Design and space vertical stirrups for an 18-ft simple span rectangular beam, b = 11 in., d = 18 in., f_c' = 3000 psi, Grade 40 steel, w_d (including beam weight) = 800 plf, and w_l = 3200 plf.
(*a*) Use of stirrup spacing curve.
(*b*) Use slide-rule method.

Prob. 4.8. The loads shown in Fig. 4.39 are at fixed points, not moving loads. Each load consists of 3000 lb of dead load and a possible 4500 lb of live load. Establish the maximum shear curve and design and space the necessary stirrups for f_c' = 3000 psi and Grade 40 steel. Consider uniform load negligible.

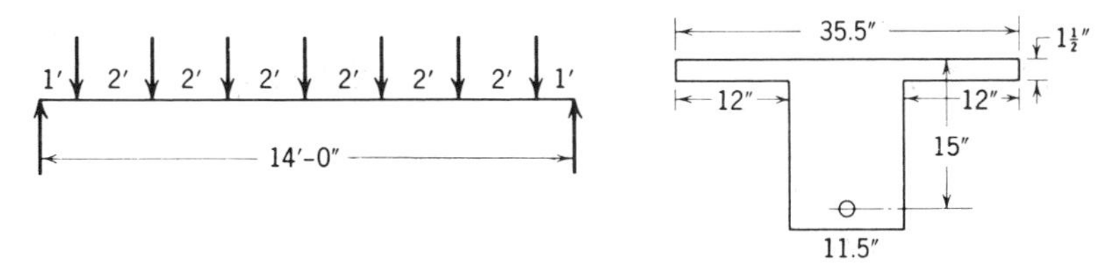

Fig. 4.39. Simple span beam for Prob. 4.8.

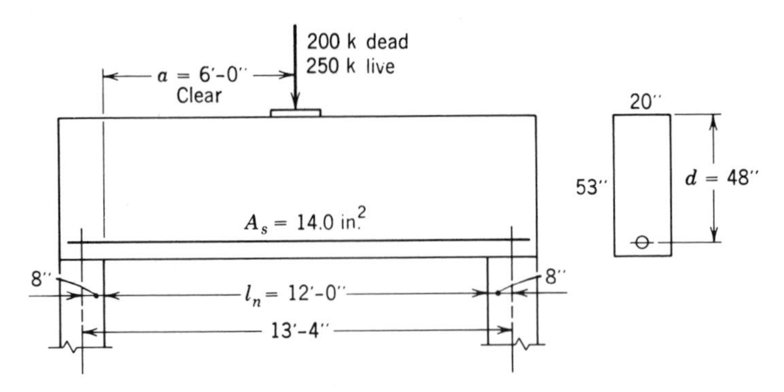

Fig. 4.40. Beam for Prob. 4.9.

Prob. 4.9. Given the essentially simple span beam of Fig. 4.40 carrying the column loads shown, $f_c' = 4000$ psi, Grade 60 steel. Design shear reinforcement required and sketch its placement on elevation of beam.

Prob. 4.10. A rectangular beam, freestanding except for being fixed at each end against any rotation, must carry a midspan live load of 35 kips which can be as much as 12 in. off the axis of the beam. Given $b = 12$ in., $d = 20$ in., $h = 23$ in., $l_n = 20$ ft, $f_c' = 4000$ psi, and Grade 60 steel. Design the necessary shear and torsion reinforcement.

Prob. 4.11. A spandrel beam 12-in. wide by 20-in. deep ($d = 17.5$ in.) with a slab 4-in. thick available (on one side) to act as a flange must carry design (factored) loads producing a shear of 50 kips and a torque of 18 k-ft. Using $f_c' = 4000$ psi and Grade 60 steel, design stirrups and longitudinal steel to be added (to the flexural demands).

5
Development and Splicing of Reinforcement

DEVELOPMENT LENGTH

5.1 EVOLUTION OF THE DEVELOPMENT LENGTH CONCEPT

Experiment shows that, if a bar has enough embedment in concrete, it cannot be pulled out. The term "bar development length" is the embedment necessary, under specific surrounding conditions, to assure that the bar can be stressed to the yield point, with some reserve to insure member toughness.

Prior to the 1971 Code, development and anchorage of bars was always treated as a subdivision of bond stress. Bond stress is the local longitudinal shearing force, between the surrounding concrete and the perimeter surface of the bar, which changes the bar stress from point to point. Studies over the past dozen years show that the bond stress calculation is not itself helpful in the present state-of-the-art. In fact bond stress calculation is not mentioned in this Code.

The reason for this change in emphasis is not that bond stress is unimportant but because:

1. Bond stress presents a very complex problem for which there is no immediate dependable solution usable in a design. The well known equation $u = V/(\Sigma o.z)$ must be supplemented by development length checks to be safe.
2. Strength over a given length appears not sensitive to local peak bond stresses, but can be based on an average value.

3. Failures observed are most commonly the result of tensile splitting stresses around the bar,* with the bond failure only a secondary effect of this collapse.
4. The development length concept summarizes or incorporates most of the present usable knowledge in this area; flexural bond would add a calculation not actually helpful in the design.

One remnant of caution still exists with regard to bond stress concentrations at the end of a bar, say, at the end of a simple span and, by analogy, at the point of inflection (P.I.). For convenience this caution has also been expressed in terms of an equivalent development length requirement (Sec. 5.13).

Hopefully, in the future, possibly even in the next Code, calculation methods for average splitting stresses or values derived therefrom may be available. It will then be apparent that bars so embedded in mass concrete as to avoid critical splitting stresses constitute a different problem, one probably discriminated against in the 1971 Code.

In such a developing and changing area it is important to understand what is known (and what is unknown) about bond stress, even though the Code now simplifies this in such a way that bond stress itself is not usually calculated. Sections 5.2 to 5.9 summarize the present state of knowledge in this basic area. The detailed Code treatment starts with Sec. 5.10, and the balance of the chapter relates more to design.

5.2 THE MECHANICS OF BOND—TENSION BARS

Tension force is put into or taken out of a reinforcing bar by bond stresses between the steel and the adjacent concrete. Only rarely is the load applied directly on the bar, as in a hanger.

The mechanism of stress transfer with deformed bars involves three primary elements, which act at a given point more in sequence than in unison.

1. Shearing resistance of the adhesion itself.
2. Frictional resistance to sliding after adhesion is broken.
3. Bearing against the lugs.

Usually, near ultimate, adhesion has been broken and sliding friction is not very important, making the bearing against the lugs the primary mechanism. This bearing stress against the lugs cannot exist by itself; it produces:

*Possibly accentuated by some dowel force at times.

4. Shearing forces tending to shear off the concrete between the lugs at the outer boundary of the lugs.
5. Less obviously, the bearing has a major radial component which produces large ring tensions around the bar.

In most bond failures the resulting longitudinal splitting stresses become the weak link in the system that leads eventually to loss of bond resistance. The crack patterns of the four beams in Fig. 5.3 which failed in bond under the loading of Fig. 5.9 demonstrate the longitudinal splitting which develops into failure.

Until very recently the internal stress distribution could only be rationalized on the basis of external evidence such as these longitudinal cracks. In 1971 Goto published[1] photographs of internal cracking patterns which resulted from pulling both ends of a tension bar embedded in a concrete prism encasement, as shown in Fig. 5.1. Axial tension cracks developed at intervals, much like the appearance of flexural cracks in a beam, and in effect created a number of small length specimens. Adjacent to each side of each crack and adjacent to each end these specimens developed the same internal crack pattern.

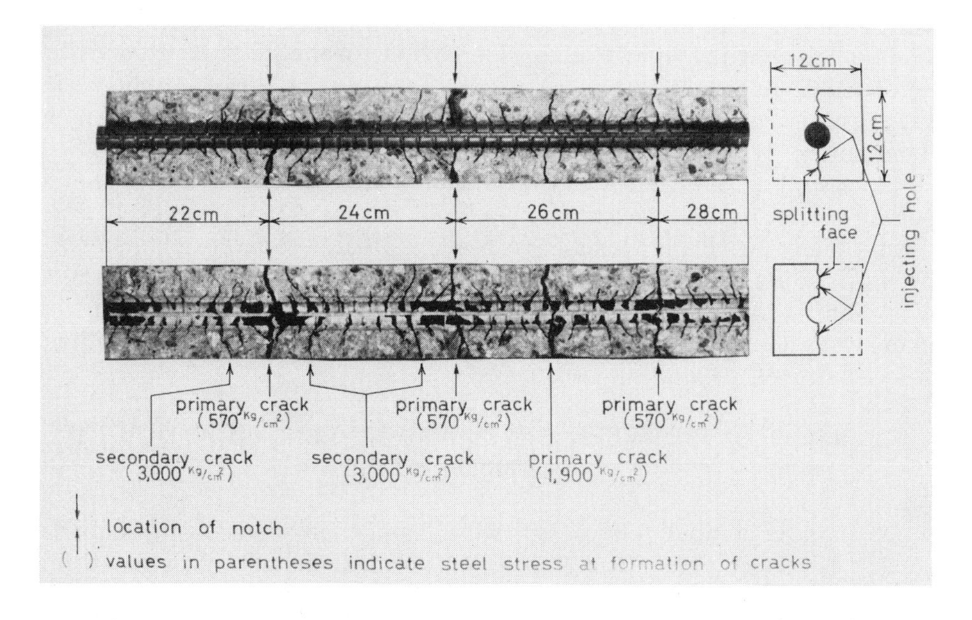

Fig. 5.1. Internal cracks created by tension pull on both ends of the bar. Cracks were dyed with an injection of red ink which, in the bar imprint below, also spread over part of the bar surface where slip occurred. (Courtesy ACI, Ref. 1.)

The systematic internal crack pattern, cracks starting just behind each lug and progressing diagonally a limited distance toward the nearest transverse crack, is impressive. Since cracking in concrete results from a principal tensile stress, it follows that there are compressive stresses parallel to these cracks, forces directed outwardly around the bar to form a hollow truncated cone of pressure. This pressure is directed inwardly against the lug and outwardly must be resisted by ring tension in the specimen.

The pictured angle between crack and bar axis is surprisingly large, considerably more than 45 degrees. This points to radial splitting forces much exceeding the parallel component. It seems rational to hypothesise that these crack directions are actually functions of the thickness of cover, more steeply angled to the bar when the cover is large enough to produce a large resisting ring tension and at flatter slopes when cover is small. In any event, the radial pressures are large and the idealized sketch of Fig. 5.2 now shows much steeper outward forces than those shown in earlier editions of this book. These outward forces, like water pressure in a pipe, lead to splitting on weak planes along the bar unless the cover is unusually large.

If Goto's work is extended to the case of bars picking up stress, as in a beam, and to splices, it may open the possibility of computing *splitting* stresses and a more rational design procedure for developing bars.

Although splitting should be considered apart from bond, at present it is not possible to separate the two. Hence splitting, such as that shown in Fig. 5.3*a*, is considered simply as the most visable sign of approaching bond failure. It occurs even with heavy stirrups, as shown in Fig. 5.3*d*.

Likewise, Fig. 5.3*c* indicates that diagonal tension (shear) cracking and possibly resulting dowel action can be closely associated with bond splitting. Traditionally bond and shear have been treated as separate phenomena. This procedure has been followed here for the most part. However, Sec. 4.4*g* points out that bond and shear are interrelated topics and that in many ways neither can be understood fully without knowledge of the other.

5.3 RELATION BETWEEN DEVELOPMENT LENGTH AND BOND STRESS

As already pointed out, a bar with enough embedment in concrete cannot be pulled out. After slip at the loaded end has progressed far enough to develop bond over a considerable length, such a bar reaches its yield strength and fails in tension; it is then described as fully anchored in concrete. The length of embedment necessary to provide against pullout failure is called the anchorage length.

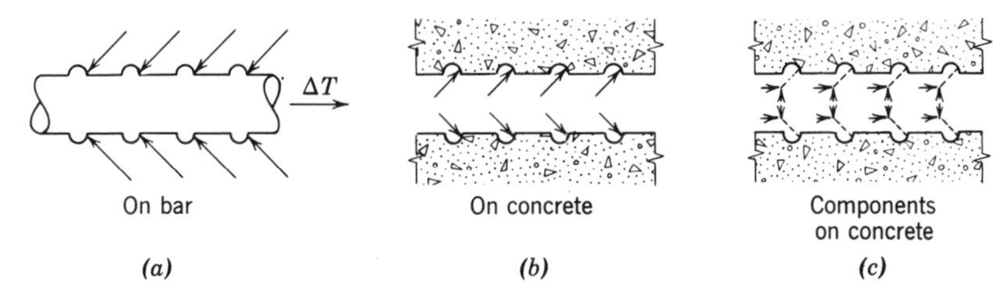

Fig. 5.2. The forces between a deformed bar and concrete which may cause splitting, as in Fig. 5.3.

A similar concept has developed to describe the development of a bar in a structure where the stress varies from zero at bar end to a maximum, usually considered f_y, at some critical cross section. The required development and anchorage lengths would be the same if the two lengths related to the same materials and surroundings. To be specific, a top bar in an overhanging (cantilever) end of a beam must be developed in the overhang. If it is anchored back into the adjacent beam span, the bar size, yield strength, and concrete strength are automatically the same. If the bar spacing, bar cover, and possibly the shear forces* are the same, the minimum anchorage and development lengths should be the same.

The basic concept of anchorage length considers a bar embedded in a mass of concrete, as in Fig. 5.4. The actual bond stress will be distributed similarly to that of the pullout test (Sec. 5.6), quite large near the surface and nearly zero at the embedded end until fairly close to failure. If the average bond stress u is limited to a permissible value determined from comparable pullout tests, safe results should be obtained. Based on this logic, at ultimate

$$A_b f_y = u l_d \pi d_b \text{ (for one bar)}$$

For round bars of diameter d_b, $A_b = \pi d_b^2/4$, giving

$$\pi d_b^2 f_y/4 = l_d \pi d_b u$$

$$l_d = \frac{f_y}{4u} d_b \tag{5.1}$$

This l_d is the minimum permissible anchorage length. However, even the deformed bar anchored in mass concrete exhibits several flaws in this logic, some on the safe side, some unsafe insofar as design is concerned.

On the safe side, the mass concrete does not fail like the cylinder or prism cast around a pullout bar, which always splits open without actually

*Interaction of shear and anchorage is mentioned in Sec. 4.4*g*.

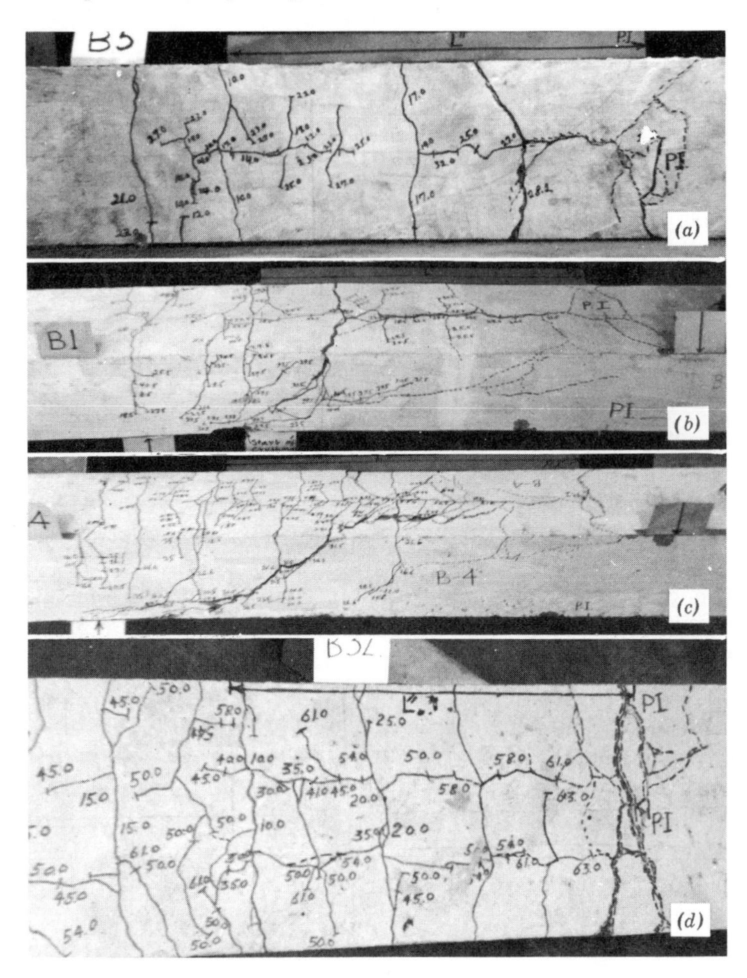

Fig. 5.3. **Bond splitting photographs of beams tested as shown in Fig. 5.9. (*a*) Splitting is directly over bar and runs lengthwise. The cross cracks are flexural cracks. (*b*) Failure chiefly from bond splitting. (*c*) Diagonal tension (inclined side crack) combined with longitudinal splitting to produce failure. (*d*) Splitting over two bars in beam with heavy stirrups. (The L'' marked is the development length l_d.)**

failing the bar (unless the bar is *very* small or the aggregate is weak against local crushing). The same anchorage length in ordinary mass concrete should thus be stronger. One must note, however, that even in mass concrete a bar parallel to the surface and with only nominal cover is as free to split off this cover as if it were in the top of a thin slab.

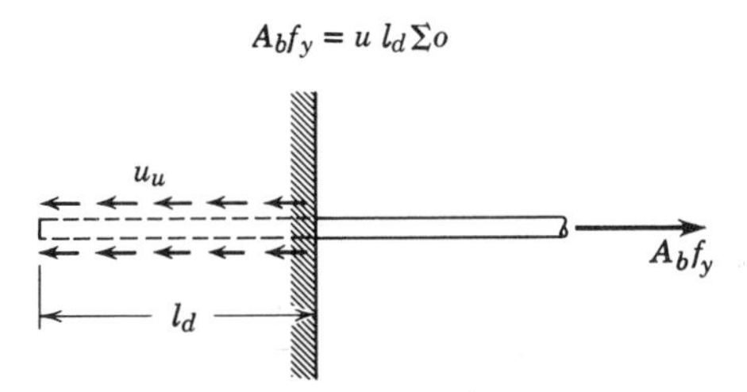

Fig. 5.4. Anchorage of a bar.

On the unsafe side, doubling the pullout embedment length will not double the pull it will resist. The longer the specimen the less efficient it becomes and linear extrapolation of data is normally on the unsafe side. Also, groups of bars are probably less effectively anchored than individual bars.

Development lengths computed on the above basis (Eq. 5.1) can be no better than the values of bond stress used in deriving them. Since the development may occur in very diverse surroundings, as to spacing, cover, edge distance, and enclosing stirrups, no single value of u or l_d is appropriate over a large range.

This discussion may appear to be an argument for calculating the bond stress itself and limiting it to an acceptable value. Earlier Codes have indicated computation of the flexural bond stress based on the necessary change in bar tension to provide for the moment change, or, in terms of the flexural bond stress u on the surface area of the bar

$$u = V/(\Sigma o \times z)$$

Here V is the shear, as a measure of the changing moment, Σo is the total perimeter of the bars at the section, and z is the arm of the internal resisting couple.* The reasons mentioned in the next paragraphs indicate design control by bond stress is no more reliable than development length.

Basically, strength is not sharply influenced by local bond stresses or local splitting. Although concrete appears to be brittle where failing in tension, it has a surprising ability to carry local overstress and readjust to where *average* stress appears to be the sounder failure criterion. The average attainable, however, is subject to variations for length, cover, spacing, and other factors. The only reliable way to determine an allowable average bond stress is to test a similar situation. Any such test establishes a maximum change in bar tension over a given development length; the allowable

*Formerly this arm z was called jd.

bond stress can be evaluated only from the development length equation rearranged to

$$u = f_s d_b / 4 l_d$$

Because a varying shear is not usually practical in a test situation, the test moment variation is usually linear; one then obtains the same answer as from the bond stress equation, $u = V/(\Sigma o \times z)$.

If one uses the flexural bond equation, the allowable stress must thus come from the development length equation. Since there is no advantage in going from test development length to allowable bond and then back to required development length, and since no improved accuracy results from the allowable bond concept, the 1971 Code no longer uses it. The development of the specific l_d values used in this Code is covered in Sec. 5.9 and Sec. 5.10.

5.4 BOND STRESS CONCENTRATIONS UNDER NORMAL CONDITIONS

Not for direct design but for a better understanding of member behavior, the very uneven distribution of bond stress at service loads and probable readjustment prior to failure are important. Consider a simple beam as shown in Fig. 5.5. Although at B in the constant moment length the nominal steel stress is constant, the bond stress is not zero but large adjacent to each flexural crack.

At the crack the steel carries most of the tension; but between cracks, tension in the concrete shares significantly. Thus the bond stress condition on each side of a crack is almost identical with that at the loaded end of a pullout specimen, with a near ultimate value of bond stress existing for a short distance adjacent to the crack,[2] of opposite sign on the two sides of the crack. The author calls these the "in-and-out" bond stresses. At a moment crack in a section carrying a shear, as at C, the bond stresses needed to build up the steel stress are superimposed on those caused by the crack itself, but some readjustment is necessary on the side of the lesser bar stress since the total bond stress at any point cannot exceed the ultimate already existing next to the crack. With #18 bars of Grade 60 steel, service loads may even lead to a little splitting over the bars on each side of the cracks, where moment is constant, and on the low moment side of the cracks elsewhere. On closely spaced bars the splitting would show on the side of the beam near the level of the steel. With #11 bars the service load will probably produce no such cracks, but some will show in a pure

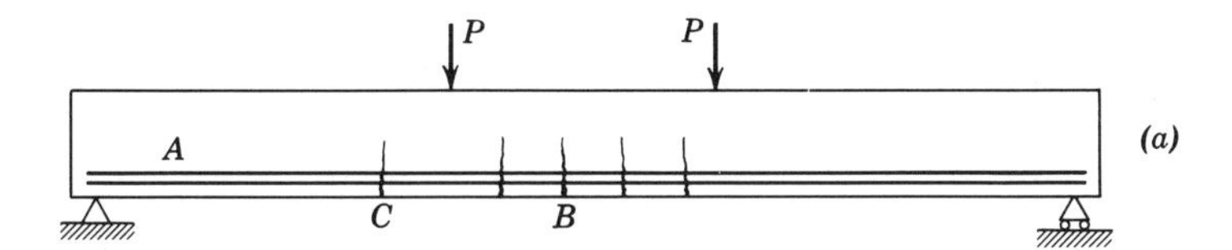

Fig. 5.5. Crack pattern in a simple span.

flexure zone before the yield stress of Grade 75 bars is reached. The probability of the splitting cracks decreases with decreasing bar size and lower steel stress.

At point *A* in Fig. 5.5 the concrete carries most of the tension and both steel stress and bond stress are low. At *C*, the first flexural crack from the reaction, the steel stress increases to carry almost all of the tension and the short length to the left of *C* must marshal bond stress to offset all the deficiency between *C* and the reaction. The smaller the percent of steel in the beam the longer the necessary section of this high bond stress and the greater the possibility of some local splitting.*

The special cases of brackets, of variable depth members, deep beams, and bar cutoff points, where special design precautions are necessary, have already been discussed in connection with shear, Secs. 4.14 to 4.18.

It appears probable that bond stresses become less variable as slip takes place and loads approach the ultimate. Goto's crack pictures seem to show an equalizing process. Near ultimate, splitting will have started and will have lowered some of the peak bond stresses. The detail of what occurs has not been mapped for development length, but for splices some conclusions are developing (Sec. 5.18).

Tests show that some of the in-and-out bond stresses between cracks relax substantially under repeated loading, but they reappear when a higher load is applied; the relaxation is local and develops only at the load level which is often repeated. With reversal of forces this decay in bond resistance is more severe. Study is continuing with reversing stresses.

5.5 THE PROBLEM OF TESTING FOR BOND STRESS OR DEVELOPMENT LENGTH

Except in relation to splice tests, a good bond test has not yet been devised. The problems are several, including:

a. Inability to scale down specimens with confidence.

*To the extent smaller ρ means smaller bars, this is an offsetting factor.

b. Some interaction between shear and bond stress.
c. Nonlinear response, either with bar size, steel stress, or development length (which in design are related quantities).
d. A wide range of failure modes which can range from crushing against the lugs or longitudinal shearing of a surface just outside the lugs to the more usual splitting of concrete, either over individual bars or through an entire layer of bars.

(*a*) Possibly because much higher bond stresses can be developed around small bars than large ones, or possibly because the spacing of flexural cracks does not scale down properly as specimen size is reduced, scaled specimens sometimes react differently from full size ones.

(*b*) As already pointed out in Sec. 4.4*g*, particular combinations of shear and bond stress can lead to lower shear strengths. The mixture of shear and bond problems is evident in Fig. 5.3*b*,*c*.

(*c*) The longer the test length of the bar in the specimen the lower the average bond stress attained at failure. A Grade 40 bar can be developed in less than two-thirds the length required for the same size of a Grade 60 bar. (In some recent splice tests, 83% of two-thirds of the length of the Grade 60 bar splice proved adequate for the Grade 40 bars.)

(*d*) The tensile splitting stress and the influence of cover thickness on ultimate strength,* neither of which can yet be evaluated with confidence, make generalizations difficult.

The only sure test at present is a full size specimen combining the given mix of variables, which is a very expensive procedure and one difficult to generalize to a different mix of variables. The next several sections discuss various specimens that have been used and their weaknesses.

5.6 BOND PULLOUT TESTS

Permissible bond stresses were formerly established largely from pullout tests with some beam tests as confirmation. In the pullout test a bar is embedded in a cylinder or rectangular block of concrete and the force required to pull it out or make it slip excessively is measured. Figure 5.6 shows such a test schematically, omitting details such as hemispherical bearing plates. Slip of the bar relative to the concrete is measured at the bottom (loaded end) and top (free end). The bond stress distribution in such a specimen is very nonuniform. Even a very small load causes some slip and develops a high bond stress near the loaded end, but leaves the

*But see Sec. 5.18 for cover effect on splices.

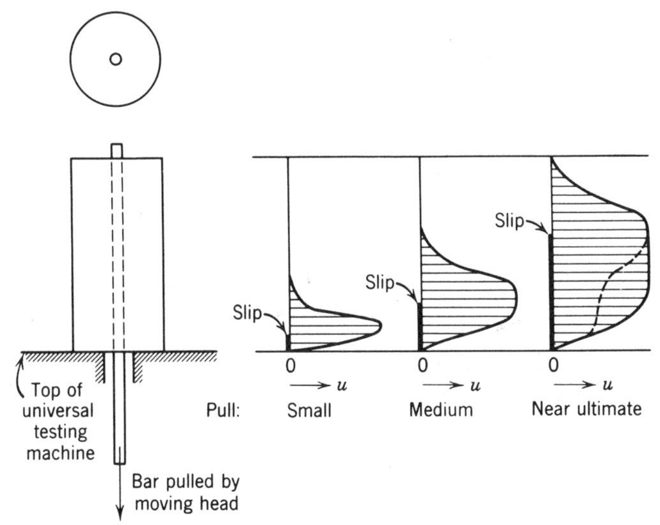

Fig. 5.6. Bond pullout test, with bond stress distributions.

upper part of the bar totally unstressed, as shown in Fig. 5.6. As more load is applied, the slip at the loaded end increases, and both the high bond stress and slip extend deeper into the specimen. With plain bars the bond stress will decrease to the friction or drag value wherever adhesion has been broken by slip, as shown dotted in the last sketch; such slip is indicated by the heavy line in the righthand sketches of Fig. 5.6. The maximum bond is somewhat idealized in these sketches; its distribution depends on the type of bar and probably varies along the bar more than is shown.

When the slip first reaches the unloaded end, the maximum resistance has nearly been reached. Failure will usually occur (1) by longitudinal splitting of the concrete in the case of deformed bars, or (2) by pulling the bar through the concrete in the cases of a soft light-weight aggregate or a very smooth bar, or (3) by breaking the bar, if the embedment is long enough.

The average bond resistance is always calculated just as though it were uniform over the bar embedment length. Actually, the bond stress distribution varies greatly as slip develops, and at any load the average is one of values quite dissimilar, an average between ultimates and smaller values. The very first slip of the loaded end of the bar represents essentially an ultimate bond stress over a short length of the bar, although the calculated average bond stress may be quite low. Attention is called to the fact that this test procedure does not attempt to measure maximum unit bond resistance.

In this test there are no tension cracks across the bar and the adjacent concrete is in compression, which tends to increase the end slip at loaded end. The friction on the base also restrains splitting of the specimen and many tests have included spirals that avoid total collapse from splitting. To the author the test appears useful chiefly where relative rather than real bond resistance is acceptable, as in comparing the slip resistance of various lug sizes and patterns. The principal problem of splitting is not realistically handled in this specimen.

Modifications of this test, called the tensile pullout specimen, have also been used. These eliminate the compression of the concrete, as shown in Fig. 5.7. In these specimens the pull at one end is on one bar in (*a*) or two bars in (*b*) while the other end is held by pulls on the remaining bars. These specimens are an improvement; but each introduces some of the special problems of spaced splices and any crack pattern is influenced by this interaction.

5.7 BOND BEAM TESTS

Pullout specimens are now considered less reliable than beam tests since it is recognized that flexural tension cracks influence bond behavior. Two types of beams have been used: that of Fig. 5.8 at the Bureau of Standards and the one of Fig. 5.9 at The University of Texas at Austin. A major consideration in each was to remove reaction restraints which might confine the concrete over the bar and thereby increase splitting resistance.

The Bureau of Standards beams were heavily reinforced with stirrups and failed as sketched in Fig. 5.8*b*, the diagonal crack creating an increased steel stress much closer to the end of the bar and concentrating bond stresses nearer the end.

The University of Texas beams (Fig. 5.9) placed the bar in a negative moment region where there was no external restraint against splitting except by the use of a rather wide concrete beam.[4] Some beams were with stirrups, some without. Figure 5.3*b* and *d* shows beams which failed by splitting out to the end of the bar, whereas the beam of Fig. 5.3*c* first developed a diagonal crack which increased the bar stress as in the Bureau of Standards tests. When two bars were placed in the same beam width, with stirrups to carry the expected shear, the bond resistance was generally not much improved by the stirrups.

While the National Bureau of Standards tests incorporated an excessive amount of stirrups to avoid diagonal tension failures, the University of Texas beams were generally excessively wide and represented bar spacings too wide to be fully practical. Both specimens were expensive to make and

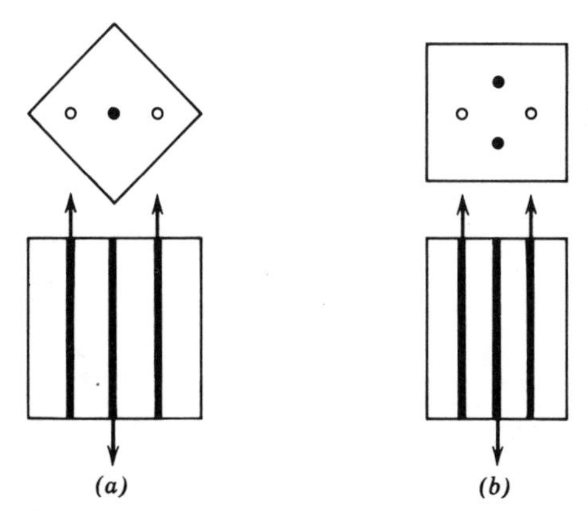

Fig. 5.7. Tension pullout tests, schematic.

heavy to handle, since both used full-size members with the larger sizes of bars, which create the most serious bond problems.

The results from these two widely different test specimens were reasonably in agreement and had much influence on the 1963 Code bond provisions, especially on fixing bond stresses varying as $\sqrt{f_c'}$ and decreasing

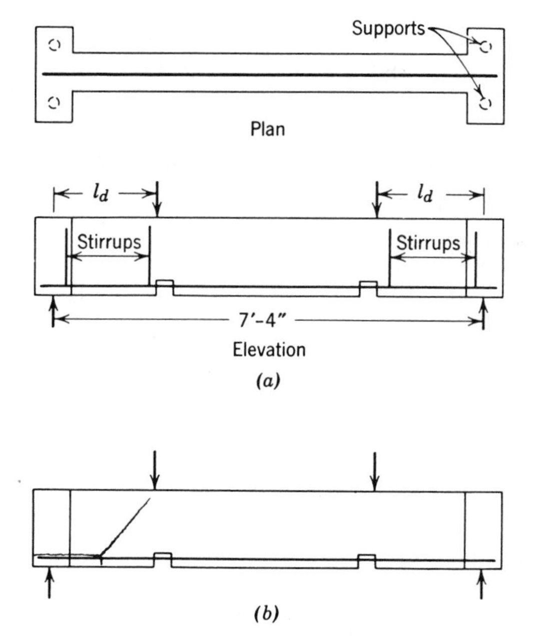

Fig. 5.8. National Bureau of Standards bond test beam. (*a*) Load and support system. (*b*) Failure pattern.

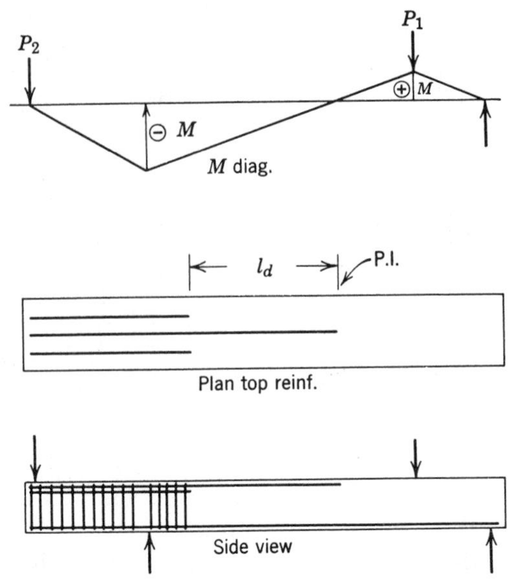

Fig. 5.9. The University of Texas beams. (Failure shown in Fig. 5.3.)

with increasing length l_d. In a following section it is pointed out that in tension members the permissible bond varies as $1/d_b$, where d_b is the bar diameter up to #11 bar size. Thus l_d lengths will actually vary as d_b^2 in-instead of the apparent d_b shown in Sec. 5.3. In other words, these beams acted as though the splitting of the concrete limited the pull that could be put into a bar in a given length to the same total number of pounds whether the bar was large or small.

For #14S and #18S bars beam tests were quite few in number, but seemed to indicate that a further decrease in bond stress for these bar diameters was not necessary.

5.8 SEMIBEAM SPECIMENS

To reduce specimen size and expense, partial beam specimens have been developed, such as the one sketched in Fig. 5.10. Although various details are used, those in this figure convery the general idea. The bottom reaction may bear against the end of the bar, may be spread such as not to be near the bar or, as sketched, the bar may be shielded and isolated by a soft covering. The test pull is put directly into the bar.

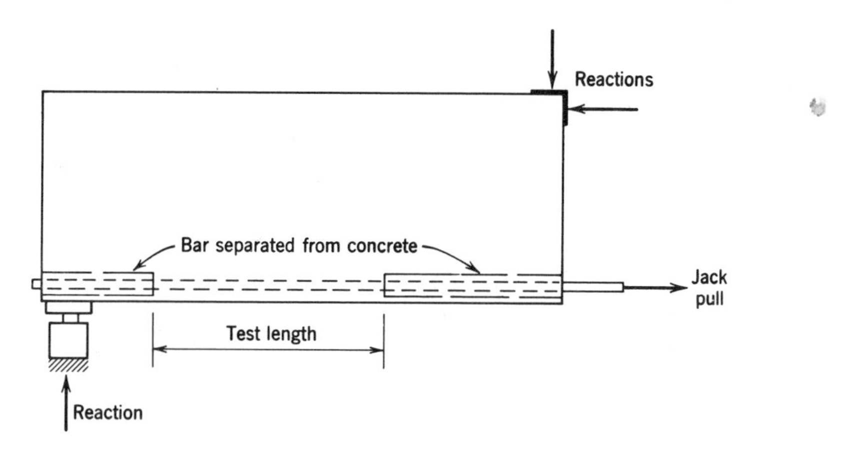

Fig. 5.10. Cantilever bond specimen.

The overall length and the part of it used for the test length can each be varied, anything from the full length of the specimen to the reduced length. One or several bars, with or without stirrups may be used.

The advantage of this specimen lies in its simplicity. The disadvantages lie in the confining pressure against the bar (if the shield is not used) and in the increased length over which splitting resistance tends to be mobilized. In the specimen in the sketch, there is a greater length of concrete subject to splitting for a given bar test-length than will exist in the actual member, making the apparent strength too large.

5.9 THE SHIFT FROM 1963 BOND STRESS TO 1971 DEVELOPMENT LENGTH

Development length or development bond was a part of the 1963 Code but more emphasis was then on flexural bond. In the interim there have been few data justifying any changes in the allowable bond values of that Code, excepting the observation that closely spaced bars were quite typical, constituted a real problem, and should be handled more like closely spaced splices. The 1963 Code in Sec. 805(b) read:

> For contact splices spaced laterally closer than 12 bar diameters or located closer than 6 in. or 6 bar diameters from an outside edge, the lap shall be increased 20 percent or stirrups . . . or closely spaced spirals shall enclose. . . .

In Sec. 1801(c) that Code specified the following allowable bond stresses written here in the 1971 Code notation:

For tension bars in sizes and deformations conforming to ASTM A305 (the standard #3 to #11 bars):

Top bars	$6.7\sqrt{f_c'}/d_b \not> 560$ psi
Bars other than top bars	$9.5\sqrt{f_c'}/d_b \not> 800$ psi

For tension bars with sizes and deformations conforming to ASTM A 408 (the #14 and #18 bars):

Top bars	$4.2\sqrt{f_c'}$
Bars other than top bars	$6\sqrt{f_c'}$

For all deformed compression bars $13\sqrt{f_c'} \not> 800$ psi

In effect the 1971 Code specifies development lengths l_d of 1.2 times those these bond stresses would require, making the required lengths more suitable for close spacing. For example, for a bar in the first group:

$$\begin{aligned}
\text{1963 allowable } u &= 9.5\sqrt{f_c'}/d_b, \quad A_b = \pi d_b^2/4 \\
\text{Required } l_d &= 1.2\, f_y d_b/4u = 1.2 f_y d_b/(4 \times 9.5\sqrt{f_c'}/d_b) \\
&= 0.0316 f_y d_b^2/\sqrt{f_c'} = 0.0316 f_y(4A_b/\pi)/\sqrt{f_c'} \\
&= 0.0403 f_y A_b/\sqrt{f_c'} = \text{say, } 0.04\, f_y A_b/\sqrt{f_c'}
\end{aligned}$$

For spacing at least 6 in. on center, a length 0.8 times this value is permitted. Since $1.2 \times 0.8 = 0.96$, this returns the development length essentially to the 1963 status for the wider bar spacings.

5.10 DEVELOPMENT LENGTHS FOR TENSION BARS

The 1971 Code specifies four basic development lengths for tension bars, corresponding to the four allowable stresses listed in the 1963 Code for bars other than top bars, each adjusted as noted in the last section; also a value for deformed wire and an overall minimum of 12 in. The version below is slightly condensed.

12.5 Development length of deformed bars and deformed wire in tension

(a) The basic development length shall be:

For #11 or smaller bars	$0.04A_b f_y/\sqrt{f_c'}$
but not less than	$0.0004 d_b f_y$
For #14 bars	$0.085 f_y/\sqrt{f_c'}$
For #18 bars	$0.11 f_y/\sqrt{f_c'}$

(b) The basic development length shall be multiplied by the applicable factors for:

Top reinforcement — 1.4

Bars with f_y greater than 60,000 psi — $2 - 60{,}000/f_y$

(c) When lightweight aggregate concrete is used, the basic development lengths in (a) shall be multiplied by 1.33 for "all lightweight" concrete and 1.18 for "sand-lightweight" concrete with linear interpolation when partial sand replacement is used, or . . . (split cylinder tests can modify this).

(d) Multiply by the applicable factor or factors for:

Reinforcement being developed in the length under consideration and spaced laterally at least 6 in. on center and at least 3 in. from the side face of the member — 0.8

Reinforcement in a flexural member in excess of that required — (required A_s)/(A_s provided)

Bars enclosed within a spiral which is not less than $\frac{1}{4}$ in. diameter and not more than 4 in. pitch — 0.75

Development lengths are in inches when f_y and f_c' are taken in psi. Dimensionally no other combination of units may be used.

The values of (a) are always applicable but are to be multiplied by all of the applicable factors from (b), (c), and (d). In (a) the second limitation of $0.0004d_bf_y$ governs only for very small bars, such as #5, or very high strength concrete.

Because of the nonlinear response (Sec. 5.5) these equations are a bit severe for Grade 40 bars, but not severe enough for Grade 75 bars. Thus in (b) the Code specifies that for $f_y > 60{,}000$ psi the lengths are to be increased by the factor $2 - 60{,}000/f_y$, in addition to basing the value on an (a) value which includes f_y as 75,000 psi, as indicated in Fig. 5.11.

For top bars, that is, horizontal reinforcement having more than 12 in. of concrete cast below the bars, 1.4 times the basic l_d must be used. An accumulation of air and water that rises beneath such bars becomes entrapped on their under sides. Such bars are not as tightly held and slip at lower loads.

As already noted in Sec. 5.9, a 0.8 factor may be used for bars spaced at least 6 in. on center (and not less than half that distance from an edge). Reduced development length is also specified when a spiral is available to control splitting of the concrete or where excess steel is available and it is not necessary to develop the full yield strength.

Particular attention is called to the increased development lengths needed with lightweight concrete, values of 1.33 for "all lightweight" (including lightweight for the fine aggregate) and 1.18 for "sand-lightweight" concrete. When the mix is well-controlled and split cylinder data are available this rule is relaxed. More data on the development length requirements in lightweight concrete are badly needed.

Table D.3 in Appendix D summarizes many of the l_d values for tension bars; Table D.4 covers compression bar l_d values (Code 12.6).

In the formulation of required l_d values, allowable bond stresses, and lap splice lengths, some protection against a brittle type of failure must always be considered. To design a beam that is somewhat ductile in flexure but allow it to fail suddenly in bond splitting at the yield stress is inconsistent. Hence reserve strength in l_d above f_y is essential to maintain member ductility. The Code values aim at making this extra strength about 25% of f_y. This should maintain the local bond strength and cause the failure section to occur at the moment section which has no corresponding large strength reserve above that calculated. However, especially in splices and possibly also in simple development lengths, the steel strain that can be accommodated is not unlimited. If the steel has a long yield plateau on its stress-strain curve the 25% extra strength may not be attainable, but the ability to accept a high strain would serve the same function. Strain limit information is very skimpy and deserves further study.

5.11 CRITICAL STRESSES AND CRITICAL SECTIONS FOR DEVELOPMENT

The proper use of development lengths requires a clear picture of how bar tensile or compressive stresses change along a beam. Bar stresses will be a maximum where moment is a maximum; obviously no maximum point should be closer than l_d to the end of a bar, in either direction; l_d must exist to put stress into a bar or to take it out.

Equally important, and easier to overlook, is the peak stress that develops in bars wherever a neighboring tension bar is cut off or bent. At that point the total tension must crowd into the (reduced number of) continuing bars. Since the cutoff for moment is usually made where the remaining bars have just enough capacity to handle the total tension, the full yield stress in the continuing bars determines their l_d requirement. The design stress just beyond the cutoff depends on what has controlled this location.

First, if the cutoff has been determined by the development length of the bar which is cut off, the cutoff is farther than needed for moment and the f_s in the remaining bars (less than f_y) determines l_d. It is safe (sometimes also wasteful) to ignore this extra extension and measure the next l_d as though the bar had been cut at the point determined for moment.

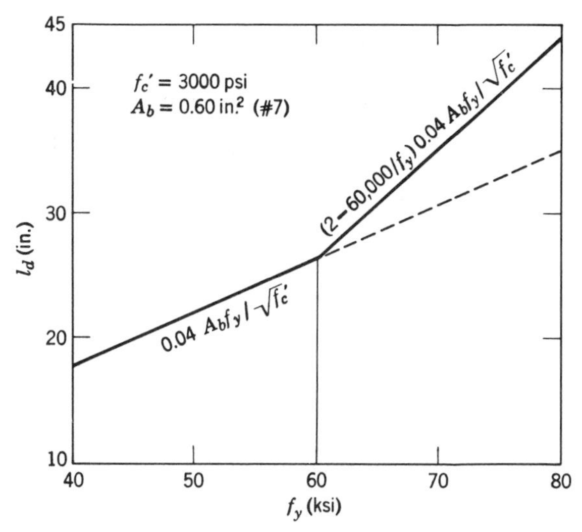

Fig. 5.11. Development length variation with f_y for No. 7 bar.

The second control is involved with design philosophy, already introduced at the close of Sec. 3.10d. If the arbitrary extension of the bars is considered entirely arbitrary, not related to stress, then stresses permit a cutoff at the theoretical moment diagram used; the development length of the continuing bars can then be considered to start there and this satisfies the Code. The author recommends instead that the arbitrary extension be considered as defining a new (shifted) moment diagram; this concept leads to l_d measured from this shifted diagram rather than the basic one. The examples show the use of both concepts, the legal Code requirement and the author's recommendation.

5.12 EXAMPLES SHOWING USE OF DEVELOPMENT LENGTHS

Examples illustrate more specifically some of the above general ideas.* The basic discussion is continued within the example format.

(a) A 12 in. concrete wall heavily loaded requires a footing 5 ft 6 in. wide as sketched in Fig. 5.12. The footing is designed for $f_c' = 3000$ psi and Grade 60 reinforcement. What maximum size bars will conform to the development length requirements?

*See also Fig. 10.11 (from the Code) showing good practice for slabs supported directly on columns rather than beams.

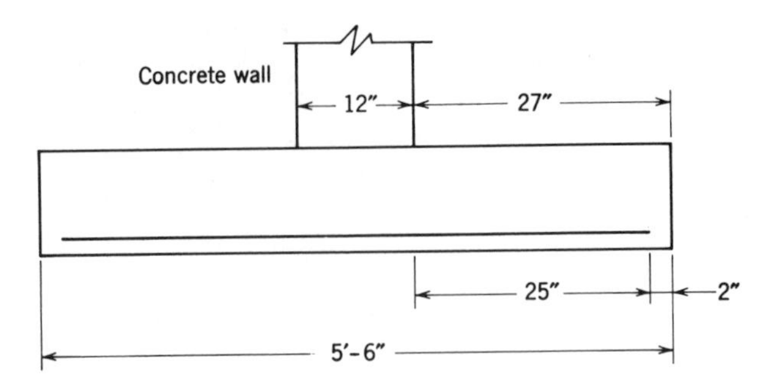

Fig. 5.12. Concrete wall footing.

Solution

With a concrete wall, the critical footing section for flexure is at the face of wall (Sec. 15.8). The bars are there stressed to maximum and project 25 in., leaving 2 in. end cover. Based on the values of Sec. 5.10, a bar must be found which has an l_d of 25 in. or less. In the absence of a table of l_d values of various bars under various conditions (which a design office will surely develop), the required

$l_d = 0.04\,A_b f_y/\sqrt{f_c'}$ but not less than

$l_d = 0.0004\,d_b f_y$.

The second value governs only with small bars and will not be checked initially.

$$0.04\,A_b\,60{,}000/\sqrt{3000} = 25 \text{ in.}, \quad A_b = 0.57 \text{ in.}^2$$

A #6 bar ($A_b = 0.44$) is the largest permissible without a hook. Checking the other limit with d_b for the #6 bar,

$$0.0004\,d_b f_y = 0.0004 \times 0.75 \times 60{,}000 = 18 \text{ in.} < 25 \text{ in.} \qquad \text{O.K.}$$

The number of bars must come out of a flexure calculation. If this number permits a 6 in. spacing or more, a larger bar may be possible because the required l_d can then be reduced by an 0.8 factor to give

$$0.8(0.04\,A_b \times 60{,}000/\sqrt{3000}) = 25 \text{ in.}$$
$$A_b = 0.71 \text{ in.}^2$$

A #7 bar is then permissible. The alternate equation need not be checked again, since the 0.8 factor also applies to it. The #7 bar possibility improves the chances of using the wider spacing.

It is noted that a hook at the end of bar would reduce the requirement for straight bar development length and permit still larger bars.

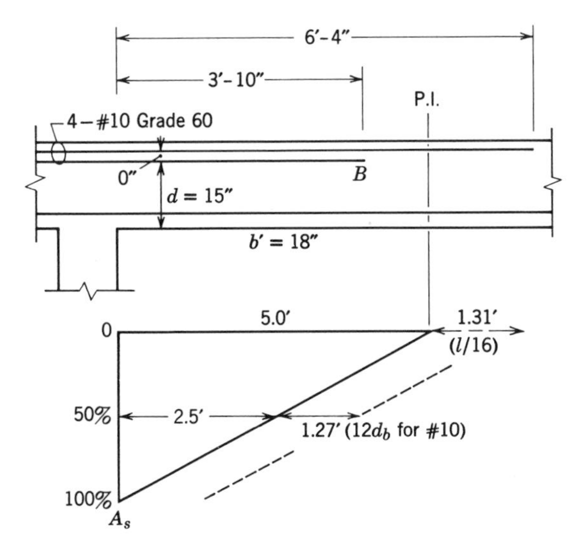

Fig. 5.13. Data for example (c).

(b) Assume the footing is made of sand-lightweight concrete but the example is otherwise the same.

Solution

Required development length is increased by a 1.18 factor when lightweight concrete with a sand fine aggregate is used. If the spacing is under 6 in.,

$1.18(0.04A_b \times 60{,}000/\sqrt{3000}) = 25$ in.

$A_b = 0.48$ in.^{2}

A #6 bar remains the largest usable bar (0.44 in.^{2}) at this spacing.

(c) The 4 – #10 bars in Fig. 5.13 are designed for the moment diagram shown. The 1.31-ft extension of one-third the bars beyond the P.I. is required by the $l/16$ rule* (Code 12.3.3). May two bars be cut off as indicated on the moment diagram? Assume that the shear requirements of Code 12.1.6 have been satisfied.

Solution

If the bars are cut as shown, the spacing of the extended bars in an 18-in. beam width will be near 9 in. Also, since these would then be developed at

*Or d or $12d_b$, if larger.

or near B, the two short bars would be contributing all the important splitting forces from B to the support and each of these could also mobilize splitting resistance over a 9-in. width. Hence, for the bars cut off the effective spacing would also be 9 in., justifying the 0.8 coefficient on l_d for $s \geqq 6$ in. As top bars they require a 1.4 coefficient.

$$\begin{aligned} \text{Reqd. } l_d &= 0.8 \times 1.4(0.04A_b f_y/\sqrt{f_c'}) \\ &= 0.8 \times 1.4 \times 0.04 \times 1.27 \times 60{,}000/\sqrt{3000} \\ &= 63 \text{ in.} = 5.25 \text{ ft} \end{aligned}$$

The "short" bars must extend 5.25 ft which is beyond the P.I., making a moment *cutoff not feasible*. The 5.25-ft cutoff also satisfies the following consideration.

The author has recommended that the dashed diagram be considered as the potential or shifted moment diagram which might result from the conditions discussed in Sec. 3.10*d*, something beyond the legal requirement of the Code. The 5.25-ft dimension would then be so near the P.I. that one would not need to worry about the capacity of the longer bars to pick up the small necessary moment at this location. Regardless of whether a designer chooses to accept the shifted moment diagram idea or not, a stagger of cutoff points is desirable, of the order of 6 in. when closely spaced or the bar spacing where wider than 6 in. If external forces produce any significant tension at the section, the discontinuity from stopping all bars at one length will produce an early crack. Even if flexural tension never occurs, shrinkage tends to open the crack at such a cutoff.

(d) If in Fig. 5.13 the moment requires only 4 – #8 bars, may the given dimensions be used with the #8 bars?

Solution

l_d in this case involves the same 0.8 and 1.4 coefficients as in (c).

$$l_d = 0.8 \times 1.4 \times 0.04 \times 0.79 \times 60{,}000/\sqrt{3000} = 39 \text{ in.}$$

The 3 ft 10in. dimension is larger and satisfactory for the shorter bars.

The development of the longer bars must also be considered. The author recommends that these bars be considered fully stressed at B, because of the shifted (dashed) moment diagram. The longer bars would then need l_d of 39 in.beyond the *required* distance to B, or a total from the column of $(2.5 + 1.25)^* + 3.25 = 7.00$. The longer bars should then be extended 7 ft instead of 6 ft 4 in.

If one totally disregards the shifted moment diagram concept, the long bars could be considered fully developed at $2.5 + 3.25 = 5.75$ ft and the 6 ft 4 in. dimension could be accepted under the Code as giving 7 in. leeway.

*The smaller #8 bars cause the extension to be governed by d rather than $12d_b$.

The extra 7 in. would make the author feel more comfortable than no extra. His concern exists because settlements do frequently occur and the usual moment diagrams, even those from moment distributions, are approximate. They neglect the cracking of the beams, the axial shortening of columns, and the differences between a joint as a point and as a substantial portion of the member lengths.

Since cutting off some bars frequently requires others to be run farther, each bar cutoff point becomes a critical section to be checked as regards the continuing bars.

The implications of a situation often encountered (which might be overlooked in the case just discussed) is emphasized in the next example.

(e) If in Fig. 5.13 the moment requires 4 – #9 bars, may the given dimensions be used with the #9 bars?

Solution

$$l_d = 0.8 \times 1.4 \times 0.04 \times 1.00 \times 60{,}000/\sqrt{3000} = 49 \text{ in.} = 4.08 \text{ ft}$$

The cutoff point B must be extended from 3 ft 10 in. to 4 ft 1 in. This throws point B beyond both the basic and the shifted moment diagram requirements for moment length. As a result, the remaining bars provide excess steel area at B and could safely be developed beyond B for a stress lower than f_y. It is simpler (and on the safe side where this calculation controls) to develop the longer bars for a full f_y, *not* from the actual point B used, but from the point where the bars could be cut for moment alone. Based on the dashed moment diagram this gives a required dimension from the column of $(2.5 + 1.25) + 4.08 = 7.83$, say 7 ft 10 in., which the author would recommend instead of the 6 ft 4 in. Considering only the basic moment requirement, a designer could legally use $2.5 + 4.08 = 6.58$ ft, say 6 ft 7 in. Either answer should alert the designer to the fact that both lengths are increased because of using the #9 bars. Construction might be cheaper with 5 – #8 bars, particularly if the author's shifted moment diagram idea with its longer requirement is used.

5.13 POSITIVE MOMENT BARS

(a) Continuous Beams*

When uniform loads are considered in simple spans or within the positive moment length of continuous spans, there is a problem in applying the basic development length concept in a meaningful way. The question finally

*The detailing of continuous beam reinforcement, including detailed examples, is covered in Secs. 8.13 to 8.16.

emerging is what is the maximum bar size where moment is nearly zero. The answer is essentially the equivalent of the old flexural bond concept, which is presented below in terms of development length and some liberalization.

Consider the case shown in Fig. 5.14*a* and a simple case where l_d is exactly l_0 to midspan. Such a bar would develop its yield strength at midspan but would be unsafe at $l_0/2$ from the P.I. There the bar would be developed for only half its f_y but would need to care for $0.75M_{u0}$, which requires $0.75f_y$. Nor is it safe to use a bar developing this $0.75f_y$ in exactly $l_0/2$, because then a somewhat similar overstress exists at points closer to the P.I. The simple answer developed below is identical with the old flexural bond demands, but encourages realistic adjustments to relax these demands.

The slope of the moment diagram at the P.I. is the shear V_u at that point. If the bars at the P.I. develop toward f_y at the rate M is initially developing, they can develop their full value f_y in the length M_t/V_u, as indicated

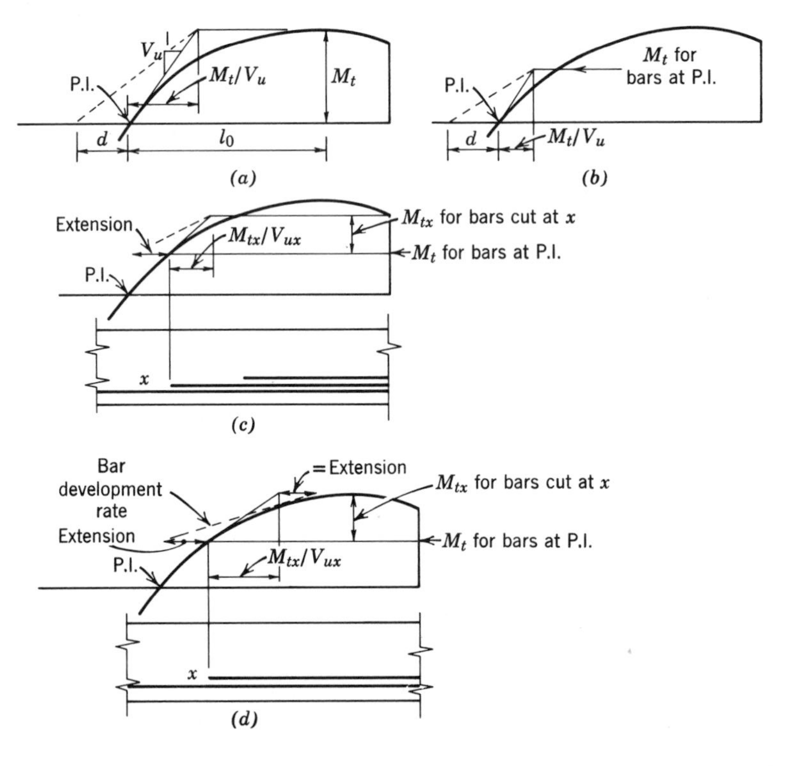

Fig. 5.14. Available development lengths for positive moment bars at a P.I. (*a*) No bars cut. (*b*) Bars at P.I. less than total A_s. (*c*) General case. (*d*) For first bars cut off (nearest point of maximum positive moment).

in Fig. 5.14*a*. If the bars extend beyond the P.I., as some must according to the Code, the development length concept implies that local "overstress" in bond at the P.I. is not a problem because it no longer occurs at the end of the bar. The extra length beyond the P.I. which may be counted is a judgement decision, but the Code limits it to l_a equal to the effective beam depth d. The criterion becomes:

At a P.I. choose bars with $l_d \overline{<} M_t/V_u + d$.

The dotted development rate line indicates this would be safe even with a slightly shifted moment diagram. The shifted moment diagram is less critical in a positive moment region, but still valid in the author's opinion.

If not all bars extend to the P.I., Fig. 5.14*b* indicates the only change needed is to include in M_t only the bars actually reaching the P.I.

Usually, if bar size at the P.I. is satisfied and other positive moment bars are not too widely different in size, the required bar lengths for flexure will control over development lengths. However, if larger bars are used away from the P.I. they could be selected by the same procedure used at the P.I. with one modification indicated in Fig. 5.14*c*. These bars must care for all or part of the moment in excess of M_t developed by the other bars at the P.I. If their moment capacity is designated M_{tx}, the local demands* indicate l_d equal to or less than M_{tx}/V_{ux}, where V_{ux} is the shear at x. Any further extension of the bar obviously adds to the available l_d to give the criterion for bars not extending to the P.I. as:

For the added bars $l_d \overline{<} M_{tx}/V_{ux}$ + extension used.

If such a criterion is used for the first bars cut off, the moment diagram will be rather flat. Any added arbitrary bar extension not only moves the critical bond stress away from the nominal cutoff point but also permits a flatter development line as sketched in Fig. 5.14*d*. It appears safe, *for a moment diagram that can shift very little*, to use the criterion:

For the added bars *first* cut off $l_d \overline{<} M_{tx}/V_{ux}$ + twice the extension.

It is noted again that the need for this approach occurs only when the moment diagram has this general shape and when larger bars are used here than at the P.I. If the moment diagram were a straight line, as for a concentrated midspan load, the procedures used for negative moment would be adequate.

*The old flexural bond requirement would not even indicate this problem unless Σo were used as only the bars being cut off, not the total number at x.

(b) Simple Spans

At the support of a simple span that is hung from above by an embedded bar hanger, the condition is exactly that at the P.I. of the continuous span, except that usually the bar ends with an extension l_a less than d (Code 12.2.1). However, more commonly the reaction is from below and provides a compressive force that restricts possible splitting of concrete caused by bond stresses. The Code recognizes this by specifying that the bars may be chosen from the criterion:

For bars confined by a reaction use $l_d \ngtr 1.3M_t/V_u + l_a$.

The 1.3 factor leaves this less strict than at a P.I. *Simple* beams can rarely be supported by framing into a girder which picks up the reaction in shear, but where such a case exists the 1.3 factor must not be used.

5.14 HOOKS AND END ANCHORAGE

Where bars must be anchored (or developed) in tension within a limited distance, a hook can add some capacity. A hook in compression is useless. Although in tension a hook is not as good as an equal length of straight bar, a tension hook can be used to advantage where no room exists for the equal length of straight bar. For example, the footing bars of the example in Sec. 5.12*a* can be made larger if hooks are added at the bar ends.

A standard hook is defined to have one of the forms shown in Fig. 5.15*a*. A general evaluation of a hook is difficult. In mass concrete or other confined conditions a hooked bar slips more than the same length of straight bar, although the ultimate strength may be reached. In a structural member where the bar is close to an exposed face the hook pressure on the concrete tends to split off the concrete face and lowered strength results. In

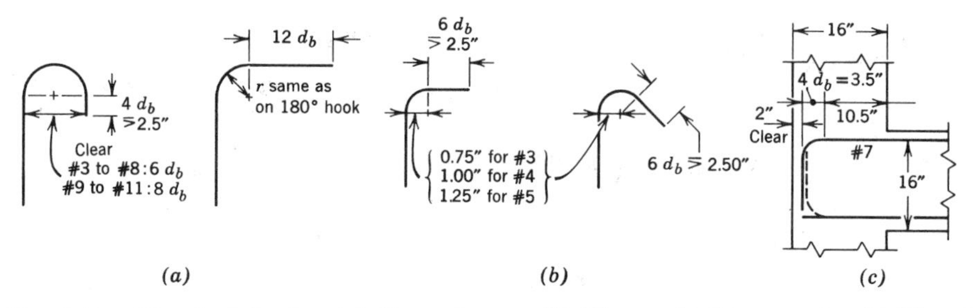

Fig. 5.15. Standard hooks. (*a*) General case. (*b*) Ties and stirrups only. (*c*) Use of hooks.

such a place the value is limited to the splitting resistance of the concrete, a value not well established for this condition but definitely lower than the hook value in mass concrete.

The Code (12.8) assumes either standard hook of Fig. 5.15*a* develops a tensile resistance $f_h = \xi\sqrt{f_c'}$, where the ξ values in Table 5.1 are those tabulated in the Code for #3 to #11 bars. A tie or stirrup hook (Fig. 5.15*b*) is valued in Code 12.13.1.1 as the equal of $0.5l_d$.

TABLE 5.1. ξ Values

	$f_y = 60$ ksi		$f_y = 40$ ksi
Bar Size	Top Bars	Other Bars	All Bars
#3 to #5	540	540	360
#6	450	540	360
#7 to #9	360	540	360
#10	360	480	360
#11	360	420	360

Analysis of the top and bottom beam bars in Fig. 5.15*c* emphasizes several differences that are important. End anchorage of such bars is not as simple as at interior columns and may be quite troublesome.

For the top bars the designer has the option of developing fully (anchoring) smaller diameter bars or of using larger bar diameters at less than 100% efficiency. The upper #7 bars must be classified as *top* bars and Table 5.1 shows ξ of 360 for either Grade 40 or Grade 60 bars. The value of the hooks, assuming normal weight concrete and $f_c' = 3000$ psi is

$$f_h = \xi\sqrt{f_c'} = 360\sqrt{3000} = 19{,}700 \text{ psi}$$

The remaining length of straight bar is 10.5 in., which can be used in the l_d relationship to evaluate the f_s obtainable.

$$l_d = 1.4(0.04A_bf_y/\sqrt{f_c'})$$
$$10.5 = 1.4(0.04 \times 0.60f_s/\sqrt{3000}),\ f_s = 17{,}100 \text{ psi}$$

The total values of the #7 bar is thus 19.7 + 17.1 = 36.8 ksi, about 92% of the normal value of Grade 40 bars or 62% of the value of Grade 60 bars. The design moment requirements for A_s must then be based on this lowered efficiency if #7 bars are used.

If the column has a dependable compression load, this might improve the value of the straight bar portion of the top bar, as discussed for bottom bars in simple spans in Sec. 5.13. For such a case the value of the 10.5-in. length might be taken as $1.3 \times 17{,}100 = 22{,}200$ psi. The hook portion is probably not improved by the vertical compression. Note that if the column ever loses its compression, as in an overturning or earthquake situation, the extra strength is *not* available.

The bottom bars might need evaluation on either of two bases. If the frame has shear wall bracing, the value of a straight #7 bar in compression would be established from the available 14-in. embedment:

$$l_d = 0.02\, f_y d_b/\sqrt{f_c'}$$

$$14 = 0.02\, f_s' \times 0.875/\sqrt{3000}, \qquad f_s' = 43{,}700 \text{ psi}$$

This would be the limiting value of such a bar used as compression steel, if f_y is assumed to be greater. For Grade 60 bars a hook would *not* improve its compression strength; even if a hook were used to meet the requirements of the next paragraph, the author would use the straight 10.5 in. portion then available to compute the usable f_s'.

The second basis of evaluation would apply if the beam were a part of a primary lateral load resisting system, that is, if external shears were not taken by a shear wall. The 1971 Code then considers the possibility of a reversing moment on the column and beam and requires (for the first time) a built-in energy absorbing detail (12.2.2). At least 25% of the bottom steel required for positive moment resistance must not only be carried *into* the column, but must also be anchored there *for its full* f_y. This requirement will fix the maximum bar size usable at the end because the substitution of excess bars less well anchored is not acceptable; the energy absorption of the yielding bars is the end objective. In this case a hook is helpful. Except in massive construction such bars must usually be smaller than the #9 bar which in Table 5.1 leads essentially to $f_s = 0.5\, f_y$ for either Grade 40 or Grade 60 bars. Precisely, for Grade 60,

$$f_h = 540\sqrt{3000} = 29{,}600 \text{ psi}$$

The straight bar section, say 10.5 in., must develop the remaining part of f_y, equal for Grade 60 to $60{,}000 - 29{,}600 = 30{,}400$ psi.

$$l_d = 0.04\, A_b f_y/\sqrt{f_c'}$$

$$10.5 = 0.04\, A_b \times 30{,}400/\sqrt{3000} \qquad \text{Max. } A_b = 0.47 \text{ in.}^2$$

The bar size is thus limited to #6, even though a #6 hook would leave 11 in. straight instead of 10.5 in. Note that a #7 bar of Grade 40 steel would be entirely acceptable, with hook value 19.7 ksi and 20.3 ksi left for the 10.5 in. straight section. This is a place where an *upper* limit on the value of f_y could sometimes be helpful, although it would be a considerable complication on the job.

5.15 MIXED BAR SIZES AND BUNDLED BARS

The variation in bond stress that elastic analysis has indicated as existing at the same cross section on bars of different diameter (a larger bond stress on the larger bars) is not believed to be significant at the ultimate stage. These inequalities can be ignored just as are various other bond stress concentrations (Sec. 5.4) on individual bars. The development length of each size bar must be maintained as separately calculated.

The Code permits up to four bars to be bundled or clustered together. In the absence of test data, it is assumed that the bars must be developed individually but that the lengths must be increased 20% for a three-bar bundle and 33% for a four-bar bundle. The increase is to cover any uncertainty as to how well mortar will penetrate into the core of the bundle. All bundled bars must be within ties or stirrups.

Bundles are limited to bar sizes #11 or smaller and a maximum of four bars. Within a span of a beam the cutoff point of individual bars must be staggered at least 40 diameters, and splices of individual bars must not overlap each other. In applying bar spacing rules, the group is treated as having a single diameter which would give the actual total area of the group.

5.16 DEVELOPMENT LENGTH IN COMPRESSION

Little recent exploration has been done on the development of compression bars. The absence of tension cracks eliminates a major weakness and permits shorter development lengths in compression than in tension. The 1963 Code allowable bond stress has been directly transformed into a development length requirement with no modification.

$$l_d = 0.02 f_y d_b / \sqrt{f_c'} \not< 0.0003 \, f_y d_b$$

The $0.0003\, f_y d_b$ controls wherever $f_c' > 4440$ psi. An absolute minimum of 8 in. is specified (Code 12.6). Inside of spirals the equation length may be reduced 25%. Reductions may be made when excess compression steel is used.

Attention is called to Code 7.10.5 which says definitely that not all column reinforcing may be considered as compression steel for the purpose of splicing (and presumably for development).

> At horizontal cross sections of columns where splices are located, a minimum tensile strength at each face equal to one-fourth the area of vertical reinforcement in that face multiplied by f_y shall be provided.

SPLICES

5.17 SPLICES IN TENSION

It is generally necessary to splice tension bars, partly because of the limited (usually 60 ft) length of commercial bars, but more because of the awkwardness of interweaving long bars on the job. Splicing may be by welding, by Cadweld (fusion type) mechanical connections, or most frequently, for bars #11 and smaller, by lapping bars as in Fig. 5.16*a*. The lapped bars are commonly tied in contact with each other, but may be spaced up to 6 in. apart, with an upper limit of one-fifth the lap length.

With development lengths already rather carefully specified, the splice lap in tension has been specified (Sec. 7.6) as a multiple of l_d, where l_d in this case must be based on the *full* f_y.* Unless this splice can be made away from a point where the bar stress is high, this multiple must be from 1.3 to 1.7, as diagramed in Fig. 5.16*b* and as outlined below. Two factors lead to the necessary increase above l_d. Obviously there is a severe stress condition at *both* ends of a lap splice. Second, the concrete at a splice must take out of the two bars a total stress of $2A_bf_y$ and this increases the splitting forces. Recent research advances in this area are discussed in the next section.

The Code has broken down splices into four classes in increasing order of severity of conditions, as follows:

For stress always less than $f_y/2$:
- Class A if not over 75% of bars are spliced within one lap length. Use $1.0l_d$.
- Class B if more than 75% are spliced within one lap length. Use $1.3l_d$.

For stress exceeding $f_y/2$:
- Class B if not over half the bars are spliced within one lap length. Use $1.3l_d$.
- Class C for more than half the bars spliced within one lap length. Use $1.7l_d$.

For a bar splice in a tension tie member (Class D) the lap must be $2.0l_d$. These splices *must* be staggered, enclosed in a spiral (without any reduction allowed in l_d), and, for bars larger than #4, must have 180-degree hooks at the bar ends.

Where important tension tie members are involved, welding or some other positive connection would be preferred, although the Class D splice appears adequately guarded.

Attention is directed to the saving in splice length if splices are staggered, a reduction from $1.7l_d$ to $1.3l_d$ for a net saving of $0.4l_d$. The wider spacing of splices should then legally qualify for the 0.8 factor on l_d in many cases, but Sec. 5.19 indicates this is a problem area and recommends otherwise.

*See Sec. 5.19 for additional recommendations.

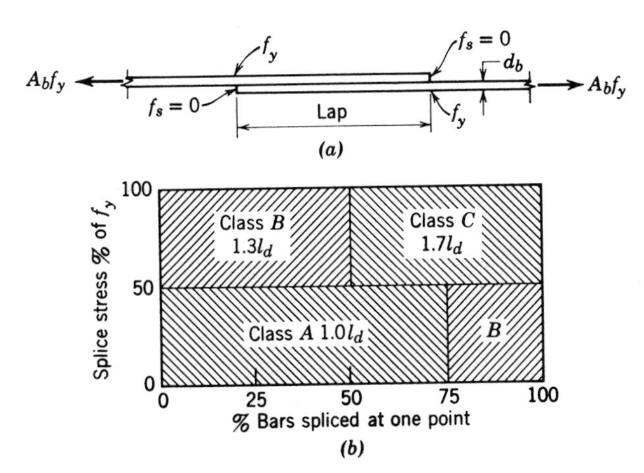

Fig. 5.16. Tension lap splices. (*a*) Diagram of stress transfer. (*b*) Requirements in terms of stress and percent of area spliced at one section.

The behavior of a spliced member is much improved if only half the bars are spliced, an improvement that automatically results from staggering them and nearly doubling the splitting width per splice.

The Code requires a splice lap of 1.7 l_d (etc.) where l_d includes a minimum value of 12 in. However, this minimum lap was not a deliberate intent in setting up this section and the Code will probably be changed* to eliminate the 1.7 × 12 in. interpretation. However, the author would recommend an arbitrary minimum of at least 12 in. for any lap.

5.18 SPLITTING FAILURE GOVERNS SPLICE STRENGTH—RECENT RESEARCH

Research under the author[5,6] the past several years for the Texas Highway Department has somewhat clarified the picture of splice behavior and ultimate strength. This section reports some of the present conclusions, and Sec. 5.19 gives some related design recommendations. The research is continuing.

(a) Failure Modes

A basic finding is that all tension lap splices fail from splitting of the surrounding concrete in one of the basic patterns of Fig. 5.17. At very close

*By the new annual revision procedures which will become effective about 1974.

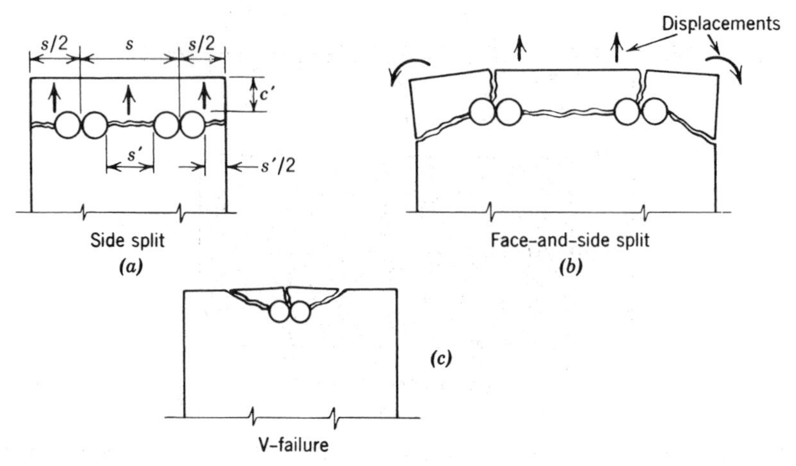

Fig. 5.17. Splitting around splices.

spacings the side split failure will occur and additional cover changes it little (Fig. 5.18*a*); lessened cover tends to change it to the face-and-side split type. In the latter the face splitting over part of the length leaves weak corner sections and finally reaches enough of the length to cause the side split failure at a lowered efficiency (Fig. 5.17*b*). Extra cover in this case increases the resistance offered to the face splitting; lesser cover tends to move the failure toward the V-type (Fig. 5.17*c*) where a splice is not seriously influenced by further spacing changes.

At failure, the concrete splitting stress over the full splice length is far from uniform. The steel stress is sharply changing in the spliced bars and just before failure becomes remarkably linear in both bars (Fig. 15.19*a*), although not totally so. Slip of one bar relative to the other must occur except near the middle of the lap where the two bar stresses are nearly equal. The slip must be large, near $0.5f_y/E_s$ times half the splice length, as shown in Fig. 5.19*b*. This slip implies loss of the "tight" shear connection in the slip plane directly between the two bars over, at least, portions of the length. At least the end section of long splices must have failed in the slip plane earlier than the time of the general failure.

An adequate splice of a large bar goes through several stages in addition to flexural cracking:

1. Splitting starts at the ends of the splice, beginning from the end flexural cracks. The splitting may be on the tension face or on the sides of the beam.

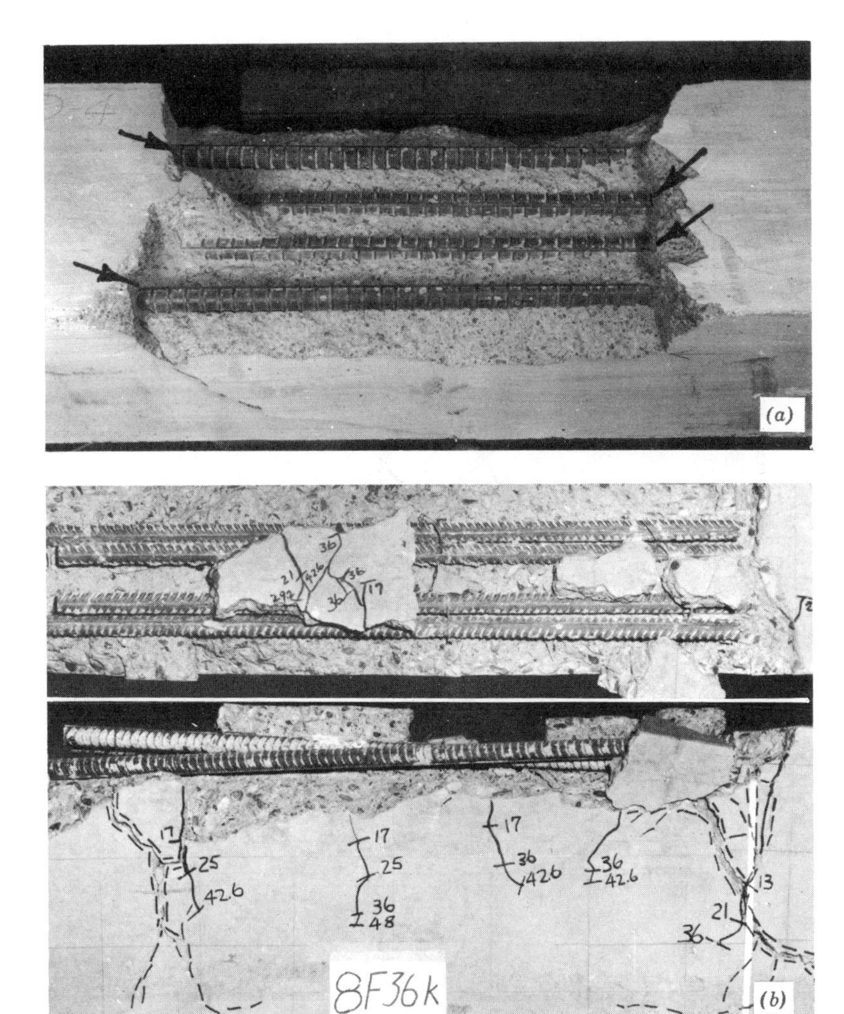

Fig. 5.18. Splice failure in test beams. (*a*) Spaced splice with side split failure. Arrows mark ends of bars where they have separated from the concrete as they slipped at failure. (*b*) A longer contact splice, also in a constant moment length; a face-and-side split failure.

2. The splitting progresses toward the middle of the splice, although not usually completely to the next crack before splitting starts there also, etc. Near failure, splitting may extend over 30 to 80% of the splice length.
3. Final failure is sudden and complete unless confining steel is present.

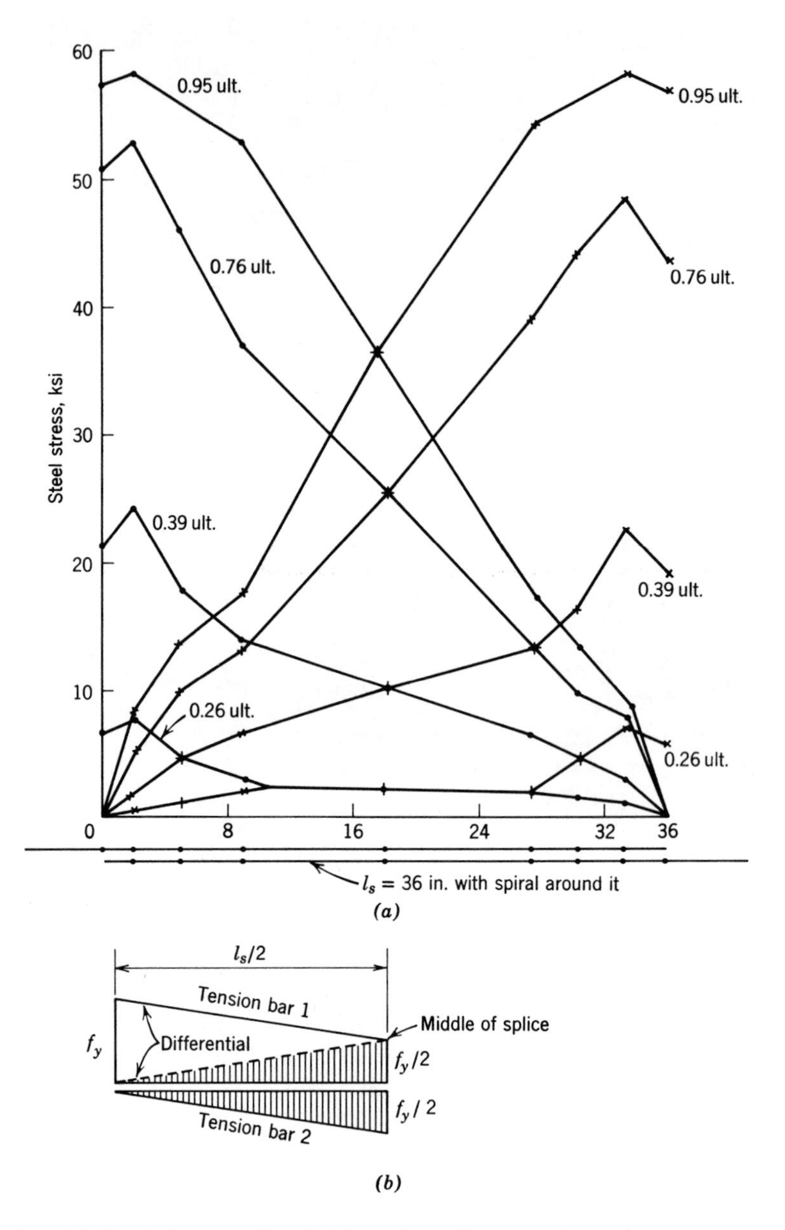

Fig. 5.19. (*a*) Typical stress distribution along No. 14 splice at various load levels. Constant moment zone; face-and side split failure. (*b*) Stress differential between two bars (roughly).

The concrete tensile splitting stresses existing just prior to failure must vary considerably over the splice length, probably being most effective near midlength were cracking is less, and least effective where splitting has started. A bar is not released by a single face split over it, but it would seem to be held less effectively.

(b) Semiempirical Analysis

An oversimplified approach to analysis is helpful, as follows.

1. Assume a uniform bond stress over each bar representing in total a pull of $A_b f_y$ as indicated in Fig. 5.20*a*.
2. Assume also a radial component equal to this bond stress, the concept of the bar pressing on the concrete just like water pressure held in a pipe (Fig. 5.20*b*).
3. On a horizontal plane through the splices calculate the total splitting per unit length as in Fig. 5.20*c*.
4. Assume a *uniform* resisting tension in the concrete across the section and calculate this tensile stress.
5. Calculate the ratio of this stress to the concrete tensile splitting strength.

There are too many approximations for this calculated stress to be realistic:

1. Bond stress is not uniform (Fig. 5.19*a*), being different near the splice ends.
2. The Goto picture (Fig. 5.1) would indicate radial forces greater than the bond stresses.
3. Tensile resistance over the member width is not uniform where the clear spacing s' is large, nor over the length (less where splitting has occurred).

The calculated stress ratios begin to take meaningful form when plotted against s'/c', where s' is the clear spacing between splices and c' is the clear cover over the splice, as shown in Fig. 5.20*d*. It was noted that the side split failures were all at s'/c' of 1.7 or less, the face-and-side split failures from 1.5 to 8, and the V-type only beyond 7.5. Also, a reasonable lower bound curve can be drawn for f_y of 60 ksi, the small cluster of points excluded being cases where yield stresses have led to large strains. Short splices leading to low steel stresses plot higher, and the points are loosely stratified in terms of stress or splice length; but the lower bound furnishes a reasonable base for design. When relevant data are replotted in terms of the inverse of the stress ratio above, an upper bound curve can be drawn,

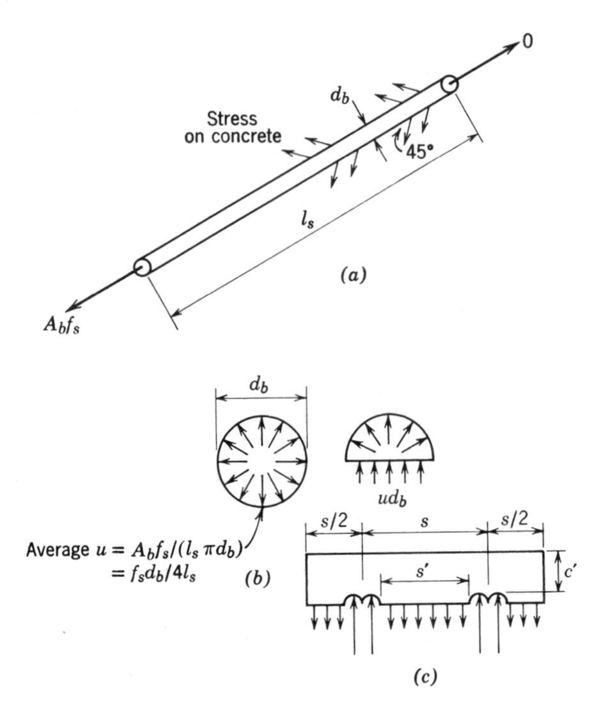

0
d_b
Stress
on concrete
45°
l_s
(a)
A_bf_s
d_b
ud_b
Average $u = A_bf_s/(l_s \pi d_b)$
$= f_sd_b/4l_s$
(b)
s/2
s
s/2
s'
c'
(c)

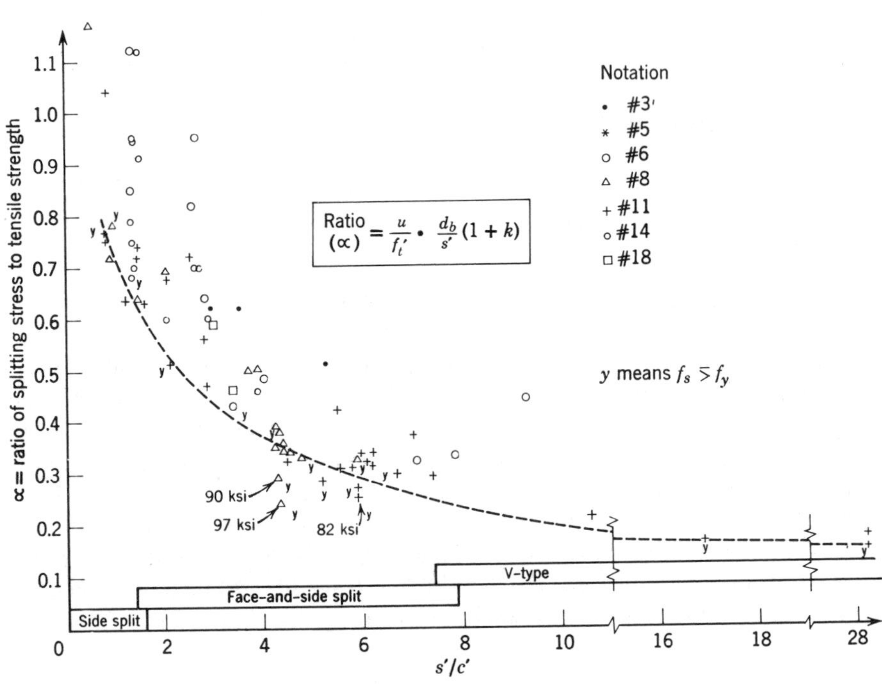

∝ = ratio of splitting stress to tensile strength
Notation
• #3
* #5
○ #6
△ #8
+ #11
○ #14
□ #18
Ratio (∝) $= \frac{u}{f_t'} \cdot \frac{d_b}{s'}(1 + k)$
y means $f_s \geq f_y$
90 ksi
97 ksi
82 ksi
Side split
Face-and-side split
V-type
s'/c'

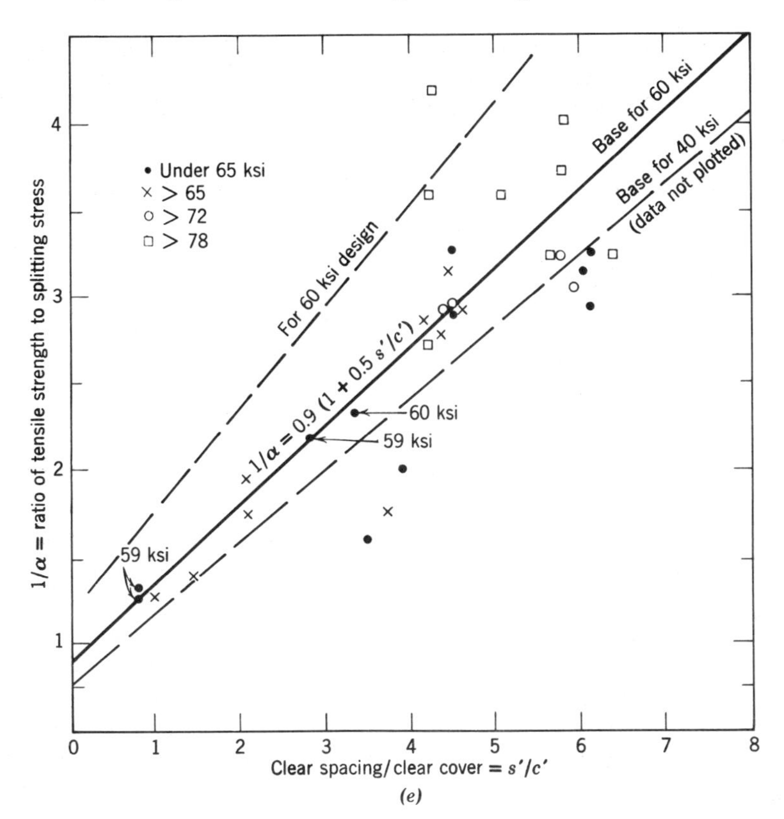

Fig. 5.20. Analysis of splice. (*a*) Forces on concrete from bond. (*b*) Equivalent loading on concrete. (*c*) Equilibrium on short length of splice. (*d*) Test data show decreasing ratio of computed stress to tensile strength of concrete as s'/c' increases. (*e*) Design limit on ratio of tensile strength to calculated stress, the inverse of (*d*) for a particular f_y.

as shown by the center diagonal line in Fig. 5.20*e*, which forms a good design base. To allow for ductility needs, this base was raised, as noted in the next subsection, and was used to set up design equations.

(c) Development of Design Equations

On the basis just discussed, for $f_y = 60{,}000$ psi and $f_c' = 3000$ psi, design equations can be set up.

Splitting force per linear inch of lap $= u2d_b = (f_s d_b/4l_s)2d_b = f_s d_b^2/2l_s$
Resisting tensile stress psi $= (f_s d_b^2/2l_s)/s' = f_s d_b^2/(2l_s s')$
Tensile strength $= 6.4\sqrt{f_c'} = 351$ psi for $f_c' = 3000$ psi

For $f_y = 60{,}000$ psi, the basic ratio $(1/\alpha)$ from tests (Fig. 5.19*e*) is 0.9 $(1 + 0.5s'/c')$, which ought to be increased for ductility in the ratio of $75/60 = 1.25$ plus a small further factor for dropping efficiency at higher stresses, to give a ratio, say,

$$1.30 \times 0.90(1 + 0.5s'/c') = \text{(tensile strength)}/\text{(splitting stress)}$$

$$= \frac{351}{60{,}000d_b^2/(2l_s s')} = \frac{351 \times 2l_s s'}{60{,}000d_b^2}$$

$$l_{s60} = \frac{1.30 \times 0.90(1 + 0.5s'/c')\,60{,}000d_b^2}{351 \times 2s'} = 100d_b^2\left(\frac{1}{s'} + \frac{1}{2c'}\right)$$

A similar basic ratio for a limiting stress of 40,000 psi leads to

$$l_{s40} = 57d_b^2\left(\frac{1}{s'} + \frac{1}{2c'}\right)$$

These appear to be equations (for $f_c' = 3000$ psi) more general than any others that have been available and for the first time point up the importance of clear cover c'. For other f_c' values l_s should be multiplied by $\sqrt{3000/f_c'}$. For lightweight concrete the multipliers of Code 12.5.1c are needed, 1.33 for "all lightweight" and 1.18 for "sand-lightweight" concrete. Top bar splices need to be lengthened by the multiplier 1.4.

The equations for l_{s60} and l_{s40} are plotted in Fig. 5.21 and 5.22 for several values of c'. The basic Code values of $1.7l_d$ and $1.3l_d$ are also shown for reference.

(d) Different Steel Stress at Two Ends of Splice

In a retaining wall stem, or a similar situation, the moment varies significantly over the splice length, and this leads to a stress at the upper end of $kf_y < f_y$ where k is a ratio <1. The derivations above still apply except that, instead of two bars at f_y giving a total stress of $2f_yA_b$, there are two bars with a total stress of $(1 + k)f_yA_b$. The l_s equations can be multiplied by $0.5(1 + k)$ for either the required l_{s60} or l_{s40}. The result is a shorter lap required.

(e) Wall Splices

In any wall it appears that the weak corner elements of the face-and-side split failure can be eliminated or changed from crack 1 in Fig. 5.23 to crack 2 or crack 3. Except at the extreme end of the wall segment, the

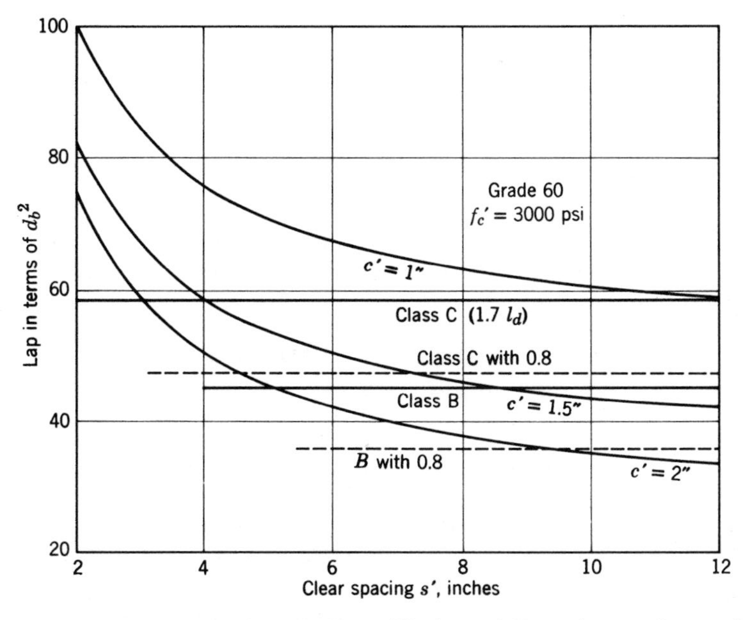

Fig. 5.21. Required lap length for splices. Horizontal lines show code requirement, for comparison.

elements splitting off are not free to rotate as in Fig. 5.17. The end bar in the wall can have a staggered splice or a longer splice (which should delay its failure) and thus can allow the interior splices to capture an extra element of strength. Work on defining the necessary conditions is continuing, but a preliminary (hopefully conservative) judgement has already been

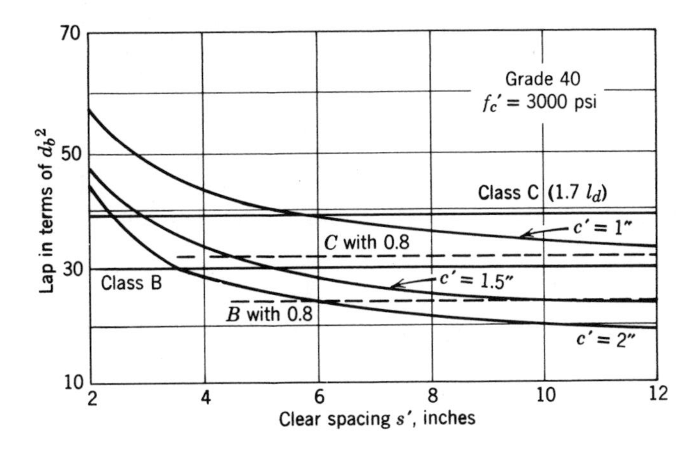

Fig. 5.22. Required lap length for splices.

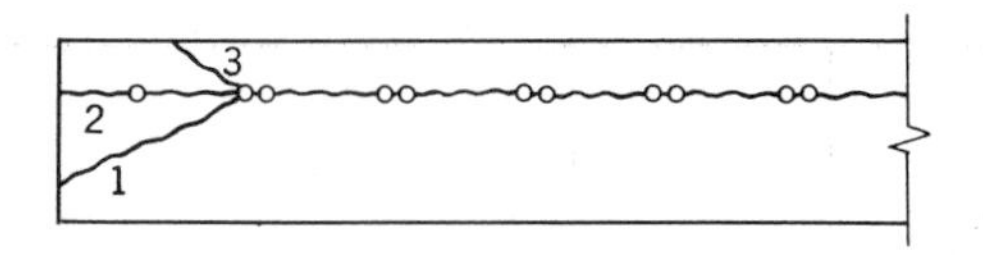

Fig. 5.23. Strengthening wall splices by elimination of end splice at weak section.

reached that a 5/6 factor may safely be applied to the required lap. For a retaining wall with $f_y = 60{,}000$ psi and $f_c' = 3000$ psi, this would give, for an *interior* splice,

$$l_{s60w} = 0.83 \times 0.5(1 + k)\, 100 d_b^2(1/s' + 1/2c')$$
$$= 42(1 + k) d_b^2(1/s' + 1/2c')$$

(f) Transverse Reinforcement Around Splice To Reduce Length

Splice length can be reduced by providing confinement reinforcement in the form of ties or spirals over the lap. Spirals are, at least, a little more effective, but they are probably more awkward in the field. Tentative design recommendations, the same for either ties or spiral, give some idea of the approach. Although transverse steel can only be fully effective after concrete splitting has lowered concrete resistance, it is convenient to proceed as if the strengths were additive. In design one can assign a stress f_{st} to transverse steel and compute the total required transverse steel A_v of yield strength f_{yt} as follows, for $k = 1$, that is, for a splice having f_y at both ends

$$\text{Reqd. } A_v f_{yt} = 0.26 f_{st} d_b^2(1 + 2s'/c')$$

The lap length l_{s60}, for example, can then be reduced (multiplied) by the factor $(f_y - f_{st})/f_y$.

(g) Staggered Splices

No change in the basic equation is needed for staggered splices, but s' then becomes $2s - 3d_b$, where s is the basic bar spacing and $2s$ the splice spacing center-to-center.

5.19 RECOMMENDATIONS FOR TENSION SPLICE DESIGN

Splices with greater cover are shown in Fig. 5.21 and 5.22 as requiring less length than those with reduced cover. The Code requirements fit best the needs with a 2 in. cover.

In view of this most recent research, the author recommends that the following further restrictions be incorporated into the use of the Code requirements for tension lap splices:

1. That for Grade 60 bars the Code rules be considered not applicable for clear covers of less than 1.5 in.; that the 0.8 factor for wide spacing (Code 12.5.*d*) not be used unless checked against Fig. 5.21 or modified in accordance therewith.
2. That for Grade 40 bars the Code rules be considered not applicable for clear covers less than 1.25 in.; that the 0.8 factor for wide spacing (Code 12.5.1*d*) not be used for Class B splices ($1.3l_d$) unless checked against Fig. 5.22 or modified in accordance therewith.
3. That, with the *minimum* covers of 1 and 2 above, the *clear* spacing between splices not be less than 4 in. for Class C ($1.7l_d$) splices or 8 in. for Class B ($1.3l_d$).

The curves in Fig. 5.21 and 5.22 reflect the best available information for splices in 3000 psi concrete. Where the Code is not mandatory the curves are recommended, with the usual factors for other strengths of concrete, lightweight concrete, and top bars to be imposed.

5.20 SPLICING OF STIRRUPS

In some congested spots the use of two U-stirrups (without hooks) turned together to form closed stirrups is feasible. Tested in either shear or torsion, members so reinforced show reasonable ductility, but final failure tends to be more sudden and complete when it occurs.

Code 12.13.4 covers the requirements:

> Pairs of U-stirrups or ties so placed as to form a closed unit shall be considered properly spliced when the laps are $1.7l_d$. In members at least 18 in. deep, such splices having A_bf_y not more than 9000 pounds per leg may be considered adequate if the legs extend the full available depth of the member.

The last sentence definitely exempts stirrup laps of #3 bars of either Grade 40 or 60 and #4 bars of Grade 40 from the 1.7 × 12 in. (minimum l_d) values. The #3 stirrup lap for Grade 60 would even be some 10% less than 1.7 times the theoretical l_d. This is possible because the lap in such a beam cannot be subject to much error in placement and it is known that the small bars are slightly penalized in fitting them into a general specification which also covers large bars.

5.21 COMPRESSION SPLICES

Where bars are required only for compression the bar ends may be cut square, then butted together, and positively held in place to transmit the stress in end bearing. The restrictions are detailed in Code 7.7.2, but Code 7.10.5 is probably more important. This requires some tensile steel on each face at column sections where splices are located, as already mentioned in Sec. 5.16.

Compression lap splices transmit a substantial portion of their load in end bearing on the concrete, as indicated by the PCA test results shown in Fig. 5.24 with the intercept high on the load axis. This bearing is evidently dependent on how well the concrete is confined, spirals showing better than ties.

The Code in effect specifies for f_c' of 3000 psi or better $20d_b$ lap for Grade 40, $30d_b$ lap for Grade 60, and $44d_b$ lap for Grade 75 bars, but in no case a lesser lap than 12 in. Within ties or spirals of specified makeup, these laps may be reduced to 0.83 or 0.75, respectively, of these amounts, but still must be not less than 12 in.

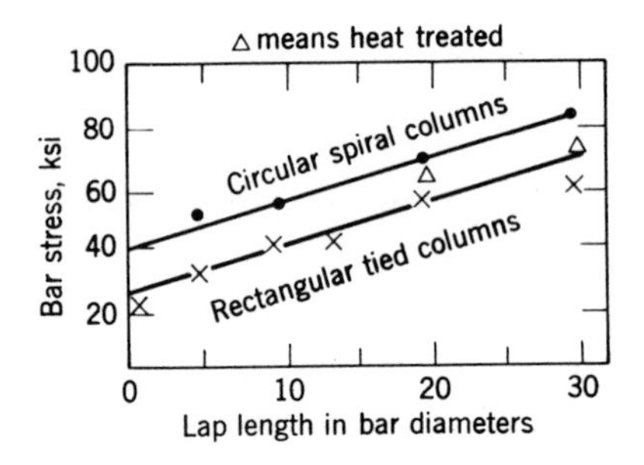

Fig. 5.24. Compression splice strengths in columns. (Modified from Ref. 2.)

SELECTED REFERENCES

1. Yakimasa Goto, "Cracks formed in Concrete Around Deformed Tension Bars," *ACI Jour.*, 68, No. 4, April 1971, p. 244.
2. ACI Comm. 408, "Bond Stress—The State of the Art," *ACI Jour.*, *63*, No. 11, Nov. 1966, p. 1161.
3. John A. Hribar and Raymond C. Vasko, "End Anchorage of High Strength Steel Reinforcing Bars," *Jour. ACI*, 66, No. 11, Nov. 1969, p. 875.

4. Phil M. Ferguson and J. Neils Thompson, "Development Length of Large High Strength Reinforcing Bars," *Jour. ACI*, 62, No. 1, Jan. 1965, p. 71.
5. C. N. Krishnaswamy, "Tensile Lap Splices in Reinforced Concrete," Ph.D. dissertation, The University of Texas at Austin, Dec. 1970.
6. Phil M. Ferguson and C. N. Krishnaswamy, "Tensile Lap Splices, Part 2: Design Recommendations for Retaining Wall Splices and Large Bar Splices," *Research Report 113-3*, Center for Highway Research, The University of Texas at Austin, April, 1971, 60 pp.

PROBLEMS

Prob. 5.1.
(a) In Fig. 4.37 with $f_c' = 4000$ psi, Grade 60 steel, $w_d = 1300$ plf, $w_l = 4500$ plf, do the #6 bars at the end satisfy development length requirements? Assume that bars extend 4 in. beyond the reaction.
(b) If the steel in (a) is changed to 2– #8 with one bar bent up, is the development length satisfactory?

Prob. 5.2. In Fig. 4.38*a*, for $f_c' = 4000$ psi, Grade 60 steel fully stressed, a triangle (assumed) for the moment diagram, find the anchorage l_d into the support. Check development length and moment length to find whether the arrangement shown can be adequate. If not, modify the arrangement of the #6 bars to make it work properly. (There are two ways other than using all straight bars.)

Prob. 5.3. Same as Prob. 5.2 but assume the beam loading changed to uniform load that gives the same maximum moment.

Prob. 5.4. How large a Grade 60 stirrup can be developed in a beam having a 24-in. overall depth?

Prob. 5.5. Some frames must have the bottom beam bars anchored into the column for their full f_y.
(a) In a 16 in.-square exterior column, how large a Grade 60 bar can be fully anchored without exceeding Code evaluations?
(b) If the vertical load from the column is assumed to increase bar resistance by 30% * as allowed at the support of a simple span, what maximum size bar is acceptable?

Prob. 5.6.
(a) If you wish to anchor Grade 60 top bars from a beam into an exterior girder 12 in. wide, what is the maximum size bar which can be used at full value?
(b) If Grade 40 top bars are used?

*No such increase should apply on a hook turned vertically, but on a hook turned horizontally it would be as logical as on the straight bar.

Prob. 5.7. A tied column 20 in. square contains 8- #9 bars of Grade 60 steel, equally on all four faces. The column design indicates no computed tension on the bars. Sketch and detail the arrangement of splices.

Prob. 5.8. The beam in Fig. 5.25 is heavily loaded with balanced overhanging end loads such that for this problem the negative moment is essentially constant across the center span and requires 8- #11 bars of Grade 60 steel arranged in two layers of 4 bars each. $f_c' = 4000$ psi. Because the maximum length of available bars is 60 ft, tension splices of all bars are necessary.
(a) Design and sketch splices, assuming splices are staggered into two groups.
(b) As in (a) with three groups of 3, 3, and 2 splices.
(c) Compare (a) and (b) with the case of all splices made at a single section. Is the latter detail permissible under the Code?

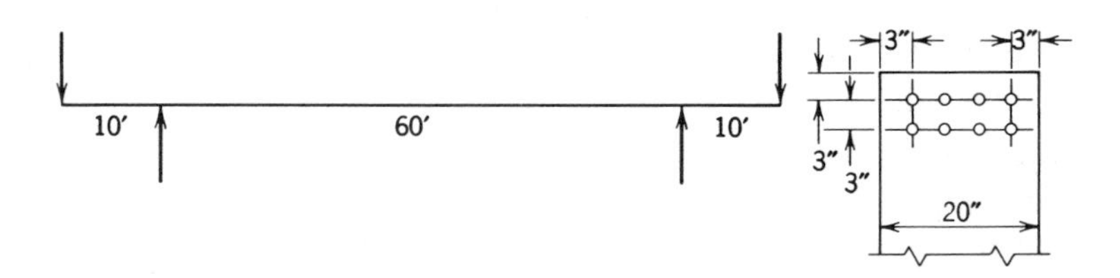

Fig. 5.25. Beam of Prob. 5.8.

6
Serviceability of Slabs and Beams

6.1 INCREASING IMPORTANCE OF SERVICEABILITY

As design procedures become more accurate and more efficient, practice tends to move into the realm of smaller members; greater deflections and greater need for attention to serviceability result.

Strength design methods were significantly introduced into practice in the United States in the middle 1950's. These improved strength evaluations made smaller columns possible. The increased column slenderness led to emphasis on column deflections and increased secondary moments. New column provisions are now in the Code and are discussed in Chapter 14.

The changes in slabs and beams were less pronounced because (1) steel costs increased substantially as member thickness was reduced, and (2) long-term deflections had already been realized as a problem, at least, in slabs. But the possibility of reduced depths without loss of strength, coupled with the possibility of offsetting increased steel costs with smaller areas of high strength steel, made desirable a more definitive statement of what constitutes poor behavior. For serviceability is almost as much a problem of proper standards as of proper calculations. What constitutes poor serviceability?

Aside from corrosion and poor weathering or wearing properties, which are normally avoided by proper controls in mixing and placing, poor serviceability usually relates to excess deflections, occasionally to extensive cracking or to excessive crack width.

6.2 THE DEFLECTION PROBLEM

Excessive deflections can lead to crushing of partitions, bulging and buckling of glass enclosure walls or metal partitions, sagging of floors, and unsightly drooping of overhanging canopies. ACI Committee 435 on deflections lists[1] four categories of deflection problems for which it discusses proper limits.

1. Sensory problems which include vibrations (a function of member stiffness) and appearance of droopy members.
2. Serviceability problems such as roofs that do not drain and floors that are not plane enough for their intended use, exemplified by the requirements for bowling alleys or for sensitive equipment installations.
3. Effects on nonstructural elements such as masonry and plaster, movable partitions. This category must include beam deflections caused by lateral building deflection and vertical movements of columns from temperature differentials.
4. Effects on structural behavior, instability, etc., which is not a serviceability but a strength problem.

The report suggests proper limits to apply for each category.*

Several factors unique to reinforced concrete make exact deflection prediction very difficult. (a) Unsymmetrical reinforcing automatically leads to shrinkage deflections as a normal phenomenon. (b) Creep of concrete under stress leads to a gradual increase in deflection of members left under loads. (c) Even the somewhat uncertain stage at which reinforced concrete begins to crack significantly adds uncertainty as to the actual deflections to be expected at service loads.

6.3 THE CODE SOLUTION FOR DEFLECTION

The Code takes an overall approach in terms of the immediate deflection plus the expected overall percentage increase with shrinkage and time effects.

The immediate deflection which is the starting point is quite sensitive to whether the member is uncracked or cracked, and if cracked how severely cracked. This severity of cracking necessarily varies along the span as the moment changes. The load-deflection curve shown in Fig. 6.1*a* is typical for a moderately reinforced slab that is simply supported and loaded for the first time. Diagrammatically it could be simplified about

*Note also the l/h controls discussed briefly at the close of Sec. 2.8.

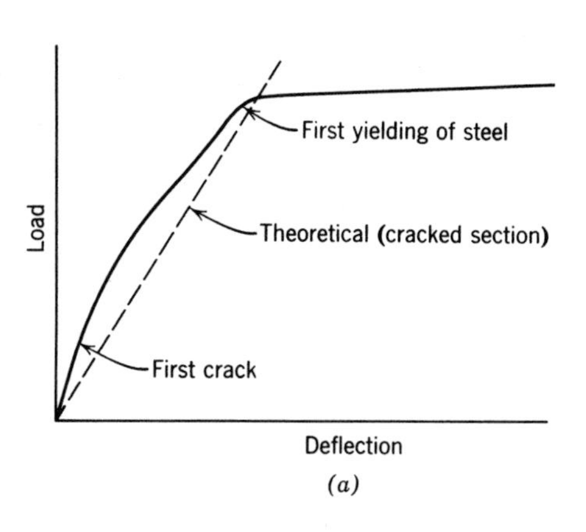

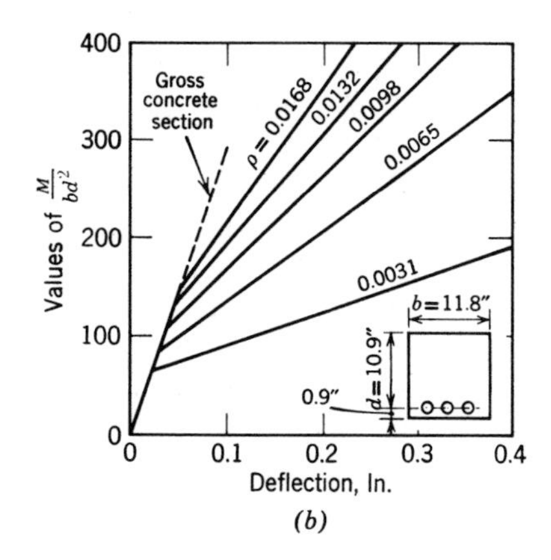

Fig. 6.1. (*a*) Simple span slab deflection under increasing load (*b*) Schematic of influence of ρ on deflection. (From Reference 4, Portland Cement Assn.)

as shown in Fig. 6.1*b*, where the importance of the percentage of tensile steel is shown. The deflections of interest are usually around a steel stress of 50 to 60% of the yield stress or less.

The Code uses an *average* as the effective moment of inertia and then applies the usual methods of elastic analysis (Appendix A) to compute the immediate deflection. The Code then requires that the deflection from the sustained portion of the load be multiplied by a factor that reflects the ratio of A_s'/A_s to determine the added long-term effects from creep and shrinkage.

The average moment of inertia used in Sec. 6.5 involves the transformed area concept which must first be explained in general and then applied for the beam. The application to deflections is picked up again in Sec. 6.5.

6.4 TRANSFORMED AREA CONCEPT—ELASTIC ANALYSIS

(a) General

Where concrete and steel are deformed to the same unit strain ϵ, their stresses vary as their modulus of elasticity.

$$f_c = \epsilon E_c \qquad \text{and} \qquad f_s = \epsilon E_s = f_c(E_s/E_c) = nf_c$$

where the modular ratio n is defined as the ratio E_s/E_c. The total stress on a bar of area A_s then can be written as

$$f_sA_s = nf_cA_s = f_c(nA_s)$$

The last form associates n with the steel area instead of with the steel stress. It shows that the steel area A_s acts identically as would a concrete area nA_s. This concrete area is called the *transformed area* of the steel. When the steel is thus replaced in a member by its transformed area, the result is a total area of homogeneous material, in this case concrete, which is relatively easy to analyze for stress or deformation. To be exact the effective added area is $(n - 1)A_s$ because A_s displaces an equal volume of concrete, but the difference between $n-1$ and n is beyond the accuracy otherwise attainable in deflection calculations.

In the column of Fig. 6.2*a* the transformed area of the column steel may be sketched as shown in Fig. 6.2*b*, *c*, or *d*, that is, at any point around the cross section provided that symmetry is maintained and only an axial load is considered. This is true only because the unit deformation ϵ is everywhere the same.

With bending, f_s again equals nf_c provided both stresses correspond to the same strain ϵ. The nA_s area must then be spread *parallel* to the neutral axis to hold to the same level of strain. In Fig. 6.2*a* for moment or eccentricity about axis x–x the transformed area would have to be sketched parallel to the axis of bending as shown in Fig. 6.2*c* and as shown in Fig. 6.2*d* for moment or eccentricity about axis y–y. Bars in different planes would then carry different stresses in accordance with their locations.

(b) Applied to Beams

With bending, any tension concrete is normally assumed as cracked and omitted from the calculations, but the tension nA_s area is assumed

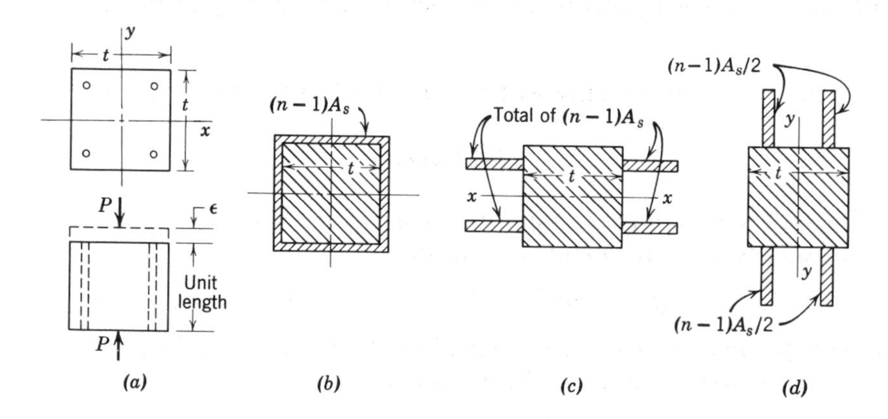

Fig. 6.2. Transformed area of a column, elastic theory.

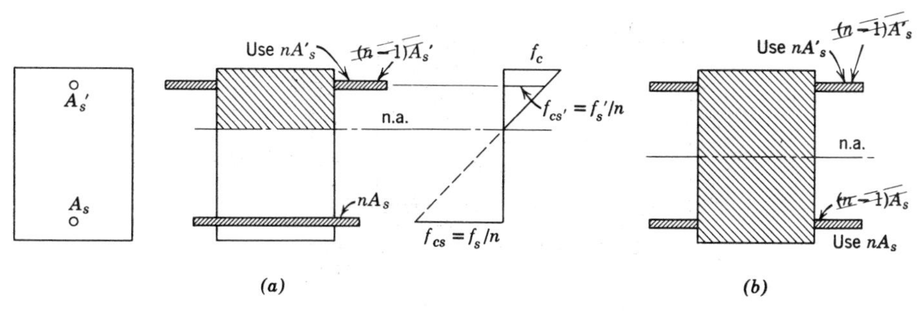

Fig. 6.3. Transformed area concept for effect of short-time loads on a beam. (*a*) Usual cracked section. (*b*) Before cracking.

active in tension to any level. The transformed area of a double-reinforced beam would be as shown in Fig. 6.3*a* with A_s' replaced by an effective transformed area of $(n - 1)A_s'$ and A_s replaced by nA_s, there being no useful concrete displaced on the tension side. Prior to cracking with the concrete still carrying tension, the area would be that of Fig. 6.3*b*.

(c) Neutral Axis of Beam—Mechanics of Elastic Analysis

The transformed area concept makes it possible to replace any reinforced concrete member (under moderate loads) with an equivalent member of homogeneous elastic material to which basic strength of materials relationships apply. For flexure without axial loading, the neutral axis must lie at the centroid of the cross section to make the total tension equal to the total compression and thus provide a resultant couple. The moment and stress are related by the equation

$$Mc = fI$$

where M = bending moment, usually in.-lb or in.-k
c = distance to extreme fiber, in.
f = bending stress, psi or ksi, at extreme fiber (at distance c from neutral axis). This term is used as s or σ in engineering mechanics notation, but f is standard notation in reinforced concrete in the United States.
I = area moment of inertia, $\int y^2\, dA$, about the neutral axis, in.^{4}

(d) Analysis Example with Transformed Area

The following example illustrates the application of the transformed area.

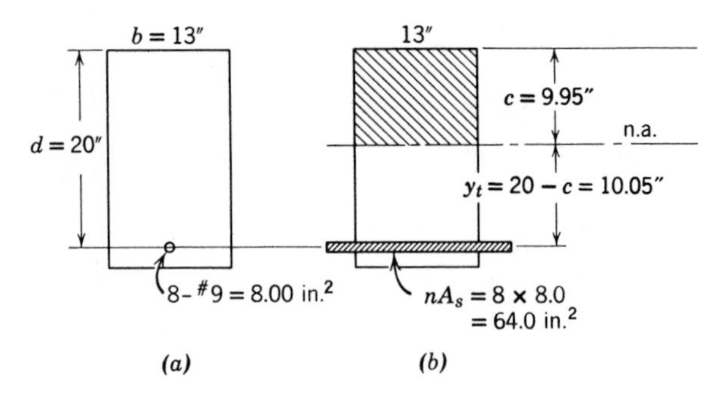

Fig. 6.4. Analysis of a rectangular beam.

Find f_c and f_s for the rectangular beam shown in Fig. 6.4*a* for $M = 100$ k–ft, $f_c' = 4000$ psi, and $E_s = 29 \times 10^6$ psi.

Solution

E_c (Code 8.3.1) $= 57{,}000 \sqrt{f_c'} = 57{,}000 \times \sqrt{4000}$
$= 3.61 \times 10^6$ psi

$n = E_s/E_c = 29 \times 10^6 \div (3.61 \times 10^6) = 8.05$, say 8

This transformed area is first sketched (Fig. 6.4*b*) and the neutral axis indicated at the unknown depth c. The neutral axis lies at the centroid of the transformed area. The moment of the areas about the neutral axis* gives a quadratic equation for which a solution by completing the squares is recommended.

$13\, c \cdot c/2 = 64(20 - c)$
$6.5\, c^2 + 64c = 1280$
$c^2 + 9.85c + (9.85/2)^2 = 196.9 + (9.85/2)^2 = 221$
$c + 4.92 = \pm 14.87$
$c = +9.95$ in. (or negative)

Sec. C.12 discusses a design aid which avoids the quadratic equation. From Fig. 6.4*b* the moment of inertia is:

$13 \times 9.95^3/12$	$= 1067$
$13 \times 9.95 \times 4.97^2$	$= 3200$
64×10.05^2	$= 6460$
Total	10,730 in.^{4}

*Although moments about an axis not yet located may seem awkward, it should be noted that this gives a simpler equation. In general, $\bar{y} \int dA = \int y\, dA$. If y is measured from the centroid, $\bar{y} = 0$ and the equation becomes simply $\int y\, dA = 0$.

The transformed area of the steel is considered negligibly thin, so that its moment of inertia about its centroidal axis can be neglected.

$$f_c = Mc/I = 100{,}000 \times 12 \times 9.95/10{,}730 = 1113 \text{ psi}$$
$$f_s/n = My_t/I = 100{,}000 \times 12 \times 10.05/10{,}730 = 1124$$
$$f_s = 8 \times 1124 = 8990 \text{ psi}$$

6.5 DEFLECTION COMPUTATIONS FOLLOWING THE CODE

(a) Algebraic Recommendations

The effective moment of inertia is specified as an *average* value to be used all across a simple span and is a weighted value dependent on the extent of probable cracking under moment:

$$I_e = (M_{cr}/M_a)^3 I_g + [1 - (M_{cr}/M_a)^3 I_{cr} \leq I_g$$

in which $M_{cr} = f_r I_g/y_t = 7.5\sqrt{f_c'}\, I_g/y_t$ except that f_r must be modified (Code 9.5.2.2*a* or *b*) for lightweight concrete.

M_a = max. M in member at stage for which the deflection is computed
I_{cr} = I based on transformed area of cracked section
I_g = I based on gross area of concrete section (without steel)

The ACI Committee 435 report noted that I_g in this relation would be more accurate if it included the transformed area of the reinforcement, especially where this was heavy.

The above equation can be more simply written as

$$I_e = I_{cr} + (M_{cr}/M_a)^3(I_g - I_{cr})$$

In this form notice that M_{cr}/M_a is most important when M_a is not too far different from M_{cr}.

The use of I_e gives the immediate deflection when used in the usual formulas for elastic deflections. Both shrinkage and creep will result in substantial increases unless the load is primarily a transient one. The Code requires the following multiplier to establish the *increase* in the deflection (added to the basic deflection) caused by the load that is usually sustained:

$$(2 - 1.2A_s'/A_s) \geq 0.6$$

For continuous beams I_e may be computed at mid-span and at supports and the average used in calculating the immediate deflection. Branson's Code discussion[2] indicates that a weighting of the two I_e values in terms of the proportion of negative to positive moments is also possible.

A later study by Committee 435[11] reports that "for controlled laboratory conditions, there is a 90% chance that the deflections of a particular beam will be within the range of 20% less to 30% more than the calculated value" as given by the 1971 Code. The corresponding range based on the 1963 Code procedure* would be 30% less and 30% more.

(b) Practical Complications

The Code procedure covers the simplest possible case, an immediate sustained load and its time effects, plus a later live load regarded as transient. The designer will recognize practical complications. How much will the immediate deflection and time effects be increased by other cracking induced by normal construction loading? Would not much of the time effect be based on cracking that goes with the normal full live load, even if such loading is itself transient? When live load is heavy, does this not imply manufacturing or storage usage where much of this live load causes creep starting *after* the early initial period?

For the heavy live load case, the graphs of Fig. 6.5 show the complications to be expected. Particularly it is important to note that in the long run the dead load deflection becomes that based on the maximum cracking condition plus any accumulated time effects. Hence early smaller calculated deflections are useful only for (1) evaluating the maximum increase in deflections which partitions must accept and (2) for evaluating time effects. As a nonexpert in this area the author wonders whether the sustained load deflection base with I_e based on M_a for that sustained load is actually the best for use in determining time effects. Early construction loads or transient live loads will induce cracking which lowers I_e even where it is not a part of the sustained load. It appears that the Code procedure might give an overly precise value of the early I_e, one which may well be too high in view of the actual physical complications.

(c) Example Based on T-Beam of Sec. 8.10 and 8.11

The cross sections of Fig. 6.6 are taken directly from Fig. 8.11*f*,*g* with steel areas added. $f_c' = 3000$ psi, $E_c = 57{,}000\sqrt{f_c'} = 3.12 \times 10^6$ psi,

*For immediate deflection the 1963 Code, 909c, stated:

> The moment of inertia shall be based on the gross section when ρf_y is equal to or less than 500 and on the transformed cracked section when ρf_y is greater. In continuous spans the moment of inertia may be taken as the average of the values obtained for the positive and negative moment regions.

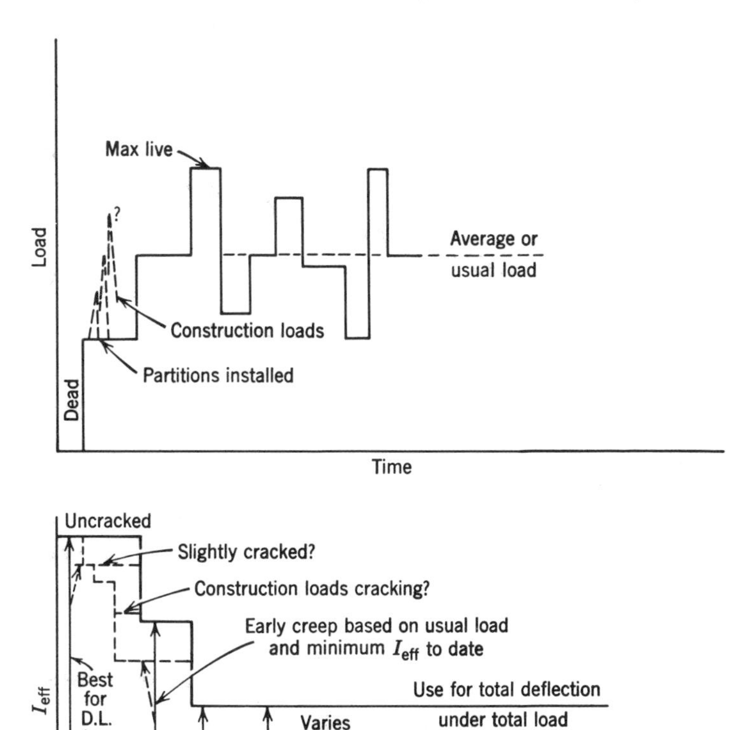

Fig. 6.5. Relation between loading and I_{eff} in diagrammatic form.

$n = 8$,* $w_d = 1156$ plf, $w_l = 2640$ plf, total service load $w = 3800$ plf, $l_n = 20$ ft, design moments of $+0.072wl_n^2$ and $-0.091wl_n^2$. Does the beam deflection satisfy the Code requirement for a beam attached to non-structural members such as partitions?

Solution

The final short time deflection, involving the total load, probably with cracked cross sections, is the most straight forward starting step. To this must be added the shrinkage and creep deflection from the sustained load, a slightly more involved consideration. From this total, insofar as

*The 8 used here more properly should be taken as 9.

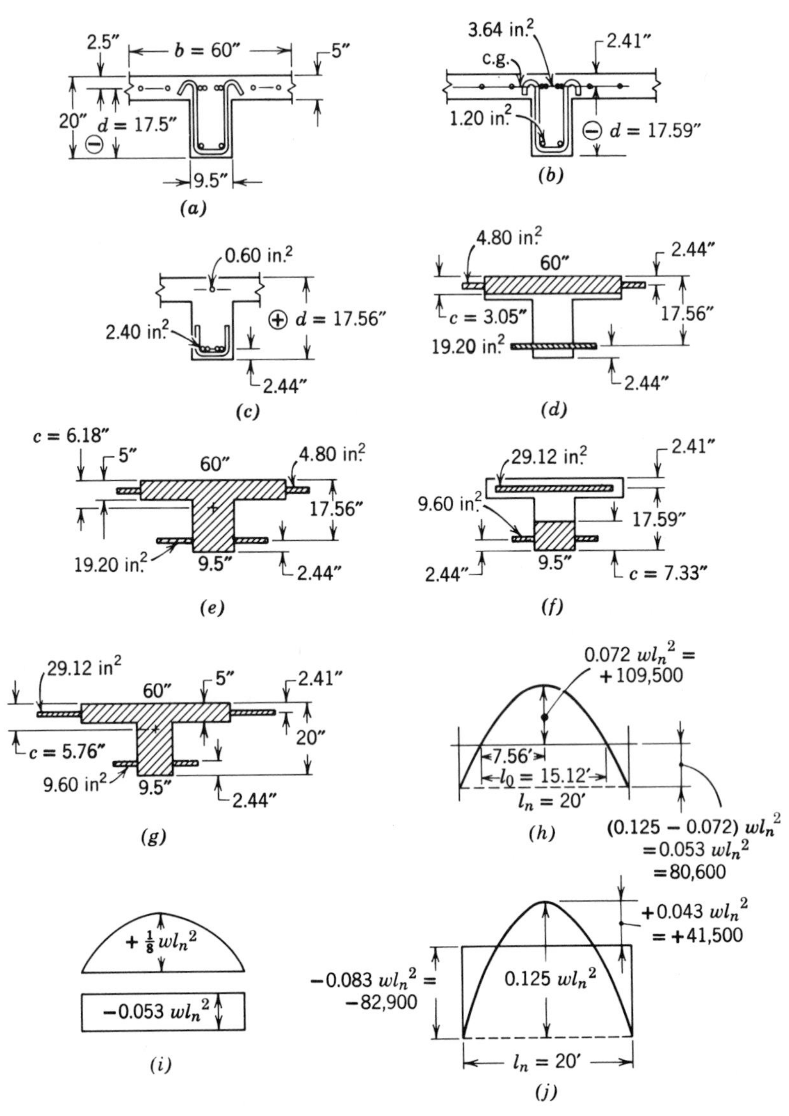

Fig. 6.6. Details considered in deflection of T-beam. (*a*) General layout, negative moment section. (*b*) Reinforcement for negative moment. (*c*) Reinforcement for positive moment. (*d*) Positive moment transformed area, cracked section. (*e*) Positive moment gross area. (*f*) Negative moment, cracked area. (*g*) Negative moment gross area. (*h*) Maximum positive moment diagram. (*i*) Equivalent moment areas for easy use. (*j*) Dead load moment diagram.

partitions are concerned, any deflection existing when they are installed is deductible. The short-time full load deflection requires values of I_{cr} and I_g both at mid-span and at the support for averaging.

Mid-span Section

$nA_s = 8 \times 2.40 = 19.20 \text{ in.}^2 \qquad nA_s' = 8 \times 0.60 = 4.80 \text{ in.}^2$

Cracked section I_{cr}

Centroid of transformed area of Fig. 6.6*d* is at c from top.

$$60c^2/2 + 4.80(c - 2.44) - 19.20(17.56 - c) = 0$$

$$c^2 + 0.80c + 0.40^2 = 11.63 + 0.40^2 = 11.79$$

$$c = \sqrt{11.79} - 0.40 = 3.45 - 0.40 = 3.05 \text{ in.}$$

$$\begin{aligned}
I_{cr}\text{: } & 60 \times 3.05^3/3 &= 566\\
& 4.80(3.05 - 2.44)^2 &= 1\\
& 19.20(17.56 - 3.05)^2 &= 4030\\
\hline
& I_{cr} &= 4600 \text{ in.}^4
\end{aligned}$$

Gross section I_g (Fig. 6.6*e*)

$$(60 - 9.5)5(c - 2.5) + 9.5 \times 20(c - 10) + 4.80\ (c - 2.44) - 19.20\ (17.56 - c) = 0$$

$466c = 2879$, $c = 6.18$ in.

$$\begin{aligned}
I_g\text{: } & (1/12)50.5 \times 5^3 &= 525\\
& + 50.5 \times 5(6.18 - 2.5)^2 &= 3330\\
& (1/12)\ 9.5 \times 20^3 &= 6330\\
& + 9.5 \times 20(10 - 6.18)^2 &= 2770\\
& 4.8(6.18 - 2.44)^2 &= 67\\
& 19.20(17.56 - 6.18)^2 &= 2470\\
\hline
& I_g &= 15{,}490, \text{ say } 15{,}500 \text{ in.}^4
\end{aligned}$$

$$\begin{aligned}
M_{cr} = 7.5\sqrt{3000}\ I_g/y_t &= 7.5 \times 54.8 \times 15{,}500/(20 - 6.18)\\
&= 460{,}000 \text{ in.-lb} = 38{,}300 \text{ ft-lb}
\end{aligned}$$

Support Section

$nA_s = 8 \times 3.64 = 29.12 \text{ in.}^2. \qquad A_s' = 8 \times 1.20 = 9.60 \text{ in.}^2$

Cracked section I_{cr} (Fig. 6.6*f*)

$$9.5c^2/2 + 9.60(c - 2.44) + 29.12(c - 17.59) = 0$$

$$c^2 + 8.15c + 4.07^2 = 112.8 + 4.07^2 = 112.8 + 16.5 = 129.3$$

$$c = \sqrt{129.3} - 4.07 = 11.40 - 4.07 = 7.33 \text{ in.}$$

$$\begin{aligned}
I_{cr}\text{: } & 9.5 \times 7.33^3/3 &= 1250\\
& 9.60(7.33 - 2.44)^2 &= 230\\
& 29.12(17.59 - 7.33)^2 &= 3070\\
\hline
& I_{cr} &= 4550 \text{ in.}^4
\end{aligned}$$

Gross section I_g (Fig. 6.6*g*)

Although slightly debatable, the tension flange in which part of A_s is placed will be counted. The use of the I_g above would not be a serious error.

$(60 - 9.5)5(c - 2.5) + 9.5 \times 20(c - 10) + 9.60(c - 17.59) + 29.12(c - 2.41) = 0$

$c = 2770/481 = 5.76$ in.

$$
\begin{aligned}
I_g\text{: } & (1/12)50.5 \times 5^3 &= \quad 525 \\
& + 50.5 \times 5(5.76 - 2.5)^2 &= \quad 2680 \\
& (1/12)9.5 \times 20^3 &= \quad 6330 \\
& + 190(5.76 - 10)^2 &= \quad 3420 \\
& 9.60(17.56 - 5.76)^2 &= \quad 1330 \\
& 29.12(5.76 - 2.41)^2 &= \quad 326 \\
& I_g &= 14{,}610, \text{ say } 14{,}600 \text{ in.}^4
\end{aligned}
$$

$$
\begin{aligned}
M_{cr} = 7.5 \times 54.8 \times 14{,}600/5.76 &= 1{,}040{,}000 \text{ lb-in.} \\
&= 86{,}800 \text{ lb-ft}
\end{aligned}
$$

Full Load Short Time Deflection

The loading that gives maximum positive moment also gives maximum deflection and the most cracking at mid-span.

$$+ M_a = +0.072wl_n^2 = 0.072 \times 3800 \times 20^2 = 109{,}500 \text{ lb-ft}$$

$$+M_{cr}/M_a = 38{,}300/109{,}500 = 0.350$$

$$+ I_{\text{eff}} = 4600 + (0.35)^3(15{,}500 - 4600) = 4600 + 465 = 5060 \text{ in.}^4$$

At the support the simultaneous negative moment (Fig. 6.6*h*) is

$$- M_a = -(0.053/0.072)109{,}500 = -80{,}600 \text{ lb-ft} < M_{cr} = 86{,}800$$

If this were an initial laboratory loading, the interpretation of $M_a < M_{cr}$ would be a negative moment region uncracked and I_{eff} for that region would be I_g. However, in an actual structure it appears sloppy reasoning to assume that a pattern loading is needed for maximum deflection, but at a support region cracking will not be worse than that caused by the reduced negative moment accompanying this particular pattern. Cracking certainly should be as great as with full load on several panels, and there is, at least, an even chance that it might correspond to the pattern for maximum negative moment. In this example the author assumes that the support region *will* have been cracked by the maximum negative moment there, at service load level.

$$-M_a = -(0.091/0.072)109{,}500 = -138{,}600 \text{ lb-ft}$$

$$M_{cr}/M_a = 86{,}800/138{,}600 = 0.625$$

$$-I_{\text{eff}} = 4550 + 0.625^3(14{,}600 - 4550) = 4550 + 2450 = 7000 \text{ in.}^4$$

$$\text{Aver. } I_{\text{eff}} = (7000 + 5060)/2 = 6030 \text{ in.}^4$$

Assuming that f_c' has been developed at time of initial loading,

$$EI_{eff} = 3.12 \times 10^6 \times 6030 = 1.88 \times 10^{10} \text{ lb-in.}^2$$

The total short time deflection is calculated as the sum of two values easily calculated* (Fig. 6.6*i*).

$$\begin{aligned} y_1 &= (5/384)wl^4/EI \\ &= (5/384)3800 \times 20^4 \times 1728/EI &&= 13.67 \times 10^9/EI \\ y_2 &= -Ml^2/8EI = -80{,}600 \times 12 \times 20^2 \times 144/8EI &&= -6.94 \times 10^9/EI \\ &\text{Short time } y &&= 6.73 \times 10^9/EI \end{aligned}$$

Without creep or shrinkage, the pattern loading maximum is:

$$y_{d+l} = 6.73 \times 10^9/(1.88 \times 10^{10}) = 0.357 \text{ in.} = l_n/670$$

Long Term Deflection

The creep and shrinkage deflection is to be taken as the multiplier $(2 - 1.2A_s'/A_s)$ times the deflection under the usual or sustained load. For this beam the sustained load is assumed to be the dead load plus half the live load. The 220 psf live load implies something like storage or light manufacturing. It seems reasonable, although only a guess, that all or part of this load will be present enough of the time to average half of the load all the time, that is, $1156 + 2640/2 = 2476$, say 2480 plf.

There is no reason to assume this usual load will be in a pattern producing maximum moment; it will be assumed over all spans and this, for uniform spans, gives the fixed end moments of Fig. 6.6*j*.

$$\text{Usual negative moment} = -(1/12)wl^2 = (1/12)\,2480 \times 20^2 = 82{,}900 \text{ lb-ft}$$

The proper value of I_{eff} is now a problem that the Code seems to answer too simply. At the very start, creep should be based on the cracking from dead load and any construction loading. After building occupancy I_{eff} must soon represent cracking from full live load. Theoretically and greatly oversimplified, one needs the sum of two creep and shrinkage calculations. First, the dead load deflection alone would be computed on the limited stage of cracking with the multiplier term for the limited time prior to occupancy (or loading) estimated from the curves of Fig. 6.7; the product then gives the increased deflection before live loads come on. Then the dead plus half live load deflection would be computed on the *final* cracked stage, with $EI = 1.88 \times 10^{10}$ as for the pattern loading, since much of this creep would occur after full live load had been experienced.

*With *unequal* end moments the average used, as here, is usually adequate, except in end-spans.

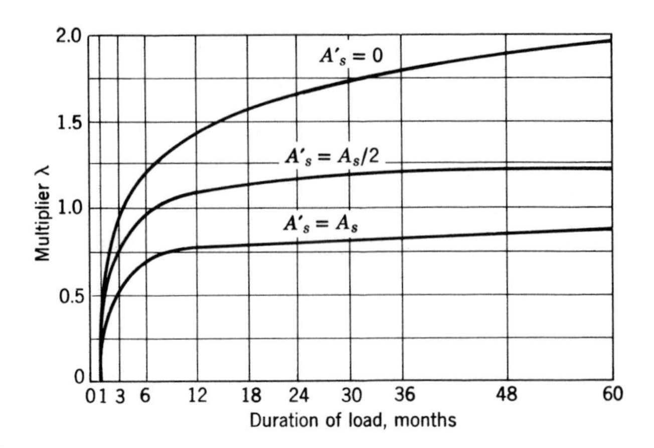

Fig. 6.7. Multipliers for long-term deflection. (From Code Commentary)

Obviously such a procedure tends to get into much detail and many arbitrary assumptions. Instead, the author will first calculate the second of these basic deflections and use it as a basis for a judgement decision about the first one. The immediate deflection for the sustained loading is as follows, using data from the previous calculations and Fig. 6.6*j*:

$$y_1 = (5/384)wl^4/EI = (13.67 \times 10^9/EI)(2480/3800) = 8.90 \times 10^9/EI$$
$$y_2 = -Ml^2/8EI = -82{,}900 \times 12 \times 20^2 \times 12^2/8EI = -7.16 \times 10^9/EI$$
$$\text{Sustained load } y = 1.74 \times 10^9/EI$$

$$EI_{\text{eff}} = 3.12 \times 10^6 \times 6030 = 1.88 \times 10^{10}, \text{ as above}$$

Without shrinkage or creep,

$$\text{Sustained load } y = 1.74 \times 10^9/(1.88 \times 10^{10}) = 0.093 \text{ in.}$$

The nominal portion of this deflection caused by dead load would be

$$y = (1156/2480)0.093 = 0.043 \text{ in.}$$

A check of the dead load moments* against the cracking moments will show the beam uncracked and a smaller deflection yet, roughly 40% of 0.043 in. or less than 0.02 in. The author notes from Fig. 6.7 that early loading for 2 or 3 months might account for half of the total creep expected (for *that* stress condition) and that the creep after a later loading (different stresses) is not really covered by the chart, but would be somewhat reduced

* $-82{,}900(1156/2480) = -38{,}500$ vs 86,800 lb-ft for M_{cr} at support. $+41{,}500(1156/2480) = +19{,}300$ vs 38,300 lb-ft for M_{cr} at mid-span.

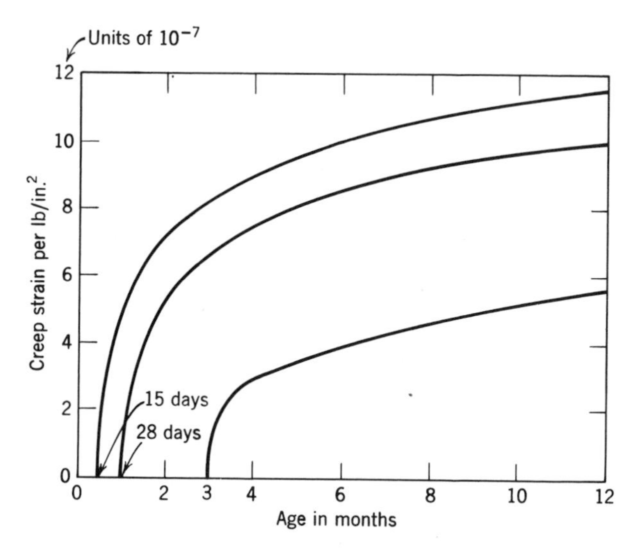

Fig. 6.8. Variation in creep with age of concrete at loading.

by the delayed loading, as in Fig. 6.8 and Fig. 1.10. The deflection increment is, therefore, rather arbitrarily computed on a multiplier of only 1.50, compared to the full computed value of 1.65 determined as follows.

$$\text{Multiplier (Code 9.5.2.3)} = 2 - 1.2(A_s'/A_s) \geqq 0.6$$
$$\text{Mid-span } A_s'/A_s = 0.60/2.40 = 0.25$$
$$\text{Support } A_s'/A_s = 1.20/3.64 = 0.33 \qquad \text{Average} = 0.29$$
$$\text{Multiplier} = 2 - 1.2 \times 0.29 = 1.65$$

On this arbitrary basis, creep deflection $= 1.50 \times 0.093 = 0.14$ in.

$$\text{Maximum deflection} = 0.352 + 0.14 = 0.49 \text{ in.} = l_n/490$$

Of this total, approximately 0.02 in. estimated above will occur prior to installation of partitions. There might also be some prior creep deflection before the installation.

$$\text{Critical deflection} = 0.47 \text{ in.} = l_n/510^*$$

This barely meets the criterion in Code Table 9.5*b* of $l_n/480$†

*The Code uses the length specified for moments, l_n in this case.

†Code limits are from $l/180$ to $l/360$ on *immediate deflection caused by live load* for members not attached or supporting "nonstructural elements likely to be damaged by large deflections." *For* members attached to or supporting such nonstructural elements the *deflection after attachment* is limited to from $l/240$ to $l/480$.

The real accuracy of deflections calculated for beams in structures is not nearly as accurate as these numbers suggest. The improvement in the deflection method of the 1971 Code over that of the 1963 Code (see footnote on Sec. 6.5*a*) is rather minor[11] in the author's opinion. It does not seem to justify the calculation detail that it adds. In practical cases the designer must make many estimates in computing a beam deflection, and the end result will only by chance be as good as indicated in Ref. 11. Some questions have already been noted concerning the loading sequence and the age at loading, neither of which is under the designer's control. It is also the *average* concrete strength that determines the E rather than the design f_c'. To the extent that curing differences do not completely obscure the picture, the average cyliuder strength is the most usable value for deflecting calculations. Much more data and study are needed in this area.

6.6 FURTHER DATA ON COMBINED CREEP AND SHRINKAGE

In 1960 Yu and Winter[5] presented Table 6.1 to show their evaluation of the proper multiplier to be used in terms of the duration of loading and the amount of compressive reinforcement. As in the Code, these multipliers lead to the *additional* (not the total) deflection when applied to an initial deflection calculated from the cracked section I_{cr}. The 1963 ACI Code used essentially the values indicated for the five-year period.

More recently ACI Committee 435 on Deflections is reported[3] as suggesting the Table 6.2 multipliers to give the additional deflection for either normal or lightweight concrete with original deflections based on I_{eff}, but adding: "For unusual cases (particularly for very shallow members such as canopies) or when very early application of load is necessary, it is suggested that shrinkage and creep deflections be considered separately and that the

TABLE 6.1. Multipliers for Additional Deflection Due to Shrinkage-plus-Creep

Duration of Loading	$A_s' = 0$	$A_s' = 0.5A_s$	$A_s' = A_s$
1 month	0.53	0.42	0.27
3 months	0.95	0.77	0.55
6 months	1.17	0.95	0.69
1 year	1.42	1.08	0.78
3 years	1.78	1.18	0.81
5 years	1.95	1.21	0.82

TABLE 6.2. Multiplier for Additional Long-Time Deflections Due to Shrinkage and Creep

Concrete Strength f_c' at 28 Days	Average Relative Humidity, Age When Loaded								
	100%			70%			50%		
	$\leq 7d$	$14d$	$\geq 28d$	$\leq 7d$	$14d$	$\geq 28d$	$\leq 7d$	$14d$	$\geq 28d$
2500 to 4000 psi	2.0	1.5	1.0	3.0	2.0	1.5	4.0	3.0	2.0
>4000 psi	1.5	1.0	0.7	2.5	1.8	1.2	3.5	2.5	1.5

For the period:	1 month or less	3 months	1 year	5 years or more
Use:	25%	50%	75%	100% of table values

The 50% humidity values may normally be used for lower relative humidities, as in a heated building, for example.

choice of the appropriate shrinkage and creep coefficients be made by the designer (preferably from local test data),"*

6.7 SEPARATE CALCULATION OF CREEP DEFLECTION

The above discussions have combined creep and shrinkage deflections because the numerical values of the variables are not easily determined in advance, and also because exact methods are rather involved to use. The age at which loading starts is very important as Table 6.2 suggests. The practical simplification is to say that the creep deflection from two cumulative loadings is the sum of what would be the separately computed deflections. But one must beware of any case where f_c becomes much in excess of $0.5f_c'$, because above $0.5f_c'$ to $0.6f_c'$ the influence of microcracking begins to build up and creep increases more rapidly than the stress increases. Fortunately most sustained loads fall below this level.

The designer in the past has frequently used a reduced modulus, sometimes called the sustained or effective modulus as a creep approximation. In effect, this is equivalent to assuming E_c effectively reduced to $E_c/2$ or $E_c/2.5$, which is a cruder approximation than those already discussed.

*Ref. 3, p. 732.

Branson[3] notes that, since creep in the compression concrete shifts the neutral axis toward the tensile steel (this shift in turn lowering f_c), the creep deflection is not as large with respect to percentage as the creep on a concrete prism under constant stress. He suggests that this reduction factor k_r be taken as

$$k_r = 0.85 - 0.45A_s'/A_s \not< 0.4$$

for creep alone. The corresponding factor for creep plus shrinkage is

$$k_r = 1 - 0.6A_s'/A_s \not< 0.4$$

The latter, with a basic factor of 2 comparable to data from Table 6.2, is the basis for the code multiplier $2 - 1.2A_s'/A_s$, although the Code uses a limiting value of 0.6 instead of 2×0.4.

6.8 SHRINKAGE DEFLECTION PHILOSOPHY

Shrinkage deflection is discussed at some length here. For cantilever slabs where deflection can be particularly serious, the empirical method of Sec. 6.9 has the best correlation according to ACI Committee 435 on Deflections. It is recommended for this particular type of slab. The semielastic method of Sec. 6.10*, although not so accurate in predicting the results from available test data[6] on 14 beams, is still reasonable, considering the accuracy of the shrinkage coefficient that must be assumed in most cases. The error in predictions may be of interest, compared in terms of the distribution in each group:

Percent Error	0–9%	10–13%	14–20%	21–25%	Over 25%	Max	Mean
Empirical	11	0	1	1	1	29% low	0.989
Semielastic	2	4	4	2	1	44% high	1.064

The unequal intervals are selected to emphasize the different groupings of the results in the two cases.

The semielastic method is flexible in use and able to handle any shape, although sections cracked to an unknown depth become a trial-and-error solution after a first estimate of the extent of the cracking. The method can handle any assumed distribution of shrinkage, although the basic estimate of shrinkage will normally be a crude one at best. The basic process is equally usable for uniform or nonuniform temperature changes.

*The method of Sec. 6.10 uses a transformed area for moment of inertia with $2n$ or $2.5n$ while the Committee formulated the comparison on the basis of an I_g using gross area of concrete alone with $E_c/2$ for the modulus. The author has not made similar comparison for the method of Sec. 6.10, but does consider it a more viable approach for varied use.

6.9 BRANSON'S EMPIRICAL METHOD FOR SHRINKAGE OF UNCRACKED SECTIONS

This method is quite suitable for slabs with or without compression steel. The cover over the bars is not one of the variables included, and the equations probably fit unusual cover conditions less accurately. However, the design value of the assumed shrinkage (Sec. 1.8) is the largest probable source of error for any method.

For ρ expressed as a percent of steel rather than a ratio, the curvature is expressed for a singly reinforced member in terms of the shrinkage strain ϵ_{cs} as

$$\theta_{cs} = 0.7 \frac{\epsilon_{cs}}{h} \sqrt[3]{\rho} = (\text{coefficient})\ \epsilon_{cs}/h$$

or

ρ	Coefficient
0.5%	0.56
1	0.70
1.5	0.80
2	0.88

With θ_{cs} known and uniform over the length, this leads to the very simple deflection calculation of $\theta_{cs}l^2/8$ for a simple span and $\theta_{cs}l^2/2$ for a cantilever.

With compression reinforcement the curvature is

$$\theta_{cs} = 0.7 \frac{\epsilon_{cs}}{h} \sqrt[3]{\rho - \rho'}\ \sqrt{(\rho - \rho')/\rho} \qquad \text{for} \qquad \rho - \rho' \lesseqgtr 3\%$$

$$\theta_{cs} = \epsilon_{cs}/h \qquad \text{for} \qquad \rho - \rho' > 3\%$$

The method gives no information as to stresses which, because of creep, are rather uncertain.

6.10 SEMIELASTIC METHOD FOR SHRINKAGE OR TEMPERATURE

(a) General Theory

The shrinkage deflection can be approximated for a given uniform shrinkage by using a reduced modulus of elasticity in lieu of creep effects, as suggested in Sec. 1.7. The usual transformed area under elastic stresses can be used in several ways under this artificial assumption of elastic behavior of the beam. A method of superposition is used here.

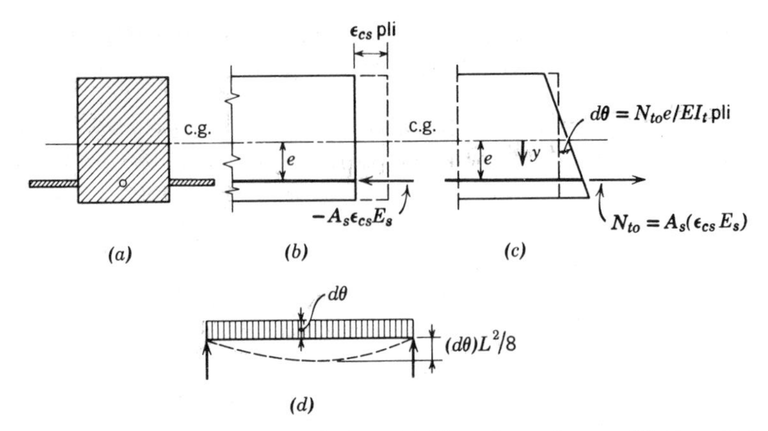

Fig. 6.9. Shrinkage deflection calculation. (*a*) Transformed area. (*b*) Free shrinkage of concrete, forced shortening of steel. (*c*) Elimination of forcing force. (*d*) Deflection from uniform shrinkage.

If a member is symmetrically reinforced, it shortens as the concrete shrinks but does not become curved or deflect laterally. Consider then a member with unsymmetrical reinforcing as shown in Fig. 6.9*a*. Assume the shrinkage ϵ_{cs} (inch per inch) shortens the concrete uniformly, without restraint from the steel,* as in Fig. 6.9*b*; this can be accomplished by applying an external compressive force $A_s(\epsilon_{cs}E_s)$ on the steel to shorten it an equal amount. This gives $f_{c1} = 0$ and $f_{s1} = -\epsilon_{cs}E_s$, with the minus sign representing compression. Then release the restraint of this unwanted external force by applying an equal and opposite (tensile) force N_{t0} to the entire transformed area (A_t, I_t) as shown in Fig. 6.9*c*. The stresses for this loading at any distance y from the centroid are given, as for any eccentrically loaded member, by

$$f_{cy} = \frac{N_{t0}}{A_t} + \frac{(N_{t0}e)y}{I_t}$$

for tension positive and y positive toward the steel. The total concrete stresses are the algebraic sum of the corresponding stresses from the two cases and the steel stress can be obtained similarly, of course, with $f_{s2} = nf_c$ at the level of the steel.

For curvature, the moment $N_{t0}e$ is the only term of significance, leading to $d\theta = M\,ds/EI_t = N_{t0}e\,ds/EI_t$ or $N_{t0}e/EI_t$ per unit length. For constant reinforcing the member shape is a circular curve which, for a simply

*The basic idea is to retain or develop a plane section, and an unrotated plane simplifies the later steps. In the given case the elongation of the concrete to match the unstressed steel would also be an easy starting step.

supported member as in Fig. 6.9*d* leads to maximum deflection $y = (d\theta)l^2/8$ or for a cantilever $y = (d\theta)l^2/2$. No significant modification of the method is needed for steel at more than one level. The uncertainties of the method come from (1) the assumption of a uniform ϵ_{cs} and its magnitude, (2) the assumption of the value of n to include the effect of creep, and (3) the necessity of assuming the member either cracked or uncracked in establishing A_t and I_t for the calculation of $d\theta$.

For beams or slabs with varying A_s over the length, deflections can be calculated by area-moment methods by noting that a $d\theta$ curve simply takes the place of an M/EI curve. The sign of $d\theta$ is positive where A_s is in the bottom of the beam, negative with top tension steel.

It should be noted that compression steel reduces shrinkage deflection, eliminating it completely for uniform shrinkage in slabs and rectangular beams when $A_s' = A_s$. Good practice calls for A_s' in *all* cantilever slabs to avoid shrinkage deflection and to reduce creep deflection from dead load.

(b) Example of Shrinkage Deflection, Uncracked Section

A canopy 4 in. thick (Fig. 6.10) extends 10 ft as a cantilever. If $E = 3.1 \times 10^6$ psi, $n = 9$, find the deflection at the outer end from uniform shrinkage $\epsilon_{cs} = 0.0004$, using $2.5n$, say, 23 to approximate the creep effects and assuming an uncracked slab section and a horizontal tangent at the support.

Solution

Transform the area as in Fig. 6.10*a* and locate the centroid:

$$y = \frac{11.5 \times 1.0}{11.5 + 12 \times 4} = 0.193 \text{ in. above middle}$$

$$N_{t0} = 0.5(0.0004 \times 29 \times 10^6) = 5800 \text{ lb at } e = 3 - 2.19 = 0.81 \text{ in.}$$

$$M = N_{t0}e = -5800 \times 0.81 = -4680 \text{ in.-lb}$$

$$I_t = 1/12 \times 12 \times 4^3 + (12 \times 4)0.193^2 + 11.5 \times 0.81^2 = 64 + 1.8 + 7.5 = 73 \text{ in.}^4$$

$$\text{Effective } E = 3.1 \times 10^6/2.5 = 1.24 \times 10^6 \text{ psi}$$

$$d\theta = M/EI_t = -4680/(1.24 \times 10^6 \times 73) = -52 \times 10^{-6} \text{ uniform over } l$$

$$y = (d\theta)l^2/2 = -52 \times 10^{-6} \times 120^2/2 = -0.38 \text{ in.}$$

More significant digits are not justified.

(c) Stresses from Shrinkage

Shrinkage stresses in structural members are rarely large unless member deformations are restrained, but shrinkage stresses add to flexural tension to induce member cracking earlier than would otherwise be expected. In investigating shrinkage stresses, it is the tensile stresses that are the more important. Hence it would be proper to consider all the transformed area shown in the middle upper sketch of Fig. 6.10. In this slab the shrinkage stresses can be found as the sum of those indicated at the bottom of Fig. 6.10*a*. In the first sketch, $f_{c1} = 0$, $f_{s1} = -5800/0.5 = -11{,}600$ psi (minus representing compression). In the second sketch, for axial load:

$$f_{c2} = N/A = +5800/(11.5 + 12 \times 4) = +97.5 \text{ psi}$$
$$f_{s2} = nf_{c2} = +23 \times 97.5 = +2240 \text{ psi}$$

and for flexure, with y positive toward the compression face:

$$f_{c3\text{top}} = My/I = -4680(-1.81)/73 = +116 \text{ psi}$$
$$f_{c3\text{bott}} = -4680 \times 2.19/73 = -140 \text{ psi}$$
$$f_{s3} = 23(-4680)(-0.81)/73 = +1190 \text{ psi}$$

$$\text{Total } f_{c,\text{top}} = 0 + 97.5 + 116 = +214 \text{ psi (tens.)} < \text{cracking}$$
$$f_{c,\text{bott}} = 0 + 97.5 - 140 = -43 \text{ psi (comp.)}$$
$$f_s = -11{,}600 + 2240 + 1190 = -8260 \text{ psi (comp.)}$$

However, if other loading has cracked the concrete, the transformed area in the middle sketch is not valid; all resulting tension areas already cracked must be omitted.

(d) Example of Shrinkage Deflection, Cracked Section

The above canopy may be considered as being cracked to a depth of 3 in. from earlier live loading. Calculate shrinkage deflection.

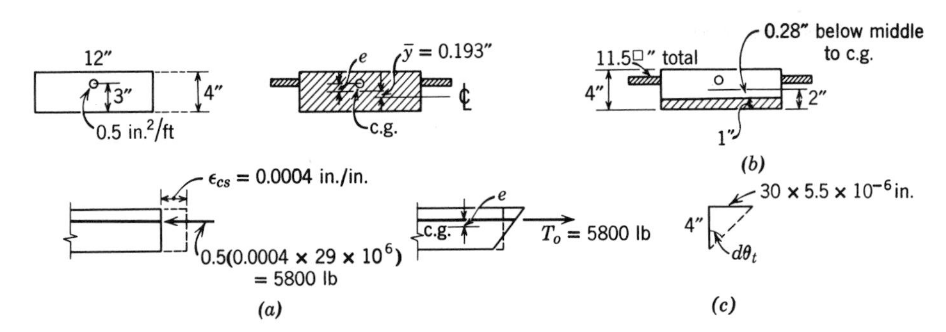

Fig. 6.10. Shrinkage deflection calculation. (*a*) With cracked section. (*b*) Transformed area of cracked section. (*c*) Temperature deformation pli.

Solution

For the transformed area of Fig. 6.10*b*, with all the concrete below the crack as effective. $A = 12 \times 1 + 11.5 = 23.5$ in.^{2}

$\bar{y}$ from mid-depth $= (12 \times 1.5 - 11.5 \times 1)/23.5 = 0.28$ in. (below)
$I = 12 \times 1^3/12 + 12\ (1.72 - 0.50)^2 + 11.5 \times 1.28^2 = 37.7$, say 38 in.^{4}
$M = N_{t0}e = -5800 \times 1.28 = -7440$ in.-lb.
Stress at bottom of original crack $= N/A + My/I$
$= +5800/23.5 - 7440 \times 0.72/38 = 247 - 141 = 106$ psi (tens.)

This stress is less than enough to crack the concrete (usually taken as $7.5\sqrt{f_c'}$), and no tension can exist at the crack itself. Assumed section is O.K. The deflection calculation could be next, but the other stresses will be calculated as a matter of possible interest.

$f_{c,\text{bott}} = 247 - 7440 \times 1.72/38 = 247 - 337 = -90$ psi (comp.)
$f_s = -5800/0.5 + 23 \times 247 + 23 \times 7440 \times 1.28/38$
$= -11{,}600 + 5670 + 5750 = -180$ psi (comp.)

Notice that with a cracked section the shrinkage stresses tend to dissipate when no A_s' is present, but this very fact means an increase in curvature.

$d\theta_{cs} = M/EI = -7440/(1.24 \times 10^6 \times 38) = -157 \times 10^{-6}$
$y = \theta_{cs}l^2/2 = -157 \times 10^{-6} \times 120^2/2 = -1.13$ in.

The probable deflection lies between -1.13 and the original -0.38 in., because the entire length of the beam will probably not be cracked from loading (probably nearer the larger value because the most influencial part of the beam is the one which load will crack).

Deflection will be further increased by a differential temperature from the sun on the top surface. For example, assume that the direct sun raises the temperature of the top to 30°F above the underside with a linear temperature gradient between top and bottom. The coefficient of expansion of steel is 6.5×10^{-6} per degree (F) and of concrete around 5.5×10^{-6}, making this difference of little importance in most cases. As shown in Fig. 6.10*c*,

$d\theta_t = 5.5 \times 10^{-6} \times 30/4 = 4.1 \times 10^{-5}$
Temperature $y = -d\theta_t l^2/2 = 4.1 \times 10^{-5} \times 120^2/2 = -295 \times 10^{-3}$
$= -0.29$ in.

The total addition to deflection from dead or live load and creep is thus $-1.13 - 0.29 = -1.42$ in.

(e) Temperature Curvature and Stresses

As a sample for discussion, consider a roof frame where the upper part of the beam is exposed above the roof for architectural reasons while the lower part is within the enclosure. With seasonal temperature changes outside and air conditioning inside, the concrete temperature shifts around. Assume an analysis is needed when the internal concrete temperature is that of Fig. 6.11. Any distribution, straight line or curving, uniform or nonuniform, is possible under the method used above.

As part of a frame, both changes in length and curvature are important; both should be relative to the basic analysis conditions. For simplicity here, but not essential to the method, consider that creep has adjusted the frame to an average 70° temperature. Step 1 would add on the transformed area the forces required to change the beam and slab section to a plane one, as if all were at 70°, that is, $f_{c1} = \epsilon_{ct}(130 - 70)E_c$ at the top. E_c is the most troublesome term, probably the elastic value for daily variations, a reduced value for seasonal variations. (In an extreme case parts of the temperature could be assigned to each.) Step 2 is to apply reversed forces (or their resultant) to the transformed area, cracked or uncracked as is realistic. For a statically determinate system, elongation, angle change, and stresses of interest are determined as above from these reversed forces. For a continuous frame one must introduce these elongations and angle changes into the system as part of the applied load. Always, for deformations, deflections, or stresses it is the sum of step 1 and step 2 values that must be considered, plus any statical redundants induced by them.

6.11 EXTRA DEFLECTION OF CANTILEVERS AND OVERHANGING ENDS

In contrast to the reduction in deflection from continuity, cantilevers and overchanging ends often have their normal deflections y_w increased by rotations of their supported ends, as indicated in Fig. 6.12.

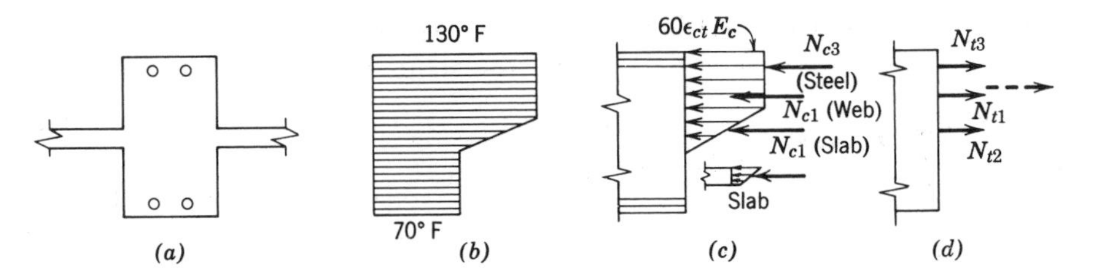

Fig. 6.11. Temperature stress example. (*a*) Beam. (*b*) Temperature in concrete. (*c*) Step 1 forces. (*d*) Step 2 forces.

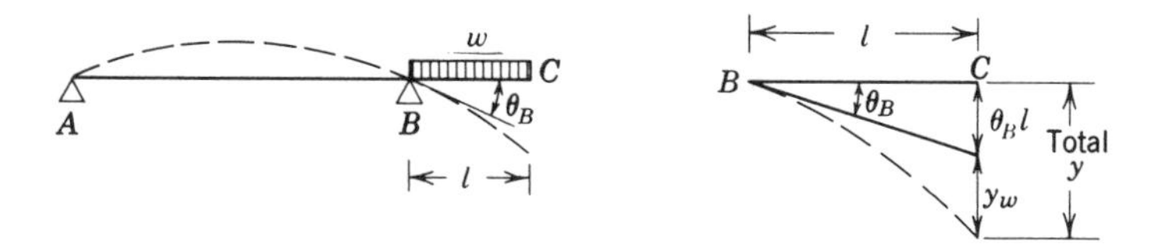

Fig. 6.12. Extra cantilever deflection caused by rotation of supports.

6.12 EFFECT OF INELASTIC ACTION ON DEFLECTION

The calculation of deflections is more complex when inelastic action occurs. If the relation between moment and angle change is known, as in Fig. 9.1, it is possible to calculate deflections by a summation process in which the real $d\theta$ (or ϕ) for each given moment is obtained from the curve. Each of these $d\theta$ values contributes to the member deflection just as the $M\ ds/EI$ increments do in ordinary area-moment procedures. The area-moment calculation simply changes from $\Sigma x(M\ ds/EI)$ to $\Sigma x\ d\theta$, as indicated in Fig. 6.13. Of course, in indeterminate structures the process is more involved because the moment itself is influenced (shifted) by inelastic deformations.

When the reinforcing steel reaches the yield point, ϵ_s and $d\theta$ increase rapidly with small changes in load. The increasing steel deformation crowds the neutral axis toward the compression face at this section and brings about increasing concrete deformations and finally a secondary compression failure. For an ordinary simple span slab the difference between the load at steel yield point and the failure load will be only a few percent, say, 5% or possibly 10%. The deflection before failure will be many times that at the first yielding of the steel. The total relative deflection will depend on the percentage of steel, the span, and the type of loading, but ten times the yield-point deflection would be quite usual. This added deflection will be almost entirely due to the large angle change θ_2 at the yielding section, the slab tending to fold as in Fig. 6.14 about this point as though it constituted a stiff hinge. This is the general basis for limit design (Chapter 9) and the yield-line method (Chapter 11).

Except for the general uncertainties as to partially elastic action prior to yielding, the total deflection is easy to calculate for any given θ_2 value. If the moment-ϕ diagram became truly horizontal, θ_2 would increase indefinitely to failure. If, however, a definite known relation exists between M and the angle change $d\theta$ or ϕ per unit length, as in Fig. 9.1 the simple beam moment diagram permits $d\theta$ to be established for each section and

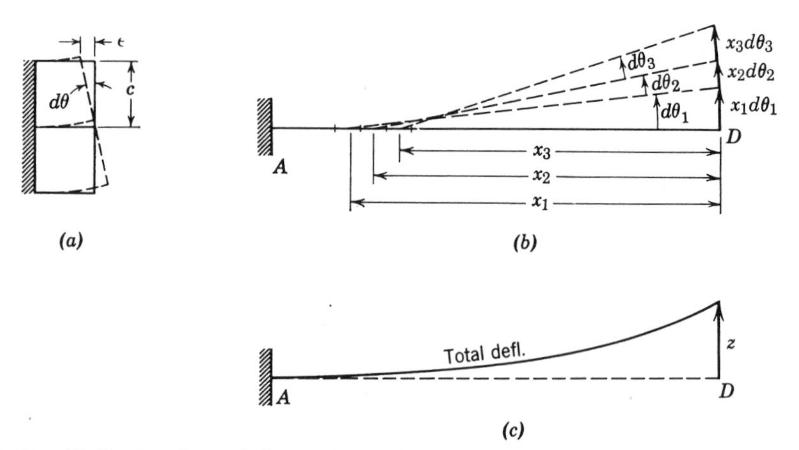

Fig. 6.13. Deflection from deformation of beam elements, positive moment.

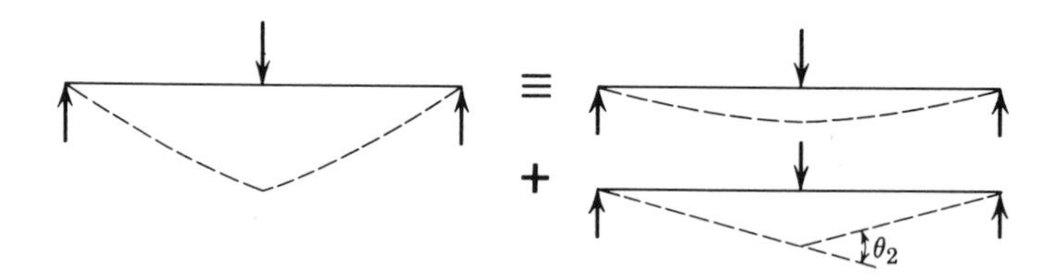

Fig. 6.14. Beam deflection after a plastic hinge forms equals elastic deflection plus hinge deflection.

particularly for those critical sections near the point of maximum moment, that is, at the so-called plastic hinge section.

The area-moment theorems are restated in Ref. 12 in a manner consistent with inelastic and plastic hinge action.

6.13 CRACK WIDTH

(a) Code Requirement Related to Crack Width

Both for appearance and for resistance to corrosion, the Code discourages large crack widths without dealing directly in terms of this width.

In 1968 Gergely and Lutz developed an equation[7] for expected maximum crack width at the tension face of a flexural member, as follows:

$$w = 0.076\beta f_s \sqrt[3]{d_c A}$$

where w = expected maximum crack width in 0.001 in. units
β = ratio of distances to the neutral axis from the extreme tension fiber and from the centroid of A_s
f_s = steel stress in ksi
d_c = cover of outermost bar of A_s measured to the center of bar
A = tension area per bar measured as in Fig. 6.15, that is, equal to 1/5 of the shaded area for the 5 bars used, centered on c.g. of bar areas.

The Code (10.6.3) uses this equation with β taken as 1.2 as the basis for computing limiting values of

$$z = f_s \sqrt[3]{d_c A}$$

which are set at 175 kips per inch for interior exposure and 145 kips per inch for exterior exposure. These correspond to w values of 16 and 13, respectively, that is, to 0.016 and 0.013 in. Since crack widths scatter over $\pm 50\%$, the object of specifying z is to push toward more and smaller bars rather than toward fewer large ones. In increasing from Grade 40 to Grade 60 bars it has been easy to keep bars as large as before and to use fewer of them. Better crack widths result if smaller bars are used, and this is the remedy where the requirement for z is not met. Both positive and negative moment bars must be checked. For mixed bar sizes use the number of bars as A_s/(area of largest bar).

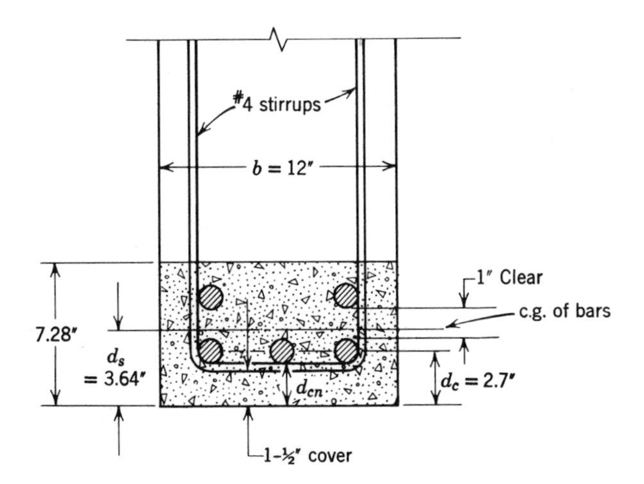

Fig. 6.15. For computing the area A, the shaded area is divided by the number of bars which is five. (Modified from ACI Commentary.)

The Code permits the use of $0.6f_y$ in lieu of a calculated f_s if the designer wishes, and it does not require this check when Grade 40 bars are used. Appendix Fig. D.9 uses the $0.6f_y$ and gives curves that fix the maximum value of the third variable when any two are known.[14]

The value of β in the Gergely-Lutz equation may logically be used in establishing the limiting z in lieu of the 1.2 value used in fixing the Code limits. The Commentary suggests for one-way slabs that 1.35 would be better than 1.2, which would mean that the permissible z given would be multiplied by the ratio 1.2/1.35, to hold the same nominal crack width. Something similarly done for a beam, using a more "exact" β, might be regarded as simply using a more refined method than that in the Code.

Research by Nawy[8,9] has indicated that cracking in two-way slabs follows a pattern somewhat different, a pattern much influenced by the spacing of the reinforcement grid in the slab. This agrees with the author's observation from research testing that, with transverse steel somewhat widely spaced, flexural cracks nearly always develop first over the stirrups or ties rather than between them. Transverse steel exerts a considerable influence on crack spacing.

A short discussion of Nawy's equation for crack width in two-way slabs is in the Commentary discussion relating to Code 10.6.5. The reader is referred to Refs. 8 and 9 for more detail.

(b) Recent Crack Research at University of Texas

The internal shape of cracks were investigated[10] by holding a beam under known loads while flexural cracks in a constant moment region were filled with a colored epoxy. The beams were then sawed such as to cut across the cracks directly over the negative moment bars and permit measurements with a microscope. Cracks varied greatly in thickness and contours, and the mean values were used for plotting. Such mean widths for six beams are superimposed in two groups (30 ksi and 20 ksi stresses) in Fig. 6.16. The surface crack width varies almost linearly with the stress and also almost linearly with the cover over the bar. The interior crack width *at the surface of the bar* is very narrow in comparison. Twenty cycles of loading made little difference in any of these findings.

Official reports are being prepared on corrosion studies related to daily salt spraying of stressed beams (outdoors), a very severe exposure. Two very interesting findings are that, with a water-cement ratio of 6.25 U.S. gallons per 94-lb sack, (1) the corrosion of a stressed length in two years was only 10 to 15% more than in an unloaded and uncracked length of the same beam; and (2) the ratio of clear cover to bar diameter was a more

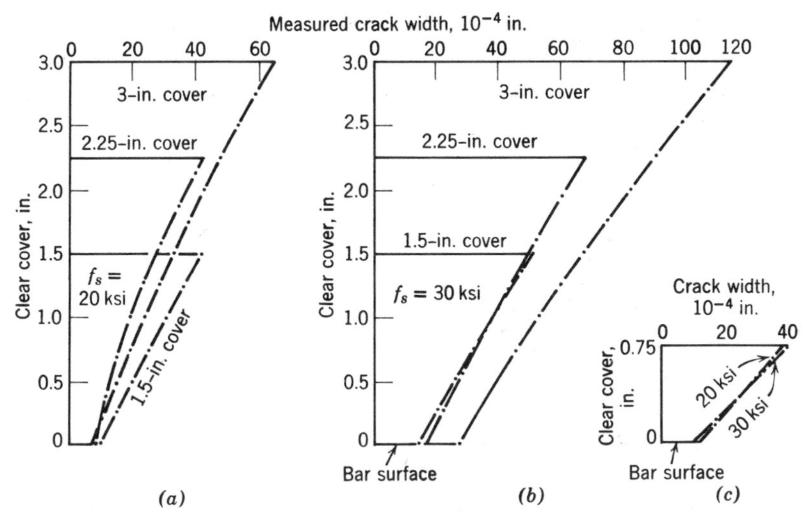

Fig. 6.16. Measured widths of cracks from bar surface to tension face of member. The shape of crack varied more than these comparisons suggest. (*a*) Beams at 20 ksi. (*b*) Beams at 30 ksi. (*c*) Thick slabs.

meaningful variable than clear cover alone, that is, the reduction of the bar diameter to half was as effective as doubling the cover. The results are necessarily limited to chlorine (saltwater) corrosion and probably do not apply to carbonation problems.

SELECTED REFERENCES

1. ACI Committee 425, "Allowable Deflections," *Jour ACI*, *65*, No. 6, June 1968, p. 433.
2. Dan E. Branson, Discussion of "Proposed Revision of the ACI 318-63: Building Code Requirements for Reinforced Concrete," by ACI Committee 318, *Jour. ACI*, *67*, No. 9, Sept. 1970, p. 692.
3. Dan E. Branson, "Design Procedures for Computing Deflections," *Jour. ACI*, *65*, No. 9, Sept. 1968, p. 730.
4. *Deflection of Reinforced Concrete Members*, Bulletin ST-70, Portland Cement Assn., 1947.
5. Wei-Wen Yu and George Winter, "Instantaneous and Long-Time Deflections of Reinforced Concrete Beams Under Working Loads," *Jour ACI*, *57*, No. 1, July 1960, p. 29.
6. ACI Committee 435, "Deflections of Reinforced Concrete Flexural Members," *Jour. ACI*, *63*, No. 6, Jan. 1966, p. 637.

7. P. Gergely and L. A. Lutz, "Maximum Crack Width in Reinforced Concrete Flexural Members," *Causes, Mechanism, and Control of Cracking in Concrete*, SP-20, American Concrete Institute, Detroit, 1968, p. 87.
8. Edward G. Nawy and G. S. Orenstein, "Crack Width Control in Reinforced Concrete Two-Way Slabs," *Proc. ASCE, 96*, ST3, Mar. 1970, p. 701.
9. Edward G. Nawy and Kenneth W. Blair, "Further Studies in Flexural Crack Control in Structural Slab Systems," *Cracking, Deflection, and Ultimate Load of Concrete Slab Systems*, SP-30, American Concrete Institute, Detroit, 1971, p. 1.
10. Syed I. Husain and Phil M. Ferguson, "Flexural Crack Width at Bars in Reinforced Concrete Beams," Research *Report 104-1F*, Center for Highway Research, The University of Texas at Austin, June 1968, 35 pp.
11. ACI Committee 435, "Variability of Deflections of Simply Supported Reinforced Concrete Beams," *Jour. ACI, 69*, No. 1, Jan. 1972, p. 29.
12. George C. Ernst, "Ultimate Slopes and Deflections—A Brief for Limit Design," *ASCE Trans., 121*, 1956, p. 605.
13. G. D. Base, J. B. Read, A. W. Beeby, and H. J. P. Taylor, "An Investigation of the Crack Control Characteristics of Various Types of Bars in Reinforced Concrete Beams," Research Report No. 18, Cement and Concrete Assn., London, 1966.
14. Peter Gergely, "Distribution of Reinforcement for Crack Control," *Jour. ACI, 69*, No. 5, May 1972, p. 275.

7

Cantilever Retaining Wall Design

7.1 TYPES OF RETAINING WALLS

Retaining walls provide soil stability at a change in ground elevation. Dead weight in such a wall is a major requirement, both to resist overturning from the lateral earth pressures and to resist horizontal sliding from the same forces. (The curved-plane sliding of soil on soil well below the retaining wall constitutes the most common kind of sliding failure,* but this is strictly a matter of soil mechanics, not of reinforced concrete. A wall can also slide *over* the soil.)

A *gravity* retaining wall (Fig. 7.1*a*) depends entirely on its own weight to provide the necessary stability. Plain concrete or even stone masonry constitutes an adequate material. Design is then concerned chiefly with keeping the thrust line within the middle third of the cross section.

The *cantilever* retaining wall (Figs. 7.1*c* and 7.2) is a reinforced concrete wall that utilizes the weight of the soil itself to provide the desired weight. Stem, toe, and heel are each designed as cantilever slabs, as indicated in Fig. 7.8.

The *semigravity* type of wall uses very light reinforcement and is intermediate between the cantilever and gravity types (Fig. 7.1*b*).

The *counterfort* retaining wall looks something like a cantilever wall and likewise uses the weight of the soil for stability. The wall and base are tied together at intervals by counterforts or bracing walls (Figs. 7.1*d* and 7.3). These act as tension ties and totally change the supports for stem

*See Fig. 24–2 of Ref. 1.

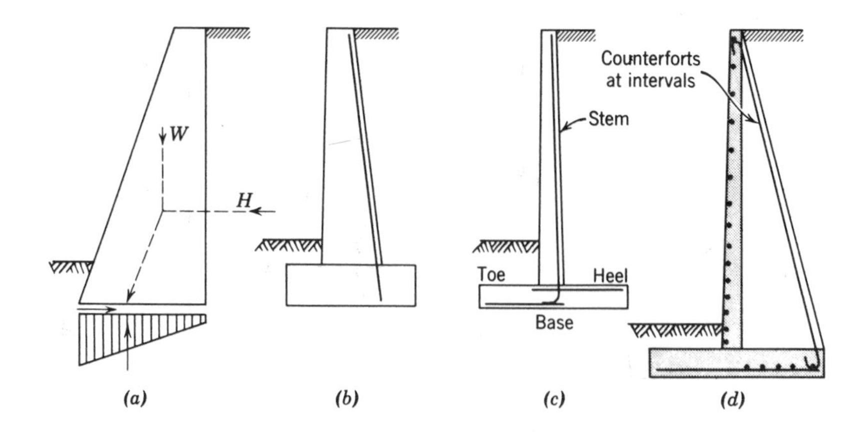

Fig. 7.1. Common types of retaining walls. (*a*) Gravity. (*b*) Semigravity. (*c*) Cantilever. (*d*) Counterfort.

Fig. 7.2. Cantilever retaining wall. (Courtesy Texas Highway Department.)

(*a*)

(*b*)

Fig. 7.3. Counterfort retaining wall. (Courtesy Texas Highway Department.) (*a*) Under construction. (*b*) Before backfilling.

and heel slabs. The stem becomes a slab spanning horizontally between counterforts and the heel becomes a slab supported on three sides. This type of wall becomes more economical than the cantilever type somewhere in the 20 to 25-ft height range.

A buttressed wall is similar to the counterfort wall except that the bracing members are on the opposite side of the wall and act in compression.

Crib-type retaining walls may be made of precast concrete, timber, or metal. The face pieces are supported by anchor pieces extending back into the soil for anchorage.

7.2 ACTIVE SOIL PRESSURE

For the purpose of this text only a general idea of the soil mechanics theory of soil pressure is needed. Only the case of cohesionless soil will be considered, that is, essentially a dry sand. Cohesion in the soil theoretically reduces the demands on a retaining wall, but cohesion generally goes with other adverse factors, such as reduced friction and expansive-type soils, which increase the total effect. An expansive type of soil, such as many clays, can introduce problems of such magnitude as to be beyond economic solution. This type of soil is totally unsuited for a backfill material behind a wall.

Dry sand left unbraced will not stand steeper than a certain slope which depends on its internal friction. When confined behind a wall, such a sand tends to slide. Closely enough, the sliding can be considered as taking place on plane surfaces, with the slope to be established.

Rankine in 1857 analyzed this condition of active soil pressure on a smooth wall as one of plastic equilibrium in the soil. In such a case sliding occurs on two sets of planes in a wedge behind the wall, as sketched in Fig. 7.4. For a smooth (frictionless) wall these planes make an angle of $45 + \phi/2$ with the horizontal, where ϕ is the friction angle for soil on soil. Slip on these planes gives the maximum pressure which can follow through against a wall. (For a wall which does not deflect at all, larger pressures can exist.) The plastic equilibrium idea of sliding planes is a fundamental of soil mechanics. However, for all conditions except those of a vertical frictionless wall and a horizontal fill, Coulomb's theory presented in 1773 gives a better result than that of Rankine's; and since for this special case the two theories give the same result, only Coulomb's will be developed.

Coulomb considered possible sliding planes at different slopes and found the one which demanded the greatest holding force on the part of the wall. When friction on the wall is neglected, the holding force is perpendicular to the wall, that is, horizontal for a vertical wall face. Neglect of wall friction is on the safe side for active pressure (but not for the passive pressures of Sec. 7.4).

Consider a wall at AB in Fig. 7.5*a*. The sand behind will tend to slide on some plane such as BC_1. If the wedge of sand ABC_1 is taken as a free body, it will be in equilibrium under three forces: (1) W_1, the weight of the soil; (2) R_1, the reaction from the soil below BC_1, which may be considered as a normal reaction N_1 and a friction force F_1 resisting sliding; (3) H_1, the holding force from the wall. For any given slope angle, θ_1, the force H_1 (and R_1 if desired) can be found from the force triangle as shown in Fig. 7.5*b*. Similar force triangles can be constructed for other potential sliding planes such as BC_2, BC_3, and so on, leading to other necessary holding forces H_2, H_3, and so forth. A number of trials will lead to the determination in Fig. 7.5*b* of the maximum possible value of H. A helpful

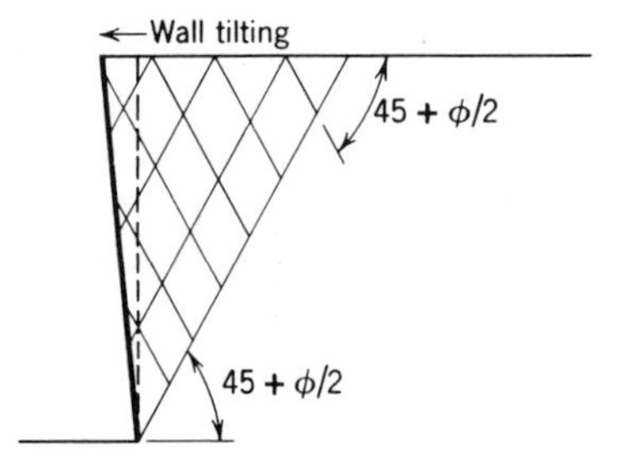

Fig. 7.4. Rankine's plastic equilibrium for active soil pressure.

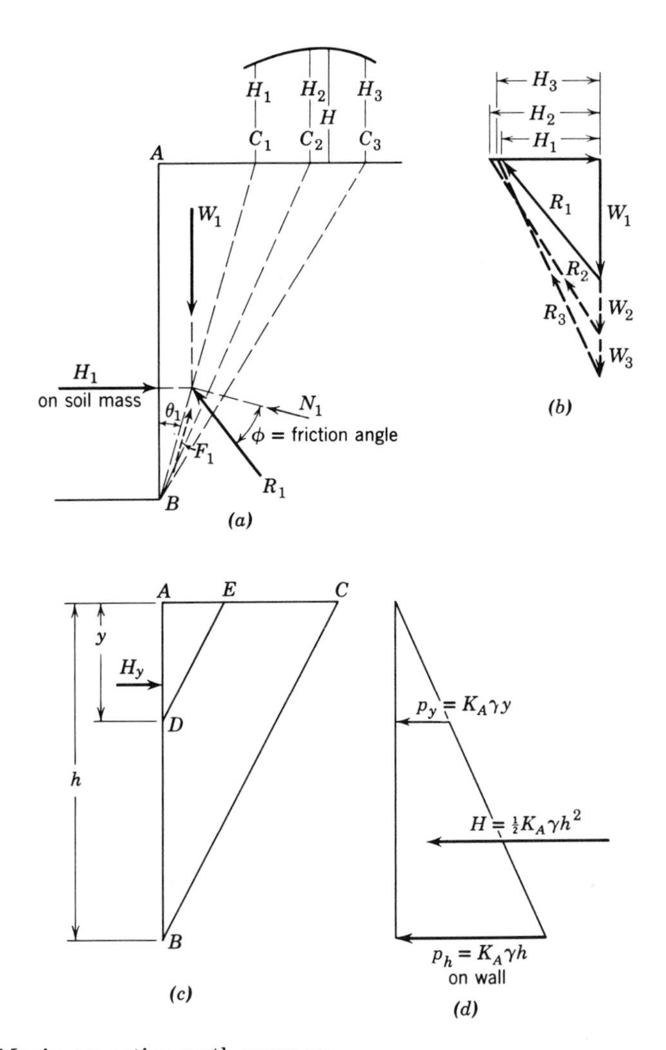

Fig. 7.5. Maximum active earth pressure.

procedure for visualization is to plot an ordinate H_1 over C_1, H_2 over C_2, and so on, and sketch a curve through the points so located. Enough ordinates can be determined to evaluate the peak value H and to locate its sliding plane, say BC in Fig. 7.5c.

A similar analysis of the soil above point D will lead to a critical sliding plane through D parallel to BC. The holding force for the depth AD will be proportional to the weight of sand in the wedge ADE, that is, proportional to y^2 and the unit weight of soil γ. If the constant of proportionality is indicated as $K_A/2$,

$$H_y = K_A \gamma y^2/2$$

The coefficient K_A is called the coefficient of active earth pressure and is usually in the order or 0.27 to 0.34, depending on the sliding friction angle ϕ. It can be shown mathematically that for this simple case

$$K_A = \frac{1 - \sqrt{1 - \cos^2 \phi}}{1 + \sqrt{1 - \cos^2 \phi}} = \frac{1 - \sin \phi}{1 + \sin \phi} = \tan^2 \left(45° - \frac{\phi}{2}\right)$$

Rankine's formula for the more general case of a surface at a slope:

$$K_A = \cos \theta \left(\frac{\cos \theta - \sqrt{\cos^2 \theta - \cos^2 \phi}}{\cos \theta + \sqrt{\cos^2 \theta - \cos^2 \phi}} \right)$$

where θ is the surface slope measured from the horizontal and ϕ is the friction angle for soil on soil.

A total pressure increasing with the square of the depth corresponds to a unit pressure increasing directly with the depth, that is,

$$p_y = K_A \gamma y$$

Thus the pressure on the wall is as shown in Fig. 7.5*d*.

Since the pressure on the wall is like that from a fluid, that is, one weighing somewhat less than water, many designers have used the term *equivalent fluid weight* or *equivalent fluid* for the term $K_A\gamma$. Call this fluid weight w_f. Then

$$p_y = w_f y \qquad H_y = w_f y^2/2$$

Equivalent fluid weight is often assumed without an adequate knowledge of the factors which enter into a calculated value of K_A. As presented here for strength design, there appears to be no great advantage from using the w_f concept.

These pressures are called active soil pressures because they can continue to act on a wall after it deflects or slides. Pressures in a confined soil may be higher, because the active earth pressure has been calculated on the favorable basis of a considerable holding force developed by friction. Some small movement on plane *BC* in Fig. 7.5c is necessary to develop this friction; without it *H* will be larger. For the triangular pressure distribution to be possible, there must be some sliding on all parallel planes, such as *DE*, above *BC*. Hence the wall must deflect more at the top than at the base, by approximately 0.001 times its height, this necessary theoretical deflection corresponding to a rotation of the wall about the base at *B*.

It should be noted that many practical constructions fail to satisfy this deflection requirement, for example, basement walls when supported at or near the ground level by the first floor framing. Another case is the usual braced trench construction, where excavation starts with a brace placed near the top of the trench. In both these cases, the total pressure is roughly the same (say 10% more, for the ideal cohesionless soil, than discussed above, or in extreme cases on an individual strut in loose sand possibly as much as 45%), but the resultant acts nearer mid-depth than at the lower third point. The distribution may vary considerably, but it may be thought of as somewhat parabolic as shown in Fig. 7.6. It might be emphasized that this entire discussion has related to cohesionless soils and has ignored the complications brought about by cohesion and swelling action. Materials which expand under increasing moisture content should not be used as backfill behind retaining walls.

7.3 SURCHARGE

Loads on the surface of the ground over a possible sliding plane, as in Fig. 7.7, increase the horizontal pressure by adding to the ordinary soil weight W in Fig. 7.5*a*. Uniform surcharge over the entire area adds the same effect as an additional height of soil weighing a like amount. Although such a surcharge can be evaluated in terms of an equivalent height of earth, direct methods are actually simpler because the load factor for live load differs from that used for dead load. The concept is convenient, however, in visualizing the effect of the surcharge, which is simply that of an added height of earth weighing the same amount. Such a surcharge adds a uniform pressure to the triangular soil pressure already discussed, as shown in Fig. 7.7*a*.

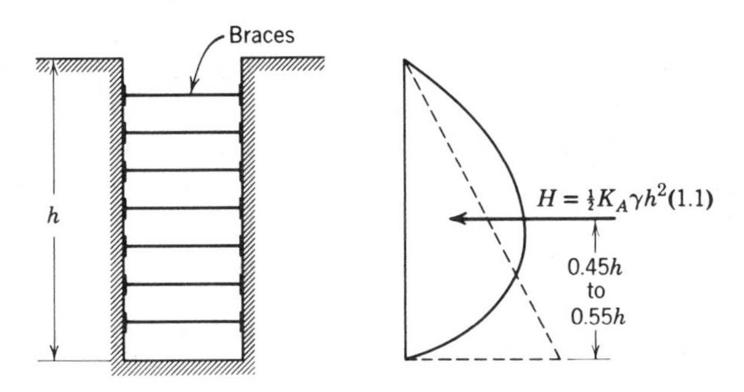

Fig. 7.6. Earth pressure in a braced trench.

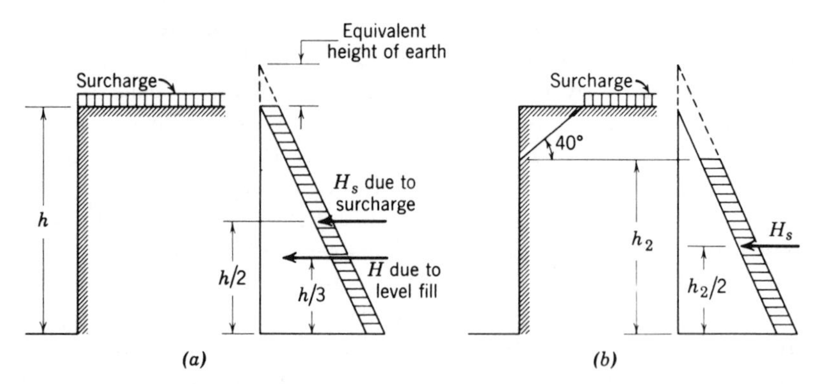

Fig. 7.7. Effect of surcharge on earth pressure.

Surcharge far enough removed from the wall causes no pressure on the wall. The presence of a surcharge well to the right of *C* in Fig. 7.5*c* cannot influence the sliding plane *BC* or the pressure *H*. A load just to the right of *C* would influence the sliding on a slightly flatter plane and might make such a plane critical.

Engineers commonly assume that a surcharge cannot influence the pressure above the point where a line sloping downward from the load intersects the wall, as shown in Fig. 7.7*b*. A slope of 40° or 45° has been used in the past, although now somewhat better methods are available in soil mechanics which recognize the extent of the surcharge. The actual pressure does not change as abruptly as shown in Fig. 7.7*b*, but this assumption is within reason and indicates a greatly reduced overturning effect compared to Fig. 7.7*a*.

7.4 PASSIVE EARTH PRESSURE

If the wall is pushed against the soil, the resistance is very much higher than the active pressure because the soil friction then resists the wall movement. In this case the sliding plane is much flatter ($45 - \phi/2$ according to Rankine) and an analysis similar to that of Fig. 7.5 involves a greatly increased *W*. The mathematical solution leads to

$$H = \tfrac{1}{2}K_p\gamma h^2$$

$$\text{where } K_g = \cos\phi\left(\frac{\cos\theta + \sqrt{\cos^2\theta - \cos^2\phi}}{\cos\theta - \sqrt{\cos^2\theta - \cos^2\phi}}\right)$$

θ = surface slope measured from the horizontal

ϕ = friction angle for soil on soil

Unfortunately, the wall friction is an important element in this case, the true failure is on a curved plane, and the actual H developed is significantly less than this solution indicates. Nevertheless, passive resistance is several times as large as active pressure.

7.5 DESIGN OF CANTILEVER RETAINING WALL

(a) Data

Overall height = 18 ft 0 in.
Level fill, 400 psf surcharge, soil weight of 100 pcf.
Horizontal pressure: $K_a = 0.31$.
Sliding friction of concrete on soil, minimum coefficient assumed as 0.5; of soil on soil, minimum coefficient assumed as 0.62.
Permissible soil pressure under toe = 3500 psf for service load.
$f_c' = 3000$ psi, Grade 60 steel.
Design wall for general provisions of ACI Building Code, using about 1.0% steel ($\rho = 0.01$).

(b) Design Sequence

The design of a cantilever wall involves the choice of heel and toe lengths and the separate design of stem, heel, and toe slabs. Each of these three slabs acts as a cantilever, as shown in Fig. 7.8.

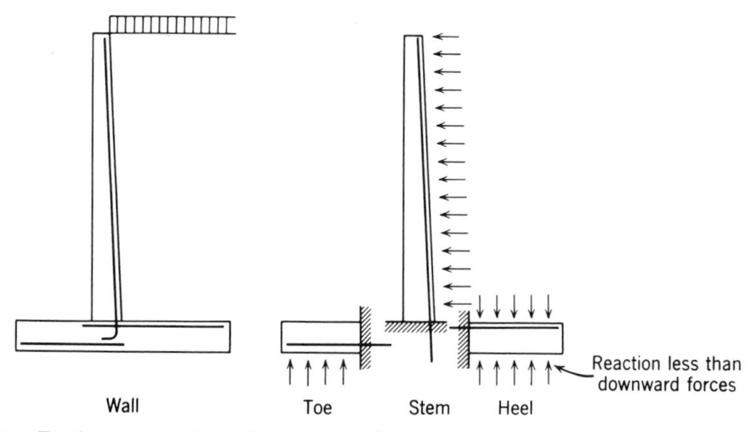

Fig. 7.8. Design parts of cantilever retaining wall.

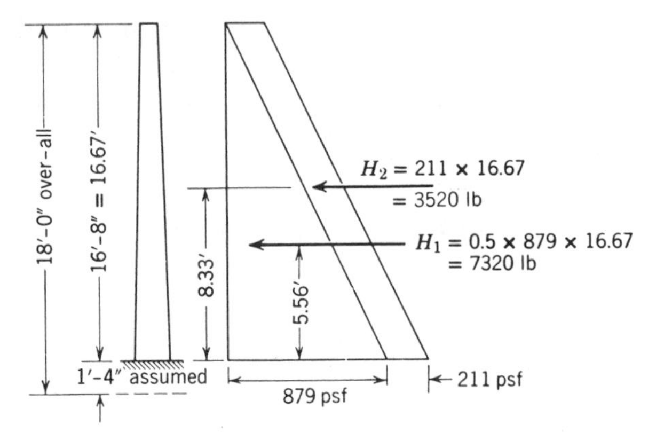

Fig. 7.9. Earth pressure on stem on a 1-ft width.

Foundation conditions determine the elevation of the bottom of the base and thus the overall height. The thickness of base must then be estimated in order to establish a design height for the stem. A tentative stem thickness can next be calculated. The necessary length of heel and toe for stability against sliding and overturning can next be established. Heel and toe can then be completely designed and the stem design can be completed on the basis of the actual base thickness chosen. The design solution of this wall follows, with discussion interspersed, especially on:

The special problem of safety as reflected in the use of load factors for retaining wall equilibrium, in (d).
Allowable soil pressure for use with involved ultimate loads, in (e).
Conventional soil pressure distribution under eccentric reactions, in (g).

(c) Stem Design

The base thickness will be roughly 7 to 10% of the over-all height with a minimum of about 12 in. Assume a 16-in. base, giving a stem height of 18.0 − 1.33 = 16.67 ft as shown in Fig. 7.9. The 400-psf surcharge will be treated as live load with a load factor of 1.7 and the soil weight as dead load with a load factor of 1.4. The Code (9.3.4) provides a separate load factor of 1.7 for the lateral earth pressure H which is to be applied to the service load values of lateral pressure.*

*The last clause of Code 9.3.4 which is related to reducing load factors for D and L in some cases is discussed in detail in the next subsection.

Because of the surcharge there is an added lateral soil pressure of $1.7 \times 0.31 \times 400 = 211$ psf from top to bottom of the wall. This adds to the basic triangular soil pressure from the weight of the soil, a pressure increasing from zero at the top to $1.7 \times 0.31 \times 16.67 \times 100 = 879$ psf at the base of the stem. The resultant lateral pressures per ft. length of wall are calculated in Fig. 7.9.

Maximum moment and maximum shear occur at the bottom of the stem and the initial design will be at this section for a 1-ft length of wall. Stem dead load creates no moment in the stem, and the small direct compression it causes, about 24 psi, would be generally ignored. Strictly under Code 9.2.1.2d the ϕ factor of 0.9 for flexure would be slightly decreased toward the column value of 0.70 by using $P_u/A_g = 24$ psi in the equation:

$$\begin{aligned}\phi &= 0.90 - 0.20P_u/(0.10f_c'A_g)\\ &= 0.90 - 0.20 \times 24/(0.10 \times 3000) = 0.884\end{aligned}$$

The reduction from 0.90 to 0.884 is *not* significant, especially if the axial load is ignored, as here.

$$\begin{aligned}M_u &= 3520 \times 8.33 + 7320 \times 5.56 = 70{,}000 \text{ ft-lb/ft width}\\ \overline{M} &= 70{,}000/0.884 = 79{,}400 \text{ ft-lb}\end{aligned}$$

For $\rho = 0.01$, $\overline{N}_t = 0.01\text{bd} \times 60{,}000 = 600\,bd$

$$\begin{aligned}a &= 600bd/(0.85 \times 3000 \times b) = 0.235d\\ z &= d - 0.118d = 0.882d\\ \overline{M} &= 0.882d \times 600\,bd = 530bd^2, \qquad \bar{k} = 530\end{aligned}$$

With $b = 12$ in., $bd^2 = 12d^2 = \overline{M}/530 = 79{,}400 \times 12/530$

$$\begin{aligned}d &= \sqrt{79{,}400/530} = 12.26 \text{ in.}\\ h &= 12.26 + 2 \text{ in. cover (Code 7.14.1.1)} + 0.5d_b \text{ (bar diam.)}\\ &= \text{say, } 14.76 \text{ in.*}\end{aligned}$$

USE $h = 15$ in., $d = 12.5$ in. (for estimated #8 bars)

The design would normally proceed to the choice of base length at this stage, but, at the risk of necessary revision later, stem design will be carried further here in order to show it as a complete unit of design.

Since d is slightly increased above the required, a is slightly reduced and z is slightly increased above $0.882d$, say, to $0.89d = 11.1$ in.

$$A_s = \frac{\overline{M}}{f_y z} = \frac{79{,}400 \times 12}{60{,}000 \times 11.1} = 1.43 \text{ in.}^2/\text{ft} = 0.119 \text{ in.}^2/\text{in.}$$

$$\begin{aligned}a &= 1.43 \times 60{,}000/(0.85 \times 3000 \times 12) = 2.81 \text{ in.}\\ z &= 12.5 - 1.40 = 11.1 \text{ in.} \qquad \text{O.K. without revision}\end{aligned}$$

*Since the 1.0% of steel is strictly a judgement decision, not a Code limit, h could be made either 14.5 in. with slightly more steel or 15 in. with slightly less steel.

Bar spacing for				
	#7	= 0.60/0.119 =	5.02 in.	d and z increased 0.06 in.
	#8	= 0.79/0.119 =	6.62	d as assumed
	#9	= 1.00/0.119 =	8.38	d and z reduced 0.06 in.
	#10	= 1.27/0.119 =	10.65	d and z reduced 0.13 in.
	#11	= 1.56/0.119 =	13.08	d and z reduced 0.20 in.

If the designer can foresee the development requirement for dowels into the base at this stage, it would be helpful. (This is one of several practical reasons for holding the choice of bars until the base design is complete.) Subject to a later check on bar development requirements, it is noted that the #7 fits nicely, although the 5-in. spacing means many bars to handle. The reduced d for #11 bars would mean a further increase in required A_s and a 12.5-in. spacing. Some excess A_s is thus required for all sizes except the #7. For simplicity of spacing

~~USE #9 at 8 in.~~ ($A_s = 1.00 \times 12/8 = 1.50$ in.^{2}/ft)
Reinf. revised in subsection (j) to #7 at 5 in.

The 0.06-in. reduction in d and z is about 0.5%, amply offset by the increased A_s that resulted from the practical bar spacing.

In slab design, shear rarely governs thickness unless the span is short or loads are usually heavy. Hence shear can usually be delayed until a late check. It is critical at a distance d from the support, but the calculation of critical shear there may be avoided if the larger (but more easily calculated) shear *at* support is below the allowable. At the support

$$V_u = H_1 + H_2 = 7320 + 3520 = 10{,}840 \text{ lb}$$
$$\overline{V} = 10{,}840/0.85 = 12{,}800 \text{ lb}$$
$$v = 12{,}800/(12 \times 12.5) = 85.5 \text{ psi} < 2\sqrt{3000} = 109$$

Development length into the stem creates no problem of consequence. Dowel development length into the base is discussed in subsection (j).

(d) Heel and Toe Length—Design Philosophy

Heel and toe length are interrelated but can be established in sequence. The heel has often been made just long enough to cause the resultant of the service load forces on the wall to strike the ground under the stem (Fig. 7.10a). In addition to necessary further provision against overturning, this can lead to complications from insufficient sliding resistance which require the use of a key into the soil to take a major part of the horizontal force. Such a key is rather uncertain to calculate and it functions

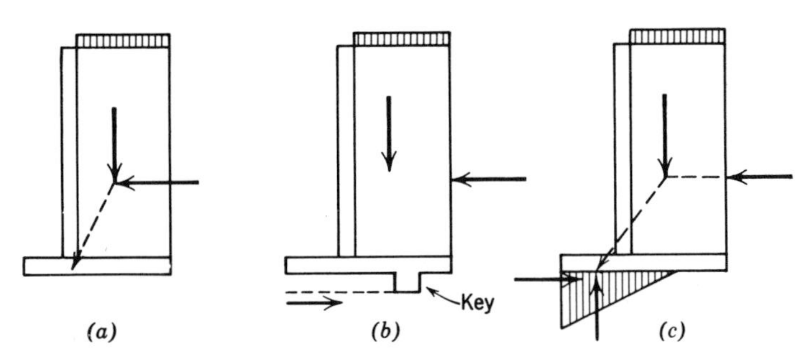

Fig. 7.10. Design criteria. (*a*) Heel length to put resultant under wall, service loads. (*b*) Heel length to limit sliding under factored loads. (*c*) Toe length to give satisfactory soil pressure under factored loads.

chiefly to change the critical friction surface from concrete resting on soil to soil resting on soil, which usually increases the available friction. Alternatively, the heel length can be lengthened to increase vertical load and frictional resistance. Such a long heel increases wall costs and sometimes excavation costs.

The method proposed in the following subsection compromises between these two approaches, basing the heel length on friction of soil-on-soil, with a key added simply to avoid sliding of the concrete at the soil-concrete interface (Fig. 7.10*b*).

The toe length must then be based on maintaining soil pressures within the allowable ultimate while avoiding overturning from overloads, Fig. 7.11. In poor soils, the minimizing of settlement problems [subsection(g)] may control the design length. The length of toe is intimately tied in with the length of heel, since a long heel is a counterbalance to overturning.

Both these equilibrium conditions, for heel and for toe, involve some loads offsetting the effects of other loads. Horizontal loads tend to cause sliding, while vertical loads mobilize more frictional resistance. Horizontal loads cause overturning moments while vertical loads develop counterbalancing moments. As a result, two serious considerations exist outside the basic areas of member design for strength and the allowable soil pressure at service load. These considerations are not, in the author's opinion, yet solved to the point of defining good practice in precise terms:

1. The open question of how best to apply a theory of safety, discussed in the next paragraphs.
2. The correlation of service load pressure with factored loads is here not as clear as in the case of column footings. Ultimate soil pressure is discussed with the choice of toe length, see subsection (f)

With regard to ϕ the Code (9.3.4) appears quite specific:

> If lateral earth pressure H must be included in design, the strength U shall be at least equal to $1.4D + 1.7L + 1.7H$, but where D or L reduce the effect of H, the corresponding coefficients shall be taken as 0.90 for D and zero for L.

This Code section, written with the design of members in mind, seems not adequate for the equilibrium problem of the wall. Two examples will be outlined to define the problem.

The author considers surcharge as a live load, since it may be present or absent as a loading. When present, it both (1) creates a lateral pressure on the wall, similar to H_2 of Fig. 7.9, and (2) adds to the frictional resistance below the base. Item 2 definitely reduces the sliding potential of item (1), but the Code seems to prescribe a zero load factor for this live load. It is illogical here to take this load factor on surcharge as zero and yet include the H it produces; the wall is acted on by both or by neither. (The Code is well designed for a different case, where, say, earth pressure increases a wall moment but a floor load introduces a counteracting moment.)

Likewise, in the case of the soil weight on the heel, the load tends to counteract the sliding tendency created by the horizontal pressure. Although here the Code gives 0.9 for the load factor (not zero) this concept is not really valid in evaluating the wall equilibrium. If D were a floor weight corresponding to the live load in the paragraph above, the 0.9 factor would be a sound value for use. Here, however, $1.7H$ will never exist, that is, have a measurable chance of occurring, if only $0.9D$ (as a weight of soil) exists. The $1.7H$ is intended to cover not only a possible increase in K_A but also a possible increase in the soil weight in $1.4D$. One might analyze the 1.7 as the result of the sum* of 1.4 times the nominal soil weight added to an 0.3 factor for variation in K_A.

The author recommends, therefore, that the last half of Code 9.3.4 be considered not applicable here except that 0.9 should be applied to any weight of concrete that tends to balance the effect of H. He further recommends *for the equilibrium equations* (not for the design of cross sections) the use of $1.7H$ for all lateral pressures with 1.4 for surcharge weight and 1.4 for soil weight.

The less satisfactory alternate discussed in the opening paragraph deals with service loads, requires investigation of overturning when horizontal loading is increased (usually considered as doubled), and may still leave sliding problems.

*Probability theory does not directly add two factors, but for the present discussion the basic concept is correct even if 0.3 is not quantitatively correct.

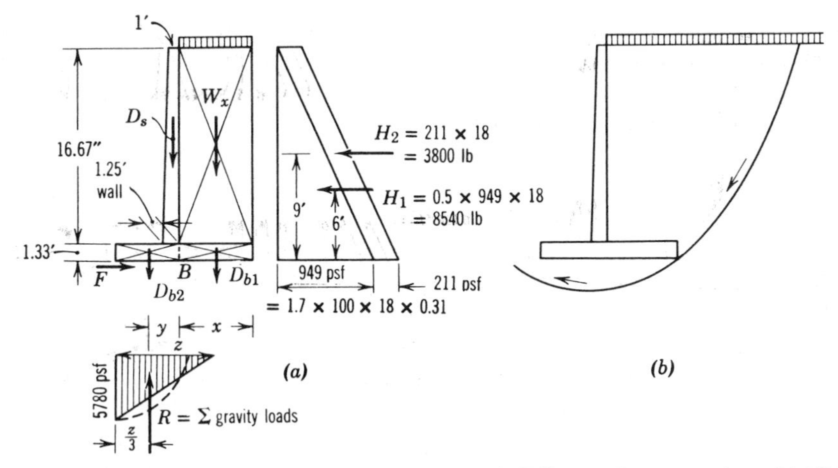

Fig. 7.11. Retaining wall stability. (*a*) Horizontal sliding and overturning. (*b*) Sliding failure in soil.

(e) Choice of Heel Length

The heel length will be chosen such that adequate friction of soil-on-soil is provided. A key will be added below the base to prevent failure from the lesser frictional resistance of concrete on soil. The method can be easily adjusted to count on either more or less help from the key. The coefficient of friction given is assumed to be a reasonable lower bound value, not a value greatly reduced to provide a large factor of safety.

In Fig. 7.11*a* the summation of horizontal forces gives the simple equation:

$$H_1 + H_2 - F = 0$$

where F is given by the coefficient of friction μ times the total weight $\Sigma(W + D)$. The given 0.62 coefficient for soil on soil will be used.

$$\begin{aligned}
W_x &= 1.4\,(400 + 100 \times 16.67)x &&= 2890x \\
D_s &= 0.9 \times 0.5(1.0 + 1.25) \times 16.67 \times 150 &&= \qquad 2530 \\
D_{b1} &= 0.9 \times 1.33 \times 150x &&= 180x \\
D_{b2} &= 0.9 \times 1.33 \times 150 \times 4 \text{ (assumed)} &&= \qquad 720
\end{aligned}$$

$$8540 + 3800 - 0.62\,(3070x + 3250) = 0$$

$$\begin{aligned}
3070x + 3250 &= 12{,}340/0.62 = 19{,}800 \\
x &= 16{,}550/3070 = 5.27 \text{ ft*}
\end{aligned}$$

Use heel projecting 5 ft 3 in. beyond B (5.25 ft)

***Arithmetic error. Correct** $x = 5.39$ **ft.**

While the distance x protects against horizontal sliding of the wall on the soil, a soil mechanics check is also needed against sliding on a curved plane such as that indicated in Fig. 7.11*b*. A complete design requires that both kinds of sliding be investigated.

(f) Choice of Toe Length

The total horizontal forces (somewhat larger than those on the stem) tend to overturn the wall and the gravity loads tend to prevent such overturning. The reaction under the base must equal the total downward load, and the necessary location of the resultant reaction can be calculated by moments about some convenient center, say B of Fig. 7.11*a*.

	Load		Arm	Moment abt. B	
W_x	$1.4(400 + 1667)5.25$	15,200	+2.63 ft	+40,000	
D_s	$0.9 \times 1 \times 150 \times 16.67$	2,250	−0.50		− 1,125
	$+0.9 \times 0.5 \times 0.25 \times 150 \times 16.67$	280	−1.08		− 300
D_{b1}	$0.9 \times 1.33 \times 150 \times 5.25$	945	+2.63	+ 2,490	
D_{b2}	$0.9 \times 1.33 \times 150 \times 4$ (est.)	720	−2.00		− 1,440
	$\Sigma\,(W + D)$ =	19,400			
	H_2 = 3,800		−9.0		−34,200
	H_1 = 8,450		−6.0		−50,700
	12,250			+42,500	−87,800
					−45,300
	R (from Σ gravity loads)	−19,400 (up)	y		$-19{,}440y$

$$M_B = -45{,}300 - 19{,}400\,y = 0$$
$$y = -2.42 \text{ ft (to left)}$$

Thus the resultant reaction acts 2.42 ft to the left of B.

The soil cannot support a concentrated reaction. It must be distributed in some fashion over some distance z. It can be asumed that the allowable ultimate soil resistance is developed at the toe before the soil fails. What should be used as the allowable at ultimate is not readily apparent and must be discussed.

In the case of a concentrically loaded footing one senses automatically that the allowable service load pressure should increase at the same rate as the factored loads increase, that is:

$$\frac{\text{Allow. } q \text{ at ultimate loads}}{\text{Equiv. } q \text{ at service loads}} = \frac{1.4D + 1.7L}{D + L} \quad \text{or} \quad \frac{1.4D + 1.7L + 1.7W}{D + L + W}$$

This relation is used in footing design (Chapter 15) to obtain the equivalent soil pressure at ultimate. For the retaining wall stability, the relationship is more difficult because the effect of the H load enters into the solution as a moment. The use of the above relation in terms of moment seems totally impractical. The permitted increase in soil pressure should probably be somewhere between the 1.4 factor used for L and soil weight W_x (in the moment equation) and the 1.7 factor used for H, since the concrete weight (0.9 factor) is relatively small. The best weighting is obscured by the fact that the moments as used are of opposite signs. Here a 1.65 factor will be used,* this being a pure judgement decision. One does not escape this problem by using service loads and allowable soil pressure in the equilibrium equation. He then faces the same decision when he picks an appropriate soil pressure for the check against overturning under overloads.

USE allowable soil p at ultimate $= 1.65 \times 3500 = 5780$ psf

Although the real reaction distribution is uncertain, designers usually assume a straight-line distribution, which in this case means a triangle of pressure as shown in Fig. 7.11*a*.

$$0.5 \times 5780z = \Sigma \text{ gravity loads above} = 19{,}400$$
$$z = 6.75 \text{ ft back from the toe}$$

The resultant vertical reaction R of 19,400 lb acts at $z/3$ from the toe, or 2.25 ft. Thus the toe extends beyond B by $y + z/3 = 2.42 + 2.25 = 4.67$ ft, or beyond the front of wall by $4.67 - 1.25 = 3.42$ ft. The effect of the weight D_{b2} is so small that the weight for the assumed 4-ft length need not be revised for these calculations.

USE toe projection = 3 ft 5 in. (3.42 ft).

USE total base length = 9 ft 11 in.

(g) Service Reaction Pressure Conditions

With the permissible soil pressures as large as 3500 psf (at service loads) settlements are probably not a serious matter. However, where settlement is a serious problem, the soil reaction under service loads should be established. This omits the load factors on all loads. The first step is to locate the resultant reaction, that is, the term y as calculated in Sec. 7.5f but based on service loads. The reaction pressure is usually assumed to vary linearly although the actual pressure will probably follow some curved

*A better solution would be to obtain from the soil mechanics specialist a *safe* ultimate soil pressure, something comparable to ϕ times the ultimate that could be expected 95% of the time. The correlation factors otherwise are simply mathematical equivalents without real physical significance, an equivalent in this particular case only roughly approximated.

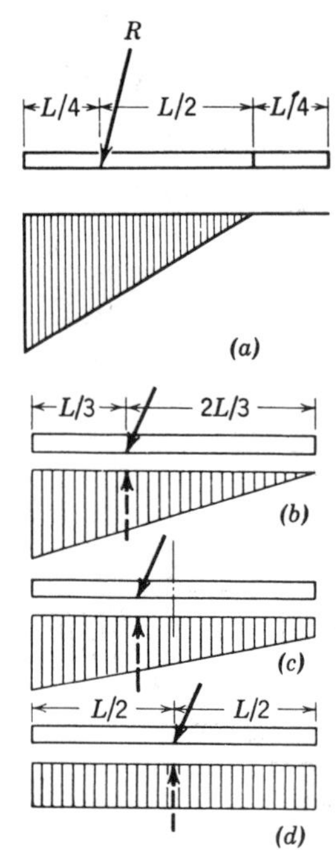

Fig. 7.12. Reaction pressure distributions for various locations of resultant of load.

line. The resulting pressure distribution will necessarily vary with the location of the resultant R. If the resultant is at the third point of the base or nearer the toe, a nominal triangular reaction will be assumed as shown in Fig. 7.12*a* or *b*. If the resultant falls within the middle third of the base as in Fig. 7.12*c*, a trapezoidal pressure results; if at the center as in Fig. 7.12*d*, a uniform pressure results.

The more serious the settlement problem the more attention must be given to the possibility of tilting; obviously, only the last case of uniform reaction pressure eliminates this problem. The position of the resultant is little changed by a change in toe length. Hence lengthening the toe will leave the resultant nearer the middle of the base. The designer can thus increase toe length as much as service load conditions seem to require. A sample of such a reaction calculation follows.

The detail calculation is not shown here, but for vertical service loads without surcharge = 13,410 lb and $y = -0.56$ ft (from B), the stress condition of Fig. 7.13 results, with R at 4.11 ft from toe. The toe and heel pressures can be found from statics (Fig. 7.13). The upward pressure can be considered as two triangles of pressure with individual resultants at their third points. Moments about the smaller resultant (on the right in Fig. 7.13) give:

$$(0.5q_t \times 9.92)3.30 - 13{,}410 \times 2.50 = 0$$
$$q_t = 2050 \text{ psf}$$

Similarly,

$$(0.5q_h \times 9.92)3.30 - 13{,}410 \times 0.80 = 0$$
$$q_h = 655 \text{ psf}$$

The pressures can also be calculated considering the vertical load applied as an eccentric load on a rectangular section 9.92 ft by 1.0 ft. The eccentricity is 0.85 ft from the center of the base. The moment of inertia of this section is $1 \times 9.92^3/12 = 81.5 \text{ ft}^4$.

$$q_t = \frac{P}{A} + \frac{Pec}{I} = \frac{13{,}410}{9.92} + \frac{13{,}410 \times 0.85 \times 4.96}{81.5}$$
$$= 1353 + 695 = 2050 \text{ psf}$$
$$q_h = \frac{P}{A} - \frac{Pec}{I} = 1353 - 695 = 660 \text{ psf}$$

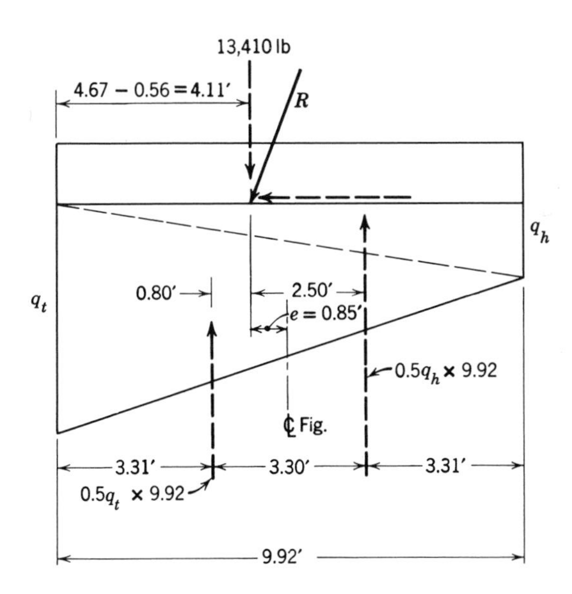

Fig. 7.13. Calculation of soil reaction at service load without surcharge.

If a simple formula is preferred, the above can be expressed as such.

$$c/I = 6/bh^2 = 6/Ah$$

$$p = \frac{P}{A} \pm \frac{Pe\,6}{Ah} = \frac{P}{A}(1 \pm 6e/h)$$

$$= \frac{13{,}410}{9.92}\left(1 \pm \frac{6 \times 0.85}{9.92}\right) = 1351\ (1 \pm 0.515)$$

$$= 2050 \text{ psf or } 655 \text{ psf}$$

(h) Design of Heel

The load on the heel is predominantly a downward load. The supporting upward reaction comes from the tension in the stem steel. Hence the effective cantilever length is 5.25 ft plus the stem cover to center of steel $= 5.25 + 2.56/12 = 5.46$ ft as shown in Fig. 7.14.

The downward load of earth and surcharge totals 3010 plf on the heel cantilever strip. The reaction was calculated in Sec. 7.5f as extending back 6.75 ft from the toe or 2.29 ft into this cantilever length with a triangle of reaction pressure as shown in Fig. 7.14*a*. In the following design this upward pressure has been omitted because (1) there is no assurance that the straight-line pressure really exists* and (2) it reduces the calculated M_u and V_u on the heel, but by such a small percentage that its omission is not really wasteful.

$$M_u = (3010 + 280) \times 5.46^2/2 = 49{,}300 \text{ ft-lb}$$

$$\overline{M} = 49{,}300/0.9 = 54{,}800 \text{ ft-lb} = \bar{k}\,bd^2$$

$$530 \times 12d^2 = 54{,}800 \times 12$$

$$d_m = \sqrt{103.4} = 10.2 \text{ in.}$$

Since shear might control under such heavy loads on a short span, the unit shear equation $v_u = \overline{V}/bd$ will be solved for $d_v = \overline{V}/v_u b$, with the v subscript on d_v added here for identification and the m subscript added to the moment depth for the same reason. Although the critical design section for shear is specified at a distance d from the support, this is not safe when the reaction is a tension "hanger" as it is here. Hence shear will be calculated essentially at the bar support (recognizing that 0.21 ft of the length is protected from vertical load by the stem)

*Note that the 2.39 ft of upward load might largely disappear if the reaction pressure were more parabolic as sketched in Fig. 7.14.

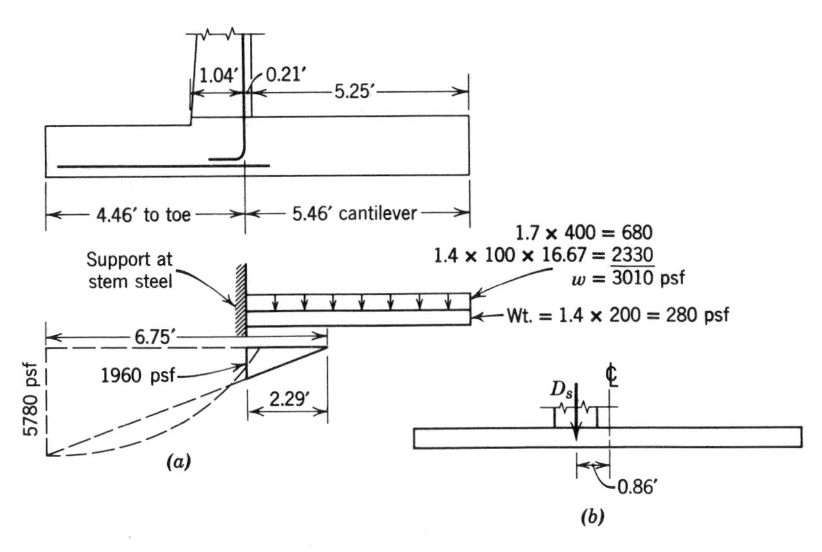

Fig. 7.14. Heel cantilever and loading. (*a*) Maximum. (*b*) Before backfilling.

$$V_u = (3010 + 280)5.25 = 17{,}400 \text{ lb}, \ \overline{V} = 17{,}400/0.85 = 20{,}500 \text{ lb}$$

$$\text{Permissible } v_c = 2\sqrt{f_c'} = 2\sqrt{3000} = 109.8 \text{ psi without stirrups}$$

$$d_v = \overline{V}/v_c b = 20{,}500/(109.8 \times 12) = 15.55 \text{ in.}$$

Since shear does govern, check the more exact permissible v_c with ρ_w assumed as 0.01 (10.2/15.55) = 0.0066 where the parenthesis is the ratio of d_m/d_v.

$$v_c = 1.9\sqrt{f_c'} + 2500(\rho_w V_u d/M_u)$$

$$v_c = 1.9\sqrt{3000} + 2500\frac{0.0066 \times 17{,}400 \times 15.5}{49{,}300 \times 12}$$

$$= 104.0 + 7.5 = 111.5 \text{ psi}$$

(For usual cases, the more exact value differs so little that it is probably more practical just to ignore it.)

$$d_v = 20{,}500/(111.5 \times 12) = 15.35 \text{ in.}$$

$$h = 15.35 + 2.00 \text{ in. cover} + d_b/2 = 17.85 \text{ in. for } d_b = 1 \text{ in.}$$

$$\text{or } 17.73 \text{ in. for } d_b = 0.75 \text{ in.}$$

$$\text{USE } h = 18 \text{ in.}, d = 15.5 \text{ in}$$

This increases the assumed heel weight by 25 × 1.4 = 35 psf, but since it also reduces the earth cover by 2 in. or 17 × 1.4 = 24 psf, it is obvious that no revision in $\overline{M}$ or $\overline{V}$ is really needed.

Since the d selected is considerably more than the required d_m, z exceeds the assumed $0.882d$ value. Assume $z = 0.94d = 0.94 \times 15.5 = 14.6$ in.

$$\text{Reqd. } A_s = \frac{\overline{M}}{f_y z} = \frac{54{,}800 \times 12}{60{,}000 \times 14.6} = 0.752 \text{ in.}^2$$

$$a = A_s f_y / 0.85 f_c' b = 0.752 \times 60/(0.85 \times 3 \times 12) = 1.47 \text{ in.}$$

$$A_s = \frac{54{,}800 \times 12}{60{,}000(15.5 - 0.74)} = 0.740 \text{ in.}^2/\text{ft} = 0.0616 \text{ in.}^2/\text{in.}$$

Since, when shear is large, it is possible for bar development length to limit the choice of bars, this requirement will be investigated before choosing bars. Since these are top bars (more than 12 in. of concrete below the bars) the required development length is 1.4 times the basic value of $0.04A_b f_y \sqrt{f_c'}$, but an 0.8 factor is also probable, since spacing (hopefully) will be 6 in. or more. The available length for development is 5.46 ft less 2 in. of end cover:

$$\text{Available } l_d = 5.46 \times 12 - 2 = 63.5 \text{ in.}$$
$$= (0.04A_b \times 60{,}000/\sqrt{3000})\, 1.4 \times 0.8$$
$$\text{Max. } A_b = 1.29 \text{ in.}^2$$

This A_b permits bars up to #10 (1.27 in.^{2} ea.)

#9 at 16.2 in.	#7 at 9.8 in.
#8 at 12.8	#6 at 7.1

The 16 in. spacing for #9 is a little large, greater than the heel d. The #6 at 7 in. is reasonable, but to reduce the number of bars to handle

USE #8 at 12.5 in. for heel (0.76 in.^{2}/ft)

With *high* walls the possible need for bottom steel in the heel before backfill is placed should be investigated. The factored weight of stem is

$$1.4[(12 + 15)/(2 \times 12)] \times 150 \times 16.67 = 3940 \text{ lb}$$

acting about 4.10 ft from toe and 5.82 ft from heel, that is, 0.86 ft off center as indicated in Fig. 7.14*b*. Using the equation at the close of Sec. 7.5*g*.

$$q = (P/A)\,(1 \pm 6e/h)$$
$$= (3940/9.92)\,(1 \pm 6 \times 0.86/9.92) = 397\,(1 \pm 0.516)$$
$$= 600 \text{ psf at toe and 192 at heel}$$
$$q \text{ at heel side of stem} = 192 + 408 \times 5.25/9.92 = 408 \text{ psf}$$

$$+M_{\text{heel}} \text{ at face of wall} = 0.5 \times 408 \times 5.25^2/3 + 0.5 \times 192 \times 5.25^2 \times 2/3$$
$$= 1870 + 1750 = 3620 \text{ ft-lb}$$
$$\overline{M} = 3620/0.65 = 5560 \text{ ft-lb (plain concrete } \phi\text{, Code 9.2.1.5)}$$

The weight of the base adds reaction pressure but no moment. As an unreinforced beam the effective thickness should neglect the 1 in. next to ground, leaving 18-1 = 17 in., the section modulus = $bt^2/6 = 12 \times 17^2/6 = 578$ in.^{3}

$$f = 5560 \times 12/578 = 116 \text{ psi}$$

Code 15.7.2 permits a flexural stress of $5\phi\sqrt{f_c'}$ with $\phi = 0.65$, to be used with M_u (not $\overline{M}$). Here $\overline{M}$ has been calculated to be consistent with reinforced concrete sections. With $\overline{M}$ the allowable flexural stress becomes $5\sqrt{f_c'} = 273$ psi $\gg 116$. Hence, footing is O.K. for this temporary loading.

(i) Design of Toe

Toe design is similar to that of heel design, except that diagonal tension is critical at a distance d from the face of the wall. Earth fill over the toe has been neglected thus far since it may not always be present. This fill adds also to the toe reaction, but this reaction and fill weight cause little change in M and V on the toe.

The maximum moment on the toe occurs where the shear changes sign. Since the large compression from the stem causes this shear reversal to occur just inside the face of the stem, the case is usually simplified by considering the cantilever support *at* the face of stem (Fig. 7.15).

The toe weight is assumed to be 200 psf, or 280 psf with the 1.4 load factor. The load factor and the footing weight is already included in the reaction pressure. Shear stress is almost sure to control the toe depth. Check d_v at distance d from support, say $d \doteq 16 - 3 - 0.5d_b = 12.5$ in. = 1.04 ft

$$V_{1.04} = 5780 \times 2.38 - 0.5(855 \times 2.38)2.38 - 280 \times 2.38 = 10{,}650 \text{ lb}$$

$$\overline{V}_u = 10{,}650/0.85 = 12{,}550 \text{ lb},$$

$$d_v = \overline{V}_u/v_u b = 12{,}550/(109.6 \times 12) = 9.55 \text{ in.}$$

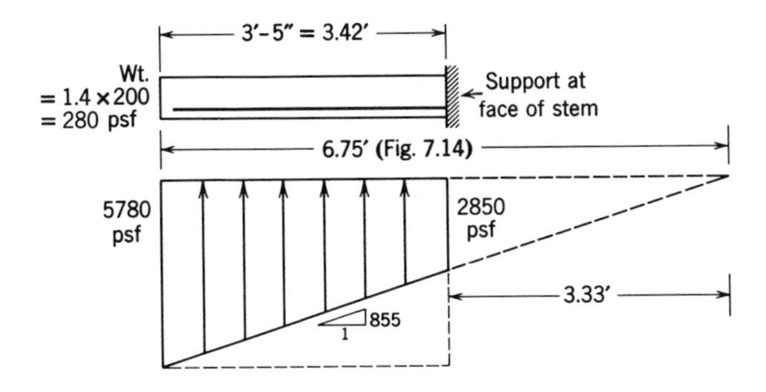

Fig. 7.15. Toe cantilever and loading.

Since this calculation is quite sensitive to variation in the assumed d, try again with assumed $d = 10$ in. $= 0.83$ ft.

$$V_{0.83} = 5780 \times 2.59 - 0.5(855 \times 2.59)2.59 - 280 \times 2.59 = 11{,}320 \text{ lb.}$$
$$\overline{V}_{0.83} = 11{,}320/0.85 = 13{,}330 \text{ lb}$$
$$d_v = 13{,}320/(109.8 \times 12) = 10.10 \text{ in. (Closely enough)}$$
$$h = 10.10 + 3.0 \text{ cover (Art. } 808a) + d_b/2 = 13.60 \text{ in. for } d_b = 1 \text{ in.}$$
$$13.48 \text{ in. for } d_b = 0.75 \text{ in.}$$

Check next for moment. The same 280 psf weight term will be used here because it is an original weight term which is now being taken out of a reaction computed on that basis. Strictly, both weight and reaction change in almost equal amounts, but the calculations based (in effect) on net pressures are ignoring the minor change in gross pressures.

$$M_u = 5780 \times 3.42^2/2 - 0.5(855 \times 3.42)3.42^2/3$$
$$- 280 \times 3.24^2/2 = 26{,}500 \text{ ft-lb}$$
$$\overline{M} = 26{,}500/0.9 = 29{,}400 \text{ ft-lb} \times 12 = \bar{k}\, bd^2$$
$$530 \times 12d^2 = 29{,}400 \times 12, \qquad d_m = \sqrt{55.5} = 7.46 \text{ in.} < d_v$$

Most designers, including the author, normally prefer to use the same toe and heel thickness, but this is not essential. A little thicker toe simplifies the shear key between wall and base; a little thinner toe complicates it slightly, as discussed in the next subsection. In this design, to introduce this complication, while not maintaining the extreme thickness difference that is theoretically possible,

USE $h = 15.0$ in., $\quad d = 11.62$ in. (based on #6 bars)

With d_m increased more than 50%, try $z = 0.95 \times 11.62 = 11.0$ in.

$$\text{Approx. } A_s = \overline{M}/(f_y z) = 29{,}400 \times 12/(60{,}000 \times 11.0)$$
$$= 0.533 \text{ in.}^2/\text{ft} = 0.0443 \text{ in.}^2/\text{in.}$$
$$\text{Min. } A_s = (200/f_y)bd = (200/60{,}000)\, 12 \times 11.62 = 0.465 \text{ in.}^2/\text{ft} < 0.53 \quad \text{O.K.}$$
$$a = 0.53 \times 60{,}000/(0.85 \times 3000 \times 12) = 1.04 \text{ in.}$$
$$z = 11.62 - 0.52 = 11.10 \text{ in.} \quad \text{Say, O.K.}$$

Try #6, spcg. $= 0.44/0.0443 = 10.2$ in.

USE #6 at 10 in. for toe (0.52 in.^{2}/ft)

$$l_d = (0.04 \times 0.44 \times 60{,}000/\sqrt{3000})\, 0.8 = 15.4 \text{ in., say, 16 in.}$$
$$\ll \text{toe length less 2 in. end cover.} \quad \text{O.K.}$$

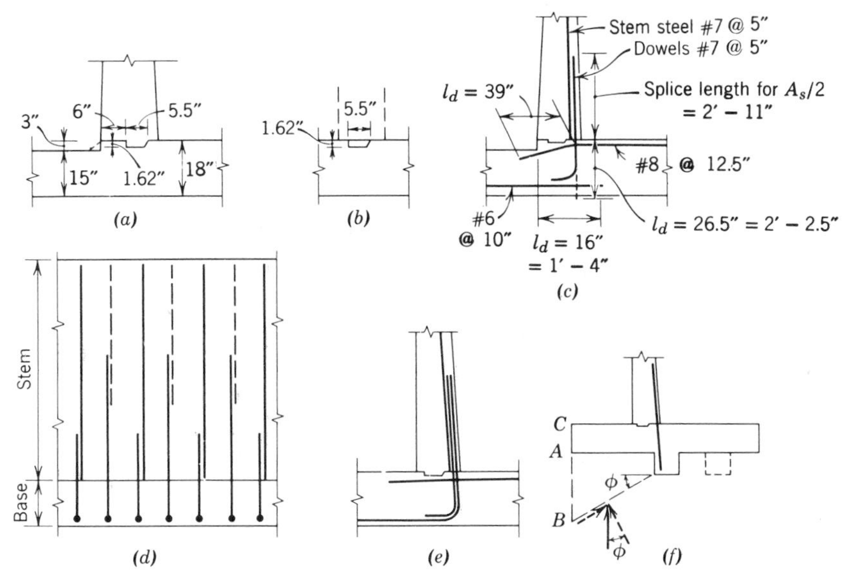

Fig. 7.16. Details at junction of stem and base.

(j) Assembly Problems at Junction of Stem, Heel, and Toe; Revision of Stem

The stem height was based on an assumed 16-in. base thickness, whereas the design above uses 18 in. for the heel and 15 in. for the toe. The way the junction between the three members is handled influences the stem height and stem design. For this reason the *final* stem thickness and stem A_s would normally be left until after the heel and toe designs were completed.

The junction proposed by the author is shown in Fig. 7.16*a* and his reasoning will be clear as the various checks are made.

First, provision is usually made for a shear key between the stem and base. The key is usually made by embedding a beveled 2-in. by 4-in. or 2-in. by 6-in. timber in the top of the footing as shown in Fig. 7.16*b*.* When the timber is removed and the stem is cast in place, the 1.62-in. vertical face provides an allowable bearing (Code 10.14.1) of $12 \times 1.62 (0.70 \times 0.85 \times 3000) = 34{,}700$ lb/ft of wall compared to the approximately 10,840 lb horizontal force on the stem. The distribution of shear force on

*Some designers prefer that the key be cast as a projection on top of the base which extends up into the stem. Some now question the need for a shear key.

the horizontal section of the key at the top of footing is uncertain. If it is taken as parabolic, as for a homogeneous rectangular beam, the equation is

$$V/\phi = (vbh)2/3$$

$$v = 1.5(10{,}840/0.85)/(12 \times 5.5) = 291 \text{ psi}$$

The allowable shear in such a case is also not too definite. It is somewhat similar to the shear permitted between stirrups, which the Code limits to roughly $10\sqrt{f_c'} = 548$ psi. As a practical matter, the bending moment on the stem gives $N_c = N_t = \phi A_s f_y = 0.9 \times 1.41 \times 60{,}000 = 76{,}000$ lb of compression pressing down on top of the base, and friction alone should be adequate to resist the 10,840 lb sliding force. The shear strength in front of the key must also be maintained, especially when the toe is of lesser thickness. Under this compression and with a little more area it seems to present no problem.

If heel and toe were of equal thickness, the cantilever stem height would clearly be measured from the top of the base. In the present case it is less specific. It will be quite safe to calculate stem thickness at the level of the top of the toe and steel area at the level of the heel steel; or one might calculate both at the top of the heel and put in a fillet between stem and toe to deepen the stem, as shown by the dotted line in Fig. 7.16*a*. The first appears the simpler from the construction viewpoint and will be used.

For compression, the height of stem is 18 ft minus 15 in. (or 1.25 ft) for a net of 16.75 ft but the loading is only on the height above the heel which is 18 ft minus 18 in. (or 1.50 ft) for a net of 16.50 ft. The stem thickness was based on 16.67 ft for both, which is close enough, representing a deficiency for maybe 1.0% in moment in an underreinforced member.

For tension, the height of stem is 18 ft minus the heel d of 15.5 in., which is 16.71 ft, nearly the same 16.67 ft used originally. No change in the stem cross section is necessary.

Since the stem steel cannot be readily supported in position while the base concrete is placed, it is customary to provide dowels (or stub bars) in the base equal to the stem steel, #9 at 8 in. These must project a full anchorage length into the base and a full splice length (Sec. 5.17) above the base, as shown in Fig. 7.16*c*. The anchorage length into the base, for a spacing $\gtrapprox$ 6 in., is

$$l_d = 0.8[0.04 \times 1.00 \times 60{,}000/\sqrt{3000}] = 0.8 \times 43.8 = 35.1 \text{ in.}^*$$

*l_d could be taken directly from Table D.3, Appendix D.

This length cannot be obtained vertically below the stem even if bars were extended into a reasonable depth key below the base. The bar could be curved into the toe, but more than half the l_d length would be beyond the bend, which the author considers undesirable. A bend can possibly be made more effective than a standard hook, but there should be some limitation on anchorage around a bend. For a #9 the limitation of 29,600 psi on a standard hook (Code 12.8.1) is not unreasonable. Anchorage length can be reduced by using more dowels of lower strength steel or by using smaller bars. For example, if #7 bars at 5.0 in. are used,

$$l_d = 0.04 \times 0.60 \times 60{,}000/\sqrt{3000} = 26.3 \text{ in.}$$

This dowel length into base is entirely satisfactory and will be used. The designer has choices here. He can use #7 and 5.0 in. for both stem steel and dowels or use alternate #6 and #7 dowels for each #9 bar of stem steel, spacing the dowel bars at 4 in. on centers to match the 8-in. spacing of #9 bars, which increases the dowel area to 1.56 in.^{2}/ft. The #9 bar would control the lap required with mixed sizes. USE #7 at 5.0 in. for stem and dowel bars

$$d = 12.56 \text{ in.}, \qquad A_s = 1.44 \text{ in.}^2/\text{ft}, \qquad l_d = 26.3 \text{ in.}$$

The splice lap in the stem must be $1.3l_d$ if not over half of the bars are spliced at one level, as in Fig. 7.16*d*, but $1.7l_d$ if all bars are spliced at one point.

$$\text{USE staggered splices (min. lap} = 1.3 \times 26.3 = 34.2 \text{ in.)}$$
$$\text{with 35-in. lap}$$

The possiblity of eliminating the splice on half the bars by extending half the dowels as high in the stem as they are needed for moment will be checked later.

The #8 bars in the heel are top bars (more than 12 in. of concrete below them) and are spaced more than 6 in. apart.

$$l_d = 1.4 \times 0.8(0.04 \times 0.79 \times 60{,}000/\sqrt{3000}) = 38.7 \text{ in.}$$

Anchor the #8 bars 39 in., as shown in Fig. 7.16*c*. The compression from the stem across part of the length merits some recognition, but what numerically this should be is unknown.

The bars in the toe must be anchored 16 in.

It should be obvious, as a matter of detailing, that some of the stem dowel steel might be replaced by toe steel bent up as shown in Fig. 7.16*e*. This would probably save some steel and certainly would tie the toe well to the wall. Figures 7.20 and 7.21 show such a design. The detailing then is facilitated by using the same spacings for toe and stem steel.

(k) Key Against Sliding

Resistance against sliding as provided in Sec. 7.5*e* assumed a key into the soil to take the difference between the coefficient of friction of 0.62 and 0.50, or $0.12\ \Sigma(W + D) = 0.12 \times 19{,}480 = 2340$ lb. This requires only a limited amount of depth into the soil, and 10 in. or 12 in. appears adequate, as sketched in Fig. 7.16*f*. The key is usually placed sufficiently below the stem in order to use it for anchorage of the stem dowels, but it may be more effective when somewhat more to the rear. The maximum effect of such a shift cannot be more than to mobilize the passive resistance of the soil over the depth AB in Fig. 7.15*f*, where ϕ is the friction angle for soil on soil. The key should be designed as a bracket and the details will not be shown here.

(l) Temperature and Shrinkage Steel

Longitudinal bars are used to space the moment bars and to provide for shrinkage and temperature stresses; both horizontal and vertical bars are needed on exposed faces. The wall should be placed in short lengths, not to exceed 20 to 30 feet, to reduce shrinkage stresses.

The Code contains the following provisions for walls, not relating particularly to retaining walls:

10.16.2. The minimum ratio of vertical reinforcement area to gross concrete area shall be:

(a) 0.0012 for deformed bars not larger than #5 and with a specified yield strength of 60,000 psi or greater, or
(b) 0.0015 for other deformed bars, or
(c) 0.0012 for welded wire fabric not larger than $\frac{5}{8}$ in. in diameter.

10.16.3. Vertical reinforcement shall be spaced not farther than three times the wall thickness nor 18 in.

10.16.5. The minimum ratio of horizontal reinforcement to gross concrete area shall be:

(a) 0.0020 for deformed bars not larger than #5 and with a specified yield strength of 60,000 psi or greater, or
(b) 0.025 for other deformed bars, or
(c) 0.0020 for welded wire fabric not larger than $\frac{5}{8}$ in. in diameter.

10.16.6. Horizontal reinforcement shall be spaced not farther apart than one and one-half times the wall thickness nor 18 in.

For Grade 60 bars #5 or smaller, this leads to areas on *each* face:

	Vertical, $0.0006bh$	Horizontal, $0.0010bh$
b = 12 in., h = 12 in.	0.086 in.²/ft	0.144 in.²/ft
15 in.	0.108	0.180
18 in.	0.130	0.216

In the exposed face of the retaining wall stem the author would prefer a little more than this and possibly less on the side against the earth. In the base that will be covered, no bars are needed on the compression faces of heel and toe, and the main function of the longitudinal bars is as spacers. The following are chosen:

Stem: exposed face, horizontal, upper 8 ft #5 at 18 in. = 0.21 in.²/ft
below 8 ft #5 at 12 in. = 0.31 in.²/ft
exposed face, vertical, #5 at 18 in. = 0.21 in.²/ft
rear face, horizontal, #5 at 18 in. = 0.21 in.²/ft
Toe: longitudinal spacers in bottom, #4 at 18 in. = 0.13 in.²/ft
Heel: longitudinal spacers in top, #5 at 18 in. = 0.21 in.²/ft

Temperature and shrinkage steel is shown in Fig. 7.17, the exposed face steel being in the foreground, as can be seen most clearly at the bottom of the illustration, adjacent to the form panel which is being lifted into place.

Fig. 7.17. Reinforcing in wall stem with the front form being lifted into place. Frequently the rear or front face form is first erected and the reinforcement is next placed. (Courtesy Texas Highway Department.)

(m) Drainage

Since walls are usually not designed against water pressure, the design must provide complete drainage of backfill by French drains, drains through the wall, or the like. A porous backfill, such as gravel, should be provided directly behind every wall to allow the water to reach these drains.

7.6 ARRANGEMENT OF STEM STEEL—STOPPING BARS

Not all of the stem steel must run full height, because the moment decreases rapidly above the base. The simple method of Sec. 3.10c must be modified here to take care of the variable depth. For the design of Sec. 7.5 the upper sketches in Fig. 7.18 show the effective depth d and the soil pressures at various depths. From these the moment and steel requirements have been determined and are tabulated in Table 7.1. For M_u the ϕ value has been taken as 0.884 as in Sec. 7.5c; it should increase to 0.9 at the top and it would be adequate in this case to use 0.9 all the way.

TABLE 7.1 Calculations for Required A_s Curve

Distance y from Top, ft	M_u, Bending Moment, ft-lb	$\overline{M}_u$	Slab Depth d, in.	Required A_s, in.^{2}/ft
0	0	0	9.56	0
5	3,740	4,230	10.46	0.09
9	14,950	16,930	11.18	0.34
12	30,500	34,500	11.72	0.65
15	53,450	60,500	12,26	1,09
16.67	70,000	79,400	12.56	1.43

These data are the basis for the plotted curve in Fig. 7.18, which is to be used for a graphical solution. A dotted line in Fig. 7.18 is plotted the (variable) distance d higher to show how far the bars really must extend.

In order to avoid some of the splicing of steel just above the base, half of the dowel bars have been made in two heights that are adequate to take care of their portion of the moment: one-fourth to be 4 ft 11 in. high and the other one-fourth 2 ft 11 in. (for simplicity the same as the normal splice lap length of Sec. 7.5j). Note that the stopping of tension bars at their *minimum* length leaves the remaining steel at maximum stress and requires that none of these remaining bars be stopped closer than l_d (=2.21 ft in

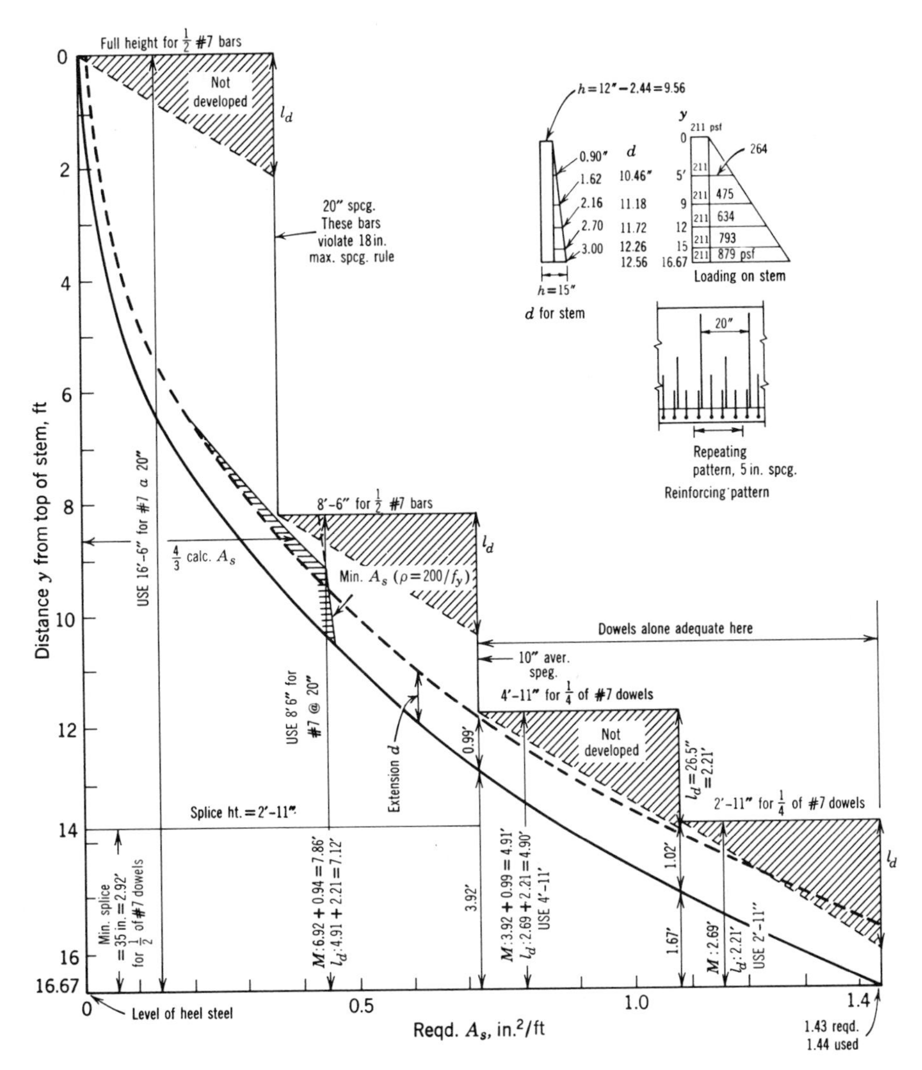

Fig. 7.18. Stopping of stem steel, based on required A, curve.

Sec. 7.5j) to the next lower cutoff point.* This l_d requirement is indicated at three places, but happens not to be governing. The insert at the right of Fig. 7.18 shows the schematic arrangement.

*For the three longer groups of bars one might logically use the 0.8 factor on l_d (for bars spaced more than 6 in. on center), but this would not be logical for the shortest bars that parallel the highly stressed splice lengths.

In the 7 ft to 10.5 ft depth range minimum Code limits for flexural A_s ($\rho = 200/f_y$) actually govern the required reinforcement, although not by a very large amount. A nearly vertical line with shading (at about the 10 ft depth) represents this minimum, but this rule is relaxed when a considerable excess of A_s is used, beyond that required for moment (Code 10.5.1). Four-thirds of the calculated A_s requirement may be substituted, which is indicated by the curved shaded line running to the left of the other limit.

In cutting off tension reinforcement, a further condition must be checked. Code Sec. 12.1.6 forbids a bar cutoff in a tension zone unless one of three

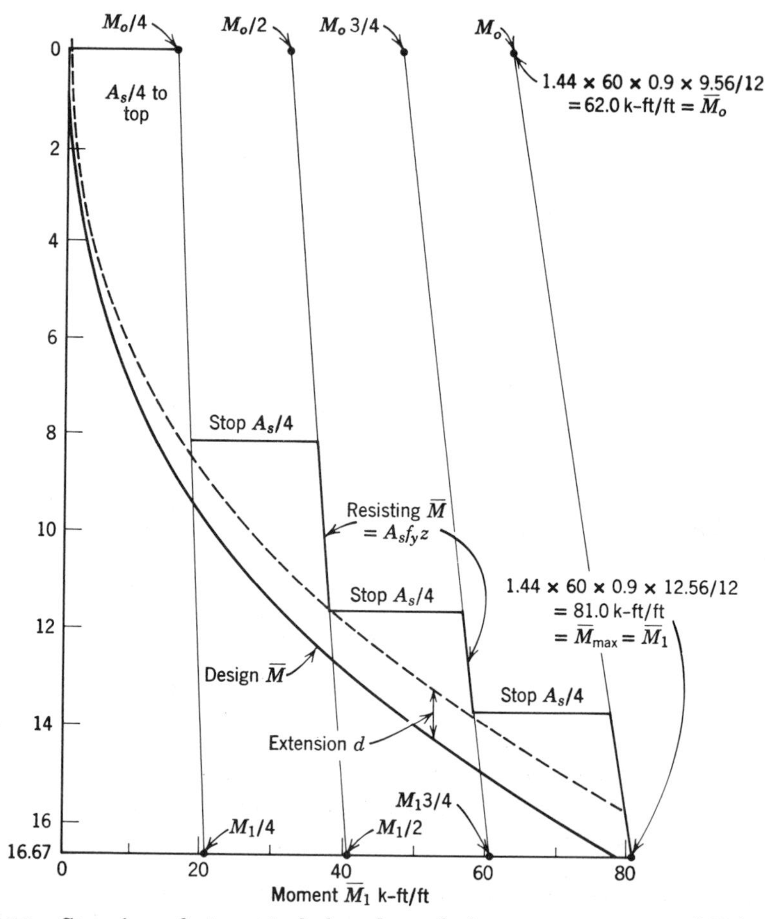

Fig. 7.19. Stopping of stem steel, based on design moment curve. (Minimum A_s limits also to be considered.)

enumerated conditions is met. Usually in slabs (other than two-way slabs) the shear is less than two-thirds the allowable, and this will satisfy the first of these conditions. The stem shear at the top of the base was only 85.5 psi, 78% of the allowable. A rough check indicates that the shear at the minimum cutoff length of 2 ft 11 in. is less than the permissible two-thirds of the allowable and thus permits the cutoff without any special provisions.

An alternate procedure for moment length which is preferred by many is to plot design moment and resisting moments as in Fig. 7.19. At the base, allowable moments are proportional to A_s (neglecting change in z) and at top, they are reduced in the ratio of reduced effective depth. The l_d lengths and minimum A_s (not shown in the graph) must be considered here as in Fig. 7.18 and the final bar lengths should be identical by the two procedures.

A design layout often leading to economy in reinforcing is based on turning up the toe steel as dowels for matching stem bars as indicated in Fig. 7.20. The toe bars must extend a splice length into the stem which, in the case of the design of Sec. 7.5, for #6 toe steel making up less than one-half of the stem steel requirements, should be:

$$\text{Lap} = 1.3\, l_d = 1.3(0.04 \times 0.44 \times 60{,}000/\sqrt{3000}) = 25.0 \text{ in.}$$

The deficit in the stem steel, after matching the #6 bars, is provided by other bars, preferably at a compatible spacing, as indicated in Fig. 7.21*b*.

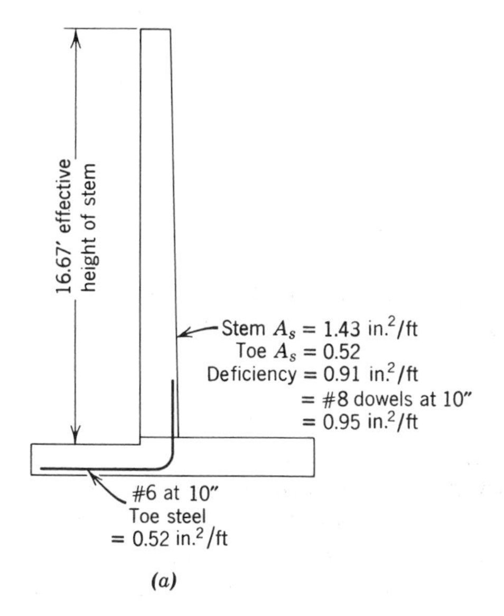

Fig. 7.20. Basis of revised steel layout to utilize toe steel as dowels for stem steel.

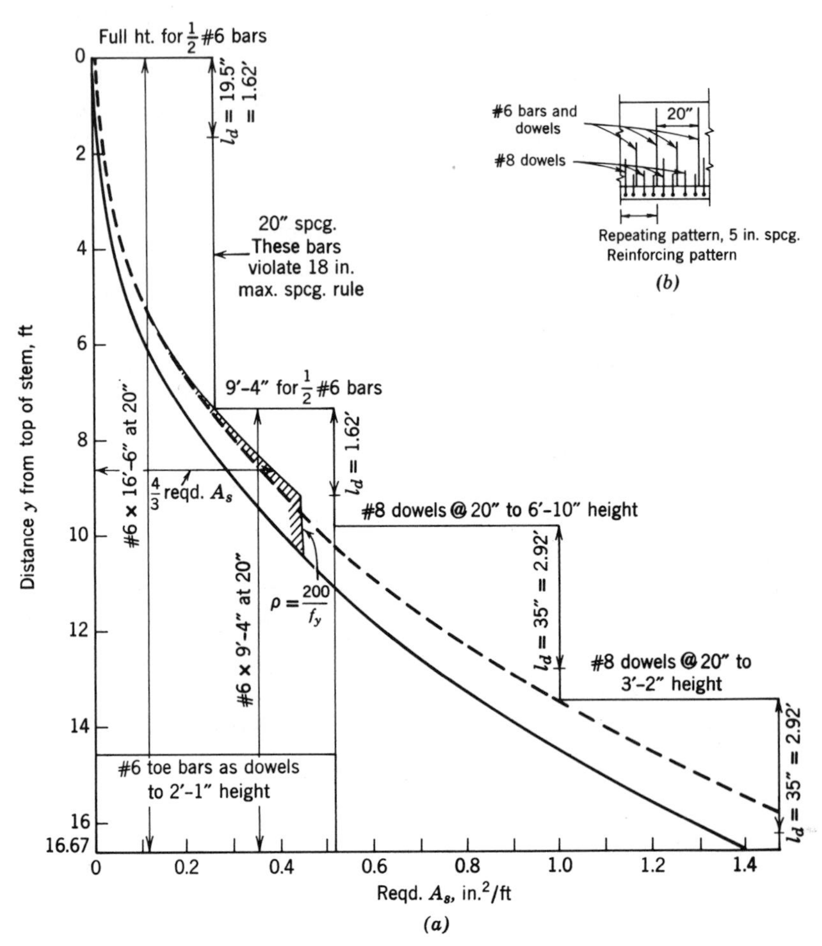

Fig. 7.21. Details of reinforcement for layout based on Fig. 7.20.

These bars can usually be short enough that they can be bent to form their own anchorage in the footing. This extension into the footing must be at least the l_d for #8 bars in this case.

$$\#8\ l_d = 0.04 \times 0.79 \times 60{,}000/\sqrt{3000} = 34.6 \text{ in.}$$

The 0.8 factor for spacing of 6 in. or more cannot be used here because the #6 toe bars must be considered part of the same group of bars performing the same function. The 35-in. length is longer than desirable in this base thickness, since more than one-half the anchorage must be beyond the bend.* Even though the 5% extra area helps a little, this is not as good a

*Even with an increased radius of bend this is of uncertain efficiency.

design in this particular case as that with the #7 bars used in the earlier layout.

SELECTED REFERENCES

1. Whitney Clark Huntington, *Earth Pressures and Retaining Walls*, John Wiley and Sons, New York, 1957.
2. Karl Terzaghi and Ralph B. Peck, *Soil Mechanics in Engineering Practice*, John Wiley and Sons, New York, 2nd ed., 1968.
3. *Standard Specifications for Highway Bridge*, AASHO, Washington, 10th ed., 1969.
4. "Retaining Walls and Abutments," *AREA Manual*, Vol. 1, Chapter 8, Part 5, Chicago, 1953.
5. B. K. Hough, *Basic Soils Engineering*, Ronald Press, New York, 1957.
6. Wayne C. Teng, *Foundation Design*, Prentice-Hall, Englewood Cliffs, New Jersey, 1962.

PROBLEMS

Prob. 7.1. (*a*) Investigate under service load conditions the stability, sliding resistance, and foundation soil pressure of the wall of Fig. 7.22. Surcharge = 200 psf, soil weight = 100 pcf, K_a = 0.30, concrete weight

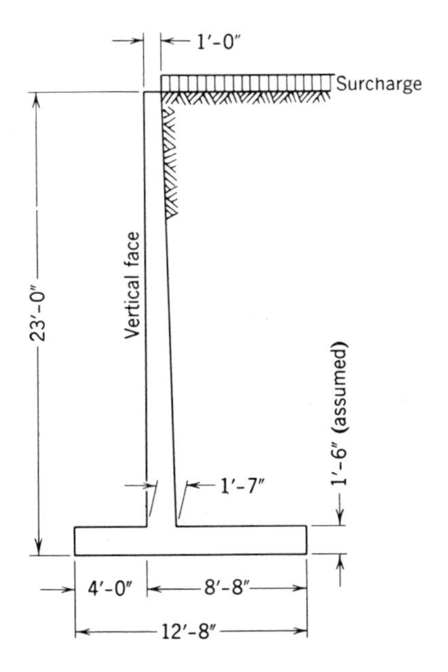

Fig. 7.22. Retaining wall for Probs. 7.1 and 7.2.

(including reinforcing) = 150 pcf, coefficient of friction of concrete on soil = 0.40, allowable soil pressure = 3000 psf.
(*b*) Same as (*a*) except use load factors of Sec. 7.5 and a permissible ultimate soil pressure of 6000 psf.

Prob. 7.2. Design the stem, heel, and toe for the wall of Fig. 7.22 assuming all the foundation soil pressures and sliding resistances found in Prob. 7.1 are used as satisfactory. Use factored loading.

Prob. 7.3. Redesign the retaining wall of Sec. 7.5 for horizontal pressure based on $K_a = 0.27$, $f_c' = 4000$ psi, allowable foundation soil pressure under factored loading = 8000 psf, and a *minimum* base length.

(*a*) Stem design.
(*b*) Minimum base length.
(*c*) Heel design.
(*d*) Toe design with thickness based on stresses, but not less than thickness of heel.
(*e*) Sketch the detail of the joint where the stem, heel, and toe come together.

Prob. 7.4. Redesign the retaining wall of Sec. 7.5 changing only the following data from those originally used:

Omit surcharge.
Limit foundation soil pressure under factored loading to 8000 psf.
Use minimum length base.
Coefficient of friction of concrete on soil = 0.40, of soil on soil = 0.55.

Prob. 7.5. Redesign the retaining wall of Sec. 7.5 on the assumption that the service load foundation soil pressure must not exceed 4000 psf, all other data remaining unchanged.

Prob. 7.6. Assume that the retaining wall of Sec. 7.5 is to be made without any toe projection, that is, as an L-shaped wall. What length of heel will be necessary to keep the service load resultant at the third point of the base? What will then be the maximum foundation pressure (ignoring the change in heel thickness which will probably be necessary): (a) At service load? (b) With load factors?

Prob. 7.7. Would it be feasible to design the retaining wall of Sec. 7.5 without any heel slab, that is, as an L-shaped wall? (Two problem aspects.)

Prob. 7.8. It is possible in Fig. 7.18 to stop all the #7 bars at the same height and splice them with #5 bars which continue upward. Where can such a splice be made and what is the length of all bars to be used, assuming half of the #5 bars are stopped as soon as the stresses permit?

Prob. 7.9. It is possible in the steel arrangement of Fig. 7.18 to stop the longest #7 bars somewhat shorter than full height and splice on smaller bars, say #5 bars. Sketch the bar arrangement and specify the bar lengths. Does this save steel?

Prob. 7.10. Assume that the #7 bars at 10-in. average spacing shown in the stem in Fig. 7.18 are to be made into three equal groups and cut off at three different levels, instead of two as shown. Sketch the reinforcing pattern and establish the length of each group of bars. (For this problem overlook the wide top spacing which results.)

8

Continuous Beams and One-Way Slabs

8.1 TYPES OF CONSTRUCTION

Most reinforced concrete members are statically indeterminate because they are parts of monolithic structures. However, it would convey the wrong impression to ignore the many forms of precast concrete construction which are taking a significant part of the present market. These vary from statically determinate floor channels, double-T sections, and other shapes to systems where precast units are combined with cast-in-place girders or a cast-in-place slab to form essentially monolithic construction or composite construction. Often the precast members are also prestressed.

The most usual form of building construction consists of a slab cast monolithically with a beam-and-girder floor framing which carries the floor load to the columns. The plan view of such a floor in Fig. 8.1*a* indicates that such slabs are supported on all four sides and should properly be designed with two-way steel by the methods of Chapter 10 (probably by Sec. 10.22).

When the slab is more than twice as long as it is wide, it is usually designed as a one-way slab continuous over the beams, but with special negative moment steel added across the girders as required by Code 8.7.5.*

*Because they find the one-way steel arrangement simpler, many engineers actually design one-way slabs for all panels except those that are nearly square.

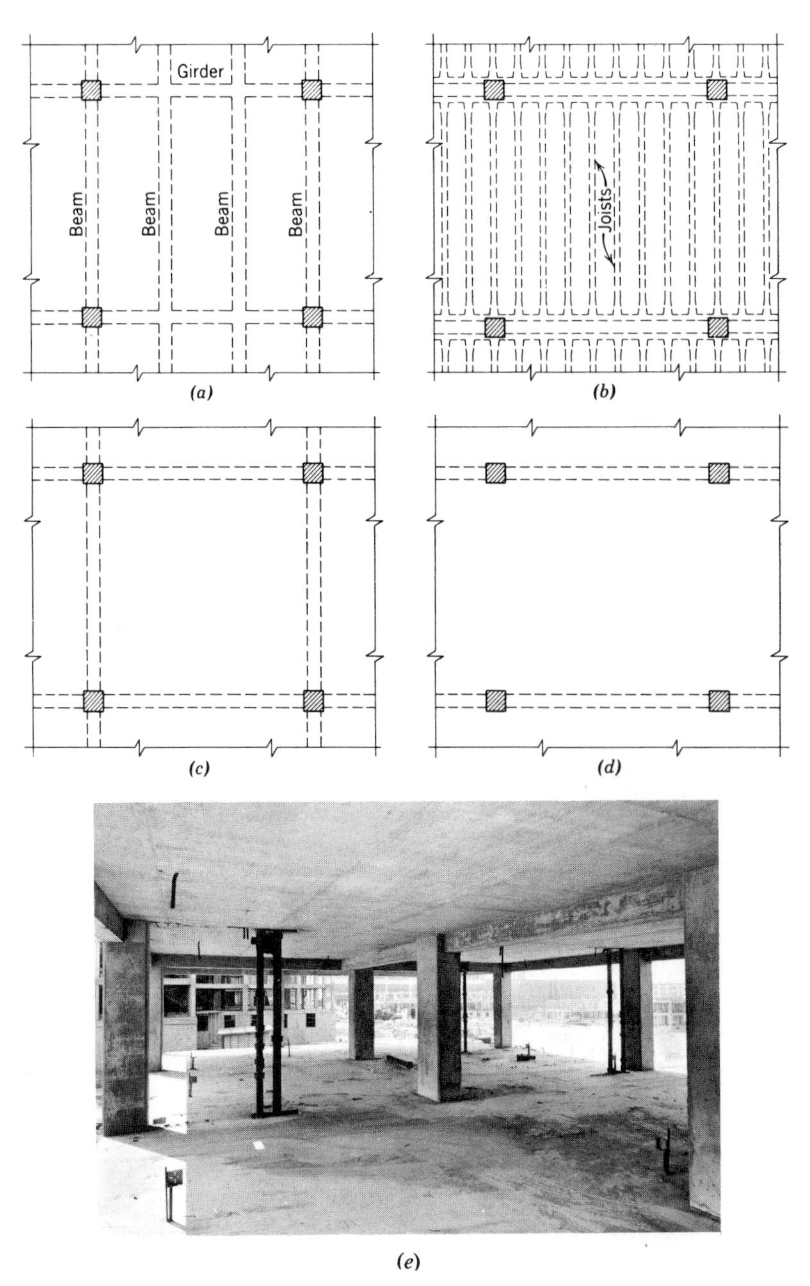

Fig. 8.1. Typical floor beam framing. (*a*) Beam-and-girder layout. (*b*) Joist construction. (*c*) Two-way slab and beams. (*d*) One-way slab and beams. (*e*) One-way construction. (Courtesy Portland Cement Assn.)

When such a slab is designed for a uniform load the supporting beams are also designed for a uniform load,* this process in effect ignoring the portion of the slab load which goes directly to the girder from the end of the panel. For the beams, the uniform load assumption is on the safe side since it overestimates the beam load. For the girders, the corresponding assumption is to place the beam reactions on the girder as concentrated loads and to add the uniform load situated directly over the girder. This assumption is not on the safe side for the girder design, particularly for maximum shear. More realistic tributary loading areas are shown in Fig. 10.20*b*. The student might investigate the difference in maximum girder shear between the two loadings for a girder simply supported.

Sometimes, for long spans, closely spaced beams with a very thin slab are used. This so-called joist construction (Fig. 8.1.*b*) is facilitated by the availability of removable pans which are used as forms between the joints.

When no beams are used except those between columns, as in Fig. 8.1*c*. the slabs are definitely supported on all four sides and both beams and slabs should be designed as discussed in Chapter 10.

In light construction the beams at times are run in only one direction, as shown in Fig. 8.1*d* and *e*. In this case true one-way slabs result, except at the beams supporting the walls at the ends. The slab band construction of Sec. 10.15 is really this same type of construction utilizing shallow beams, often quite wide.†

Two slab-type floors with beams only at outside walls and around large openings are economical, with the flat plate floor used for light loads and the flat slab floor almost always used for very heavy loads. The (patented) lift slab construction is a flat plate floor without any beams, which is cast at grade and provided with special collars around the columns for lifting into place. Flat slabs and flat plates are presented in detail in Chapter 10, including mention of the two-way joist slab commonly referred to as the waffle slab (Fig. 10.6*a*).

8.2 INTERACTION BETWEEN PARTS OF THE STRUCTURE

In all these various types of construction the designer is faced with a highly indeterminate type of structure, a structure in three dimensions which cannot be precisely analyzed as a planar structure.[1–3] More specifically, the intermediate beams of Fig. 8.1*a* cannot be analyzed exactly

*As computers come more into the design process, such oversimplifications are less often used.

†The spandrel beams on the outside are often omitted with light live loads.

without considering the vertical deflection and torsional stiffness of the girders as well as the stiffness of the columns. Likewise, the beams framing into the columns have moments which are influenced by the column joint rotations and hence by any torsion present in the girders.* The Code (8.5.3.1) permits "any reasonable assumptions for computing relative flexural and torsional stiffnesses," but the assumptions must be "consistent throughout the analysis." This may include ignoring torsion in many situations, but Chapter 10 points out the importance of torsion in designing two-way slabs and flat slabs and plates.

A designer commonly utilizes approximate methods of analysis, but he becomes a better designer when he understands the nature of his approximations and how exact or how crude these may be. This chapter assumes that the reader is familiar with ordinary moment distribution procedures (Appendix B) and the plotting of continuous member moment and shear diagrams (Sec. A.5 in Appendix A).

The ordinary conventional assumption is that analysis in two dimensions is adequate for most designs. This assuption fits in better with structures designed for uniform floor loads than for those with moving concentrated or wheel loads. It may be reasonably assumed that, in carrying a uniformly distributed load, any one beam gets little help from its neighbor, because this neighboring beam is probably also fully loaded.† Likewise, in a one-way slab each strip may be assumed to carry the load directly above it when the entire slab is loaded.

When moving concentrated loads are involved, each slab strip may carry a different moment and the load on a single beam may depend greatly upon the stiffness of the slab and the adjoining beams. The problem of moving concentrated loads is discussed in Chapter 13.

Moment diagrams for continuous beams normally show negative moments over supports and positive moments near mid-span, as indicated in Fig. 8.2*a*. The presence of columns changes the picture only slightly, typically by making the moment at opposite faces of the column slightly different. The moment diagram in any span may be considered as the sum of two parts, one the simple beam moment diagram for that span, and the

*Torque stiffness drops sharply after first diagonal cracking occurs. See Sec. 4.21 discussion.

†This statement is not strictly true, of course. In Fig. 8.1*a*, with equal spacing of beams, the beam framing directly into the column is stiffer because its end joints are stiffer, whereas the girder provides only a yielding support for the neighboring beam. The beam framing into the column thus tends to carry the greater load and to some extent relieves the intermediate beam. It might be noted, however, that at the ultimate or collapse load, for beams of equal size, each beam would probably be carrying almost exactly the same load, since the load distribution changes after some yielding takes place.

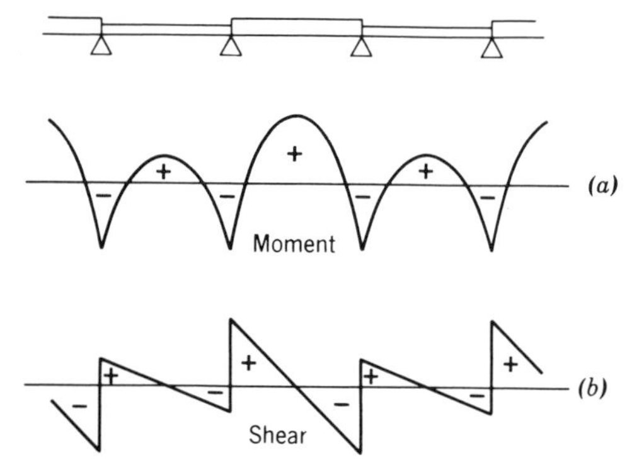

Fig. 8.2. Typical M and V diagrams for a continuous beam.

other the moments across the span due to the negative support moments, as outlined in Sec. A.5. Since these negative moments at supports are influenced by loads on *any* span, it follows that the moment at each point is influenced by every load on every span. Loading patterns for maximum moment are thus a major consideration.

The typical shear diagram (Fig. 8.2*b*) differs only slightly from that for a series of simple spans. In any span the shear diagram for a given loading consists of the simple beam shear plus a constant correction which will be designated as the continuity shear V_c. The continuity shear is usually relatively small except in end spans.

8.3 THE GENERAL DESIGN PROBLEM FOR CONTINUOUS MEMBERS

Each span of a continuous beam or slab requires a separate design for negative and positive moment conditions, even though practice in the United States usually uses a constant depth member for any given span, often for all spans. Figure 8.3 shows the design conditions for continuous slabs, rectangular beams, and T-beams, in diagrammatic fashion. To indicate clearly that at this stage no consideration is being given to the arrangement of steel, the reinforcement is indicated only in the vicinity of maximum moment zones. Moment at A (on the left) is assumed to be the governing (maximum) negative moment in all cases. The letter t designates the tension face.

In the thinner slabs compression steel is not desirable because it would lie so close to the neutral axis that a slight displacement would make it

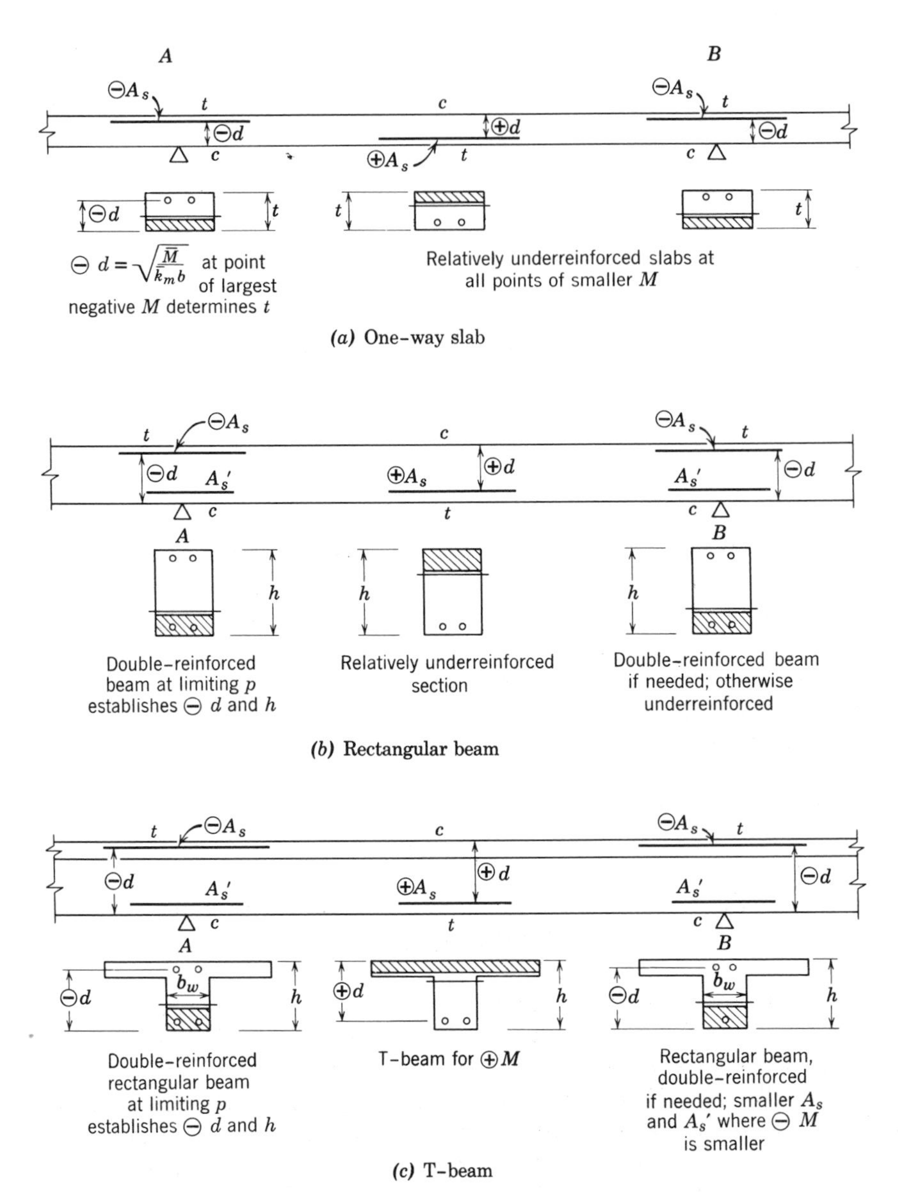

Fig. 8.3. Strength design procedures for continuous slabs and beams.

ineffective. Hence Fig. 8.3*a* shows the maximum negative moment at A and a limiting percentage of steel determining the slab depth, with relatively underreinforced sections elsewhere. Thick slabs can be designed with compression steel, similar to rectangular beams, if desired, but this does not reduce slab cost.

Rectangular beams may be designed with compression steel at supports, as in Fig. 8.3*b*, or without it. In the latter case the design is similar to that shown for slabs.

T-beams become, in effect, inverted rectangular beams at the supports with only the web width b_w effective in compression, as indicated in Fig. 8.3*c*. This restricted compression zone is improved by using a double-reinforced section; and such is recommended for ordinary use.

Shear (diagonal tension) rarely governs in ordinary one-way slab design, in spite of the fact that the allowable v_c is only $2\sqrt{f_c'}$. Stirrups in slabs would be awkward even if needed. The Code permits them now, with $v_c = 2\sqrt{f_c'}$ a change from the 1963 Code which counted them not at all in slabs thinner than 10 in. Continuous rectangular beams and T-beams normally require stirrups, but the web size for the continuous T-beams very rarely will be governed by shear instead of moment and only occasionally by the size needed for placing the bars.

8.4 LOADING PATTERNS FOR MAXIMUM MOMENTS

(a) Maximum Positive Moment

Elastic theory is specified for calculation of design moments except that some limit design modification is permitted (Chapter 9).

Influence lines might be used to determine the loading arrangement, but the critical patterns for an elastic analysis are easy to deduce from a single load-deflection sketch, such as that in Fig. 8.4*a* in which deflections are greatly exaggerated. The usual carry-over moment idea leads directly to the moment diagram in Fig. 8.4*b*. It is observed that this loading produces positive moment near mid-span in all even-numbered spans. One concludes that if all even-numbered spans were loaded, each such load would increase the positive moments in the other loaded spans. This loading pattern is usually stated as follows:

> For maximum positive moment near the middle of a span, load that span and alternate spans on each side, as shown in Fig. 8.5*a*.

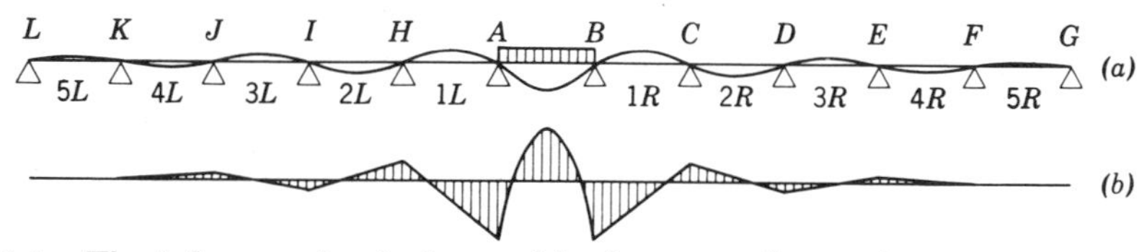

Fig. 8.4. The influence of a single panel load on a continuous beam.

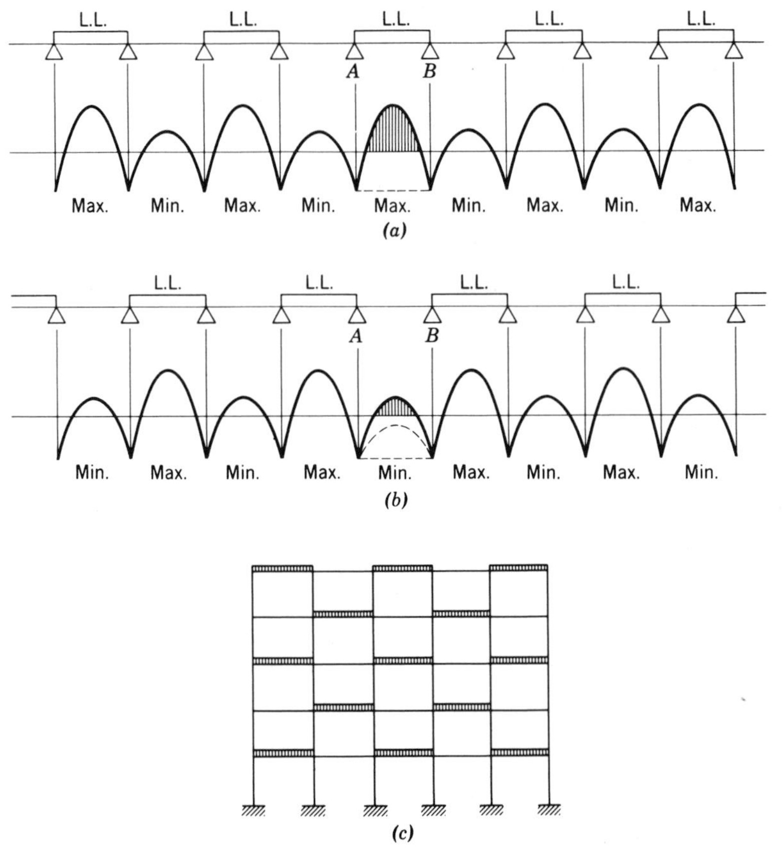

Fig. 8.5. Loading for (*a*) maximum positive moment and (*b*) minimum positive moment at mid-span. (*c*) Checkerboard loading for multiple stories.

Load factors are applied to all loads. The dead load with its load factor is always considered to act* and is included on the moment diagrams shown. This one loading arrangement gives the maximum positive moments on all the loaded spans. The maximum positive moment on all other spans is given by loading only all those spans, as in Fig. 8.5*b*, taking off the live loads shown in Fig. 8.5*a*. Thus all maximum positive moments on all spans are determined by two load arrangements and two moment distributions. Extended to a multistory frame, these patterns become checkerboard arrangements of loading (Fig. 8.5*c*), still with only two patterns needed for the complete analysis.

*Except that when it offsets the effect of wind or earthquake loading, the load factor is reduced.

(b) Minimum Positive Moment (or Maximum Negative Moment) Near Mid-span

Since the minimum positive moment is simply the opposite extreme of loading from that for maximum positive moment on beam AB, all the live loads in Fig. 8.5*a* must be removed and live loads with their load factor must be placed on the other spans as in Fig. 8.5*b*. It should be noted that the two loadings already discussed for maximum positive moment, Fig. 8.5*a* and *b*, give all the necessary minimum positive moments or maximum negative moments near mid-span. The loading criterion is:

> Omit live load on span considered; load adjacent spans and alternate spans beyond.

With smaller dead load moment, the mid-span moment in AB could be negative, as indicated by the dashed curve in Fig. 8.5*b*, especially since the load factor for dead load is smaller than that for live load.

(c) Maximum Negative Moment at Left Support

The single panel loading of Fig. 8.4. shows that negative M is produced at

B by loading adjacent span to left
D by loading third span to left
F by loading fifth span to left
A by loading adjacent span to right
I by loading third span to right
K by loading fifth span to right

This can be summarized by the criterion:

> For maximum negative moment at a given support, load adjacent spans on each side and alternate spans beyond.

Load factors must be applied to the loads. Applied to maximum moment at the left support A, this criterion results in the loading and moment diagram shown in Fig. 8.6*a*. Only the resulting moments adjacent to A are significant, since no others are either maximum or minimum moments. These moments are critical at the face of support, as will be discussed in Sec. 8.4g.

Adjustment of these moments from elastic analysis in the direction of limit design is considered in Chapter 9.

(d) Maximum Negative Moment at Right Support

The same criterion applies to this maximum moment as in the preceding case. The loadings are placed on spans adjacent to B, and alternate spans

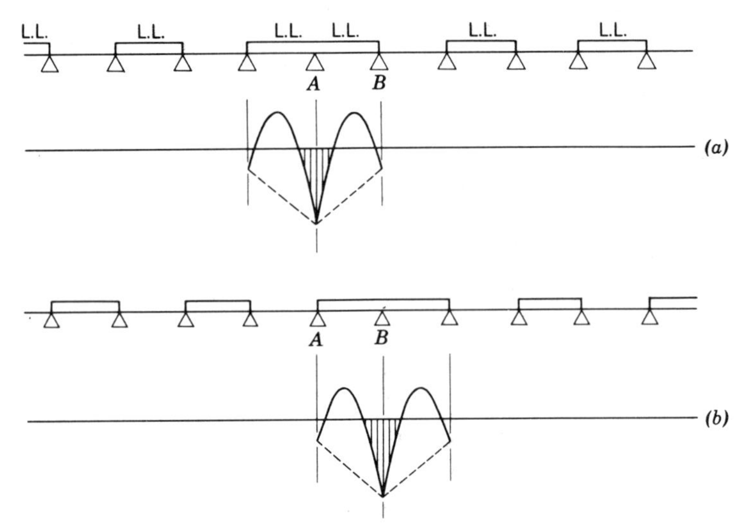

Fig. 8.6. Loadings for maximum negative moment: (*a*) at left support A; (*b*) at right support B.

beyond, as shown in Fig. 8.6*b*. Unfortunately, the loading gives maximums only near the one support at B.

(e) Partial Span Loadings

It should be noted that both maximum positive and maximum negative moment conditions call for full load on the span under consideration. Maximum negative moment occurs with a very unsymmetricl moment diagram in the span involved. On the other hand, maximum positive moment results in a moment diagram nearly symmetrical about the middle of the span. Minimum positive moment at mid-span (or maximum negative there when dead load is small) also gives a moment diagram nearly symmetrical, in this case a moment diagram based on dead load alone on the span in question.

In building frame design it is not customary to deal with partial span loads since they do not increase the principal design moments. In highway structures partial span loads would have a considerable influence on detailing the steel. Partial span loads are mentioned briefly in Sec. 8.6 in connection with maximum moment diagrams.

Some codes specify a single moving load to be placed anywhere on the structure as an alternate loading to care for special cases.

(f) Permissible Simplifications

The ACI Code encourages the use of a reduced or simplified frame in the analysis of buildings (Code 8.5.1.1) and specifies that moments be calculated by elastic analysis. One floor may be analyzed at a time with the far ends of columns taken as fixed. In calculating maximum negative moments the live load may be applied on only the two adjacent spans. This permits the use of simplified moment distribution procedures such as the two-cycle procedure given in Ref. 1. These methods, in the author's opinion, are quite reasonable for ordinary structures, except that they give beam and column moments at the exterior columns which tend to be considerably too small. The load factors, on both dead and live load, must not be overlooked for any design.

(g) Moment at Face of Support

Moment distribution generally implies moments calculated on the basis of spans taken center to center of supports. Since such calculations treat the support reaction as though it were concentrated at a point, the resulting moment diagram within the support width is entirely imaginary. This is of no concern since both theory and tests show that the critical moment is at the *face* of the support (Code 8.5.2.2).

Many engineers establish the design moment at the face of support from the calculated moment at the center line of support, simply by deducting $Va/2$, where V is the shear and a is the column width, as illustrated in Fig. 8.7*a*. This is almost the same as scaling the moment value from the moment diagram, because the small length of uniform load within the column width (Fig. 8.7*b*) would cause little change in moment.

The author prefers the more conservative correction recommended by the old Joint Committee Specification, namely, a reduction taken as $Va/3$. The reasoning behind this value is as follows. The support stiffens the end of the beam much as would a haunch. A calculation considering this increased end stiffness would lead to an increased negative moment at the center of the column, as indicated by M_1' in Fig. 8.7*c*. Instead of calculating M_1' (which might be roughly $Va/6$ larger than M_1), an approximate equivalent is obtained by applying a smaller correction, $Va/3$, to the original calculated M_1 value. The design moment is then $M_1 - Va/3$, as in Fig. 8.7*c*.

The extra stiffness at the support also leads to smaller positive moments. However, the correction would be only about $Va/6$. Most engineers consider this correction less certain than that to the face of the column. They

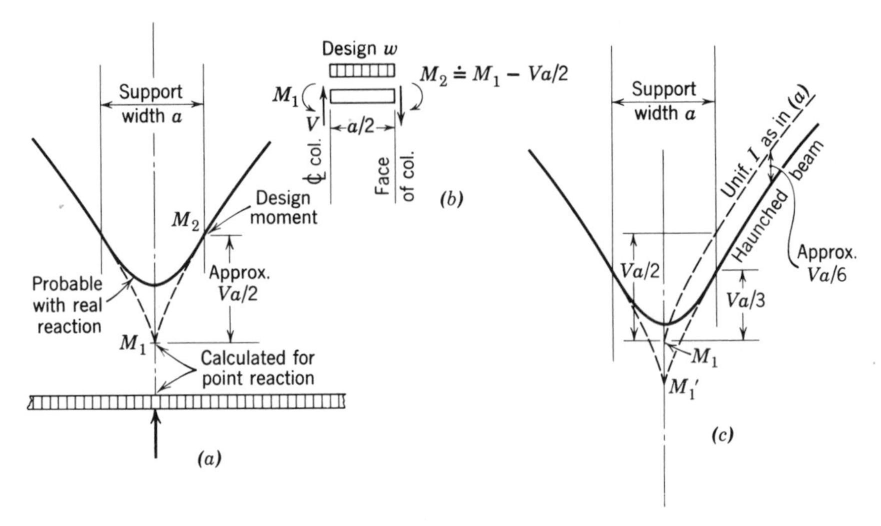

Fig. 8.7. Moment at face of support. (*a*) Without correction for increased stiffness at support. (*b*) Free-body diagram establishing procedure in (*a*). (*c*) Recommended procedure recognizing increased stiffness at support.

recognize the lesser accuracy of positive moment calculations, always sensitive to relative column stiffness values, and simply use the original positive moments without any correction.

(h) Validity of Elastic Analysis in Strength Design

Philosophically, the use of elastic theory for moments and the use of the very nonelastic methods of section analysis are not consistent. This leads design thinking toward some kind of inelastic frame analysis as discussed in Chapter 9. Code 8.6 permits some redistribution in this direction. Although the elastic theory may seem incompatible with strength design, it errs only by giving moments that are too large and hence too safe. This chapter is based wholly on moments obtained by the elastic theory.

8.5 MOMENT COEFFICIENTS

Any given moment can be expressed as a moment coefficient times wl_n^2, where w is the total design load (including load factors) and l_n is the clear span. The maximum moment coefficients will be largest when the ratio of live load to dead load is large and when the column or other joint restraint is relatively small. Negative moment coefficients may also be large when adjacent spans are longer or more heavily loaded than the span in question.

Based on uniform live loads not greater than three times the dead load and on span lengths "approximately equal (the larger of two adjacent spans not exceeding the shorter by more than 20%)," the Code committee has established by analysis certain reasonable moment coefficients to use for maximum moment calculations. Code 8.4.2 tabulates these coefficients and Fig. 8.8 presents them in diagrammatic fashion.

The same Code section lists two shear values:

In end members at face of first interior support	$1.15wl_n/2$
At face of all other supports	$wl_n/2$

When the ratio of dead to live load is particularly large, some economy can be achieved by calculating the moments by more accurate methods. Also some economy can be achieved by less approximate analysis and the redistribution which is discussed in Chapter 9.

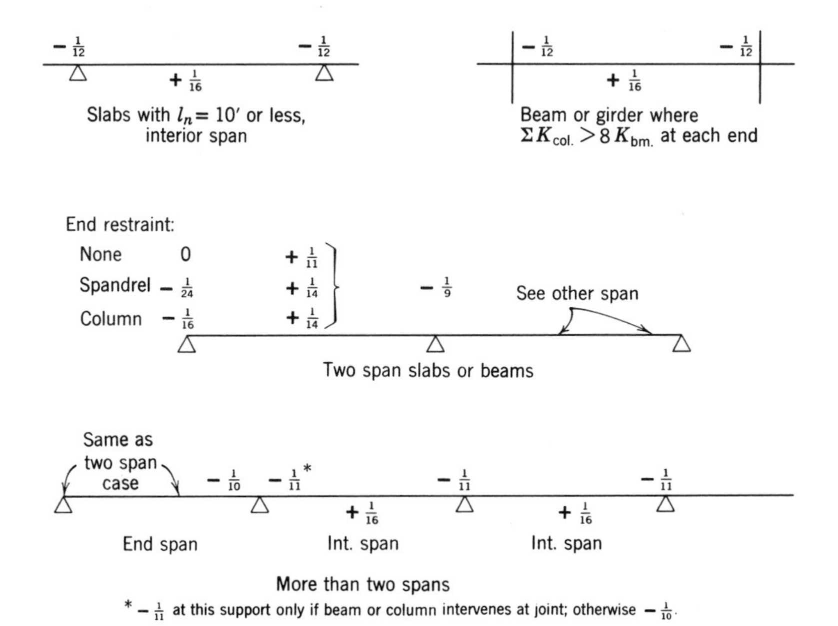

Fig. 8.8. ACI Code (8.4.2) moment coefficients for nearly equal spans and live load less than three times the dead load. $M = (\text{coef.})\ wl_n^2$.

The use of moment coefficients should be restricted to rather standard conditions unless one uses general tables such as those in the Appendix of the ACI *Reinforced Concrete Design Handbook.*[4]

8.6 MAXIMUM MOMENT DIAGRAMS

To determine the best arrangement of the reinforcing it is necessary for the designer to have in mind a clear picture of the extreme range of moments all along the beam or slab. The major part of such a diagram can be assembled from the several moment diagrams for critical maximum moments already illustrated in Figs. 8.5, 8.6, and 8.7. For a typical interior span with equal spans and equal live loads, these moment curves are drawn to larger scale in Fig. 8.9*a–d* and grouped together in Fig. 8.9*e*.

Recent frame studies have pointed out a moment condition not shown in Fig. 8.9 and one that is easy to overlook. When the floor system is light and the lower-story columns (rather than a shear wall) resist high wind or earthquake shears, the sum of the moments from the column above and the column below add at the joint and rotate it and the attached beams. This increases the negative moment in one of the beams enough to make it the weakest member of the system. Hence the designer should be alert to combinations of live load with wind or earthquake to produce the governing *beam* negative moment in spite of the appropriate reduced load factors.

By loadings exactly opposite to those for maximum negative moment, some small positive moment can often be obtained over supports as suggested by the dashed lines. Some partial span loadings may also increase the negative moments slightly, as indicated by dashed lines, but these are not very significant changes.

A very significant fact about the maximum moment diagrams is that, over a considerable portion of the beam, the moment may change from positive to negative, or the reverse, as the loading in adjacent spans is varied. Hence the designer must provide tension steel in both top and bottom over this zone.

Definite locations for some of the inflection points on Fig. 8.9 are calculated in connection with the bending of bars in Fig. 8.11 and in Sec. 8.14.

8.7 MAXIMUM SHEARS

The loading for maximum shear at a support is the same as for maximum negative moment there. Hence the end shear can always exceed the simple beam shear by an amount equal to the continuity shear V_c. On

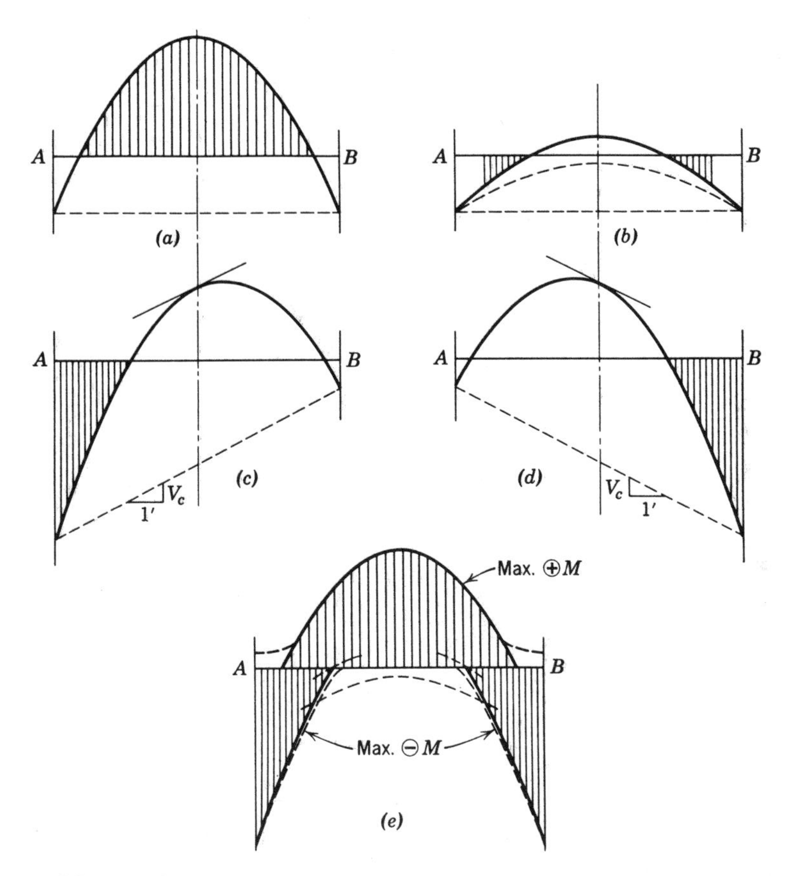

Fig. 8.9. Moment diagrams for maximum moment. (*a*) Positive moment at mid-span. (*b*) Negative moment near mid-span; the dotted curve would show with a relatively large live load or long adjacent span. (*c*) Negative moment at left support. (*d*) Negative moment at right support. (*e*) Composite maximum moment diagrams.

interior spans the value of V_c will be in the order of 3 to 12% of the simple beam shear, with 8% a fair average value. On end spans it can run as large as 20% or more of the simple beam shear. This large V_c on the end span is additive only on the half beam adjacent to the first interior support. For nearly equal spans, the Code (8.4.2) suggests 15% extra shear in end members at the first interior support and no addition elsewhere. The author prefers to make a nominal addition in interior spans as well as the substantial addition in the end span.

8.8 DESIGN OF CONTINUOUS ONE-WAY SLAB

Design a continuous one-way slab supported on beams at 12 ft 0 in. on centers, ACI Code moment coefficients (Sec. 8.5 above), dead load of 22 psf (plus slab weight), live load of 220 psf, $f_c' = 3000$ psi, Grade 60 steel with ρ limited to $0.18f_c'/f_y$. Assume the beam stem is 12 in. wide.

No particular merit is associated with ρ of $0.18f_c'/f_y$ except that (1) design constants have already been developed in Fig. 2.8, and (2) anything near this value will be more economical than the use of the maximum ρ from Table 2.1. The maximum ρ is uneconomical for slabs or beams and will frequently lead to thin sections with deflection problems; it should be avoided except where conditions do demand an extreme design.

Solution

The moment coefficients shown in Fig. 8.8 indicate that the negative moment at the first interior support is the maximum, at $0.10wl_n^2$. The slab at this point will be designed for the arbitrary limiting steel ratio, which leads to $\bar{k} = 0.161f_c'$ (Fig. 2.8) $= 483$. Slab weight will be based on assumed h of 6 in.

$w_l = 220 \times 1.7$ (L.F.) $= 374$ psf

$w_d = 22 \times 1.4$ (L.F.) $= 31$

Slab wt. $= 75 \times 1.4 =$ ~~105~~ (estimate) 87 (revised)

Total $w =$ ~~510~~ psf 492

$l_n = 12.0 - 1.0 = 11.0$-ft clear span

$M_u = 0.10 \times 510 \times 11^2 = 6170$ ft-lb/ft width of slab

$\bar{M} = M_u/\phi = 6170/0.9 = 6880$ ft-lb/ft

$\bar{k}bd^2 = 483 \times 12d^2 = 6880 \times 12$

$d = \sqrt{14.2} = 3.77$

Code 7.14.1.1 specifies 0.75-in. clear cover, if not exposed to weather.

$h = d + d_b/2 + 0.75 = 3.77 + 0.25 + 0.75 = 4.77$ in. for #4 bars

This leads to $h = 5$ in.; it could be 4.5 in. only if a very gross overestimate of weight had been included. The orginal 75 psf assumption corresponded to $h = 6$ in.

Try $h = 5$ in., wt $= \frac{5}{12}(150) = 63$ psf $\times$ 1.4 L.F. $= 87$ psf. This indicates a decrease of 18 psf, which is 3.5% of the total load, and causes a 3.5% change in moment and about $3.5/2 = 1.7\%$ change in required d, since d varies as $\sqrt{\bar{M}}$. (If 4.5 in. had been tried, the change in h could not have been as much as the 5.6% required to change 4.77 to 4.5 in.)

$$\begin{aligned}
\overline{M} &= -0.10 \times 492 \times 11^2/0.9 = 6600 \text{ ft-lb/ft} \\
d &= \sqrt{6600 \times 12/(483 \times 12)} = \sqrt{13.62} = 3.69 \\
h &= 3.69 + 0.25 + 0.75 = 4.69 \text{ in., say 5 in. as estimated} \\
\text{USE } h &= 5 \text{ in., } d = 5 - 0.25 - 0.75 = 4.00 \text{ in. for \#4 bars}
\end{aligned}$$

Code 9.5 gives a deflection warning (Table 2.2, Sec. 2.8) if $h < l/28 = 11 \times 12/28 = 4.72$ in. $<$ 5in. O.K. This design cannot use the savings from the limit design idea of Code 8.6 because approximate moment coefficients have been used.

All other sections have less moment and hence all steel can be designed assuming z slightly greater than the $0.894d$ from Fig. 2.8, say $z = 0.90d$. (For very small moments it may be desirable to check a and z for a more economical value.) Noting that A_s and $\overline{M}$ are each per foot of slab width,

$$A_s = \frac{\overline{M}}{f_y z} = \frac{(\overline{M} \text{ in ft-lb})12}{60{,}000 \times 0.9 \times 4.00} = \frac{\overline{M} \text{ in ft-lb}}{18{,}000}$$

It is convenient to tabulate A_s per inch of slab width if the bar spacing is to be determined without the use of tables.

$$A_s/\text{in.} = \frac{\overline{M} \text{ in ft-lb}}{18{,}000 \times 12} = \frac{\overline{M} \text{ in ft-lb}}{216{,}000}$$

where $\overline{M} = \text{coef.}\ (492 \times 11^2/0.9) = 66{,}000 \times \text{coef.}$ The calculations are tabulated in Table 8.1. Code 7.13 for Grade 60 bars requires $\rho = 0.0018$ for shrinkage and temperature, as a minimum. (For beams Code 10.5 makes the minimum $\rho = 200/f_y$ which is $\rho = 0.0033$ for Grade 60 bars.) The required areas and the spacing of bars are given in Table 8.1 as though top and bottom steel were to be totally separate as they are in Fig. 8.10*a*.

For the layout of Fig. 8.10*a* the requirements are well matched by the bars specified at the bottom of the table.

Some designers prefer to use some bent-up bars, feeling that this positions the top steel more exactly. The arrangement of Fig. 8.10*b* is then the most common bending pattern. The top bars are actually all in one layer and

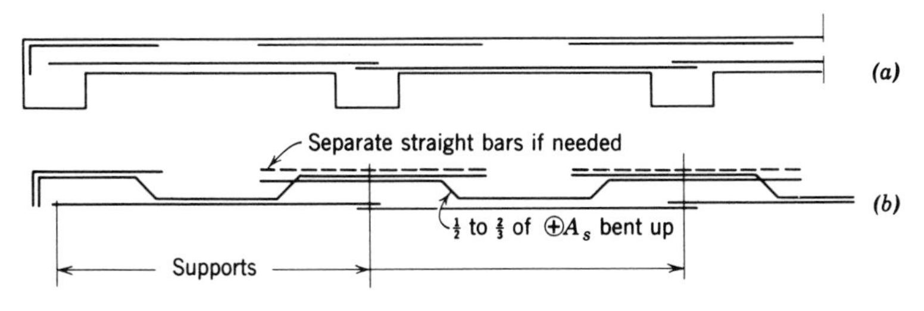

Fig. 8.10. Arrangement of slab steel. (*a*) Straight bars alone. (*b*) Some bars bent.

TABLE 8.1. Calculation of Slab Steel

	Exterior Span		First Int.	Typical Interior	
	Ext. End	Middle	Sup.	Middle	Support
M coef.	$-\frac{1}{24}$	$+\frac{1}{14}$	$-\frac{1}{10}$	$+\frac{1}{16}$	$-\frac{1}{11}$
$\overline{M}$ = coef. × 66,000	−2750 ft-lb/ft	+4720	−6600	+4130	−6000
A_s/ft = $\overline{M}$/18,000	0.153 in.^{2}/ft	0.262	0.367	0.230	0.333
A_s/in. = $\overline{M}$/216,000	0.0127 in.^{2}/in.	0.0218	0.0305	0.0191	0.0277
Min + A_s = 0.0018bh	–	0.009 in.^{2}/in.	—	0.009	—
Spcg. #3 bars	8.65 in.	5.04	3.60	5.78	3.98
Spcg. #4 bars	15.7	9.16	6.56	10.46	7.23
USE #3 at:	8.5 in.	5	—	5.5	—
Plus #4 at:	—	—	6.5 in.	—	7

the bottom bars in one layer, but they are sketched separately to indicate the patterns more clearly. Bending up half the bars provides negative moment reinforcement equal to the average positive moment reinforcement in adjacent spans and added straight bars can make up the deficit. This requires #3 at 12 in. over the first interior support and #3 at 14 in. at all typical interior supports. In this particular case it would be simpler not to bend up at the outer end; otherwise add #3 at 15 in.

The bend points and stop points for bars in slabs involve almost exactly the same considerations as for any other continuous member. These considerations are discussed in considerable detail in Secs. 8.13 through 8.16 in connection with the design of a continuous T-beam.

The student should note that temperature steel is required parallel to the beams, at least in the amount specified in Code 7.13. The temperature bars are used as spacers for both top and bottom steel, tied to the underside of top bars and to the top of bottom bars. In addition, bar supports or chairs should be provided to hold the steel at proper levels.

Slab shear stresses could easily have been computed, but they do not control on one-way slabs of ordinary span.

The design of rectangular beams will be discussed in Sec. 8.20.

8.9 CONTINUOUS T-BEAM DESIGN—GENERAL

The principles of design for a continuous T-beam can be discussed most easily in terms of a numerical example. Much of the remainder of this chapter (through Sec. 8.20) is devoted to various aspects of such design in terms of a typical interior span as follows:

> A typical interior panel of a continuous T-beam of 20-ft clear span with 15-in. square columns is to be designed to carry the slab designed in Sec. 8.8. Beams are 12 ft 0 in. on centers, $h_f = 5$ in., $w_l = 220$ psf, $w_d = 84$ psf (including slab weight but excluding stem weight), $f_c' = 3000$ psi, Grade 60 bars, and moment coefficients calculated by moment distribution, when expressed in terms of total load and clear span (wl_n^2), are as follows: -0.091, $+0.072$, and for minimum positive moment at midspan, -0.010.

There are many different choices of b_w and d for the web of a T-beam, any one of which might be a good design for specific conditions. A choice of d might be on the basis of stiffness against deflection, after the fashion of Table 2.2, especially for long spans. This would set $l/21$ as a probable minimum overall depth. The depth may be chosen to match some other member which is more critical in strength or stiffness or it may be related to depths needed for ducts. This member, for example, might well match the depth required for the exterior span with its larger moments and shears, if of the same span.

It would seem logical to choose depth on the basis of economy, remembering that A_s decreases with depth, whereas concrete and formwork are more expensive as depth increases. However, the economy of the structural frame may rarely agree with over-all building economy since greater depths for beams mean heavier beams (usually) and hence heavier columns and footings, higher exterior walls with expensive finish, more steps in a story height and hence larger stair wells, higher elevator lifts, etc. The stem width may on occasion be chosen to fit into a wall.

One can almost say that the choice of web b_w and d is an arbitrary choice. Four primary conditions must be satisfied. (1) The capacity of the inverted rectangular section which carries negative moment (Fig. 8.3*c*) must be adequate for flexure with a "reasonable amount" of compression steel. (2) The shear capacity must be adequate, but such a wide range can be carried by stirrups that this provision is rarely restrictive. (3) There must be width for the required number of reinforcing bars, but the possibility of using higher strength steels, of bundling the bars, and of spreading the negative moment steel into the flange makes this much less restrictive than formerly. (4) There must be enough stiffness to keep deflections within proper limits, but the use of compression steel can stiffen considerably a shallow beam (Code 9.5.2.3).

In this example, deflection is assumed not to be critical on a heavily loaded 20-ft span.* Depth will be considered on the basis of some arbitrary requirements for negative moment. The given moment coefficients indicate that positive moment steel will be approximately 75 to 80% of the negative moment tension steel, since z will not be greatly different for positive and negative moment. If half of the positive moment steel is bent up in the fashion indicated in Fig. 8.13*b*, the other half will continue in the bottom part of the beam into the column and thereby almost automatically provide compression steel in the amount $A_s' = 0.5 \times 0.8A_s = 0.4A_s$, where A_s is the negative moment tension steel. By lapping this compression steel from two adjacent spans as in Fig. 8.17*b*, A_s' could become approximately 0.8 A_s.

A new provision of the 1971 Code is also pertinent here. A long-standing rule in continuous beam design has been to run, at least, one-fourth the positive moment steel into the support. Now it is further specified (12.2.2) that, if the beam is part of a frame resisting lateral load, this one-fourth be anchored such that its full f_y in tension be available at the face of support. (The objective is greater ductility or energy absorption in case of moment reversal from unexpected loadings such as explosion, earthquake, unusual settlement, etc.) This reinforcement is obviously the minimum automatically available for A_s' at face of support.

The designer thus has considerable freedom in selecting the approximate A_s' to use, and estimates such as the first above can only be approximate at this stage. (For example, the positive moment steel may work out to be five bars, which means either 40 or 60% bent up, instead of half.) Here, for the first step, it is assumed that deflection probably does not control and that A_s' will be $0.4A_s$. The possible range of $\bar{k}$ values for moment will be explored.

For $f_c' = 3000$ psi, $f_y = 60{,}000$ psi, Table 2.1 shows $\bar{k} = 783$, $\rho = 0.0161$, and $a/d = 0.378$. For the present ignore the complication about not counting A_s' at full value (Sec. 2.14 and 3.5*a*) and assume that $A_{s2} = A_s'$.

$$A_{s2} = A_s' = 0.4\ A_s = 0.4(A_{s1} + A_{s2}) = 0.4(0.0161bd + A_s')$$
$$A_s' - 0.4A_s' = 0.00644bd, \qquad A_s' = 0.0107bd$$

With $d - d'$ estimated as $0.88d$,

$$\bar{M} = 783bd^2 + 0.0107bd \times 60{,}000 \times 0.88d$$
$$= 783\ bd^2 + 565bd^2 = 1348bd^2$$

*The deflection calculation of Sec. 6.5c indicates a possible need for more consideration of deflection here. If the beam supports or is attached to partitions, the resulting deflection is barely satisfactory.

Thus bd^2 can be reduced by, at least, 50% with this small A_s', if one desires. The resultant small d would certainly not be economical if A_s' ran full length of a member, but here the need is only at extreme ends of the span. This small section might also get more into deflection problems or, with its larger A_s, into bar spacing problems.

This short analysis confirms the earlier statement that the choice of web can be quite arbitrary, with experience and judgment quite helpful in making the choice. On this basis, and recalling that minimum size is rarely the best solution, the author arbitrarily picks an intermediate value of $\bar{k} = 1050$ (near the middle of the range) for his starting point. This may prove either a wise or unwise decision; the choice can only be evaluated after the design is further advanced.

8.10 CONTINUOUS T-BEAM DESIGN—NEGATIVE MOMENT SECTION

Size will first be established for moment as a double-reinforced rectangular beam.

$$LL = 220 \times 12 \times 1.7 = 4500 \text{ plf}$$
$$\text{Slab} + DL = 84 \times 12 \times 1.4 = 1410$$
$$5910$$
$$\text{Estimated stem wt.} = 200 \times 1.4 = \cancel{280}\ 207$$
$$w = \cancel{6190}\ 6120 \text{ plf}$$

$$M_u = -0.091 \times 6190 \times 20^2 = 225{,}000 \text{ ft-lb},$$
$$\overline{M} = 225{,}000/0.9 = 250{,}000 \text{ ft-lb}$$

This moment could be reduced by the redistribution provisions of Code 8.6 and Chapter 9, but will first be used as would be necessary when only approximate moments are available.

$$\text{Reqd. } b'd^2 = \overline{M}/\bar{k} = 250{,}000 \times 12/1050 = 2850$$
$$\text{If } b' = 11.5 \text{ in.},\quad d = \sqrt{2850/11.5} = \sqrt{248} = 15.8 \text{ in.}$$
$$b' = 9.4 \text{ in.},\quad d = \sqrt{300} = 17.32 \text{ in.}$$

The latter is nearer the usual d/b ratio of 1.5 to 2.5, although the narrow width may give a tight detailing situation. Add 1.5-in. cover + 0.5-in. stirrup + 0.5-in. $(= d_b/2) = 2.5$ in.

Min $h = 17.3 + 2.5 = 19.8$ in., say 20 in. for over-all beam height.

USE ~~Try~~ $b' = 9.5''$, $d = 20 - 2.5 = 17.5$ in.

Check shear before calculating A_s. The shear is critical at d from support and an extra 8% of the simple beam shear* will be allowed for the continuity shear. (More generally, to fix the stem size, the *end* span shear $1.15wl_n/2$ would be used to permit uniformity in size for exterior and interior spans.)

$$V_u = 6190(10 \times 1.08 - 1.46) = 57{,}700 \text{ lb}$$
$$\overline{V} = 57{,}700/0.85 = 68{,}000 \text{ lb}$$

$$v = \overline{V}/b'd = 68{,}000/(9.5 \times 17.5)$$
$$= 408 \text{ psi} < 10\sqrt{f_c'} = 548 \text{ psi} \qquad \text{O.K.}$$

$$\text{Stem wt. (Fig. 8.11}b\text{)},\ w_s = 9.5(20 - 5) \times 150/144$$
$$= 148 \times 1.4LF = 207 \text{ plf}$$

$$\text{Revised } w = 6120 \text{ plf}$$

The 1.1% drop in w would reduce the depth required for M by about 0.5%, which means no change in d, but M_u will be recomputed for steel calculations.

$$M_u = -0.091 \times 6120 \times 20^2 = 222{,}000 \text{ ft-lb}$$
$$\overline{M} = 222{,}000/0.90 = 247{,}000 \text{ ft-lb}$$

From Table 2.1, $\bar{k} = 783$, $p = 0.0161$, $a/d = 0.378$.

$$\overline{M}_1 = 783 \times 9.5 \times 17.5^2/12 = 189{,}000 \text{ ft-lb}$$
$$\overline{M}_2 = 247{,}000 - 189{,}000 = 58{,}000 \text{ ft-lb} \qquad d - d' = 15.0 \text{ in.}$$
$$A_{s1} = 0.0161 \times 9.5 \times 17.5 = 2.68 \text{ in.}^2$$
$$A_{s2} = 58{,}000 \times 12/(60{,}000 \times 15) = 0.77$$
$$A_s = 3.45 \text{ in.}^2$$

Although A_s' will be near the magnitude of A_{s2} and apparently not critical, it will be calculated exactly to indicate the method in detail. First check the stress on A_s', starting with the depth of stress block, as in Fig. 8.11*c*.

$$a = 0.378 \times 17.5 = 6.64 \text{ in.} \qquad c = a/0.85 = 7.80 \text{ in.}$$
$$\epsilon_s' = 0.003(7.80 - 2.5)/7.80 = 0.00204$$
$$f_s' = 0.00204 \times 29 \times 10^6 = 59{,}300 \text{ psi}$$

*In Fig. 8.11*h*,

$$V_c = \frac{(0.091 - 0.048)wl_n^2}{l_n} = 0.043wl_n$$

which is 8.6% of the simple beam shear at the end and 10 to 11% of that at a distance d from the support.

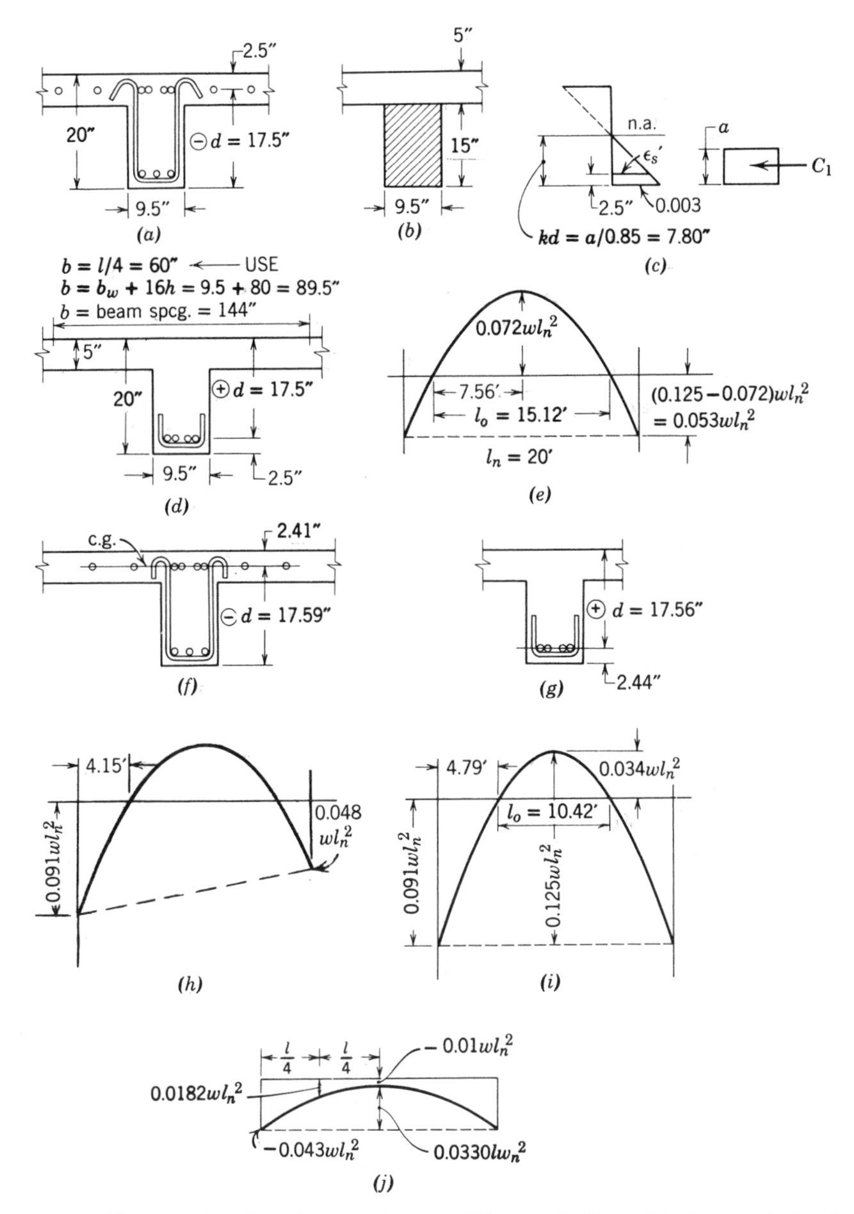

Fig. 8.11. Design sketches for continuous T-beam design. (a) Assumed depth for negative moment. (b) Estimated weight, shaded area. (c) Strain triangle for f_s' calculation. (d) Assumed depth for positive moment. (e) Location of P.I. with maximum positive M. (f) Actual depth to top steel as selected. (g) Actual depth to bottom steel as selected. (h) "Exact" location of P.I. with maximum negative M. (i) Approximate location of P.I. with maximum negative M. (j) Minimum positive moment near mid-span.

Considering displaced concrete, effective $f_s' = 59{,}300 - 0.85 \times 3000$

$$= 59{,}300 - 2550 = 56{,}700 \text{ psi}$$

This compares to a maximum possible of $60{,}000 - 2550 = 57{,}400$ psi

$$A_s' = 58{,}000 \times 12/(56{,}700 \times 15) = 0.82 \text{ in.}^2 \text{ for stress.}$$

To keep the design compression within 0.75 of the balanced beam compression (with A_s') requires that A_{s2} be increased to $(4/3)0.82(56{,}700/57{,}400) = 1.08$ in.^{2} The author would make this correction only where legally required, since it is not really needed for ductility; he would then probably ignore the last ratio which is there only because f_s' at the balanced condition would be at f_y rather than 59,300 psi.

At mid-span the minimum positive moment is actually negative, $-0.010wl_n^2$, requiring, closely enough by proportion,

$$A_s = 3.45(0.010/0.091) = 0.38 \text{ in.}^2$$

with 1 − #7 adequate (0.60 in.^{2}). At the quarter point of span, $M = -0.0182wl_n^2$ and

$$A_s = 3.45(0.0182/0.091) = 0.69 \text{ in.}^2$$

The 1 − #7 is adequate back to about 6 ft from the support.

8.11 CONTINUOUS T-BEAM DESIGN—POSITIVE MOMENT DESIGN

The overall depth of 20 in. already chosen for negative moment fixes the effective depth at mid-span. For one layer of steel (Fig. 8.11d),

Positive $d = 20$ in. − 1.5-in. cover − 0.5-in. stirrup − 0.5 in. (for $d_b/2$) = 17.5 in.

Trial $z = 17.5 - h_f/2 = 15.0$ in., or $0.9d = 15.75$ in. Use larger value.

$$M = +0.072 \times 6120 \times 20^2 = 176{,}000 \text{ ft-lb,}$$
$$\overline{M} = 176{,}000/0.9 = 195{,}500 \text{ ft-lb}$$

$$\text{Trial } A_s = \frac{\overline{M}}{f_y z} = \frac{195{,}500 \times 12}{60{,}000 \times 15.75} = 2.48 \text{ in.}^2$$

The compression area is probably limited to less than the depth of the 60-in. wide flange and thus acts as in a wide rectangular beam.

$$a = 2.48 \times 60{,}000/(0.85 \times 3000 \times 60) = 0.97 \text{ in.}$$
$$z = 17.5 - 0.97/2 = 17.02 \text{ in.}$$
$$A_s = 195{,}500 \times 12/(60{,}000 \times 17.02) = 2.30 \text{ in.}^2$$

A further cycle could change only the last digit. Since a is so small, the beam is obviously in the underreinforced classification.

This steel should be checked against the minimum for positive moment steel as required by Code 10.5, that is, against $200\, b_w d/f_y = 200 \times 9.5 \times 17.5/60{,}000 = 0.56$ in.^{2} Bars must be so arranged to maintain this steel area throughout the entire positive moment length, but this is no problem since it is here essentially the one-quarter of the positive moment steel which the Code requires to be continued into the column in all continuous beams.

8.12 CONTINUOUS T-BEAM DESIGN—CHOICE OF BARS

The size of bars used for positive moment may be limited by the specified l_d at the point of inflection (Sec. 5.13a, Code 12.2.3) at least as regards the bars continuing toward the column. For these n bars of individual area A_b the available $l_d = M_t/V_u + d$. At the P.I. which is located in Fig. 8.11e at 7.56 ft from mid-span,

$$M_t/V_u = nA_b\, 60{,}000 \times 17/(6120 \times 7.56) = 22.0\, nA_b$$

If nA_b is half of the maximum required positive moment steel (required steel being 2.30 in.^{2}),

$$M_t/V_u = 1.15 \times 22.0 = 25.3 \text{ in., available } l_d = 25.3 + 17.5 = 42.8 \text{ in.}$$

For #8, $l_d = 0.04 A_b f_y/\sqrt{f_c'} = 0.04 \times 0.79 \times 60{,}000/54.8 = 34.6$ in. This permits the use of #8 bars, possibly #9 if needed (and checked).

Bars will now be selected and detailed, using bent bars. The detailing will be revised in Sec. 8.15 e,f for straight bars.

	$-A_s$	A_s'	$+A_s$
Reqd, in.^{2}	3.45	1.08	2.30
USE$_u$ 4- #7 Bent	= 2.40 in.^{2} ←		2- #7 Bent = 1.20 in.^{2}
4- #5 Straight	= 1.24	2- #7 str.←	2- #7 Straight = 1.20
	3.64	= 1.20	2.40

The schematic bending arrangement of the steel selected is the same as that shown for the slab in Fig. 8.10b.

For the positive moment steel, 2- #7 bars will be bundled* in each corner of the stirrups since there is not room to space four bars singly

*Note that bundled bars require the use of stirrups around them. If shear did not call for stirrups all the way across the span as happens to be the case in Sec. 8.17 the bundling here would have required them anyway.

in the clear space between stirrup legs which is 9.5 − 2 × 1.5 cover −2 × 0.5 stirrup = 5.5 in. With bars bundled as above, the clear space between bundles is 5.5 − 2 × 0.88 − 2 × 1.00 = 1.75 in., actually nearer 1.4 in. because the diameter over the lugs is nearly $\frac{1}{8}$ in. extra for each bar. The spacing for bundled bars in Code 7.4.2 is the same as for a single bar of the *combined* area. In this case the area is 2 × 0.60 = 1.20 in.^{2} with equivalent diameter of 1.24 in.; hence the space between bars is ample, since aggregate will probably be 0.75 in. maximum size. Notice that in this case the slight excess A_s would also permit the 2- #7 bars to be bundled one above the other.

The crack control provisions should be considered here, but this will be delayed until after all the bars are tentatively selected in order not to confuse the basic selection process.

The negative moment steel can be placed in one layer by bundling the four bent-up #7 bars into two bundles adjacent to the stirrups and placing the other four straight bars in the slab, say at 10 in. spacing. The use of #7 bars increases the estimated d for all steel by 0.06 in., a little more with the #5 bars at the top, but the change appears negligible (Fig. 8.11f,g).

The compression steel (at support) might be detailed as 2- #7 bars from each side, each caring for the stresses on the entering face of the column. Since this means developing all four bars into (and actually beyond) the far face of the column, it appears simpler to run 1- #7 from each side far enough to make it also effective on the far face of the column. This leaves 1- #7 bottom bar to be cut off short of the column on each side.

The crack control provision of Sec. 6.13 (Code 10.6.3-4) is

$$z = f_s\sqrt[3]{d_c A} \not> 175 \text{ (for interior exposure)}$$

The positive moment bars will be considered first. This equation was not designed for bundled bars. Whether to count each bundle as one or two bars is not definite; the use here as one bar is on the conservative side. Try the two bars stacked vertically as suggested above, instead of horizontally as shown in Fig. 8.11g, since this is on the safe side with A larger (Fig. 6.15). Although it may at first sound contradictory, d_c will be taken from the center of outermost bar (2.44 in.) because crack width spreads from the outermost bar (Fig. 6.16) to a width dependent upon the free cover. The suggested $0.6f_y$ will be used for f_s.

$$A = 2(1.5 \text{ cover} + 0.5 \text{ stirrup} + 0.875)9.5/2 = 27.3 \text{ in.}^2$$

$$z = 0.6 \times 60\sqrt[3]{2.44 \times 27.3} = 36 \times 4.07 = 147 < 175$$

The bars are thus satisfactory for either orientation of the bundle.

For the negative moment bars, with the bundles in the horizontal plane and the bars spread over the flange width of 60 in., $d_c = 2.44$ in., $A = 2 \times 2.44 \times 60/6 = 48.8$ in.^{2}, again counting a bundle as one bar.

$$z = 0.6 \times 60\sqrt[3]{2.44 \times 48.8} = 36\sqrt[3]{119} = 178 > 175.$$

In consideration of the conservative treatment of the bundled bars, the arrangement is approved.

Although the area around the individual bar is not called for in the equation for z, the designer might be concerned about the #5 bars at 10 in. spacing near the edges of the effective flange where the individual bar A is apparently $2 \times 10 \times 2.42 = 48.4$ in.^{2}

$$z = 36\sqrt[3]{2.42 \times 48.4} = 36 \times 4.90 = 177 > 175$$

This is about the same as for the entire group. Three #4 bars at 7 in. spacing could be substituted on each side for the 2- #5 at 10 in., but this is probably using the equation improperly and the author would not make the change unless this individual bar calculation was far out of line.

It is not uncommon for the designer to increase b_w arbitrarily or even change d to make the steel used space properly or to modify the required A_s to fit available bar sizes closer.

A cheaper beam would result if advantage were taken of the redistribution provisions of Code 8.6. Where the moments have been established by elastic theory this is recommended.

8.13 CONTINUOUS T-BEAM DESIGN—GENERAL REQUIREMENTS FOR BENDING BARS

The fabricator must detail each bar, but the designer can be satisfied to locate bend points and cutoff points reasonably accurately. Many offices use rules such as "bend up half the bottom steel at the quarter point of clear span," but more exact procedures are presented here to indicate more adaptable methods needed in many cases.

The author recommends a very conservative attitude toward bar detailing. There has been much ineffective detailing of what would otherwise have been good designs. The final member is no better than its details.

Bend points may be governed by:

1. Moment requirements.
2. Development lengths.
3. "Arbitrary" requirements of Code 12.1.4, 12.2.1, 12.2.2 (Sec. 3.9*d*)
4. Use of bent bars as stirrups.

The several "arbitrary" specification requirements will first be summarized. Code 12.3.3 requires that at least one-third of the total reinforcement for negative moment be extended beyond the extreme location of the point of inflection by the greater of: (1) $l_c/16$; (2) beam depth d; (3) $12d_b$. Code 12.2.1 requires that at least one-foruth of the positive reinforcement in continuous beams shall extend 6 in. into the support. Both of these, in the words of the old Joint Committee Specification, are "to provide for contingencies arising from unanticipated distribution of loads, yielding of supports, shifting of points of inflection, or other lack of agreement with assumed conditions governing the design of elastic structures." Code 12.2.2 goes further to add that when the member is part of a "primary lateral load resisting system" the extension of the bottom steel covered in 12.2.1 must anchor it to develop its full-yield stress in tension at the face of the support, that is, be much more than 6 in. This requirement is to provide some ductility in the event of stress reversal from wind, earthquake, or explosion.

Code 12.1.4 also requires that every bar, whether required for positive or negative reinforcement, be extended the depth of the beam or 12 bar diameters beyond the point at which it is no longer needed to resist flexure. This requirement, discussed in Sec. 5.11, the author designates as a and uses as a shifted moment diagram, but the Code itself seems to be less strict and to leave much to the designer's discretion. Also, since the Code uses simply the word "extended," it seems to imply that the extension might exist entirely as a bent bar. The author prefers to consider at least half of it ($a/2$) as a relaxed requirement for a shifted moment diagram requiring one-half of the extension to remain straight, but relaxed because a bent bar remains partially effective in flexure while in the tension half of the effective depth.

Arrangements satisfying moment requirements, development lengths, and the "arbitrary" requirements may sometimes be varied slightly to help out in development length demands elsewhere or in web reinforcement. For example, if development of the bottom steel at the point of inflection proves difficult, it may be possible to shift the bend-up points toward the column and to keep more steel available at the P.I.; or the spacing of bend points may be shifted to make bent bars more useful for web steel.

8.14 CONTINUOUS T-BEAM DESIGN—MOMENT DIAGRAMS GOVERNING BAR BENDS

The positive moment diagram has already been established from Fig. 8.11*e*.

For maximum negative moment the actual moment diagram is unsymmetrical and the maximum moment calculation alone does not provide sufficient data for the entire diagram. The diagram can be reasonably approximated by using a negative moment at the far end as slightly less than that accompanying the maximum positive moment, in this case say $-0.048wl_n^2$ instead of the $-0.053wl_n^2$ value shown in Fig. 8.11*e*. Figure 8.11*h* would be almost exact insofar as the location of P.I (point of inflection) on the left is concerned. The method of Sec. A.5 can then be used to locate the P.I. at 4.15 ft from the face of column.

More commonly the approximate P.I. is calculated for the assumed symmetrical diagram of Fig. 8.11*i*, which will be used here.

$$\tfrac{1}{8}wl_0^2 = 0.035w \times 20^2$$
$$l_0 = 20\sqrt{0.272} = 10.42 \text{ ft}$$

Hence the P.I. is 4.79 ft from the column. The futher approximation of using a triangle for the negative moment section of the parabola is reasonable when the P.I. is not too near mid-span, say, is outside the middle third of the span.

The minimum positive moment at mid-span is also part of a symmetrical diagram, but in this case the simple beam moment diagram is that for dead load alone, which can be expressed as follows in terms of total load w, that is, 1620 plf out of 6120 plf total.

$$\text{Simple beam } M_s = 0.125w_d l_n^2 = 0.125\,\frac{1620}{6120}\,wl_n^2 = 0.0330wl_n^2$$

The diagram is sketched in Fig. 8.11*j*. The maximum moment diagrams are assembled in Fig. 8.12*a*.

As is usual in the office unless it is a special case, the A_s required diagram is based on the total steel used.

8.15 CONTINUOUS T-BEAM DESIGN—BENDS AND STOP POINTS FOR TENSION BARS

(a) Purpose of Discussion

It is not the intent of this section to consider simply the details of bar detailing. Rather, experience has shown that no other type of problem so clearly brings to the fore the fundamentals of how reinforced concrete works and how both flexure and bar development length affect each other.

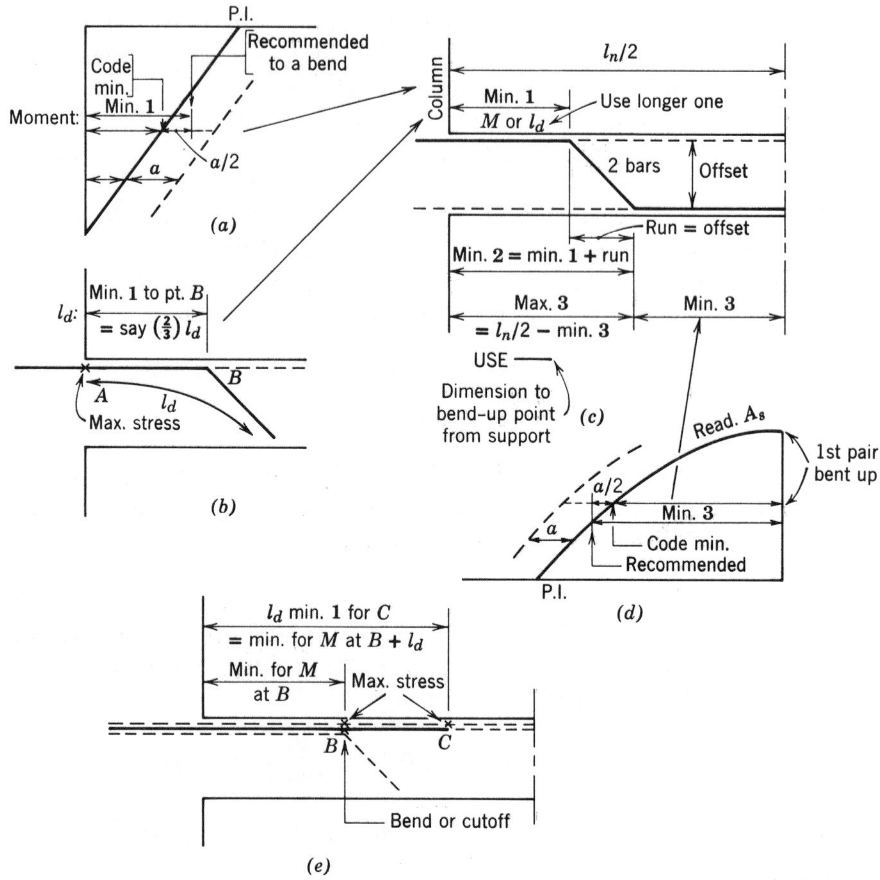

Fig. 8.12. Schematic for data assembly.

(b) General Assembly of Data

The data can be arranged in many possible patterns, but several pieces of data are interrelated in each decision, and a regular repeating pattern of presentation is desirable. The author's form is shown schematically in Fig. 8.12, for one pair of bars bent first.* The minimum distance to the bend-down point is determined by the larger of two requirements:

*The student might note that "first" relates to the nearness to the point of maximum moment for that steel. If only one pair is bent, the first bar bent down is also the first bar bent up. If two bars are bent singly, the first bent down matches the last one bent up and the second bent down matches the first bent up.

1. Moment length plus $a/2$, as shown in Fig. 8.12*a*, the second $a/2$ being available in the bend length itself.
2. The development needs, shown in Fig. 8.12*b*. The author assumes two-thirds of l_d is permissible here because the bent portion also helps to develop the bar.

These moment length and development length values are tabulated in Fig. 8.12*c*, immediately above the bar sketch; the larger is "Min.1." Since bend points are usually dimensioned at the level of the bottom bars, "Min.2" is next found from "Min.1" by adding the run of the bar, equal to the offset of the bar. The bottom bar can be bent up at "Min.3" (Fig. 8.12*d*) given by the moment length plus $a/2$ from mid-span or at "Max.3" from the column $= l_n/2 -$ Min.3. The values of "Max.3" and "Min.2" determine whether the bend can be made and the leeway open to the designer in choosing the final dimension.

If Fig. 8.12 were based on a first cutoff point instead of bend point, only three changes would be necessary (but the stirrups of Code 12.1.6 would be necessary; see Sec. 4.14).

1. The moment length in Fig. 8.12*a*,*d* would include the full *a* length (not $a/2$), as a Code requirement.
2. The development length would be the full l_d.
3. Only "Min.1" would determine the top cutoff and only "Min.3" the bottom cutoff.

For any succeeding bar cutoffs one other change is involved, as in Fig. 8.12*e*. The l_d value is to be measured from point B, the *theoretical* cutoff for the last bar stopped or bent. The point B does not shift if the bar is run farther because of l_d or some other reason. (If the bar runs on to *C*, the maximum stress point is eliminated and the column face takes its place.)

This form is expandable to as many bars or cutoffs as are necessary. Development lengths are obviously needed for this example as follows.

$$\text{Bottom } \#7,\ l_d = 0.04A_b f_y/\sqrt{f_c'} = 0.04 \times 0.60 \times 60{,}000/54.8 = 26.3 \text{ in.} = 2.19 \text{ ft}$$

$$\text{Top } \#7,\ l_d = 1.4 \times 26.3 = 36.8 \text{ in.} = 3.07 \text{ ft.}$$

$$\text{Top } \#5,\ l_d = 1.4 \times 0.04 \times 0.31 \times 60{,}000/54.8 = 1.4 \times 13.6 = 19.1 \text{ in.} = 1.59 \text{ ft}$$

(c) Author's Preferred Arrangement of Bars

The author's preferred arrangement consists of 2 – #7 bars bent as a pair (or singly) and all other bars straight and continuing at least to the

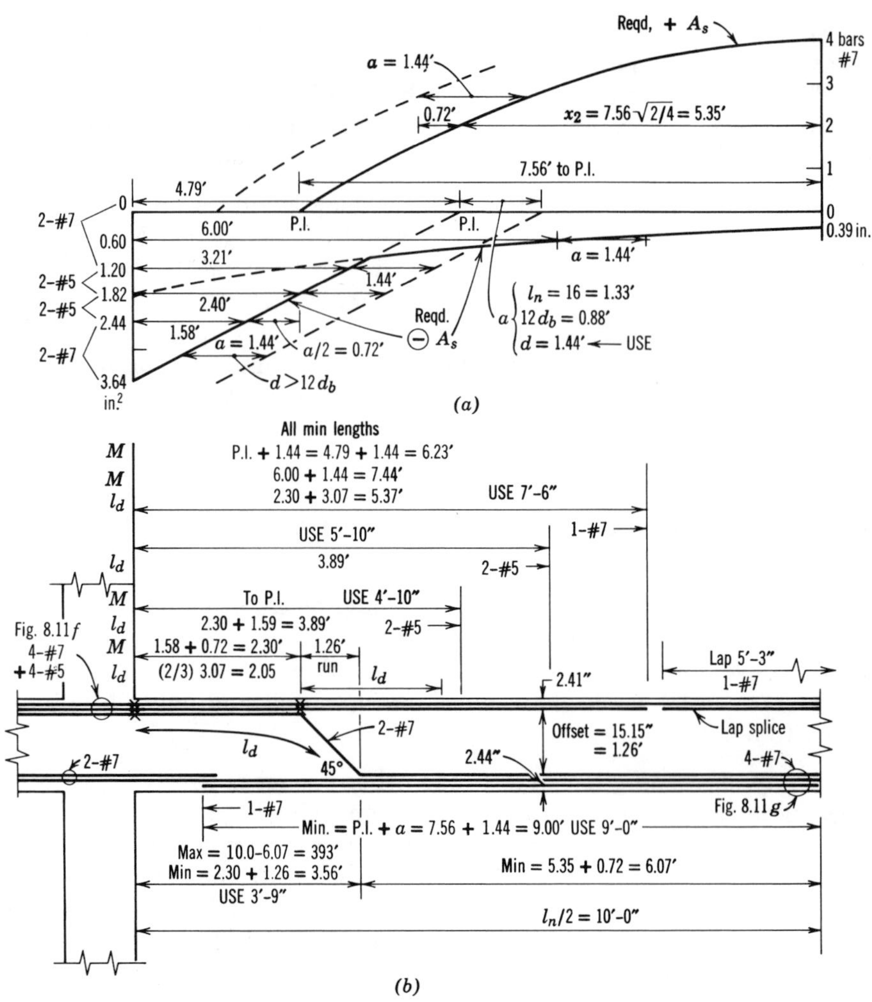

Fig. 8.13. Detailing for bent bars and shifted moment diagram.

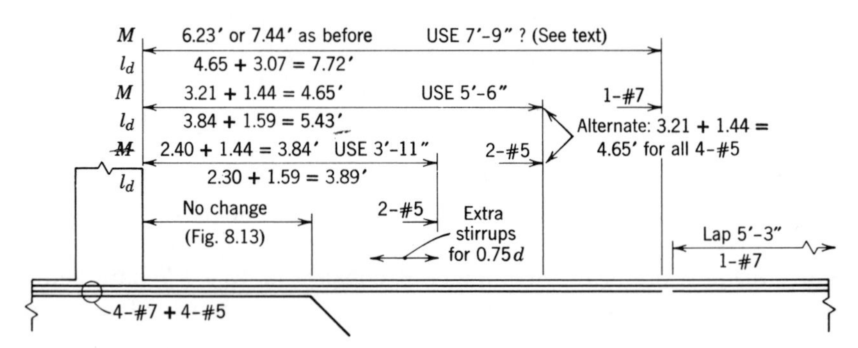

Fig. 8.14. Detailing with shifted moment diagram but with minimum cutoff points.

nominal P.I. as sketched in Fig. 8.13*b*. Since no bars are cut off inside the P.I., the extra stirrups of Code 12.1.6 are not needed.

The bent bars are calculated in Fig. 8.13 (based on the ideas of Fig. 8.12*a* to *d*); and at the bottom bar level a minimum of 3.56 ft and a maximum of 3.93 ft result, with only about a 4-in. leeway. The next bottom stop point is for the third #7 bar, established solely from the minimum from mid-span. Since the size of this bar was originally selected in Sec. 8.12 to match development length needs, it is automatically satisfactory if extended d beyond the P.I. Since a #8 or larger bar size was found permissible there, the distance d could be reduced if one wished to check it more closely.

The top straight bars are next considered. One-third of the total 3.64 in.^{2}, say 2 − #7 = 1.20 in.^{2}, must extend past the P.I. (which goes with that negative moment diagram) at least $l_n/16$, d, or $12d_b$, in this case 1.44 ft which is d. One of these will necessarily be spliced at mid-span with a lap of $1.7l_d = 1.7 \times 3.07 = 5.22$, say 5 ft-3in., the 1.3 factor not being acceptable because more than 50% of the bar is needed at mid-span, that is, $f_s > f_y/2$. The other #7 bar must wait until the development length requirements are fixed by what is done with neighboring bars.

The initial decision to carry straight bars to the nominal P.I. automatically goes beyond any moment requirement. Hence moment lengths are not all shown. For this case, arrangement becomes primarily a matter of staggering the bar cutoffs. For the first 2- #5 bars the critical stress point is at 2.30 ft where the 2- #7 are bent down and the #5 l_d of 1.59 ft is added to give an l_d requirement of 3.89, well inside the P.I. These bars are stopped at the nominal P.I., at 4 ft 10 in. The second pair of 2- #5 bars has the same requirements, and the cutoff is simply staggered one foot to 5 ft 10 in. The 1- #7 must go to the nominal P.I. plus 1.44 ft to satisfy the arbitrary extension, or $4.79 + 1.44 = 6.23$ ft; or for the dead load moment diagram $6.00 + 1.44 = 7.44$ ft. The small stress peak where the last #5 bars are cut off can be ignored. The flat dead load moment diagram demands also the extension and center splicing of the last 1- #7 bar.

(d) Bent Bars with Straight Bars Cut Off by Shifted Moment Diagram

This arrangement will require extra stirrups under Code 12.1.6 over the bars cut off inside the P.I. These stirrups are not computed here.

The bend point (Fig. 8.14) for the 2- #7 bars is the same as in *c* above, at 3 ft 9 in. measured at the bottom steel level. The first 2- #5 have the same 3.89 ft l_d requirement as in Fig. 8.13 and the moment requirement is unchanged at 3.84 ft. The second pair of 2- #5 have an l_d requirement of the 3.84 ft moment length just noted for the bars cut plus an l_d of 1.59 ft beyond, a total of 5.43, say 5 ft 6 in., with the moment requirements much less. The separate cutting of the two #5 pairs is costing extra length of the

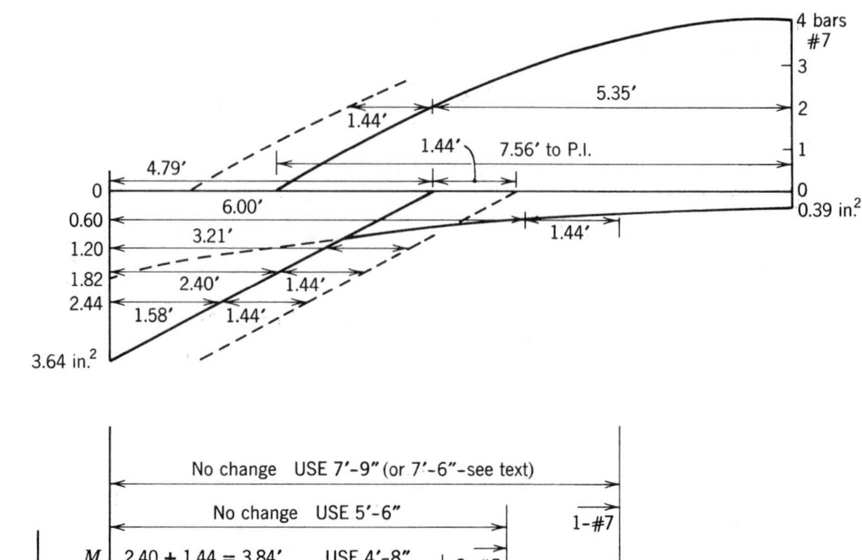

Fig. 8.15. Detailing straight bars with shifted moment diagram.

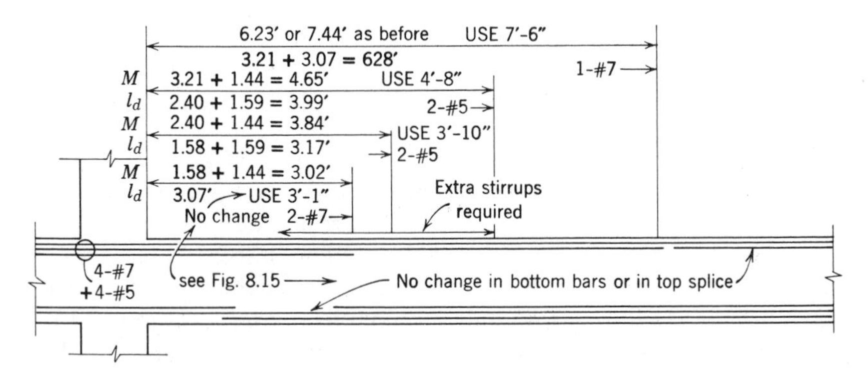

Fig. 8.16. Detailing straight bars without shifted moment diagram.

second pair. Both bars together could stop at 4.65 ft for moment,, since l_d then remains 3.89 ft.

The 7.72-ft development length of the 1- #7 is overly large because the bar stress where the #5 bars are acutally cut is much less than this calculation assumes. If this stress at the actual cutoff point were calculated, the length of the #7 could be reduced, almost surely enough to make the moment length of 7.44 ft control; say use 7 ft 6 in.

(

(e) All Bars Straight and Cut Off by Shifted Moment Diagram

The final results are tabulated in Fig. 8.15 with data from the moment diagrams of Fig. 8.14. Only the pair of #7 bottom bars, the pair of #7 top bars cut off first, and the pair of #5 bars cut off first are changed from Fig. 8.14. The third #7 bottom bar cutoff might be shortened, but to do so one must recalculate as indicated in discussing the layout in (c) above.

(f) All Bars Straight Without Shifted Moment Diagram Idea

The change which occurs in this case compared to (e) is that all lengths based on l_d are relaxed by assuming that the bar stress is maximum at the basic moment diagram rather than the shifted diagram. This would make the moment requirement of Fig. 8.15 the larger requirement in Fig. 8.16 and thus lower the lengths for the #5 bars to 3 ft 10 in. and 4 ft 8 in., respectively.

(g) Bent Versus Straight Bars

A major advantage of bent or offset bars is that they do not lower the shear strength of the member as do bars simply cut off and stopped. They also help to keep top steel at the proper level and probably tend to reduce placement errors. However, practice has moved largely toward straight bars, probably because of cost considerations in placement. If bars are bent, it is easier to satisfy moment requirements with bars bent singly; but it is easier to satisfy development lengths needed with bars bent as pairs.

8.16 CONTINUOUS T-BEAM DESIGN—STOP POINTS FOR COMPRESSION STEEL

The compression length of the #7 straight bottom bars is also established by both a moment and a development requirement.

Since only two bars are required at each column face, bars from *both* beams are not required (Fig. 8.17*c*), although it proved convenient in this design to lap one bar from each beam. Since both bars are then needed at both sides of the column, the extension must go through the column and then beyond the far side enough to develop the bar there in compression, as in Fig. 8.17*b*; the length for moment may also control.

For compression development length Sec. 5.16 gave the relation for $f_c' < 4440$ psi

$$l_d = 0.02 f_y d_b / \sqrt{f_c'} \ngtr 8 \text{ in.}$$

$$\text{For } \#7,\ l_d = 0.02 \times 60{,}000 \times 0.875/54.8 = 19.12 \text{ in.} = 1.59 \text{ ft}$$

For moment the length needing some help from M_2 is easily established in Fig. 8.17*a* from the negative moment diagram. In this case, $\overline{M}_2$ was found in Sec. 8.10 to be 58,000 ft-lb out of a total of 247,000 ft-lb, or $M_2 = (58{,}000/247{,}000)M_u = 0.235 M_u$. This indicates *all* compression steel could theoretically be omitted at 0.235×4.79 (distance to P.I.) $= 1.13$ ft

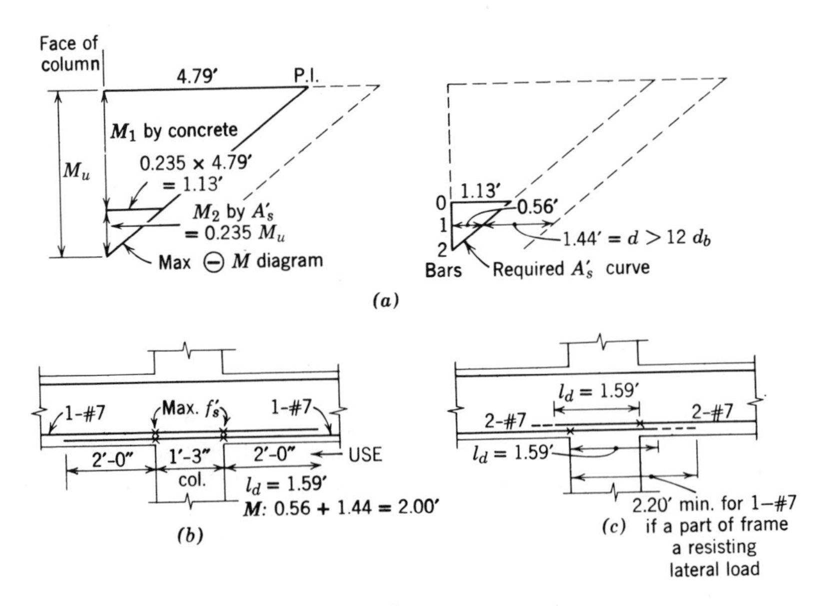

Fig. 8.17. Detailing of compression steel.

from the column. Half the steel could be stopped at 1.13/2 = 0.56 ft *plus* the arbitrary requirement of d or $12d_b$ (1.44 ft controlling) for a total of 2.00 ft. This exceeds l_d and controls the bar length, as sketched in Fig. 8.17*b*.

The detail of A_s' if both bars required are furnished from the adjacent beam is shown in Fig. 8.17*c*. Since compression steel in this case extends as far as negative moment, there is no need to consider the moment diagram at all. (Within the column the column compression provides good development conditions and even if the column were very wide the negative moment in the middle of the column would in effect be in a deeper beam.) The l_d already calculated gives the distance the bars must extend *into* the column, which in this case happens to go a little beyond the far face of column.

Finally, attention is called to the special requirements of Code 12.2.2, already mentioned in Sec. 8.13. If the beam is part of a primary lateral load resisting frame, that is, if the beam must hold a column carrying lateral shear from, say, a wind loading on the frame, one-fourth of the required positive moment steel (1- #7 bar in this beam) must be anchored into the column for its full f_y in tension. For tension

$$l_d = 0.04\, A_b f_y/\sqrt{f_c'} = 0.04 \times 0.60 \times 60{,}000/\sqrt{3000} = 26.3 \text{ in.} = 2.20 \text{ ft}$$

This would call for a change in Fig. 8.17*c* but would not change Fig. 8.17*b*. However, notice the special limitation this places on bottom bar sizes in the case of a beam stopping at an exterior column, already discussed in Sec. 5.14.

8.17 CONTINUOUS T-BEAM DESIGN—STIRRUPS

Since bundled bars were selected in Sec. 8.12, stirrups are required all across the middle of the beam to satisfy Code 7.4.2. However, stirrups are usually considered a matter of shear design.

Under maximum moment loading the beam is subject to an end shear equal to the simple beam shear $wl_n/2 = 61{,}200$ lb plus a continuity shear, again assumed as $0.08wl_n/2$ or 4890 lb, acting both at the end and at mid-span as shown dashed in Fig. 8.18*a*. The shear at mid-span will be greater when live load is removed from the left half of the span. This loading will produce a smaller continuity shear, which will be neglected. This simple beam shear at mid-span is $4500 \times 10 \times \frac{5}{20} = 11{,}200$ lb. The solid line in Fig. 8.13*a* will be used as a maximum shear diagram for stirrup design.

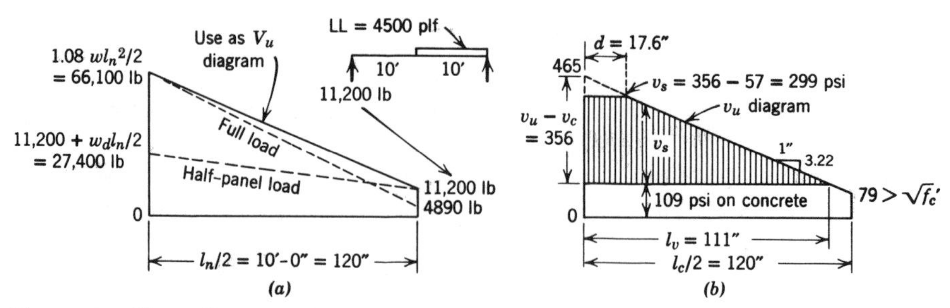

Fig. 8.18. Shear diagrams for stirrup calculations. (*a*) V_u diagram. (*b*) v_r diagram and v_s (shaded).

With $b' = 9.5$ in. and $d = 17.59$ in. at the end, Fig. 8.11*f*, the critical shear is at the distance d from support.

$$\text{At support} \quad v_0 = \frac{66{,}100/0.85}{9.5 \times 17.59} = 465 \text{ psi}$$

$$\text{At mid-span } v_{10} = \frac{11{,}200/0.85}{9.5 \times 17.59} = 79 \text{ psi}$$

The slope of the v diagram is $(465 - 79)/120 = 3.22$ psi/in. The critical shear at 17.6 in. from support is $465 - 3.22 \times 17.6 = 408$ psi. The unit shear curve is plotted Fig. 8.18*b*. Stirrups are needed for the l_v length where v is in excess of $2\sqrt{f_c'}$; also for all the remainder of the length since everywhere $v > v_c/2 = \sqrt{f_c'} = 55$ psi. The stirrup requirement over bundled bars is thus automatically satisfied; also the requirement for stirrups over A_s' bars under Code 7.12.5.

If bars were cut off in a tension zone, *extra* stirrups would be needed for $0.75d$ over each cut bar as indicated in Figs. 8.14, 8.15, and 8.16. Code 12.1.6 waives these stirrups only where shear is equal to or less than two-thirds the allowable. These stirrups must provide an extra $\rho_n f_y$ of 60 psi with a resulting spacing not exceeding $d/8\beta_b$, where β_b is the proportionate part of the bars cut off at the particular section.

Except for such special extra stirrups, a design with similar data is worked out in detail in Sec. 4.22*c*.

8.18 CONTINUOUS T-BEAM DESIGN—PLACEMENT OF STIRRUPS

The proper placement of stirrups in continuous beams presents a problem. The best anchorage of stirrups would call for the hooks to be in compression concrete, which near the supports would be the bottom concrete. Since construction is simpler with the open end of the stirrup turned up, the matter of anchorage has generally been ignored. Furthering this easier placement is the requirement that the compression (bottom) steel be tied as specified in Code 7.12.5. One stirrup must extend completely around all longitudinal A_s' bars. The ties, over the length where A_s' is needed must satisfy the requirements for a column, that is, be "so arranged that every corner (bar) and alternate . . . bar shall have lateral support provided by the corner of a tie . . . . and no bar shall be further than 6 in. clear on either side of such a laterally supported bar." Closed stirrups, like column ties, would be excellent but these have limited usage except where torsion may be present because they complicate bar placement. The chief objection to these relates to difficulty in dropping reinforcement into place inside the ties.

8.19 CONTINUOUS T-BEAM DESIGN—DEFLECTION

Deflection under service load is the critical case and this calculation is somewhat toward an elastic analysis. The deflection calculations for the particular design of this chapter are given in Chapter 6.

8.20 CONTINUOUS RECTANGULAR BEAMS

The design differences between continuous rectangular beams and continuous T-beams are all minor. The foregoing design of the T-beam can also serve as a model for continuous rectangular beam design. Other than the obvious minor difference in designing the positive moment steel for a rectangular beam and a somewhat greater congestion of negative moment steel (possibly two layers required), attention must be called to the special requirements for lateral supports as given in Code 10.4. Freestanding rectangular beams lack the extra stiffness and strength of the T-beam in resisting torsion.

8.21 SPANDREL AND OTHER L-BEAMS

A spandrel beam is one supporting an exterior wall, although occasionally the problems (without the name) may be associated with an interior beam adjacent to a stairwell, shaft, or other interior opening. The spandrel carries more dead load because of the exterior wall weight, but often is lighter than an interior beam because of live load coming from a reduced floor area.

The L-beam, with flange on only one side, is an unsymmetrical section. As such, Code 8.7.4 limits the flange counted for flexure more strictly than in a T-beam, that is, to the least of $l_n/12$, $6h_f$, or half the clear distance to the next beam. However, in flexure, an L-beam cannot act as an unsymmetrical section because the slab prevents the normal lateral deflection which would occur when freestanding.

The lateral deflection is actually prevented only at the level of the slab. With the slab loaded, a secondary rotation about an axis on the intersection of the center lines of slab and web does occur and causes some lateral deflection of the web. This results from the slab deflection and rotation at the beam junction, and it introduces torsion into the beam.

There is a current trend, with which the author disagrees, to consider only the rectangular web of the L-beam in computing torsional resistance (author's comments in Sec. 4.21). The slab is both a loading and a resisting portion of the system; a strength calculation that ignores the strength aspect is oversimplified and wasteful.

How to establish the design torsion is still the major uncertainty, in the author's opinion. Elastic analysis tends to overemphasize torsional demands. After cracking, a rapid decrease in torsional stiffness occurs. For the additional rotation necessary to develop the resisting stirrups, torsion will build up at only 5 to 10% of the rate it does under initial rotations. It is, therefore, difficult to project the initial elastic calculations to a reasonable design basis.

8.22 END SPANS AND IRREGULAR SPANS

End spans always involve negative moments smaller at the outer end and larger at the inner end, with the points of inflection and point of maximum positive moment shifted toward the outer support. Irregular spans and loads also result in maximum moment diagrams which are less symmetrical than those used in this chapter.

Proper detailing of end spans and irregular spans requires a better knowledge of unsymmetrical moment diagrams, but no additional reinforced concrete theory. Where approximations are deemed proper, the designer should be more conservative than where more exact moment requirements are known.

SELECTED REFERENCES

1. "Continuity in Concrete Building Frames," Portland Cement Association, Chicago, 3rd ed.
2. Phil M. Ferguson, "Analysis of Three-Dimensional Beam-and-Girder Framing," *Jour. ACI*, 22, Sept. 1950; *Proc.*, 47, p. 61.
3. R. H. Wood, "Studies in Composite Construction: Part I, The Composite Action of Brick Panel Walls Supported on Reinforced Concrete Beams; Part II, the interaction of Floors and Beams in Multi-Story Buildings," National Building Studies, *Research Papers No. 13* (1952) and *22* (1955), Her Majesty's Stationery Office, London.
4. *Reinforced Concrete Design Handbook, SP-3,* ACI, Detroit, 3rd ed., 1965.
5. *Notes on ACI 318-71 Building Code Requirements with Design Applications.* Portland Cement Association, Skokie, Illinois, 1972.

PROBLEMS

Prob. 8.1. The reduced frame of Fig. 8.19*b* should be used for the analysis of the second floor of the bent of Fig. 8.19*a* since standard coefficients are not applicable. Assume all the beams have $I = 25{,}000$ in.^{4}, the 16-in. columns below the floor have $I = 5450$ in.^{4}, and the 14-in. column above the floor have $I = 3200$ in.^{4} Each beam carries a dead load of 1050 plf and a live load of 2150 plf. With factored loads:

(*a*) Calculate the maximum negative moment for the beam *CD* at *C* and correct this to the design moment at the face of the column. (Note that symmetry about *C* is equivalent to a fixed end for moment distribution purposes.)

(*b*) Calculate the maximum negative design moment at *D* of beam *CD*.

(*c*) Calculate the maximum positive moment for *CD*; also the minimum positive moment.

(*d*) Locate the several points of inflection that are useful in detailing steel.

(*e*) Compare the points of inflection in (*d*) with those which would be obtained if each moment diagram were assumed to be symmetrical about mid-span, as in Fig. 8.11*i*.

Prob. 8.2. In Prob. 8.1 calculate the corresponding maximum moments in span *DF* and the points of inflection needed. (For this end span the use of a symmetrical moment diagram is scarcely valid.)

Prob. 8.3. In the frame of Prob. 8.1 calculate the maximum bending moment on column *D*, using factored loads.

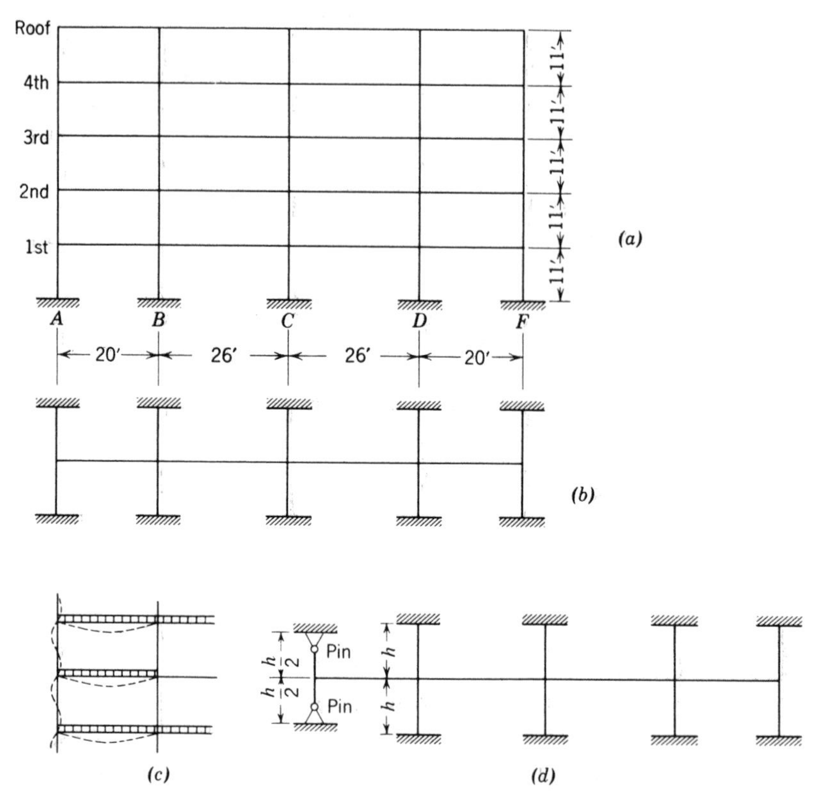

Fig. 8.19. Analysis of building frame. (*a*) The frame considered. (*b*) Ordinary reduced frame for calculation of moments on a single floor. (*c*) Loading for maximum negative moment at exterior end of beam; also nearly maximum for exterior column moment. (*d*) Improved form of reduced frame for maximum moment at exterior joint.

Prob. 8.4. Repeat Prob. 8.3 for the exterior column *F*, using the reduced frame of Fig. 8.19*b*.

In addition, note that loads on successive floors as in Fig. 8.19*c* give a reverse bending condition in the exterior columns which is almost equivalent to a reduced frame similar to that of Fig. 8.19*d*. Recalculate column *D* moments and negative moment at *D* in beam *CD*, using factored loads and this reduced frame.

Prob. 8.5. Design an interior span of a continuous one-way slab supported on beams 14 ft 0 in. on centers using moment coefficients of Sec. 8.5 (Code 8.4.2), live load of 150 psf, no dead except slab weight, $f_c' = 3000$ psi, Grade 40 steel, beam stems 11 in. wide, cover over center line of steel of

1.12 in. (as a simplification). Carry the design through the choice and spacing of bars (same size bars for both positive and negative moment). Draw a lengthwise section of slab and sketch the bars in place, assuming no bars bent up.

Prob. 8.6. Assume the steel found in Table 8.1 for a typical interior span of the slab of Sec. 8.8 is to be arranged as shown in Fig. 8.10*b*, with half of the bottom bars bent up and extra straight top bars added over the support. (This involves a new choice of steel at the support.) Calculate all bend points and stop points (for a typical interior span).

Prob. 8.7. A continuous rectangular beam is to carry a uniform live load of 2000 plf plus 2000 plf dead load (including its weight) over equal 22-ft spans. Supports may be considered of negligible width (knife edges) and moment coefficients of Fig. 8.8. may be used. Design a typical interior span through the choice of beam size and the choice of reinforcing. Assume maximum $V = 1.05wl_n/2$, $f_c' = 4000$ psi, Grade 60 steel. Limit $\bar{k}$ to 1300, which means some compression steel at the worst section. Show a cross section at support and at mid-span with detailed spacing of bars; also an elevation of beam showing the schematic arrangement (bending) of bars. Exact bend points, and so forth, are not a part of this problem.

Prob. 8.8. Design an intermediate span of one of a series of continuous T-beams spaced 11 ft 0 in. on centers to carry a 5-in. slab, live load of 175 psf, and its own weight over a 22-ft clear span, using $f_c' = 3000$ psi, Grade 40 steel, ACI moment coefficients, $V = 1.08wl_n/2$. It is suggested that the stem size be based on using $\bar{k} = 1100$ at the support and a stem width of 11.5 in. be first tried. Stem weight may be assumed 250 plf (without revision). Sketch the cross section at the support and at mid-span and also show the trial arrangement of longitudinal steel.

Prob. 8.9. Assume that an interior span of a continuous beam with uniform live load has been designed for moment coefficients of -0.093 and $+0.067$, based on using the clear span of 23 ft. The steel used is 10- #7 for negative moment, 7- #7 for positive moment, and 4- #7 for compression steel at support. Assume columns 16 in. square, $b_w = 16$ in., $d = 19$ in., $d - d' = 16.5$ in., $f_c' = 4000$ psi, $f_y = 40{,}000$ psi, and $f_s' = 36{,}000$ psi. The M_2 couple represents 0.33 of the maximum negative moment. Arrange the bars to be bent up, using a single bent bar nearest mid-span and then a pair of bent bars nearer the support. For location of points of inflection, the simplifying assumption of Fig. 8.11*i* may be used. Sketch the arrangement of steel and detail all bend and stop points. (If any bends fail to work out satisfactorily, note this fact and compromise lengths as seems best, but do not revise the bending scheme.) Show the results on a sketch similar to Fig. 8.13. Assume minimum positive moment does not control bar lengths. Assume the beam is cast with a 4 in. slab and is part of a frame braced by a shear wall.

Prob. 8.10. A typical interior span of a continuous beams of 21-ft clear span supported by 18-in. square columns in a braced frame and designed for uniform load moment coefficients of $-\frac{1}{12}$ and $+\frac{1}{16}$ (based on clear span) requires 9- #8 for negative moment, 6- #8 for positive moment, and 6- #8 for compression steel at the support. If $M_2 = 0.54$ of maximum negative moment, $f_c' = 3000$ psi, $f_s = 40{,}000$ psi, $f_s' = 28{,}000$ psi, and the offset distance between top and bottom steel is 14 in., sketch the steel arrangement using straight bars and locate stop points for bars. For detailing bar lengths take d as 17 in. Assume mid-span minimum positive moment as $+0.015wl_n^2$ with the factored dead load equal to $w/3$. Record the results as in Fig. 8.15. (A 5-in. slab may be assumed with beam $b_w = 14$ in.)

Prob. 8.11. An interior 19 ft 6 in. clear span of continuous T-beam is loaded at its third points with concentrated loads P such that the uniform load may be neglected in establishing the shape of the moment diagram. Columns are 16 in. square and a primary part of the lateral load resisting system. The maximum moments, including an allowance for uniform load, are $-0.25Pl_n$ and $+0.17Pl_n$. The negative moment is broken down into $M_1 = 0.42M$ and $M_2 = 0.58M$ for purposes of design. The minimum mid-span positive moment is $0.050Pl_n$. Factored dead load is $P/3$. For bar lengths, d may be taken as 18 in. and $d - d'$ as 15 in. The required number of #7 bars is 8 for negative moment A_s, 5 for compression steel A_s', and 5 for positive moment steel. $f_c' = 4000$ psi, $f_y = 60{,}000$ psi, $f_s' = 50{,}000$ psi. Arrange and detail the steel attempting to bend at least two bars up from the bottom. Record the results as in Fig. 8.13. (If any bends fail to work out satisfactorily, note the fact and compromise lengths as seems best, but for this problem stay with two bars bent up in a web 15 in. wide.)

Prob. 8.12. Same as Prob. 8.11 except straight bars are to be used and detailed.

9
Limit Design

9.1 TERMINOLOGY

Limit design and plastic design are often considered as synonymous terms. A distinction is made here and in most discussions of reinforced concrete. In steel design, the term *plastic design* includes not only the change in the pattern of moments beyond the yield point but also the increased resistance of a cross section after its extreme fiber reaches the yield point. In reinforced concrete, *limit design* is used only to refer to the changing moment pattern; ultimate strength design already includes the increase in strength of a given cross section after some stress reaches the yield point.

For reinforced concrete, the frame can be designed under any *fixed* loading to have strengths closely matching the moment diagram, with the concrete tailored to fit by cutting off and the bending of bars. In such a case the limit design concept adds nothing because capacity is reached simultaneously at a number of points and the elastic theory analysis is fully adequate to predict collapse.

However, when several pattern loadings are considered possible on a continuous flexural member, the sum of the maximum positive moment and the average of the two maximum negative end moments will total more than the simple beam moment. In such a case proper limit design might reduce the member cost. Slab design, on the basis of extensive testing, uses the limit design idea in Chapter 10 without the name or the usual limit design calculations.

9.2 LIMIT DESIGN FOR REINFORCED CONCRETE

When yielding starts, deflections increase sharply and repeated loading would introduce an element of fatigue. Hence under working load conditions yielding is certainly rather undesirable. On the other hand, as a reserve against final failure or collapse, the strength between yielding and failure has proved to be quite significant for structural steel, and the fact that this stage involves larger deflections does not seem too important.

Limit design is significant because it points out that (1) a statically indeterminate member or frame cannot collapse as the result of a single yielding section, and (2) between first yielding and final frame failure there normally exists a large reserve of strength.

Limit design was not recognized in the ACI Code until 1963 and then only in a rather minor way. In the 1971 Code under the title of redistribution of negative moments (Code 8.6) the provisions of 1963 are relaxed somewhat, but not expanded. (The beams carrying two-way slabs, discussed in Chapter 10, might be considered as now under a special form of limit design, but it is not so designated.) Nonelastic design conditions have long been recognized in averaging the moments across a footing or across fairly wide strips of two-ways slabs, flat plates, and flat slabs. This is a kind of limit design applied transversely, but it is not the usual limit design concept applied lengthwise to change the given moment diagram itself.

There are several reasons why the ACI Code can afford to appear even to be a little timid in the area of limit design.

First, in terms of economics, the situations for steel and for concrete are not comparable. With steel the efficient wide flange section has only a nominal increase in strength between M_y and M_p, these being the moment at first yield of steel and the fully plastic moment. With essentially the constant strength of a steel beam along the span, a change which equalizes several maximum moments is particularly helpful for most types of loading.*

It is only because reinforced concrete frames are designed for many alternate loadings that reserve strength exists somewhere under almost any specific loading. Although such reserve strength can very often be utilized to advantage by limit design, it is not proportionately as important as in steel construction.

A second reason why limit design is not pushed so strongly for reinforced concrete is that there is more uncertainty as to how concrete acts under all conditions of overload. It does not have the uniform ductile character

*Not, for instance, with a fixed mid-span load on a fixed-end beam which under elastic conditions has equal positive and negative moments.

exhibited by steel—unless certain design restrictions are imposed. Some of these behavior patterns are not yet sufficiently documented to make complete limit design acceptable to all designers.

9.3 MOMENT-CURVATURE RELATIONS

The basis for limit design lies in the inelastic behavior of materials at high stresses, that is, their ability to sustain a given yield moment while a considerable increase in local curvature develops. In a statically indeterminate frame this means that a given local section which tends to be overloaded yields and, in effect, refuses to accept more moment, but does not fail. Instead it forms what is called a "hinge" and thereby forces responses to further loading onto sections less fully stressed. A study of limit design must start by considering the relation between bending moment and the resulting curvature of a short length of the member, as shown in Fig. 9.1*a* for steel and 9.1*b* for reinforced concrete. The curve for steel can be approximated with considerable accuracy by two straight lines, as shown in Fig. 9.2*a*. The curves for reinforced concrete can, a little less accurately, be approximated by two straight lines or three straight lines as in Fig. 9.2*b*. For small percentages of sharply yielding steels, such as ordinary Grade 40 steel, the approximation is quite satisfactory. With a large percentage of the same steel, the yielding actually usable would be too short (before failure) to modify the ultimate behavior of the frame significantly. With a balanced beam, such as shown in the upper curve of Fig. 9.2, the inelastic behavior of the concrete in compression makes for some small nonproportional curvature increases at high loads, but does not provide the ductility necessary for significant modification of moments in the frame.

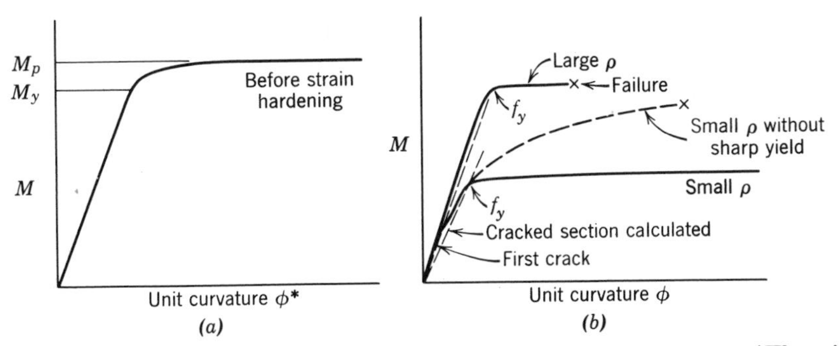

Fig. 9.1. Moment curvature. (*a*) Structural steel. (*b*) Reinforced concrete. (*There is no real distinction between ϕ as used here and $d\theta$ used in Chapter 6.)

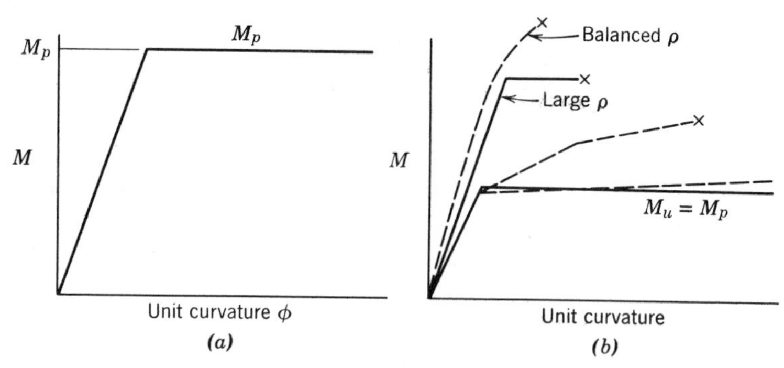

Fig. 9.2. Conventional moment-curvature relations. (*a*) Structural steel. (*b*) Reinforced concrete. Compare Fig. 9.1*b*.

For reinforcing steels without a sharp yield point, typical of some of the present-day high strength steels, the behavior might call for three straight lines as shown in Fig. 9.2*b*. Although behavior with such steel may be quite favorable, it is much more difficult to handle this case mathematically and it will not be considered further in this discussion.

It should be noted that either axial tension or axial compression in the member substantially modifies the M–ϕ curve for either steel or reinforced concrete, but more so for the latter.

9.4 FIXED-END BEAM WITH HINGES, AS AN EXAMPLE

The principal behavior which makes limit design interesting can be illustrated with a fixed-end beam under increasing uniform load as shown in Fig. 9.3. This will be discussed for the simplified two line curves of

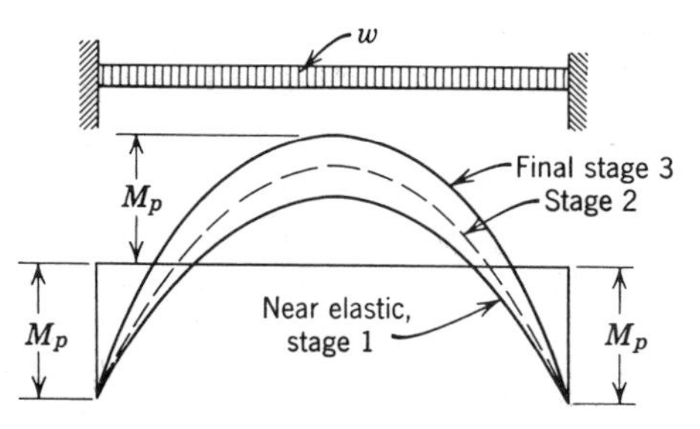

Fig. 9.3. Development of moment with increasing load, for idealized moment curvature.

Fig. 9.2. Moments build up proportionately until moment diagram 1 shows the negative moment as M_p and the positive moment roughly half as large. Under further loading M_p cannot increase, but the end element is forced to turn through a larger angle ϕ to permit the beam to deflect more and build up the necessary added resistance elsewhere. Thus the moment diagram would go through increased stages such as 2 until it finally developed a positive moment M_p as shown by diagram 3. At this point it could carry no more added load because it would be on the verge of developing into a mechanism, with hinges at each end and at the middle of the span.

The increase in load between diagrams 1 and 3 is 33% if M_p is the same for positive and negative moments, as in a rolled steel beam. However in reinforced concrete the M_p for positive moment would usually be much less than for negative moment. Thus this redistribution of moments which constitutes limit design as it relates to reinforced concrete, is of less economic significance for reinforced concrete than for steel. Nevertheless, a reinforced concrete beam designed under elastic theory normally is not fixed ended and is designed separately for different loads to produce maximum positive moment and maximum negative moment, as illustrated by Fig. 8.9*e*. Under either loading there is surplus moment capacity, in the one case at the supports, in the other at the mid-point. Thus, even in reinforced concrete, there is some room for limit design concepts.

9.5 CODE PROVISIONS FOR LIMIT DESIGN

Code provisions for limit design are under the title "redistribution of negative moments in continuous . . . members," Code 8.6. Redistribution is permitted *only* if the moments at supports are calculated by elastic analysis and if the ratio of steel ρ (or $\rho - \rho'$) is less than $0.5\rho_b$ where

$$\rho_b = \frac{0.85\beta_1 f_c'}{f_y} \times \frac{87{,}000}{87{,}000 + f_y}$$

Thus ρ_b is defined such that it is independent of any A_s' which may be present.*

The negative moments may be decreased (or increased) by a percentage defined as

$$20\,[1 - (\rho - \rho')/\rho_b]$$

*This definition avoids the complications found in Sec. 2.14 and Sec. 3.5 in fixing maximum allowable A_s in a beam in combination with A_s' under Code 10.3.2 and 10.3.3.

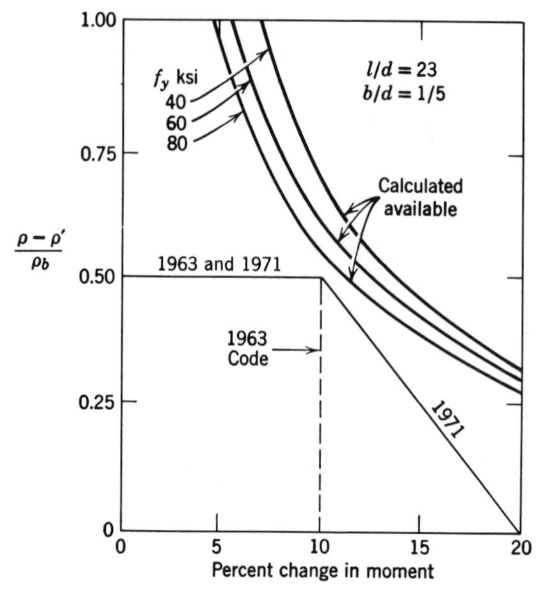

Fig. 9.4. Moment redistribution allowed and studies of minimum rotation capacities available. (Modified from ACI Code Commentary.)

Restricted in this way the problem of necessary hinge rotation at the ends of the beam is very easily handled by the very long flat M–ϕ curve which applies to the beam with small percentages of reinforcement. The curves in Fig. 9.4 for various grades of steel in the beam (indicated in the upper left of the figure) relate the theoretical possible change in moment to the steel used in the beam. The permissible change in moment under the Code is bounded by the two straight lines.

The next section reconsiders the continuous beam design of Sec. 8.10 and changes that might be made by redistribution of its moments.

9.6 REDISTRIBUTION APPLIED TO BEAM OF CHAPTER 8

The beam chosen in Chapter 8 starting at Sec. 8.10 is compact and rather small (9.5-in. web and h of 20 in. overall), just stiff enough for deflection and, simply as a structural member, on the expensive side because of the area of steel required. If beam weight is not critical, this member could be made eligible for redistribution of moments simply by widening its web to 14 or 15 in. Beam weight would go up 50% to 70% and steel would not be greatly reduced by this process, but ρ would be reduced by the extra width to around the $0.5\rho_b$ threshold required for redistribution.

If clearances permit, it would be better to increase b_w and d without changing their ratio so much. To this end, one could note that $\bar{k}$ (or k_u) in Table 2.1 is based on a ρ of 0.75 ρ_b. The use of $\bar{k}$ about 70% to 72% of the tabulated value (allowing for increased internal lever arm) would lead to ρ approximately 0.5 ρ_b. Alternatively, consideration of the usual required presence of 25% of the positive moment reinforcement at the support would mean a ρ' of roughly 20% of ρ and would permit a $\bar{k}$ of around 85% of the tabulated value, say, 650 for $\bar{k}$ instead of the 783 tabulated and the 1050 used in the earlier design. The first decision toward redistribution would be to accept the size this $\bar{k}$ leads to as an appropriate beam.

Balanced reinforcement, needed in assessing redistribution, is accurately available from Table 2.1, simply as 4/3 of the maximum ρ permitted there.

For a member chosen on the above basis (with ρ and ρ' to be verified later), at least a 10% reduction in negative moments is permissible by redistribution, a little more if $\rho - \rho'$ proves to be less than 0.5 ρ_b. From the negative moment diagram of Fig. 8.11*h* and the positive moment diagram of Fig. 8.11*e*, the modified diagrams of Fig. 9.5 (moment coefficients shown but omitting the wl_n^2 which goes with each) are obtained. In Fig. 9.5*a* the negative moment at the left is decreased to $-0.082wl_n^2$, approximately 90% of the original $-0.091wl_n^2$. At the same time the moment on the right is similarly increased to $-0.053wl_n^2$, leaving a positive moment at midspan almost unchanged, $+0.053wl_n^2$, and less than the positive moment in Fig. 9.5*b*. The same diagram except opposite hand would be appropriate for the right support of the beam.

In a similar fashion the negative moment in Fig. 9.5*b* can be increased 10% to $-0.058wl_n^2$, which reduces the positive moment coefficient from 0.072 to 0.067. This remains larger than the coefficient in Fig. 9.5*a* and hence is permissible; in fact there is still the spread between 0.067 and

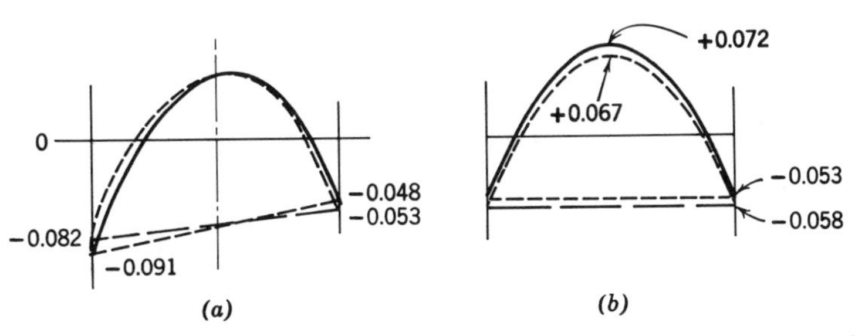

Fig. 9.5. Design moments (solid lines) redistributed (dashed lines). (*a*) Maximum negative moment at left. (*b*) Maximum positive moment.

0.053 which could be reduced if the redistribution could exceed the 10% used here.

The result of these moment changes is nearly a 10% reduction of all A_s values and the elimination of any A_s' requirement as such. Should the resulting $\rho - \rho'$ values at the support be less than $0.5\rho_b$, the reduction could be greater, as limited by the spread in positive moment values in the above paragraph. In that connection some allowance should be made for the slightly higher positive moment off center in Fig. 9.5*a*.

The positive moment reinforcement ρ is subject to the $0.5\rho_b$ limit just discussed. This ρ should be lower than at the support unless A_s' there was large.

9.7 BENDING AND STOPPING BARS

The Code, in taking the first step toward limit design, has not been specific with regard to details. Hopefully, the joint ACI-ASCE Committee studying limit design will recommend details with regard to bar lengths and development needed as the moment diagrams shift.

It is not the Code intent, as the author visualizes it, to reduce the steel requirements all across the span. Its intent is to reduce the peak moments and to reduce steel concentrations across columns where beam and girder steels must cross each other. The reduced steel of Sec. 9.6 will not modify the moments of Fig. 8.11*e* and *h* except in the last 10% increment of load. Cracking away from points of maximum moment should not be encouraged by reducing bar lengths in addition to total steel area. Hence for bar bending the author recommends using the basic moment diagrams of Fig. 8.11*e* and *h* with only the minor modifications shown in Fig. 9.6*a* and *d*. The required A_s curves then become those of Fig. 9.6*b* and *d* and their use involves little difficulty since the peak dotted values are easily calculated, either in terms of bars or bar areas, from the ratios 0.072/0.067 and 0.091/0.082. These required A_s curves are directly comparable to those of Fig. 8.13*a*.

9.8 PRESENT STATUS OF LIMIT DESIGN THEORY

Aside from bounding the redistribution problem more closely and liberalizing it within these bounds, there has been little progress toward general agreement on the way limit design for reinforced concrete should develop. In 1972 a general system of limit design appears farther away than it did

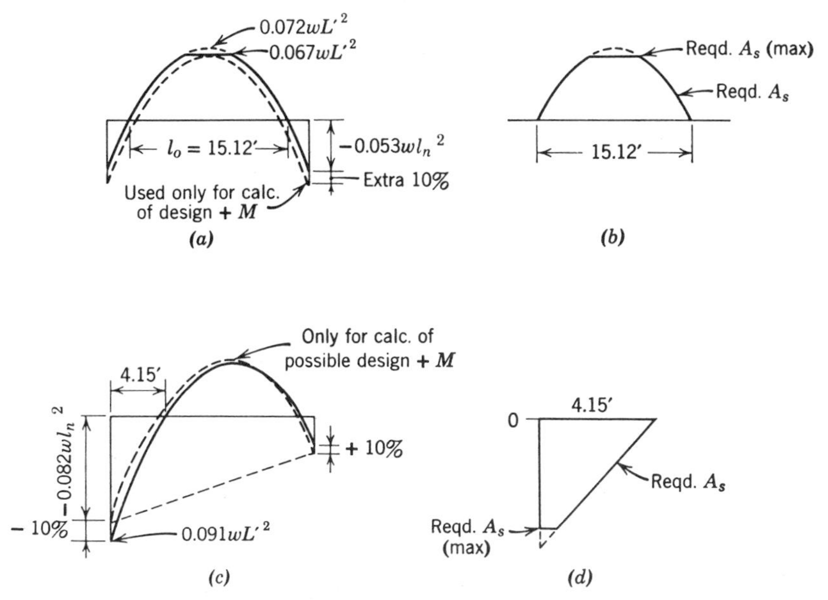

Fig. 9.6. Modification of design moments toward limit design. (*a*) Reduced positive moment for design. (*b*) Required A_s diagram for positive moment reinforcement. (*c*) Reduced negative moment for design. (*d*) Required A_s diagram for negative moment reinforcement.

in 1964.* There has been some considerable progress in basic concepts, but limit design in unbraced frames still appears improbable in the near future.

If one considers only frames sufficiently braced (without restricting normal joint rotation) that lateral deflection of the joints is eliminated, this restricted limit design may be reasonably near practical application.[5] Such frame restrictions make frame failure as a whole meaningless and failure of individual members the basic concept. The continuity which surrounds the member at service load gradually fades away as hinges form (Sec. 9.9) first at one end and later at the other end of the member (Sec. 9.10). Thus this problem gradually reduces to one of individual member collapse. Frame action after first yielding is discussed in Sec. 9.11 where it is pointed out that it is not necessary to trace these intermediate steps even where complete frame collapse is involved.

*When the second edition of this text was being prepared.

9.9 THE PLASTIC HINGE IDEA

If ϕ designates the angle change which develops over a unit length of the member, the M–ϕ curve or the relation between M and ϕ is shown in the simplest or ideal case in Fig. 9.1*a* and idealized in Fig. 9.2*a*. It can be similarly idealized for reinforced concrete members as in Fig. 9.2*b*.

The significant part of these M–ϕ curves for limit design lies beyond the first yield moment, in the zone where the plastic moment M_p remains unchanged over a very large range in ϕ values. When a point along the beam develops a moment M_p, it will act as a plastic hinge. This means it will continue to resist the moment M_p, but further loading will result only in an increased angle change ϕ rather than an increased moment. If an ordinary structural hinge is though of as frictionless, a plastic hinge may be considered as a "rusty hinge" having a definite, but limited, resistance to rotation. Thus, upon further loading of the member, the moment *changes* produced elsewhere on the member are the same as though a real hinge existed at the plastic hinge point.

The concept of a hinge at a point is mathematically convenient, but in real frames (steel or concrete) a plastic hinge is spread over some length of top and bottom beam surfaces, a length normally in the order of the depth of the member.[6] Occasionally, as in a symmetrical third-point loading of a beam, the yielding region can occur over all the middle third of the beam, a very generalized form not too closely resembling the simple plastic hinge idea.

9.10 THE COLLAPSE MECHANISM

In limit design as developed for steel frames, the ultimate strength considered is that which brings either the frame or the individual member to the verge of failure or collapse, assuming perfect plastic hinge action. To fit the present restricted idea of braced frames of reinforced concrete, consideration here is first directed to the simpler individual beam members.

A member cannot collapse under moment loading (except by buckling) until enough actual hinges or plastic hinges form to transform it into a mechanism. A cantilever beam thus collapses when a single hinge forms at the support, but a simple or continuous beam span must have three hinges to collapse, as indicated in Fig. 9.7. For the simple beam this requires the formation of only one plastic hinge because the supports furnish two real hinges. The entire moment diagram is statically determinate for this case. The total deflection consists of two parts, as indicated in Fig. 6.14 with the

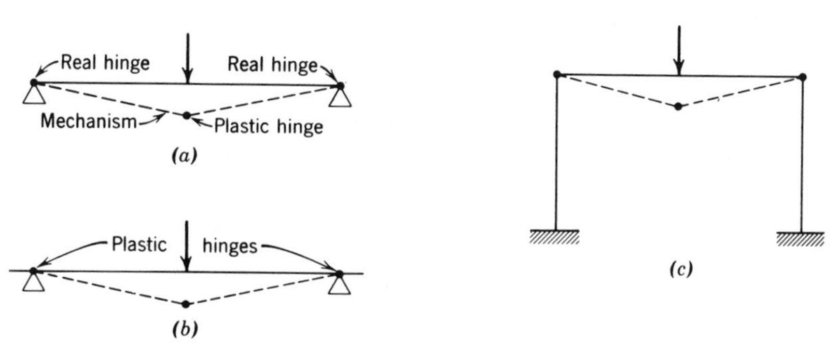

Fig. 9.7. Local collapse mechanism. (*a*) Simple span. (*b*) Continuous span. (*c*) Frame member.

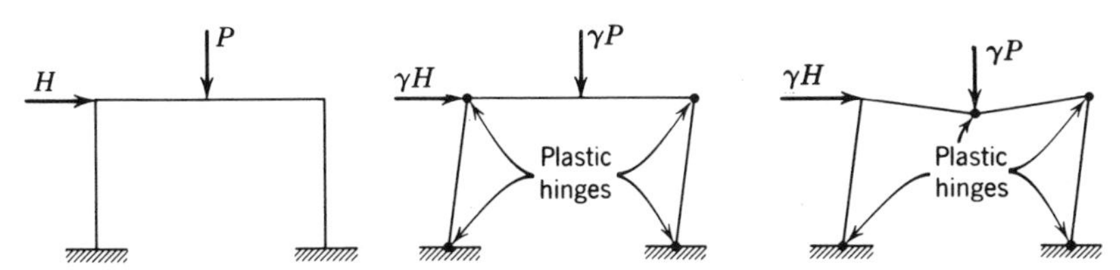

Fig. 9.8. Mechanisms for collapse of frame as a whole, under load factor γ.

added deflection which occurs after M_p develops consisting of a simple triangle, just as though the two segments were plane and not curved by the earlier loading.

For a single span of the usual continuous beam three plastic hinges are required for collapse; and these hinges can greatly modify the moments from their elastic analysis values.

A frame may collapse as a whole without any individual member developing the three hinges, provided the frame as a whole develops enough hinges to act as a mechanism, as in the middle sketch of Fig. 9.8.

9.11 FRAME ACTION BETWEEN FIRST YIELDING AND COLLAPSE

In Sec. 9.2 it was stated that yielding under working loads was objectionable, although this does not mean that such a condition is totally prohibited under all circumstances. Generally, however, a study under working loads involves analysis on an elastic basis.

As overloads are applied, one or more of the moments might be expected to pass the M_y value and reach the M_p value.* However, it would be quite an unusual frame which would have its maximum moments so equalized that all the plastic hinges required to form a mechanism would develop at any one loading stage and thus lead to an immediate collapse. The usual pattern is for a single highly stressed section (or several) to develop first M_y and then M_p, with more load necessary before some other plastic hinge forms; a number of such loading increments might be required to produce all the plastic hinges necessary for a mechanism.

Although it might be interesting to trace out this sequence of loads and moment diagrams, it would be lengthy process. Fortunately it is an unnecessary process. For any given loading on a given frame, the collapse pattern can be established entirely independently of the original elastic moment pattern and of the intervening elastic-plastic stages.

The ratio of this collapse loading to the service loading is the available safety factor γ, or in reinforced concrete it might be better to say the available load factor γ. This brings out one limitation inherent in the usual limit design approach. *This available load factor is meaningful only as it relates to that particular loading used* and it implies that all portions of that loading are considered only as increasing proportionately.

This limitation is important enough for elaboration through a specific example. If the frame of Fig. 9.8 were analysed and γ were established as 3, it would mean that if all loads were increased in proportion, the frame would not collapse until it carried loads of $3H$ and $3P$. This would not mean it could carry $3H$ alone or $3P$ alone or $3H$ along with P or any other combination except $3H$ and $3P$. If the load factor for H alone is needed, this requires a separate analysis.

The student should also note that the collapse load is far beyond the elastic range; hence the effect of the two loads H and P may be far different from the sum of their individual effects.

9.12 LOWER BOUND OR LIMIT ON LOAD FACTOR

A lower bound or lower limit on the real load factor is determined from a moment diagram, *any* moment diagram that is consistent with the given loads, that is, *any* moment diagram that satisfies statics.

The fundamental idea may be very simply indicated in terms of the elastic moment diagram. Let M_p represent the plastic moment (different

*In steel construction M_p is definitely greater than M_y; in reinforced concrete M_p is definitely M_u and M_y may not be differentiated in the calculations.

values for different members or for different parts of members) and M_s represent the corresponding service load moments from the moment diagram. If M_p/M_s is determined for every member, the smallest ratio found is a lower bound on the true load factor. Why? Up to value M_p the idealized M–ϕ diagram assumes elastic action, which means the moment is directly proportional to the load. Hence the ratio M_p/M_s indicates how much the load may be increased without inelastic action and before any readjustment of the moment diagram shape occurs. This is a *lower* limit on how much the load may be increased because the moment diagram will probably shift to a more favorable shape as this critical section begins to act as a plastic hinge. For further loading the critical moment at the hinge remains constant at M_p, whereas other parts of the moment diagram increase toward the M_p values.

Greenberg and Prager have proved mathematically that any moment diagram satisfying statics may be used as the basis of a lower bound. Normally it is easier to develop an "arbitrary" moment diagram than to establish the elastic moment values.

As one tries different possible moment diagrams, he discovers that higher values of this *lower* bound are obtained when more individual ratios of M_p/M_s become equal to this bound. The moment diagram can thus be "equalized" toward values establishing the true load factor. In complex cases it is desirable finally to assume plastic hinges at these high moment points and check such a solution by the energy-mechanism procedure (Sec. 9.13) which fixes an *upper* bound, just to verify the solution.

If this approach is applied to the beam of Fig. 9.9*a*, with all M_p values 100 k-ft, one might start with a simple beam moment diagram as in (*b*) with

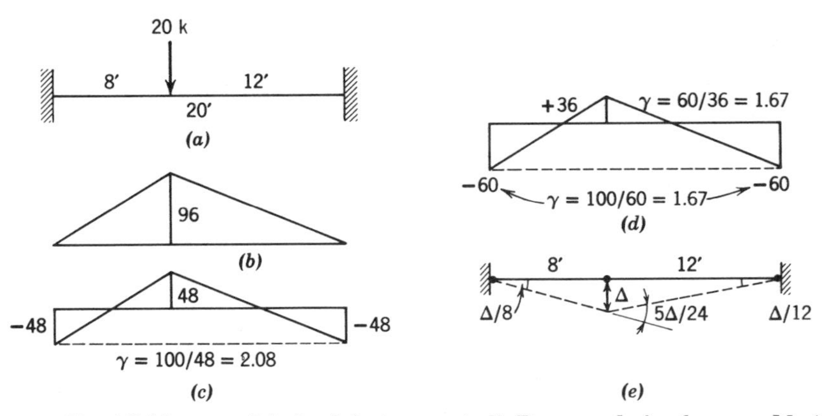

Fig. 9.9. Establishing possible load factor, γ. (*a-b*) Beam and simple span M. (*c-d*) Adjusting M diagram. (*e*) Energy approach.

$M_{max} = (20 \times 12/20)8 = 96$ k-ft and then try to develop three peaks of moment having the same ratio of M_p/M_s. This is easily accomplished by shifting the axis as in (*c*), making M_p/M_s uniformly $100/48 = 2.08$ at each peak moment value. This 2.08 is the true load factor possible. If M_p for positive moment were 60 k-ft and M_p for negative moment remained 100 k-ft, the moment diagram of (*d*) would be the proper one and the load factor would be 1.67.

9.13 UPPER BOUND OR UPPER LIMIT ON LOAD FACTOR

If a frame is on the verge of collapse, it must have developed a sufficient number of plastic hinges to change it into a mechanism. As the mechanism starts to collapse, the loads move in such a way as to contribute energy to the system while rotations occur at the plastic hinges which absorb energy. For a specific given or assumed mechanism a token deflection establishes specific angle changes at all plastic hinges. The required external load to maintain this movement can be established by equating internal energy absorbed at the plastic hinges to the external energy input by the loads.

The true load factor for the frame can be no larger than the ratio of *P* (collapse) obtained for this particular mechanism to *P* (working load); it may be smaller for some other (more probable) mechanism. The mechanism-energy approach thus establishes an upper bound or upper limit on the value of the true load factor.

If all possible mechanisms are investigated the lowest load factor will be the true one; but it is desirable to check this load factor against the lower bound discussed in the Sec. 9.12. For this purpose it should be noted that the moments are all statistically determinate when M_p values are used at each plastic hinge in a mechanism.

Consider again the beam of Fig. 9.9*a*, with positive moment $M_p = 60$ k-ft, negative moment $M_p = 100$ k-ft. From the discussion of Sec. 9.12 the necessary hinge locations here are obviously at the three possible peaks on the moment diagram, as in Fig. 9.9*e*. The energy imput from the load is 20Δ. Energy absorption at hinges $= 100(\Delta/12 + \Delta/8) + 60(5\Delta/24) = 20.8\Delta + 12.5\Delta = 33.3\Delta$. This leads to the load factor possible as $33.3\Delta/20\Delta = 1.67$. It is not necessary to check other solutions even though this is an upper bound theorem, because the pattern is unique.

In entire frames the hinge location is not so obvious. The student might wish to imagine the hinge at midspan instead of the load and see how it leads to a higher and hence invalid γ.

9.14 BASIC CONCEPTS ESSENTIAL TO LIMIT DESIGN AS SUMMARIZED BY FURLONG

(a) Assigned Limit Moments

Furlong proposes[5] limits within which he feels it feasible to design totally without an elastic frame analysis. He limits himself to frames braced to avoid joint deflection laterally. His paper has taken the limit design concept out of the realm of involved theory and injected a practical design approach. Whether the profession is ready to depart this far from elastic analysis remains to be seen. Some such approach appears necessary if a later Code is to go beyond the moment redistribution now authorized. The direct approach appeals to the author as having much merit compared to the present "patch-up" of elastic analysis. Furlong's proposal at least presents pertinent data and a challenge and encourages examination of the basic problem of whether limit design for reinforced concrete can be made worthwhile.

(b) Flexural Ductility

All limit design analyses require a material that will sustain M_p through a large angle of rotation in order that the simple hinge concept be feasible. Solutions are more complex if a sloping line, as shown dotted near the bottom of Fig. 9.2b, is required and quite complex if the three-line-dotted shape in the middle of that figure is required. Ductility here is defined as the ratio of the ultimate ϕ_u at the end of the horizontal line to the initial ϕ_e at the start of the horizontal section. Values of 4 to 6 are desirable and Ref. 5 relates available ductility to variation in either $\rho - \rho'$ or $M_u/(\phi bd^2)$. The method of A. L. L. Baker[3] in general frames uses a group of simultaneous equations to find the related values of needed hinge rotation at each joint to assure that local failure of a hinge will not lower the frame capacity as usually computed.

(c) Ratio of M_y to M_p

The ratio of M_y/M_p lies between 0.94 and unity for singly reinforced members made with reinforcement having a long flat yield section in their stress-strain curve. This implies to the author that the use of $M_p = M_u$ will usually be feasible without involving M_y much, if at all, in the calculations.

(d) Yielding at Service Loads

Furlong has shown that reasonable limits are possible to assure no yield under service loads in load patterns which elastic analyses usually recognize.

SELECTED REFERENCES

1. Lynn S. Beedle, *Plastic Design of Steel Frames*, John Wiley and Sons, New York, 1958.
2. Michael Rex Horne, "A Moment Distribution Method for the Analysis and Design of Structures by the Plastic Theory," *Jour. Inst. Civil Engrs.*, 3, Part III, Apr. 1954, p. 51.
3. Report of Institution Research Committee on Ultimate Load Design, A. L. L. Baker, "Utlimate Load Design of Concrete Structures," *Proc. Inst. Civil Engineers (London)*, 21, Feb. 1962, p. 399
4. M. Z. Cohn, "Rotational Compatibility in the Limit Design of Reinforced Concrete Continuous Beams," *Flexural Mechanics of Reinforced Concrete, SP-12*, ACI/ASCE, Detroit, 1965, p. 359.
5. Richard W. Furlong, "Design of Concrete Frames by Assigned Limit Moments," *Jour. ACI*, 67, No. 4, April 1970, p. 341.
6. Alan H. Mattock, "Rotational Capacity of Hinging Regions in Reinforced Concrete Beams," *Flexural Mechanics of Reinforced Concrete, SP-12*, ACI/ASCE, Detroit, 1965, p. 143.

10
Two-Way Slabs

10.1 ONE-WAY AND TWO-WAY SLABS

In Chapter 8 one-way slabs were discussed in connection with continuous beams. These differ only in the large width-to-depth ratio of the slabs. This large ratio results in shear stresses that are less important in one-way slabs than in beams. Such slabs need transverse steel, usually an arbitrary amount, to care for shrinkage and temperature stresses. Transverse steel also assists in distributing concentrated loads which may come on the slab.

Although true one-way slabs are sometimes used, Sec. 8.1 has already pointed out the many two-way slab variations which are frequently economical and often used. Chapter 10 deals with the behavior and design of slabs which should *not* be designed as one-way slabs. All these can logically be called two-way slabs although common usage has preempted this title generally to mean slabs well supported on beams on all four sides.

This presentation is organized in the following sequence:

1. Identification, behavior, and analysis
   a. Two-way slabs on beams, Sec. 10.2
   b. Flat slabs and flat plates, Sec. 10.3
2. Flat slabs and flat plates – basic direct design rules, Sec. 10.4
3. Flat slabs and flat plates–design examples
   a. Interior panel of flat plate, Sec. 10.5.

b. Special aspects of design.

   Minimum stiffness of column, Sec. 10.6.
   Effective column stiffness in a slab system, Sec. 10.7.
   Concentration of reinforcement over the column, Sec. 10.8.
   Code thickness equations, Sec. 10.9

c. Exterior panel of flat plate without edge beam, Sec. 10.10 and 10.11
d. Interior panel of flat slab with drop panel, Sec. 10.12

4. Slabs on beams—design examples.
   a. General, Sec. 10.13.
   b. Two-way slabs on stiff beams, Sec. 10.14.
   c. Two-way slabs with stiffened strips, Sec. 10.15.
5. Equivalent frame method.
   a. Use and frame constants, Sec. 10.16 and 10.17.
   b. Design example, Sec. 10.18 and 10.19
6. Designs outside the Code slab system rules, Sec. 10.20, 10.21, and 10.22.
7. Special conditions.
   a. Shear reinforcement, Sec. 10.23
   b. Lift slabs, Sec. 10.24.
   c. Cantilevers from slabs, Sec. 10.25.
   d. Freely supported slab corners, Sec. 10.26.

IDENTIFICATION, BEHAVIOR, AND ANALYSIS

10.2 TWO-WAY SLABS ON BEAMS

(a) General

In the 1963 Code two-way slabs were treated entirely separately from flat slabs, both numerically and in basic philosophy of design. The usual two-way slab is square or rectangular, monolithic with relatively stiff beams; only this case was specifically treated in the 1963 Code. In the 1971 Code this case is treated as a special case of the general slab solution which considers beams of varying stiffness as well as no beams at all. The 1971 Code covers only slabs on beams that frame directly into columns, not slab panels away from columns.

(b) Elastic Analysis—Mathematical Approach

Two-way slabs, even single panels simply supported, require a three-dimensional approach for analysis. Such slabs are rarely statically determinate in their internal moments and shears. They are the most highly indeterminate of all ordinary structures.

Usually slabs have been analyzed as flat thin plates made of a homogeneous elastic material which has equal strength and stiffness in every direction, that is, an isotropic material. On this basis solutions for simple cases can be established from partial differential equations by the use of advanced mathematics. Westergaard[1] was a pioneer in such analysis in this country.

With differential equations, approximations must generally be introduced to handle practical cases. The use of difference equations and various computer techniques has led to reasonably accurate solutions for some problems. Theoretical boundaries thus established are valuable in assessing the merits of shorter and less accurate analyses. Today computer solutions can handle most cases, but these solutions are not ready for everyday design.

(c) Typical Moment Patterns

Since the student cannot really analyze a slab except by arbitrary code provisions, it is desirable that he have a clear picture of the physical action of the slab. With a little imagination he should be able to visualize the general deflected shape taken by a uniformly loaded slab. A simply supported square slab will deflect into a saucerlike shape; and unless the corners are held down they will actually rise a little off the supports. An oblong slab will take a platterlike shape. A very long narrow slab will take a troughlike shape except near the ends. For fixed edges, there must also be a transition zone around the edges in which the slope gradually turns downward from the horizontal edge tangents. If contours are roughly sketched, as in Fig. 10.1 for the simple supports, they give more than a clue to the moment pattern.

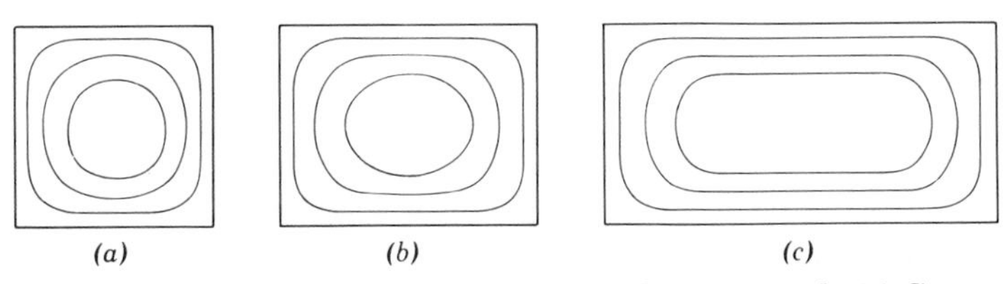

Fig. 10.1. Approximate contours for two-way simply supported. (*a*) Square slab. (*b*) Oblong slab. (*c*) Long rectangular slab.

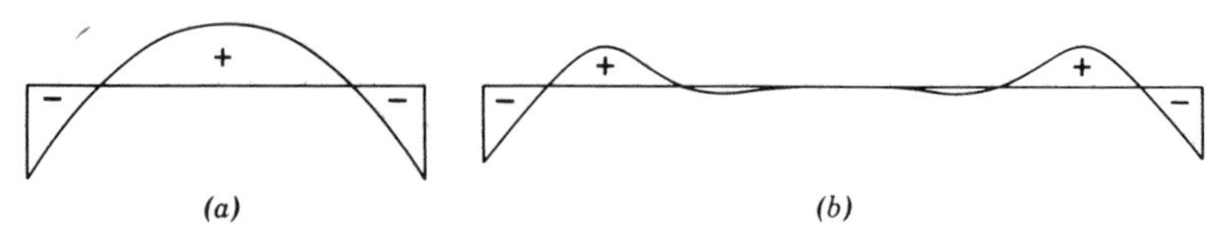

Fig. 10.2. Typical moment diagrams for continuous slabs. (*a*) Short-span strip. (*b*) A very long strip, say, length three times width.

In a square slab, simply supported, a strip across the middle cuts the greatest number of contours and has the sharpest curvature and largest moment. In a long narrow slab only the short strips have significant curvature and bending over most of the panel length. The long center strip is essentially flat and without moment except near its ends.

In the continuous slab all slab strips in each direction have negative moments near the supports and positive moments near mid-span, as shown in Fig. 10.2*a*. The long center strip in a long narrow slab is an exception, or a special case, in that its positive moment occurs not at the center but at the point where the strip starts to curve upward; the moment over its mid-span length is small or zero (Fig. 10.2*b*) and the short center strips act almost exactly as one-way slabs.

Mathematical analysis shows that the negative moment on such a long strip is nearly the same regardless of the long span length. It is almost the same as it would be for a square panel having the short span dimensions. One method in the 1963 Code* found it appropriate to express both long span and short span moments in terms of coefficients times the *short span* squared.

(d) Approximate Analyses

Approximate analyses of slabs are usually somewhat crude, certainly so from the standpoint of the theoretical man or mathematician. Subsection (c) has indicated that the slab in each direction acts somewhat as a one-way slab. But the perpendicular slab strips of Fig. 10.3*a* are not independent in action. They share in carrying the load and thus each has a smaller moment than a one-way slab. They must deflect the same total amount; hence their relative stiffness becomes a factor in establishing the load and the moment each must carry. With slab thickness a common factor, the longer span is the more flexible and carries the smaller moment and load.

*See text Sec. 10.22.

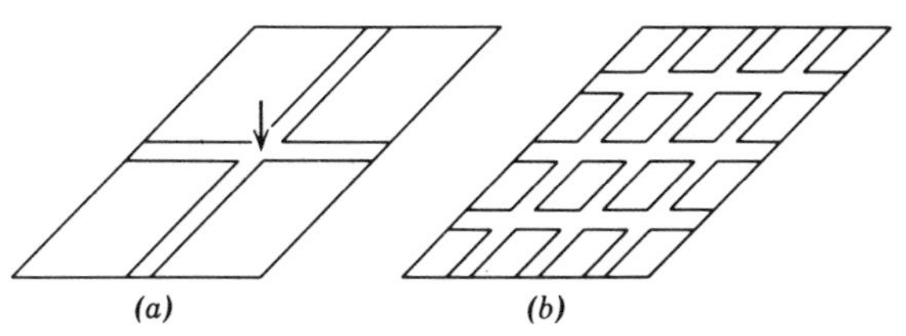

Fig. 10.3. The strip idea for two-way slabs. (*a*) Single intersecting strips. (*b*) Multiple intersecting strips.

The use of a *single* slab strip in each direction is obviously a crude analogy. The short-span element across the middle of the panel deflects a much greater amount that a parallel strip alongside the edge beam; the edge strip can deflect little more than the beam. A better analysis would result if the slab were considered as several strips in each direction, held to common deflections at their intersections, as in Fig. 10.3*b*. A still better approximation would consider torsional stiffness of these strips as well as their bending stiffness. Since three strips each way would give nine intersections, it should be noted that the labor of such an analysis increases at least with the square of the number of strips considered. Many simultaneous equations must be solved, one for each intersection, or many successive approximations must be tried for a solution. With enough strips, results closely equivalent to those obtained from the partial differential equation solution are found, but the labor is excessive except on the computer.

Although elastic analyses of homogeneous isotropic plates provide much information that is helpful, they are exact solutions only for the assumed conditions. They usually do not recognize the fact that reinforcing steel is made lighter near the panel edges than at the center and that short span steel is heavier than long span steel. Nor do they usually recognize the changes in relative stiffness which result as cracked sections develop from the bending moment.

(e) Inelastic Considerations in Slab Design

In a 1926 paper Westergaard recommended moment coefficients which gave considerable weight to the nonelastic readjustments in slab moments which take place before failure. In recognition of these favorable readjustments, his recommended coefficients were established at 28% below strictly elastic values. The percentage reduction corresponds in a way to a similar reduction which had been accepted for flat slabs, but for two-way slabs it gave more recognition to maximum moment loadings. The 1963 ACI Code requirements fundamentally stemmed from Westergaard's recommendations, although work by Marcus and by vanBuren and di Stassio was also recognized.

Design practice and codes by no means attempt to design for the real distribution of bending moment existing across the slab under elastic conditions. This moment varies from element to element across a strip. If a square slab is considered as subdivided into 1-ft strips in a given direction, the center strip obviously has the sharpest curvature and largest moment. Curvature and moment on adjacent strips decrease gradually until alongside the edge beam the deflection, curvature, and moment all approach zero.

Even if this were a one-way slab under variable loading it would not collapse when the maximum moment raised the steel stress in a single narrow strip beyond its yield point. This one strip would then simply become more flexible and leave more of any added load to be carried by adjacent elements less highly stressed. Before real failure could occur, all elements would have to be yielding in some fashion. (Chapter 11 discusses the yield-line analysis and collapse conditions for the ultimate strength of slabs.)

A two-way slab is highly redundant both in the slabs and the beams. Careful consideration of statics will show that any load, say, near the middle of the panel, must be carried in full to reactions at the four corners of this slab panel. It is often overlooked that this means that *all the load* must be carried in (say) the north-south direction to these reactions and *in addition* that *all the load* must be carried also in the east-west direction to these reactions. In fulfilling these two requirements slab and beams work as a team. Only part of the load is carried by the slab in the north-south direction; the remainder is carried in the east-west direction by the slab and then picked up by the north-south beams, which in turn carry it north-south. Thus one portion is carried directly north-south by the slab and the remainder must eventually be carried north-south by the beams. A similar condition exists in an east-west direction.

When a given slab strip in a two-way slab is overstressed, it not only can shift added load to neighboring parallel strips (as in a one-way slab) but it can also shift load to perpendicular strips if these are not overloaded and if their supporting beam is not overloaded. Thus the distribution of load support between the slab in one direction and its parallel beams is very redundant, but the total load on the two is fixed.

This readjustment in stiffness and bending moment indicates that the steel does not have to be placed exactly in accordance with the elastic moment requirements. With its highly redundant system of supporting the load, the slab will, before failure, go far towards shifting the moments to the sections capable of resisting them. One can say that *almost* any arbitrary arrangements of steel, in sufficient quantity to carry the total load, will be developed before a slab completely fails. Chapter 12 design moves in this direction.

At working loads, however, a poor distribution of slab steel may result in local yielding of the steel, large cracks, and increased deflection. For desirable action under working loads, steel should be placed at least roughly in accord with the moments existing at the service load; and these moments are more nearly the elastic analysis moments.

Engineering practice uses a uniform spacing of steel over the center strip (one-half the panel width in a square panel) and reduces the steel towards the edges of the panel. This uniform center strip steel is designed to take the *average* rather than the maximum moment on the center strip, as shown in Fig. 10.4 for a simply supported slab. Thus the method takes considerable cognizance of nonelastic behavior as failure approaches.

Strictly, slab and beam design in two-way slabs should be a joint matter. The 1971 Code for the first time recognizes this joint behavior, now usable in design for both flexible and stiff beams.

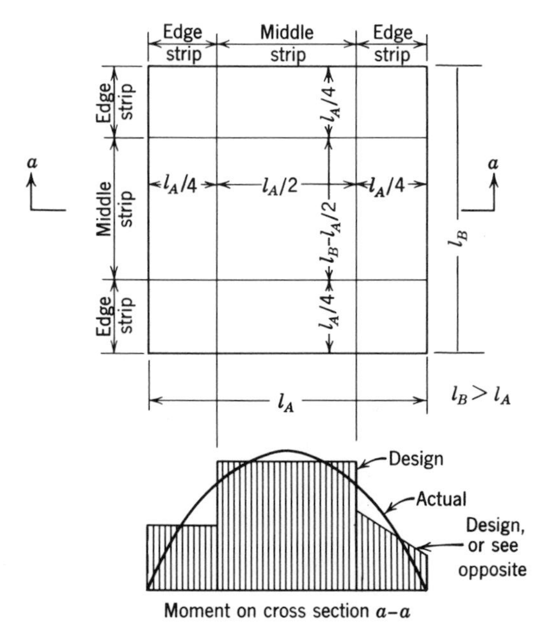

Fig. 10.4. Middle strip positive moment, slab simply supported. Comparison of actual and design moments.

Despite the fact that in practice approximate moments are used and the steel is designed for average moments rather than maximum, slabs have proved one of the most trustworthy structural elements and have stood up well under great abuse and overload.

10.3 FLAT SLABS AND FLAT PLATES

(a) Identification

A flat slab is a concrete slab so reinforced in two (or more) directions as to bring its load directly to supporting columns, generally without the help of any beams or girders, as in Fig. 10.5. Beams are used where the slab is interrupted, as around stair walls, and at the discontinuous edges of the slab.

The supporting columns may be increased in size near the top to form a column head or column capital, as shown in Fig. 10.5. In addition, the slab may be thickened by a drop panel around the column, as shown in Fig. 10.5, but many slabs are constructed without the drop panel.

The ACI Code also considers "slabs with recesses or pockets made by permanent or removable fillers between ribs or joists" as flat slabs. This includes two-way joist systems and the so-called waffle slab (Fig. 10.6*a*).

The flat plate floor is a flat slab having neither drop panel nor column capital, as in Fig. 10.6*b*.

The slab band type of floor thickens the slab into bands of greater depth, as in Fig. 10.7. Although the most usual pattern results in one-way transverse slabs of variable depth supported on wide shallow beams or bands in the longitudinal direction, some engineers have used such thickened sec-

Fig. 10.5. Typical flat slab construction, with drop panel. (Courtesy Portland Cement Assn.)

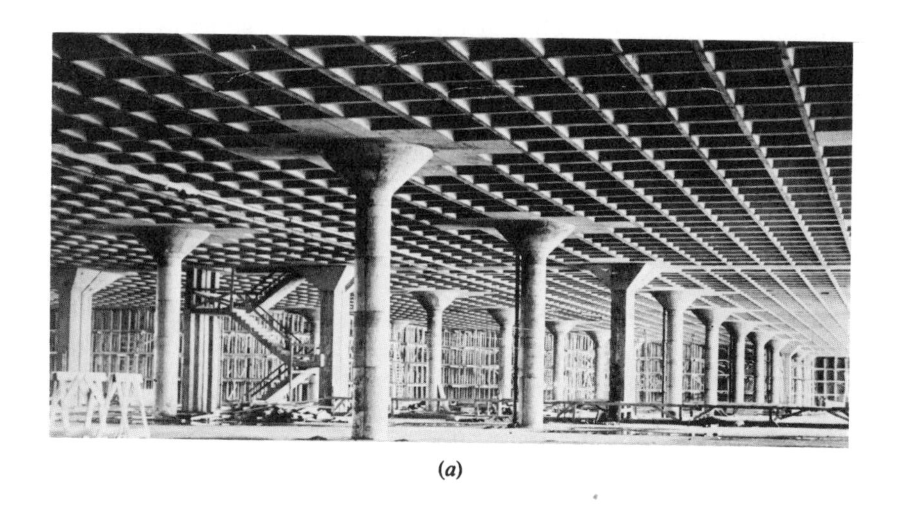

(*a*)

(*b*)

Fig. 10.6. Variations of flat slab form. (Courtesy Portland Cement Assn.) (*a*) Waffle slab. (*b*) Flat plate.

tions in end panels much like a drop panel around a flat slab column. Such construction partakes somewhat of the nature of flat slabs.

Two-way slabs supported on all sides by wide shallow beams also involve some flat slab action. When the slab proper occupies only the middle half of the panel, the Code recognizes it as a paneled ceiling form of flat slab.

Flat slabs, being thin members, are not economical of steel, but they are economical in their formwork. Since formwork represents over half the cost of reinforced concrete, economy of formwork often means over-all economy.

For heavy live loads, that is, over 100 psf, flat slabs have long been recognized as the most economical construction. From the 1950's flat plate floors has proved economical in tall apartment house construction. Reduced story height resulting from the thin floor, the smooth ceiling, and the possibility of slightly shifting column locations to fit the room arrangement all seem to be factors in the over -all economy.

(b) Early Slabs as Empirical Construction

Flat slabs were built and sold many years before an adequate analysis was available.[2] The originator was an American, C. A. P. Turner. He built from intuition as to how slabs would function and applied a proof load to the completed slab to satisfy the owner.

Fig. 10.7. Slab band construction. (Courtesy Portland Cement Assn.)

Numerous flat slab structures were load-tested during the period from 1910 to 1920.* These slabs performed well under test loads. The difficulty in closely correlating test results with the moments which statics shows must be present is mentioned in (d).

(c) The Statics of a Flat Slab

In 1914 J. R. Nichols[3] showed that statics required a total of positive and negative moments equal to

$$M_0 = \frac{1}{8} Wl \left(1 - \frac{2}{3}\frac{c}{l}\right)^2$$

where W is the total uniform panel load, l is the span, and c is the diameter of the column capital. This conclusion follows from an analysis of a half panel.

Consider the interior panel of Fig. 10.8a, loaded with a unit load w and surrounded with similar panels equally loaded. The straight boundaries of the slab are all lines of symmetry, indicating that they are free from shear and torsion. Hence all the shear and torsion must be carried around the curved corner sections which follow the column capital.

Likewise, if the slab is subdivided along the middle of the panel, this line is a line of zero shear and torsion. Thus the free body of Fig. 10.8b,c is subject to a downward load W_1 acting at the centroid of the loaded area, an equal upward shear W_1 acting on the curved quadrants, the total positive moment M_1 acting on the middle section ab, and the total negative moment M_2 acting about the y-axis on section $cdef$. This assumes the bending moment around the column capital is uniformly distributed, which means no torsional moments exist there. Moments also exist about the x-ais on efa and dcb, but these do not enter into $\Sigma M_y = 0$.

$$W_1 = w(l^2/2 - \pi c^2/8) = (l^2 - \pi c^2/4)w/2$$

The moment of this load about axis y-y is:

$$\frac{wl^2}{2} \times \frac{l}{4} - \frac{\pi c^2 w}{8} \times \frac{2c}{3\pi} = \frac{wl^3}{8} - \frac{wc^3}{12}$$

*In 1947 Professor J. Neils Thompson and the author ran such a test on a flat slab constructed about 1912. Although it cracked badly, its strength was surprising.

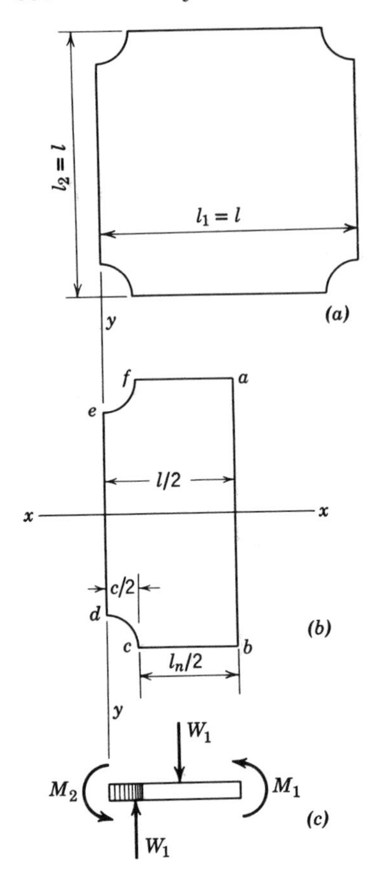

Fig. 10.8. Equilibrium conditions indicate $M_1 + M_2$ is established by statics.

If the upward shear W_1 is considered uniformly distributed around the quadrants *cd* and *ef*, the resultant acts at a distance c/π from the *y*-axis. Equilibrium of moments about the *y*-axis then gives:

$$-M_1 - M_2 + \frac{wl^3}{8} - \frac{wc^3}{12} - \frac{w}{2}\left(l^2 - \frac{\pi c^2}{4}\right)\frac{c}{\pi} = 0$$

$$M_1 + M_2 = M_0 = \frac{wl^3}{8} - \frac{wc^3}{12} - \frac{wcl^2}{2\pi} + \frac{wc^3}{8} = \frac{wl^3}{8}\left(1 + \frac{c^3}{3l^3} - \frac{4c}{\pi l}\right)$$

$$M_0 = \frac{wl^3}{8}\left(1 - \frac{2}{3}\frac{c}{l}\right)^2$$

The value of $[1 - \frac{2}{3}(c/l)]^2$ approximates the longer parenthesis reasonably well.*

*The error is 0.5% low for $c/l = 0.1$, 0.5% high for $c/l = 0.2$, 1.3% high for $c/l = 0.25$, and 2.0% high for $c/l = 0.30$.

This statics solution tells nothing about how this total moment is distributed between positive or negative moment or how either varies along the slab width. The solution also neglects possible torsional moments around the column capital which will act if the tangential bending moments there are not uniform in their distribution.

When flat slab panels are not square, the long span produces the larger moment. The student should contrast this with the two-way slab supported on all four sides where the short slab strips carry the larger load and larger moment. In such a slab the long slab strips carry less load and less moment, but it should be noted that the long beams have to carry heavy slab reactions and large moments. Two-way slab moments in both directions are reduced because the beams help to carry the moment. In the flat slab, in contrast, the *full* load must be carried in *both* directions by the slab alone.

(d) Correlation of Statics with Tests

In early tests (mechanical) strain measurements were necessarily over considerable lengths and indicated average rather than maximum steel stresses. This led to an overly optimistic evaluation of the reserve strength, even when attempts were made to allow for this problem. Also tests were not to failure and thus reflected less tension cracking than would develop at ultimate. Thus early test evaluations led to a total M_o of $0.09wl_1^2l_2[1 - (2/3)(c/l_1)]^2$, which was accepted until the change in the present Code, in spite of the apparent conflict with statics.

The behavior of large areas of this type of construction has proven that the static analysis fails to give the total picture. There is a horizontal thrust, a form of arching which, under favorable conditions, greatly strengthens interior panels and even adds some strength to edge panels. Tests in the 1960s have also shown this reserve strength, to such an extent that the 1971 Code uses $M_o = wl_2l_n^2/8$, which bases the moment on clear span l_n between square supports*. This looks in Fig. 10.10*a* like true statics and it is an adaptation (reduction) of Nichol's value. It is admittedly still short of the full static moment, since all of the reaction is not on the near face of the column, but this M_o is larger than in earlier codes.

(e) Theoretical Distribution of Moment

Westergaard developed a theoretical slab analysis which accompanied Slater's test analyses.[1] He established the distribution of positive and negative moments which would exist throughout a flat slab when it was

*If supports are not square replace them (for this value of l_n) with square columns of equal area.

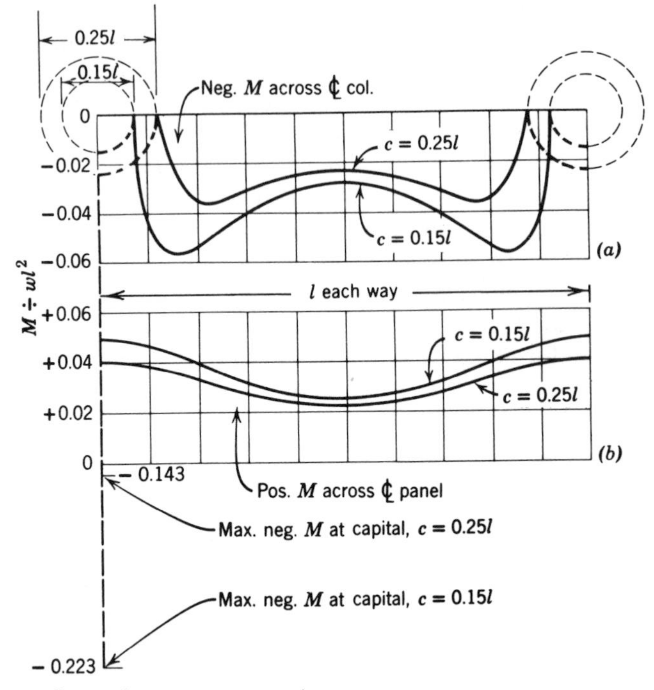

Fig. 10.9 (a, b). Legend appears on next page.

considered as an isotropic plate. His work, roughly checked by Slater's test analyses, formed the basis for the specified subdivision of the total M_0 into positive and negative moments and for the further subdivision of these into moments on the two design strips set up in the specifications. Figure 10.9a shows his calculated distribution of negative moments on strips crossing the column center line for $c/l = 0.15$ and 0.25. Around the column capitals the moment in a radial direction is considered constant at $-0.223\ wl^2$ for $c/l = 0.15$ and at $-0.143wl^2$ for $c/l = 0.25$. These values are nearly four times the maximums shown between columns. The positive moments on strips crossing the middle of the span are shown in Fig. 10.9*b*. For a strip running along the center line between columns, the moment diagram is shown in Fig. 10.9*c*, and for a strip running along the middle of the panel in Fig. 10.9*d*. It will be noted that both these strips have a form of moment diagram very similar to that of a one-way slab, with positive moment in the middle of the span and negative moments at the ends.

As a result of extensive analytical work and quarter scale tests on multipanel slabs[4,5,6] at the University of Illinois, and a three-quarter scale test by the Portland Cement Association,[7] the distribution of moments

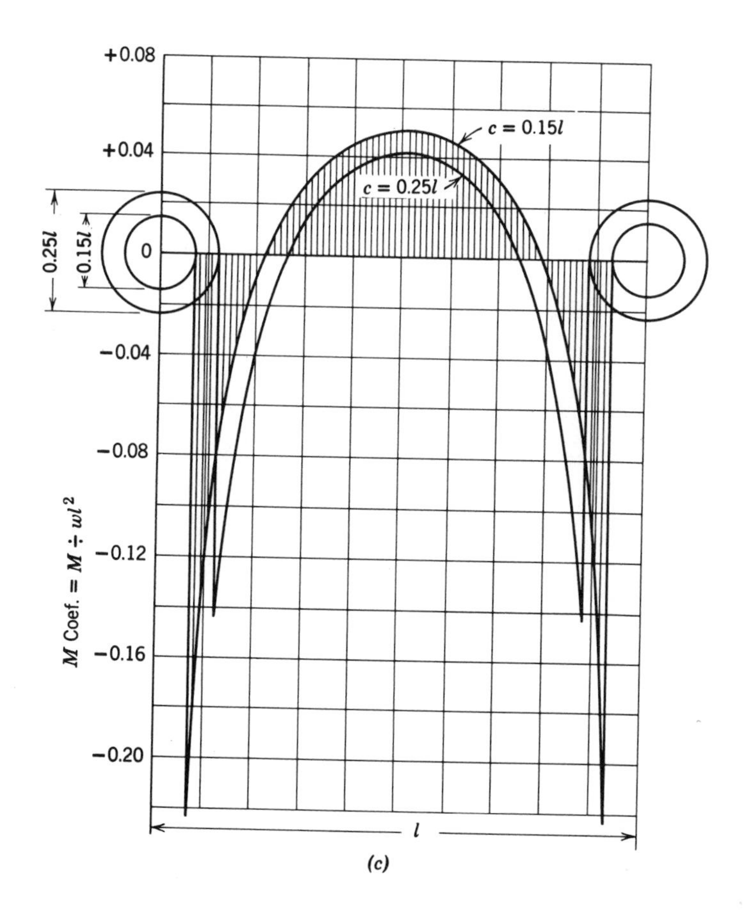

(c)

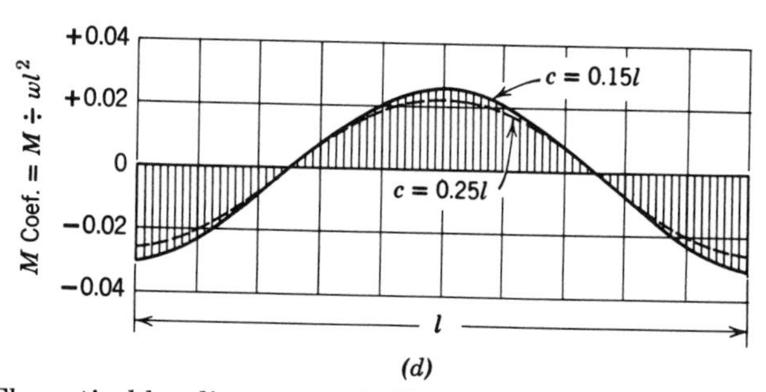

(d)

Fig. 10.9. Theoretical bending moments (Poisson's ratio zero). (Adapted from Ref. 3, ACI.) (*a*) Negative moments on strips crossing the center line of column and the column capital. (*b*) Positive moments on strips crossing the center line of span. (*c*) Moment diagram for strip along center line of columns. (*d*) Moment diagram for strip along middle of panel.

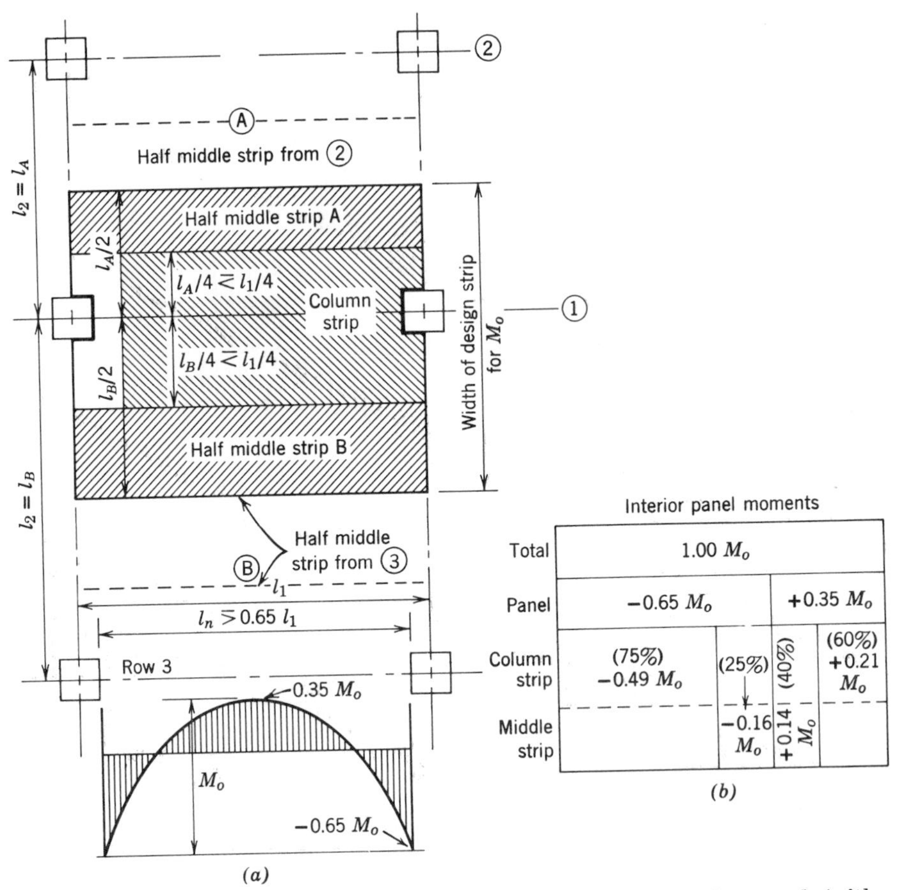

Fig. 10.10. Calculation of M_o and its subdivision, for an interior panel (without beams) in x-direction; similar in y-direction except for half column strip at exterior.

actually used in the 1971 Code (Fig. 10.10*b*) has been slightly modified from the 1963 Code values. The 1971 values now apply to cases of slabs with monolithic cast beams (if framing directly into columns) as well as to flat slabs.

FLAT SLABS AND FLAT PLATES—BASIC DIRECT DESIGN RULES

10.4 DIRECT DESIGN—PRIMARY RULES

(a) Code Direct Design Requirements for Flexure

When layouts are relatively simple a direct design procedure is given in Code 13.3, limited to slabs meeting these limitations:

1. A minimum of three spans each way, directly supported on columns.
2. Rectangular panels with the long span not more than twice the short span.
3. Successive spans not differing by more than 1/3 of the longer span.
4. Live load not more than 3 times dead load.

Columns may be offset from either axis by up to 10% of the span in the direction of the offset.

The minimum thickness rules of Code 9.5.3.1 appear very complex but, for the case of no beams, the interior panels are always governed by the minimum of 5 in. when without drops, 4 in. with drops, or by Code Eq. 9.6 and 9.8 which then become the same* and specify the thickness as:

$$h = l_n(800 + 0.005f_y)/36{,}000.$$

This equation reduces to $h = l_n/36$ for Grade 40 bars and $h = l_n/32.7$ for Grade 60 bars, where l_n is the clear span.

For exterior and corner panels *without* edge beams the thickness must be increased 10% above those just stated (Code 9.5.3.3).†

Only bending moments for the interior panels will be discussed first, to show the basic procedures without the slight complications arising from evaluation of the effective restraints at the exterior columns (discussed in Sec. 10.7).

The basic moment calculation, $M_o = (\frac{1}{8})wl_2l_n^2$, is for a panel centered on a column line, that is, for a half panel from panel A and a half panel from panel B in Fig. 10.10*a*. This moment calculation, centering on the column lines in each direction, uses moments based on the clear span l_n and the average width of $0.5(l_{2A} + l_{2B})$. For interior panels (with or without beams) $-0.65M_0$ is considered negative moment and $+0.35M_0$ is positive moment, as in Fig. 10.10*a*.

The moments, as already noted earlier, are larger close to the columns than elsewhere. While they vary greatly, it is safe to average them over a considerable width for a design. For this purpose the Code defines a column strip and a middle strip.

A column strip is a width (Fig. 10.10*a*) of $l_{2A}/4$ plus $l_{2B}/4$ (but not more than $l_1/2$ total) adjacent to the centerline of the columns. The remainder of the area used in Fig. 10.10 makes up two half middle strips, one for panel A and one for panel B. In design the two adjacent half strips at the column are treated as one design strip and the two adjacent half middle strips as the other design strip. In uniform panels each strip is $l_2/2$ in width.

*$\alpha_m = 0$ when no beams; $\beta_t = 1$ for interior panel. See Sec. 10.9 for the general discussion of the h equations.

†As used in Sec. 10.11*a*.

Without beams (which means $\alpha_1 = 0$ in the Code notation), the column strip negative moment is 75% of the $-0.65M_o$, i.e., $-0.49M_o$, the other $-0.16M_o$ going to the two half middle strips, as indicated in Fig. 10.10*b*. Likewise the column strip is assigned 60% of the $+0.35M_o$ moment, or $+0.21M_o$, and the $+0.14M_o$ remaining goes to the two half middle strips. For uniform panels, the two half middle strips are alike, giving $-0.16M_o$ and $+0.14M_o$ for the total middle strip moments.

The designer is permitted to modify any design moment by 10%, if all of M_o is still assigned.

The system for exterior panel moments is quite similar, except that less negative moment goes to the exterior, and more to the interior column line; the positive moment is correspondingly increased (Sec. 10.10), much as for typical exterior beam panels. The detailed discussion of this operation is delayed because the exterior slab restraint involves not only the exterior column but also torsion in the edge beams or, in their absence, in the slab strip. The details of this now would tend to cloud the simple structure of the direct design method.

(b) Shear Requirements

Although shear requirements were adequately discussed in Chapter 4, shear is introduced here again for emphasis. While the yield line theory shows that flexural capacity can be obtained by several arrangements of reinforcing, no such freedom exists in caring for shear around columns in the case of flat plates and flat slabs. In either, shear is *the critical element* of design. Nearly any test slab, if loaded to collapse, will show a final shear failure or, around an exterior column, a combined shear and torsion failure.

On interior columns it is possible to use crossed steel shearheads to strengthen the slab (up to 75% more v_u) in shear (Code 11.11.2); but a shearhead modification of an exterior column to the same specification *does not function* and is wasted. Around interior columns it is also possible to use closed stirrups effectively (up to 50% more v_u but with v_c decreased to $2\sqrt{f_c'}$), if the bars or stirrups are carefully developed.

Around exterior columns, especially when spandrel beams (edge beams) are omitted or when the slab extends only to the outside of the column (less than all the way *around* the column), careful design for shear *and torsion* is absolutely essential. An overhang of the slab provides a great improvement. Many flat plate structures have been built without extending the slab beyond the column, but the author is still prejudiced against this construction detail. A very conservative approach is fully warranted. Attention is specifically directed to Code 13.3.4.10:

Edge beams or the edges of slabs shall be proportioned to resist in torsion their share of the exterior negative design moments.

The provisions of Code 11.13.2 about the transfer of part of the moment by an eccentricity of shear are also in the right direction. The designer must be alert to the fact that the shear transfer to the exterior column is (nearly always) the weakest point in the slab.

(c) Development of Reinforcement

The Code includes Fig. 10.11 which indicates the requirements for extending some of the positive moment bars into the support area, to within 3 in. of the centerline of supports. The left half of this figure is for slabs of uniform thickness; the right half is for slabs which are thickened around the column by a drop panel (Fig. 10.5 and Sec. 10.12). Of course bars must be of such a size that they can be developed within the lengths shown (or else be extended farther), but this is not commonly a serious problem.

FLAT SLABS AND FLAT PLATES—DESIGN EXAMPLES

10.5 DESIGN OF FLAT PLATE—INTERIOR PANEL

Design by the direct method an 18 ft by 20 ft flat plate interior panel for a service dead load, including its own weight, of 100 psf* and a live load of 60 psf. Assume 15 in. square columns, $f_c' = 4000$ psi, Grade 60 steel, 0.75 in. cover because not exposed to the weather.

Solution

Normally the slab thickness used would be determined by the thickness of an exterior panel. However, since a transition in depth can often be accommodated, the design will be developed here as if this panel stood alone in the design.

Depth probably will be determined by shear around the column or by the thickness needed against deflection. The minimum thickness (Code 9.5.3.1) is 5 in. or $h = l_n(800 + 0.005f_y)/36{,}000 = l_n/32.7 = (20 - 1.25) \times 12/32.7 = 6.87$ in., say 7 in.

*This should include floor finish, partition allowance, and the like.

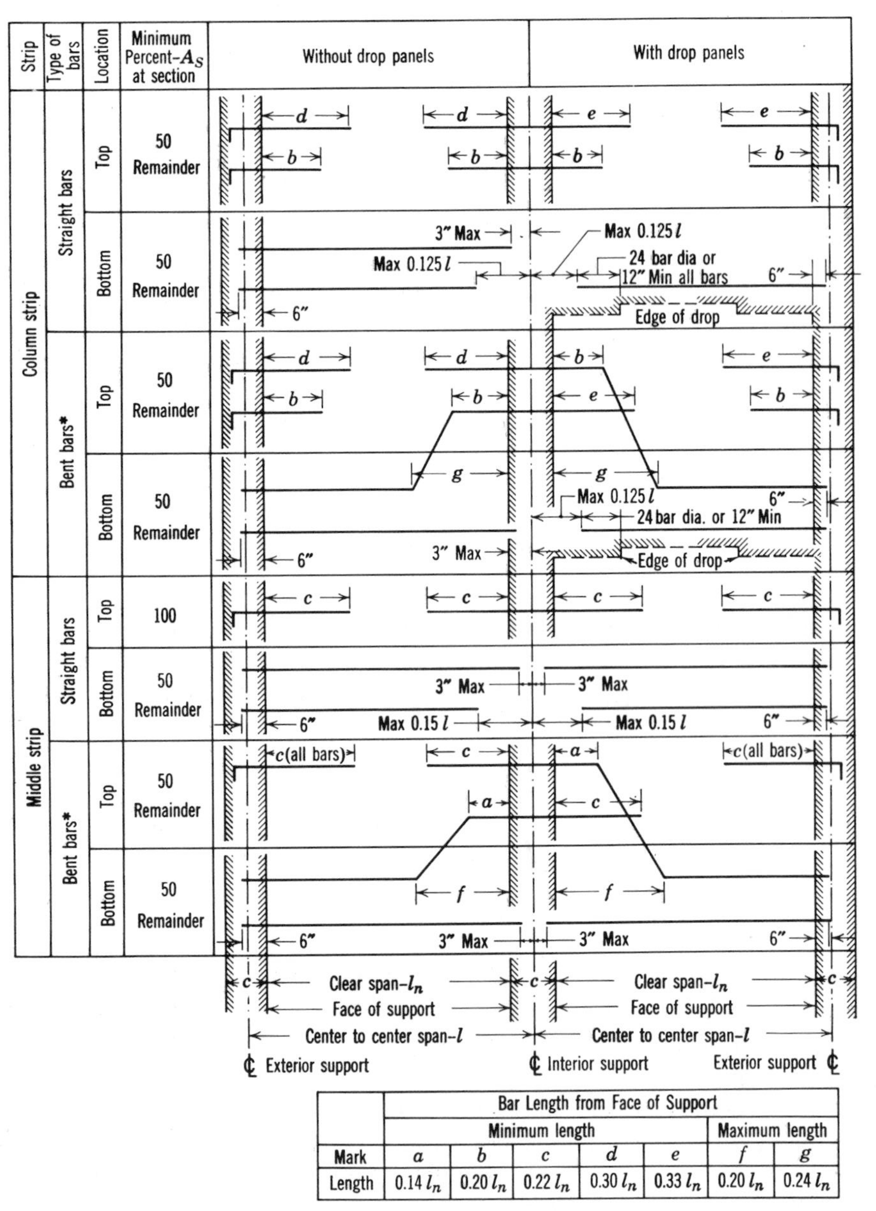

	Bar Length from Face of Support						
	Minimum length					Maximum length	
Mark	*a*	*b*	*c*	*d*	*e*	*f*	*g*
Length	$0.14\ l_n$	$0.20\ l_n$	$0.22\ l_n$	$0.30\ l_n$	$0.33\ l_n$	$0.20\ l_n$	$0.24\ l_n$

Fig. 10.11. Minimum length of slab reinforcement, slabs without beams (From ACI Code, p. 50). (See Code 12.2.1 regarding extending reinforcing into supports.) *Bent bars at exterior supports may be used if a general analysis is made.

Dead load = 1.4 × 100 = 140 psf
Live load = 1.7 × 60 = 102

w = 242 psf

Shear must be checked on a section enclosing the column at $d/2$ from the column face. For shear, the average depth of the two steel layers seems logical and safe, say, $d = 7 - 0.75 - 0.50$ bar = 5.75 in. The width of each side of this section thus becomes 15 + 5.75 = 20.75 in. = 1.73 ft.

$V_u = 242(20 \times 18 - 1.73^2) = 86{,}500$ lb
$\overline{V} = V_u/\phi = 86{,}500/0.85 = 102{,}000$ lb
Allowable $v_c = 4\sqrt{f_c'} = 4\sqrt{4000} = 253$ psi
$d_v = \overline{V}/v_c b = 102{,}000/(253 \times 4 \times 20.75) = 4.83$ in. $< h = 7$in. above
USE $h = 7$ in.

For moment the steel in the long direction will be nearest the top and bottom faces of the slab, that for the short span next inside, making:

Trial $d_l = 7.0 - 0.75 - d_b/2 = 6.00$ in. for long span
$d_s = 7.0 - 0.75 - 1.5d_b = 5.50$ in. for short span

both assuming #4 bars. Note that such round numbers for the resulting d occur only for a few bar sizes; also that in a square panel the smaller d would control, *not* the average d, for steel in *both* directions, unless supervision were close enough to justify different steel in the two directions. The small bar size was selected here because the depth for deflection seems to be governing over stress criteria and this points to a lightly reinforced slab.

M_{ol} (long span) $= 0.125wl_2l_n^2$
$= 0.125 \times 242 \times 18\ (20 - 1.25)^2/1000$
$= 192$ k-ft
M_{os} (short span) $0.125 \times 242 \times 20\ (18 - 1.25)^2/1000$
$= 170$ k-ft

Next check depth for moment by comparing $k_m = M_u/bd^2$ to the allowable of 937 from Table 2.1. The highest moment is the long span negative moment in the column strip (over the column).

$-M_u = -0.75(0.65M_o) = -0.75 \times 0.65 \times 192 = -93.5$ k-ft
$-\overline{M} = -104.4$ k-ft

The strip width is half the transverse panel length, 9 ft or 108 in.

Reqd. $k_m = M_u/bd^2 = 93.5 \times 12{,}000/(108 \times 6.00^2) = 287 \ll 937$

This indicates such a greatly underreinforced slab that a check in the short direction is not needed.

A tabular form, as in Table 10.1, expedites this type design and organizes the results in a manner easily available to the detailer. The author finds the use of $\overline{M}$ and "ideal" materials convenient since it eliminates ϕ from all subsequent calculations, but this is optional. The lever arm z of the internal couple will be estimated as $0.95d$, a little high in the usual range of $0.90d$ to $0.95d$ because the slab is much thicker than moment requires. This value typically need not be modified for the various moments; but z could be checked where any close decision is to be made in fixing the bar spacing. Here, since the student probably has little "feel" for the proper z, a check will be made at the worst section, that just used in checking k_m.

$$A_s = \overline{M}/f_y z = (104.4 \times 12)/(60 \times 0.95 \times 6.0) = 3.65 \text{ in.}^2$$

$$a = \overline{N}_t/(0.85 f_c' b) = (3.65 \times 60)/(0.85 \times 4 \times 108) = 0.598 \text{ in.}$$

$$z = 6.0 - 0.598/2 = 5.70 \text{ in.} = 0.95d$$

Thus $0.95d$ is satisfactory here and will be on the safe side elsewhere (where smaller moments lead to still smaller a values).

This area of steel also indicates 19- #4 bars (3.80 in.^{2}) at a spacing of $108/19 = 5.65$ in., which is reasonable (possibly a trifle close) for the worst spot. Bars will usually be specified by total number in the strip rather than by exact spacing, as 19- #4 at 5.5 in. $\pm$, thus also letting the field man know that a uniform spacing close to 5.5 in. will need little rearrangement to cover the strip properly.

In Table 10.1 the cross-section sketches are helpful to emphasize the steel arrangement. Solid circles represent the bars being designed. The zf_y constant used has also incorporated the 12 factor needed to change the moment from k-ft to k-in. Minimum temperature and spacing steel (Code 7.13) is also shown, with the ratio for Grade 60 bars being 0.0018 based on the area of the *total slab thickness*. This steel includes both top and bottom steel where both exist, as in a negative moment region. Although in such a region the main steel is top steel for negative moment, half of the positive moment steel is required to be run into the support region and stays effective for temperature cracking resistance. If the temperature needs *at* the support appear critical these bottom bars can be lapped instead of stopping them 3 in. short of the column center line as indicated in Fig. 10.11. Finally, the maximum spacing for slabs is fixed at $2h$ which is 14 in. and this governs in several places.

Two arrangements of bars are shown in the table, the first for all bars straight as shown schematically in Fig. 10.12*a*. These would usually be shown on the drawings as bands, somewhat as in Fig. 10.12*b*. There for simplification the listings have been shown almost on two separate panels, one for the steel in each direction. Where spans are unequal and steel must

TABLE 10.1. Steel Calculations for Flat Plate Panel of Sec. 10.5.

	Long Span				Short Span			
	Column Strip (9′)		Middle Strip (9′)		Column Strip (9′)		Middle Strip (11′)	
	Negative	Positive	Negative	Positive	Negative	Positive	Negative	Positive
For #4 bars 7″	6.00″	6.00″	6.00″	6.00″	5.50″	5.50″	5.50″	5.50″
$\overline{M}_o = M_o/0.9$	213 k-ft				189 k-ft			
$\overline{M}$, k-ft	−104.4	+44.8	−34.2	+29.9	−92.8	+39.7	−30.3	+26.5
d, in.	6.00	6.00	6.00	6.00	5.50	5.50	5.50	5.50
$zf_y/12$	28.5	28.5	28.5	28.5	26.2	26.2	26.2	26.2
A_s reqd., in.^{2}	3.65	1.57←	1.20←	1.05	3.53	1.51←	1.16←[a]	1.01
Min. $A_s = 0.0018 \times 7b$	—	1.36	1.36	1.36←	—	1.36	1.66	1.66←
Straight Bars								
No. of #4	19	8	~~6~~		18	8	~~6~~	
Aver. spcg., in.	5.5 ±	13.5 ±	~~18~~ Use #3		6.0	13.5	~~24~~ Use #3	
Max. spcg., in.	—	14	14		—	14	14	
No. of #3			11	13			11	15
Aver. spcg., in.			10 ±	8.5 ±			12	9 ±
Alternate-Bent & Str.								
Bent	8- #4	4- #4	12- #3	6- #3	8- #4	4- #4	12- #3	6- #3
Straight	11- #4	4- #4	—	7- #3	10- #4	4- #4	—	9- #3

[a] Neg. A_s + half pos. A_s can be made effective for temp.

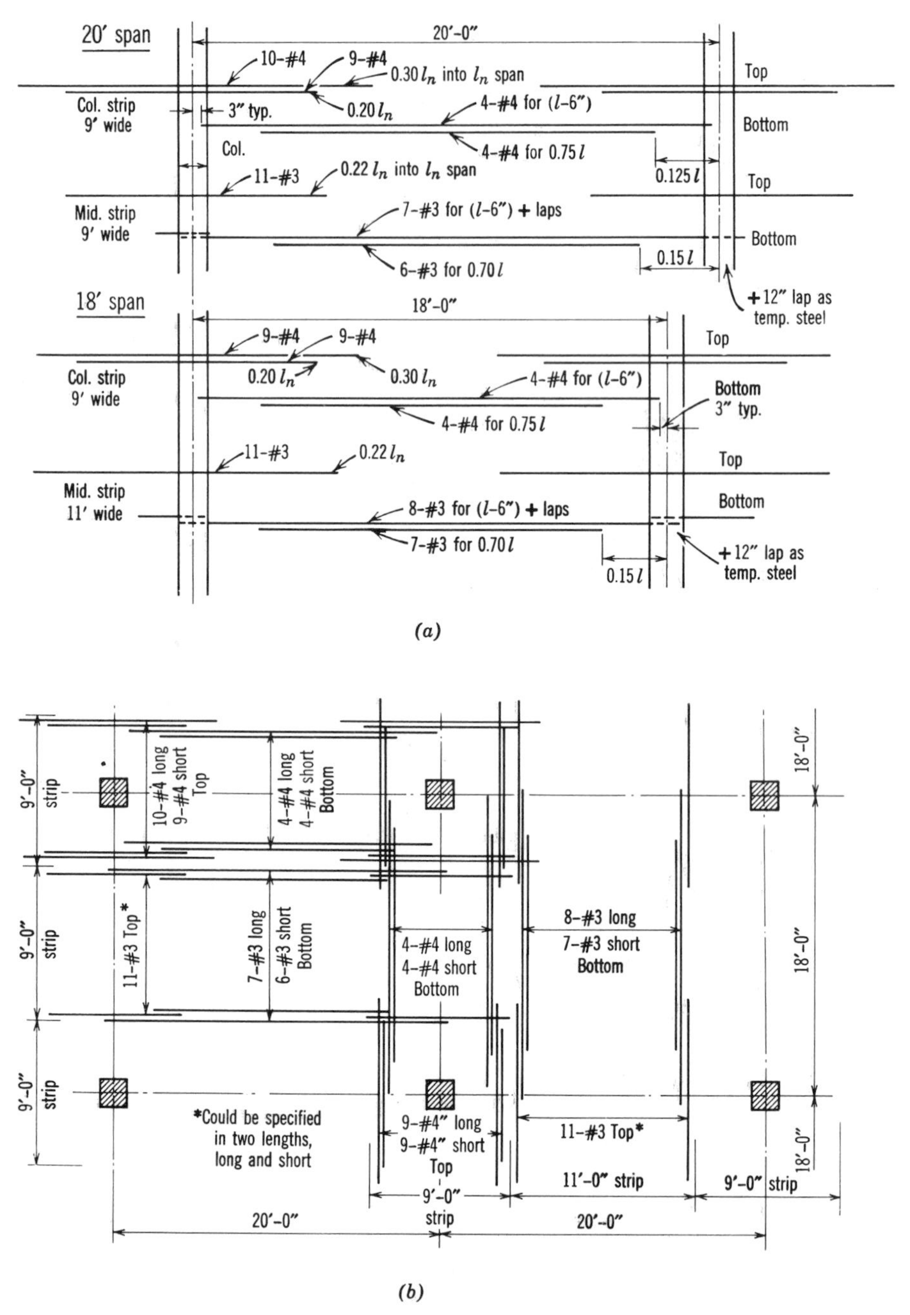

Fig. 10.12. Steel for flat plate floor. (*a*) Symmetrical schematic arrangement as in elevation. (*b*) Plan showing bands of reinforcing. Lengths of bars would often be listed instead of simply "long" or "short."

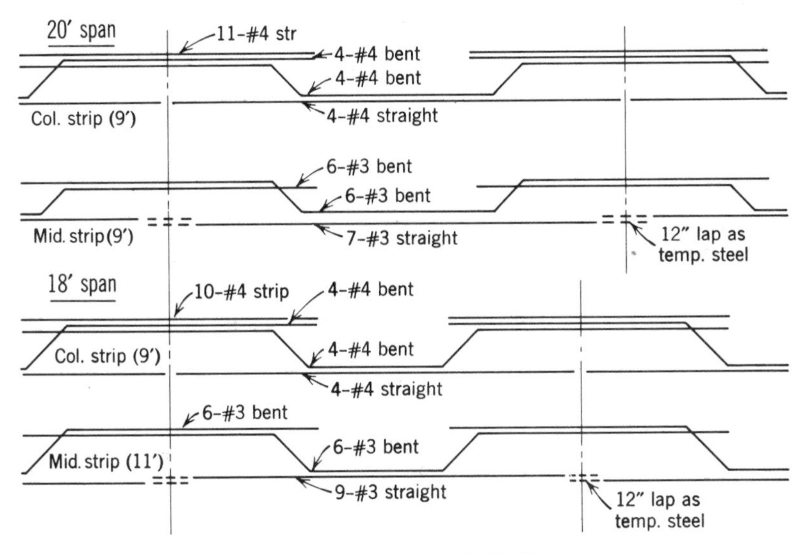

Fig. 10.13. Pattern for bent bars, the alternate in Table 10.1.

be shown in both directions on many panels, it is possible to superimpose as Fig. 10.12*b* shows along one column centerline.

The second arrangement in Table 10.1 is for bent bars, as sketched in Fig. 10.13.

The foregoing concludes the primary or main design calculations for this example. However, there are four special aspects of design which might influence or change the decisions already made. The following aspects deserve more generalized comment than would have been appropriate within the design itself:

Sec. 10.6 Minimum stiffness of columns.
Sec. 10.7 Effective column stiffness in a slab system.
Sec. 10.8 Concentration of reinforcement over the column.
Sec. 10.9 Code thickness equations – comments on their use

10.6 MINIMUM STIFFNESS OF COLUMNS

Slab moments, especially for slabs without beams, are sensitive to pattern loadings.[5] Provision for the total M_o moment based on full loading of all panels still leaves overstress possible under pattern loadings. The Code accepts a possible overstress to 33% under these conditions. To stay within this limit when the unfactored dead load is less than twice the unfactored live load, the system requires either a specific minimum size column or, with a lesser column, an increase in the positive moments above those indicated by M_o.

The minimum column is defined by a minimum required ratio

$$\alpha_{min} = \Sigma K_c / \Sigma(K_s + K_b)$$

as tabulated in Table 10.2 for various values of β_a, the ratio of dead to live load (at service load level), panel proportions l_2/l_1, and relative beam stiffness values α. For dead load of twice the live load, α_{min} is zero, but α_{min} increases as β_a decreases.

TABLE 10.2.[a] Minimum Column, α_{min} Values

		Relative beam stiffness, α				
β_a	Aspect ratio l_2/l_1	0	0.5	1.0	2.0	4.0
2.0	0.5–2.0	0	0	0	0	0
1.0	0.5	0.6	0	0	0	0
	0.8	0.7	0	0	0	0
	1.0	0.7	0.1	0	0	0
	1.25	0.8	0.4	0	0	0
	2.0	1.2	0.5	0.2	0	0
0.5	0.5	1.3	0.3	0	0	0
	0.8	1.5	0.5	0.2	0	0
	1.0	1.6	0.6	0.2	0	0
	1.25	1.9	1.0	0.5	0	0
	2.0	4.9	1.6	0.8	0.3	0
0.33	0.5	1.8	0.5	0.1	0	0
	0.8	2.0	0.9	0.3	0	0
	1.0	2.3	0.9	0.4	0	0
	1.25	2.8	1.5	0.8	0.2	0
	2.0	13.0	2.6	1.2	0.5	0.3

[a] Copy of Code Table 13.3.6.1

As an example, consider the slab design of Sec. 10.5 where $\beta_a = 100/60 = 1.67$, which is less than the specified ratio of 2 and requires investigation. $l_2/l_1 = 20/18 = 1.11$. $\alpha = 0$ (no beams). From the table:

$$\beta_a = 2.0, \text{ any } l_2/l_1, \alpha = 0, \alpha_{min} = 0$$

$$\beta_a = 1.0, l_2/l_1 = 1.11, \alpha = 0, \alpha_{min} = 0.74$$

Interpolating: $\beta_a = 1.67$, $l_2/l_1 = 1.11$, $\alpha = 0$, $\alpha_{\min} = 0.24+$, say 0.25

$\alpha_{\min} = 2K_c/2K_s = K_c/K_s = 0.25$
The smaller $I_s = (18 \times 12)7^3/12 = 6180$ in.^{4}
For a slab span of 20 ft, $K_s = 4EI_s/l = 4E\ 6180/240 = 103E$

The varying I of Code 13.4.1.4, is not used although this would be appropriate (not required) if one had the related curves convenient.

$$0.25 = K_c/K_s = K_c/103E,\ K_c = 25.7E$$

For a story height of 10 ft, $K_c = 25.7E = 4EI_c/120 = EI_c/30$

$$\text{Min. } I_c = 25.7E \times 30/\text{E} = 771 \text{ in.}^4 = t^4/12,\ t = 9.85 \text{ in.}$$

The 15 in. column used is thus more than adequate with this 7 in. slab. The design positive moments used remain acceptable without penalty.

If $\alpha_{\min}$ could not be satisfied, an increased positive moment would be required as given by multiplying the nominal moment values by δ_s:

$$\delta_s = 1 + \frac{2 - \beta_a}{4 + \beta_a}(1 - \alpha_c/\alpha_{\min})$$

where all terms have already been defined except α_c, the actual column value.

10.7 EFFECTIVE COLUMN STIFFNESS IN A SLAB SYSTEM

Even if a flat plate were supported on very stiff columns, the slab spans would *not* be really fixed at the column. The slab strip framing directly into the column would be fixed over the column width, but the parallel strips farthest from the column would be restrained only by the next slab span, practically none at all by the column itself. The torsional restraint that the slab could transfer laterally to this strip would be nearly negligible. With a heavy transverse beam there could be more restraint, but on the average across the slab the restraint would still be less than directly at the column. This is a way of saying that insofar as the slab is concerned the slab restraint averages less than the column stiffness would suggest. Only a transverse beam infinitely stiff in torsion between two columns could make this difference negligible.

This reduced column effectiveness is a new concept insofar as the Code is concerned. The Code uses the term "the equivalent column" and describes its behavior in terms of flexibilities (the inverse of stiffnesses). For the complete exterior column (above and below) the specified flexibility is

$$1/K_{ec} = 1/(\Sigma K_c) + 1/K_t$$

where ΣK_c represents the sum of the stiffnesses of column above and below the joint and K_t represents the sum of the torsional stiffness of the slab or beam on each side of the column. The reason K_t and K_{ec} do not carry a Σ is that the Σ shows earlier as part of the K_t equation and K_{ec} definition.

The author prefers to think in terms of stiffness rather than flexibility, which would make the equation

$$K_{ec} = \frac{1}{1/(\Sigma K_c) + 1/K_t}$$

The effective column K_{ec} taking the place of ΣK_c is either more flexible or less stiff than the column would be alone. If K_t is small, say, only a slab, K_{ec} is *much less* stiff than ΣK_c.

The evaluation of K_t is based on simple (rather crude) assumptions and basic theory, with final adjustment by an empirical factor to fit test results. The final evaluation is taken as $\frac{1}{3}$ as stiff as Fig. 10.14 from the Code Commentary would suggest. This figure shows the transverse beam (or slab) loaded in torque by unbalanced slab moments which vary from zero at midspan to a maximum at the centerline of the column. The total torque (area of the load curve) is 0.5 for each half span and the shape of the torque curve is a second power parabola with an ordinate at the face of column

$$T_u = \tfrac{1}{2}\left(1 - \frac{c_2}{l_2}\right)^2$$

The unit rotation angle at any point is the torque divided by both G, the shear modulus, and C, the torque stiffness of the cross section. The integral of the area under this curve measures the difference between column rotation and midspan beam rotation,

$$\begin{aligned} \theta_t &= (1/3)\,[(l_2/2)(1 - c_2/l_2)0.5(1 - c_2/l_2)^2]/CG \\ &= (1/12)l_2(1 - c_2/l_2)^3/CG \\ &= (1/6)l_2(1 - c_2/l_2)^3/CE, \text{ assuming } G = 0.5E. \end{aligned}$$

One may think of the average θ for the slab as $\theta_t/3$ or he may think of using a factor of 1/3 to fit the equation to the test data. Fitting could be necessary to represent a nonlinear torque load, the difference between torque at the column face and at the center of column, etc.,

$$\theta_t = (1/18)l_2(1 - c_2/l_2)^3/EC^*$$

The stiffness is the ratio of total torque (used as 0.5 per arm) to this θ_t, or

$$K_t = 0.5/\theta_t = 9\,EC/[l_2(1 - c_2/l_2)^3]$$

*C is similar in use to the polar moment of intertia used for a circular elastic member under torsion.

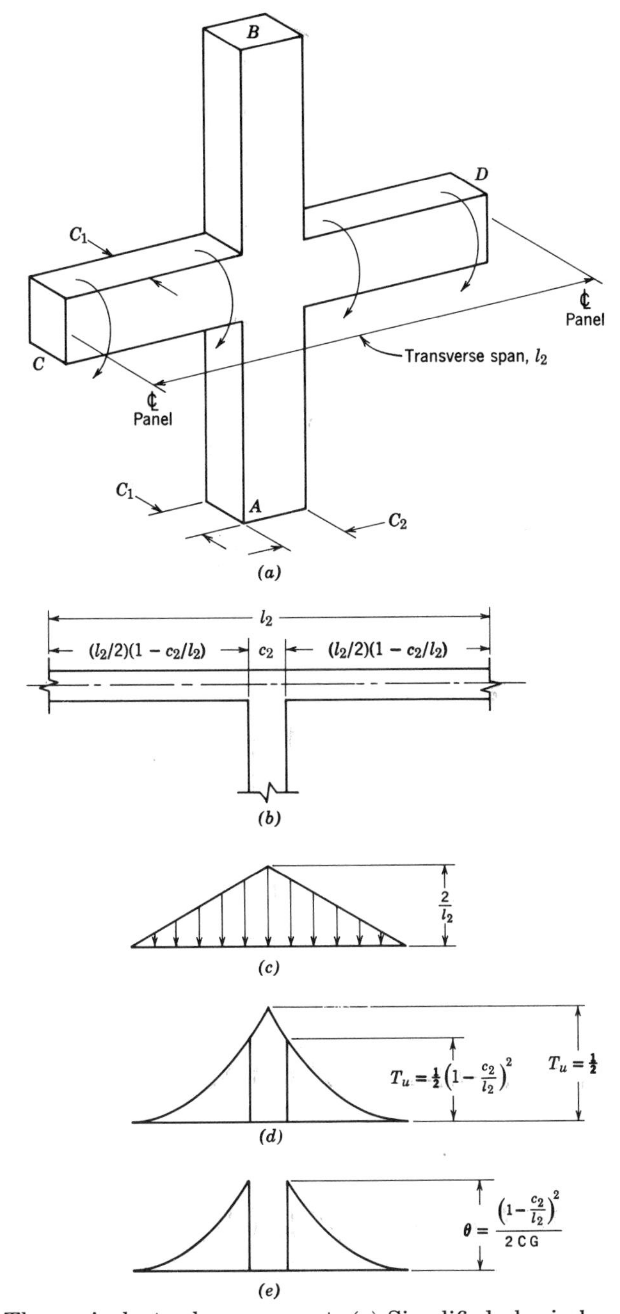

Fig. 10.14. The equivalent column concept. (*a*) Simplified physical model. (*b*) Beam-column combination. (*c*) Distribution of torque or twist load, based on a total of unity. (*d*) Twisting moment or torque diagram. (*e*) Distribution of twist angle per unit length. (Modified from ACI Commentary on Building Code.)

for one arm, or twice this for the two arms in the symmetrical case. The Code indicates the doubling by writing the Σ sign as part of the definition of K_t

$$K_t = \Sigma 9EC/[l_2(1 - c_2/l_2)^3]$$

(Clarity, to the author, would be improved if the Σ were moved into the K_{ec} equation to show ΣK_t.)

The value of C in these equations is the necessary term to replace the polar moment of inertia when considering the torque of noncircular members:

$$C = \Sigma(1 - 0.63x/y)x^3y/3$$

where x and y are the small and large dimensions of the various rectangles making up the cross section, as shown in Fig. 10.15*a*. Where no beam stem is present, Code 13.1.5 defines the beam as a width of slab equal to the column width. Where a beam stem is present, the beam includes the stem plus the adjoining slab on *each* side (one side if a spandrel or edge beam) of width equal to $4h_s$ or, if smaller, the stem projection above or below the slab, as indicated in Fig. 10.15*b*.

For moment distribution under the equivalent frame method, or in moment formula evaluation under the direct design method, K_{ec} is the column stiffness used. It reflects the reduced fixity of slab because some middle strips of the slab are less restrained by abutting slabs than are those close by the column. Where this creates a torque in the slab this torque is carried back to the column and the effect on the column is a bending moment at the joint equal to the sum of moment transferred directly from the slab to the column and that transferred indirectly through torque. But this total is less than if all the slab were directly framing into the column, which is what K_{ec} is intended to indicate.

10.8 CONCENTRATION OF REINFORCEMENT OVER THE COLUMN

Where moment must be transferred from slab to column, as at an exterior column, or from column to slab, as in resistance to wind or earthquake moments, it is desirable that this be accomplished as directly as possible.

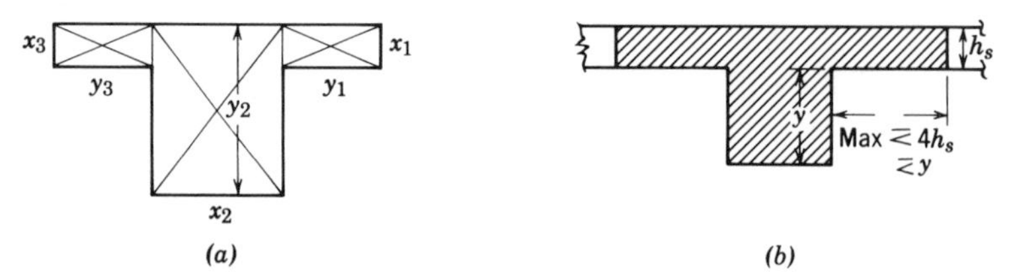

Fig. 10.15. (*a*) Rectangles used in calculation of C for torsion. (*b*) Flange in slab assumed as part of beam.

The Code (13.2.4) considers a direct transfer of moment possible only through reinforcement that passes through the column itself or within a width $h_s/2$ on each side.* This width is nearly the same as the critical width for shear, which adds $d/2$ on each side. The difference is outside the accuracy of the state-of-the art. Since the shear requirement is of longer standing, it will be used here for *both* types of calculation and no further mention will be made of the $h/2$ limit for direct moment transfer.

Tests have shown that the shear stresses are shifted by the general moment transfer taking place at the column, somewhat as indicated in Sec. 4.6c and as calculated in Sec. 10.11f. The Code Commentary suggests that 40% of the moment transferred should be represented by this eccentric shear (or torque) and that 60% should be transferred directly. If the column depth c_1 (parallel to l_1) is greater than the transverse column width, Code 11.13.2 gives a formula that increases the percent to be transferred by eccentric shear or torque.

Some concentration of the column strip negative moment steel within the specified width may be necessary for the direct transfer of moment to the columns. If wind load moment is being transferred to the slab, extra moment steel may be required. Live load pattern moments or unequal spans also involve these transfer moments in interior panels. For equal spans the usual steel arrangement for the column strip will usually be adequate, as now will be demonstrated.

The Code (13.3.5.2) gives an equation defining the moment to be transferred to columns above and below the joint. For equal spans this equation simplifies to:

$$M = 0.08(0.5w_l l_2 l_n^2)/(1 + 1/\alpha_{ec})$$

where w_l is the live load and α_{ec} comes from K_{ec} just discussed in Sec. 10.7:

$$\alpha_{ec} = K_{ec}/\Sigma(K_s + K_b)$$

where subscript s is for slab and b for beam.

Thus far there has been no need to calculate K_{ec} numerically. Rather than do so now where emphasis is on moment transfer rather than the equivalent column, the present calculation will be based on data developed in Sec. 10.11b,c where, for an exterior column of the same size, K_{ec} was $108E$, K_s for an 8 in. slab was $154E$, K_b was zero, α_{ec} was 0.700, and $1 + 1/\alpha_{ec} = 2.42$. At the typical interior column of Sec. 10.5 K_s must be based on a 7-in. slab and be doubled for two slabs instead of one at the exterior column. $K_s = 154E(7/8)^3 = 103\text{E}$.

$$\alpha_{ec} = 108E/(2 \times 103E) = 0.525$$
$$1 + 1/\alpha_{ec} = 1 + 1/0.525 = 2.80$$
$$M = 0.08 \times 0.5(1.7 \times 60) \times 18 \times 18.75^2/(2.80 \times 1000) = 9.2 \text{ k-ft}$$

*1973 Supplement to Code says 1.5h on each side.

Some 60% of this M should be carried within a column width plus d* which totals 1.75 ft, or $M = 0.60 \times 9.2 = 5.5$ k-ft. The original column strip design M was 104.4 k-ft on a 9-ft width and this has already provided a capacity in the required width of $(1.75/9)104.4 = 20.3$ k-ft. No concentration of reinforcing is necessary or even nearly so.

For unequal spans the Code formula adds to the numerator already used the difference between $0.08w_d l_2 l_n^2$ for the long span and the corresponding quantity for the short span.

For the exterior column this requirement is more serious, as in Sec. 10.11(f). There 60% of the total negative moment must be carried within this section close to the column and special provisions must be made for bringing in the torque.

There will be no need to check the concrete in compression for this direct moment transfer if the designer is careful to keep the steel ratio less than $0.75\rho_b$ locally. The slab thickness provided against deflection should usually result in lower steel ratios.

10.9 CODE THICKNESS EQUATIONS—COMMENTS ON THEIR USE

The Code establishes minimum thicknesses of slab in two-way construction primarily by formula values, but with certain limiting considerations, especially minimum values (9.5.3.1):

For slabs without beams or drop panels 5 in.

For slabs without beams, but with drop panels satisfying Code 9.5.3.2 4 in.

For slabs having beams on all four edges with a value of α_m at least equal to 2 3.5 in.

The three Code equations, minimum values in Eq. 9–6 and 9–7 and an upper limit in Eq. 9–8, are:

$$\text{Eq. 9-6:} \quad h \geqslant \frac{l_n(800 + 0.005f_y)}{36{,}000 + 5000\,\beta[\alpha_m - 0.5(1 - \beta_s)(1 + 1/\beta)]}$$

$$\text{Eq. 9-7:} \quad h \geqslant \frac{l_n(800 + 0.005f_y)}{36{,}000 + 5000\,\beta(1 + \beta_s)}$$

$$\text{Eq. 9-8:} \quad h \ngtr \frac{l_n(800 + 0.005f_y)}{36{,}000}, \text{ permissive, not mandatory}$$

A few comments on the use of such complex formulas are warranted.

*As noted above the Code says h_s.

All three equations have the common numerator which reduces to $1000l_n$ for Grade 40 steel and $1100l_n$ for Grade 60 steel.* The denominators include three ratios:

β = ratio of long to short *clear* spans = 1 for a square panel; above a ratio of 2, based on spans center-to-center of columns, the slab is to be designed as a one way slab.

β_s = ratio of length of continuous edges to total perimeter of a slab panel = 1 for an interior panel and never less than 0.5 for the direct design method.

α_m = average ratio of flexural stiffness of beam section to that of the slab. Slab width to the center of adjacent panels is used, that is, for uniform panels l_2. The subscript m calls for the average for all beams around the panel. The value of α varies from 0 for an interior flat plate upward.

A combination of $\alpha\beta = \alpha l_2/l_1 = 1$ represents a beam stiff enough to mark a breakpoint in the Code recommendations. Above that $\alpha l_2/l_1$ value, moment coefficients vary only with l_2/l_1. In effect all such beams furnish adequate supports for the slab and contribute little to slab deflection.

To establish a feel for the deflection equations, consider a square panel, which makes $\beta = 1$. The denominators then reduce to:

$$\text{Eq. 9-6:} \quad 31{,}000 + 5000\alpha_m + 5000\beta_s$$
$$\text{Eq. 9-7:} \quad 41{,}000 + 5000\beta_s$$

Unless α_m exceeds 2 (rather stiff beams) the second equation will control the depth. The last equation of the group, Eq. 9-8, can govern, as an upper limit on h, only when the denominator of Eq. 9-6 drops below 36,000.

The Portland Cement Association[13] treatment of these equations appears to offer better visualization of the thickness requirements. The three equations above are first restated with both numerator and denominator divided by 1000, as follows:

$$\text{Eq. 9-6:} \quad h \gtrless \frac{l_n(0.8 + f_y/200{,}000)}{36 + 5\beta[\alpha_m - 0.5(1 - \beta_s)(1 + 1/\beta)]}$$

$$\text{Eq. 9-7:} \quad h \gtrless \frac{l_n(0.8 + f_y/200{,}000)}{36 + 5\beta(1 + \beta_s)}$$

*Note that this difference of 10% between the two grades of steel is smaller than the 20% difference for beams suggested with Code Table 9.5*a* reprinted here as Table 2.2 (Sec. 2.8). Slabs crack less than beams under service conditions where deflections are of greatest interest; the grade and quantity of steel makes little difference in the uncracked portions.

$$\text{Eq. 9-8:}\quad h \not> \frac{l_n(0.8 + f_y/200{,}000)}{36}, \text{ permissive, not mandatory}$$

The numerators then all reduce to l_n for Grade 40 bars and 1.1 l_n for Grade 60 bars. The last equation, for example, can be restated for Grade 40 bars as a required $l_n/h \not< 36$, that is, l_n/h need not be lower than 36; or for Grade 60 bars that l_n/h need not be lower than $36/1.1 \doteq 33$. Likewise the denominator of the other equations establish similar requirements. These denominators, directly suitable for Grade 40 bars, are plotted in Fig. 10.16. For any given α_m one can tell at a glance what equation governs; or one can interpolate with design values of β and β_s for the governing design ratio. For Grade 60 bars the graph value of the governing ratio is simply divided by 1.1. Although Fig. 10.16 is based directly upon a PCA graph, it has been recast for use here.

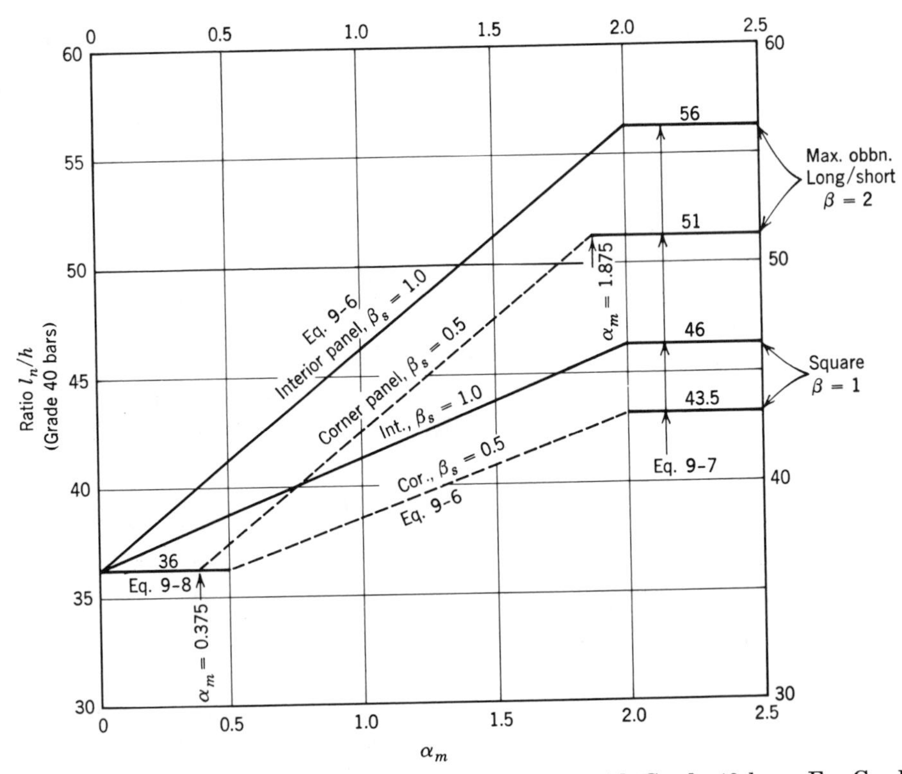

Fig. 10.16. Required l_n/h ratio for deflection control with Grade 40 bars. For Grade 60 divide ratio by 1.1, for Grade 50 by 1.05. (Recast from Ref. 13, PCA.)

Unfortunately, the publication schedule of this book does not permit the recasting of the design problems, in this chapter, to the use of this version of the minimum thickness equations. The author believes these provide an improved approach.

Two special cases are treated as modifications of the Code values of thickness. With drop panels at the columns meeting the requirements illustrated in Sec. 10.12, a thickness 10% smaller than the equation values may be used. In edge panels the formula thickness is too shallow unless edge beams having a stiffness ratio $\alpha \geq 0.80$ are provided; otherwise the required thickness is 10% more than the formula values, or 10% more than the special drop panel provision requires.

Finally, it is permitted, in lieu of the formulas, that deflections may be calculated and the immediate deflection under the live load then be limited to maximum values given in Code Table 9.5*b* which is not reproduced here. The listed permissible deflection limits vary with structure usage and are strictest where attached nonstructural elements may be damaged by deflections.

10.10 FLAT PLATES AND FLAT SLABS—EXTERIOR SPANS

For strips parallel to the exterior wall, the methods for interior panels are available with little change. For the perpendicular strips running to the exterior a number of problems arise: the unsymmetrical negative moments, the combined shear and torsion problem at the exterior column, the torsion reinforcement often required in the slab or spandrel beam, and the direct transfer of moment to the exterior column (Sec. 10.8).

The Code gives formulas for the assignment of M_o into positive and negative moments, using functions of $\alpha_{ec} = K_{ec}/[\Sigma(K_s + K_b)]$, the ratio of equivalent stiffness of exterior column to the stiffness of the slab (or slab and beam). These multipliers for M_o are:

Interior negative moment	$0.75 - 0.10/(1 + 1/\alpha_{ec})$
Positive moment	$0.63 - 0.28/(1 + 1/\alpha_{ec})$
Exterior negative moment	$0.65/(1 + 1/\alpha_{ec})$

Note that the denominator $(1 + 1/\alpha_{ec})$ shows in each case. The sum of the average negative moment and the positive moment is still M_o.

The break of moments into column strip and middle strip moments at the interior support is the same as for interior spans, 75% and 25% for negative moment and 60% and 40% for positive moment.

At the exterior support, for the same concrete in beam and column ($E_{cb}/E_{cs} = 1$), Table 10.3 shows the portion going to the column strip is a

TABLE 10.3. Distribution of Column Strip Negative Moment at Exterior. (Reproduced from Code 13.3.4.2)

l_2/l_1		0.5	1.0	2.0
$(\alpha_1/l_2l_1) = 0$	$\beta_t = 0$	100	100	100
	$\beta_t \geq 2.5$	75	75	75
$(\alpha_1 l_2/l_1) \geq 1.0$	$\beta_t = 0$	100	100	100
	$\beta_t \geq 2.5$	90	75	45

function of $\beta_t = C/2I_s$, where C is the torsional value defining the edge beam and I_s is for a slab width equal to the beam span center-to-center of supports. The 2 factor comes from a ratio of E/G and the assumption that $G = 0.5E$. It might be debated whether to use $C = 0$ when no exterior spandrel beam is present. When calculating C in the determination of K_{ec}, a slab strip the width of the column is specified to be used. It would appear consistent to do the same here. Actually, unless C is large, nearly all the moment finally goes to the column strip and it is usually just as practical an answer to assume 100% to the column strip and skip the calculation unless a real beam is present.

The following section covers the basic calculations for an exterior panel matching the interior panel of Sec. 10.5.

10.11 DESIGN OF FLAT PLATE—EXTERIOR PANEL

Design an exterior panel flat plate slab 18-ft wide parallel to the free edge and 20-ft center to center of columns perpendicular to this edge. Assume no beam,* no wall load (an omission for simplicity in the design example), and a slab extending to the outside face of the column, as in Fig. 10.17*a*. Use $f_c' = 4000$ psi for both slab and columns, Grade 60 steel, dead load (including slab weight) of 100 psf and live load of 60 psf. Assume all columns are 15 in. square with story heights of 10 ft.

Solution

(a) Slab Thickness

Thickness for deflection will be considered first, following the comments of Sec. 10.9.

$$l_n = 20 - 1.25 = 18.75 \text{ ft} \qquad f_y = 60{,}000 \text{ psi}$$

*The influence of an edge beam is discussed in Sec. 10.11*g*, at the close of this section.

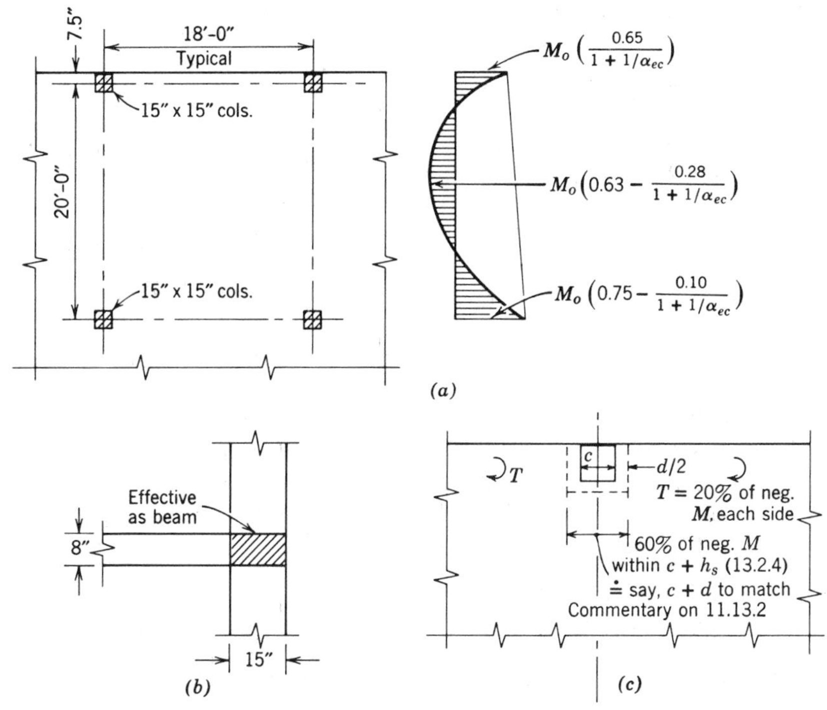

Fig. 10.17. Exterior slab panel. (*a*) Layout and bending moments. (*b*) Slab section acting as beam in torsion. (*c*) Moment concentration over column. The 1973 Supplement to Code changes width to $c + 3h_s$.

β = ratio of *clear* spans = $(20 - 1.25)/(18 - 1.25) = 1.12$

$\alpha_m = 0$ (no beams)

β_s = ratio of length of continuous edge to total perimeter of panel = $(2 \times 20 + 18)/(2 \times 20 + 2 \times 18) = 0.763$

The denominator of Code Eq. 9-6 becomes:

$$36{,}000 + 5000\beta[\alpha_m - 0.5(1 - \beta_s)(1 + 1/\beta)]$$
$$= 36{,}000 + 5000 \times 1.12[0 - 0.5 \times (1 - 0.763)(1 + 1/1.12)]$$
$$= 36{,}000 + 5000 \times 1.12\,(-0.225) < 36{,}000 \text{ of Eq. 9-8}$$

Equation 9-8 cuts off any greater requirement and Eq. 9-6 permits no less; Eq. 9-7 is surplus in this case.

$$h \nless l_n(800 + 0.005f_y)/36{,}000 = 18.75 \times 12 \times 1100/36{,}000 = 6.88 \text{ in.}$$

Code 9.5.3.3 requires either a spandrel beam or an increase of 10% in these formula values.

$h = 1.10 \times 6.88 = 7.57$ in., say 8 in.*

The most critical design provisions are those for shear around the columns, but these cannot be calculated before the bending moments in the exterior panel perpendicular to the free edge are known. (An interior panel under the same conditions was found in Sec. 10.5 to require a d_v of only 4.83 in.)

(b) Effective Exterior Column Stiffness

The equivalent exterior column stiffness must be found, which requires that the slab strip between exterior columns first be considered as a beam carrying torsion, as discussed in Sec. 10.7. This slab strip (Fig. 10.17*b*) is 15 in. wide by 8 in. deep, leading to the effective torsional inertia term C, or use Table 10.11 in Sec. 10.17.

$$C = (1 - 0.63\, x/y)(x^3y/3)$$
$$= (1 - 0.63 \times 8/15)(8^3 \times 15/3) = 0.664 \times 2570 = 1700 \text{ in.}^4$$

The torsional stiffness for the beam to one side of the column is

$$K_t = 9E_{cs}C/[l_2(1 - c_2/l_2)^3]$$

where E_{cs} is simply E_c for the slab. With column width c_2 in feet,

$$K_t = 9E_c \times 1700/[18 \times 12(1 - 1.25/18)^3]$$
$$= 70.8\, E_c/0.931^3 = 88E_c$$

For the two adjacent slabs, acting as transverse beams, $K_t = 2 \times 88E_c = 176E_c$.

The column above or below gives, for a *uniform* cross section:†

$$K_c = 4EI/l = 4E_c(15^4/12)/(10 \times 12) = 141E_c$$
$$K_c = 2 \times 141E_c = 282E_c$$

This leads to the equivalent column stiffness:

$$K_{ec} = 1/[(1/282E_c) + (1/176E_c)] = E_c/(0.00355 + 0.00568)$$
$$= 108E_c$$

*Some designers would accept 7.75 in. for a slab.

†In the equivalent frame method the extra stiffness within the slab thicknesses is recognized.

(c) Slab Design Moments

For strips running perpendicular to the exterior edge, the design moments of Sec. 10.10 each contain a fractional multiplier that includes the denominator $1 + 1/\alpha_{ec}$, where α_{ec} by definition is $K_{ec}/\Sigma(K_s + K_b) = K_{ec}/K_s$ in this case.

$I_s = 18 \times 12 \times 8^3/12 = 9240 \text{ in.}^4$
$K_s = 4E_cI_s/l_1 = 4E_c \times 9240/(20 \times 12) = 154E_c$
$\alpha_{ec} = 108E_c/154E_c = 0.700$
$1 + 1/\alpha_{ec} = 1 + 1/0.700 = 2.42$

The design moments can now be established from the design load.

$w = 1.4 \times 100 + 1.7 \times 60 = 242 \text{ psf}$

20 ft Strips

$M_o = 0.125 \times 242 \times 18\,(20 - 1.25)^2/1000 = 192 \text{ k-ft}$
Total neg. M at int. col. $= [-0.75 - 0.10/(1 + 1/\alpha_{ec})]\, M_o$
$= -(0.75 - 0.10/2.42)\, 192 = -136.0 \text{ k-ft}$
Total positive $M = [0.63 - 0.28/(1 + 1/\alpha_{ec})]M_o$
$= (0.63 - 0.28/2.42)\, 192 = +98.5 \text{ k-ft}$
Total neg. M at ext. col. $= [0.65/(1 + 1/\alpha_{ec})]\, M_o$
$= -(0.65/2.42)\, 192 = -51.6 \text{ k-ft}$

The interior negative moment is assigned to column strip (75%) and middle strip (25%) just as in an interior panel.

Int. neg. M on column strip $= -0.75 \times 136.0 = -102 \text{ k-ft}$
Int. neg. M on middle strip $= -136.0 + 102 = -34.0 \text{ k-ft}$

Likewise the positive moment is assigned 60% to the column strip and 40% to the middle strip, just as in an interior panel.

Positive M on column strip $= 0.60 \times 98.5 = 59.1 \text{ k-ft}$
Positive M on middle strip $= 98.5 - 59.1 = +39.4 \text{ k-ft}$

For the exterior negative moment, the proportion going to the column strip (Table 10.3) depends upon $\beta_t = E_{cb}C/(2E_{cs}I_s)$.

$\beta_t = 1700\, E_c/(2E_c \times 9240) = 0.0920$

Interpolating in Table 10.3:

For $\beta_t = 0$, 100% goes to column strip
$\beta_t = 2.5$, 75% goes to column strip

$$\beta_t = 0.09,\ 100 - \frac{0.09}{2.5}(100 - 75) = 100 - 0.9 = 99.1\%$$

The difference between this and 100% is negligible and this is the reason for stating earlier that the difference could be ignored.

Exterior neg. M on column strip = −51.6 k-ft
Exterior neg. M on middle strip = 0

This completes the design moments in the one direction.

Where unequal negative moments occur at a column line, as at the first interior column line, Code 13.3.3.4 requires the slab design be for the larger moment unless analysis is made to distribute the moments. The interior panel design in Sec. 10.5 found column strip negative moment of −104.4 k-ft against −102 k-ft here and middle strip negative moment of −34.4 k-ft against −34.0 k-ft here. Design for the larger is no real penalty in this case.* Where significant, the unbalanced moment may be distributed to the slabs and column (using the K_{ec} values).

The strips parallel to the exterior are not different from an interior panel except that a spandrel or edge beam is more often present. Also this beam, or in this case the slab since no beam is present, must be checked for torsion.

The design of reinforcement for these strips is no serious problem and the student may follow the method of Table 10.1.

(d) Concentration of Negative Moment Reinforcement at Exterior Column

At the exterior column there is a necessary concentration of the column strip negative moment[8] as already discussed in Sec. 10.8. This requires a concentration of flexural reinforcement within a width equal to the column face plus $h_s/2$ (say $d/2$) on each side for all moment to be transferred directly to the column; the remainder of the slab moment is transferred by torque, causing a resultant eccentricity of shear around the column on a width equal to the column face plus d, as indicated in Fig. 10.17*c*.

*Note that the simplified "total dead load" used here has neglected the increased weight of the thicker exterior panel slab, in order to simplify the example; this is not a g ood practice.

For a square column 60% of the exterior negative moment is considered transferred by moment, 40% by eccentric shear. (Note Code 11.13.2 for rectangular columns.)

Direct moment transfer $= 0.60 \times 51.6 = 31.0$ k-ft

Assuming $d = 7.00$ in., $A_s = (31.0 \times 12/0.9)/(60 \times 0.9 \times 7) = 1.10$ in.^{2} This area, equivalent to 6- #4 or 4- #5, must be placed within a width of $15 + 7 = 22$ in.* This presents no serious problem, but will require a special note on the drawing as to this closer spacing over the column.

(e) Minimum Column Stiffness

For the interior panel, minimum column stiffness was checked in Sec. 10.6. Since it is relative stiffness of column to slab which is checked, even a smaller column would satisfy the needed restraint at the exterior column. Parallel to the exterior face only a half panel width of slab is involved, which nearly doubles the relative column stiffness. In the other direction there is slab on only one face of the column, which has the same effect. The 15 in. square column is heavier than need be for this purpose.

(f) Shear at Exterior Column

Normally there would be some wall load, although for simplicity none is included in this text example; this would add a substantial extra shear. From Fig. 10.18*a*:

$$\begin{aligned} V_u &= 242(0.5 \times 20 \times 18 - 1.80 \times 0.92 + 0.62 \times 16.20)/1000 \\ &= 45.7\ k, \qquad \overline{V} = 45.7/0.85 = 53.5k \end{aligned}$$

From the basic ideas of Sec. 4.6(c), the centroid of shear resistance is at the centroid of the shaded areas of Fig. 10.18*b*.

$$A_c = 2 \times 18.5 \times 6.5 + 21.5 \times 7 = 241 + 150 = 391 \text{ in.}^2$$

From center of column,

$$\bar{x} = (2 \times 18.5 \times 6.5 \times 1.75 + 21.5 \times 7 \times 11)/391 = 5.33 \text{ in.}$$

or from center of area A of Fig. 10.18*b* = 3.58 in.

The torque from both sides of the column is the 40% of the column strip negative moment not carried directly into the column.

$$T = 0.40 \times 51.6 = 20.6 \text{ k-ft}, \ \overline{T} = 20.6/0.85 = 24.3 \text{ k-ft}$$

*Actually $c + 3h_s \gg 1.92$ ft, but see footnote on p. 357.

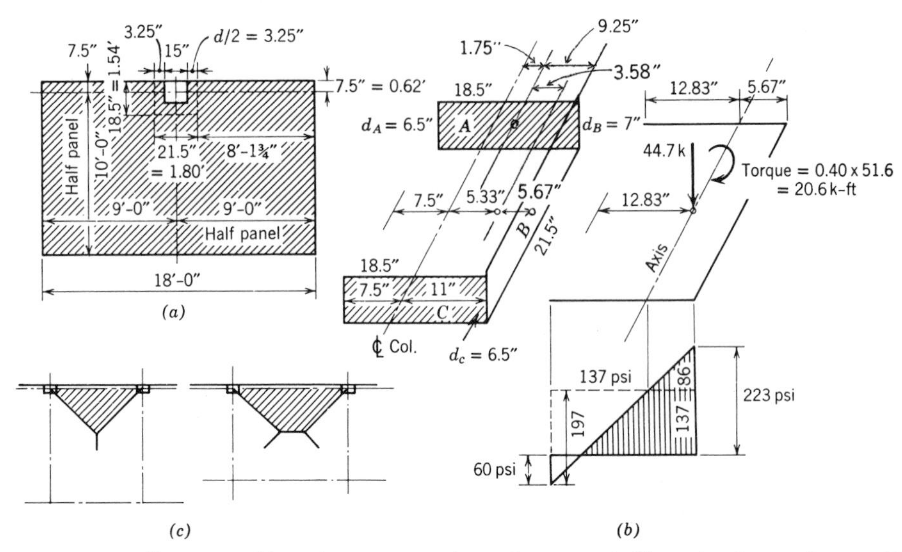

Fig. 10.18. Shear considerations at exterior columns. (*a*) Shear load on column. (*b*) Shear and torque loading with resultant stresses. (*c*) Load to edge beam, when present.

The polar moment of inertia of these areas about a horizontal axis through their centroid and parallel to the center line of column is required. Using the area designations in Fig. 10.18*b*:

$$J_C = J_A = I_{xA} + I_{yA}$$
$$I_{xA} = 18.5 \times 6.5^3/12 = 424$$
$$I_{yA} = 6.5 \times 18.5^3/12 = 3440$$
$$+18.5 \times 6.5 \times 5.33^2 {}^* = 3420$$
$$J_A = 7280 \text{ in.}^4$$

$$J_B = A_B(18.5 - 7.5 - 5.33)^2$$
$$= 5.67^2 A_B$$
$$J_B = 150 \times 5.67^2 = 4830 \text{ in.}^4$$

$$J = 2 \times 7280 + 4830 = 19{,}300 \text{ in.}^4$$
$$v_{\max} = \overline{V}/A_c + \overline{T}x_B/J$$
$$= 53{,}500/391 + 24{,}300 \times 12 \times 5.67/19{,}300 = 137 + 86 = 223 \text{ psi}$$
$$v_{\min} = 137 - 86 \times 12.83/5.67 = 137 - 197 = -60 \text{ psi}$$

The allowable maximum is $4\sqrt{f_c'} = 4\sqrt{4000} = 253$ psi > 220 O.K.

If a wall had been included, as would be normal, it would have added substantially to the average shear but would have decreased the torque some since its weight would be to the left of the centroid.

*Error: Change 5.33^2 to $(12.83\text{-}0.5 \times 18.5)^2 = 3.58^2$ modifying J and V values.

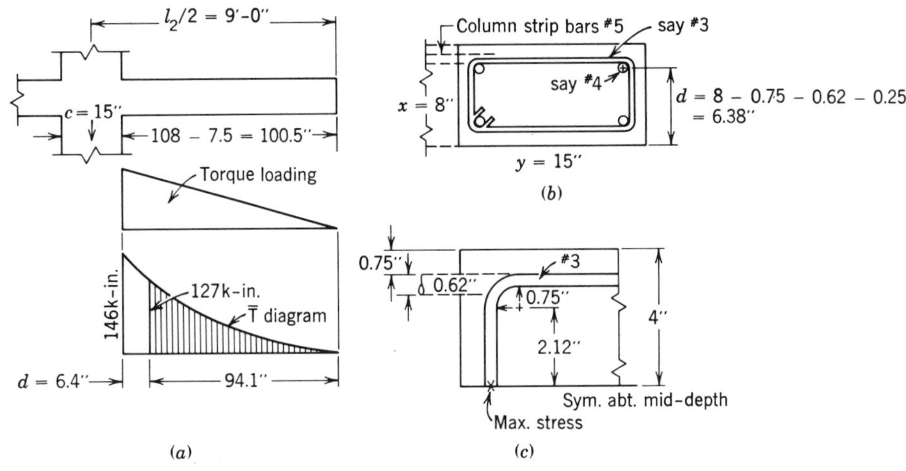

Fig. 10.19. Torque provisions in beam element. (*a*) $\overline{T}$ diagram. (*b*) Cross section. (*c*) Development in vertical leg of closed tie.

(g) Torque on Slab

Torque might be calculated at a distance d from the column (11.2.2) but will first be checked for the total torque to see if torque is serious.

$\overline{T}$ from above $= 24.3/2 = 12.2$ k-ft $= 146$ k-in.

For the specified slab strip equal to the column in width (Fig. 10.19*b*) and with torsion always on gross area (Sec. 4.19d), $\Sigma x^2 y/3 = 8^2 \times 15/3 = 320$ in.^{3}

$v_{tu} = \overline{T}/320 = 146{,}000/320 = 457$ psi

The problem is serious. If one considers a parabolic torque diagram* similar to the one used in establishing K_t for the moment calculation, but with maximum at the *face* of column as shown in Fig. 10.19*a*, and takes $d = 8 - 0.75 - 0.62$(col. strip bar) $- 0.25 = 6.38$ in., $\overline{T}$ at a distance d from the column is:

$\overline{T} = 146(94.1/100.5)^2 = 127$ k-in.
$v_t = 127{,}000/320 = 398$ psi $= 6.3\sqrt{f_c'}$

*This is not quite consistent with the 99.1% of negative moment to the column strip (subsection c), but is simple and on the safe side.

The large torque shear stress indicates the basic problem of the slab without an edge beam.* Nor does the Code speak directly to the problem of how to design for torque quite close to a concentrated slab load, in this case the column reaction. Some two-way action exists on the inner face of the column but primarily one-way action exists near the outside column face. In this situation decisions must be somewhat arbitrary and there is no practical meaning in writing a third significant digit, although habit may make it easier that way.

If a beam existed in fact, one would here be calculating v_c and v_{tc} at a distance d from the column face, as here, but the criteria would be those for beam shear, critical on the inside face of the beam.

Since the vertical slab shear was calculated as some 87% of the permissible, the interaction of shear with torque will be considered relatively a normal one. The permissible v_{tc} will be crudely estimated at about 50% of the pure torque allowable,† $v_{tc} = 0.5 \times 2.4\sqrt{4000} = 76$ psi.

$v_{tu} - v_{tc} = 398 - 76 = 320$ psi

$A_t = [(v_{tu} - v_{tc})s\Sigma x^2y/3]/(\alpha_t x_1 y_1 f_y)$

$\alpha_t = 0.66 + 0.33(y_1/x_1) = 0.66 + 0.33(13.12/5.62) = 1.44$

Reqd. $A_t = 320 \times 320s/(1.44 \times 5.62 \times 13.12 \times 60{,}000) = 0.016s$

Max. spcg. in Code $= (x_1 + y_1)/4 = (5.62 + 13.12)/4 = 4.7$ in.

If $A_t = \#3 = 0.11$ in.^{2}, $s = 0.11/0.016 = 6.9$ in. $>$ max. spcg.

USE #3 closed ties at 4.5 in.

Reqd. extra longitudinal steel $= 2A_t(x_1 + y_1)/s$

$A_l = 2 \times 0.016s(5.62 + 13.12)/s = 0.60$ in.^{2}

$= 0.30$ in.^{2} top and bottom

With the above assumed v_{tc}, ties are needed wherever v_{tu} exceeds 76 psi or where $\bar{T}$ exceeds $76 \times 320/1000 = 24.4$ k-in. With a parabolic T distribution rising to 146 k-in. at face of column

$x^2/100.5^2 = 24.4/146$, $x = 41$ in. from midspan or 59.5 in from col.

USE 14- #3 closed ties at 4.5 in.

3- #3 top and bottom bars extending 5 ft from face of column

The latter steel is in addition to the usual flexural reinforcement.

*This paragraph reflects the author's bias against slabs with neither an overhanging cantilever nor an edge beam. He feels that the 1971 Code does well to point to the problem. The state-of-the-art, however, does not provide a documented design procedure. As in many design situations the engineer must visualize the slab behavior (deformation) and be overconservative in meeting it until more tests have been made of torsion reinforced slabs.

†The interaction with $2\sqrt{f_c'}$ for v_c seems inappropriate here with two-way bending and quite near a $4\sqrt{f_c'}$ allowable shear zone.

The #3 tie in an 8-in. slab is on the verge of doubtful efficiency because of under development of its vertical leg. The hook (around the corner) is good (Code 12.8.1) for $540\sqrt{f_c'} = 34{,}200$ psi. The clear vertical length between the start of hooks or bends (0.75 in. inside radius) is $8 - 2(0.75 + 0.38 + 0.75) = 4.25$ in. If half of this 4.25 in. is available for development, the critical section being at midheight, as indicated in Fig. 10.19*c*, the straight development length is:

$$2.12 = 0.0004 d_b f_s = 0.004 \times 0.375 f_s \qquad f_s = 14{,}100 \text{ psi.}$$

This gives a total capacity of $34{,}200 + 14{,}100 = 48{,}300$ psi if one ignores the arbitrary 12 in.* minimum l_d. The use of 4.5 in. spacing here instead of a theoretical 6.9 in. seems to give the required margin in strength required.

(h) Modifications in Design Where Edge Beam Exists

A separate example will not be developed for an exterior panel with an edge beam, but it is well to note the essential design differences in such a case.

The formula values for slab thickness will not then need the 10% increase if the edge beam has an α of at least 0.80.

The edge strip (half column strip) would be made up of a beam plus a slab, instead of a slab alone. The methods used in Sec. 10.14 and 10.15 are available for establishing the distribution of M_o in this half panel.

The equivalent exterior column stiffness found in (*b*) above is influenced by increased torsional stiffness of the beam, C in the edge beam case being based on the sum of two areas making up the inverted L section. The flange counted is limited to the smaller of $4h_s$ or the projection of the beam above or below the slab. The increased K_t results in a higher K_{ec}, α_{ec}, and negative moment assigned to the exterior column line, with minor changes in other moments. The relative torsional stiffness β_t of the beam would also modify (lower) the distribution of negative moment to the column strip, while increasing that to the middle strip.

The shear in this case would, in part, be brought into the column by the beam and, in part, directly by the slab. The load on the beams, including any direct wall load, could be assumed that bounded by 45 degree diagonals in the panel, as in Fig. 10.18*c*, plus a parallel centerline of panel if the exterior edge is the longer edge. Code 13.3.4.7 suggests that where $\alpha_1 l_2/l_1$ for the edge beam is less than unity this proportional part $(\alpha_1 l_2/l_1)$ of the shaded area be used. The beam torque and shear should be analyzed (the

*Since stirrups cannot be carelessly displaced in the direction of l_d, the need for the 12 in. minimum is less here than for A_s bars.

shear at $2\sqrt{f_c'}$ allowable in combination with the allowable v_{tc}) at a distance d from the column, with closed stirrups provided for any excess.

The remainder of the column shear might, in the author's opinion, be assumed carried around the column outside the limits of the beam web, at a distance $d/2$ from the column faces. This shear would be limited to $4\sqrt{f_c'}$ unless the beam framed flush with the loaded face of the column and turned this into a one-way shear problem, critical at a distance d from the face. It would not be necessary to worry about a nonuniform distribution of this shear (in the nonflush case), although some variation undoubtedly would result.

10.12 FLAT SLAB WITH DROP PANEL—DESIGN EXAMPLE

Design a typical interior bay of a flat slab to carry a dead load (including its own weight) of 100 psf and a live load of 200 psf over an 18 ft by 19 ft 6 in. panel, center-to-center of columns. $f_c' = 3000$ psi, Grade 60 steel. Use a story height of 11 ft, a column capital, and a drop panel.

Solution

A flat slab of this type (Fig. 10.20a) is particularly good for heavy manufacturing or warehouse loads. Shear can be controlled by the size of capital used and strength is more apt to control slab thickness than is deflection. However, deflection can be simply checked by calculating the required thickness h from the Code equations as soon as l_n, the net span, is fixed.

(a) Thickness for Deflection Control

The equations discussed in Sec. 10.9 will be compared by evaluating their denominators.

$$\text{Eq. 9-6: } 36{,}000 + 5000\beta[\alpha_m - 0.5(1 - \beta_s)(1 + 1/\beta)]$$
$$= 36{,}000 + 5000(19.5/18)(0 - 0) = 36{,}000$$

Eq. 9-8 cuts off any greater requirement and Eq. 9-6 permits no less; Eq. 9-7 is surplus in this case.

$$h = l_n(800 + 0.005f_y)/36{,}000 = 1100l_n/36{,}000$$

The net length must be estimated. Assume a column with a 3 ft 6 in. capital with an area of $\pi \times 3.5^2/4 = 9.63$ ft^2. The equivalent side of a square having the same area $= \sqrt{9.63} = 3.11$ ft.

$$l_n = 19.5 - 3.11 = 16.39 \text{ ft}$$
$$h = 1100 \times 16.39 \times 12/36{,}000 = 6 \text{ in.}$$

With the drop panels specified in 13.3.5 (see (b) below) the Code lowers the requirement 10% to 5.4 in., say 5.5 in.

(b) Depth for Flexure

$$
\begin{aligned}
\text{Dead load} &= 1.40 \times \overset{106}{\cancel{100}} = \cancel{140} \qquad 148 \text{ (final)} \\
\text{Live load} &= 1.70 \times 200 = 340 \\
w &= \cancel{480}\ \text{psf} \qquad 488 \text{ (final)}
\end{aligned}
$$

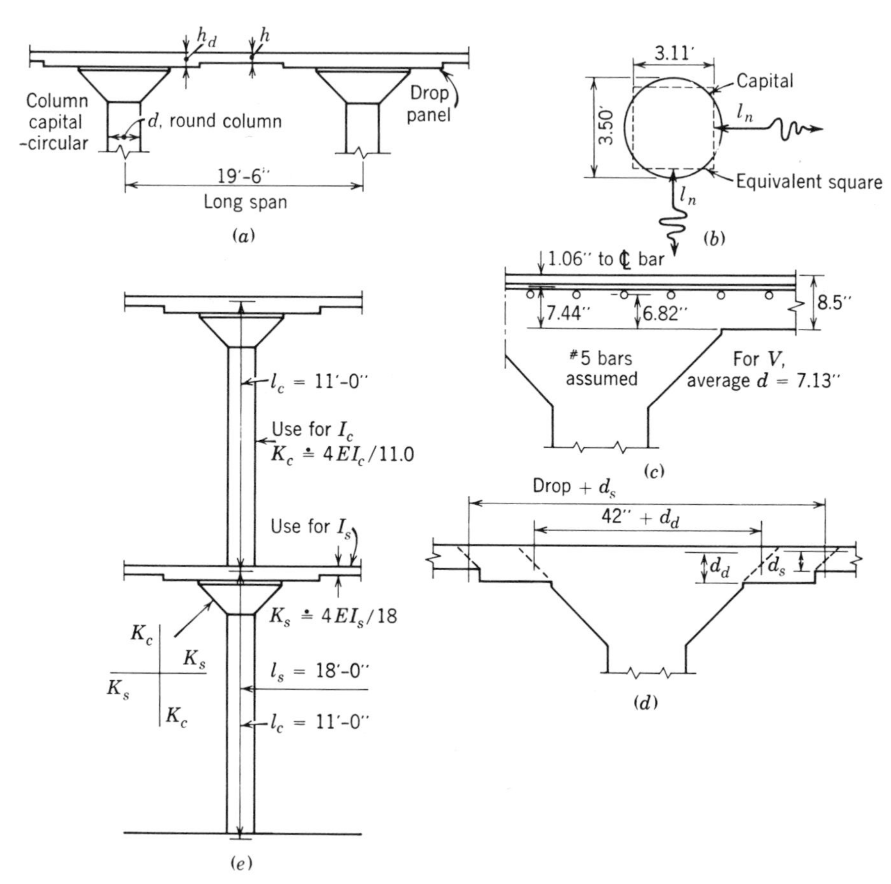

Fig. 10.20. Flat slab construction with drop. (*a*) Nomenclature. (*b*) Net span l_n to face of equivalent square support. (*c*) Depth for shear. (*d*) Critical shear sections. (*e*) Calculation of minimum column, about center joint.

Long span $M_o = 0.125 \times 480 \times 18 \times 16.39^2/1000 = 291$ k-ft
Total negative moment $= 0.65M_o = 0.65 \times 291 = 189$ k-ft
Col. strip neg. $M = 0.75 \times 189 = 142$ k-ft

The drop panel (13.5.5) should be at least $l_1/3$ by $l_2/3$, that is, in the long span direction 6 ft 6 in. and transversely 6 ft 0 in. Its thickness h_d must be at least $(5/4)h_s$, say in this case 7.0 in., which seems light for this heavy a load. For strength design it is desirable to keep the steel ratio around $0.5\rho_b$* which means k_m slightly over two-thirds the value in Table 2.1 (for $0.75\rho_b$) which shows as 705. Try $k_m = 500$. The width in compression is that of the drop panel, 72 in.

Reqd. $d_m = \sqrt{146 \times 12{,}000/(0.9 \times 500 \times 72)} = \sqrt{52.5} = 7.23$ in.
$h_d = 7.23 + 0.75$ cover $+ 0.31$ for $d_b/2 = 8.29$ in.
USE drop $h_d = 8.5$ in., subject to shear check,
in panel 6 ft 0 in. by 6 ft 6 in. Revised in (e)
to 6 ft 6 in. square

Dead load must increase to 106 psf, as corrected at start of (b), to give total $w = 488$ psf.

(c) Shear at Column Capital

Use the assumed capital diameter of 3.5 ft $+ d = 3.5 + 0.6 = 4.1$ ft to calculate V_u and the required capital for shear.

$v_u = 488(18 \times 19.5 - \pi \times 4.1^2/4) = 169{,}000$ lb

Assume average $d = 8.5 - 0.75 - 0.62\ (=d_b) = 7.13$ in., as shown in Fig. 10.20*c*.

Shear perimeter (Fig. 10.20*d*) $= 2\pi(r + 0.5 \times 7.13) = 2\pi r + 22.4$

$$2\pi r + 22.4 = V/(\phi v_c d)$$
$$= 169{,}000/(0.85 \times 4\sqrt{3000} \times 7.13) = 127$$
$$r = (127 - 22.4)/2\pi = 16.7 \text{ in.}$$
$$2r = 33.4 \text{ in. vs. 42 in. assumed}$$

USE 36 in. diameter column capital

The equivalent square (same area) $= 32$ in. $= 2.67$ ft for use in l_n.

*For some extra ductility and usually also for overall economy.

(d) Shear at Drop

This is critical at $d/2$ outside of the drop.

$d = 5.5 - 0.75 - 0.62\ (= d_b) = 4.13$ in. average

Perimeter of shear section if keep drop dimensions one-third of center-to-center span each way and add $d/2$ each way to the critical section $= 2(78 + 4.13) + 2(72 + 4.13) = 317$ in.

$V_u = (19.5 \times 18 - 82.13 \times 76.13/144)488/1000 = 150\ k$
$v_u = 150{,}000/(0.85 \times 317 \times 4.13) = 135$ psi
Allowable $v_c = 4\sqrt{3000} = 220$ psi O.K.

(e) Critical Moments

These must be recalculated since the size of capital has been changed.

Long span $M_o = 0.125 \times 488 \times 18(19.5 - 2.67)^2/1000 = 312$ k-ft
Short span $M_o = 0.125 \times 488 \times 19.5(18 - 2.67)^2/1000 = 280$ k-ft

The moment coefficients of Fig. 10.10*b* will be entered directly into Table 10.4.

Although the short strip moment is smaller than for the long strip, the effective depth is also necessarily smaller by one bar diameter (because of crossing bars) than the depth with the long strip moment. Both depth requirements will be rechecked.

Long span neg. $M = 0.65 \times 0.75 \times 312 = 153$ k-ft
Short span neg. $M = 0.65 \times 0.75 \times 280 = 137$ k-ft
Reqd. long span $d = \sqrt{153{,}000 \times 12/(0.9 \times 500 \times 72)} = 7.55$ in.
Min. $h_d = 7.55 + 0.75 + 0.31$ (for #5) $= 8.61 > 8.5$ in above.
Reqd. short span $d = \sqrt{137{,}000 \times 12/(0.9 \times 500 \times 78)} = 6.85$ in.
Min. $h_d = 6.85 + 0.75 + 1.5 \times 0.62 = 8.53$ in. > 8.5 in., say O.K.

The latter is certainly satisfactory because the k_u value of 500 was a judgement decision and overrun is negligible. The long span requirement may be handled in any of four ways:

1. Accept k_u several percent greater than 500.
2. Increase the column capital, to lower M_o.
3. Widen the drop panel to 78 in. square.
4. Change depth, weight, and moments.

For this example the drop panel will be made 6 ft 6 in. square which will obviously provide the width to maintain k_u below 500 while lowering the required h_d to 8.5 in., as earlier assumed.

USE drop panel 6 ft 6 in. square

(f) Reinforcement Calculations

The remainder of the reinforcement calculations are made in Table 10.4. Since heavy loads and moments are involved the internal lever arm z is taken initially as $0.90d$ (compared to 0.95 d in the earlier examples). The required A_s at the column in the long strip, the heaviest required, is 5.07 in.^{2} and will be used to check z. The concrete width used is the revised drop panel width of 78 in.

$$a = 5.07 \times 60{,}000/(0.85 \times 3000 \times 78) = 1.53 \text{ in.}$$
$$z = 7.44 - 1.53/2 = 6.67 \text{ in.} = 0.90d$$

The assumed z is (accidentally) almost precise.

In the table, rather than estimate z individually for each case, the $0.90d$ was first used throughout, leading to tentative bar selections. Next minor changes in d resulting from the use of #4 bars were noted and the whole inspected for possible weakness. In the second column, z for positive moment looked close but the positive moment under the 6th sketch showed as more critical. The z for this short column strip positive moment is therefore checked:

$$a = 3.72 \times 60/(0.85 \times 3 \times 108) = 0.814 \text{ in.}$$
$$z = 3.81 - 0.41 = 3.40 \text{ in.} = 0.89d$$

which would give a required $A_s = 3.78 \times 0.90/0.89 = 3.83$ in.^{2} compared to 3.72 in.^{2} used. This would not be acceptable, but it is noted that the A_s for negative moment in this strip is 5.27 in.^{2} compared to 4.99 in.^{2} required. The *two* sets of bars *can be accepted* if the designer chooses to assume a little less positive moment and a little more negative moment in the strip. The Code (13.3.3.6) provides that a design moment may be modified 10% provided that all of M_o is taken care of. The second column by comparison has a satisfactory z since some 2% excess A_s was actually used anyway.

The short column middle strip positive moment reinforcement appears to be 2% short, permissible because this small A_s obviously leads to $z > 0.90d$ which was used originally. It is possible that more detailed study of z might lead to economy elsewhere, but this refined calculation would be

TABLE 10.4. Reinforcement for Flat Slab of Sec. 10.12.

	Long Span (19′-6″)				Short Span (18′-0″)			
	M_o = 312 k-ft				M_o = 280 k-ft			
	Column Strip		Middle Strip		Column Strip		Middle Strip	
#5 bars assumed Strip width	8.5″, 7.44″, 5.5″ 9′	4.44″	4.44″ 9′	5.5″, 4.44″	8.5″, 6.81″, 5.5″ 9′	3.81″	5.5″, 4.44″ 10″-6′	5.5″, 3.81″
	Neg.	Pos.	Neg.	Pos.	Neg.	Pos.	Neg.	Pos.
M coef.	0.49	0.21	0.16	0.14	0.49	0.21	0.16	0.14
$\overline{M}$ = (coef.) M_o /0.90 (k-ft)	−170	+73	−56	+49	−153	+65	−50	+44
d, in.	7.44	4.44	4.44	4.44	6.82	3.82	4.44	~~3.82~~ 4.00
$A_s = \overline{M} \times 12/(60 \times 0.9d)$ $= \overline{M}/(4.5d)$, in.^{2}	5.07	3.65	2.80	2.45	4.99	3.78	2.50	~~2.57~~ 2.45
Min. no. bars for $s = 2h$	10	10	10	10	10	10	12	12
No. of #5 Area, in.^{2}	17- #5@6½±″ (5.27)	12- #5@9″ (3.72)			17- #5@6½±″ (5.27) ←	12- #5@9″ (3.72) ↑ Pair		
No. of #4 Area, in.^{2}			14- #4@7.5±″ (2.80)	12- #4@9″ (2.40)[a]			13- #4@9.5±″ (2.60)	12- #4@10½″←USE ~~13- #4@9½±″~~ (~~2.60~~) (2.40)[a]
Revised d, in.	O.K.	O.K.	4.50	4.50	O.K.	3.94	4.50	4.00

[a] O.K. because $z > 0.9d$ with these low M values.

worth while only if many like panels were involved. When tables are available, one can go from M to calculated actual k_m to a table value of ρ suitable for that k_u. This eliminates checks on z.

Temperature steel calculations are not shown since a quick mental check of $0.0018bh$ in the short (wide) span middle strip indicated it could not control.

(g) Column Stiffness

Ratio of dead to live load, $\beta_a = 100/200 = 0.5$, is much below the factor of 2 required to make the above calculations acceptable without a special check on the column stiffness. $l_2/l_1 = 19.5/18 = 1.08$. Table 10.2 shows, fortunately with only the interpolation for this particular ratio and none for β_a, that $\alpha_{\min} = 1.7$.

The real column stiffness with the capital is greater than $4EI/l$. If I is based on minimum cross section, as usual, the coefficient for K of the column above is larger than 4 and in the column below (with the capital at the joint) much larger than 4, making more than half of the unbalanced moment at the joint go to the column below. The Commentary suggests more approximate methods are appropriate to the direct design method, both for the columns and slabs, but that similar simplifications should be used for both columns and slabs.

The slab stiffness is modified by both the drop panel and the column capital. Here the minimum I_s will be used, with the short span which is more restrictive in this case.

$$I_s = (19.5 \times 12)5.5^3/12 = 3250 \text{ in.}^4$$

$$K_s = 4E \times 3250/(18 \times 12) = 60E$$

$$\alpha_{\min} = 1.7 = K_c/K_s = (2 \times 4EI_c/l_c)/(2 \times 60E)$$

$$= [2 \times 4EI_c/(11 \times 12)]/(2 \times 60E) = 5.03 \times 10^{-4} I_c$$

For a circular column of diameter d,

$$I_c = \pi d^4/64 = 1.7/(5.03 \times 10^{-4}) = 3380 \text{ in.}^4$$

$$d^4 = 3380 \times 64/\pi = 68{,}800$$

$$d = 16.2 \text{ in.}$$

This calls for a minimum 17 in. diameter column (or an increase in the positive moments already used for the slab design). In the strength design of this column the moments of Code 13.3.5.2 must be considered.

SLABS ON BEAMS—DESIGN EXAMPLES

10.13 TWO-WAY SLABS ON BEAMS—GENERAL

Code Chapter 13 on slab systems covers two-way slabs on beams that frame directly into columns, as well as flat plates and flat slabs. For the first time it provides a transition between the two extremes that will care for the range from slight stiffening of the slab to the provision of rather rigid beam supports. Being so broad in their coverage, the requirements are not simple to visualize in all their ramifications. This presentation introduces this design area by stages such as those already used for the flat plate and flat slab portions.

At the beginning this type of slab will be introduced by a layout with beams stiff enough to minimize their contribution to slab deflection. The moment assignment tables from Code 13.3.4.1 and 13.3.4.3 are shown for reference as Table 10.5. For interior spans these assignment percentages indicate that $\alpha_1 l_2/l_1 \gtrsim 1$ is used as a break point. One might think of this as the point where beams carry essentially all the load the slab tends to deliver to them; below this $\alpha_1 l_2/l_1$ value lie small beams or deepened slab sections which simply stiffen the slab as an inseparable combination. Of course, where $\alpha_1 = 0$, these are no beams and a flat plate results.

TABLE 10.5. Percentages of Moment Assigned to Column Strip.

	Negative Moment			Positive Moment		
l_2/l_1	0.5	1.0	2.0	0.5	1.0	2.0
$\alpha_1 l_2/l_1 = 0$	75%	75%	75%	60%	60%	60%
$\alpha_1 l_2/l_1 \geq 1.0$	90%	75%	45%	90%	75%	45%

For convenience where $\alpha_1 l_2/l_1 \gtrsim 1$, the term stiff beams is used here, although this stiffness is relative to the slab and not to beams in general. The beam values in the example of Sec. 10.14 indicate that the resulting "stiff" beam, viewed separately from a slab problem, might be considered just a usual beam.

Whatever the concept, at $\alpha_1 l_2/l_1 \gtrsim 1$ the slabs and beams can be almost independently designed.* Where $\alpha_1 l_2/l_1 < 1$, the design must be a joint slab

*α_1 is relative beam stiffness, discussed in the next paragraph. Since l_2/l_1 is limited to the range 0.5 to 2, an $\alpha_1 = 2$ will always give $\alpha_1 l_2/l_1 \gtrsim 1$.

and beam problem starting from a trial combination of slab and stiffening beam, a problem where there are many independent design possibilities. For such combination design the designer must start with some assumed stiffening of the slab suiting the architectural demands or the designer's interest and verify its adequacy and the necessary reinforcement to be used, as is done in Sec. 10.15.

The ratio $\alpha_1 = I_b/I_s$ needs some special notice since I_b is for an arbitrary definition of a beam and I_s is for an entire panel width (not what is left after the beam is separated). The beam includes projections above or below the slab plus a flange equal on each side to $4h_s$ or the beam projection below (or above) the slab, whichever is smaller (Fig. 10.15*b*). The value of I_s is that of all the slab, a width l_2, as though no beam were present. Thus the same beam cross section on adjacent sides of a slab would lead to different α_1 values unless the slab were square ($l_1 = l_2$).

The next section designs a slab on stiff beams and Sec. 10.15 one employing small slab bands as stiffening.

10.14 TWO-WAY SLABS ON STIFF BEAMS—DESIGN EXAMPLE

Design an interior bay two-way slab on *stiff* beams for a bay 16 ft by 20 ft (to column centers) with service dead load (including slab weight) of 75 psf and live load of 100 psf, $f_c' = 3000$ psi, Grade 60 bars, 0.75 in. clear cover on slabs, 1.5 in. clear cover on beams, columns 15 in. square. Assume beams stiff enough to give α = beam I/slab I = 1.00 in the short direction and 1.25 in the long direction.

Solution

(a) Relative α_1 Values

For a given size of beam on all four sides such that α in the short direction is 1.0, the ratio $\alpha_1 l_2/l_1 = 1.0 \times 20/16 = 1.25$ in the short direction. In the other direction I_s is only 16/20 as large, α_1 becomes 1.25, and $\alpha_1 l_2/l_1 = 1.25 \times 16/20 = 1.0$. These α_1 values of 1.0 and 1.25 will be assumed here as the minimum to make $\alpha_1 l_2/l_1 \geqq 1$, although stiffer beams in either direction will not influence the slab part of the design.

(b) Slab Thickness

Based on the comments in Sec. 10.9:

$l_n = 20 - 1.25 = 18.75$ ft in longer direction

β = long/short *clear* span ratio = 18.75/14.75 = 1.27
β_s = 1, continuous all around
α_m = average α for all beams around panel = 2(1.0 + 1.25)/4 = 1.12

Compare the denominators of the h equations:

Eq. 9-6: $36{,}000 + 5000\beta[\alpha_m - 0.5(1 - \beta_s)(1 + 1/\beta)]$
$= 36{,}000 + 5000 \times 1.27(1.12) = 43{,}120 > 36{,}000$

Eq. 9-7: $36{,}000 + 5000\,(1 + \beta_s)$
$= 36{,}000 + 5000 \times 1.27(1 + 1) >$ above; cannot govern.

Eq. 9-8 is an upper limit of no interest here, since Eq. 9-6 will call for a smaller thickness.

Reqd. $h = (800 + 0.005 \times 60{,}000)l_n/43{,}120$
$= 1100 \times 18.75 \times 12/43{,}120 = 5.75$ in., say 6 in.

This will be checked later for moment. Shear with this type slab with stiff beams is rarely critical.

(c) Design Moments

Dead load = 1.4 × 75 = 105
Live load = 1.7 × 100 = 170

w = 275 psf

This does not include the stem weight of the beams.

Long Span

$M_o = 0.125 \times 275 \times 16 \times 18.75^2/1000 = 193.0$ k-ft
Total negative moment = $-0.65 \times 193.0 = -125.4$ k-ft
Total positive moment = $0.35 \times 193.0 = 67.6$ k-ft

Column strip:

$\alpha_1 = 1.25$ assumed, $l_2/l_1 = 16/20 = 0.8$, $\alpha_1 l_2/l_1 = 1.0$

For this $\alpha_1 l_2/l_1$, Table 10.5 (from Code 13.3.4.1 and .3) shows moment assignment percentages to the column strip the same for negative and positive moment. Interpolations for l_2/l_1 of 0.8 and 1.25 give 81% and 67.5%:

l_2/l_1	0.5	0.8	1.0	1.25	2.00
$\alpha_1 l_2/l_1$	90%	81%	75%	67.5%	45%

Code 13.3.4.4 assigns 85% of the column strip moment to the beam for these stiff beams.

Col. strip neg. $M = 0.81(-125.4) = -101.3$ k-ft
Beam neg. $M = 0.85(-101.3) = -86.0$ k-ft
Slab neg. $M = -101.3 + 86.0 -15.3$ k-ft $= -7.65$ k-ft/half strip
Col. strip pos. $M = 0.81 \times 67.6 = +54.8$ k-ft
Beam pos. $M = 0.85 \times 54.8 = +46.5$ k-ft
Slab pos. $M = 54.8 - 46.5 = +8.3$ k-ft $= +4.15$ k-ft/half strip
Middle strip (takes the remaining moments)
Mid. strip neg. $M = -125.4 + 101.3 = -24.1$ k-ft
Mid. strip pos. $M = 67.6 - 54.8 = +12.8$ k-ft

Short Span

$M_o = 0.125 \times 275 \times 20 \times 14.75^2/1000 = 149.5$ k-ft
Total neg. $M = -0.65 \times 149.5 = -97.2$ k-ft
Total pos. $M = 0.35 \times 149.5 = +52.3$ k-ft

Column strip:
$\alpha_1 = 1.0$ assumed, $l_2/l_1 = 20/16 = 1.25$, $\alpha_1 l_2/l_1 = 1.25$

From the earlier interpolation in Table 10.5:

Col. strip neg. $M = 0.675(-97.2) = -65.6$ k-ft
Beam neg. $M = 0.85(-65.6) = -55.7$ k-ft
Slab neg. $M = -65.6 + 55.7 = +9.9$ k-ft $= 4.95$ k-ft/half strip
Col. strip pos. $M = 0.675 \times 52.3 = +35.2$ k-ft
Beam pos. $M = 0.85 \times 35.2 = +29.9$ k-ft
Slab pos. $M = 35.2 - 29.9 = +5.3$ k-ft $= +2.65$ k-ft/half strip

Middle strip
Mid. strip neg. $M = -97.2 + 65.6 = -31.6$ k-ft
Mid. strip pos. $M = 52.3 - 35.2 = +17.1$ k-ft

These moments are shown in Table 10.6 and on the slab layout in Fig. 10.21.

The middle strip negative moment on the short span is normally considered the most critical with regard to required slab depth for flexure. In this example the total negative moment on this strip is -31.6 compared to -24.1 on the long strip, but the respective widths (Fig. 10.20*a*) are 12 ft and 8 ft, making the worse moment per foot width that on the long span ($24.1/8 = 3.01$ vs. $32.1/12 = 2.68$).

$k_u = M/bd^2 = 3.01 \times 12{,}000/(12 \times 5^2) = 121$

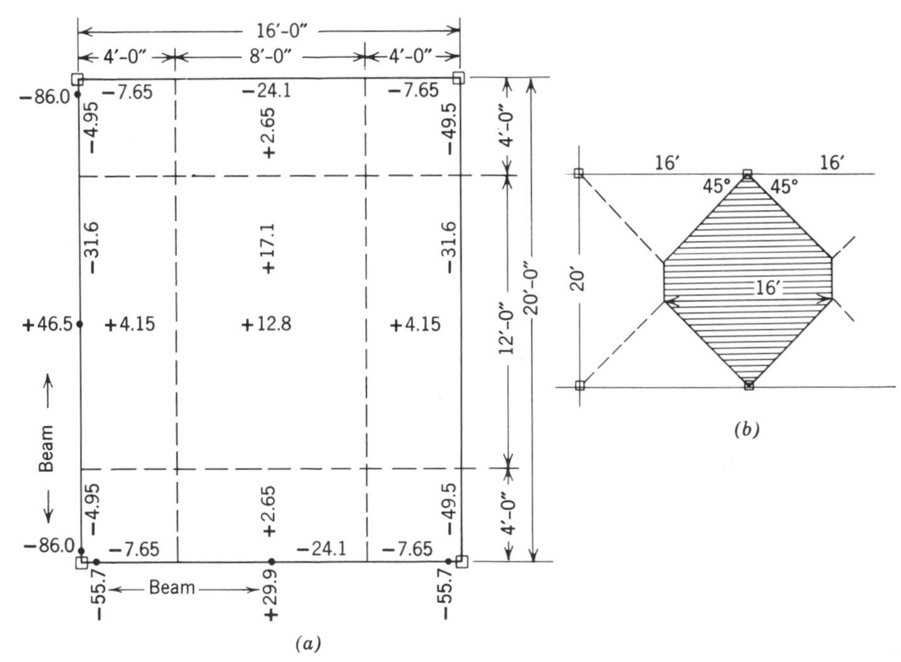

Fig. 10.21. Slab on beams. (*a*) Distribution of moments to the strips and beams. (*b*) Load distribution to beams.

This is very much under the allowable k_u of 705 in Table 2.1; the 6 in. slab is obviously satisfactoy for moment.

USE slab $h = 6$ in.

TABLE 10.6. Moment Assignment for Sec. 10.14.

	Long Span				Short Span			
	Neg. M		Pos. M		Neg. M		Pos. M	
Total	$0.65M_o$ $= -125.4$ ↓		$0.35M_o$ $= +67.6$ ↓		$0.65M_o$ $= -97.2$ ↓		$0.35M_o$ $= +52.3$ ↓	
Col. strip	-101.3 ↙ ↘		$+54.8$ ↙ ↘		-65.6 ↙ ↘		$+35.2$ ↙ ↘	
Beam	-86.0		$+46.5$		-55.7		$+29.9$	
Slab		-15.3		$+8.3$		-9.9		$+5.3$
Mid. strip		-24.1		$+12.8$		-31.6		$+17.1$

(d) Check Minimum Column Stiffness for 10 ft Story Height

The column stiffness requirement is in the Code to protect slabs from pattern loading. The requirement should not be a problem where stiff beams are being used, but it will be checked anyway. The ratio of service dead to live load is $\beta_a = 75/100 = 0.75$, less than the criterion of 2, which causes the designer to check α_{min} from Table 10.2. In the two directions $l_2/l_1 = 20/16 = 1.25$ with $\alpha = 1.0$ and $l_2/l_1 = 16/20 = 0.8$ with $\alpha = 1.25$. Values of α_{min} increase as l_2/l_1 increases and decrease as the beam α increases. Hence the short span with $l_2/l_1 = 1.25$ and $\alpha = 1.0$ is the more critical for this check. Interpolating with $l_2/l_1 = 1.25$ and $\alpha = 1.0$ fixed:

$$\beta_a = 1.0 \qquad \alpha_{min} = 0$$
$$\beta_a = 0.5 \qquad \alpha_{min} = 0.5 \qquad \beta_a = 0.75 \qquad \alpha_{min} = 0.25$$

To use this, the stiffness of slab and beam must be developed for the short span.

$$I_s = 20 \times 12 \times 6^3/12 = 4320 \text{ in.}^4$$
$$I_b = \alpha I_s = 1.0\, I_s = 4320 \text{ in.}^4$$

With slab and beams on both sides of the column and the column both above and below the joint:

$$\alpha_{min} = \Sigma K_c/\Sigma(K_s + K_b) = K_c/(K_s + K_b)$$
$$0.25 = (I_c/10)/[(4320 + 4320)/16] = I_c/5400$$
$$\text{Min. } I_c = 0.25 \times 5400 = 1350 \text{ in.}^4 = h^4/12$$
$$\text{Min. } h = 11.3 \text{ in.}$$

The column size has not been specified, but should easily satisfy this minimum thickness.

(e) Slab Reinforcement

The positive moment bars in the middle strips do cross each other and require design for different depths. With moments (per foot width) so nearly the same it does not matter much whether long span or short span bars are given preferential location. Contrary to recommendations made earlier in connection with the 1963 Code, the author will here place the long bars on the bottom, making:

$$d_l = 6 - 0.75 - 0.25 = 5.00 \text{ in. (for \#4 bars)}$$
$$d_s = 6 - 0.75 - 0.50 - .025 = 4.50 \text{ in. (for \#4 bars)}$$

Middle strip top bars in both directions may be in the top with $d = 5.00$ in., since the only interference is possibly near the quarter points of the sides.

Column strip bottom bars will have to match what was done for middle strip bottom bars. Negative moments in the perpendicular column strips occur in the same area. Use long strip bars on top with $d = 5.00$ in. and short strip bars with $d = 4.50$ in. The sketches and calculations in Table 10.7 complete the basic design of the slab.

The maximum bar spacing (13.5.1) is $2h_s = 12$ in. The steel must also be adequate for temperature steel requirements, that is, $0.0018\ bh$.

TABLE 10.7. Reinforcement for Slab of Sec. 10.14.

	Long Span		Short Span	
Assume #4	6" 5.00"	5.00"	5.00"	4.50" 6"
	Neg. M	Pos. M	Neg. M	Pos. M
Middle Strips				
Strip width	8 ft		12 ft	
M_u, k-ft	−24.1	+12.8	−31.6	+17.1
$\overline{M} = M_u/0.9$, k-ft	−26.8	+14.2	−35.0	+19.0
d, in.	5.00	5.00	5.00	4.50
$A_s = \overline{M} \times 12/(60 \times 0.95d)$				
$= \overline{M}/(4.75d)$, in.^{2}	1.13	0.60	1.47	0.89
$0.0018bh$, in.^{2}	1.04	1.04	1.56	1.56
Min. no. bars, $s = 2h$	8	8	12	12
#3 bars (0.11 in.^{2})	11- #3@8.5±"	10- #3@9.5"	14- #3@10.5±"	15- #3@9½±"
	(1.21)	(1.10)	(1.54)	(1.65)
Revised d, in.	5.06	5.06	5.06	4.69
Column Strips				
Strip width	Half strip = 4 ft (incl. bm.) = 3.5 ft net		Half strip = 4 ft (bm.) = 3.5 ft net	
M_u, k-ft	−7.65/half	+4.15/half	−4.95/half	+2.65/half
$\overline{M} = M_u/0.9$, k-ft	− 8.50	+ 4.60	− 5.50	+ 2.94
d, in.	5.06	5.06	4.69	4.69
$A_s = \overline{M}/(4.75d)$, in.^{2}	0.35	0.19	0.25	0.14
$0.0018\ bh$, in.^{2}	0.46	0.46	0.46	0.46
Min. no. bars, $s = 2h$	4	4	4	4
#3 bars (0.11 in.^{2})	4- #3@10±"		Same for each	
	(0.44)			

(f) Beams

The moments are available for the beam designs except that the weight of stem must still be added. Although beams may be designed for strength, unless they provide $\alpha_1 l_2/l_1 \geqq 1.0$, the slab computations already made must be revised. This is equivalent to stating that the minimum stiffness represented by the assumed α values should be met. In this case this only means I_b in the short span of 4320 in.[4] or 3460 in.[4] for the long span or, say, a beam vertical projection below the 6 in. slab of about 9 in. over a 12 in. width.

The author calls attention to the fact that beam moments obtained in this way are lower than would be developed from conventional analysis. In effect this is a mild form of limit design for the beams comparable to the general slab design rules. As such, the effective percentage of reinforcement, in the author's opinion, should preferably meet the requirements of Code 8.6, that is, ρ or $\rho\text{-}\rho'$ not exceeding $0.5\rho_b$, where ρ_b is that for a balanced member; this is lower than the $0.75\rho_b$ upon which Table 2.1 values of k_u are based. Code Chapter 13 is not written such as to require this limitation.

The shear load taken by the beams in this case must be that from the shaded area of Fig. 10.20*b* and any directly applied loads. If $\alpha_1 l_2/l_1 < 1.0$, the shear would be reduced proportionately with this ratio, the remainder being carried by the slab around the column.

(g) Exterior Panels

Design procedures for exterior panels parallel those of Sec. 10.11 for the flat plate, the column strip simply including beams and slab as above. There is one special provision (13.5.4) which requires special reinforcement, both top and bottom, in any *exterior* corner of the slab system to take care of a moment equal to the maximum positive moment per foot of slab. The critical top reinforcement is parallel to the diagonal of the slab, and that on the bottom is at 90 degrees to that diagonal. Either diagonal bars or a rectangular grid of bars may be used, with the effective A_s/ft the maximum required in the slab for postive moment.

10.15 TWO-WAY SLAB WITH STIFFENED STRIPS—DESIGN EXAMPLE

Design an alternate interior panel slab for the conditions of Sec. 10.14, in this case considering the slab as stiffened only by strips 15 in. wide and 3 in. deeper than the slab running along the column centerlines each way.

Solution

(a) Slab Thickness

The α value for each beam can be established as soon as the slab thickness has been fixed. On the basis of h values already used in this chapter for deflection (5.75 in. in Sec. 10.14 for slab with stiff beams and 6.9 in. for a flat plate in Sec. 10.5 which happened to be 2 ft wider than the present slab), 6.5 in. looks reasonable as an initial trial. Note that this requires an increase in dead load to 81 psf, the initially specified 75 psf not being adequate.

With a trial $h = 6.5$ in. the beam is then that of Fig. 10.22*a*. From the middle of the slab, $\bar{y}$ to the centroid becomes:

$$\bar{y} = -15 \times 3 \times 4.75/(21 \times 6.5 + 15 \times 3) = -1.18 \text{ in.}$$

$$\begin{aligned} I_b: \quad & 21 \times 6.5^3/12 && = 482 \\ & +21 \times 6.5 \times 1.18^2 && = 190 \\ & 15 \times 3^3/12 && = 34 \\ & +45(4.75 - 1.18)^2 && = 574 \\ & \qquad\qquad I_b && = 1280 \text{ in.}^4 \end{aligned}$$

With $l_1 = 20$ ft, $I_{sl} = 16 \times 12 \times 6.5^3/12 = 4400$ in.^{4}
With $l_1 = 16$ ft, $I_{ss} = 20 \times 12 \times 6.5^3/12 = 5500$ in.^{4}

$$\alpha_{1l} = 1280/4400 = 0.29 \text{ for long span}$$
$$\alpha_{1s} = 1280/5500 = 0.23 \text{ for short span}$$

Average $\alpha_1 = 0.26$

The Code method is usable only if the relative beam stiffness in the two directions is within the ratio of 0.2 to 5.0 (13.3.1.6).

$$(I_{bl}/l_l)/(I_{bs}/l_s) = (\alpha_{1l}/\alpha_{1s})(l_s/l_l)^2 = (0.29/0.23)(16/20)^2 = 0.81 \qquad \text{O.K.}$$

$l_n = 20 - 1.25 = 18.75$ ft in the longer direction
$\beta =$ (long/short) clear span ratio $= 18.75/14.75 = 1.27$
$\beta_s = 1$, since continuous all around the panel

Compare the denominators of the h equations.

Eq. 9-6: $36{,}000 + 5000\beta[\alpha_m - 0.5(1 - \beta_s)(1 + 1/\beta)]$
$= 36{,}000 + 5000 \times 1.27(0.26) = 37{,}700 > 36{,}000$

The upper limit equation (9-8) is thus not useful here.

Eq. 9-7: $36{,}000 + 5000\beta(1 + \beta_s)$
$= 36{,}000 + 5000 \times 1.27(1 + 1) = \gg 37{,}700$

Reqd. $h = [(800 + 0.005 \times 60{,}000)18.75 \times 12]/37{,}700 = 6.57$ in.
Accept this 1.1% > 6.5 in. USE $h = 6.5$ in.

The alternate, of increasing the denominator 1.1% by increasing α_m(i.e., I_b) by 25% could also be considered rather than increase the entire slab a half inch.

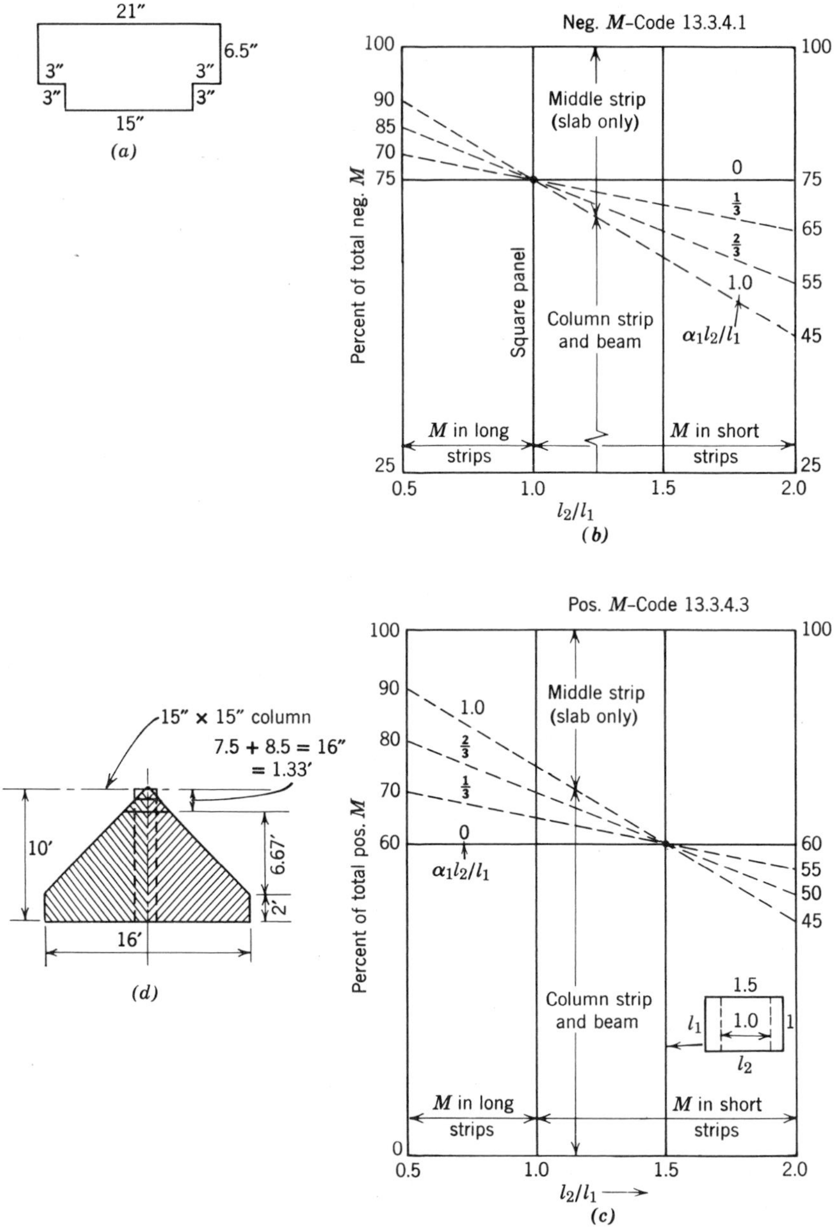

Fig. 10.22. Slab of Sec. 10.15. (*a*) Beam used. (*b*) Moment assignment for negative moment. (*c*) Moment assignment for positive moment. (*d*) Shear load on beam.

(b) Shear Check

At this stage one can check the shear only in a rough fashion. If the slab is considered simply a flat plate, the critical shear section is a square around the column at $d/2$ beyond each face, with each side $15 + d = 15 + 5.25 = 20.25$ in. $= 1.69$ ft wide on each side of the column.

$$\begin{aligned} \text{Dead load} &= 1.4 \times 81 = 113 \\ \text{Live load} &= 1.7 \times 100 = 170 \\ w &= 283 \text{ psf} \end{aligned}$$

$$\text{Design stem weight} = 1.4(3 \times 15/144)150 = 66 \text{ plf}$$

$$\begin{aligned} V_u &\doteq 283(20 \times 16 - 1.69^2) + 66 \times 2(20 + 16 - 2 \times 1.69) \\ &= 90{,}000 + 4300 = 94{,}300 \text{ lb} \\ v_u &= V_u/(\phi b d) = 94{,}300/(0.85 \times 4 \times 20.25 \times 5.25) = 262 \text{ psi} \end{aligned}$$

The shear exceeds the allowable of $4\sqrt{3000} = 220$ psi, on the basis of the oversimplified concept of the flat plate. The designer has several choices. First the slab might be thickened, although it has not been proved that the beams with stirrups, in spite of their lower v_c, cannot handle v_u; the beams are rather shallow for effective stirrups. Second, the concrete strength might be increased to $f_c' = 4000$ psi which gives allowable $v_c = 253$ psi and might be close enough. Third, the beams might be widened to make the 9.5 in. thickness effective all around the column, roughly a width of $15 + 7 = 22$ in. plus a little to keep the failure plane in the deeper elements, say 26 in. This same result could be accomplished by adding a triangle of concrete in each corner between beams to build up a kind of octagonal drop panel effect to resist shear. Fourth, a steel shearhead might be included within the slab depth. Tentatively, the design will proceed with f_c' to be changed to 4000 psi if that becomes necessary.

(c) Design Moments

Long Span

$$\begin{aligned} M_o &= 0.125 \times 283 \times 16 \times 18.75^2/1000 = 199 \text{ k-ft} \\ \text{Total negative moment} &= -0.65 \times 199 = -129 \text{ k-ft} \\ \text{Total positive moment} &= 0.35 \times 199 = 70 \text{ k-ft} \end{aligned}$$

Coefficients for moment distribution or assignment are given in Table 10.5 but for visualization have also been plotted in Fig. 10.22*b*,*c*.* For use the following constants are required:

$$\alpha_1 = 0.29, \; l_2/l_1 = 16/20 = 0.80, \; \alpha_1 l_2/l_1 = 0.23$$

*See also further detailed discussion in Sec. 10.21.

Negative moment to the column strip is then 75 + (0.23/0.333) 0.4 × 5 = 75 + 1.4 = 76%* and positive moment to column strip is 60 + (0.23/0.333) 0.7 × 10 = 65%. Code 13.3.4.4 for $\alpha_1 l_2/l_1 < 1.0$ assigns $85\alpha_1 l_2/l_1$ percent of the negative or positive moment in the column strip to the beam. These two conditions lead to:

Column strip negative moment = −0.76 × 129 = −98 k-ft

Beam neg. M = −0.85 × 0.23 × 98 = −19 k-ft

Slab neg. M = −98 + 19 = −79 k-ft = −39.5 k-ft/half strip

Column strip positive M = 0.65 × 70 = 46 k-ft

Beam pos. M = +0.85 × 0.23 × 46 = 9 k-ft

Slab pos. M = 46 − 9 = +37 k-ft = +18.5 k-ft/half strip

Middle strip neg. M = −129 = 98 = −31 k-ft

Middle strip pos. M = 70 − 46 = +24 k-ft

The author notes that with this very light beam a distribution in proprotion to web width alone would have given −14 k-ft to the beam, which is over two-thirds of the −19 k-ft found above; with the stiffer beam of Sec. 10.14 the 85% assignment led to −86.0 k-ft to the beam.

Short span

M_{os} = 0.125 × 283 × 20 × 14.75^2/1000 = 154 k-ft

Total negative moment = −0.65 × 154 = −100 k-ft

Total positive moment = 0.35 × 154 = +54 k-ft

$\alpha_1 = 0.23,\ l_2/l_1 = 1.25,\ \alpha_1 l_2/l_1 = 0.29$

For negative moment to the column strip Fig. 10.22*b* shows 75 − 2 = 73% and for positive moment 60 + (0.29/0.333) 0.7 × 10 = 66% (Fig. 10.22*c*).

Column strip neg. M = −0.73 × 100 = −73 k-ft

Beam neg. M = −0.85 × 0.29 × 73 = −18 k-ft

Slab neg. M = −73 + 18 = −55 k-ft = −27.5 k-ft/half strip

Column strip pos. M = +0.66 × 54 = +36 k-ft

Beam pos. M = +0.85 × 0.29 × 36 = +9 k-ft

Slab pos. M = +36 − 9 = +27 k-ft = 13.5 k-ft/half strip

Middle strip neg. M = −100 + 73 = −27 k-ft

Middle strip pos. M = +54 − 36 = +18 k-ft

*This calculation, in effect, modifies the line for $\alpha_1 l_2/l_1$ of ⅓ in Fig. 10.22*b* to fit $\alpha_1 l_2/l_1$ of 0.26. A value closer than a full percentage point is unrealistic.

(d) Depth for Moment

The worst slab moment is -39.5 k-ft on a half column strip of width $(8 - 1.25)/2 = 3.37$ ft in the long span where $d = 5.5$ in.

$k_u = 39.5 \times 12{,}000/(3.37 \times 12 \times 5.5^2) = 387$

The allowable from Table 2.1 is 705.

The worst beam section is 15 in. wide and carries $M = -19$ k-ft with d about $9.5 - 0.75$ top cover $-$ 0.375 stir. $- 0.25 = 8.13$ in. If beam moments were more important the normal beam cover of 1.5 in. would be more appropriate.

$k_u = 19 \times 12{,}000/(15 \times 8.13^2) = 232 \ll 705$

(e) Recheck Shear

A recheck is needed because the preliminary estimate left questions to be settled later. The load area contributing to the beam load is that shown shaded in Fig. 10.22*d*, but Code 13.3.4.7 reduces the load to $\alpha_1 l_2/l_1$ parts where this ratio is less than 1.0. On the long beam, the critical shear is at d from face of column, $0.5 \times 18.75 - 8.5/12 = 8.67$ ft from midspan.

Gross V at critical section

$= (8.67 \times 16 - 2 \times 0.5 \times 6.67^2)\ 281 = 26{,}500$ lb + bm. wt.

Effective $V_u = (\alpha_1 l_2/l_1)26{,}500 + 8.67 \times 66$

$= 0.23 \times 26{,}500 + 572 = 6100 + 572 = 6670$ lb

$v_u = 6670/(0.85 \times 15 \times 8.50) = 62 \text{ psi} < 2\sqrt{3000} = 109$ O.K.

A quick mental check on the short beam with $\alpha_1 l_2/l_1 = 0.29$ and less gross load shows it also has a low v_u. However, the remaining shear, around 70,000 lb, cannot be carried by the slab in the corners between the beams. (Allocating the low shear to the beams has increased the shear left for the remaining slab.)

With the beams working so inefficiently under the Code assignments, the check must return to the flat plate concept. The shear capacity, as a matter of logic, is not automatically reduced by deepening some portions 3 in. The original calculation of $v_u = 262$ psi > 220 psi allowable still stands under this assumption.

The author is personally confident that the excess 42 psi could be summed all around the critical section, $94{,}300(42/262) = 15{,}100$ lb, and this excess assigned to stirrups in the four beams. Minimum stirrups ($\rho_n f_y$ of 50 psi) would be ample, but stirrups in a 9.5 in. thickness lead to questions about stirrup development. If the irregular depth was created by steps on the tension face, instead of on the compression face, the greater effective depth would be available. The author is hesitant to count this extra depth here since diagonal cracks might break through the 6.5 in. depth before the 9.5 in. slab became fully effective. (Such a minimized resistance must be accepted where tests of unusual situations are not available.) The assumption is made here that specification of $f_c' = 4000$ psi is more acceptable than a change in slab thickness or stiffening thickness. With the change in concrete strength the author accepts the remaining 9 psi overstress $(262 - 4\sqrt{4000})$ as undoubtedly offset by the increased depth at the beams.

USE $f_c' = 4000$ psi.

(f) Check on Minimum Column Size

A check on column stiffness is required where the ratio β_a, the ratio of service dead to live load, is less than 2. Here $\beta_a = 81/100$. Relative beam stiffness is 0.23 in the short direction where $l_2/l_1 = 1.25$, and 0.29 in the long direction where $l_2/l_1 = 0.8$. Table 10.2 shows that $\alpha_{\min}$ increases as l_2/l_1 increases and also as α decreases. Therefore the short span controls for a square column. Assume a column height of 10 ft. With a beam stiffness ratio α of 0.23 and a length ratio of 1.25, Table 10.2 leads to:

$\beta_a = 1.0 \quad \alpha_{\min} \quad 0.6$

$\beta_a = 0.5 \quad \alpha_{\min} = 1.5$ $\qquad \beta_a = 0.81 \quad \alpha_{\min} = 1.0$

$$\alpha_{\min} = \Sigma K_c/(\Sigma K_s + \Sigma K_b) = I_c/l_c/(I_s/l_s + I_b/l_b)$$

The values of I_s and I_b were calculated in (a) above.

$$1.0 = (2I_c/10)/(2 \times 5500/16 + 2 \times 1280/16) = 2I_c/(10 \times 846)$$

Min. $I_c = 4230$ in.^{4} $\qquad$ Min. $h = 15.0$ in.

The column is barely adequate. If it had failed this test, it could have been enlarged or the positive moments could have been increased under Code 13.3.6.1b by the ratio:

$$\delta_s = 1 + [(2 - \beta_a)/(4 + \beta_a)](1 - \alpha_c/\alpha_{\min})$$

where δ_s is the multiplier for the positive moment

α_c is the actual column stiffness ratio

The α ratio can be replaced with the ratio of actual I_c to required I_c since other parts of α are the same. When α_c equals $\alpha_{\min}$ the multiplier δ_s becomes unity.

(g) Reinforcement design

The calculations are summarized in Table 10.8 with no new problems except the extra calculation of beam steel and a possible alternate decision as to whether the 3 in. deepening of the slab really makes a beam with 1.5 in. cover required or just a deepened slab with 0.75 in. cover. The assigned moments are so low for positive moment it, in fact, makes no difference here. The table treats it as a slab type of element for positive and negative moment.

TABLE 10.8. Reinforcement for Slab of Sec. 10.15.

h = 6.5 in. generally, with 15 in. wide by 9.5 in. between columns

Middle Strips	Long Span		Short Span	
Strip width	8 ft		12 ft	
	Neg. M	Pos. M	Neg. M	Pos. M
M_u, k-ft	−31.0	+24.0	−27.0	+18.0
$\overline{M} = M_u/\phi$, k-ft	−34.5	+26.7	−30.0	+20.0
d, in.	5.5	5.5	5.5	5.0
$A_s = \overline{M} \times 12/(60 \times 0.95d)$				
$= \overline{M}/(4.75d)$, in.^{2}	1.32	1.02	1.15	0.84
$0.0018bh$	1.12	1.12	1.69[a]	1.69
Min. no. bars, $s = 2h$	8	8	11	11
Use #3 (0.11 in.^{2})	12- #3@8″	11- #3@9 ±″	11- #3@13 ±″	16- #3@4″
	(1.32)	(1.21)	(1.21)	(1.76)
Revised d	5.06	5.06	5.06	5.19
Column Strip				
Long span	Slab (each half strip)		Beam	
Strip width	Half strip = 4 ft (incl. bm.) = 3.37 ft = 40.5 in. net		b_w = 15″, h = 9.5″	
	Neg. M	Pos. M	Neg. M	Pos. M
M_u, k-ft	−39.5	+18.5	−19	+ 9
$\overline{M} = M_u/0.9$, k-ft	−43.8	+20.5	−21	+10
d, in.	5.5	5.5	8.5	7.75
$A_s = \overline{M}/(4.75d)$, in.^{2}	1.68	0.79	0.52	0.27
$0.0018\ bh$	0.48	0.48	0.26	0.26
Min. no. bars, $s = 2h$	4	4	1	1
USE bars	6- #5@6.5 ±″	8- #3@5 ±″	2- #5@7″	3- #3@5″
	(1.86)	(0.88)	(0.62)	(0.33)
Revised d	5.44	5.56	8.44	7.81
Column strip				
Short span	Slab (each half strip)		Beam	
Strip width	Half strip = 4 ft (incl. bm.) = 3.38 ft = 40.5 in. net		b_w = 15 in.	
	Neg. M	Pos. M	Neg. M	Pos. M
M_u, k-ft	−27.5	+13.5	−18	+ 9
$\overline{M}$, k-ft	−30.6	+15.0	−20.0	+10.0
d, in.	4.81	5.0	8.0	7.75
$A_s = \overline{M}/(4.75d$, in.^{2}	1.34	0.63	0.53	0.27
$0.0018\ bh$	0.48	0.48	0.26	0.26
Min. no. bars, $s = 2h$	4	4	1	1
USE bars	5- #5@8 ±″	6- #3@7 ±″	3- #4@5″	3- #3@5″
	(1.55)	(0.66)	(0.60)	(0.33)
Revised d	4.81	5.06	8.00	8.44

[a] Half of bottom bars count on this.

EQUIVALENT FRAME METHOD

10.16 USE OF EQUIVALENT FRAME METHOD OF SLAB ANALYSIS

For slab systems not meeting the direct method requirements of Sec. 10.4 (Code 13.3.1) the design moments must be computed by the equivalent frame method; elsewhere it is an optional method.

As in an analysis of beams and columns, the structure is subdivided into bents running each way through the frame. The entire slab width, that is, one-half panel width on each side of the column, is considered in establishing the slab load and slab stiffness. Stiffness is based on gross concrete area, but considers a variable moment of inertia because the stiffness at column or drop panel becomes part of the problem.

If the service live load is not more than 75% of the dead load, the frame analysis is made with full design (factored) load on all spans instead of a pattern loading. For higher live load ratios pattern loads must be used (Sec. 10.19) but, as in frame analysis for beam design, certain simplifications are permissible:

a. Columns fixed at their far end (at floor above and below).
b. Two adjacent spans loaded with live load for maximum negative moment.
c. Alternate spans loaded with live load for maximum positive moment.
d. Slab-beams may be considered fixed two panels away from the point where moment is being computed, provided the slab continues further.

In addition the pattern loading may be applied with only 75% of the design live load.

Although spans to centerline of columns are used in analysis, the critical negative moments are considered those at the face of any rectangular column or, if not rectangular, at the face of the equivalent (equal area) square column.

The chief complication, a new one in this Code, is the use of the equivalent column stiffness K_{ec}, recognizing the part that the slab or transverse beam torsional stiffness plays in the system. Equivalent stiffness has already been discussed in Sec. 10.7 and calculated for an exterior column in Sec. 10.10b (but there using a simplified version of K_c for the column).

10.17 FRAME CONSTANTS

Since no member of the equivalent frame has a uniform I, the stiffness is not $4EI/l$ but something more than 4 for the constant and something more than 0.5 for the carry-over factor, C_f* or C.O.F. Tables can tabulate k for use in kEI/l and C_f, although C_f is directional, not the same from each end except with symmetry of crosssection. Likewise the fixed-end moment is normally greater than for uniform I.

Table 10.9, slightly modified from Ref. 10, lists for flat plates the slab stiffness factors, moment factors, carry-over factors, and at the very bottom the useful† quantity $(1 - c_2/l_2)^3$. The EI variation used in preparing this table is sketched alongside. With a drop panel the EI for the drop length is that of a very flat wide T-shape of width l_2; this requires a separate table. Reference 10 has similar tables for drop panel thicknesses of $1.25h_s$ and $1.50h_s$ in symmetrical panels. The Commentary has similar tables for flat plates with $c_1/l_1 = c_2l_2$ but accommodating unequal ratios at the two slab ends; also a similar table for a drop panel slab with thickness $1.25h_s$.

Values of column K_{ec} cannot be readily tabulated, but K_c values for the column are easily obtained from the coefficients in Table 10.10, also slightly modified from Ref. 10. The EI variations used are indicated on the sketch with the table, $1/EI$ as zero for the thickness of the slab (and of the drop if any) and for the column capital height. Since such a stiffness is different at the two ends (if there is either drop or capital or both), care must be taken to use a/l_c for the *near end* and b/l_c for the *far end* in every case. In this way the one table works equally well for the column above or below a joint.

Alternatively, for K_c the curve data of Fig. 10.23*a* should give the same values plus the carry-over factors, *provided one used the column length* l_c *reduced* to the distance to the far fixed end created by the face of the slab above or below (or the start of the capital above where a capital is used).‡ Since both stiffness and carry-over factor are directional, the first subscript designates the joint which is rotating. Thus k_{TB} applies for a joint rotation at the top and C_{TB} gives the C.O.F. from top to bottom of the column; k_{BT} and C_{BT} are in the opposite direction.

The torsional stiffness of the transverse beam or slab (see Sec. 10.10b) is $K_t = 9EC/[l_2(1 - c_2/l_2)^3]$ for the transverse span *on one side* of the column. Values of $(1 - c_2/l_2)^3$ have already been tabulated at the bottom of Table 10.9. Values of $C = \Sigma(1 - 0.63x/y)x^3y/3$ have been tabulated for *individual* rectangles in Table 10.11, also adapted slightly from Ref. 10. The values for individual rectangles must be summed to care for an L- or T-section.

*The new symbol C_f (or other subscript) is used since the Code has preempted C for a torsional constant.

†Useful in the torque stiffness calculation.

‡The moment then carried over acts at the face of slab (the assumed far fixed end).

TABLE 10.9. Moment Distribution Factors for Slab-Beam Elements (Flat Plate with or without Column Capital)

c_1/l_1 \ c_2/l_2		0.00	0.05	0.10	0.15	0.20	0.25	0.30
0.00	M	0.083	0.083	0.083	0.083	0.083	0.083	0.083
	k	4.000	4.000	4.000	4.000	4.000	4.000	4.000
	C_f	0.500	0.500	0.500	0.500	0.500	0.500	0.500
0.05	M	0.083	0.084	0.084	0.084	0.085	0.085	0.085
	k	4.000	4.047	4.093	4.138	4.181	4.222	4.261
	C_f	0.500	0.503	0.507	0.510	0.513	0.516	0.518
0.10	M	0.083	0.084	0.085	0.085	0.086	0.087	0.087
	k	4.000	4.091	4.182	4.272	4.362	4.449	4.535
	C_f	0.500	0.506	0.513	0.519	0.524	0.530	0.535
0.15	M	0.083	0.084	0.085	0.086	0.087	0.088	0.089
	k	4.000	4.132	4.276	4.403	4.541	4.680	4.818
	C_f	0.500	0.509	0.517	0.526	0.534	0.543	0.550
0.20	M	0.083	0.085	0.086	0.087	0.088	0.089	0.090
	k	4.000	4.170	4.346	4.529	4.717	4.910	5.108
	C_f	0.500	0.511	0.522	0.532	0.543	0.554	0.564
0.25	M	0.083	0.085	0.086	0.087	0.089	0.090	0.091
	k	4.000	4.204	4.420	4.648	4.887	5.138	5.401
	C_f	0.500	0.512	0.525	0.538	0.550	0.563	0.576
0.30	M	0.083	0.085	0.086	0.088	0.089	0.091	0.092
	k	4.000	4.235	4.488	4.760	5.050	5.361	5.692
	C_f	0.500	0.514	0.527	0.542	0.556	0.571	0.585
$X = (1 - c_2/l_2)^3$		1.000	0.856	0.729	0.613	0.512	0.421	0.343

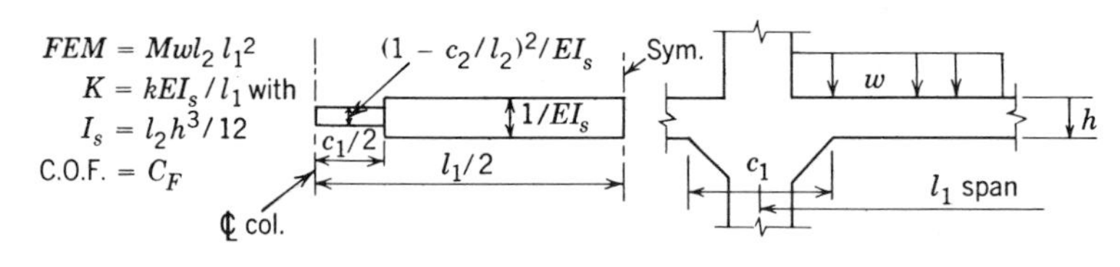

The use of these tables and the calculation of K_{ec} and the distribution factors make up a rather involved process which needs some formalizing to maintain clarity of presentation. An example is developed in the next section using a columnar form (Fig. 10.24) suggested by Ref. 10, but paralleling this with the detailed number calculations.

TABLE 10.10. Column Stiffness Coefficients, k_c[a]

a/l_c \ b/l_c	0.00	0.02	0.04	0.06	0.08	0.10	0.12	0.14	0.16	0.18	0.20	0.22	0.24
0.00	4.000	4.082	4.167	4.255	4.348	4.444	4.545	4.651	4.762	4.878	5.000	4.128	5.263
0.02	4.337	4.433	4.533	4.638	4.747	4.862	4.983	5.110	5.244	5.384	5.533	5.690	5.856
0.04	4.709	4.882	4.940	5.063	5.193	5.330	5.475	5.627	5.787	5.958	6.138	6.329	6.533
0.06	5.122	5.252	5.393	5.539	5.693	5.855	6.027	6.209	6.403	6.608	6.827	7.060	7.310
0.08	5.581	5.735	5.898	6.070	6.252	6.445	6.650	6.868	7.100	7.348	7.613	7.897	8.203
0.10	6.091	6.271	6.462	6.665	6.880	7.109	7.353	7.614	7.893	8.192	8.513	8.859	9.233
0.12	6.659	6.870	7.094	7.333	7.587	7.859	8.150	8.461	8.796	9.157	9.546	9.967	10.430
0.14	7.292	7.540	7.803	8.084	8.385	8.708	9.054	9.426	9.829	10.260	10.740	11.250	11.810
0.16	8.001	8.291	8.600	8.931	9.287	9.670	10.080	10.530	11.010	11.540	12.110	12.740	13.420
0.18	8.796	9.134	9.498	9.888	10.310	10.760	11.260	11.790	12.370	13.010	13.700	14.470	15.310
0.20	9.687	10.080	10.510	10.970	11.470	12.010	12.600	13.240	13.940	14.710	15.560	16.490	17.530
0.22	10.690	11.160	11.660	12.200	12.800	13.440	14.140	14.910	15.760	16.690	17.210	18.870	20.150
0.24	11.820	12.370	12.960	13.610	14.310	15.080	15.920	16.840	17.870	19.000	20.260	21.650	23.260

[a] $K_c = \dfrac{kcEcIc}{l_c}$

TABLE 10.11. Values of Torsion Constant, C[a]

y \ x	4	5	6	7	8	9	10	12	14	16
12	202	369	592	868	1,188	1,538	1,900	2,557	—	—
14	245	452	736	1,096	1,529	2,024	2,566	3,709	4,738	—
16	388	534	880	1,325	1,871	2,510	3,233	4,861	6,567	8,083
18	330	619	1,024	1,554	2,212	2,996	3,900	6,013	8,397	10,813
20	373	702	1,167	1,782	2,553	3,482	4,567	7,165	10,226	13,544
22	416	785	1,312	2,011	2,895	3,968	5,233	8,317	12,055	16,275
24	548	869	1,456	2,240	3,236	4,454	5,900	9,469	13,885	19,005
27	522	994	1,672	2,583	3,748	5,183	6,900	11,197	16,628	23,101
30	586	1,119	1,888	2,926	4,260	5,912	7,900	12,925	19,373	27,197
33	650	1,243	2,104	3,269	4,772	6,641	8,900	14,653	22,117	31,293
36	714	1,369	2,320	3,612	5,284	7,370	9,900	16,381	24,860	35,389
42	842	1,619	2,752	4,298	6,308	8,828	11,900	19,837	30,349	43,581
48	970	1,869	3,184	4,984	7,332	10,286	13,900	23,293	35,836	51,773
54	1,098	2,119	3,616	5,670	8,356	11,744	15,900	26,749	41,325	59,965
60	1,226	2,369	4,048	6,356	9,380	13,202	17,900	30,205	46,813	68,157

[a] $C = (1 - 0.63\,x/y)\,\dfrac{x^3/y}{3}$

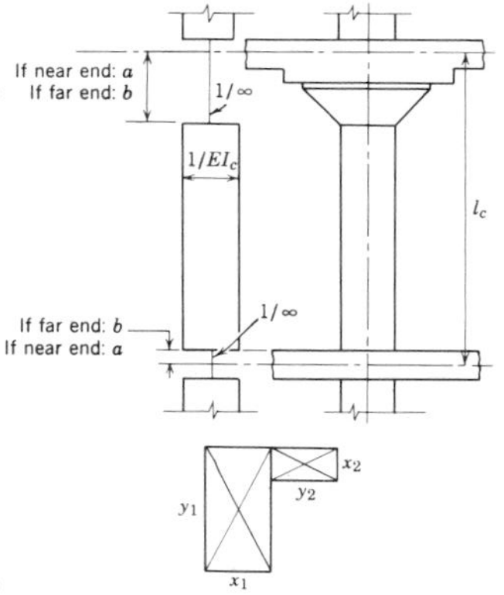

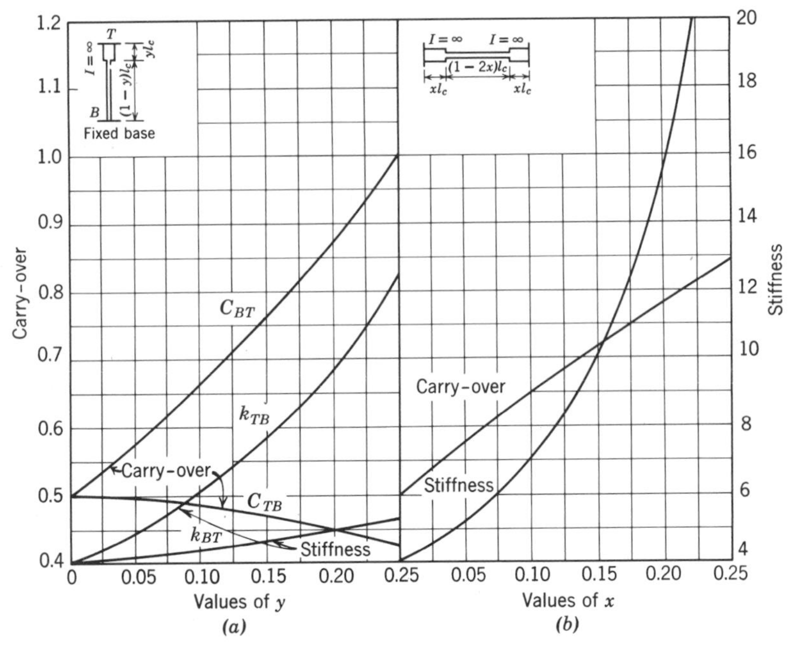

Fig. 10.23. Carry-over factors and stiffness factors for member with infinite I over part of length. (Adapted from Ref. 4, Portland Cement Assn.)

10.18 EQUIVALENT FRAME METHOD—DESIGN EXAMPLE

Although neither the spans nor the loading require it, use the data of Sec. 10.5 (interior panel) and 10.11 (exterior panel) to work out corresponding design moments for the 20 ft span by the equivalent frame method and compare the results.

Panels are 18 ft by 20 ft (center to center of columns) with the 18 ft length parallel to the exterior edge. Columns are 15 in. square, both interior and exterior, with a 10 ft story height, and the slab runs to the exterior face of the outside column without a spandrel beam. For simplicity any wall load has been excluded from the example. Live load is given as 60 psf and total dead weight 100 psf (including slab weight). Slab thickness for interior panel is 7 in., with 8 in. for the exterior panel. The difference in slab weight has been ignored (100 psf used as the total in both panels), making the design dead load 140 psf and the design live load 102 psf. Here the break in slab stiffness will be assumed on the centerline of the first interior column row. $f_c' = 4000$ psi for all columns and slabs.

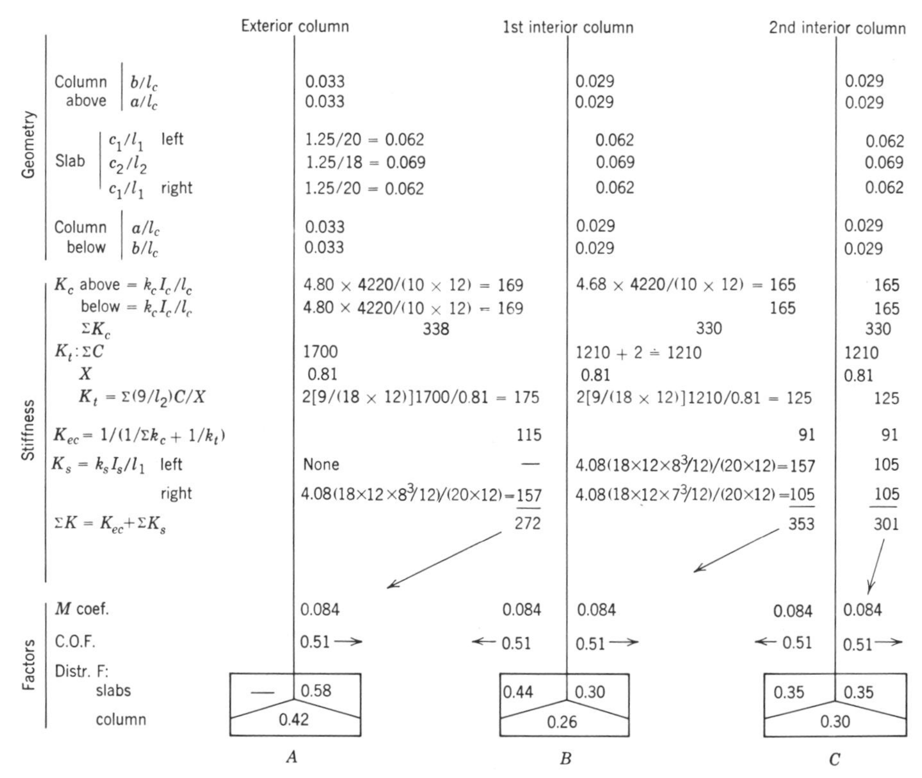

		Exterior column (A)	1st interior column (B)	2nd interior column (C)
Geometry	Column above b/l_c	0.033	0.029	0.029
	Column above a/l_c	0.033	0.029	0.029
	Slab c_1/l_1 left	1.25/20 = 0.062	0.062	0.062
	Slab c_2/l_2	1.25/18 = 0.069	0.069	0.069
	Slab c_1/l_1 right	1.25/20 = 0.062	0.062	0.062
	Column below a/l_c	0.033	0.029	0.029
	Column below b/l_c	0.033	0.029	0.029
Stiffness	K_c above = $k_c I_c/l_c$	4.80 × 4220/(10 × 12) = 169	4.68 × 4220/(10 × 12) = 165	165
	below = $k_c I_c/l_c$	4.80 × 4220/(10 × 12) = 169	165	165
	ΣK_c	338	330	330
	K_t: ΣC	1700	1210 + 2 ≐ 1210	1210
	X	0.81	0.81	0.81
	$K_t = \Sigma(9/l_2)C/X$	2[9/(18 × 12)]1700/0.81 = 175	2[9/(18 × 12)]1210/0.81 = 125	125
	$K_{ec} = 1/(1/\Sigma k_c + 1/k_t)$	115	91	91
	$K_s = k_s I_s/l_1$ left	None —	4.08(18×12×8³/12)/(20×12)=157	105
	right	4.08(18×12×8³/12)/(20×12)=157	4.08(18×12×7³/12)/(20×12)=105	105
	$\Sigma K = K_{ec} + \Sigma K_s$	272	353	301
Factors	M coef.	0.084 (right)	0.084 (left), 0.084 (right)	0.084 (left), 0.084 (right)
	C.O.F.	0.51 →	← 0.51, 0.51 →	← 0.51, 0.51 →
	Distr. F.: slabs	— (left), 0.58 (right)	0.44 (left), 0.30 (right)	0.35 (left), 0.35 (right)
	column	0.42	0.26	0.30

Fig. 10.24. Factors needed for moment distribution, flat plate.

Solution

The sequence of tabulation in Fig. 10.24 will be used as a guide for discussion, followed by items e and f.

a. Geometry
b. Stiffnesses
c. Factors (FEM, COF, distribution factors)
d. Distribution (moment distribution for each loading)
e. Design moments from each loading
f. Possible moment reductions because the slab qualifies for the direct design method.

(a) Geometry

Geometry is simplified by the uniform spans and equal column sizes. For the exterior column the half-depth of slab gives $a/l_c = b/l_c = 4/(10 \times$

12) = 0.033. For the typical interior columns $a/l_c = b/l_c = 3.5/(10 \times 12) = 0.029$. Where the break in slab thickness occurs, one might use average thickness of slab to fix a and b, but that much accuracy is not really inherent in the process; use value for typical interior column, 0.029.

For the slab $c_1/l_1 = 1.25/20 = 0.062$ at each end and c_2/l_2 for width is $1.25/18 = 0.069$.

(b) Stiffnesses

For K_c the a/l_c and b/l_c are available above. For the main part of the column above and below the joint, I_c is $15^4/12 = 4220$, E_c may be omitted (that is, used as 1 since all concrete is alike in the frame and only relative stiffnesses are finally involved). From Table 10.10 the value of k_c for exterior column ($a/l_c = b/l_c = 0.033$) is 4.80, based on interpolation.

$$K_c = 4.80 \times 4220/(10 \times 12) = 169$$
$$\Sigma K_c = 2 \times 169 = 338$$

The approximate calculation in Sec. 10.11*b* gave $282E$.

For the transverse beam at the column line the Code assumes a constant cross section, using the largest of:

1. That part of the slab having a width equal to the width of the column, bracket, or capital, in this case at the exterior column a member 15 in. by 8 in.
2. The slab just described plus that part of any transverse beam above and below the slab, in this case none.
3. The beam of Fig. 10.15*b*, in this case none.

From Table 10.11, C for the 15 in. by 8 in. "beam" at the exterior column line is 1700. (Note that interpolations in the y direction are nearly linear, but not in the x direction; any reasonable interpolation is probably as accurate as the method justifies.) Since there is a beam of this type on each side of the column,

$$K_t = 2[9EC/l_2(1 - c_2/l_2)^3] = 18\ C/(l_2X)$$

where X is the bottom line of Table 10.9 and shows as 0.81 for c_2/l_2 of 0.069.

$$K_t = 18 \times 1700/(18 \times 12 \times 0.81) = 175$$

This compares to 176 found in Sec. 10.11b without the use of the table.

For the exterior column,

$K_{ec} = 1/(1/\Sigma K_c + 1/K_t) = 1/(1/338 + 1/175) = 115$, with E of 1.

This compares to $108E$ found in Sec. 10.11b, using an approximate K_c.

For the slab in the exterior panel, $c_2/l_2 = 0.069$ and $c_1/l_1 = 0.062$, which in Table 10.9 leads to

M coef. $= 0.084$, $k = 4.08$, $C_f = 0.51$
$K_s = 4.08(18 \times 12 \times 8^3/12)/(20 \times 12) = 157$, with E of 1.

For the exterior joint, $\Sigma K = 115 + 157 = 272$.
The short $c_1/2$ length at each end of slab (which is taken stiffer), is so short that the earlier solution from a uniform I does not differ much. The earlier solution gave $K_s = 154E$.

For the first interior column line with a/l_c and b/l_c reflecting a 7 in. slab, K_c is changed to 330 and K_t is changed by the transition between the 8 in. and 7 in. slab shown in Fig. 10.25.

For C the beam shown can be divided into whatever rectangles give a maximum total, which appears to be the 15 in. by 7 in. area plus a 7.5 in. by 1 in. area. From Table 10.11

$C_1 = 1210$
$C_2 = (1 - 0.63 \times 1/7.5)1^3 \times 7.5/3 = 2$ (negligible)
Use $C = 1210$

Since all other factors aside from C remain unchanged, $K_t = (1210/1700) \times 175 = 125$. For the interior column, equivalent

$K_{ec} = 1/(1/338 + 1/125) = 91$

With the slab changing only in thickness from 8 in. to 7 in.,

$K_s = 157(7/8)^3 = 105$ for an interior slab.
ΣK for the joint $= (105 + 157) + 91 = 353$

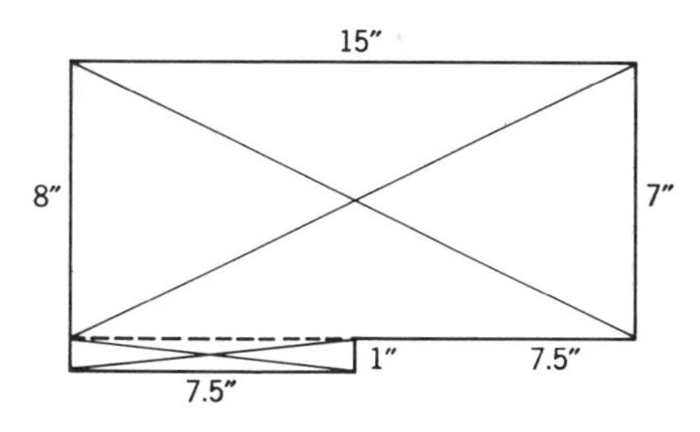

Fig. 10.25. "Beam" cross section at first interior column.

(c) Fixed-end Moments

Since $w_l/w_d = 60/100 = 0.60 < 0.75$, the Code accepts the use of *full* design load on all the frame with no pattern loadings required. For the end span, Table 10.9 shows the moment factor as 0.084 for fixed-end moment, just a trifle more than the nominal 1/12.

$$\text{End span fixed-end } M = M^F = 0.084(140 + 102)18 \times 20^2/1000$$
$$= 146.0 \text{ k-ft}$$
$$\text{Interior span fixed-end } M = M^F = 146.0 \text{ k-ft}$$

At the exterior column, there is a 7.5 in. width of slab which is beyond the column center line and adds a negative $M^F = -0.5 \times 242(18 - 1.25) \times 0.62^2/100 = 0.79$. say 0.8 k-ft. This is almost negligible but is mentioned here since there would usually be some wall weight and the slab often cantilevers significantly beyond the column itself.

(d) Moment Distribution

Consider, closely enough, that the frame system, Fig. 10.26, is fixed (roughly symmetrical) about the fourth column line. It is fully loaded, has uniform spans, and should have very little rotation several joints away from the exterior column. The alternate joint sequence is used for the moment distribution.

(e) Design Moments

The moments thus found by moment distribution are at the center of column lines and should be corrected to the face of support for rectangular (or equivalent square supports), but in no case is the correction to be farther than $0.175l_1$ from the column center line. In case of an exterior column with a bracket or capital, the correction must be reduced to that for only half the actual projection used at the column. In this example *all* corrections are to face of column, 7.5 in. = 0.62 ft from the column center line.

On the basis of the beam free bodies sketched in Fig. 10.26*b* the simple beam shear as half of the beam load, and the continuity shear as the difference between the end moments divided by the span have been calculated. Summation of moments about the face of column then leads to the following design moments.

$$\text{Ext. } M_{AB} = -66.2 + 38.3 \times 0.62 - 4.36 \times 0.62^2/2 = -43.3 \text{ k-ft}$$
$$\text{1st Int. } M_{BA} = -171.2 + 48.9 \times 0.62 - 0.8 \text{ (as for } AB) = -141.7 \text{ k-ft}$$

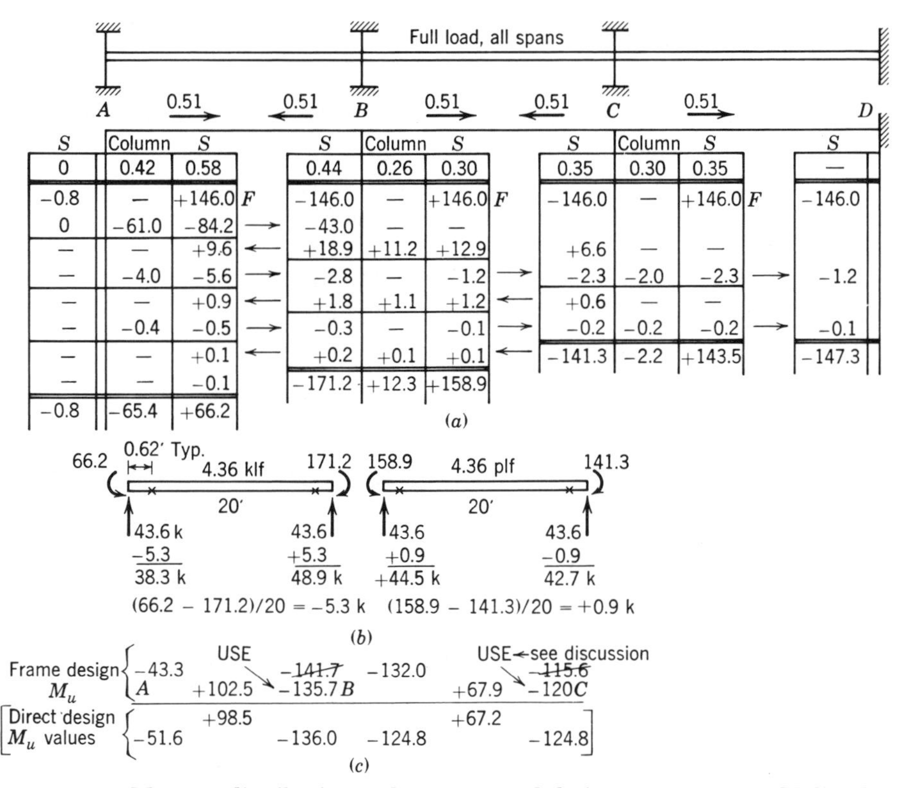

S	Column	S		S	Column	S		S	Column	S		S
0	0.42	0.58		0.44	0.26	0.30		0.35	0.30	0.35		—
-0.8	—	+146.0 F		-146.0	—	+146.0 F		-146.0	—	+146.0 F		-146.0
0	-61.0	-84.2	→	-43.0	—	—						
—	—	+9.6	←	+18.9	+11.2	+12.9		+6.6	—	—		
—	-4.0	-5.6	→	-2.8	—	-1.2	→	-2.3	-2.0	-2.3	→	-1.2
—	—	+0.9	←	+1.8	+1.1	+1.2	←	+0.6	—	—		
—	-0.4	-0.5	→	-0.3	—	-0.1	→	-0.2	-0.2	-0.2	→	-0.1
—	—	+0.1	←	+0.2	+0.1	+0.1	←	-141.3	-2.2	+143.5		-147.3
—	—	-0.1		-171.2	+12.3	+158.9						
-0.8	-65.4	+66.2										

Fig. 10.26. Moment distribution and steps toward design moments. (*a*) *M* distribution. (*b*) Shears. (*c*) Design moments.

M at midspan $= +0.125 \times 4.36 \times 20^2 - 0.5(66.2 + 171.2) = +99.3$ k-ft
V at midspan $= 38.3 - 4.36 \times 10 = -5.3$ k
For max. M, e from midspan $= -5.3/4.36 = -1.22$ ft (to left)
M = area of V diagram over length $e = +0.5 \times 5.3 \times 1.22 = +3.2$ k-ft
Max. pos. $M = +99.3 + 3.2 = +102.5$ k-ft

In similar fashion for span BC:

$M_{BC} = -158.9 + 44.5 \times 0.62 - 0.8 = -132.0$ k-ft
$M_{CB} = -141.3 + 42.7 \times 0.62 - 0.8 = -115.6$ k-ft
Pos. midspan $M = +0.125 \times 4.36 \times 20^2 - 0.5(158.9 + 141.3) = +67.9$ k-ft

The slight shift in the positive moment maximum point is negligible in this case with the end moments differing only 12%.

(f) Check Against M_o For Possible Economy

Since this slab is eligible for the direct design method, check the total moment indicated above against the M_o total that would be required by the direct design; any surplus here is unnecessary. For the exterior span, M_o in Sec. 10.11 was 192 k-ft. The total which is comparable is the sum of the positive moment plus the average end moment, which is the simple beam moment.

Simple beam M, ext. span $= 102.5 + 0.5(43.3 + 141.5) = 195.0$ k-ft This indicates that all moments for design might be reduced proportionately about 1.5%, to make up the 3.0 k-ft difference just found. Theoretically this would be the neatest solution of the problem. Alternatively, however, the author would feel free to reduce any one moment by the total correction, where this is not more than 10% of the computed value. This 10% provision is stated in the Code only for the direct design method, but it is consistent with the redistribution of moments in beam design procedures where ρ is low, as it is here. Accordingly, reduce the negative moment M_{BA} by $2 \times 3.0 = 6.0$ k-ft to $-141.7 + 6.0 = -135.7$ k-ft. This reduces the total simple span moment to match M_o and reduces the total steel required at the first interior column line, since the negative moment in the next span tends to be smaller anyway.

On the first interior span the M_o value is also 192.0 k-ft, from Sec. 10.5. For the design moments,

Simple beam $M = +67.9 + 0.5(132.0 + 115.6) = 191.7$ k-ft

This is a trifle *less* than M_o; no change is warranted and the apparent difference is probably not a significant number anyway.

It is noted that the negative moments at the joints on the third span are higher than M_{CB}(at the joint) by 2.2 k-ft and 6.0 k-ft and corrections to face of column will be similar. The author feels it would be good judgement to design at all interior column lines (except the first) for a moment of something like 5.0 k-ft more than recorded for M_{CB}, say, a total of -120 or -121 k-ft. This change is indicated on Fig. 10.26*c*.

The distribution in Fig. 10.26 has not been extended as many bays as the Code seems to require (two spans beyond the critical moment computed), but inspection indicates that under full load on all spans another span would make no noticeable difference.

The final moments for design are the modified frame design moments indicated in Fig. 10.26*c*. The distribution to middle and column strips is to follow the same procedure used in earlier examples. In this simple case it means 75% of the negative moment and 60% of the positive moments assigned to the column strip.

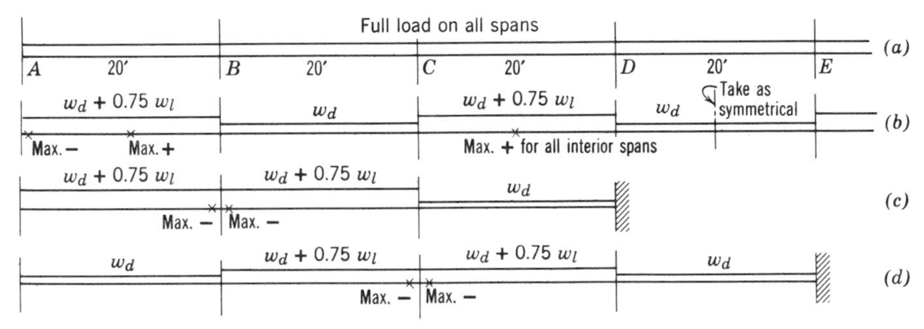

Fig. 10.27. Loadings necessary when $w_l > 0.75w_d$. Full design load required in (*a*).

10.19 PROCEDURE WHERE LIVE LOAD IS LARGER

If live load exceeds 75% of dead load (at service load stage), more loadings in pattern form are necessary to obtain design moments. In each such case the pattern live load is at the lower level of 75% of the design live load. The cases to be analyzed are sketched in Fig. 10.27, where all frame constants and all correction procedures to the critical section would be handled exactly as in the example just given. The pattern loading moments in Fig. 10.27 may not be taken smaller than those of full live load on all spans. (Where the loading qualifies for the direct design method there is no corresponding provision; the M_o calculation takes its place.)

DESIGNS OUTSIDE THE CODE SLAB SYSTEM RULES

10.20 THE GENERAL PROBLEM

(a) Where Rules Are Not Applicable

For flat slab, flat plate, and slab band types of beams, a system failing to qualify for the direct design method *must* be designed by the equivalent frame method, as the only acceptable procedure. This text section relates only to two-way slabs supported on walls, beams, or other substantial supports (not columns). The Code chapter does not cover slabs on beams that lack a supporting column at each slab corner.

For two-way slabs supported on all sides, there are many cases not qualifying for the direct design method and not involved with, or related closely enough to, a frame to justify the use of the equivalent frame method. Such slabs may be supported on masonry walls (with a single span or several continuous) or on steel beams, without composite construction. They may serve as roofs over tanks or as side walls for structures below grade. They may be supported on three sides and unsupported on the fourth side. Slabs in such borderline cases are very common and very versatile structural elements.

(b) Design Options for Nonqualifying Two-way Slabs

The yield line method of Chapter 11 is a respected tool for analyzing special cases, although not recognized in the United States as much as elsewhere. The strip method of Chapter 12 is a Swedish formalization of a method engineers everywhere have used in a rough, very conservative manner for many years, simply providing some possible or reasonable route for loads to be taken to supports. The essential basis for designing special slabs is a visualization of the deflection pattern under load, and possibly its failure pattern, and the fitting of reinforcement to the curvatures which must result. This is a rough technique and usually means an over-designed slab; but it is a highly useful technique when other methods are not available.

If the author were to criticize the Code slab systems chapter, it would be largely on the basis that it is not framed to give the designer a feel for what he is designing; the designer must simply follow by rote certain rules and like a computer come up with the indicated answers; it would take considerable experience with this method to give the designer a clear feel for what happens as he changes a dimension. For a system to handle so many varieties of slabs, this is probably a built-in problem, an unavoidable one.

The author suggests that there is still considerable merit to occasional use of the less economical design methods of the 1963 Code for design of special two-way slabs on stiff beams. Hence, Sec. 10.22 is a restatement of the simplest of these methods, Method 2 which has been in Codes for at least 30 years. Its table of moment coefficients assists the designer in developing better judgements as to what he can do in special cases. Because it contains no limit design aspects, its moments average probably 25% higher than those of the 1971 Code. The moment redistribution procedures of Sec. 9.6 are fully justified in the final choice of required reinforcement.

The author also suggests that for the case of two-way slabs on stiff beams, the designer should attempt to get more of a feel for the 1971 Code moments from the table developed in Sec. 10.21, even though these are expressed as percentages of M_o, which is the new format. It should be easier to follow trends in this table than directly from numbers such as used in Sec. 10.14. This table cannot be extrapolated to very different cases because, as a limit design concept, it will not be safe for greater live load ratios, fewer spans, and the like. However, it is a kind of lower bound which can be compared to the upper bound of the 1963 Method 2.

(c) Visualizing the Effect of Loss of Restraint

Designers of beams are accustomed to the exterior support situation where a light end support results in a large distribution of the fixed end moment back to the beam, with probably half of this distribution carried back to the first interior column joint. A slab support, even with little restraint, is rarely such that it can rotate uniformly over a considerable width;[12] mathematically this would have to be discussed in terms of summing some very unequal restraints and rotations. This is a problem of accuracy, but not one of visualization. One can accept a distribution of fixed end moment back to the slab as realistic. However, slabs differ greatly from beams in their carry-over moments. In a square panel *essentially none of this distribution carries back to the opposite beam support.* In a square panel the moment distributed back to the slab goes roughly equally to the two adjacent supports (90 degrees from the rotating edge); not only does it go to the perpendicular slab elements, but most of it goes to those parallel elements in the near half of the slab. Distribution to the slab is thus a more localized operation and this means pattern loadings are important chiefly where they are local patterns. The distribution is roughly indicated in Fig. 10.28a for a square panel and assumes *very stiff* beams.

Increasing dimension a to make a rectangular slab influences the distribution very little. If a is reduced to $b/2$, a three way distribution results roughly equal to all three beams, again for very stiff beams.

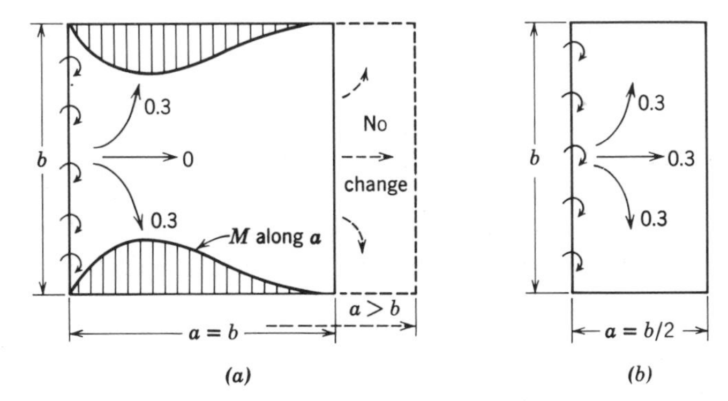

Fig. 10.28. Carry-over moment factors in continuous slabs on beams.

This three-dimensional behavior is significant because it means even free rotation of one edge is a very local matter, not very important except to the positive moments on slabs in both directions and to the negative moments added on the parallel elements nearby the rotating edge. Carry-over moments do not reach into more remote slabs anything like they do in beams. This makes pattern load moments very much less important in slabs.

10.21 SLAB MOMENT PERCENTAGES IN DIRECT DESIGN METHOD—STIFF BEAMS

For an interior panel, as in Sec. 10.14, M_o is divided into total negative moment, 65%, and total positive moment, 35%. For stiff beams (such that $\alpha_1 l_2/l_1 \geqq 1.0$), Fig. 10.22*b*,*c* shows that the amount of this negative or positive moment going to the column strip varies from 90% for $l_2/l_1 = 0.5$ to 45% for $l_2/l_1 = 2$; or this percentage can be stated algebraically as $105 - 30l_2/l_1$ percent which goes to the column strip. This is further subdivided into 85% to the beam and 15% to the column strip slab. All moment not thus assigned to the column strip becomes middle strip moment. Below, the various percentages are expressed algebraically.

$$\begin{aligned}
\text{Col. strip } \textit{slab} \text{ neg. } M &= -0.15 \times 0.65(105 - 30l_2/l_1)M_o \\
&= -(10.25 - 2.93l_2/l_1)M_o, \text{ as a percentage.} \\
\text{Col. strip } \textit{slab} \text{ pos. } M &= 0.15 \times 0.35(105 - 30l_2/l_1)M_o \\
&= +(5.53 - 1.58l_2/l_1)M_o, \text{ as a percentage.} \\
\text{Mid. strip neg. } M &= -0.65(100 - 105 + 30l_2/l_1)M_o \\
&= -(19.5l_2/l_1 - 3.25)M_o, \text{ as a percentage.} \\
\text{Mid. strip pos. } M &= +0.35(100 - 105 + 30l_2/l_1)M_o \\
&= +(10.5l_2/l_1 - 1.65)M_o, \text{ as a percentage.}
\end{aligned}$$

The beam moments (not shown) make up the missing 0.85 of the total column strip moments.

Several cases of l_2/l_1 are tabulated in Table 10.12. The widths of strips are added since otherwise the size of moment per ft width would be difficult to visualize. This table is really divided vertically at the middle between coefficients on the left to be used with M_o for the long span and its strips and on the right with M_o for the short span and its strips. The picture is improved when one pairs two columns to make one complete set of data:

l_2/l_1 of 0.5 and 2.0 for 2:1 slab proportions;
l_2/l_1 of 0.75 and 1.33 for 4:3 slab proportions;
l_2/l_1 of 1.00 alone for a square slab.

TABLE 10.12. Slab Moments for Interior Spans with Stiff[a] Beams on All Sides and at Least Three Continuous Spans

		M_o for long span			M_o for short span	
l_2/l_1:		0.50	0.75	1.00	1.33	2.00
Middle Strip	Neg.	$-6.50\%\ M_o$	$-11.40\%\ M_o$	$-16.25\%\ M_o$	$-22.75\%\ M_o$	$-35.75\%\ M_o$
	Pos.	$+3.60\%\ M_o$	$+\ 6.21\%\ M$	$+\ 8.85\%\ M_o$	$+12.35\%\ M_o$	$+19.35\%\ M_o$
	Width	$0.5l_2$	$0.5l_2$	$0.5l_2 = 0.5l_1$	$0.83l_1$	$1.50l_1$
Column Strip	Neg.	$-8.79\%\ M_o$	$-\ 8.05\%\ M_o$	$-\ 7.33\%\ M_o$	$-\ 6.33\%\ M_o$	$-\ 4.39\%\ M_o$
	Pos.	$+4.74\%$	$+\ 4.34\%\ M_o$	$+\ 3.95\%\ M_o$	$+\ 3.43\%\ M_o$	$+\ 2.37\%\ M_o$
	Width	$0.5l_2$	$0.5l_2$	$0.5l_2 = 0.5l_1$	$0.5l_1$	$0.5l_1$

[a] Technically, with $\alpha_1 l_2/l_1 \geqslant 1.0$

This table can give some idea of how moments change with panel proportions. But these percentages are rather minimum values for interior panels only. They should obviously change, somewhat upward, as boundary restraints are relaxed. A corresponding table for an exterior panel is much more difficult since these moments involve K_{ec} and α_{ec}, values neither simple to obtain nor easy to tabulate.

Thus Table 10.12 still leaves the designer with inadequate guidance when he comes to the numerous small slabs which do not qualify for the direct design method. The author has concluded that the designer still needs the guidance of a broader tabulation of slabs with various edge conditions, possible simply the moment coefficients of the 1963 Code, which said a minimum about the beams but gave detailed moment coefficients for a wide range of slab conditions.

10.22 METHOD 2 FROM THE 1963 CODE

Method 2 of the 1963 Code is the simplest and possibly the most approximate of the three 1963 methods for designing two-way slabs supported on beams. It should still be useful in estimating what should be done with all those slabs not really large enough to justify a slab systems approach. For this purpose Table 10.13 is reproduced here from the 1963 Code (Appendix A), with 1971 notation, along with about a page of that Code to indicate its limitations and its general use:

(a) *Limitations*—These recommendations are intended to apply to slabs (solid or ribbed), isolated or continuous, supported on all four sides by walls or beams, in either case built monolithically with the slabs.

TABLE 10.13. Method 2 Moment Coefficients

Moments	Short span						Long span, all span ratios
	Span ratio, short/long						
	1.0	0.9	0.8	0.7	0.6	0.5 and less	
Case 1—Interior panels							
Negative moment at—							
Continuous edge	0.033	0.040	0.048	0.055	0.063	0.083	0.033
Discontinuous edge	—	—	—	—	—	—	—
Positive moment at midspan	0.025	0.030	0.036	0.041	0.047	0.062	0.025
Case 2—One edge discontinuous							
Negative moment at—							
Continuous edge	0.041	0.048	0.055	0.062	0.069	0.085	0.041
Discontinuous edge	0.021	0.024	0.027	0.031	0.035	0.042	0.021
Positive moment at midspan	0.031	0.036	0.041	0.047	0.052	0.064	0.031
Case 3—Two edges discontinuous							
Negative moment at—							
Continuous edge	0.049	0.057	0.064	0.071	0.078	0.090	0.049
Discontinuous edge	0.025	0.028	0.032	0.036	0.039	0.045	0.025
Positive moment at midspan	0.037	0.043	0.048	0.054	0.059	0.068	0.037
Case 4—Three edges discontinuous							
Negative moment at—							
Continuous edge	0.058	0.066	0.074	0.082	0.090	0.098	0.058
Discontinuous edge	0.029	0.033	0.037	0.041	0.045	0.049	0.029
Positive moment at midspan	0.044	0.050	0.056	0.062	0.068	0.074	0.044
Case 5—Four edges discontinuous							
Negative moment at—							
Continuous edge	—	—	—	—	—	—	—
Discontinuous edge	0.033	0.038	0.043	0.047	0.053	0.055	0.033
Positive moment at midspan	0.050	0.057	0.064	0.072	0.080	0.083	0.050

l_s = length of short span for two-way slabs. The span shall be considered as the center-to-center distance between supports or the clear span plus twice the thickness of slab, whichever value is the smaller.

w = total uniform load per sq ft

A two-way slab small be considered as consisting of strips in each direction as follows:

A middle strip one-half panel in width, symmetrical about panel center line and extending through the panel in the direction in which moments are considered.

A column strip one-half panel in width, occupying the two quarter-panel areas outside the middle strip.

Where the ratio of short to long span is less than 0.5, the middle strip in the short direction shall be considered as having a width equal to the difference between the long and short span, the remaining area representing the two column strips.

The critical sections for moment calculations are referred to as principal design sections and are located as follows:

For negative moment, along the edges of the panel at the faces of the supporting beams.

For positive moment, along the center lines of the panels.

(b) *Bending moments*—The bending moments for the middle strips shall be computed from the formula

$$M = (\text{Coef.})wl_s^2$$

The average moments per foot of width in the column strip shall be two-thirds of the corresponding moments in the middle strip. In determining the spacing of the reinforcement in the column strip, the moment may be assumed to vary from a maximum at the edge of the middle strip to a minimum at the edge of the panel.

Where the negative moment on one side of a support is less than 80 percent of that on the other side, two-thirds of the difference shall be distributed in proportion to the relative stiffnesses of the slabs. (This ends the extract from Code.)

Note first that all coefficients imply application to one foot wide slab strips and these can lead to the moments per foot of width or per strip width as the designer chooses; the coefficients represent the average moment over a strip width. Second, note that long strip moments defined in terms of the coefficients are to be multiplied by the square of the *short* span. Third, the coefficients specifically assume the supporting beams or walls are "built monolithically"; if not so built on one edge, that edge provides less restraining moment and positive moments in both directions should increase a bit.

The method envisions a pattern type of loading; the coefficients thus provide more moment resistance than the M_o of the 1971 Code would require.

Slab thickness should meet the requirements already discussed in Sec. 10.9. Once the thickness and moments are established, a form similar to Table 10.7 is useful as a book-keeping operation.

For exterior slab corners the same special corner reinforcement specified in Code 13.5.4 and discussed in Sec. 10.14g is required.

SPECIAL CONDITIONS

10.23 SHEAR REINFORCEMENT IN SLABS

The limited depth of slab makes the anchorage of shear reinforcement difficult and the anchorage requirements of Code 12.13 must be closely observed. The older ring type of wire reinforcement with inclined and inverted V-shaped wires welded around the perimeter has come into disfavor because its ability to pick up or develop the necessary stresses and again properly anchor them has been seriously questioned. As a result, the 1963 Code discontinued the crediting of shear reinforcement of bars, rods, or wires in slabs having a total depth of less than 10 in. Since 1963, better tests have shown that *well anchored* bent bars and closed ties with bars in each corner can be made effective. The 1971 Code permits such shear reinforcement provided it takes care of all shear in excess of $2\sqrt{f_c'}$, thus leaving for concrete (in conjunction with shear reinforcement) only half of the usual v_c normally permitted around columns. Careful detailing and careful placement are both essential to this usage.

Because of these problems, shearheads of structural steel have also been further developed for slabs at interior columns.[11] Note that these are now only for interior columns. As now specified, *they do not work adequately at exterior columns*, but research in this area is making progress.

Shearheads consist of four crossing steel arms, welded together at a common level, to pick up both some shear and moment load from the concrete. Each shearhead arm may consist of a small wide flange beam or of two small channels turned back to back but spaced apart as much as the channel depth or more. These arms (totally within the slab thickness) pick up shear and moment beyond the column and bring the load to bearing on the column. The bottom flanges of the steel shapes are extended beyond the top flanges (with a sloping cut through the steel web) to pick up shear load which will exist low in the slab.

The critical section for shear on the concrete is thus removed to a larger perimeter farther than $d/2$ away from the column. The steel cantilever is considered effective in moving the critical shear section out from the column to a point three-fourths of its length from the column centerline. Between arms this critical section is considered a straight line joining the several three-quarter points on the arms, but never closer than $d/2$ to the face of the column.

The shearhead is a special type of construction which permits the use of a thinner slab where shear controls slab thickness at an interior column. It may be economical in some cases. Its detailed design is considered too specialized a problem for the scope of this text.

Finally a repeated warning: shearheads are *not* for exterior columns at the present state-of-the-art.

10.24 LIFT SLABS

Lift slabs are a flat plate type of slab which is cast at grade level embedding steel shoes or collars which fit loosely around the columns.After the slabs are cured they are lifted by a patented jack system to the proper level where the shoes are welded to the steel columns. Although the Code makes no special attempt to provide for this type of design, the elastic method of flat slab analysis would appear to be proper for lift slabs. Since column stiffness is small, there seems to be no slab section which deserves to be treated as having a greatly increased moment of inertia. Loadings for maximum moment are also more significant in this case. A rigid joint between the collar and the column is an essential condition.

Since lift slabs do not qualify for empirical design procedures, total moment cannot be reduced to the M_o value. The collar stiffness may determine whether the critical section is at the center of the column or farther out.

The designer must consider carefully stresses caused by differential jack movements and make certain that in the field these do not exceed the limits considered in the office design.

10.25 CANTILEVERS FROM SLAB CONSTRUCTION

Lift slabs generally, and flat slabs and flat plates frequently, have the columns set back from the outside wall, thus causing the outermost section of the slab to act as a cantilever beyond the exterior columns. This is quite favorable to regular slab action. A proper overhang of the slab can provide a total negative moment which can almost eliminate the special problems associated with exterior slab panels.

The cantilever provides a total negative moment about the exterior columns which is statically determinate, but its distribution is not uniform. It is suggested that this distribution might be taken the same as that used to distribute negative moment between the column and middle strips in the elastic analysis method. If the cantilever projects the ideal distance, it can thus balance the typical interior slab negative moments. Longitudinal steel, as in any column strip, is needed to deliver the cantilever reaction ultimately to the columns.

It might also be noted that cantilever slabs have large deflections which become conspicuous after creep of concrete has taken place. The use of some compression steel solely to reduce deflections is often justified, as noted in Chapter 6.

10.26 FREELY SUPPORTED SLAB CORNERS

Where a slab is simply supported on masonry walls or steel beams, a problem arises at the corners if the slab is not securely fastened down. Such corners rise when the slab is loaded, rising far enough to create a conspicuous horizontal crack in the masonry walls in which they may be embedded. This is especially the case with roof slabs carrying parapet walls. Such slabs should be anchored down (to a substantial mass of masonry, for example) and have reinforcing steel provided for the restraining moment thus developed.

The exterior corner of any slab needs special treatment if the support is stiff, either a beam or wall. Code 13.5.4 gives a detailed specification covering slabs monolithic with moderately stiff concrete beams. See also Sec. 10.14*g*.

SELECTED REFERENCES

1. H. M. Westergaard and W. A. Slater, "Moments and Stresses in Slabs," *ACI Proc.*, 17, 1921, p. 415.
2. Mete A. Sozen and Chester P. Siess, "Investigation of Multi-Panel Reinforced Concrete Floor Slabs: Design Methods—Their Evolution and Comparison," *Jour. ACI*, 60, No 8, Aug. 1963, p. 999.
3. J. R. Nichols, "Statical Limitations Upon the Steel Requirement in Reinforced Concrete Flat Slab Floors," *ASCE Trans.*, 77, 1914, p. 1670.
4. D. S. Hatcher, M. A. Sozen, and C. P. Siess, "Test of a Reinforced Concrete Flat Slab," *Proc. ASCE*, 95, ST6, June 1969, p. 1051.
5. J. O. Jirsa, M. A. Sozen, and C. P. Siess, "Pattern Loadings on Reinforced Concrete Floor Slabs," *Proc. ASCE*, 95, ST6, June 1969, p. 1117.
6. W. L. Gamble, M. A. Sozen, and C. P. Siess, "Test of a Two-Way Reinforced Floor Slab," *Proc. ASCE*, 95, ST6, June 1969, p. 1073.
7. S. A. Guralnick and R. W. Fraugh, "Laboratory Study of a Forty-Five-Foot Square Flat Plate Structure," *Jour. ACI*, 69, No. 9, Sept. 1963, p. 1107.
8. N. W. Hanson and J. M. Hanson, "Shear and Moment Transfer Between Concrete Slabs and Columns," *Jour. PCA Research and Development Laboratories*, 10, No. 1, Jan. 1968, p. 2.
9. W. G. Corley and J. O. Jirsa, "Equivalent Frame Analysis for Slab Design," *Jour. ACI*, 67, No. 11, Nov. 1970, p. 875.
10. Sidney H. Simmonds and Janko Misic, "Design Factors for the Equivalent Frame Method," *Jour ACI*, 68, No. 11, Nov. 1971, p. 825.
11. W. G. Corley and N. M. Hawkins, "Shearhead Reinforcement for Slabs," *Jour. ACI*, 65, No. 10, Oct. 1968, p. 811.
12. C. P. Siess and N. W. Newmark, "Rational Analysis and Design of Two-Way Concrete Slabs," *Jour. ACI*, 20, Dec. 1948; Proc. 45, p. 273.
13. *Notes on ACI 318-71 Building Code Requirements with Design Applications*, Portland Cement Association, Skokie, Illinois, 1972.

PROBLEMS

Prob. 10.1. Using $f_c' = 3000$ psi, Grade 60 steel, and no shear reinforcement, design (direct design method) an interior panel of a flat plate floor supported on columns 18 in. square spaced 18 ft on centers each way. The slab is to carry a total dead load including slab weight) of 100 psf and live load of 50 psf. Check minimum column stiffness for a story height of 9 ft floor to floor.

Prob. 10.2. Redesign the slab of Prob. 10.1 for the same data except with 20 ft square panels. (For this problem neglect change in slab weight.)

Prob. 10.3. Under the direct design method design a flat slab without a drop panel for a 20 ft square interior panel using $f_c' = 4000$ psi and Grade 60 steel. Assume total dead load of 125 psf and live load of 150 psf. For the initial trial assume column capital $c = 4$ ft 6 in. Story height is 10 ft.

Prob. 10.4. Redesign the flat slab of Prob. 10.3 using a drop panel, without changing the assumed design dead load.

Prob. 10.5. Design a 19 ft by 22 ft interior panel of a flat slab without a drop panel, using the direct design method, $f_c' = 3000$ psi, Grade 60, a total dead load of 100 psf, and a live load of 125 psf. For the first trial use $c = 4$ ft 0 in. unless otherwise instructed.

Prob. 10.6. Design the flat slab of Prob. 10.5 with a drop panel.

Prob. 10.7. Assume the flat plate designed in Sec. 10.5 has equal columns above and below with a 10 ft story height. Use the equivalent frame method of design to establish the following for the *short* strips:

(*a*) Maximum negative design moment and its distribution to column and middle strips.

(*b*) Maximum positive moment for design and its distribution to design strips.

(*c*) Maximum column design moment, about this axis, at face of slab.

(*d*) Compare these moments with those from the direct design method used in Sec. 10.5.

Prob. 10.8.

(*a*) Design an exterior panel (not a corner panel) of the flat plate described in Prob. 10.1, using the direct design method.

(*b*) What moment should be included in the exterior column design for this loading?

Prob. 10.9.

(*a*) Pick a slab thickness based on the Code deflection requirements for the following two-way slab supported by tie-beams on top of masonry walls (assumed stiff enough to qualify for $\alpha_m = 2$). The panel is isolated, 16 ft by 18 ft center-to-center of beams. $f_c' = 4000$ psi, Grade 60 steel, $w_d = 100$ psf (including slab weight), $w_l = 40$ psf.

(*b*) Check the depth for moment and design the reinforcement on the basis of Table 10.13.

Prob. 10.10.

(*a*) Pick a slab thickness based on Code deflection requirements for a two-way slab, interior panel 19 ft by 22 ft (to centers of supporting beams and columns), assuming beam stems are 12 in. wide, columns 16 in. square, $f_c' = 3000$ psi, Grade 60 bars, dead load (including slab weight) = 100 psf, live load = 125 psf. Assume $\alpha_m = 2$.

(*b*) Check the direct design moment requirements against the strength of this thickness and design the slab reinforcement.

Prob. 10.11. A 20 ft square interior panel is supported on beams 18 in. deep overall (and assumed 10 in. width) which are carried by columns 14 in. square and 10 ft high story-to-story. $f_c' = 4000$ psi, Grade 60 steel, $w_d = 90$ psf (including slab weight), $w_l = 80$ psf. Find slab thickness for deflection; check slab and beam sizes for moment; and check beam size for shear. Then design the reinforcement for the slab.

Prob. 10.12. Assume the same data as in Prob. 10.11 to be used for an exterior panel (not a corner panel).

(*a*) Design the slab.

(*b*) Design the beam running perpendicular to the wall line for flexure and check it for shear.

(*c*) Check the spandrel beam for shear and torsion, assuming a wall load of 400 lb plf of beam, applied on the axis of the spandrel beam and column center line. What stirrups are required at *d* from the face of column?

11

Yield-Line Theory for Slabs

11.1 YIELD-LINE THEORY AS A DESIGN GUIDE

The yield-line theory is not recognized by the ACI Code and is little used in this country except in research. It is introduced here because it helps the engineer to think about failure patterns and to visualize the ultimate behavior of slabs made with simple reinforcement patterns. For these the method is straightforward, reasonably simple in its concepts, and emphasizes the lines of highest stress. It is adaptable to irregular cases. From it an engineer can trace slab behavior and visualize some of the effects of moving to less simple reinforcement patterns, cutting off bars, and the like. The strip method discussed in the next chapter is even more usable as a design method where slabs are not supported directly on columns, but the designer there is apt to be embarassed by his freedoms unless he has some background such as the yield-line method presents. Yield-line analysis is a limit design method. The designer must assume a liberal load factor until he has a code to guide him. He must keep in mind that slab deflection is quite large before failure. It is possible that when and if the method is written into codes in this country, some of the ultimate slab strength will be discounted because of large deflections, just as engineers discount that portion of structural steel strength which lies between the yield point and the ultimate. This is equivalent to saying that larger load factors may be specified for such slab design. The yield-line analysis deals with moment alone; it does not assure adequate strength in diagonal tension, but diagonal tension in two-way slabs is a problem only in special cases. On the other hand, in flat plates the diagonal tension (punching shear) around the columns is quite often the weakest element of strength.

Tests have closely verified the yield-line analysis. The calculated load normally underestimates the actual test results. Whether the excess arises from the flat arch action or membrane action, it appears that engineers may use the yield-line method with confidence; it gives a conservative estimate of strength in moment.

The real concern of the designer using this method will be to establish that his slabs will be entirely satisfactory at working loads. The Danish code has for some time permitted a form of ultimate strength design for slabs. It is the author's understanding that design for ultimate strength alone does not necessarily lead to acceptable slabs. The deflection and stiffness of such slabs may not be satisfactory. Evidently these specific matters must be investigated or limited by establishing maximum l/h ratios.

11.2 BASIC IDEAS OF YIELD-LINE THEORY

(a) Angle Changes at Yielding

The yield-line theory is a form of limit design (Chapter 9) and yields an ultimate strength. Slabs are normally underreinforced, with much less than balanced steel on a strength basis. As a result, on progressive loading the steel reaches its yield-point stress before the slab reaches its ultimate strength. As the steel yields, the center of compression on the cross section moves nearer the face of the slab until finally a secondary failure in compression takes place, at a moment only slightly greater than the yield-point moment.

For a one-way simple span slab, the increase in moment after the steel starts to yield is not large, in the order of 5 to 10%, but the further angle change ϕ occurring at the point of maximum movement is quite large in comparison with the "elastic" angle change. Relative values can be shown as an M–ϕ curve, as in Fig. 9.1*b*.

In a statically indeterminate slab these extra angle changes permit (or cause) significant modifications of the resisting moments and shears. Yielding at one point in such a slab marks the gradual beginning of larger deflections but by no means marks the end of reliable load capacity.

(b) Yield Lines as Axes of Rotation

Yielding under increasing load progresses to form lines of yielding. Until yield lines are formed in sufficient numbers to break up the slab into segments which can form a collapse mechanism, additional load can be supported. To act as hinges for a collapse mechanism, yield lines must usually be straight lines although the fan pattern discussed under (c) has one curved yield-line boundary.

It would be a more accurate description to say that yielding zones develop on the tension face over narrow bands as in Fig. 11.1*a* and the yield line is an idealization in which all the angle change is considered on a line at the center of the yielding bands, as in Fig. 11.1*b*. Yield lines are axes of rotation for the movements of the several parts of the final mechanism. Yield lines form at lines of maximum moment, but this action is not restricted to those maximum moment lines originally developed under initial "elastic" conditions.

(c) Interrelationships Between Axes of Rotation

The supports of a slab determine some of the axes of rotation of the several slab segments. In general, each support line constitutes an axis of rotation and each separate column support constitutes a pivot point, that is, a point on an axis of rotation. For example, a one-way continuous slab in a given span must fail by the development of yield lines at each support acting together with an intermediate yield line which is dependent on the loading, as indicated in Fig. 11.2*a*. This might be compared to the local collapse mechanism of Fig. 9.7 in limit design discussion.

Consider a continuous slab with a skewed support *b* as shown in Fig. 11.2b. With a yield line over two adjacent supports such as *a* and *b*, one segment rotates about *a* and one about *b*; and some common yield line *between a* and *b* joins the two plate segments rotating about these axes. Hence the common yield line must lie on an axis through O_1 at the intersection point of *a* and *b* extended. Which of the possible dashed axes through O_1 will develop depends on the reinforcement and type of loading on the span. Likewise, for collapse in span *bc*, the third yield line must pass through O_2.

In Fig. 11.2*c* a slab is shown supported on two adjacent sides and a column. The general pattern of failure will be as indicated, but the axis through the column is at an unknown angle and the yield-line intersection point in the slab depends on the loading and on whether the supported edges are simple supports or are lines of negative moment resistance.

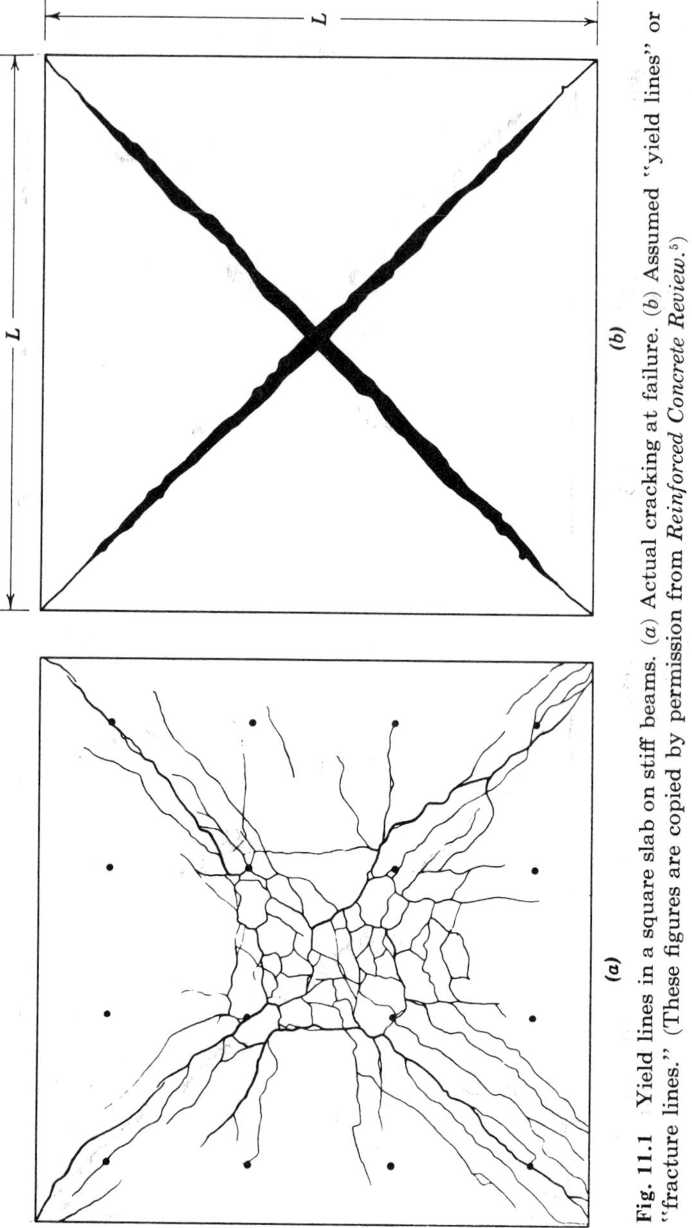

Fig. 11.1 Yield lines in a square slab on stiff beams. (*a*) Actual cracking at failure. (*b*) Assumed "yield lines" or "fracture lines." (These figures are copied by permission from *Reinforced Concrete Review*.[5])

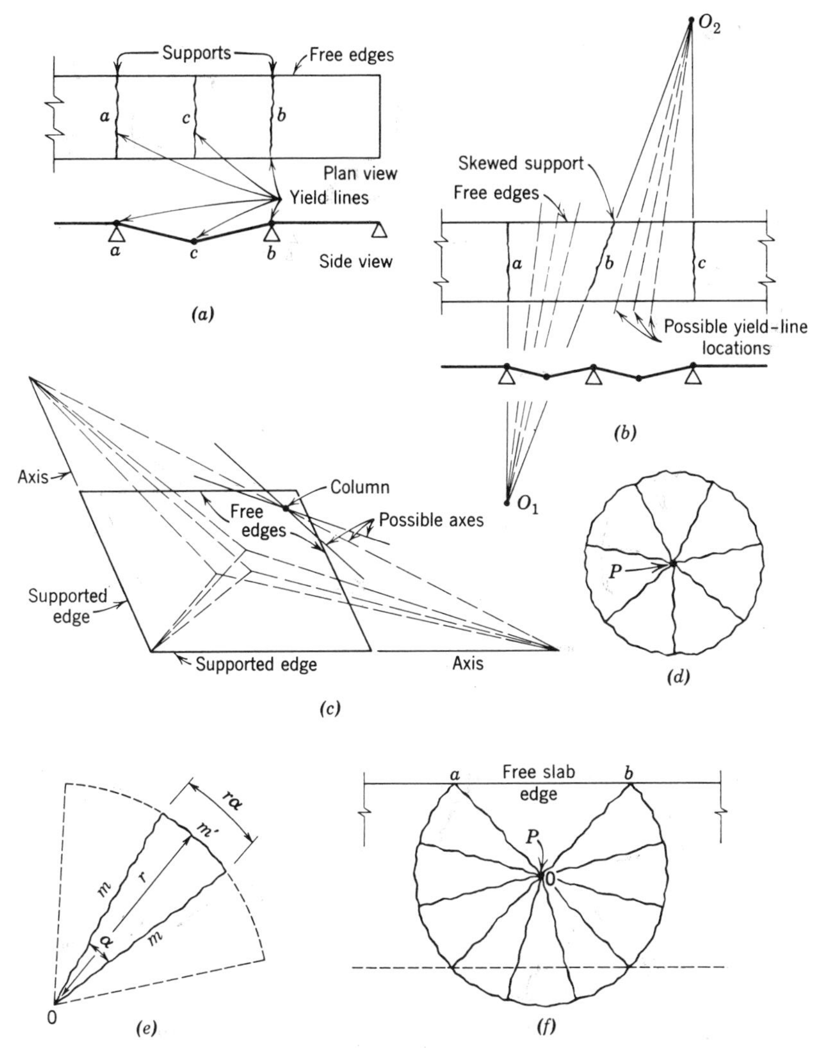

Fig. 11.2. Yield-line patterns. (*a*) Continuous one-way slab, right supports at *a* and *b*. (*b*) Continuous slab with a skewed support at *b*. (*c*) Slab supported on two adjacent edges and a column. (*d*) Fan pattern. (*e*) A fan segment. (*f*) Fan pattern near free edge.

Around a concentrated load a circular yield line of negative moment (Fig. 11.2*d*) tends to form, with radial yield lines of positive moment, like spokes of a wheel. When supports are nearby, only circular segments instead of full circles may form, as in Fig. 11.2*f*. These portions are called fans and consist of multiple segments similar to the one shown in Fig. 11.2*e*. These are discussed in Sec. 11.7.

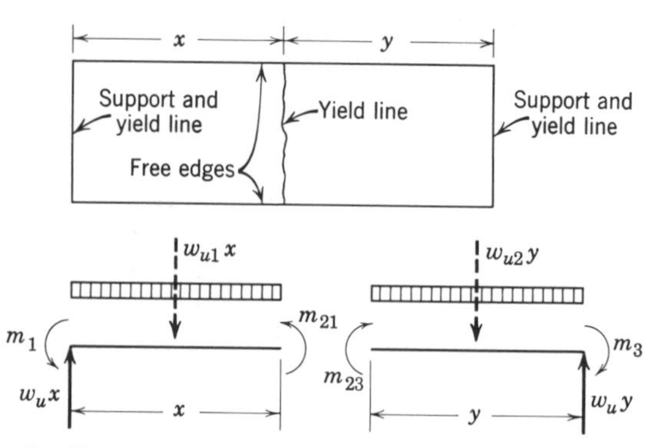

Fig. 11.3. Free body diagrams for verification of yield-line location.

(d) Segments as Free Bodies To Verify Yield-Line Locations

The free body represented by each collapsing segment must be in equilibrium under (1) its applied loads, (2) the yield moments on each yield line, (3) the reaction or shear on support lines, and at times (4) correction forces. Since the yield lines form at lines of maximum moment, neither shear nor torsion is typically present at positive moment yield lines.* Torsion can be present in special cases and then it must be represented by correction forces called nodal or knot forces, as discussed in Sec. 11.2g. The yield-line moments thus usually establish the load the segments can support, as indicated in Fig. 11.3. For a uniformly loaded slab, the location of the intermediate yield lines must subdivide the slab into segments each of which support the same uniform ultimate load w_u. If, for given values of yield moments m_1, m_2, and m_3 and a given trial location of the yield line, the calculated w_u values for the separate segments differ, the yield line must lie in a different position. The real location must be such as to reduce the segment size where the calculated w_u is smaller and to increase segment size where the calculated w_u is larger. The correct location is a unique location for a given loading and type of collapse pattern.†

In terms of design it may be convenient to take the known quantity as the load times the desired load factor. Then, with the segments as the free bodies, the required yield moments may be calculated. At a given

*Lenschow and Sozen[12] show that both torsion and a lowered moment resistance may exist at a yield line when $m_x \neq m_y$.

†However, if a different pattern is possible, this solution says nothing as to which pattern is the governing one. In complicated cases the latter demands the energy approach, at least for verification of findings.

yield line the moment thus found for one segment must match that from the adjoining segment, that is, m_{21} must be the same as m_{23}; otherwise the yield line is incorrectly located.

When only one yield line remains to be located, an algebraic equation can be solved for a governing dimension, but in more complex cases this may not be as simple as a cut-and-try procedure.

Wood[7] has raised some serious questions about the universal applicability of the correction forces discussed in Sec. 11.2g for use in equilibrium solutions. This clouds the use of the equilibrium process in novel cases, but the virtual work procedure presented in Sec. 11.2e is not subject to any limitations and may be confidently used in all cases.

(e) Virtual Work or Energy-Mechanism Criterion

If a slab has been reduced to a mechanism by yield lines acting as hinges, a known additional deflection of any specific point in the mechanism (by geometry) establishes the additional deflections at all points, along with the additional angle changes at the yield lines or hinges. For such a slab deflection, the loads also deflect and thereby contribute energy to the mechanism, whereas the hinges resist movement and absorb energy from the system. Thus, for a given mechanism, a given loading imparts enough energy to develop specific yield moments in the hinges. No smaller yield moment will be adequate to resist these loads. Some other mechanism may represent a more probable failure pattern; if so, it demands a larger yield moment to balance the given load. Thus the worst mechanism, and the real mechanism which actually forms, is the one which requires the largest resisting yield moment. Any trial mechanism or yield-line pattern establishes a lower bound or lower limit on the required yield moment.

The same procedure can be used with given yield moments to establish the ultimate load. Any given mechanism establishes an upper bound or upper limit on the collapse load. The real collapse load is the smallest that can be found from all possible mechanisms.

The energy calculations and the segment equilibrium conditions are alternative procedures. In ordinary cases either can be used to check the other. Each establishes the critical dimensions for a given pattern. Neither automatically compares this pattern with others which could be more critical. Hence, neither provides a lower bound on the load capacity (nor an upper bound on the needed moment capacity) and Wood emphasizes the need for these bounds which are available only for a few cases.

Each method has its advantages. The equilibrium method is efficient and points the way to the most critical dimensions; the energy method requires more groping. But in novel situations, especially around openings, the equilibrium method demands some correction forces [see subsection (g)] which may not be available from simple methods or may even be overlooked. Hence only the energy method is always dependable, as already mentioned a the close of subsection (d).

The energy-mechanism approach indicates that a solution is not very sensitive to small changes in the same yield-line pattern, but one must be very careful not to overlook the critical form of the pattern.

(f) Yield Moment on Axes Not Perpendicular to Reinforcing

When the yield moments in two perpendicular directions are equal and no twisting moment (torque) exists, the yield moments in all directions are equal. This condition of isotropic reinforcement simplifies the problem very considerably.

Hognestad[2] has shown (following Johansen's demonstration[1,2,4]) that, when the reinforcement in one direction differs from that in the perpendicular direction by some constant ratio, the slab dimensions can be modified to permit analysis as an isotropic slab. Such cases will not be considered here.

If m_x represents the yield moment about the x-axis and m_y about the y-axis, the yield moment about an axis at angle α with the x-axis (Fig. 11.4) will be

$$m_\alpha = m_x \cos^2 \alpha + m_y \cos^2 (90 - \alpha) = m_x \cos^2 \alpha + m_y \sin^2 \alpha$$

when $m_x = m_y = m$, this leads to $m_\alpha = m$ as stated at the beginning of this section.

Wood[7] refers to this relationship as the "square yield criterion" and states that it can be true only when applied to a point where there is a state of "all-round uniform moment," that is, uniform is all directions.

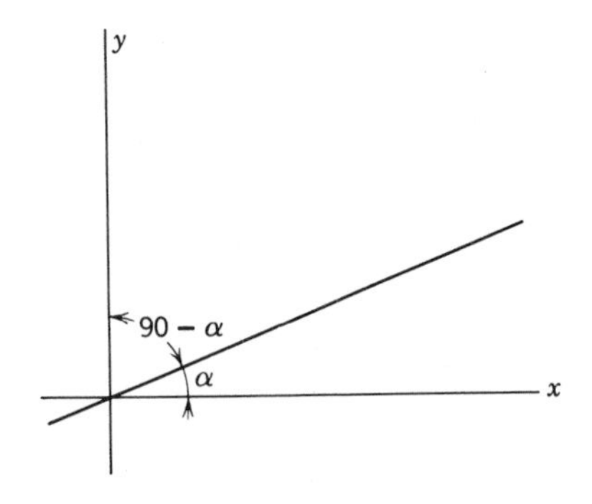

Fig. 11.4. Moment axes.

He reports that experiments indicate the slab load capacity may be increased as much as 16% by the increased m capacity on a diagonal. He concludes that the square yield criterion normally used is only an approximation for the more complicated actual criterion; but this simple criterion is the one generally used.

On the other hand, Lenschow and Sozen[12] rather conclusively prove by test and theory that there is no appreciable change in the resisting m in the case just discussed. The increased capacity observed in the English tests must have an explanation different from the one Wood suggests.

(g) Correction Forces Where Yield Line Intersects a Free Edge

Correction forces, usually called edge forces, nodal forces, or knot forces, are a necessity for general solutions by equilibrium methods. They appear most certainly when a yield line intersects a free edge at other than a 90° angle. These forces are substitutes for either torsion or twisting moments along assumed yield lines and for normal shears existing at negative moment yield lines. The correction forces are applied in the corner of segments, normal to the plane of the slab, in the magnitude $m_t = m \cot \alpha$, as shown in Fig. 11.6. Since these correction forces act between segments or between a segment and its reaction, they are of the nature of internal forces except when a segment is considered separately as a free body. Hence they completely disappear from any virtual work or energy equations applied to the entire slab. These forces can be totally ignored except in equilibrium calculations (and in reaction calculations).

These forces have been "established" in different manners by different authors. The writer agrees with Wood's evaluation that each of these proofs is lacking in rigor and open to some question. Yet, as Wood shows by various examples, the method works successfully in many instances and it avoids the necessity of differentiation for a minimum value. Hence it has merit even though it is an uncertain tool in new situations. It would seem wise to say that no equilibrium solution can now be considered entirely satisfactory by itself without a check by the energy method.

One approach to the value of the correction force is to say that the use of a straight line as the yield line can only be an approximation at an edge. The reasoning is that a single yield line, to satisfy statics, must intersect a free edge or a simply supported edge at 90° in order to avoid the necessity of torsional or twisting moments on the yield line. Such twisting moments are not normal on a true maximum moment line. Hence, near an edge, yield lines must turn as shown by the dotted lines of Fig. 11.5. If one insists on using the false "straight-line" yield line, he must consider the torsional

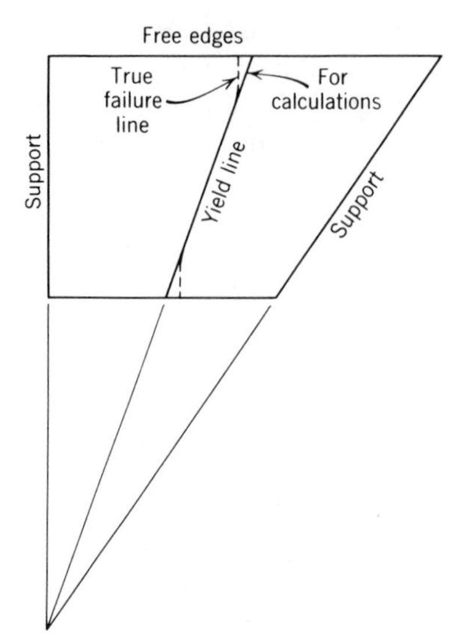

Fig. 11.5. The curved yield line at a free boundary.

moment $m_t = m \cot\alpha$ in his equilibrium equations, and this is most simply done by using correction forces at the tip of the segments. The following "proof" and Fig. 11.6 are taken directly from Hognestad's ACI paper[3]:

> "The magnitude of m_t may be established by considering the equilibrium of the infinitesimal triangle AOB shown in Fig. 12.6 in which AO is a finite length and AB is infinitesimal. Neglecting differentials of higher order, the moment in the section OB must equal the moment m in the yield line OA as m is a maximum value. Since the bending moment is zero along $AB = ds$, the total moment acting on the triangle AOB is found by vector addition
>
> $$m(\overline{AO} + \overline{OB}) = m\overline{AB} = m\,\overline{ds}$$
>
> Equilibrium of moments about OB then gives
>
> $$m\,ds \cos\alpha = m_t\,ds \sin\alpha, \qquad \text{or } m_t = m \cot\alpha$$
>
> differentials of higher order again being neglected. It should be noted that m_t acts down in the acute corner. These boundary conditions were first introduced into the yield-line theory by Johansen in 1931."

Other more detailed "proofs" start from a consideration of intersecting yield lines anywhere. These deduce the same result plus two more useful, conditions:

1. If all intersecting yield lines have positive moments (or all negative with no positive), the number of intersecting lines is unlimited, as the

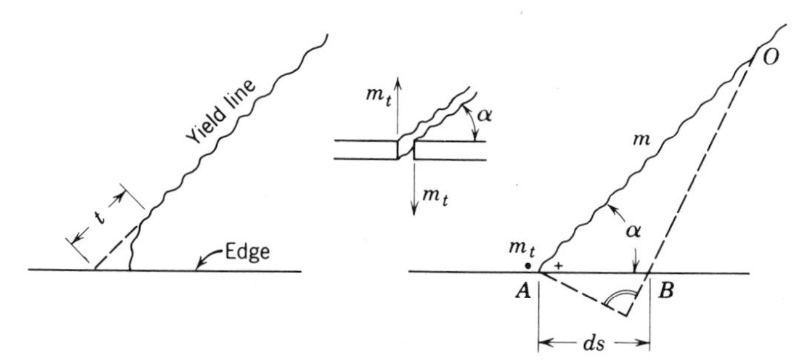

Fig. 11.6. The shear load m_t at boundary when straight yield lines are used. (Reprinted from *Jour. ACI.*[3])

"fan" of Fig. 11.2*d*, and no knot force (correction force) may be needed at the intersection where only three such yield lines intersect; but knot forces will sometimes be needed when *more than three intersect.*

2. If all intersecting lines are not of one kind, only three different directions are possible. Of the six half lines resulting, any one half line may be omitted in a particular case, *or* two half lines may be omitted if they are either halves that form a single line or are not adjacent half lines.

Wood notes that the value of $m_t = m \cot\alpha$ exceeds m when the angle α is less than 45° and that such a value would be impossible. However he points out that, in accepting an admittedly approximate or simplified yield line, one may not be limited to exactness as to such an angle. Some solutions giving essentially true results involve smaller angles.

When an entire slab is considered, this downward shear on one segment and the upward shear on the adjacent segment are internal and mutually offsetting forces, and hence do not show in the over-all energy equation.

(h) Corner Pivots and Corner Yield Lines

Supported slab corners, as in Fig. 11.11 and 11.12, introduce the problem of what might be called localized yield-line patterns. Hognestad[3] reports: "According to Johansen it is most expedient in practical design to disregard the corner levers and then later apply corrections, for which he has developed general equations and tabulated the most common cases." In this brief treatment of the subject, corner patterns will be evaluated as any other regular pattern. Such discussion will be deferred until an example not needing this corner analysis has been considered.

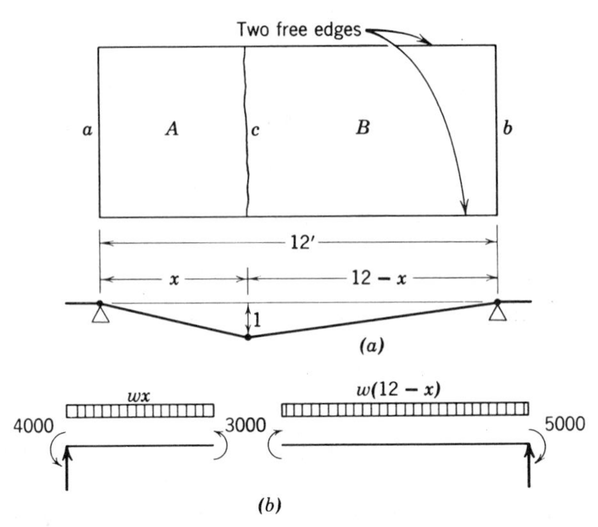

Fig. 11.7. One-way slab example. (a) Failure mechanism. (b) Free body diagrams.

11.3 SLAB EXAMPLE, ONE-WAY STEEL

In Fig. 11.7*a* consider that the yield moment at *a* is $m_a = -4000$ ft-lb/ft width, at *b* is $m_b = -5000$ ft-lb/ft, and for positive moment is $m_c = -3000$ ft-lb/ft. By the yield-line method, calculate the ultimate uniform load the slab will support on a 12-ft span.

Solution

The slab capacity is independent of adjacent panel conditions except as these are reflected in the values of m_a and m_b. The one-way slab can be very simply solved by limit design procedures, with identical results, but the object here is to introduce yield-line procedures.

Because only one yield-line pattern is possible, an algebraic solution is simple, based on the free body diagrams of Fig. 11.7*b*.

$$\Sigma M_A = -4000 - 3000 + wx^2/2 = 0, \qquad w_A = 7000 \times 2/x^2 = 14{,}000/x^2$$
$$\Sigma M_B = +3000 + 5000 - w(12 - x)^2/2 = 0, \quad w_B = 8000 \times 2/(12 - x)^2$$

Since w_A is to be equal to w_B,

$$14{,}000/x^2 = 16{,}000/(12 - x)^2$$

The quadratic could be solved algebraically, but a solution more typical of the general possibilities of this method would be by trial.

If $x = 6$ ft, $388 < 444$
$x = 5.5$ $463 > 378$
$x = 5.8$ $415 \doteq 417$, say 416
Ultimate $w = 416$ psf

The virtual work or energy-mechanism solution is also entirely feasible, using Fig. 11.7*a*. For a 1-ft strip the energy input for a unit deflection at the center yield line is

$$0.5wx + 0.5w(12 - x)$$

The work done on the slab at the yield lines or hinges is $m\theta$, where θ is the angle of rotation. At the center yield line the total angle change depends upon the combined rotation of A and B. However, the most convenient treatment is to calculate the work there as the separate amounts due to A and B respectively. The work at the yield lines is then:

Due to rotation of A: $(4000 + 3000)(1/x)$
Due to rotation of B: $(3000 + 5000)[1/(12 - x)]$

Equating energy input to energy consumption,

$$0.5wx + 0.5w(12 - x) = 7000/x + 8000/(12 - x)$$
$$6w = 7000/x + 8000/(12 - x)$$

For the needed minimum w, $dw/dx = 0$

$$6(dw/dx) = -7000/x^2 - 8000(-1)/(12 - x)^2 = 0$$

This is the same algebraic equation solved in the other approach, which gave $x = 5.8$ ft. If this value is substituted in the energy equation,

$$6w = 7000/5.8 + 8000/6.2 = 2498$$
$$w = 416 \text{ psf}$$

For the more usual slab with a more complex yield-line pattern, a group of partial derivatives would replace dw/dx. This type of solution might not be practical. However, it is also practical to try different patterns, in this case different x values, in the energy equation: $6w = 7000/x + 8000/(12 - x)$.

Try $x = 6$, $w = 1167/6 + 1333/6 = 417$ psf
$x = 5$, $w = 1167/5 + 1333/7 = 424$ psf
$x = 5.5$, $w = 1167/5.5 + 1333/6.5 = 417$ psf
$x = 5.7$, $w = 1167/5.7 + 1333/6.3 = 417$ psf
$x = 5.85$, $w = 1167/5.85 + 1333/6.15 = 416$ psf
$x = 5.9$, $w = 1167/5.9 + 1333/6.1 = 417$ psf > 416

This solution indicates, as is usually the case, that the calculated w by the virtual work or energy approach is not too sensitive to the exact yield-line location. A cut-and-try approach to the problem is thus feasible.

The calculation of w for the several segments by the equilibrium method is more sensitive, but it has the advantage of indicating more definitely the needed shift in assumed yield-line location. (The equilibrium equation for a segment is actually an identity with an energy equation written for that *single* segment.) The shifts with the energy method are more in the nature of groping one's way towards the correct solution rather than observing the needed change.

11.4 SLAB ON NONPARALLEL SUPPORTS, TWO-WAY STEEL

The slab of Fig. 11.2*b* illustrates two further points of procedure and brings up a possible practical complication. In addition, it serves to emphasize that moments are vector quantities and as such may need to be resolved into components. The ultimate uniform load will be calculated on the basis of the dimensions of Fig. 11.8*a*, $m_a = -4000$ ft-lb/ft, $m_b = -5000$ ft-lb/ft, and $m_c = 3000$ ft-lb/ft both longitudinally and transversely.

Solution

Since yield lines for negative moment will occur over each support, the corresponding axes of rotation intersect at O. The intermediate positive moment yield line (extended) must also pass through O and can be defined in terms of an unknown angle at O or by the dimension x, the length of one side of the A segment.

This solution requires the yield moment m_c along the inclined yield line between A and B. With equal m_α values longitudinally and transversely, m_c is the same for all orientations. This follows from the equation of Sec. 11.2*f*:

$$m_\alpha = m_x \cos^2 \alpha + m_y \sin^2 \alpha = m_c(\cos^2 \alpha + \sin^2 \alpha) = m_c$$

If the transverse reinforcement were lighter, as would often be the case, the yield moment would have a specific m_α value for each assumed slope of yield

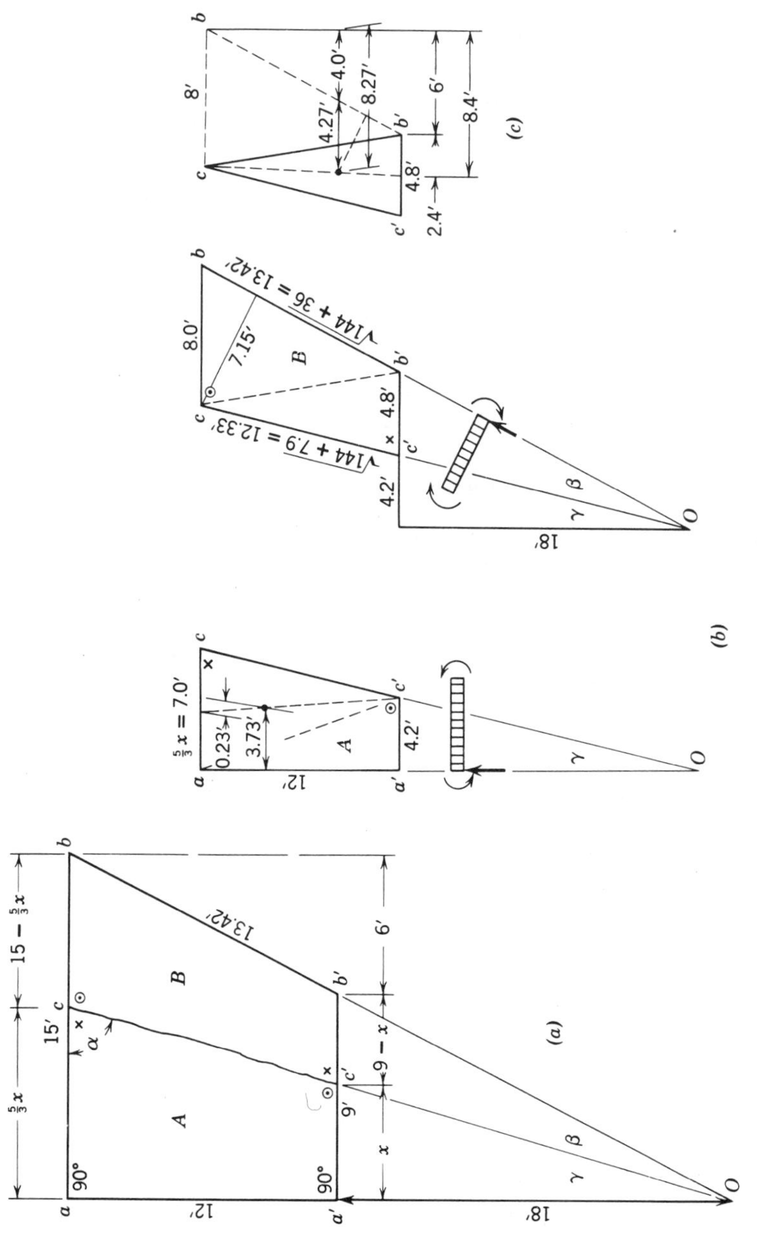

Fig. 11.8. Two-way slab on nonparallel supports. (*a*) Slab layout. (*b*) Detailed dimensions of segments *A* and *B*. (*c*) Subdivision of Segment *B*.

line. The variable m_x would constitute an extra complication* in the solution, which would make general equations rather difficult. However, successive trials for different values of x would be feasible. Solution by trial may constitute the simpler approach even for the given isotropic case because of the somewhat involved geometry for moment arms and angles.

The second new procedure arises from the fact that the middle yield line crosses two free edges at other than a 90° angle. This requires a correction (Sec. 11.2g) in the form of a downward shear m_t at the acute angles and an upward shear m_t at the obtuse angles.

$$m_t = m \cot \alpha = m(\mathrm{x}/18) = mx/18$$

These are marked on the plan view in Fig. 11.8*a* by an x for a downward force and by a dot within a small circle for an upward force. These shear forces influence an analysis by segments but not an energy equation. The first trial will be made with segments.

Assume $x = 4.2$ ft, giving the dimensions in Fig. 11.8*b*. The moment about aa' of the yield moments acting along cc' is

$$m_c(\text{length } cc')\cos \gamma = m \times 12$$

$$m_t = m_c x/18 = 3000 \times 4.2/18 = 700$$

$$\begin{aligned} \Sigma M_{aa}' = & -4000 \times 12 - 3000 \times 12 - 700 \times 4.2 + 700 \times 7 \\ & + 0.5 \times 12w \times 4.2 \times 1.4 + 0.5 \times 12w \times 7.0 \times 3.73 = 0 \\ & - 48{,}000 - 36{,}000 - 2940 + 4900 + 35.3w + 156.6w = 0 \end{aligned}$$

$$w_A = 82{,}040/191.9 = 427 \text{ psf}$$

The yield moment along $cc' = m(\text{length } cc')$, which gives a moment about axis $bb' = m(\text{length } cc')\cos \beta$. The angle β in triangle $c'Ob'$ is

$$\beta = \tan^{-1} 9/18 - \tan^{-1} 4.2/18 = 26°34' - 13°08' = 13°26'$$

$$\cos \beta = 0.973 \qquad \cos (\gamma + \beta) = 0.895$$

An alternate procedure would be to obtain the moments for cc' as components about the x- and y-axes and then to take the components of these about bb'.

*Hognestad presents Johansen's proof showing that the slab dimensions can be modified to give a solution based on the simpler isotropic case. Assume the reinforcement in the traverse direction leads to a value of μm across longitudinal sections compared to m for the longitudinal strips. The simpler isotropic case ($m_x = m_y = m$) can be used if the length in the transverse direction and the size of any concentrated loads are first divided by $\sqrt{\mu}$, any uniform load w remaining unchanged. This provides a relatively simple solution for the problem but it appears that in a general case the ratio μ would have to be the same for both positive and negative moments in a given direction.

The slab load on segment B will be subdivided into triangular areas of load as in Fig. 11.8b,c. Triangle $bb'c$ has an area $8 \times \frac{12}{2} = 48$ ft^2 which indicates an altitude perpendicular to bb' of $48 \times 2/13.42 = 7.15$ ft. For triangle $cc'b'$, the centroid will be on the median at 8.27 ft to the left of b. Thus the horizontal distance between the centroid and bb' is 4.27 ft. The arm about $bb' = 4.27 \cos (\gamma + \beta) = 4.27 \times 12/13.42 = 4.27 \times 0.895 = 3.82$ ft.

$$
\begin{aligned}
M_{bb'} &= 3000 \times 12.33 \times 0.973 + 5000 \times 13.42 - 700 \times 4.8 \times 0.895 \\
&\quad + 700 \times 8.0 \times 0.895 - 48w \times 7.15/3 - 4.8 \times 12w \times 0.5 \times 3.82 = 0 \\
&\quad + 36{,}000 + 67{,}100 - 3000 + 5000 - 114.3w - 110w = 0 \\
w_B &= 105{,}100/224.3 = 470 \text{ psf} > w_A = 427 \text{ psf}
\end{aligned}
$$

The indication is that x was taken a little too large, but the difference between w_A and w_B is quite small and an answer of ultimate $w = 445$ psf would be quite close. Normally one would not expect results of a trial to be this near to a correct answer; this choice of x benefited from the somewhat similar analysis of the slab in Sec. 11.3. On the other hand, the individual trial calculation should not be as long as here shown. The rather involved geometry was worked out in detail; graphical determination of some of the dimensions might be preferable, especially for initial trials.

The same problem will be solved from the energy-mechanism or virtual work approach, still using Fig. 11.8. Here also the geometry and angles must be carefully determined. Use $x = 4.2$ ft as before and assume a unit vertical movement at c'. Point c then deflects $30/18 = 1.667$ units (in proportion to the distance from O). Part B rotates through an angle $1/(4.8 \times 0.895) = 0.233$, and the yield hinge at bb' absorbs energy of

$$E_1 = 5000 \times 13.42 \times 0.233 = 15{,}650$$

For the yield hinge at cc', it is convenient to work with the x- and y-components of these moments and the respective components of the rotation angle. For rotation of B:

$$E_2 = 3000 \times 12 \times 1/4.8 + 3000 \times 2.8 \times \tfrac{1}{18} = 7970$$

For the rotation of A, the energy at cc' is:

$$E_3 = 3000 \times 12 \times 1/4.2 = 8570$$

At aa' $$E_4 = 4000 \times 12 \times 1/4.2 = 11{,}420$$

The total energy absorbed is $15{,}650 + 7970 + 8570 + 11{,}420 = 43{,}610$.

The same load triangles will be used as before. Triangles $bb'c$ at its centroid deflects $\frac{1}{3}(1.667) = 0.555$.

$$E_5 = (12w \times 8/2)0.555 = 26.7w$$

Triangle $cc'b$, $E_6 = 12w \times 4.8/2)(4.27/4.8)1.0 = 25.6w$

Triangle $aa'c$, $E_7 = (12w \times 4.2/2)\frac{1}{3} = 8.4w$

Triangle acc', $E_8 = (12 \times 7.0/2)(3.73/4.20)1.0 = 37.3w$

The total energy available from loads is $26.7w + 25.6w + 8.4w + 37.3w = 98.0w$.

$$98.0w = 43{,}610$$
$$w = 445 \text{ psf}$$

An ultimate load of 445 psf is a good answer, the reliability of this answer being judged more on the basis of the equilibrium calculation for parts A and B than on the energy equation. It must be kept in mind that the energy equation always gives loads at least equal to and generally greater than the true ultimate load; that is, errors are always on the unsafe side.

To indicate the effect of a small error in locating the center yield line, the energy calculation will be repeated for $x = 4.5$ ft instead of the 4.2 ft used above. (The equilibrium calculation indicated the true value was *less* than 4.2 ft) The dimensions are shown in Fig. 11.9.

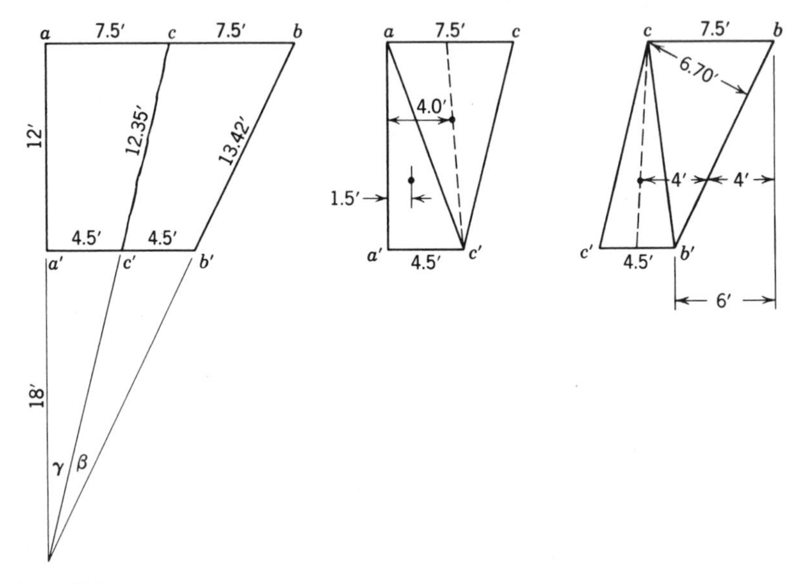

Fig. 11.9. Dimensions for another trial solution of slab of Fig. 11.8.

$\cos(\gamma + \beta) = 12/13.42 = 0.895$

Energy absorbed:

$$
\begin{aligned}
5000 \times 13.42 \times 1/(4.5 \times 0.895) &= 16{,}700 \\
3000 \times 12 \times 1/4.5 + 3000 \times 3.0 \times \tfrac{1}{18} &= 8{,}500 \\
3000 \times 12 \times 1/4.5 + 3000 \times 3.0 \times \tfrac{1}{18} &= 8{,}500 \\
4000 \times 12 \times 1/4.5 &= 10{,}660 \\
&\overline{44{,}360}
\end{aligned}
$$

Energy from loads:

$$
\begin{aligned}
bb'c\text{:}\quad 0.5 \times 12w \times 7.5 \times \tfrac{1}{3} \times 1.667 &= 25.0w \\
cc'b\text{:}\quad 0.5 \times 12w \times 4.5 \times (4/4.5) \times 1.00 &= 24.0w \\
aa'c\text{:}\quad 0.5 \times 12w \times 4.5 \times \tfrac{1}{3} &= 9.0w \\
acc'\text{:}\quad 0.5 \times 12w \times 7.5 \times (4/4.5) \times 1.00 &= 40.0w \\
&\overline{98.0w}
\end{aligned}
$$

$w = 44{,}360/98.0 = 453$ psf

This differs very lttle from the better solution above.

11.5 SQUARE PANEL, IGNORING CORNER EFFECT

Find the ultimate uniform load that a continuous two-way slab 16 ft square can carry if the yield moment is 3000 ft-lb/ft for positive moment and 4000 ft-lb/ft for negative moment, equal in both directions.

Solution

The yield pattern is extablished by symmetry in this case, with triangular segments rotating about each edge, as shown in Fig. 11.10.

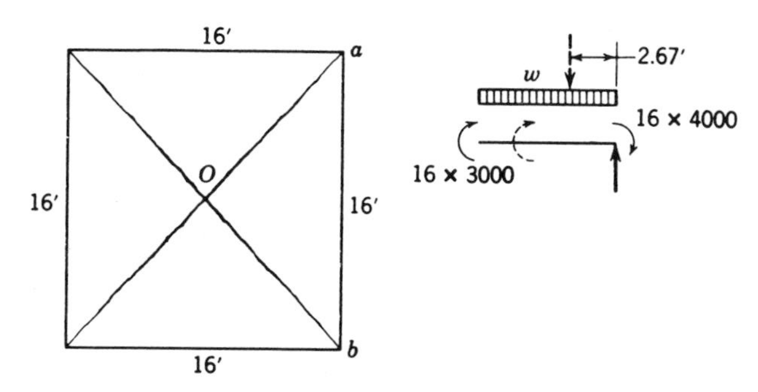

Fig. 11.10. Two-way slab supported on all four sides.

When triangle *abo* is considered, each diagonal carries a positive moment of 3000 ft-lb/ft, since the slab is isotropic with equal resistance at any angle. The component about *ab* of the moments on the diagonals is equal to the moment *m* times the projected length *ab*. Hence the equilibrium equation for this segment becomes:

$$16(4000 + 3000) - 0.5 \times 16 \times 8w \times \tfrac{8}{3} = 0$$
$$w = 112{,}000/171 = 655 \text{ psf}$$

Except for possible corner effects (Sec. 11.6), this is an exact solution and there is no need for trial solutions. The energy equation would serve just as well. For a unit deflection at *o*:

$$16(4000 + 3000)\tfrac{1}{8} = 0.5 \times 16 \times 8w \times \tfrac{1}{3}$$
$$14{,}000 = 21.3w$$
$$w = 655 \text{ psf}$$

11.6 CORNER EFFECTS

A simply supported square slab at a corner may not follow the simple yield-line pattern used in Sec. 11.5. If the corners are not fastened down, they will rise off the supports as the slab is loaded and the diagonal yield line will split or divide into two branches to form a Y, as in Fig. 11.11*a*. This forms an additional corner segment with yield moments on only two faces and with support only on the two points where the yield lines cross the boundary. This condition is called a corner pivot, since the segment pivots about these two points.

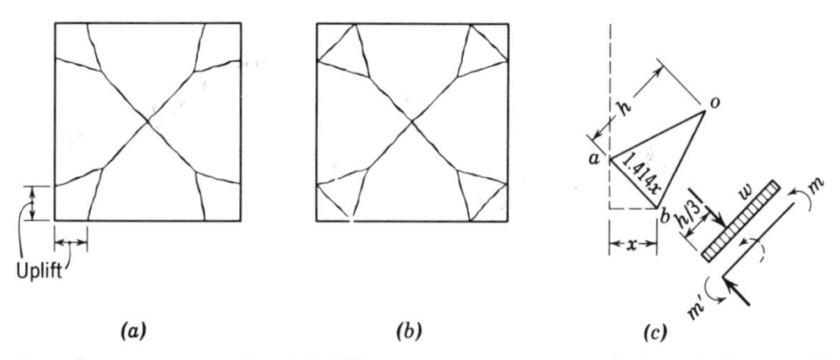

Fig. 11.11. Corner segments. (*a*) When corners are not fastened down. (*b*) When corners are held down but not reinforced for negative moment. (*c*) Equilibrium of corner segment.

If the corners are held down, but are not specially reinforced, similar yield lines form but in addition a corner crack opens along the pivot line as in Fig. 11.11*b*. If the corner is specially reinforced for negative moment, this adds a yield moment m' across this boundary of the triangular segment which increases its capacity and moves the junction of the Y farther from the corner. With a large enough m', the triangular segment fails to form and the simple diagonal yield line into the corner is correct without modification.

In continuous slabs, the corners have top steel which provides an m' for the corner segment. Whether the Y-pattern or the straight diagonal yield line forms depends upon the amount of this reinforcement.

Consider the free body formed by one of these corner segments, as in Fig. 11.11*c*. The positive moment m along sides ao and ob adds vectorially to give a positive moment m on the width ab, when m is the same in each direction. If no knot force acts at the apex of segment,

$$\Sigma M_{ab} = -1.414xm' - 1.414xm + 0.5(1.414xhwh)/3 = 0$$

$$wh^2/6 = m + m'$$

$$h = \sqrt{6(m + m')/w}$$

The length h is not dependent on x or the width of the segment and increases with the ratio of the sum of positive and negative yield moments to w. If, in a given slab, the moment-load ratio establishes a distance h much less than the length of the diagonal along which it lies, this corner pattern will control; that is, it does reduce the slab capacity. If the calculated h is large enough to push the Y-intersection beyond the limits of its particular diagonal, it means that the corner segment does not form. Intermediate values of h indicate that the corner element controls but has a smaller effect on the ultimate load or required moment strength of the slab.

With a simple supported square slab of side dimension a the ultimate moment is $wa^2/24$ if no corner segment forms. This is the true condition with the corner held down and with m' made equal to m. With $m' = 0$, or with the corners free, the corner pivots produce corner segments which increase the ultimate moment to $wa^2/22$. In the square panel the maximum effect of the corner segments is thus slightly less than 9%, increasing the required resisting moment or decreasing the allowable w.

11.7 CIRCULAR SEGMENTS OR FANS

Under a concentrated load the negative moment yield line tends to form somewhat in a circle around the load with positive moment yield lines as spokes or radial lines, as shown in Fig. 11.12*d*. If a segment of this circular area is isolated as in Fig. 11.12*e*, the total load P at O results in a load $(\alpha/2\pi)P$ to be carried by the particular segment from O to its circular boundary. The segment will be small enough to be considered as a triangle, which is similar to the corner segment of Fig. 11.11*c*, except for the loading. Then ΣM about the (circular) chord becomes:

$$(m + m')r\alpha - (\alpha/2\pi)Pr = 0$$
$$P = 2\pi(m + m')$$

It is interesting to note that the load P is thus not a function of r and hence the slab is equally subject to failure at any surrounding circle. This is not quite true when the load is applied over a finite area and the weight of the slab is considered, but it points out the need for both positive and negative moment resistance over the full area of a slab subjected to heavy concentrated loads.

Generally, the segments tend to have a large radius, and consideration of the equilibrium equation indicates that the radius might vary from segment to segment with only minor changes in the strength, resulting in a form like that of Fig. 11.15*g*. When a slab has a free edge nearby, as in Fig. 11.2*f*, it is weaker on the triangle abO because this has no m' and the circle will develop such as to make this triangle large. If a second free edge were on the other side of the load, as shown dotted, a similar triangle would also develop on that side instead of the segments shown. On the other hand, if the edges were restrained, providing an m', the failure might be similar to Fig. 11.15*h*, although the corner segments might be more like Fig. 11.12*e*.

Wood[7] points out an interesting case, that of a simple cantilever slab with a concentrated load in the corner. The failure section is more probable as a segmental fan plus edge triangles as at A in Fig. 11.15*i*. These indicate the need for m' resistance in both directions and without reduction near the corner, since any radius gives the same resistance. Even without the fan concept, as at B, a similar conclusion follows for negative moment (alone) since the moment zP must be carried by a strip $2z$ wide. The fan, however, indicates the need also for positive moment steel.

11.8 SQUARE PANEL, CONSIDERING CORNER EFFECTS

The effect of the slab corners on the ultimate moment on the square slab of Sec. 11.5 will now be investigated by basic principles.*

*Reference 3 develops three simultaneous equations for the square slab case which establish the corner segment dimensions algebraically.

In Sec. 11.5 w was found to be 655 psf, and this will not be seriously changed, say, not below 600 psf based on the 9% correction of Sec. 11.6. With this assumed w, summation of moments for the corner segment, as above, gives:

$$m' + m = wh^2/6$$
$$4000 + 3000 = 600h^2/6$$
$$h = \sqrt{7000/100} = 8.35 \text{ ft}$$

The full diagonal length is $1.414 \times \frac{16}{2} = 11.3$ ft. Hence it is probable that the corner segment does form. Since the edge does not have $M = 0$, M_t edge shears are not necessary in this example.

Try the failure pattern of Fig. 11.12*a*, using the energy method with a center deflection of one unit. Since the pattern repeats, only one set of areas A and B will be included.

The deflection of point c is $6.3/8 = 0.787$. Energy input from the load is:

$$E_A = 0.5 \times 16w \times 8 \times \tfrac{1}{3} - 2 \times 0.5 \times 0.707w \times 6.30 \times 0.787/3 = 20.1w$$
$$E_B = 0.5 \times 1w \times 8.4 \times 0.787/3 = 1.10w$$

Energy absorbed in hinges:

$$E_A = (4000 + 3000)(16 - 2 \times 0.707)\tfrac{1}{8} = 12{,}760$$
$$E_B = (4000 + 3000) \times 1 \times 0.787/8.4 = 657$$
$$20.1w + 1.10w = 12{,}760 + 657$$
$$w = 13{,}420/21.2 = 633 \text{ psf} < 655 \text{ psf of Sec. 11.5}$$

This comparison proves that the corner segment does form.

A larger corner segment may drop the ultimate w lower. Try the increased segment of Fig. 11.12*b*. The deflection of c is $6.65/8 = 0.835$. As for the above trial,

$$18.7w + 2.34w = 11{,}550 + 1390$$
$$w = 12{,}940/21.04 = 615 \text{ psf} < 643$$

Since this governs over the preceding calculation, further trials were made for the conditions of Fig. 11.12*c* and *d* which gave 607 psf and 605 psf respectively.

The corner triangles might be replaced by a small fan as shown in Fig. 11.12*e*. Then for the quarter panel the energy imput is:

$$E_A = 0.5 \times 16w \times 8 \times \tfrac{1}{3} - 2 \times 0.5 \times 2.83w \times 7.15 \times 0.894/3 = 15.30w$$
$$E_B = (w \times 0.462 \times 8.4^2/6)0.894 = 4.85w$$

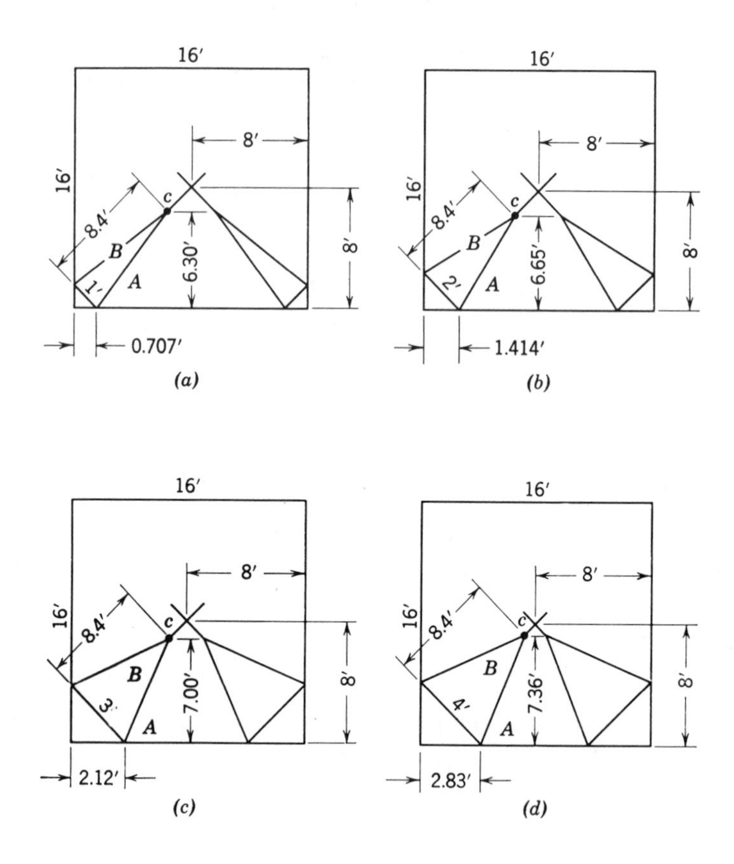

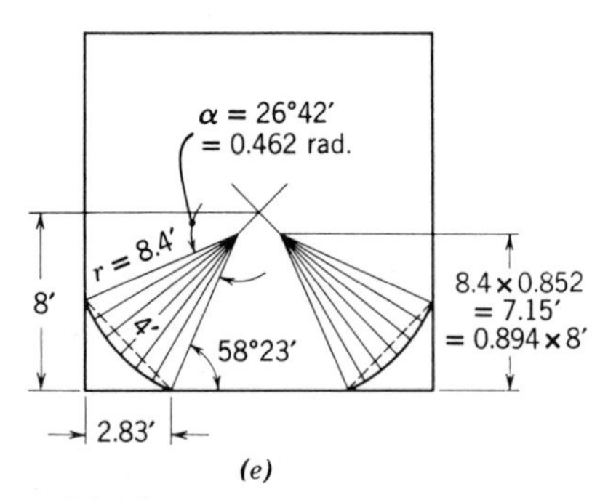

Fig. 11.12. Corner segment trials.

Energy absorbed in hinges:

$$E_A = (4000 + 3000)(16 - 2 \times 2.83)\tfrac{1}{8} = 9000$$
$$E_B = (m + m')\,0.894\,\alpha = (4000 + 3000)0.894 \times 0.462 = 2880$$
$$15.30w + 4.85w = 9000 + 2880$$
$$w = 11{,}880/20.1 = 590 \text{ psf}$$

This is another 2% lower and means w is probably slightly under 590 psi. The original h calculation is still about right. This has not exhausted all the possibilities, but when one considers the basic approximation shown in Fig. 11.1, great mathematical refinement seems out of place.

11.9 RECTANGULAR SLABS

Investigate the ultimate load capacity of 12-ft by 20-ft slab continuous on all edges, which has a yield moment of 3000 ft-lb/ft for positive moment and 4000 ft-lb/ft for negative moment, both uniform in each direction.

Solution

Symmetry dictates a yield line at the middle, parallel to the long side, which merges with corner diagonals symmetrical about unknown points O and O' as shown in Fig. 11.13*a*. The corner effect with the Y-form on the corner diagonals is also possible, but this will be treated as a later modification of the simple pattern.

The first trial dimensions shown in Fig. 11.13*a* will be considered first in terms of the equilibrium of segments A and B.

Segment A:

$$-12(4000 + 3000) + 0.5 \times 12w \times 8 \times 2.67 = 0$$
$$w_A = 84{,}000/128 = 655 \text{ psf}$$

Segment B:

$$-20(4000 + 3000) + 4w \times 6 \times 3 + 2 \times 0.5 \times 8w \times 6 \times 2 = 0$$
$$w_B = 140{,}000/168 = 834 \text{ psf} \gg w_A$$

Try a smaller A segment as noted in Fig. 11.13*a* for the second trial. These dimensions lead to $w_A = 858$ psf and $w_B = 730$ psf. A third trial with point O at 7.5 ft from the end gives $w_A = 745$ psf and $w_B = 775$ psf.

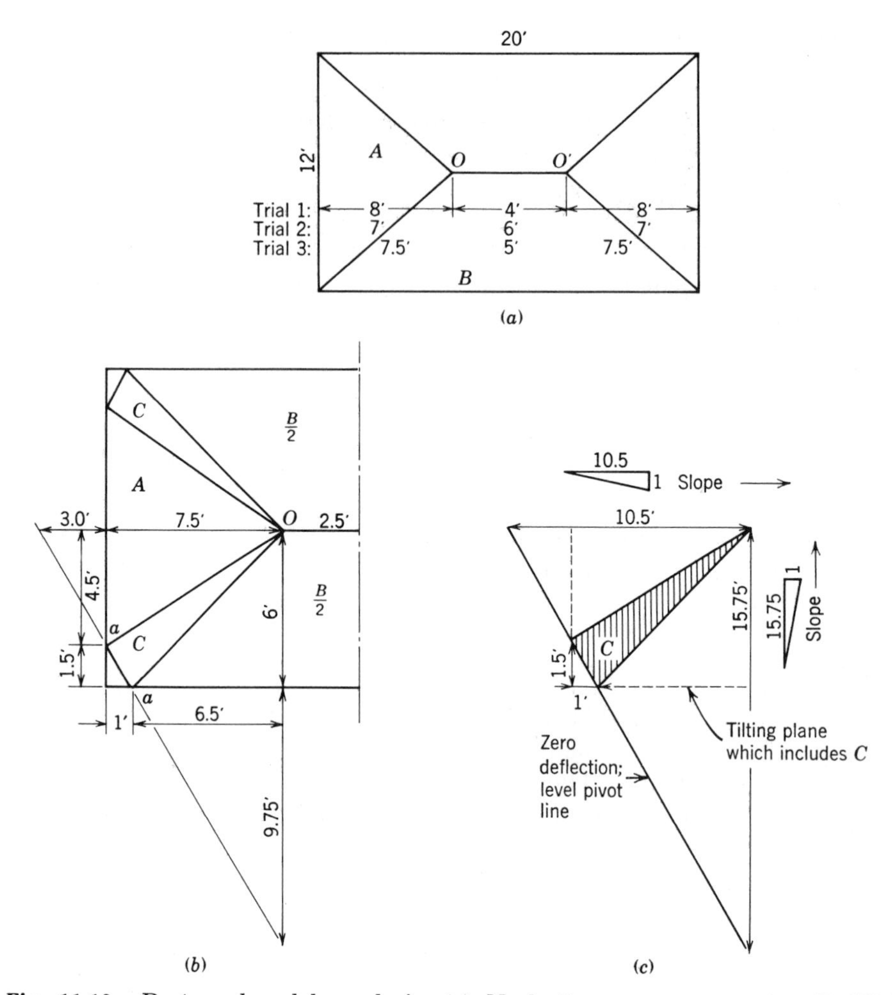

Fig. 11.13. Rectangular slab analysis. (*a*) Neglecting corner segments. (*b*) With trial corner segments. (*c*) The rotation of corner segment.

This is reasonably close and will be checked by the energy relationship, using half of the panel for convenience and a unit deflection along OO'.

$$A: 0.5 \times 12w \times 7.5 \times \tfrac{1}{3} = 15w$$

$$B: 5w \times 6 \times \tfrac{1}{2} = 15w$$

$$+2 \times 0.5 \times 7.5w \times 6 \times \tfrac{1}{3} = 15w$$

$$\overline{45w}$$

Alternatively, visualize the total slab deflections as a volume somewhat similar to an inverted pyramid. This volume times w measures the energy imput.

$$
\begin{aligned}
&7.5' \text{ end as half pyramid: } 7.5 \times 12w \times \tfrac{1}{3} &= 30w\\
&2.5' \text{ center as wedge: } \quad 2.5 \times 12w \times \tfrac{1}{2} &= 15w\\
& &\overline{45w}
\end{aligned}
$$

Energy absorbed:

$$
\begin{aligned}
&A\text{: } 12(4000 + 3000)1/7.5 &= 11{,}200\\
&B\text{: } = 20(4000 + 3000)\tfrac{1}{6} &= 23{,}300\\
& &\overline{34{,}500}
\end{aligned}
$$

$$w = 34{,}500/45 = 767 \text{ psf}$$

For the location of O the third trial dimensions will be assumed correct enough.

The corner segment situation will now be investigated. For equilibrium of the corner segment,

$$-4000 - 3000 + 0.5wh^2/3 = 0, \qquad h = \sqrt{42{,}000/w}$$

If $w = 750$ psf, $h = 7.46$ ft

This compares with the diagonal length of $\sqrt{7.5^2 + 6.0^2} = 9.60$ ft. It appears that there will be a significant corner segment, probably one extending most of the way to point O.

The corner segment shown in Fig. 11.13*b* will be analyzed by the energy approach, using the deflection at O as unity. At *aa* the x- and y-components of the moments and rotations, as diagramed in Fig. 11.13*c*, are used for calculating the energy absorbed by the hinges. Only half the panel (one segment each of A, B, and two of C) will be used.

Energy input by loads:

$$
\begin{aligned}
&B\text{: } 5w \times 6 \times \tfrac{1}{2} & & &= 15.0w\\
&\quad +2 \times 0.5 \times 6.5w \times 6 \times \tfrac{1}{3} & & &= 13.0w\\
&A\text{: } 0.5 \times 9 \times 7.5w \times \tfrac{1}{3} & & &= 11.25w\\
&C\text{: } 2 \times 0.5 \times 1w \times 6.0 &= &6.0w & &\\
&\quad +2 \times 0.5 \times 1.5w \times 7.5 &= &11.25w & &\\
&\quad -2 \times 0.5 \times 1w \times 1.5 &= &-1.5w & &\\
& & &\overline{15.75w} \times \tfrac{1}{3} &= &5.25w\\
& & & & &\overline{44.50w}
\end{aligned}
$$

Somewhat simpler, energy imput from the deflection volume:

7.5′ end as half pyramid:*	$(7.5 \times 12 - 2 \times 0.5 \times 1 \times 1.5)w \times \frac{1}{3}$	=	$29.5w$
2.5′ center as wedge:	$(2.5 \times 12)w \times \frac{1}{2}$	=	$15.0w$
			$44.5w$

Energy absorbed by hinges:

$$B{:}\ 18(4000 + 3000)\tfrac{1}{6} = 21{,}000$$
$$A{:}\ 9(4000 + 3000)1/7.5 = 8{,}400$$
$$C{:}\ 2 \times 1(4000 + 3000)1/15.75 = 890$$
$$+2 \times 1(4000 + 3000)1.5/10.5 = 2{,}000$$
$$32{,}290$$

$$w = 32{,}290/44.5 = 726 \text{ psf} < 767$$

A number of variations were computed by the energy method, partly to be certain the worst case was found, partly to study the rather minor differences in w which resulted. The results of these calculations are shown just to the right of each illustration in Fig. 11.14. Where several values of w are shown, they come from analyses of the separate segments. It appears that an ultimate load of about 710 psf is correct, possibly 690 or 700 psf with corner fans.

11.10 INFLUENCE OF STRENGTH OF EDGE BEAMS

Tests of slabs in England have indicated that the failure pattern of Fig. 11.10 or 11.11 occurs with stiff supporting beams but the pattern can be altered by varying the size of beams. If the beams in one direction are very flexible, the slab fails almost as does a one-way slab and the yield line lies directly across the middle of the span and continues through the light beams. Then the strengths of the beams and the slab add in the evaluation of the m and m' moments. Obviously, at some particular beam stiffness, it would be a matter of chance whether the diagonal failure or the mid-span failure line develops.

In tests of multiple panels of two-way slabs at the University of Illinois[10] a positive moment yield line developed near mid-span in the exterior panels parallel to the outside edges and the failure extended across the

*Where the corner elements go to a common intersection, the net area of the base as used here is convenient. Where the corner elements stop short, the "gross" pyramid for the total height less small corner triangular pyramids of lesser height is convenient as a concept.

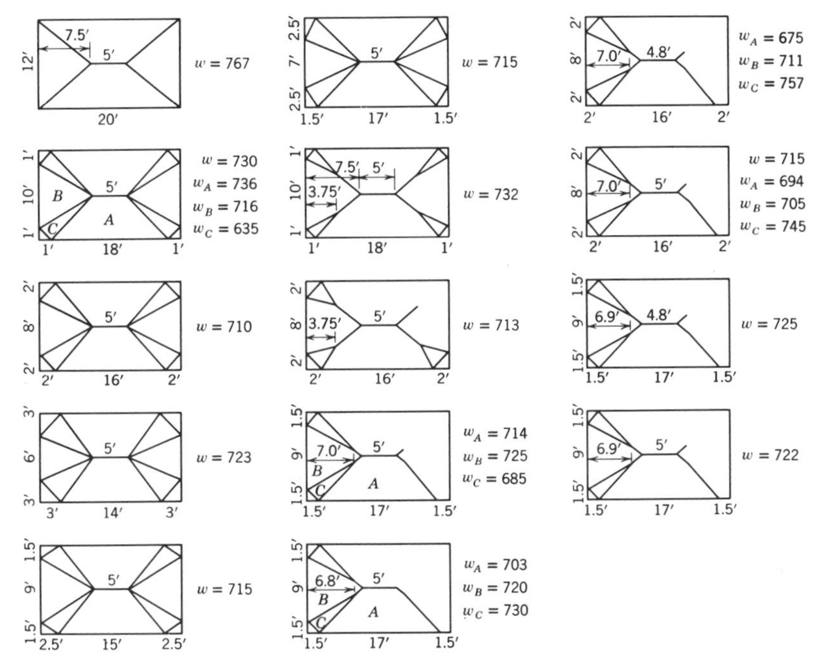

Fig. 11.14. Summary of calculations indicating the limited effect of corner segments.

interior and spandrel beams which framed perpendicular to the outside edge. In such a case slab and beam strengths are additive in establishing the maximum load capacity. However, it should be pointed out that, although the calculated failure load for this mode of failure closely matched the actual failure load, the structure did not fail in the mode the yield-line method would have predicted. Failure of the individual slabs was indicated by theory at some 25% lower load.

The example problems in this chapter have assumed that beams furnished stiff supports of such strength that the slab would fail first.

11.11 OTHER YIELD-LINE ANALYSES

A few yield-line patterns for other cases may be helpful as suggestions. Figure 11.15*a* shows a triangular slab simply supported and carrying uniform load. Johansen[4] shows for isotropic conditions that the yield lines intersect at the center of the inscribed circle and result in $m = wr^2/6$, where r is the radius of the circle. He extends this case to show the same

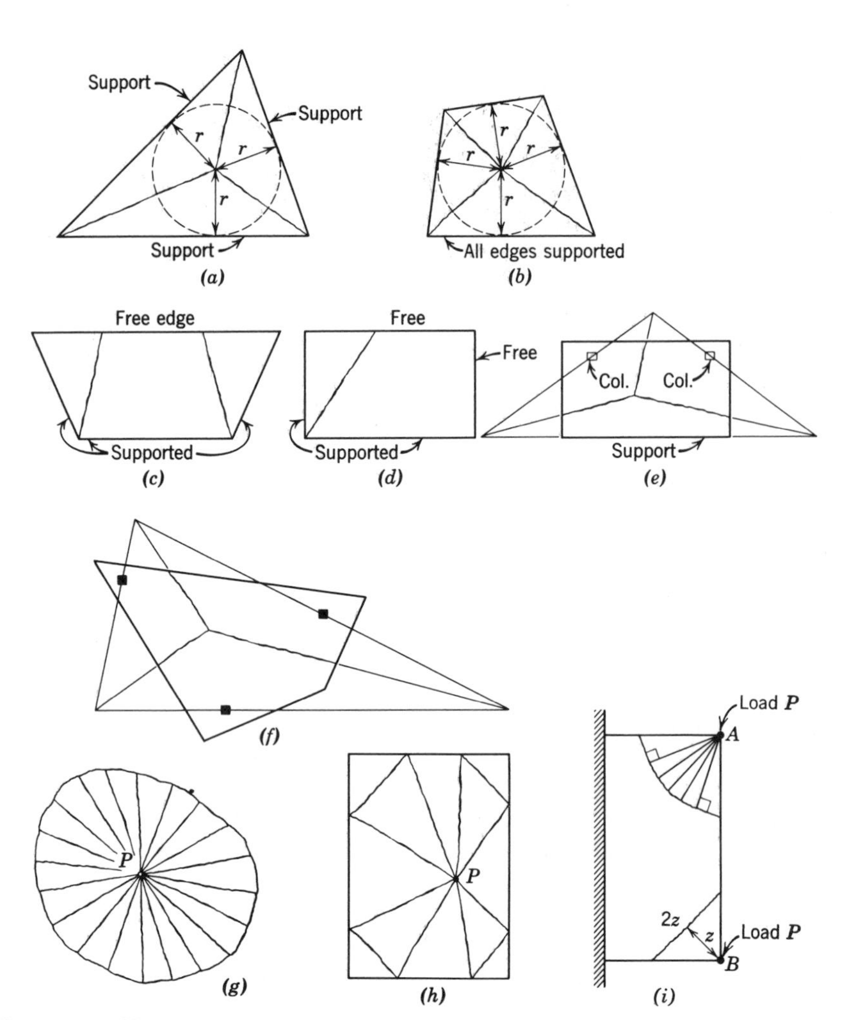

Fig. 11.15. Yield-line patterns. (*a*) Triangular slab supported on all sides. (*b*) Polygonal slab circumscribed around a circle, all edges supported. (*c*) Slab supported on three of four sides. (*d*) Slab supported only on adjacent sides. (*e*) Slab supported on one side and two columns. [(*a*) to (*e*) adapted from Ref. 4.] (*f*) Slab supported on three columns. (*g*) Yield pattern around a concentrated load. [(*f*) and (*g*) adapted from Ref. 3, ACI.] (*h*) Yield pattern from an unsymmetrical concentrated load on a rectangular slab. (Adapted from Ref. 6, Chamecki.) (*i*) Corner load on cantilever slab.

moment in any polygon shape which circumscribes a circle, as in Fig. 11.15*b*. It is assumed that any of these might need to be investigated for a possible corner pivot or segment. Johansen also shows free edge slabs as shown in Fig. 11.15*c*, *d*, *e*.

Hognestad[3] shows the slab of Fig. 11.15*f*. He also analyzes a radial pattern which at times develops under concentrated loads, Fig. 11.15*g*, as already discussed in Sec. 11.7.

With a single concentrated load in a rectangular panel Chamecki[2] shows eight triangular segments radiating from the load P whether P is centered or off center from both axes as in Fig. 11.15*h*. He develops equations to locate all key dimensions for the simple support case when the corners are anchored down and when they are free to rise. He also shows the condition necessary to eliminate the corner segments.

Elstner and Hognestad[9] show the yield patterns of Fig. 11.16 for slabs carrying a center load in the form of a column stub cast monolithically with the slab. Figure 11.16*a* shows a slab simply supported and corners free; Fig. 11.16*b* is the same except that an eccentric column load is considered as a line load. The case of simple supports on two opposite sides only is shown in Fig. 11.16c and simple support on four corners only in Fig. 11.16*d*; in the latter case the dashed lines represent an alternate yield-line pattern.

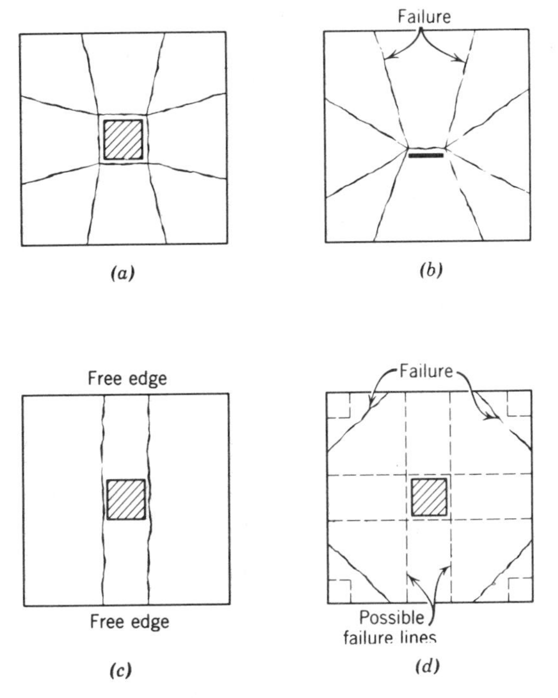

Fig. 11.16. Slabs supporting column loads. (From Ref. 9, ACI.) (*a*) Simply supported on all edges, corners free. (*b*) Eccentric column load treated as a line load, simply supported on all edges, corners free. (*c*) Simply supported on two opposite edges. (*d*) Supported at corners only.

11.12 SLABS WITH OPENINGS

One advantage of the yield-line method is that slabs with openings can be analyzed. The designer must, however, be alert to limitations on the effective usage of the equilibrium method.

When two positive moment yield lines meet at the corner of an opening, the knot forces are not zero. In fact, shearing forces will often be transferred from one element to another.

For example, the load of 668 psf and the yield lines shown in Fig. 11.17*a* were calculated as the governing case from the energy approach using $m = 3000$ and $m' = 4000$ ft-lb/ft and neglecting corner effects. The equilibrium method would lead to the idea that the height of triangle *A* should be 7.8 ft on the basis of the equation for *h* in Sec. 11.6. However, a net downward knot force at the apex makes a shorter height of triangle the correct one. The student can use equilibrium of each segment to find these knot forces, after having *w* established. For this slab anyone visualizing the deflection of segment *B* would quickly see from the way a slab deflects that some support for *B* is required at the two corners of the opening. This is what causes the knot forces there.

A second problem also involves the equilibrium method. The discontinuity at the 90° corner of the opening prevents any effective use of the ordinary edge forces m_t. It will be obvious that a single yield line going symmetrically into the corner, as in Fig. 11.17*b*, has two obtuse angles and that the nominal upward forces in each corner could not represent real

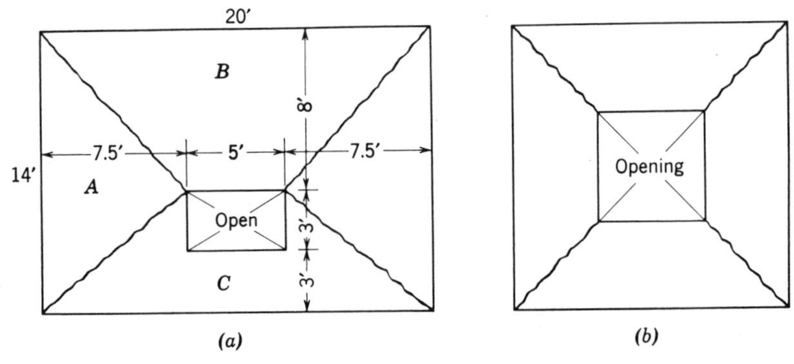

Fig. 11.17. Yield-line patterns in slabs containing openings. (*a*) Large knot forces present. (*b*) No knot forces.

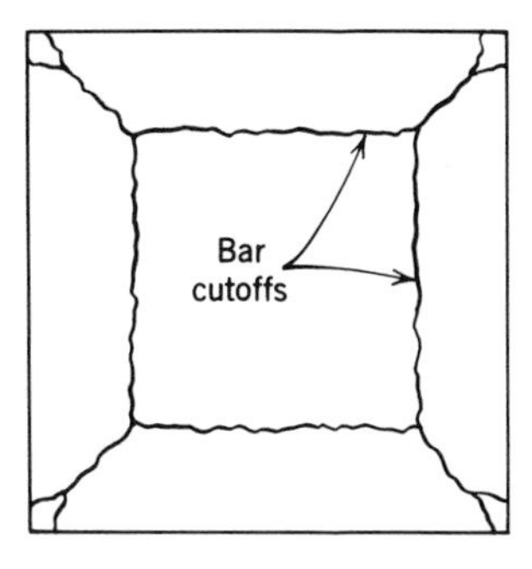

Fig. 11.18. Test yield line pattern on simply supported slabs having bars cut off according to moment.

internal forces (because they would not be "equal and *opposite*"). Symmetry considerations indicate that this must be the correct yield-line pattern (ignoring corner effects) and that there can be no knot forces at the opening. Thus one must conclude that knot forces around an opening may be absent or may be present (as in Fig. 11.17*a*), but are not likely to be of the nominal magnitude m_t. Thus the energy method is essentially mandatory for such cases.

11.13 REINFORCEMENT AT VARIABLE SPACING AND/OR WITH BARS CUT OFF

Taylor *et al.*,[13] tested a series of simply supported square slabs of several thicknesses but with the chief variable as the arrangement of reinforcement. Specimens included uniform spacing as the yield line theory uses most conveniently, variable spacing strips as most commonly used in practice, and 50 to 70% of the bars cut off to the moment requirements.

Variable spacing showed only a minimal advantage, if any, since the larger area of reinforcement near the middle decreased the internal lever arm. Variable spacing gave a slightly stiffer slab, with the cracks appearing first near the corners rather than near the center; it did not increase strength.

Some economy of material results from stopping some bars short of the supports, and ultimate design loads were still exceeded. However, the excess load capacity was less than when bars were full length and these slabs deteriorated rapidly after forming a center square of positive moment yield lines at the points where bars were cut off, as in Fig. 11.18.

SELECTED REFERENCES

1. K. W. Johansen, *Pladeformler*, Polyteknish Forening, Copenhagen, 2nd ed., 1949.
2. K. W. Johansen, *Paldeformier*; *Formelsamling*, Polyteknish Forening, Copenhagen, 2nd ed., 1954.
3. Eivind Hognestad, "Yield-Line Theory for the Ultimate Flexural Strength of Reinforced Concrete Slabs," *Jour. ACI*, *24*, No. 7, Mar. 1953; *Proc.*, 49, p. 637.
4. K. W. Johansen, "Yield-Line Theory," English translation, Cement and Concrete Assn., London, 1962.
5. F. E. Thomas, "Load Factor Methods of Designing Reinforced Concrete," *Reinf. Conc. Review*, 3, No. 8, 1955, pp. 540, 544.
6. Samuel Chamecki, *Calculo No Regime de Ruptura, Das Lajes de Concreto Armadas em Cruz*, Curitiba, Parana, Brazil, 1948, 107 pages.
7. R. H. Wood, *Plastic and Elastic Design of Slabs and Plates*, Ronald Press, New York, 1961.
8. L. L. Jones, *Ultimate Load Analysis of Reinforced Concrete Structures*, Interscience Publishers, New York, 1962.
9. Richard C. Elstner and Eivind Hognestad, "Shearing Strength of Reinforced Concrete Slabs, Appendix 1," *Jour. ACI*, 28, No. 1, July 1956; *Proc.*, 53, p. 55.
10. W. L. Gamble, M. A. Sozen, and C. P. Siess, "Measured and Theoretical Bending Moments in Reinforced Concrete Floor Slabs," Univ. of Illinois *Civil Engineering Studies Structural Research Series* No. 246, June 1962.
11. Telemaco van Langendonck, *Charneiras Plásticas em Lajes de Edifićios*, Associac̆ao Brasileira de Cimento Portland, São Paulo, Brazil, 1966, 81 pages.
12. Rolf Lenschow and Mete A. Sozen, "A Yield Criterion for Reinforced Concrete Slabs," *Jour. ACI*, 64, No. 5, May 1967, p. 266.
13. R. Taylor, D. R. H. Maher, and B. Hayes, "Effect of the Arrangement of Reinforcement on the Behavior of Reinforced Concrete Slabs," *Magazine of Concrete Research*, 18, No. 55, June 1966, p. 85.

PROBLEMS

Prob. 11.1. In Figs. 11.2*a* and 11.7 calculate the ultimate load by yield-line procedures if:

(*a*) The positive moment capacity m_c is increased to $+3800$ ft-lb/ft, m_a and m_b remaining unchanged at -4000 and -5000 ft-lb/ft respectively.

(*b*) The negative moment capacity m_a is decreased to -2800 ft-lb/ft, m_b and m_c remaining at -5000 and $+3000$ ft-lb/ft respectively.

Prob. 11.2. Recalculate the ultimate load for the slab of Sec. 11.4 and Fig. 11.8 if the right support has only a 4-ft skew, that is, if b' in Fig. 11.8*a* is moved 2 ft to the right.

Prob. 11.3. If $m = +2800$ ft-lb/ft and $m' = -3500$ ft-lb/ft, find the ultimate load which can be carried by an 20-ft square slab continuous on all sides. Consider first without corner effect and then establish effect of corner segments.

Prob. 11.4. Investigate the ultimate load capacity of a 15-ft by a 20-ft slab continuous on all sides and having a yield moment of 3000 ft-lb/ft for positive moment and 4500 ft-lb/ft for negative moment.

Prob. 11.5. Assume the rectangular slab of Sec. 11.9 had a 3-ft square hole in the middle of the span (centered each way). Find the ultimate uniform load by yield-line method: (*a*) neglecting corner effect; (*b*) with corner effects.

Prob. 11.6. Assume the rectangular slab of Sec. 11.9 has no support at the upper 20-ft side (Fig. 11.13). Find the ultimate uniform load by yield-line method.

12
Strip Method for Slab Design

12.1 DEVELOPMENT OF STRIP METHOD

A method for *designing* slabs, published by Hillerborg in 1956 and 1959 in Swedish, has received more attention in recent years because of study and tests at the Building Research Station in England. Wood and Armer[1,2,3] in 1968 reported a critical analysis of the method and their own tests of typical slabs designed by this method. They found (mathematically) that a design made by the strip method and reinforced exactly according to the moments found (without averaging across a band, but with reduction of steel as moment decreased) was an *exact* solution rather than just a lower-bound solution of the problem.* Such a slab did develop the load capacity introduced into the design.

The strip method of design gives the designer wide freedom of choice in his design approach. Hence many different solutions for a given slab design are possible. Obviously not all solutions will be of equal economy. Wood† points out that a design using moments approaching those from elastic analysis is an efficient design and to be preferred.

The strip method is simplest for slabs on simple supports, but continuity can be handled on a basis similar to limit design. The suitability

*Note that the yield-line method is an *upper* bound procedure.

†Wood's name will be used in this chapter for easy reference to the joint work of Wood and Armer.

of the method to slabs with openings is a strong point in its favor.* The most difficult slabs for this method are slabs supported on columns (Sec. 12.8). For such a case, Hillerborg developed what Crawford[4] calls the advanced strip method, using a rectangular element carrying load in two directions to a support at one corner of the element. Although the mathematical basis for this type of element could not be proved by Wood, a test result for a slab design using this method was satisfactory. Hillerborg states that the advanced strip method leads to a more economical design with a simple reinforcement pattern than does the substitute type of element Wood suggests. The discussion here will stay with the simpler strips which Wood endorsed. (The author leans heavily on work by Wood in this area.)

Wood calls attention[1] to the normal use of A_s varying in accord with the strip moments, in contrast to the complexities inherent in a yield-line analysis when m (or A_s) is not constant. If the reinforcement is cut off where it is not needed, the slab designed by the strip method does not fail by sharply developed yield lines, but by "yield in nearly all directions at failure, . . . somewhat like a plastic hammock." Hence yield-line concepts are only marginally useful with this method. The yield-line method deals with "rigid plate" rotations; the strip method tends to eliminate rigid plate failure at ultimate.

12.2 THEORETICAL BASIS

The equilibrium equation for slabs (for elastic plate analysis) is

$$\frac{\partial^2 M_x}{\partial x^2} + \frac{\partial^2 M_y}{\partial y^2} - 2\frac{\partial^2 M_{xy}}{\partial x \partial y} = -w$$

where the bending moments, M_x and M_y, and the twist moment M_{xy} follow Timoshenko's notation and w is the load per unit area on the slab. Hillerborg designs the slab to make M_{xy} unnecessary, that is, he assumes $M_{xy} = 0$ and then apportions the load to $\partial^2 M_x/\partial x^2$ and to $\partial^2 M_y/\partial y^2$ as he wishes, usually at a particular spot wholly to one or to the other. This particular apportionment is more of a convenience than a necessity, however.

Loads in a particular area are assigned to particular slab strips, as the next section will illustrate, and continuity of the resulting moments and

*The author accepts the strip method with some enthusiasm because it formalizes a very approximate method which designers have been using for many years, that of designing by their "feel" for the way the load was most apt to be transferred to the supports.

shears must be carefully maintained. Apparent discontinuity in torque or deflection may be disregarded, but a discontinuity in moment or shear is not permitted. Both elastic and plastic analysis concepts are permissible in evaluating moments on strips, but both Wood and Hillerborg note that under service load the slab behavior is more nearly in the elastic range. Hence elastic concepts are valid and acceptable even though the uses suggested below ignore relative deflections with apparent unconcern.

12.3 SIMPLY SUPPORTED RECTANGULAR SLAB

A rectangular slab is adequately represented by the simplified concept of a grid of strips in the x- and y-directions, that is, 90° apart.

Consider that boundaries, called lines of stress discontinuity (or discontinuity lines), are set up as indicated in Fig. 12.1*a*, creating an area 1 and two areas 2. These discontinuity lines indicate the designer's decision to carry all the load in areas 2 in the x-direction on x-strips and all load in area 1 in the y-direction on y-strips. The discontinuity lines are *not* yield lines, and the designer is free to choose the angle θ. If he chooses 90°, he will design a one-way slab; the result will be adequate strength, early cracking along the y-supports, and an overall solution approaching the absurd.

Wood suggests that discontinuity lines might be taken as sketched in Fig. 12.1*c* to suit bands of reinforcement, since in limit analysis one is not restricted to the use of a single straight line in a quadrant.

As in any flexural member, a load anywhere on a strip produces a shear along the entire strip. However, it is convenient to add a zero shear line on the x-axis as shown in Fig. 12.1*b* and *d* and think of the y-strip as carrying all the load above this zero shear line to the upper slab support as indicated by the arrows.

In either layout the central y-strips are simple one-way slab strips under a uniform load or such other distribution of load as may exist. The y-strips running through an area 2 are unloaded in that area and loaded only in the two area 1 end portions, as indicated by the shaded areas. Likewise x-strips in Fig. 12.1*b* are all unloaded except near the supports, but this is not quite true in Fig. 12.1*d*. With square discontinuity lines the x-strips near the top and bottom are totally unloaded, and the load pattern changes by substantial steps where the strips enter area 2.

The moment in each strip defines the necessary steel and even indicates where some may be cut off. In Fig. 12.1*b* the required steel requires a variable spacing; in Fig. 12.1*d* bands of steel are indicated. For the variable spacing Hillerborg actually substituted a weighted average; Wood shows

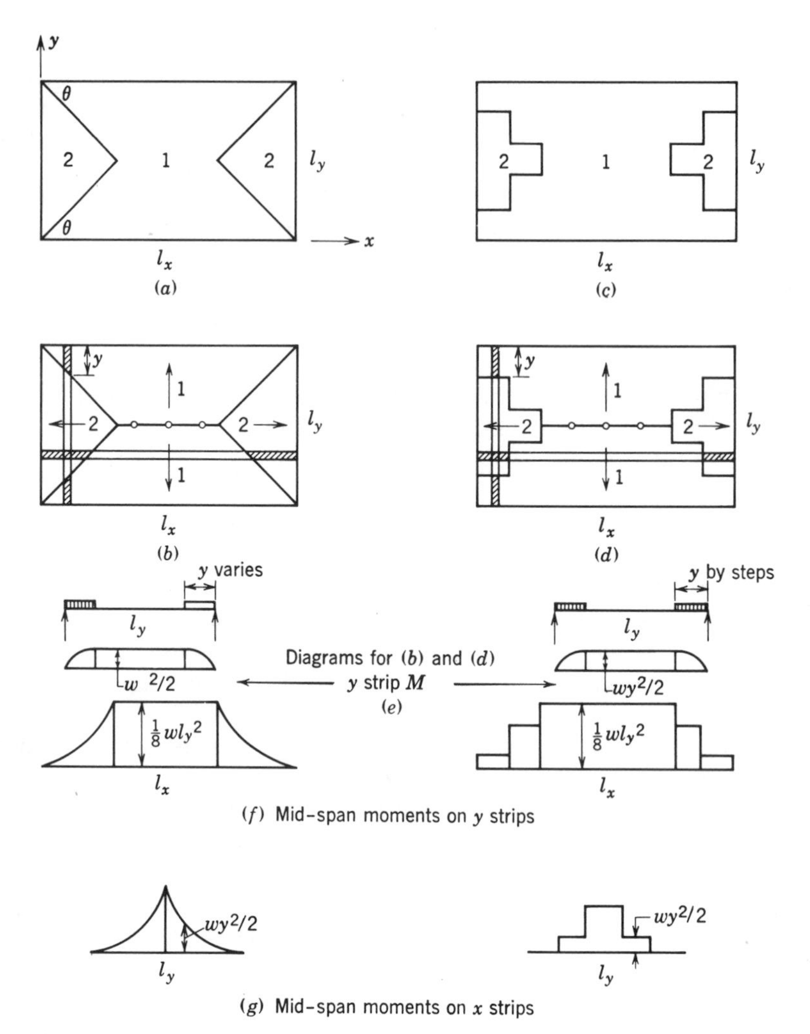

Fig. 12.1. Layouts and details for two possible strip designs of rectangular slab simply supported.

that the lower bound concept is then not satisfied, although membrane action may bridge the gap.

12.4 CONTINUITY IN RECTANGULAR SLABS

Since slab design by the strip method is a form of limit design, the ratio of negative to positive moment on a strip is not rigidly fixed. The use of moments approximating elastic analysis assures a satisfactory service load response, but this is not mandatory for strength. Wood suggests the assumption of a point of inflection (P.I.) for the middle y-strips in Fig. 12.2*a* at about $0.2l_y$ from the support, which is close to the elastic condition. Nearly the same result would be obtained by using the negative moment as $-(1/12)wl_y^2$; but for skewed slabs Hillerborg and Wood use the initial selection of the P.I. locations as the key to the desirable slab strips (Sec. 12.7).

For y strips that pass through area 2 and thus are loaded symmetrically over the y lengths only on their ends, the best assumption for the negative moment or P.I. is not so clear. Wood, following Hillerborg, breaks the P.I. trace and takes it diagonally to the corner, as shown at the top of Fig. 12.2*a*, making the distance from support to P.I. $0.4y$. If one considers the simple span moment on these strips, as in Fig. 12.2*b*, the maximum simple beam moment is $wy^2/2$, with this moment constant across area 1. The use

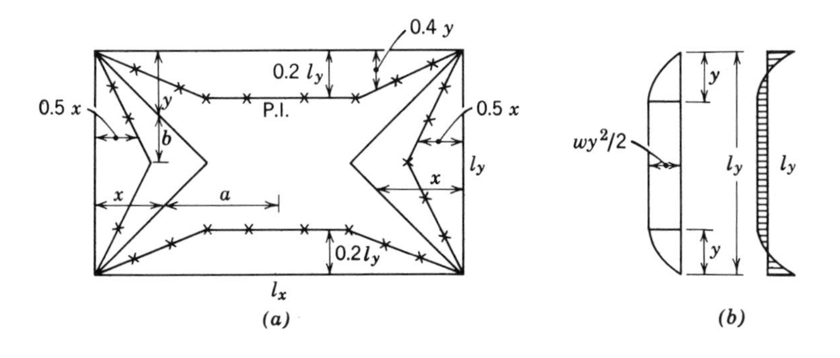

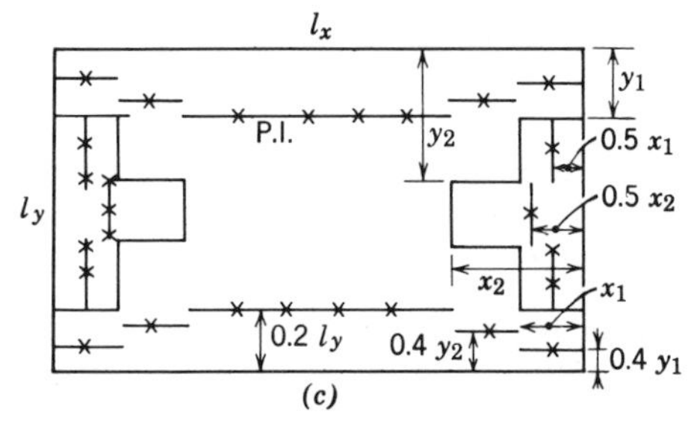

Fig. 12.2 Points of inflection for continuity. (*a*) Typical layout. (*b*) M on y strip, simple span and continuous span. (*c*) Proposed points of inflection.

of the diagonal P.I. line leads to negative moments here that are substantially less (down 50% or more*) compared to the elastic analysis values for fixed ends, but this is permissible with limit design concepts.

For the x-strips a similar treatment of P.I. lines at $0.4x$ from the support establishes negative moments relatively *much* less than elastic analysis values (in some cases down by 70%). Although the matter deserves further study in terms of the implied degree of hinging and relative costs, the author (in the absence of specifications by Wood) recommends the use of the P.I. line at $0.5x$ as possibly more realistic, Fig. 12.2*c*.

12.5 DESIGN OF RECTANGULAR SLAB

A 15 ft by 20 ft rectangular slab with restrained edges at beams is designed below for a factored load w of 300 psf. The design moments are computed without the ϕ factor included. Once the slab thickness is determined for moment or deflection (shear rarely controls a slab on beams), the required A_s calculated from $M/(\phi f_y z)$ takes the same distribution as the moments plotted in Fig. 12.3.

Discontinuity lines are arbitrarily chosen as shown in Fig. 12.3*a* and P.I. locations are selected as discussed in the section above.

Moments will be calculated first for each y-strip. The loadings are sketched to the right in Fig. 12.3*b*. Here the moments will be calculated as for a simple span between P.I. points and as a cantilever carrying these reactions plus its own uniform load. The moments on the x-strips are sketched in Fig. 12.3*c* and on y-strips in *b*.

Strip 1–1 No loading

Strip 2–2 Pos $M = 300 \times 1.5^2/2 = +337$ lb-ft

Neg $M = -300 \times 1.5 \times 1.5 - 300 \times 1.5^2/2$

$= -675 - 337 = -1012$ lb-ft

Strip 3–3 Pos $M = 300 \times 3.5^2/2 = +1840$

Neg $M = -300 \times 3.5 \times 3.5 - 300 \times 3.5^2/2$

$= -3675 - 1840 = -5520$ lb-ft

*The values vary strip to strip and this is near that for the middle strip of the group. The elastic analysis for fixed ends shows that the P.I. distance from support becomes $y(1 - 0.82\sqrt{y/l_y})$ and the negative moment is $M_s(0.67 + b/y)/(1 + b/y)$, where M_s is the maximum simple span moment, y is the distance to the discontinuity line, and b is the remaining distance to mid-span, that is, $b = (l_y/2) - y$. In these terms the same expressions can be applied to the x-strips, with distances x, a, and l_x replacing y, b, and l_y.

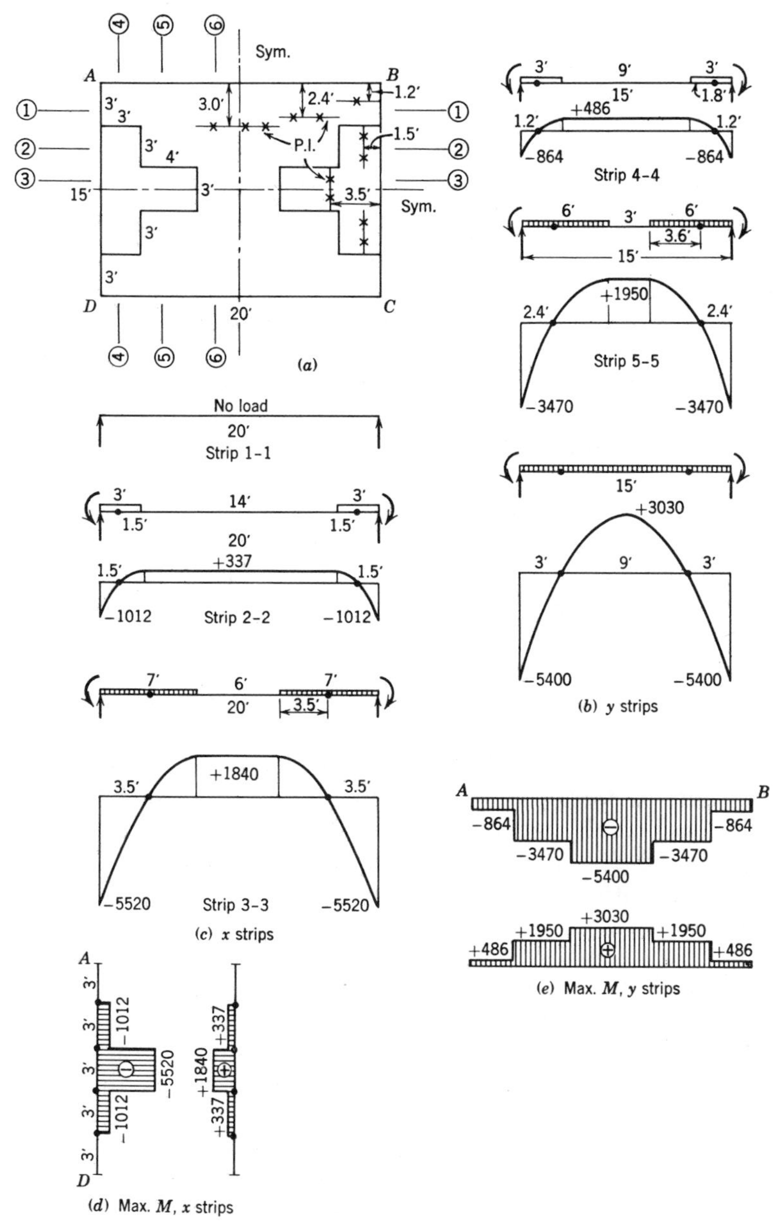

Fig. 12.3. Design moments for a rectangular slab.

Strip 4–4 Pos $M = 300 \times 1.8^2/2 = +486$ lb-ft
Neg $M = -300 \times 1.8 \times 1.2 - 300 \times 1.2^2/2$
$= -648 - 216 = -864$ lb-ft

Strip 5–5 Pos $M = 300 \times 3.6^2/2 = +1950$ lb-ft
Neg $M = -300 \times 3.6 \times 2.4 - 300 \times 2.4^2/2$
$= -2600 - 865 = -3470$ lb-ft

Strip 6–6 Pos $M = (\frac{1}{8})\, 300 \times 9^2 = +3030$ lb-ft
Neg $M = -300 \times 4.5 \times 3 - 300 \times 3^2/2$
$= -4050 - 1350 = -5400$ lb-ft

These moments are summarized in Fig. 12.3*d* and *e* into diagrams of negative moment crossing the face of support and positive moment crossing mid-span, which determine the amount and distribution of reinforcing steel. If this arrangement appears awkward or seems to have excessive local demands, the designer can modify the discontinuity lines in such a way as to shift the requirement in the desired direction. The several moment diagrams sketched in Fig. 12.3*b* and *c* are useful for establishing bar cutoff points.

The minimum temperature steel may well control where computed moments are small, certainly in strip 1-1 close to the long edge beams.

The loadings on the beams supporting the slabs are most logically established from the strip reactions.

12.6 DESIGN OF A RECTANGULAR SLAB WITH HOLE

The symmetrical slab of Fig. 12.4*a*, restrained at all edge beams illustrates the rerouting of loads when strips are interrupted. Since slabs are normally considerably under-reinforced, it is possible to use certain strips near the opening (Fig. 12.4*b*) as small beams simply by increasing the local reinforcement. If the opening is so large that even extra slab steel is inadequate to care for the moment, a real beam is needed around one or more sides of the opening, quite probably spanning to the edge beams.

The assumed "beam strips" are shown dotted around the opening and the assumed discontinuity lines and P.I. lines are added in Fig. 12.4*c*. Only the P.I. line between the long side of the opening and the long support beam requires special discussion. This 4-ft-long strip in the *y*-direction could be considered as a cantilever from *DC*, but it appears that this might be considering excessive deflection at the opening. Consider instead a discontinuity line at 3 ft from the support, which will terminate the cantilever and leave a 1-ft strip 1-1 alongside the opening in the *x*-direction. Strip 2-2 in the *y*-direction will pick up the reaction from strip 1-1. Consider this strip 2-2 to be 2 ft. wide, since strips 1-1 alone would add 50% to the

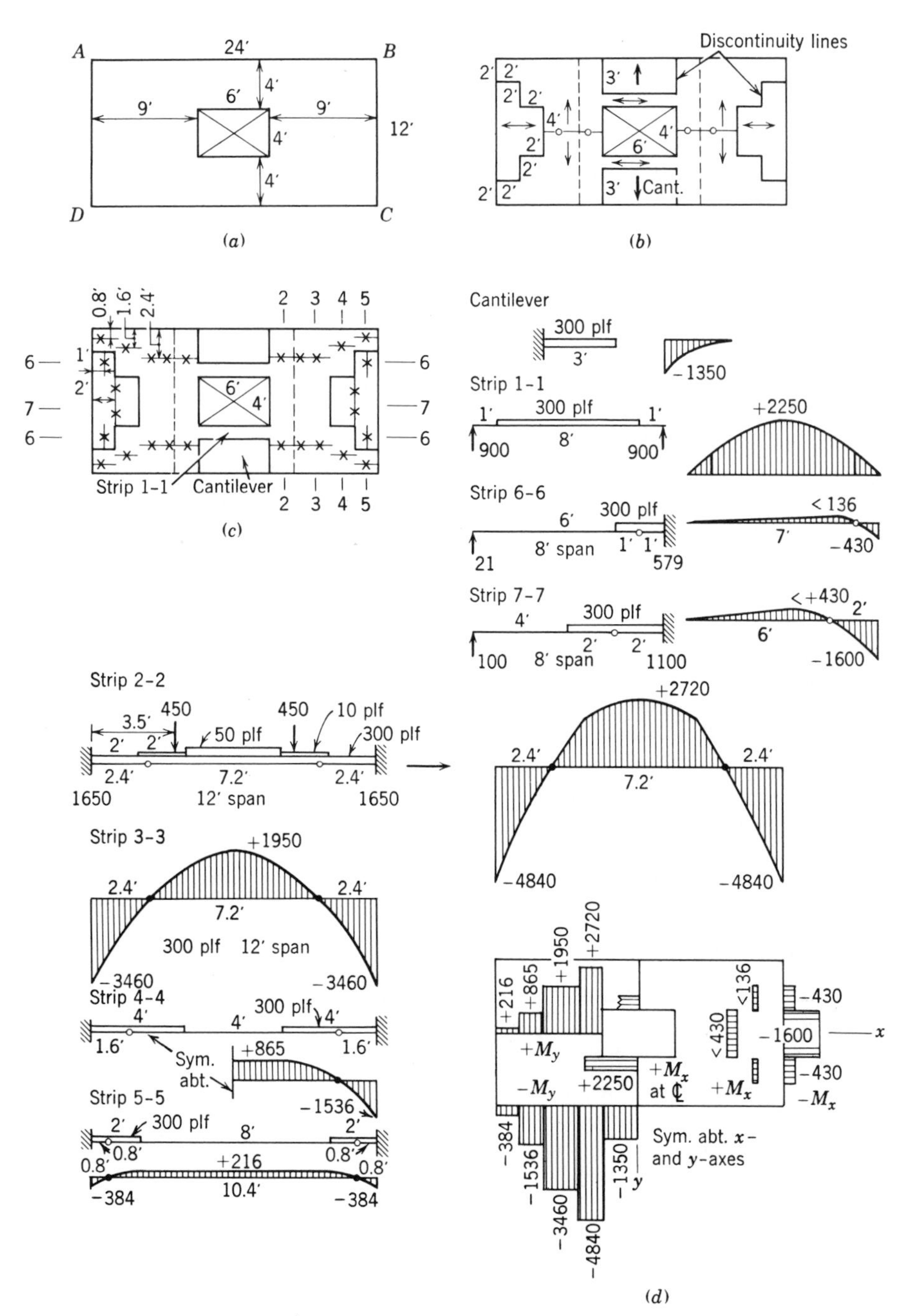

Fig. 12.4. Design based on strip method. (*a*) Layout. (*b*) Discontinuity lines. (*c*) Inflection point lines. (*d*) Design moments.

strip 2.2 load if only a 1-ft strip were used, and additional loads come from strips 6-6 and 7-7. Strip 7-7, because of the opening, is terminated on strip 2-2; it seems logical to treat trip 6-6 in the same manner.

The design moments will be calculated for a total design (factored) load of 300 psf. The strip loadings and moment diagrams are shown in Fig. 12.4*d*, along with a final sketch that assembles the maximum negative and positive design moments on both x- and y-strips. The supporting calculations are as follows.

Cantilever strips near strip 1–1, 3 ft length

Neg $M = -300 \times 3^2/2 = -1350$ lb-ft

Strip 1–1 with simple span of 8 ft to center of strip 2–2

Pos $M = +3 \times 300 \times 4 - 300 \times 3^2/2 = +2250$ lb-ft

Reaction on strip 2–2 $= 3 \times 300 = 900$ lb

Strip 6–6 spanning 8 ft to center of strip 2–2

Reaction* on strip 2–2 $= 300 \times 1 \times 0.5/7 = 21$ lb

Pos $M < 21 \times 6.5 = 136$ lb-ft

Neg $M = -(300-21) \times 1 - 300 \times 1 \times 0.5 = -279 - 150 = -430$ lb-ft

Strip 7–7 spanning 8 ft to center of strip 2–2

Reaction on strip 2–2 $= 300 \times 2 \times \frac{1}{6} = 100$ lb

Shear at P.I. $= 500$ lb

Pos $M < 100 \times 4.3 = 430$ lb-ft

Neg $M = -500 \times 2 - 300 \times 2 \times 1 = -1000 - 600 = -1600$ lb-ft

Strip 2–2 spanning 12 ft

The reactions from 1–1, 6–6, and 7–7 are shared between the two strips making up the 2 ft width.

Shear at P.I. $= 300 \times 7.2/2 + 50 \times 2 + 10 \times 1.6 + 450 = 1650$ lb

Pos $M = 1650 \times 3.6 - 300 \times 3.6^2/2 - 50 \times 2 \times 1 - 10 \times 1.6 \times 2.8$
$- 450 \times 2.5 = +5940 - 1950 - 100 - 45 - 1125 = +2720$ lb-ft

Neg $M = -1650 \times 2.4 - 300 \times 2.4^2/2 - 10 \times 0.4 \times 2.2$
$= -3960 - 865 - 10 = -4840$ lb-ft

Strip 3–3 spanning 12 ft

Pos $M = (\frac{1}{8})300 \times 7.2^2 = +1950$ lb-ft

Neg $M = -300 \times 3.6 \times 2.4 - 300 \times 2.4^2/2 = -2590 - 865 = -3460$ lb-ft

Strip 4–4 spanning 12 ft

Shear at P.I. $= 300 \times 2.4 = 720$ lb

Pos $M = 720 \times 2.4 - 300 \times 2.4^2/2 = 1730 - 865 = 865$ lb-ft

Neg $M = -720 \times 1.6 - 300 \times 1.6^2/2 = -1152 - 384 = -1536$ lb-ft

Strip 5–5 spanning 12 ft

Shear at P.I. $= 300 \times 1.2 = 360$ lb

Pos $M = +360 \times 1.2 - 300 \times 1.2^2/2 = 432 - 216 = +216$ lb-ft

Neg $M = -360 \times 0.8 - 300 \times 0.8^2/2 = -288 - 96 = -384$ lb-ft

*From ΣM about P.I. using "simple span" as freebody.

The maximum moments assembled in Fig. 12.4*d* look reasonable in distribution and no modification of the strip arrangement is indicated as desirable. The maximum moments fix the slab depth and shear may be checked from the strip loads if that appears possibly to control. The reinforcement should be arranged in bands corresponding to the strips used and this calculation is simple. The short y-strip steel should be placed at the greater depth and for the x-strips this d must be reduced by one bar diameter. Minimum temperature or spacing steel will control where the indicated moments are very small. The moment diagrams for the strips give the basic data for required length of bars, subject to the usual Code requirements for arbitrary extensions beyond the theoretical cutoff points.

Deflection *at service load* must be considered in checking serviceability. In any actual design the service load is available, and it should be on the safe side to use the strip service load moments with EI based on the cracked section. The more detailed calculations of Sec. 6.5 may be used to interpolate between cracked and uncracked sections. Although long, that method considers the usual intermediate slab condition and is less severe than the use of the cracked section.

12.7 NONRECTANGULAR SLABS

The strip method is not limited to rectangular strips supported on beams on all four sides. A few layouts, largely those already in the literature, should indicate the possibilities.

A free edge, as in Fig. 12.5*a*, can be considered as a strong strip with extra reinforcement and can even pick up a limited amount of load from cross strips.

Where triangular slabs exist with a free edge opposite an acute angle, strips can span the short way between supported sides as in Fig. 12.5*b*. The same can be done if the supported edges are restrained, but Hillerborg suggests strips that change direction, as in Fig. 12.5*c*. With an assumed P.I. line, the mid-length is like a simple span, and the negative moment region acts like a cantilever to pick up the simple span reaction plus its own load. The obvious torsion implications are not a concern in this method. However, the cantilever and mid-strip must have different widths to care for the geometry of the layout. A similar idea shows in Fig. 12. 5 *d*.

When fixed and simply supported edges occur nearby, elastic concepts are helpful. In Fig. 12.5*e*, for example, area 2 should be smaller than area 3.

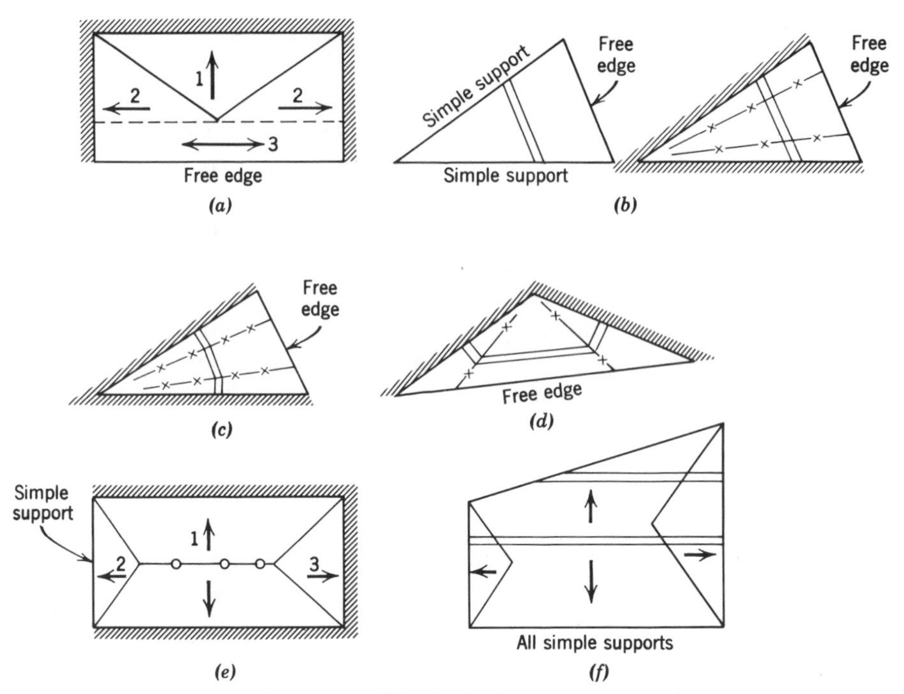

Fig. 12.5 Varied types of supports. Hatched edges are restrained.

12.8 LAYOUT AROUND COLUMNS SUPPORTING THE SLAB

The treatment of flat plates around the columns makes a difficult problem when one uses the strip method. Hillerborg's 1959 publication dealt particularly with this problem and resulted in a recommended type-3 rectangular element carrying load in two directions from zero shear lines and delivering the reaction to one corner (the column). This is obviously a concept much more involved than that of the strip. Wood was not able to prove the complete mathematical basis for this element, but he reported the resulting reinforcement looked reasonable. Armer reported[2] a test that was quite satisfactory.

Wood[1] developed a substitute procedure using the type-3 rectangular element with computer calculated tables leading to the moment coefficients. Armer designed one test specimen that substituted for the type-3 elements strong slab strips in the x- and y-directions to carry load from slab strips directly to the column. Test results on all of these variations proved satisfactory. Wood has also suggested a wide "beam band" over the columns that could be strengthened at the column by local (short) strong bands across the column, these apparently acting in flexure much like steel shearhead reinforcement in flat plates.

Developments in this particular area are intriguing and appear hopeful, but the author feels they may not at present be quite ready for everyday design.

SELECTED REFERENCES

1. R. H. Wood and G. S. T. Armer, "The Theory of the Strip Method for Design of Slabs", *Proc. Institution of Civil Engineers,* 1968, Vol. 41 (October), p. 287.
2. G. S. T. Armer, "Ultimate Load Tests of Slabs Designed by the Strip Method", *Proc. Institution of Civil Engineers,* 1968, Vol. 41 (October), p. 313.
3. G. S. T. Armer, "The Strip Method: A New Approach to the Design of Slabs", *Concrete,* Vol. 2, No. 9, Sept. 1968, p. 358.
4. Robt. L. Crawford, "Limit Design of Reinforced Concrete Slabs", *Proc. ASCE, Jour. Eng. Mech. Div.,* Oct. 1964, EM5, p. 321.

13
Distribution of Concentrated Loads and Other Special Problems

13.1 CONCRETE STRUCTURES DISTRIBUTE CONCENTRATED LOADS

The ordinary reinforced concrete structure is either monolithic or is tied together to act as a unit. Although parallel members of the structure may be analyzed somewhat independently of each other under uniform live loads, actually the entire structure is a three-dimensional frame. When moving concentrated loads are considered, their spacing and their number suggest that all parallel slab strips and all neighboring beams will not be equally loaded. The interaction of the several slab strips and beams is usually such as to make the effective slab loading less severe than if each set of loads acted separately on the individual members.

When a heavy wheel rolls over a plank floor, each plank in turn must support the total load. In contrast, when a wheel moves over a concrete slab the wheel deflects the slab locally into a saucerlike pattern and this depression moves with the wheel across or along the slab. Thus a slab strip is deflected (and must be loaded) without a wheel actually resting on it. As the wheel passes over a particular strip the deflection increases, but the single 1-ft strip of slab never carries the entire wheel load unassisted. The designer describes this by saying the wheel load in Fig. 13.1*a* is distributed over an effective width E (Fig. 13.1*b*), meaning that the moment

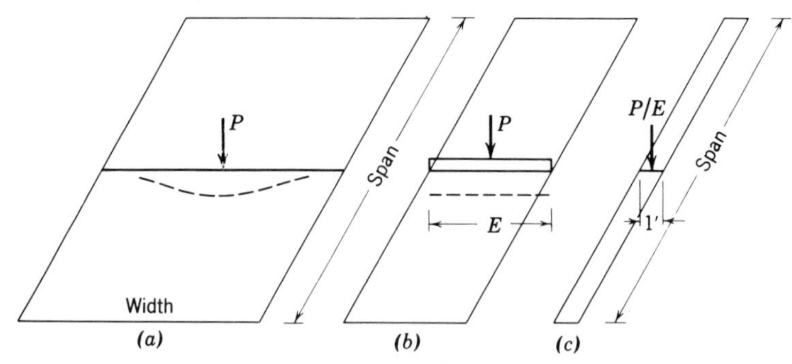

Fig. 13.1 Effective width under a single concetrated load. (*a*) Cross section through mid-span showing unequal deflections. (*b*) Design assumption of a width E with uniform deflections. (*c*) Equivalent 1-ft strip.

on the most heavily loaded 1-ft strip is that which would be produced by $1/E$ parts of the total load, as in Fig. 13.1*c*. Likewise, closely spaced beams share in carrying concentrated loads when the beams are connected by stiff floor slabs or stiff diaphragms.

The result of a theoretical study as to how a single wheel load is carried by a simple girder highway span is shown in Fig. 13.2. The load is applied

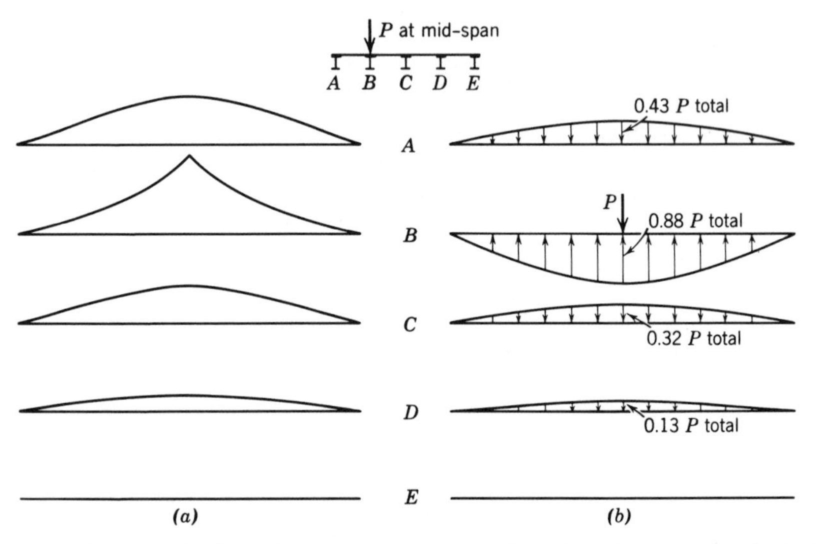

Fig. 13.2. The distribution of a single concentrated load to the girders of a bridge span, girder spacing of 0.1 span. Reproduced from Ref. 1, Highway Research Board. (*a*) Moment diagrams. (*b*) Approximate load distributions and the resulting total loads from the slab.

to mid-span, directly over beam B, and the girder stiffness assumed is five times that of the slab for a width equal to the girder span. Girder B then deflects more than its neighbors A and C. The slab (attached to the beams) is pulled down by beam B, but it resists this movement and exerts upward forces on the beam, as shown by the shaded ordinates on the right of the figure. These upward forces on beam B total $0.88P$, leaving the net downward load only $0.12P$. The resulting moment diagram on the beam is shown to the left. The neighboring beams deflect less and carry less moment than beam B. Actually, the slab imposes heavier net loads on these beams than on beam B, but the load of $0.43P$ on A and $0.32P$ on C is better distributed and produces less moment. The load on beam D is only $0.13P$ and the load on beam E is negligible.

Although such interaction of members can be approximately evaluated on a theoretical basis, design rules depend equally on field tests. This chapter will not attempt to demonstrate how load distribution factors are established nor to tabulate them for the many possible conditions. Rather its objective is to call attention to the problem of load distribution and illustrate how it can be handled in a few typical cases.

13.2 LOAD DISTRIBUTION IN A CONCRETE SLAB

Load distribution in a slab can be approached on two different bases which may be described as (1) the service load or deflection basis and (2) the ultimate strength or yield-line basis. The distribution at service loads is the one more commonly considered. For a wide slab having a 10-ft simple span, the effective width thus determined for a simple load is between 6 and 8 ft, depending somewhat on the size of the load contact area and the particular algebraic formula[2] used. For comparison, Johansen has shown[3] that at ultimate load, the effective width would be twice the span multiplied by $\sqrt{\mu}$, where μ is the ratio of the perpendicular top steel to the longitudinal bottom steel. If μ is about one-third, this would make the effective width $2 \times 10\sqrt{0.33} = 11.5$ ft.

One complication which may make calculations at ultimate strength uncertain is the shear capacity of the slab around the load. Richart and Kluge[4] found shear failures occurring from diagonal tension, with a truncated cone of concrete punched out below the load. When those shear stresses were calculated on a surface at a distance d beyond the load, the unit shear stress was low, in one series from $0.044f_c'$ to $0.057f_c'$. Since the shear failures came at loads 50% greater than those producing local yielding of the steel, these low shear stresses were not considered serious. For a yield-line analysis shear stresses around the load might be more significant.

It appears that the distribution based on elastic conditions, as commonly used, is on the safe side. Its use also tends to reduce crack size at working loads. For elastic conditions, Westergaard[5] established an extreme value of maximum positive moment on a slab as $0.315P$ for any simple span when P is distributed over a circular area with the diameter equal to one-tenth of the span, the slab thickness is one-twelfth of the span, and Poisson's ratio is 0.15. (This local moment is quite sensitive to the size of the bearing area.) The corresponding transverse moment is $0.248P$. Jensen[6] extended these results to show the effect of a rigid beam support at right angles, that is, an effect like that in a two-way slab. At this crossbeam the maximum negative moment is $-P/2\pi = -0.159P$ and it occurs with the wheel quite close to the beam.

When closely spaced multiple wheels occur, an extra slab width acts, but the effective width per wheel is reduced. The AASHO *Standard Specifications for Highway Bridges*[7] specify such an effective width E for a slab carrying a single wheel (traveling in the direction of the span) that the resultant design will be safe for multiple wheels without further calculations. Special transverse distribution steel is also specified as a percentage of the positive moment steel, in the amount $100/\sqrt{S}$ but not over 50%, where S is the span in feet.* When the wheels travel perpendicular to the span, the distribution steel is a larger percentage, $220\sqrt{S}$, but not over 67%.

13.3 CALCULATION FOR CONCENTRATED LOAD ON SLAB

A specification calls for a live load of 60 psf or a moving concentrated load of 2000 lb. Determine which loading controls for a continuous 10-ft span where the moment coefficients for uniform load are $+\frac{1}{16}$ and $-\frac{1}{12}$, and for a concentrated load $+\frac{1}{8}$ and $-\frac{1}{8}$. Consider $E = 0.68S + 2c$, where S is the span and c the diameter of the loaded area.[2]

Solution

For uniform live load, $M_{ul} = 1.7(-60 \times 10^2/12) = -850$ lb-ft.
For concentrated load, with $c = 0$ in the absence of better data,

$$E = 0.68 \times 10 + 0 = 6.8 \text{ ft}$$

Effective load, $P_e = 2000/6.8 = 294$ lb/ft strip

$$M_{ul} = 1.7(-P_eS/8) = 1.7(-294 \times \tfrac{10}{8}) = -624 \text{ lb-ft} < 850 \text{ ft-lb}$$

Therefore the uniform load moment governs the slab design.

*The AASHO notations S and E will be retained in this chapter.

It might be noted that the effective width for shear would call for the concentrated load near the support which would give less slab deflection and a much reduced effective width. The AASHO specification says that slabs designed for moment will be considered safe in bond (development length) and shear. As an extreme assumption, consider the entire load resisted by a 1-ft strip and the design shear existing at a distance *d* from the support. Assume also that a slab having $h = 4.5$ in., $d = 3.5$ in., and a weight of 56 psf.

$$V_u = 1.7 \times 2000 \times 9.71/10 + 1.4 \times 56(5 - 0.29) = 3670 \text{ lb}$$

$$v = V/bd^* = 3670/(12 \times 3.5) = 88 \text{ psi} < 2\sqrt{f_c'} = 2\sqrt{3000} = 110 \text{ psi}$$

Although there are limited data as to reaction distribution, it is difficult to imagine a diagonal tension failure which would involve less than a width of four to five times the slab thickness, in the assumed case at least 18 to 22 in.

13.4 HIGHWAY BRIDGE LOADINGS

The basic units of loading for highway bridges are the H truck, a two axle loading, and the H-S truck with trailer, a three-axle loading. The 20-ton truck is designated as H20-44, the last number denoting the year 1944 when this loading was established. Figure 13.3 shows the distribution of the load between the various wheels. The corresponding truck and trailer combination HS 20-44 represents the standard 20-ton truck plus a second 16-ton axle, as shown in Fig. 13.4. For either type of truck an alternate uniform load of 640 plf for each lane plus a concentrated load of 18,000 lb for moment or 26,000 lb for shear is to be used wherever it gives larger values, which will be the case on longer spans. The lane load occupies a 10 ft width and is to be placed in the worst position when actual lanes are wider. The reader is referred to the AASHO specification[1] for details such as the number of loaded lanes and proper reduction factors, impact allowance, and distribution of loads. In the following examples, only the governing portion of the specification will be mentioned. The AASHO notation has been retained where no equivalent has yet been set in the new ACI notation.

*The AASHO specification for service loads still uses the older equation $v = V/(bz)$ and an allowable of $0.03f_c'$. The ACI Code used here is considered the more modern in this area.

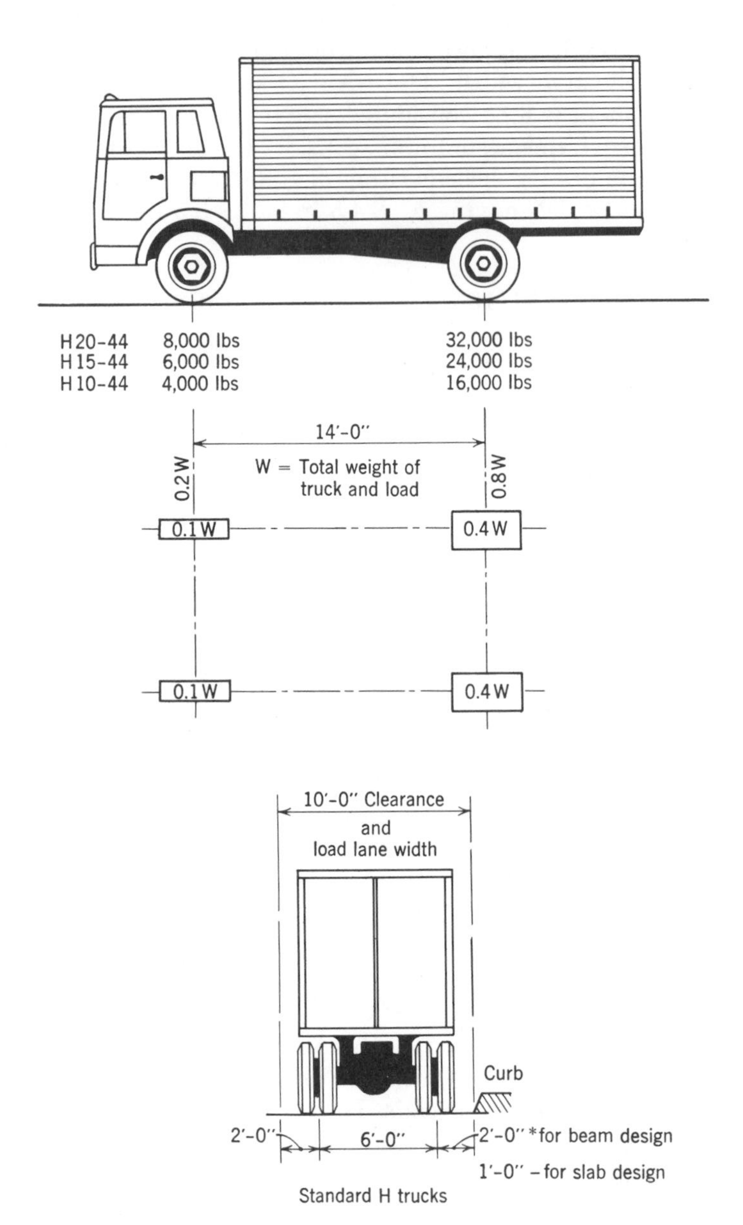

Fig. 13.3 The H truck of the AASHO *Standard Specifications for Highway Bridges*. For slab design the center line of wheels shall be assumed to be 1 ft from face of curb.

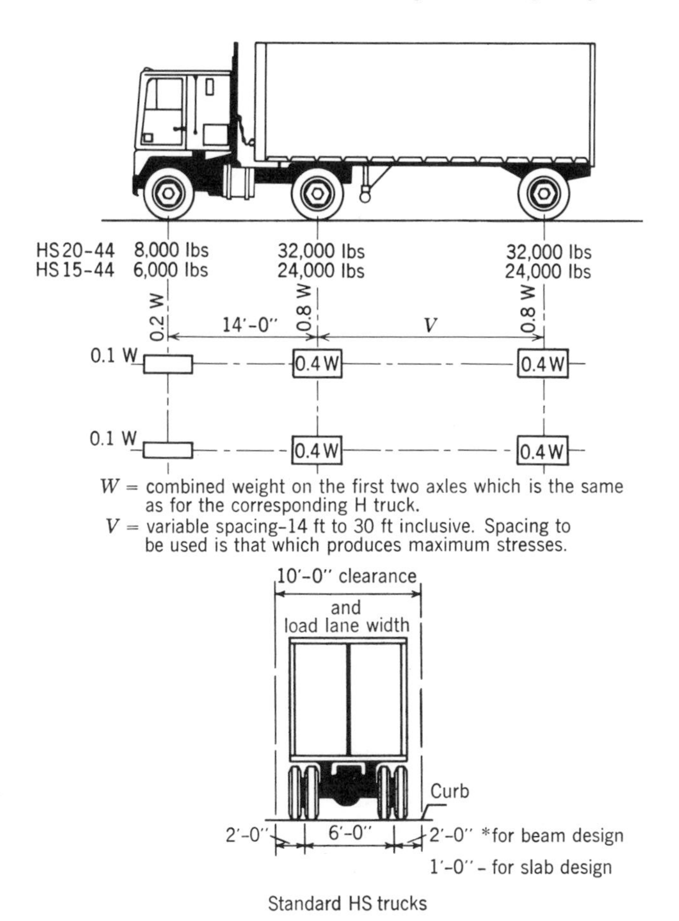

Fig. 13.4. The H-S truck of the AASHO *Standard Specifications for Highway Bridges*. For slab design the center line of wheels shall be assumed to be 1 ft from face of curb.

13.5 DESIGN OF A HIGHWAY SLAB SPAN

Design an interior panel of an 8 ft 6 in. span continuous slab for H20-44 loading, $f_c' = 3000$ psi, Grade 40 steel, AASHO specification. Consider that this slab spans longitudinally between transverse girders.* Use service load design method of Appendix C (Sec. C.9).

*More typically, highway slabs span transversely between longitudinal girders. Unfortunately for the purpose of this text (showing generally how to handle concentrated loads) AASHO for that case gives resultant moment formulas that do not show the effective width E used. That E appears to lie in the range of 6.3 to 7 ft for the two wheels (full-axle load) on the strip.

Solution

The AASHO specification allowable stresses are $f_c = 0.40f_c' = 1200$ psi, $n = 10$, $f_s = 20{,}000$ psi.

$$c_b = \frac{1200}{1200 + 20{,}000/10} d = 0.375d, \qquad z_b = d - c_b/3 = 0.875d$$

$$k_b = (0.5f_c bc)z/bd^2 = 0.5 \times 1200 \times 0.375d \times 0.875d/d^2 = 197$$

Since the reinforcement is parallel to traffic, this slab falls under Case B in the specification with an effective width $E = 4 + 0.06S$ where S is the span in ft.

$$E = 4 + 0.06S = 4 + 0.06 \times 8.5 = 4.51 \text{ ft} < 7.0 \text{ ft max.}$$
$$P = 0.4 \times 20 = 8 \text{ tons} = 16{,}000 \text{ lb}$$
$$P_e = P/E = 16{,}000/4.51 = 3530 \text{ lb}$$

If the moment for the concentrated load is taken as 80% of the simple span moment,

$$M_L = 0.8P_e S/4 = -0.20 \times 3530 \times 8.5 = -6000 \text{ ft-lb*}$$

The impact fraction is

$$I = \frac{50}{S + 125} = \frac{50}{8.5 + 125} = 0.375$$

but not to exceed 0.30, which governs here.

Impact $M_I = -0.30 \times 6000 = -1800$ ft-lb

Assume dead load $w_d = 88$ psf and moment coefficients of $-1/12$ and $+1/16$.

$$M_D = -88 \times 8.5^2/12 = -528 \text{ ft-lb}$$
$$M_T = -6000 - 1800 - 528 = -8330 \text{ ft-lb}$$
$$d = \sqrt{M/k_b b} = \sqrt{8330 \times 12/(197 \times 12)} = \sqrt{42.2} = 6.50 \text{ in.}$$

Since ASSHO specifies a minimum of 1.00 in. of cover

$$h = 6.50 + d_b/2 + 1.00 = 7.50 + 0.38 \text{ (say)} = 7.88 \text{ in.}$$

USE $h = 8$ in., $d = 8.0 - 1.00 - d_b/2 =$ say 6.56 in. (for #7 bars)

$$\text{Revised } M_D = -100 \times 8.5^2/12 = -602 \text{ ft-lb}$$

*The specification gives the approximate value of live load M for a simple span as $900S = 900 \times 8.5 = 7650$ ft-lb for a simple span which would be 6120 ft-lb on the 80% basis.

Neg. $M_T = -6000 - 1800 - 602 = -8400$ ft-lb

Pos. $M_T = +6000 + 1800 + 602 \times 12/16 = +8250$ ft-lb

$$\text{Neg. } A_s = \frac{8400 \times 12}{20{,}000 \times 0.875 \times 6.56} = 0.875 \text{ in.}^2/\text{ft} = \#7 \text{ at 8 in. } (0.90)$$

or #6 at 6 in. (0.88)

Pos. $A_s = 0.875 \times 8250/8400 = 0.858$ in.^{2}/ft $=$ #6 st 6 in.

USE #6 at 6 in. for both positive and negative M.

This design ignored the possibility of using compression steel which would be available at the supports but possibly not at mid-span. AASHO also says slabs designed for moment "shall be considered satisfactory in bond and shear."

The concentrated load requires distribution steel perpendicular to the moment steel. The amount is specified as a percentage of the positive moment steel given by $100/\sqrt{S} = 100/\sqrt{8.5} = 34.3\% < 50\%$ maximum. Transverse steel must be at least $0.343 \times 0.858 = 0.294$ in.^{2}/ft $=$ #5 at 12 in. (0.310).

13.6 DESIGN MOMENTS AND SHEARS FOR HIGHWAY GIRDER

Calculate the design live load and impact load moments and shears for a 40-ft simple span girder built integrally with a slab. Consider both H20-44 and H20-S16-44 loadings on the center girder of the cross section shown in Fig. 13.5*a*.

Solution

For shear calculations the AASHO says there shall be no lateral or longitudinal distribution of the wheel load at the end of the span. For other wheels the distribution shall be that applying for moment. The provision for moment when two or more lanes of traffic are involved calls for the wheel loads to each stringer to be $S/6.0 = 9.75/6.0 = 1.63$ of the full load.*

The longitudinal arrangement of wheels for the truck and uniform lane loads are shown in Fig. 13.5*b*, ignoring any complication which the end diaphragm may cause. Figure 13.5*c* shows the lateral arrangement of the loads on the cross section, two trucks assumed to be passing. For the truck the simple beam slab reactions would give for the end wheels:

$$(2 \times 2 \times 4.75/9.75)16 = 1.949 \times 16\text{k} = 31.2\text{k}$$

*For $S > 10$ ft, the load to the beam is the reaction computed from "simple span" slabs.

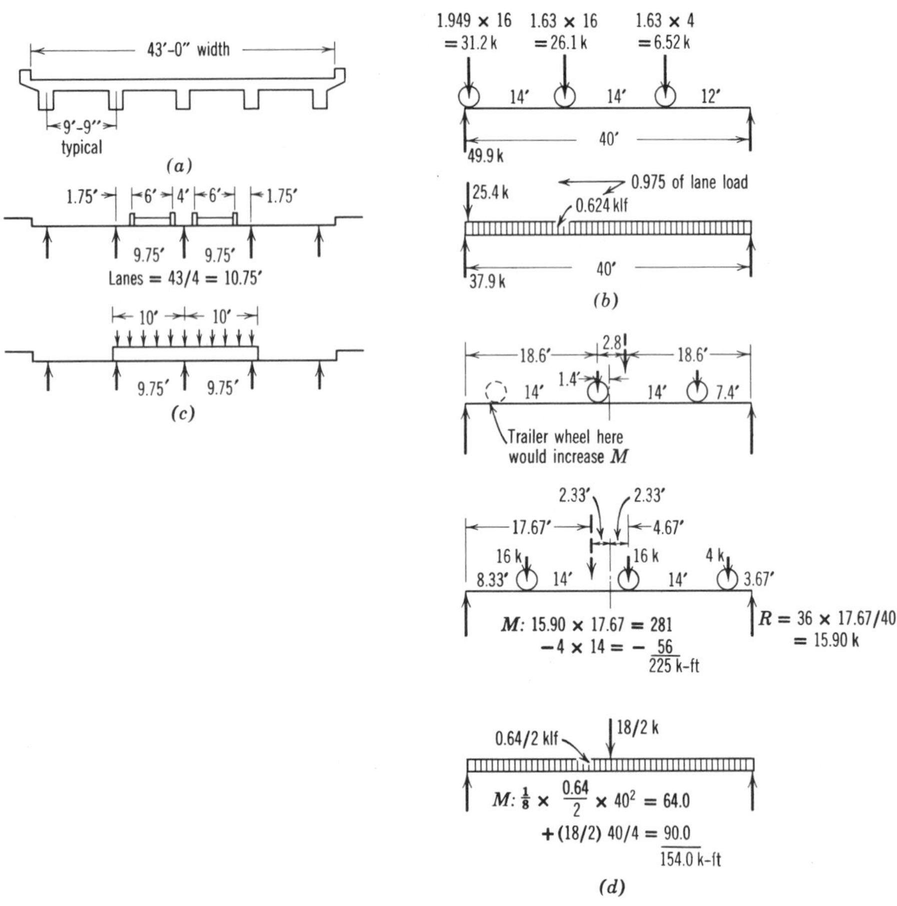

Fig. 13.5. Highway girder bridge. (*a*) Bridge cross section. (*b*) Location of wheels and lane load for maximum shear. (*c*) Position of loads laterally for maximum stresses in center girder. (*d*) Location of wheels and lane load for maximum moment.

It is only by chance that this multiplier of 1.949 is so nearly identical with the 1.95 factor based on the distribution, for moment.

For the lane load of Sec. 13.4, the end concentrated load for shear becomes $2 \times 0.5 \times 26 \times 9.75/10 = 25.4$ k and the uniform load $0.64 \times 9.75/10.0 = 0.624$ klf. For the trucks, maximum $V = 31.2 + 26.1 \times 26/40 + 6.52 \times 12/40 = 50.2$ k. For the lane load, maximum $V = 25.4 + 0.624 \times 40/2 = 37.9$ k.

Max $V_L = 50.2$ k*

*AASHO designs for shear *at* the face of support rather than *d* from the support.

By specification, L is from the shear point to the far reaction, that is, the full 40-ft span.

$$I = \frac{50}{40 + 125} = 0.303$$

Since $I = 0.30$ governs, $V_I = 0.30 \times 50.2 = 15.1$ k

For moment three loadings must be investigated, as worked out in Fig. 13.5*d*. For simplicity these loadings are compared on the basis of one line of wheels or one-half lane load. The HS 20-44 loading governs maximum moment. Since the number of lines of wheels is $S/b = 1.63$,

$$M_L = 1.63 \times 225 = 367 \text{ k-ft}$$

The loaded length for moment, by definition, is the full length, $L = 40$ ft.

$$I = \frac{50}{40 + 125} = 0.303$$

Since $I = 0.30$ governs, $M_I = 0.30 \times 367 = 110$ k-ft

The above illustrates the manner specified by AASHO for handling load distribution on stringers or longitudinal girders. The remainder of the design of the T-beams would be very similar to that of a building T-beam with these minor differences, which are usually more restrictive:

1. Different allowable unit stresses for f_c, v, and a bond calculation as in 1963 ACI Code.
2. Different rules on effective flange width.
3. Different maximum moment diagram, due to type of loading.
4. Different maximum shear diagram, due to type of loading.
5. Webs of T-beams must carry stirrups full length, at a maximum spacing not to exceed 0.75*d* where not required for stress.
6. Slightly different bar extensions.

All of these are the small differences which can be expected in changing from one specification to another.

13.7 OPENINGS IN SLABS

When openings in slabs are large, as for stairs or elevators, beams must be used around the openings. Good practice usually requires that such beams be framed into columns sufficiently to provide a stable unit without the slab.

Small openings such as pipe sleeves, if not too numerous, can be made almost anywhere in a slab, except adjacent to the columns in flat slab construction. What can be done about larger openings may require at least some rough calculations. Obviously, openings are least dangerous where shear stresses are small and bending moments are below maximum.

Electric conduits, unless closely spaced or crossing at small angles, can be included without considering any loss in moment strength. The detailed requirements of Code 6.3 might be noted in this connection.

The effect on shear of openings around columns in flat plate construction was discussed briefly in Sec. 4.6b and illustrated in Fig. 4.6. These shear provisions recognize the coexistence of large bending moments.

An equally common and troublesome problem is that of relatively large openings which interrupt the normal flexural action. Openings of any size are permitted by Code 13.6.1 if analysis shows that both strength and deflection are still acceptable. For strength the methods of Chapter 12 on the strip method seem appropriate even if they might appear rather arbitrary.

Without analysis considerable leeway is still given by Code 13.6.2:

> **13.6.2**—Openings conforming to the following requirements may be provided in slab systems not having beams without special analysis as required in Section 13.6.1.
>
> (a) Openings of any size may be placed in the area within the middle half of the span in each direction, provided the total amount of reinforcement required for the panel without the opening is maintained.
>
> (b) In the area common to two column strips, not more than one-eighth of the width of strip in either span shall be interrupted by the openings. The equivalent of reinforcement interrupted shall be added on all sides of the openings.
>
> (c) In the area common to one column strip and one middle strip, not more than one-quarter of the reinforcement in either strip shall be interrupted by the opening. The equivalent of reinforcement interrupted shall be added on all sides of the openings.
>
> (d) . . . . (refers to shear)

In two-way slabs supported on all four sides, openings in the corners of the slab are least damaging. Since for architectural reasons openings for ducts frequently need to be near the columns, this is a definite advantage of this type of slab. The negative moment zone of the short middle strips (near mid-span of the longer beams) is the least favorable zone for openings.

Often rough checks can be made on the strength of the construction after the openings are located. If openings should reduce a critical design section for moment, the required bd^2 for moment must be maintained by providing extra depth to offset the reduced width. The steel may be more

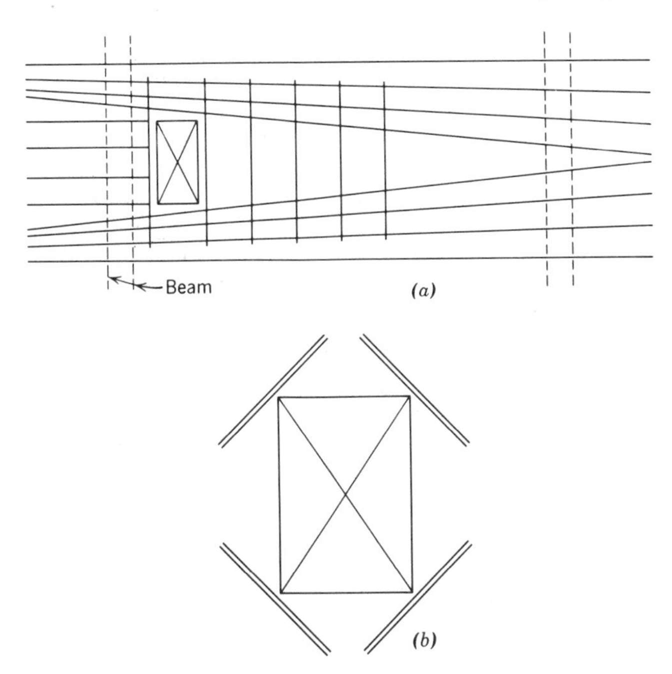

Fig. 13.6. Bars at openings in slabs. (*a*) Bars fanned out to miss opening. (*b*) Corner reinforcement to reduce shrinkage cracking.

closely spaced on each side of the opening to maintain the necessary A_s. Often it will be possible to locate openings where moment is well below the compression capacity of the slab, thereby leaving the arrangement of reinforcement as the only problem. Of course, shear strength must be maintained, but this is rarely a problem except near the columns in flat slab types, as noted above.

The arrangement of bars around any but minor openings can constitute a real problem. Bars running perpendicular to the face of an opening are not fully effective when simply cut off at the opening. This would be all right if there were a beam at the opening to act as a reaction for the slab. If there is no beam it is better to fan the bars out or splay them to go around the opening, as shown in Fig. 13.6*a*. If this leaves too wide an area without steel, extra bars can be placed parallel to the side of the opening, as indicated.

If minor cracking at the corners of an opening is objectionable, it is always well to add one or two diagonal bars at each corner, especially at large openings (Fig. 13.6*b*). This is always desirable around window and door openings in concrete wall slabs. Such reinforcement helps to take care of shrinkage stresses.

13.8 OPENINGS IN BEAMS

Openings through slabs which encroach on the flange width of T-beams require a check on the bending strength remaining.

Openings through a beam web are increasingly important for installation of building services. In a region of small shear, as near the middle of a beam span, a horizontal pipe sleeve should not be serious. Elsewhere, shear strength must be closely watched and in many places bending strength as well. Large openings in beams are particularly weakening. They destroy beam action and force this reduced section to act much as a Vierendeel truss (a truss without diagonals). In such a truss the average bending moment over the length of the opening is resisted by axial compression in one chord and tension in the other, these two forming a couple in the case of pure flexure. Where shear is present the change in the moment over the length of the opening superimposes a reversed bending resistance in each chord, the total of the four end moments on the chords equaling the external shear times the length of the opening. How the shear and these reversed moments are shared by the two chords depends on the relative chord stiffness. The designer in such a case should note that for members with significant axial tension Code 11.4.4 requires all shear to be resisted by stirrups (none assigned to the concrete).

SELECTED REFERENCES

1. C. P. Siess and A. S. Veletos, "Distribution of Loads to Girders in Slab-and-Girder Bridges: Theoretical Analyses and Their Relation to Field Tests," Highway Research Board *Report 14-B*, Washington, 1953, p. 58.
2. Clyde T. Morris, "Concentrated Loads on Slabs," Ohio State Univ. Eng. Exp. Sta. *Bull. No. 80*, 1933.
3. K. W. Johansen, "Bruchomente Der Kreuzweise Bewehrten Platten" (Moments of Rupture in Cross-Reinforced Slabs), International Association for Bridge and Structural Engineering, Liége, Vol. 1, 1932, p. 277.
4. Frank E. Richart and Ralph W. Kluge, "Tests of Reinforced Concrete Slabs Subjected to Concentrated Loads," Univ. of Ill. Eng. Exp. Sta. *Bull. No. 314*, 1949.
5. H. M. Westergaard, "Computation of Stresses in Bridge Slabs Due to Wheel Loads." *Public Roads*, 11, No. 1, Mar, 1930, p. 1.

6. Vernon P. Jensen, "Solutions for Certain Rectangular Slabs Continuous Over Flexible Supports," Univ. of Ill. Eng. Exp. Sta. *Bull. No. 303*, 1938.
7. *Standard Specifications for Highway Bridges*, AASHO, Washington, 10th ed., 1969.

PROBLEMS

NOTE: *The service load method of Appendix C is to be used for all these highway bridge type problems.*

Prob. 13.1. Design a simple span slab to carry its own weight and a single 20-k concentrated load over a 20-ft span. Assume the impact as 25% of live load and an effective width $E = 5.6$ ft. Allowable $f_c = 1200$ psi, $f_s = 20{,}000$ psi, $n = 10$, cover of 1.25 in. to center of steel. Assume the load is distributed over a sufficient bearing area to avoid a punching shear failure.

Prob. 13.2. An interior span of a continuous slab with a clear span of 13 ft is to be designed for a live load of 75 psf or a single concentrated live load of 3 k, whichever is worse. If the effective width assumed for the concentrated load is 6 ft when the load is at its worst position for moment, design the slab using the moment coefficients of Sec. 8.5 for uniform load and $\pm 0.20\ Pl$ for the concentrated load. Assume zero impact.

Prob. 13.3. Design a T-beam for a 28-ft simple span assuming a 6-in. slab, beams 7 ft 6 in. on centers, $f_c' = 3000$ psi, allowable $f_c = 1200$ psi, $f_s = 20{,}000$ psi, $n = 10$, 2.5 in. to center of steel (if in one layer). Each beam may be assumed to carry 1.50 lines of wheels for moment and 1.67* lines for shear. Use H20-44 loading with 30% of live load for impact. The stem weight below the slab may be assumed as 250 plf and this need not be revised for this problem. Sketch the beam cross section showing the steel arrangement.

Prob. 13.4. Same as Prob. 13.3 but with a 40-ft span. In this case assume the stem weight below the slab as 400 plf.

*The AASHO *Standard Specifications for Highway Bridges*[7] uses a different factor E for load on the end of span from that used for loads out on the span. The student may ignore this and, for these problems, use this factor for each wheel.

14
Columns — Axial Load Plus Bending

14.1 THE PRACTICAL COLUMN PROBLEM

All practical columns are members subject to both direct load and moment. This chapter covers the range from columns having moments relatively small to members with moment predominant and axial load relatively small.

Creep and shrinkage strains are important in column behavior and actual stresses under service conditions can only be estimated. Nevertheless the strength of a given cross section at failure is rather definite. Column length effects complicate the design moment to be used and this accounts for part of the lower ϕ factor (capacity reduction factor) which is specified, 0.70 or 0.75 as noted below. The other basic reason for the low factors lies in the structural importance of a column; it carries a larger floor area than does a beam.

14.2 TYPES OF COLUMNS

Plain concrete is not used for columns, but may be used for pedestals in which the height does not exceed three times the least lateral dimension (Fig. 14.1*a*).

Reinforced concrete columns normally contain longitudinal steel bars and are designated by the type of lateral bracing provided for these bars.

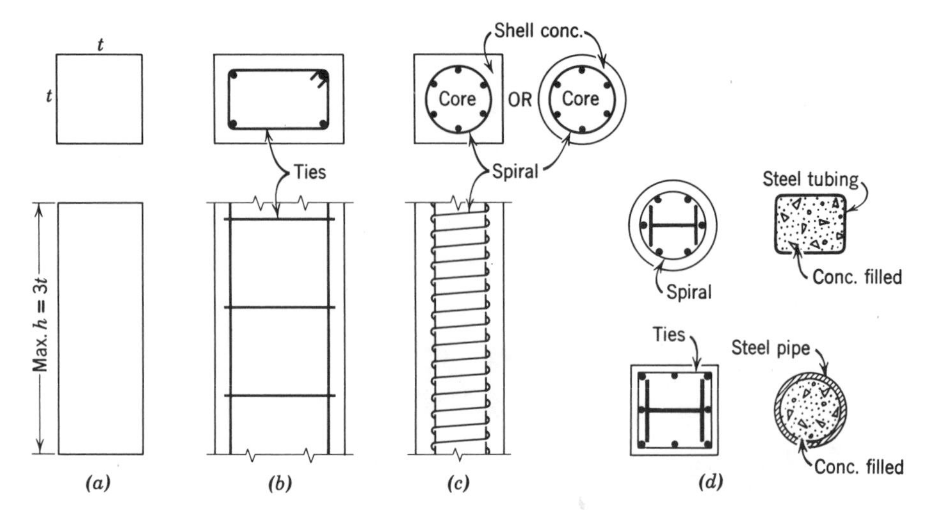

Fig. 14.1. Types of columns. (*a*) Plain concrete pedestal. (*b*) Tied column. (*c*) Spiral column. (*d*) Composite columns, four types.

Tied columns (Fig. 14.1*b*) have the bars braced or tied at intervals by closed loops called ties. *Spiral columns* have the bars (and the core concrete) wrapped with a closely spaced helix or spiral of small diameter wire or rod (Fig. 14.1*c*).

Composite columns may contain a structural steel shape surrounded by longitudinal bars with ties or spirals or may consist of high strength steel tubing filled with concrete or a steel pipe so filled.*

Tied and spiral columns are the most common forms. Either may be made circular, octagonal, square, or rectangular in cross section, as desired. Tied columns may also be of L shape.

14.3 COLUMN TESTS

For at least 40 years it has been evident that in a reinforced concrete column under sustained axial load one could not calculate f_c, the actual unit stress on the concrete, nor f_s, the actual unit stress on the steel. If the materials were really elastic it would be possible to use the transformed area (Chapter 6) to establish these stresses. However, actual observations show that the steel stress is much larger than this calculation would indicate, because of both shrinkage and creep of the concrete under load.

*The 1971 Code no longer assigns different names to the several kinds of composite columns.

Starting about 1930, a very large research project on columns was carried out at the University of Illinois[1] and at Lehigh University.[2] These tests indicated clearly that even under axial load alone there was no fixed ratio of steel stress to concrete stress in the ordinary column. The ratio of these stresses depended on the amount of shrinkage, which in turn depended on the age of the concrete and the method of curing. It also depended on the amount of creep in the concrete. Creep is greater when the load is applied at an early stage of the hardening or curing process. The amount of creep is influenced by any of the factors which determine the quality of concrete, such as cement content, water content, curing, and even type of aggregate used.

A load applied for only a short time, such as the ordinary live load, causes very little creep, especially after the concrete is well cured. The usual live load thus produces an increment or increase of stress in steel and concrete which can be calculated reasonably well by transformed area methods. However, stresses produced by dead load or any permanent or semipermanent load depend on the entire history of the column. It is even possible to have a loaded column with tension in the concrete and compression in the steel under very special circumstances (such as a large percentage of steel and a heavy initial loading which is later greatly reduced in amount).

Historically, in the United States, these tests initiated the slow switch in emphasis from service load to ultimate strength for columns. They showed that *ultimate* column strength did *not* vary appreciably with the history of loading. If, as loading was increased, the steel reached its elastic limit first, the increased deformation then occurring built up stress in the concrete until its ultimate strength was reached. If the concrete approached its ultimate strength before the steel reached its elastic limit, the increased deformation of the concrete near its maximum stress forced the steel stress to build up more rapidly. Thus, regardless of loading history, a column reached what might be called its yield point only when the load became equal to approximately 85% of the ultimate strength of the concrete (as measured by standard cyclinder tests) plus the yield-point strength of the longitudinal steel. The 85% factor for the concrete is probably due to less ideal compaction of concrete in columns (around the steel) than in cylinders, together with the reduction in apparent strength caused by the slower application of load and the longer specimen.

Up to the column yield point, tied columns and spiral columns act almost identically and the spiral adds nothing measurable to the yield-point strength. The stress-strain curves for the tied columns and the spiral column up to this point are essentially identical, similar to Fig. 14.2.

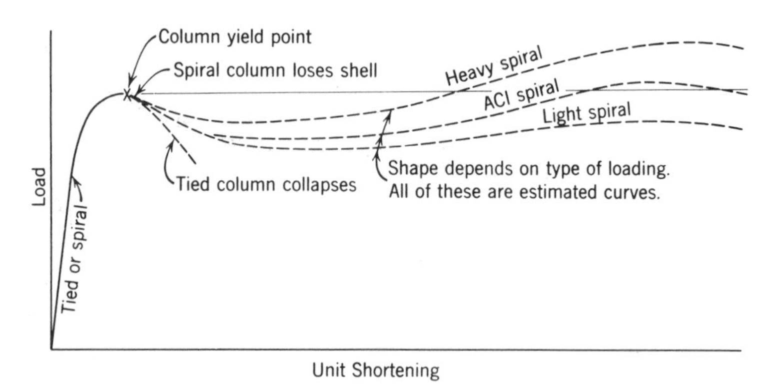

Fig. 14.2. Comparison of strains in tied and spiral columns.

After the yield-point load is reached, a tied column immediately fails with a shearing diagonal failure of the concrete (as in a test cylinder) and a buckling failure of the column steel between ties as shown in Fig. 14.3. The yield point and ultimate strength of a tied column are thus the same thing. In a spiral column, the yield-point load results in cracking or complete destruction of the shell of concrete outside the spiral (Fig. 14.4). The spiral comes into effective action only with the large increased deformation which follows yielding of the column and loss of the shell concrete. At the stage shown in the Fig. 14.4 columns, the spiral provides radial compressive forces on the concrete within the core of the column and these confining stresses add significantly to the load the core concrete can carry. Pound for pound, steel in the spiral has been found to be from 2.0 to 2.4 times as effective as longitudinal steel in contributing to the ultimate strength of the column. The spiral steel never becomes significantly effective until after the destruction of the shell concrete which covers it. Moreover, excessive longitudinal column shortening is then involved (Fig. 14.2) which makes this spiral steel of questionable value except as a factor of safety against complete collapse.

A heavy spiral can add more strength to the column than that lost in the spalling or failure of the shell, in which case the column will carry an ultimate load greater than the yield-point load, but with unsuitable shortening. If too light a spiral is used, the column will continue to carry some load beyond the column yield point, but not as much as that which caused the spalling of the shell. The ACI Building Code specifies that amount of spiral steel which will just replace* the strength lost when the

*Actually, an estimated 10% in excess of the shell strength is used just to be sure the strength after spalling is not less than before. See Sec. 14.28 for the design of a spiral.

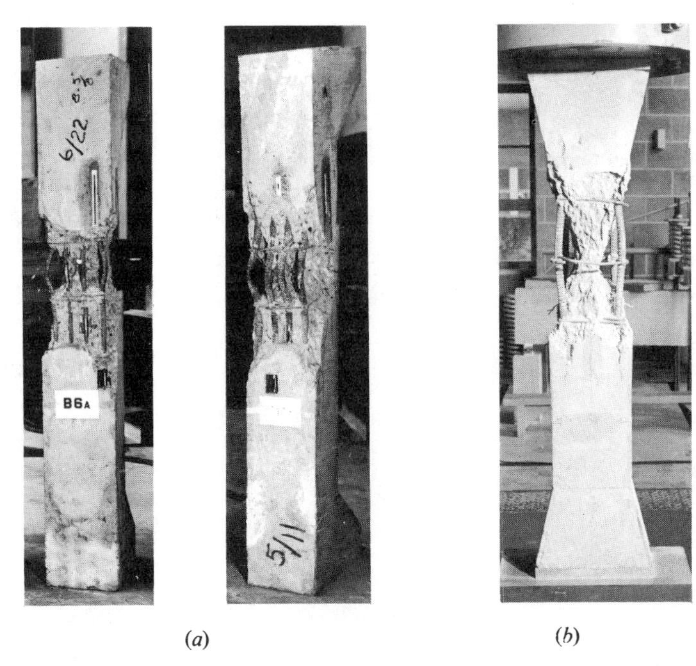

Fig. 14.3. Failure of tied columns. (*a*) Note the bars buckled between ties. Column height-thickness ratio $l_u/h = 7.5$. Special ends were cast to permit comparative tests with eccentric loads. (From Ref. 3, Univ. of Ill.) (*b*) Column in which a tie seems to have failed after yield point of column was reached. (Courtesy Portland Cement Assn.)

shell concrete spalls. The initial cracking of the shell gives some warning of overload prior to failure. The spiral also adds a considerable element of toughness to the column. Toughness is valuable in resisting explosion or blast, since it measures the energy that can be absorbed. Two columns from the same story of a building severely damaged by a recent strong earthquake (Fig. 14.5) show that only the heavily damaged spiral columns prevented a total collapse of this story.

More recently emphasis has been given to the fact that to some extent column ties also confine the concrete, although their shape makes them much less efficient than spirals. Heavy ties on columns or around compression steel in beams can establish a considerable degree of toughness in these members. Hence ties are an important requirement in earthquake resistance and limit design where members are expected to maintain their peak resistance while forming the so-called plastic hinges (Chapter 9). Code Appendix A on seismic resistance calls for heavy ties in columns where they go through the beams, which would have helped the joint above the spiral column of Fig. 14.5.

(a) (b) (c)

Fig. 14.4. Spiral column tests under concentric loads. (From Univ. of Illinois tests, Refs. 3, 4, and 5.) (*a*) Failure of 32-in. diameter column; $l_u/h = 6.6$. (*b*) Failure of 12-in. diameter column; $l_u/h = 7.3$. Shell has completely spalled off. (The special ends were cast to permit comparative tests with eccentric loads.) (*c*) Failure of column with thin cover or shell; $l_u/h = 10.0$.

Fig. 14.5. Columns almost destroyed in severe earthquake. Notice that spiral column still has its core acting.

INTERACTION OF AXIAL LOAD AND MOMENT IN SHORT COLUMNS

14.4 DEFINITIONS AND BASIC LIMITATIONS

A short column is one where the length effect or deflection response under load is a very small and is considered negligible. The ACI Column Committee estimates that 90% of all braced columns and 40% of all unbraced columns can be designed as short columns. The maximum length that can be considered a short column depends on its deflected shape and can be defined in a more meaningful way after the discussion of long columns (see Sec. 14.19). The general discussion of long columns starts with Sec. 14.12.

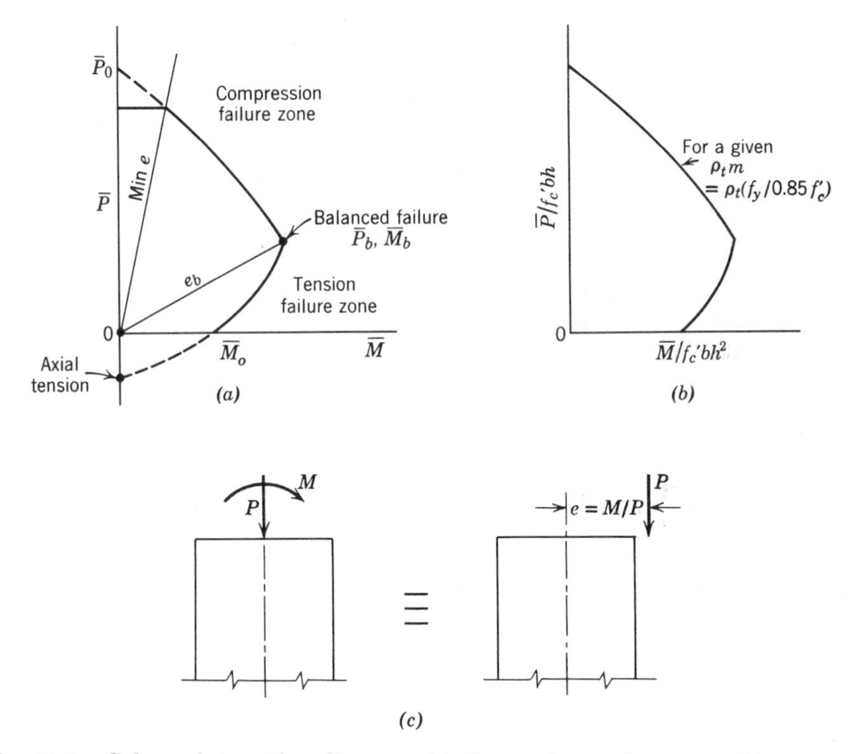

Fig. 14.6. Column interaction diagram. (*a*) For a given column. (*b*) Dimensionless form. (*c*) Equivalent eccentricity.

The axial load capacity decreases when moment is also present. A plot of the column load capacity against the moment it can carry is called a column interaction diagram. Schematically, Fig. 14.6*a* shows such a diagram, with key points and areas noted, for later discussion. Any loading which plots within this area is a possible loading; any combination outside the area represents a failure combination. A radial line from Point *O* represents a constant ratio of moment to load, that is, a constant eccentricity of load.

The interaction diagram can be put into dimensionless form as indicated in Fig. 14.6*b* for a given steel ratio and arrangement of steel, which facilitates assembly of graphs into groups for design charts. The 1970 ACI charts[11] also included a ϕ of 0.70 (or 0.75), which can be confusing since ϕ increases above these values at small axial loads (Fig. 14.9).

Since all concrete columns are subject to some moment, the Code requires all columns to be designed for a minimum eccentricity even if the computed moment is less. The minimum eccentricity is to be at least 1 in.,* or 0.10 of the column thickness about each axis (separately) for tied columns, or 0.05 of the column thickness for spiral columns. Thus all column design includes moment. The actual maximum $e = M/P$ (Fig. 14.6*c*) may exceed these minima and control the design.

All columns are required (Code 10.9.1) to contain longitudinal bars sufficient to make the steel ratio, $\rho_g = A_s/hb$, at least 0.01, because of the shrinkage and creep stresses on smaller areas, and ρ_g must not exceed 0.08. At 0.08 crowding in the member is very severe. At least 6 bars must be used when in a circular arrangement or 4 bars in a rectangular arrangement.

Ties (Code 7.12.3) must be at least #3 size for #10 bars or smaller and #4 for #11 or larger and for all bundled bars. Their spacing shall not exceed 16 bar diameters, 48 tie diameters, or the least column dimension. Every corner bar and every alternate bar must be braced by the tie and no bar shall be more than 6 in. from such a laterally supported bar.

Spirals must fully replace the strength lost when the shell of concrete outside the spiral spalls off. Spiral design is covered in Sec. 14.28 of this chapter.

14.5 AXIAL LOAD CAPACITY, $\overline{P}_0$

Although in design, axial load without moment is not a practical case, $\overline{P}_0$ is a convenient theoretical limit and one well documented experimentally.

The tests discussed in Sec. 14.3 established the ultimate strength of either a tied or a spiral column, axially loaded, as:

$$\begin{aligned}\overline{P}_0 &= 0.85\, f_c' A_n + f_y A_{st} \\ &= 0.85\, f_c'(A_g - A_{st}) + f_y A_{st} \qquad (14.1)\end{aligned}$$

where $\overline{P}_0$ = ultimate load capacity (yield-point strength) of tied or spiral column when eccentricity is zero (for ideal materials and dimensions)

A_n = net area of concrete = $A_g - A_{st}$

A_g = gross area of concrete, in.^{2}

A_{st} = area of vertical column steel, in.^{2}

f_c' = standard cylinder strength of concrete, psi

f_y = yield point stress for steel, psi

This is the ideal strength and the Code would consider the dependable strength P_0 equal to $\phi\overline{P}_0$, where ϕ would be 0.70 for a tied column and 0.75 for a spiral column.

*Code 10.3.6 permits a minimum of 0.6 in. for precast members provided that manufacturing and erection tolerances are limited to one-third as much.

Another extreme limit, generally only of theoretical interest, is the axial tension limit, which is $A_{st}f_y$ in magnitude, the concrete being fully cracked at such a limit.

14.6 BALANCED LOADING, $\overline{P}_b$, $\overline{M}_b$

A balanced section for a beam was defined in Chapter 2 as one developing simultaneously a concrete compression strain of 0.003 and a steel tension strain of f_y/E_s. For this condition a unique area of tension reinforcement was required in the beam.

For *any* column the same definition of balanced strains holds. *Any* column, regardless of its reinforcement, will reach its balanced ultimate load when the load is so placed as to maintain the eccentricity $e_b = \overline{M}_b/\overline{P}_b$. Balance in a column is a matter of *loading*, and it is more descriptive to speak of *balanced loading* rather than of a balanced column. Furthermore, although it is possible to avoid balanced beams in order to avoid compression failures and thus obtain ductility, it is not possible to avoid either compression failures or balanced failures in columns; these are primarily compression members.

For a given column it is very easy to establish the balanced load $\overline{P}_b$ for ideal conditions, and the accompanying e_b, after the fashion of Fig. 14.7. The tension steel A_s and the compression steel A_s' are each 3- #9 = 3.00 in.^{2}, $f_c' = 3000$ psi, $f_y = 60$ ksi. The maximum strain of 0.003 in compression and f_y/E_s give c from similar triangles, most simply by thinking of the large dotted triangle.

$$c_b = \frac{0.003}{0.003 + 0.00207} \times 17.5^* = 0.592 \times 17.5 = 10.35 \text{ in.}$$

As for beams, for $f_c' \lesseqgtr 4$ ksi, $a_b = 0.85c_b = 8.80$ in.

$$\overline{N}_{c1} = 0.85 \times 3000 \times 8.80 \times 18 = 403 \text{ k}$$

$$\epsilon_s' = \frac{10.35 - 2.50}{10.35} \times 0.003 = 0.00227 > \epsilon_y$$

$$\overline{N}_{c2} = 3.0(60 - 0.85 \times 3) = 172 \text{ k}$$

This value takes account of concrete displaced by steel. This correction is often omitted, but there seems to be no reason to do so when numbers are being used; algebraically, the correction is more awkward.

$$\overline{N}_t = 3.0 \times 60 = 180 \text{ k}$$

*Or this equation could be used in the form $c_b = \dfrac{87000}{87000 + f_y} d$

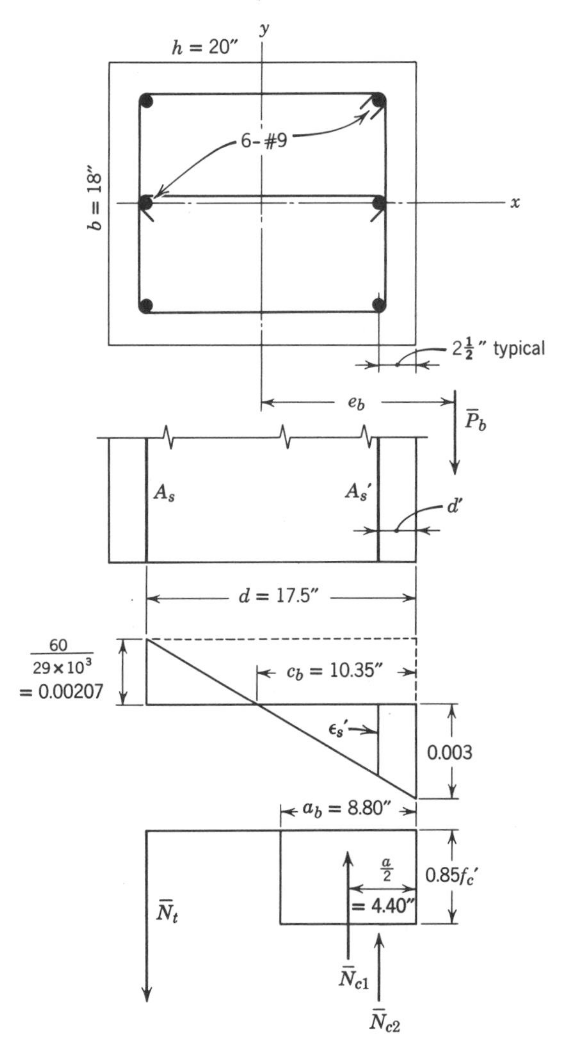

Fig. 14.7. Balanced column load.

These three forces are in equilibrium with $\overline{P}_b$.

$$\Sigma F_y = 0 = \overline{P}_b + \overline{N}_t - \overline{N}_{c1} - \overline{N}_{c2} = \overline{P}_b + 180 - 403 - 172$$

$$\overline{P}_b = 395 \text{ k}, \qquad P_b = \phi\overline{P}_b = 0.7 \times 395 = 277 \text{ k}$$

$$\Sigma M \text{ about center of column} = 0$$

$$\overline{N}_t \times 7.50 + \overline{N}_{c1}(10 - 4.40) + \overline{N}_{c2} \times 7.5 - \overline{P}_b e_b = 0$$

$$395 e_b = 180 \times 7.5 + 403 \times 5.60 + 172 \times 7.5 = 4890$$

$$e_b = 12.40 \text{ in.}$$

$$\overline{M}_b = \overline{P}_b e_b = 395 \times 12.40 = 4890 \text{ k-in.}$$

$$M_b = \phi\overline{M}_b = 0.7 \times 4890 = 3420 \text{ k-in.}$$

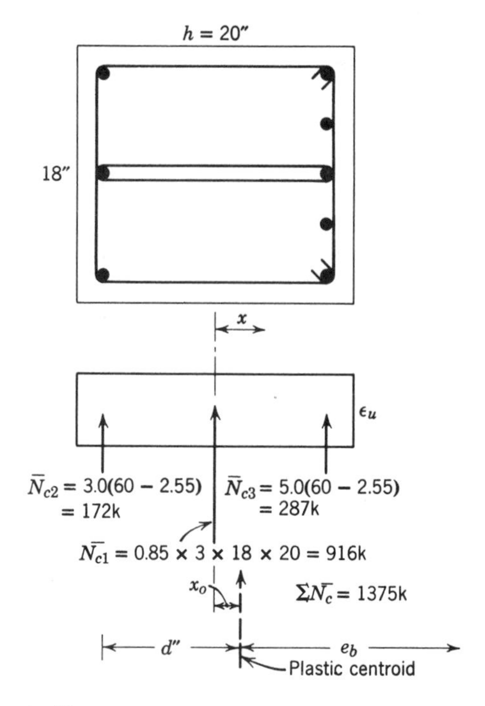

Fig. 14.8. Plastic centroid.

When a column is unsymmetrical as shown in Fig. 14.8, with A_s' of 5-#9 = 5.00 in.^{2}, the reference line for eccentricity is taken from the *plastic centroid*, which is simply the location of the resultant load which would give a uniform strain all across the column. The plastic centroid falls a distance x_0 off the center of column:

Force	x	M abt. center line	
$\overline{N}_{c1}$ = 916 k	0	0	
$\overline{N}_{c3}$ = 287	+7.5 in.	+2150	
$\overline{N}_{c2}$ = 172	−7.5	−1290	
$\Sigma\overline{N}_c$ = 1375		+ 860	x_0 = +860/1375 = +0.62 in.

The method for finding P_b and e_b is then not basically changed, but from a nomenclature point of view it is convenient to define by d'' the distance from this plastic centroid to the centroid of the tension steel.

If column steel were distributed along all four column faces, the numbers used in finding P_b would be increased by an additional term for each group of bars falling at different distances from the neutral axis, but no other complication would exist.

14.7 OTHER INTERACTION POINTS AND SPECIAL ϕ VALUES

For flexure without axial load, the point at the moment axis of the interaction diagram, the procedure used for the balanced load can be used unchanged except (1) that $\epsilon_s > \epsilon_y$, and (2) the neutral axis is unknown and must be located by trial-and-error such that the total tension equals the total compression. For curve plotting, this M_0 point need not be exactly established numerically because a solution with a very small tension resultant and another with a very small compression resultant establishes the curve across the moment axis. Since this is pure flexure, the ϕ here is 0.90.

Other points below the balanced loading also represent primary failures in tension with the tension steel past the yield strain and ϵ_c still 0.003. Plotting points are easily found with assumed c values $< c_b$.

Curve points above $\bar{P}_b$ can be established by using $c > c_b$ with $\epsilon_c = 0.003$. As c increases the tensile steel stress must drop and the failure is in primary compression. The rectangular stress block is valid for this column up to $c = 16$ in. For larger values it becomes approximate because it is based on a full triangle of strain.

As eccentricity beyond $\bar{P}_b, \bar{M}_b$ increases, the member becomes more like a beam, justifying $\phi = 0.90$ when $\bar{P}$ is zero. When longer columns in this region are considered, it is found that buckling problems can exist much below $\bar{P}_b$, indicating that the value of 0.70 should not be discarded or increased too soon. How to make the transition best is not yet well documented, and Code 9.2.1.2c is rather arbitrary, more conservative than the 1963 Code in some cases, and decidedly easier to apply. It states, in slightly reworded form to stand by itself:

> 9.2.1.2c. With f_y not exceeding 60,000 psi the values of 0.70 and 0.75 may be increased linearly to 0.90 as P_u decreases from $0.10f_c'A_g$ to zero for sections with symmetrical reinforcement and $(h - d' - d_s)/h$ not less than 0.70.

In this, h is the overall column thickness and d' and d_s are the covers to the centroid of compression and tension steel, respectively. The relationship is plotted in Fig. 14.9.

When f_y is larger than 60,000 psi or the depth ratio smaller than the limit of 0.70, the balanced load condition moves closer to the moment axis and in some cases it is not possible to secure a tension failure when a compression load exists. Hence the next paragraph in the Code substitutes "the smaller of P_b or $0.10f_c'A_g$" for the simpler reference point $(0.10f_c'A_g)$ used in Fig. 14.9.

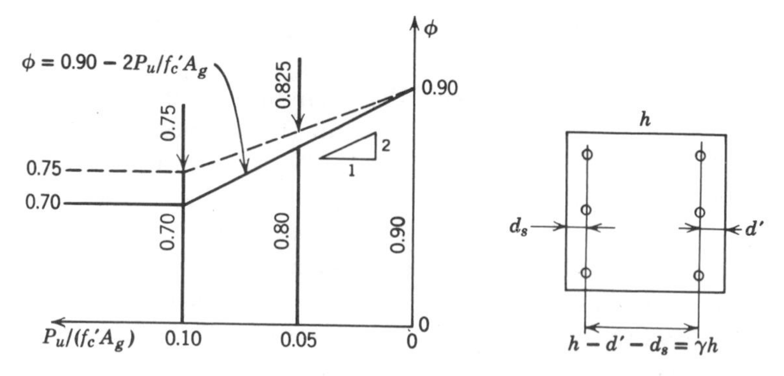

Fig. 14.9. Increase in ϕ as P_u approaches zero. Notice limitations as stated in Code 9.2.1.2c.

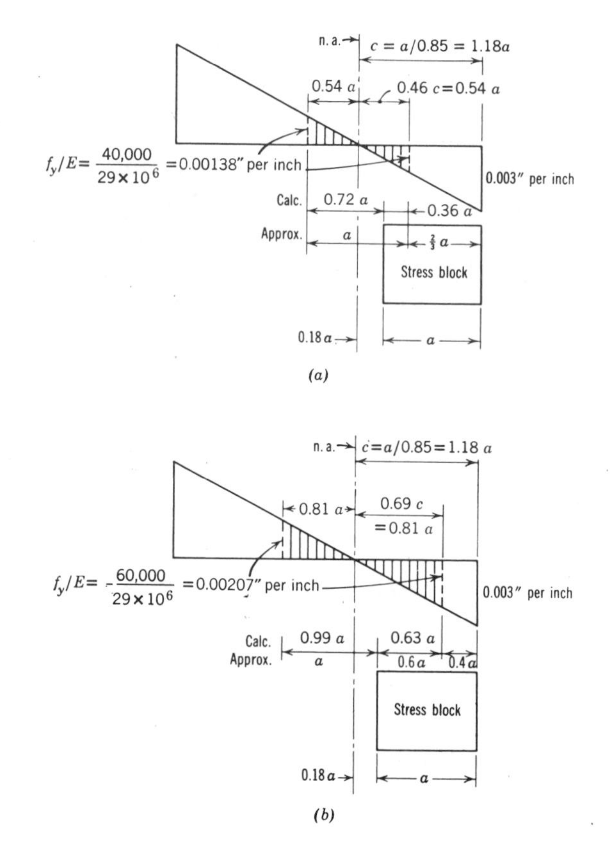

Fig. 14.10. Deformation studies for $f_c' \lessgtr 4$ ksi. (*a*) For $f_y = 40$ ksi. (*b*) For $f_y = 60$ ksi.

14.8 ACCOUNTING FOR BARS NEAR NEUTRAL AXIS

Bars very near the neutral axis will not be effective in carrying stress; for that combination of M and P any bars near the axis will have stresses lower than the yield stress. For any given neutral axis one should sketch the unit deformations to establish the status of nearby bars, starting with $\epsilon_c = 0.003$ as in Fig. 14.10a. To make $f_s = f_y = 40$ ksi, the steel deformation must be at least 0.00138. With the stress block depth as $a = 0.85c$ for f_c' not over 4000 psi, bars falling in the shaded zone will have f_s less than f_y. A rough rule would be to say that f_s is less than an f_y of 40 ksi whenever the bar falls at a depth between $2a/3$ and $5a/3$. For $f_y = 60$ ksi the corresponding zones (Fig. 14.10b) would be roughly between depths of $0.4a$ and $2.0a$.

The deformation sketch is quite simple to use whenever c is known or assumed. It also facilitates the inclusion in an analysis of bars with f_s or f_s' values less than f_y.

The placement of bars near the neutral axis modifies the shape of the interaction diagram considerably, rounding it upward or outward and making the balanced loading point less conspicuous because extreme bars yield first* and other bars later. Typical interaction diagrams are developed in Fig. 14.11, and their different shapes should be noted.

The ACI design aids have been prepared in similar fashion except that for the steel an equivalent ring or shell of steel was used instead of the discrete bar locations.

ANALYSES OF GIVEN SECTIONS—SHORT COLUMNS

14.9 COMPRESSION FAILURES

(a) Equations for Rectangular Members

Although interaction charts are the most practical tool for analysis or design (see Sec. 14.20), sometimes a direct approach, even if slightly more approximate, is handy. In originally promoting (ultimate) strength design in this country, Whitney proposed the use of an equation roughly equivalent† to the straight line of Fig. 14.11a, a line running from $\overline{P}_0$ to that point

*The Code (10.3.3) is not specific in defining how many bars must yield at the balanced condition. The author uses the yielding of the outermost bar as his criterion.

†Actually, Eq. 14.2 defines a slightly curved line which passes through P_0 and near point P_b, M_b, but exactly through P_b, M_b only for specific d'/d values. The deviation is not considered significant.

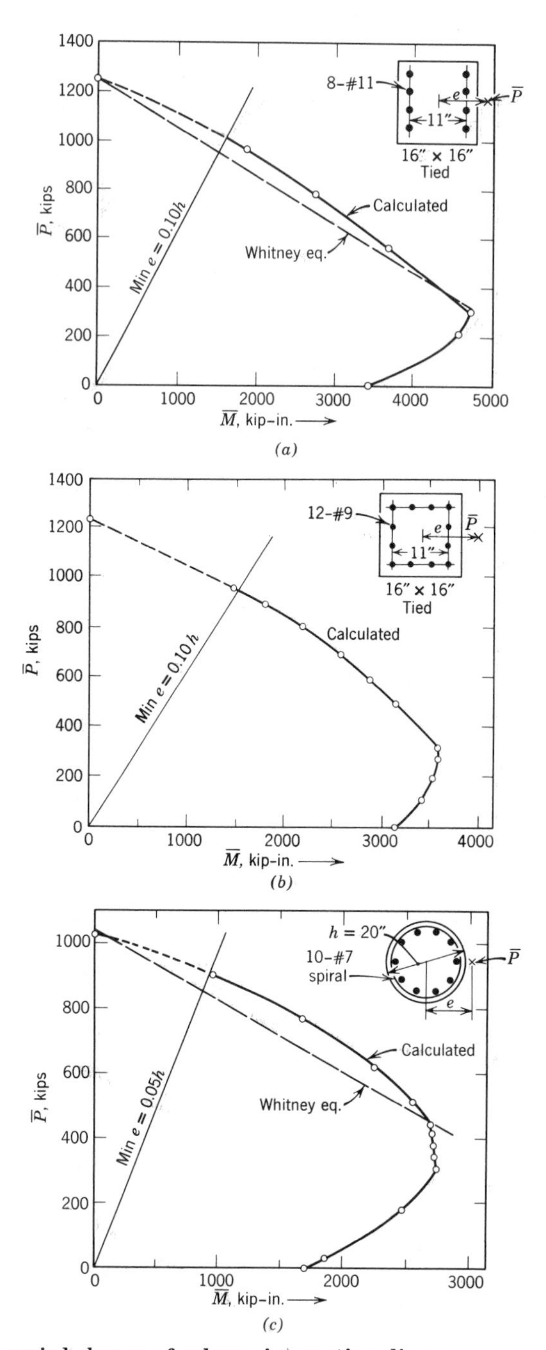

Fig. 14.11. The varied shapes of column interaction diagrams.

on the moment axis which would correspond to an imagined compression failure, as though one has a larger A_s or a greater f_y in tension, a kind of overreinforced section. His equation for rectangular columns with single layers of A_s and A_s', with ϕ added, becomes:

$$P_u = \phi\left[\frac{A_s'f_y}{e/(d-d') + \frac{1}{2}} + \frac{bhf_c'}{(3he/d^2) + 1.18}\right]$$

$$= \phi\left[\frac{A_s'f_y}{e'/(d-d')} + \frac{bhf_c'}{(3he/d^2) + 1.18}\right] \tag{14.2}$$

The notation is that of Fig. 14.7 with e measured from the column centerline and e' from the center of tension steel. When $e = 0$, this equation gives ϕ times the $\bar{P}_0$ of Eq. 14.1 except that $2A_s'$ shows in place of A_{st} and bh shows instead of A_n, thus ignoring concrete displaced by the steel; when e' becomes infinite, the equation leads to $M_{u0} = \phi[f_c'bd^2/3 + A_s'f_y(d - d')]$ which is approximately the pure moment resistance if the column could be made to fail in compression without yielding on the tension side.

The equation is not adequate in Fig. 14.11*b* unless one uses A_s and A_s' at the centroid of these areas; also the accuracy drops when the interaction diagram is so curved in the compression failure region.

The equation of the interaction diagram joining P_0 and P_b can also be solved approximately (as a straight line) to give P_u in either of the following forms:

$$P_u = \frac{P_0}{1 + [(P_0/P_b) - 1]e/e_b} \tag{14.3}$$

$$P_u = P_0 - (P_0 - P_b)M_u/M_b \tag{14.4}$$

(b) Example—Compression Failure

The short column of Fig. 14.12 will be analyzed to establish the load it may be assigned at an eccentricity of 3.0 in. above the x-axis or the Code minimum, $0.1h = 1.8$ in., about the y-axis. $f_c' = 3000$ psi, $f_y = 60$ ksi.

Solution

The cover required is 1.5 in. clear + 0.375 in. ties + 1.13/2 in. for the half bar = 2.44 in. < 2.50 in. shown. O.K.

Tie arrangement is adequate since no bar is more than 6 in. from a support. Minimum tie size must be #3 (for bars smaller than #11). Minimum tie spacing (Code 7.12.3):

16 × 1.13 = 18 in. for #9 bars

48 × 0.375 = 18 in. for #3 ties

18 in. for minimum column thickness

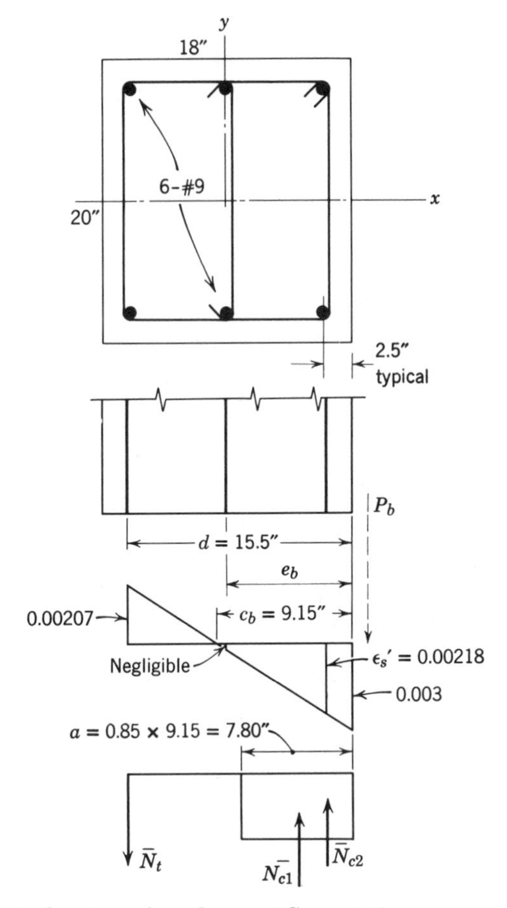

Fig. 14.12. Moment about y-axis, column of Sec. 14.9b.

The closest requirement would govern, but it happens all are the same in this case.

The check on strength about the y-axis will be investigated last, since the arrangement of steel requires more calculation about that axis. (If the e about the x-axis were not more than $0.1h = 2$ in., it would be obvious that calculations about the y-axis would actually control and should come first.)

A solution by Eq. 14.3 will first be used about the x-axis.

$$\bar{P}_0 = 0.85f_c'(A_g - A_s) + f_yA_{st}$$
$$= 0.85 \times 3(18 \times 20 - 6) + 60 \times 6 = 900 + 360 = 1260 \text{ k}$$
$$\bar{P}_b \text{ (from Sec. 14.6)} = 395 \text{ k with } e_b = 12.40 \text{ in.}$$

With e here of 3.0 in., will have compression failure with $\bar{P} > \bar{P}_b$.

$$\bar{P}_x = \frac{\bar{P}_0}{1 + [(\bar{P}_0/\bar{P}_b) - 1]e/e_b} = \frac{1260}{1 + [(1260/395) - 1]\, 3.0/12.40}$$

$$= \frac{1260}{1 + 2.19 \times 0.242} = 825 \text{ k}$$

$$P_x = \phi P_x = 0.70 \times 825 = 577\ k$$

About the y-axis a minimum e of $0.1 \times 18 = 1.8$ in. must be considered. For steel spread over the depth, the "Whitney formula" is not applicable and $\bar{P}_b$ must first be found, using the sketches of Fig. 14.12 with the advantage of certain numbers which carry over from Fig. 14.7.

The middle bars at balanced load are almost on the neutral axis and will be considered at zero strain and zero load.

$$\bar{N}_t = 2 \times 60 = 120 \text{ k}$$

$$\bar{N}_{c2} = 2(60 - 2.55) = 115 \text{ k}$$

$$\bar{N}_{c1} = 0.85 \times 3 \times 20 \times 7.80 = 398 \text{ k}$$

$$\Sigma\bar{N} = \bar{P}_b = 398 + 115 - 120 = 393 \text{ k}$$

$$\bar{M}_b = \bar{P}_b e_b = 398(9 - 3.90) + 115 \times 6.5 + 120 \times 6.5$$

$$= 2030 + 748 + 780 = 3560 \text{ k-in.}$$

$$e_b = 3560/393 = 9.04 \text{ in.}$$

$$\bar{P}_0 = 1260 \text{ k, as before}$$

$$\bar{P}_y = \frac{\bar{P}_0}{1 + [(\bar{P}_0/\bar{P}_b - 1]e/e_b} = \frac{1260}{1 + [1260/393 - 1]\, 1.80/9.04}$$

$$= \frac{1260}{1 + 2.19 \times 0.199} = 878 \text{ k} > 825 \text{ for } \bar{P}_x.$$

Permissible $P_u = P_x = 577$ k.

The Whitney equation, Eq. 14.2, is applicable to the eccentricity about the x-axis since all the steel is on opposite faces. This equation will be used to compare its simplicity to the above.

$$A_s' = 3\text{-}\#9 = 3.00 \text{ in.}^2$$

For a short column

$$\bar{P}_x = P_u/\phi = \frac{A_s' f_y}{e/(d - d') + \frac{1}{2}} + \frac{bhf_c'}{3he/d^2 + 1.18}$$

$$= \frac{3 \times 60}{3.0/15 + 0.5} + \frac{18 \times 20 \times 3}{3 \times 20 \times 3.0/17.5^2 + 1.18}$$

$$= \frac{180}{0.70} + \frac{1080}{0.588 + 1.18}$$

$$= 257 + 610 = 867 \text{ k} > \bar{P}_b = 395 \text{ k from Sec. 14.6}$$

$$P_u = \phi\bar{P} = 0.70 \times 867 = 608 \text{ k, subject to check about } y\text{-axis.}$$

The Whitney equation is thus quicker in use when the value of P_b is not already available. This P_{ux} of 608 k compares with P_u of 577 k which is about the accuracy to be expected from these formulas. The more exact solution from curves[6] which do not use the straight line for the interaction diagram gives $P_y = 637$ k. The use of curves (Sec. 14.20) is undoubtedly the practical answer for standardized situations and in this case gives a permissible load 5 to 10% higher.

14.10 TENSION FAILURES

(a) Analysis Procedures

Although compression failures have to be handled on the basis of nearly empirical formulas or strain analyses, tension failures of short columns of ideal materials are as easily analyzed as beams, from basic principles, when e is given.

When a tension failure occurs, the yielding of the tension steel causes the neutral axis to move toward the compression face, reduces the compression area, and finally brings about a secondary compression failure. Although this shift of the neutral axis lowers the strain on A_s' and makes it necessary finally to check whether $f_s' = f_y$, it is convenient to start by assuming f_y and neglecting the concrete displaced by A_s'. With these simplifications equilibrium of forces is very simply established from Fig. 14.13. For $A_{s1} = A_s/2 = A_s'$, $\overline{N}_t = \overline{N}_{c2}$, and it follows from $\Sigma F_y = 0$ that $\overline{N}_{c1} = \overline{P}$. For $\Sigma M = 0$, the forces divide into two couples, one composed of $\overline{N}_t$ and $\overline{N}_{c2}$, the other of $\overline{N}_{c1}$ and $\overline{P}$.

$$\overline{N}_{c2}(d - d') = \overline{P}e_{c1}$$

where $e_{c1} = e + d'' - d + 0.5a$ which for symmetrical sections becomes $e_{c1} = e - 0.5h + 0.5a$. These are powerful and simple equations and are adequate for many problems.

If displaced concrete is considered, the effective $\overline{N}_{c1}$ is a little greater than $\overline{P}$, and the first equation is replaced by ΣM about some convenient center.* Unsymmetrical columns can be treated similarly.

If a is small, the strain on compression steel should be checked using $c = a/k_1$ and $\epsilon_c = 0.003$. It may be necessary to substitute $\epsilon_s'E_s$ in place of f_y on A_s'. Also, if a is large, it could be that the tension f_s is less than f_y; if this occurs it is proof that $\overline{P}$ is really greater than P_b and the compression failure solution of Sec. 14.9 is needed.

*A center on A_s' is often convenient.

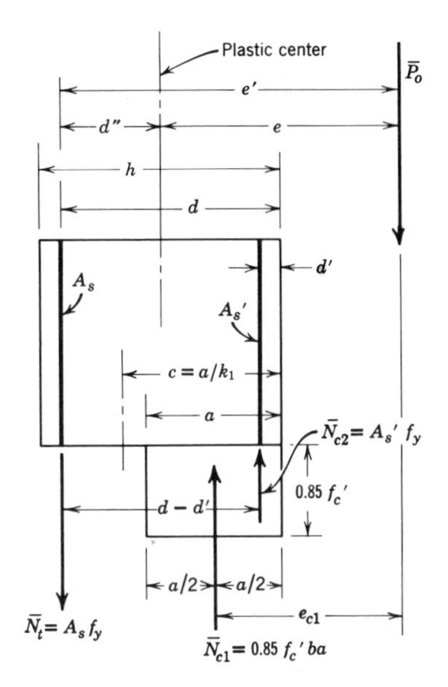

Fig. 14.13. Equilibrium of forces with tension failure, ultimate strength method.

If bars are not grouped in single layers near opposite faces, but are at varying distances from the column centerline, the same two equations $\Sigma F_y = 0$ and $\Sigma M = 0$, must be satisfied, probably by trial and error methods starting with an assumed neutral axis location.

(b) Example—Tension Failure

The short column of Fig. 14.7 will be analyzed to establish the load it may be assigned if placed on the x-axis at an eccentricity of 15 in. about the y-axis. $f_c' = 3000$ psi, $f_y = 60$ ksi.

Solution

Cover of 2.5 in. to center of bar provides more than 1.5 in. clear. O.K. First proceed (approximately and simply) by assuming $f_s' = f_y$ and neglecting displaced concrete. Find the short column strength as an ideal material.

$\bar{N}_t = \bar{N}_{c2} = 3.0 \times 60 = 180$ k

$\bar{M}_1 = \bar{N}_{c2}(d - d') = 180 \times 15 = 2700$ k-in.

Try $a = 10$ in., $e_{c1} = 15 - 10 + 10/2 = 10$ in. (compare Fig. 14.14)

$e_{c1}\,\bar{N}_{c1} = 10\bar{N}_{c1} = 2700$, $\bar{N}_{c1} = 270$ k

$a = \bar{N}_{c1}/(0.85 f_c' b) = 270/(0.85 \times 3 \times 18) = 5.88$ in.

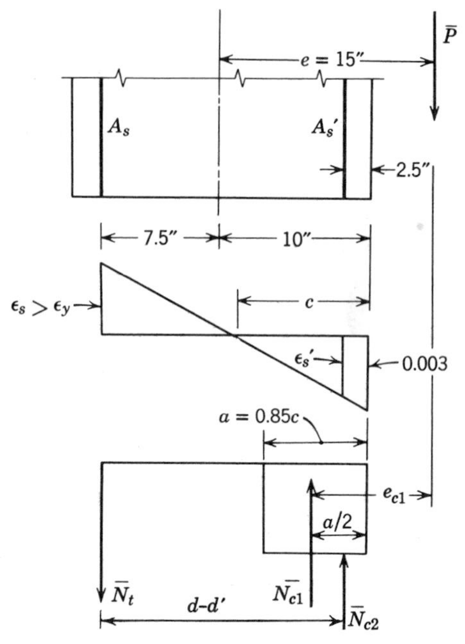

Fig. 14.14. Column of Sec. 14.10b.

Try $a = 7$ in., $e_{c1} = 15 - 10 + 7/2 = 8.5$ in. (Fig. 14.14)

$8.5\ \overline{N}_{c1} = 2700,\ \overline{N}_{c1} = 318\text{ k} = \overline{P}$

$a = 318/(0.85 \times 3 \times 18) = 6.95$ in. say O.K.

Check f_s': $c = a/0.85 = 7/0.85 = 8.25$ in.

$$\epsilon_s' = \frac{(8.25 - 2.5)0.003}{8.25} = 0.00209$$

$$f_s' = \epsilon_s' E_s = 0.00209 \times 29 \times 10^3$$

$$= 60.6 \text{ ksi} > f_y \qquad \therefore f_s' = f_y$$

$$\overline{P} = 318 \text{ k, subject to reduction for } \phi$$

The more exact solution would consider the displaced concrete. In effect, the 3.0 in.^{2} of A_s' steel displaces concrete counted at $0.85f_c' = 2.55$ ksi, or a total of about 8 k, which must come from $\overline{N}_{c1}$ (and $\overline{P}$ since $\overline{P}$ is used here as $\overline{N}_{c1}$), The second trial value of $a = 7$ in. above is a good starting point, giving

$c = 8.25$ in. as above

$\epsilon_s = 0.003(17.5 - 8.25)/8.25 = 0.00336 \gg \epsilon_y$

This proves the failure is in tension. (Alternate: Compare $\overline{P}$ with $\overline{P}_b$.) Take moments about the center of the column:

$$\overline{N}_{c1} = 0.85 \times 3 \times 7 \times 18 = 321 \text{ k}, \ \overline{N}_{c2} = 3.0(60 - 0.85 \times 3) = 172 \text{ k}$$

$$\Sigma M = 180 \times 7.5 + 172 \times 7.5 + 321(10 - 3.5) - \overline{P} \times 15 = 0$$

$$1350 + 1290 + 2080 - 15\overline{P} = 0$$

$$\overline{P} = 4720/15 = 314 \text{ k}$$

$$\Sigma F_y = 180 - 172 - \overline{N}_{c1} + 314 = 0, \ \overline{N}_{c1} = 322 \text{ k}$$

This $\overline{N}_{c1} = 322$ k $>$ 321 k assumed, but negligibly so. Also this exact $\overline{P}$ value of 314 k differs so little from the earlier value of 318 k that the exact solution will not be pursued further.

$$\text{Short column } P_u = \overline{P}\phi = 318 \times 0.7 = 223 \text{ k}$$

$$0.1f_c'\, bh = 0.1 \times 3 \times 20 \times 18 = 108 \text{ k} < 223 \text{ k}, \ \phi = 0.7 \text{ is O.K.}$$

$$M_u = P_u e = 223 \times 15 = 3350 \text{ k-in.}$$

14.11 CIRCULAR COLUMNS—ANALYSIS

(a) Approaches

Whitney also proposed an allowable P_u formula as follows for circular columns of diameter h failing in compression.

$$P_u = \phi \left\{ \frac{A_{st} f_y}{\dfrac{3e}{(h - 2d')} + 1} + \frac{A_g f_c'}{\dfrac{9.6\, he}{[0.8h + 0.67\,(h - 2d')]^2} + 1.18} \right\}$$

Essentially this is the formula for rectangular sections (Eq. 14.2) modified by considering compression steel and tension steel at approximate centroidal locations and with adjustments for the circular area. This is nearly a straight-line equation for the compression failure portion of the diagram (Fig. 14.11c). It is usable for analysis just as was the similar equation for the rectangular column in Sec. 14.9b.

It is entirely proper and more accurate for either tension or compression failure to use an assumed neutral axis and the strains procedure of Sec. 14.7. Unfortunately it is rather lengthy for the circular steel arrangement, as the following example indicates. Research[13] shows that, although circular segments are awkward, the use of $a = 0.85c$ when $f_c' \lesseqgtr 4000$ psi, as for beams, is still valid here.

(b) Example of Strain Analysis

The column of Fig. 14.15 will be analyzed to establish the ultimate load capacity for a load at a 12-in. eccentricity, assuming $f_c' = 3000$ psi, Grade 40 steel, and a short column length.

Solution

The geometrical coefficients shown in Fig. D.7 in Appendix D make the analysis of the curved shape feasible. A trial-and-error solution starting from assumed c is practical.

Try $c = 8$ in., $a = 0.85c = 6.80$ in., $a/h = 6.80/20 = 0.340 = x/h$ of Fig. D.7 and corresponds to $B = 0.24$, $A = 0.60$.* Hence, $\bar{x}$ of the segment from the center of the column $= Ar = 0.60 \times 10 = 6.00$ in. and the area $= Bh^2 = 0.24 \times 20^2 = 96$ in.^{2}

$$\text{Permissible } \overline{N}_{c1} = 0.85 \times 3 \times 96 = 245 \text{ k}$$

The assumed bar locations are shown in Fig. 14.15*b*. One pair of bars is so close to the assumed neutral axis as to be only slightly stressed. This pair is totally neglected in this calculation; the problem has been discussed in Sec. 14.8. All other bars carry the yield stress, as shown by the strain triangles of Fig. 14.15*c* which makes it a tension failure case and results in the forces of Fig. 14.15*d* (which here neglect displaced concrete). Moments of these forces about the center of the column must balance the moment $\overline{P}e$:

$$\begin{aligned}
2 \times 24 \times 7.67 &= 368 \\
2 \times 48 \times 6.23 &= 600 \\
48 \times 2.37 &= 114 \\
\overline{N}_{c1}: 245 \times 6.00 &= 1470 \\
\hline
2552 &= 12\overline{P} \\
\overline{P} &= 213 \text{ k} \\
\overline{N}_t - \overline{N}_{c2} &= 48 \\
\hline
\text{Reqd. } \overline{N}_{c1} &= 261 \text{ k}
\end{aligned}$$

This exceeds the permissible $\overline{N}_{c1}$ of 245 k for the assumed c.

For the second trial, take $c = 8.5$ in., $a = 0.85 \times 8.5 = 7.22$ in., $a/h = 7.22/20 = 0.361$. Figure D.7 for $x/h = 0.361$ shows $B = 0.26$, $A = 0.572$.

*The ACI standard notation would use lower-case Greek letters for ratios A, B, and C, but for this limited use here a change from Professor Shank's notation seems unnecessary.

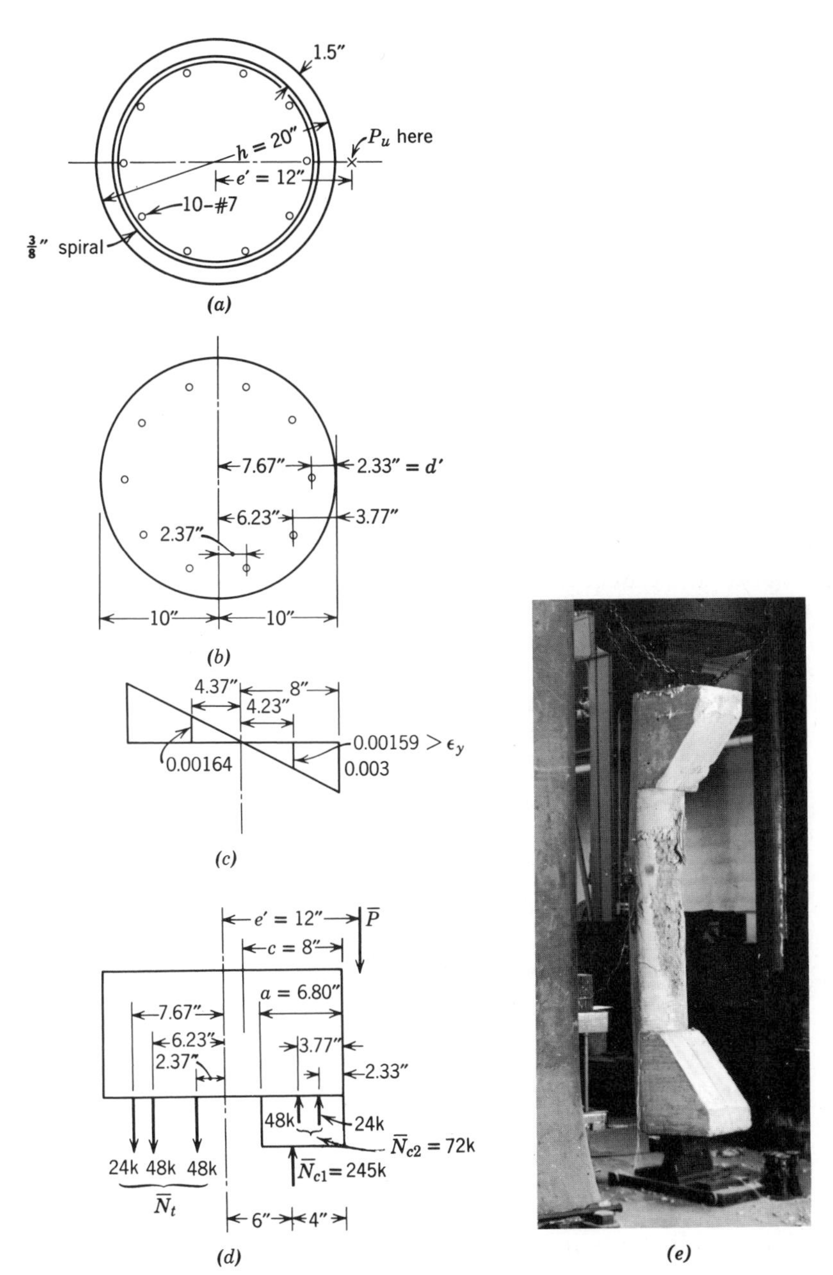

Fig. 14.15. Analysis of strength of circular column. (*a*) Column cross section. (*b*) Bar locations. (*c*) Strain triangles. (*d*) Forces in equilibrium. (*e*) The failure of an eccentrically loaded spiral column. (Courtesy University of Illinois.)

$$
\begin{aligned}
\text{Permissible } \overline{N}_{c1} &= (0.26 \times 20^2)(0.85 \times 3) = 265 \text{ k} \\
\bar{x} &= 0.572 \times 10 = 5.72 \text{ in. from center} \\
\Sigma M: \quad \text{As above} &= 368 \\
\text{As above} &= 600 \\
\text{As above} &= 114 \\
\overline{N}_{c1}: 265 \times 5.72 &= 1518 \\
& \overline{\quad 2600 = 12\overline{P}} \\
\overline{P} &= 217 \text{ k} \\
\overline{N}_t - \overline{N}_{c2} &= 48 \\
\text{Reqd. } \overline{N}_{c1} &= 265\ k \text{ as assumed} \qquad \text{O.K.}
\end{aligned}
$$

Hence the ultimate load $P_u = \phi\overline{P} = 0.75 \times 217 = 162$ k.

The failure of such a column is shown in Fig. 14.15*e*.

This type of analysis is suitable for either a tension or compression failure, with some question remaining when the neutral axis is moved just outside of the cross section or (practically) when the stress block just covers the entire depth of member h. This questionable point would arise close to the case of minimum $0.10h$ eccentricity for a rectangular column but probably not for the lower $0.05h$ minimum permitted for a spiral column. For the extreme cases the author would assume N_{c1} on the concrete uniform over the total area at $0.85f_c'$ and the steel stress varying as the strains indicated. The author would also recognize that the rectangular stress block is inherently less accurate when it covers the entire member as here than in a beam where the detailed shape of the stress-strain curve makes little numerical difference.

CRITICAL COLUMN MOMENTS AND THE INFLUENCE OF LENGTH

14.12 THE NATURE OF THE LONG COLUMN EFFECT

Columns always carry moment in the practical case, both by virtue of slight initial crookedness and by the nature of the loading from beams and slabs which causes rotation of frame joints and thus imposes curvatures on the columns. Moments in columns also produce curvatures which lead to deflections of columns between joints and sometimes sidesway deflections of the joints (similar to that occurring under wind load) even when no external horizontal loads exist.

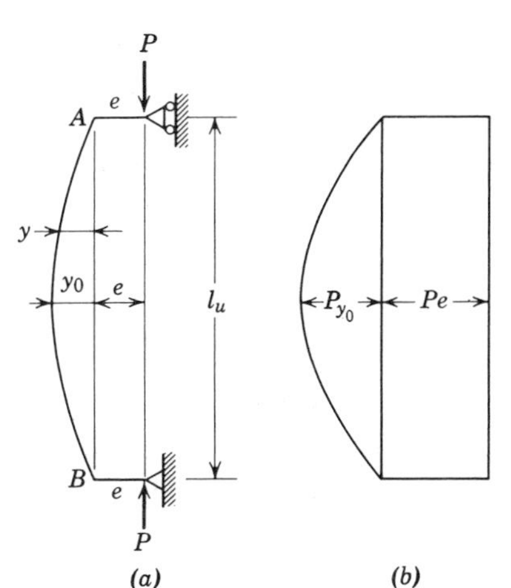

Fig. 14.16. Single curvature column. (a) Curvature. (b) Moment.

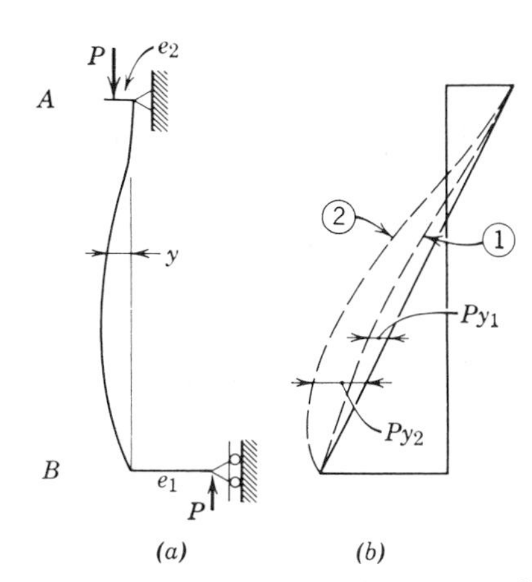

Fig. 14.17. Column with end moments of opposite kind. (a) Curvature. (b) Moment.

Columns in reinforced concrete frames are rarely loaded in a statically determinate manner, but the isolated and statically loaded column is helpful as a starting point for discussion.

In Fig. 14.16a column AB, with equal end eccentricities, is subject to a constant moment Pe from external load, as represented by the rectangular moment diagram Pe in Fig. 14.16b. Because the column deflects, each point except A and B has an extra moment Py which adds to the curvature and increases the y which would be caused by the Pe moment diagram alone. Since y depends on column length, the Py moment could be called a "long column effect" in that this extra moment reduces the capacity of the column to carry axial load. Evidently, y would be small if l_u were small; but there is no length which would not have some y under this type of loading and hence some Py moment and some length effect. However, for a very short l_u, one to two times the column thickness, the Py moment could be negligible in the computations. This general type of loading will be referred to as the single curvature case.*

Consider now the column of Fig. 14.17. Here the opposite signs of the two end moments create a point of inflection. This column also deflects and develops Py moments. However, if the length and eccentricities are such that the deflection adds a moment of the magnitude sketched as Py_1 in Fig. 14.17b, the maximum moment on the column remains at B where $Py = 0$ and there is no ill effect on column strength. However, if Py_2 exists, the maximum moment moves from the end B and there is a

*In this text the term "single curvature" is used without any added statement to designate equal top and bottom end moments, as shown in Fig. 14.16. As a general term, single curvature also includes unequal end moments; but it is not so-used herein.

small ill effect which would be called a long column effect. Practice in effect uses the term "long column effect" only when there is an ill effect, that is, a loss of strength. In a case such as this, the ratio of height to radius of gyration l_u/r has to be very large before there is an ill effect.

14.13 INCREASED COLUMN DEFLECTION UNDER SHEAR LOADING

A major case is that of a column subjected to a shear load, such as a column resisting wind or earthquake forces. Consider first the situation where the floor system is very heavy, that is, for the purpose of this initial discussion, infinitely stiff. The column deflects as in Fig. 14.18*a* and can be analyzed by considering the half-length shown in Fig. 14.18*b* which shows that the moment is increased at A, much like it is at mid-height of the column of Fig. 14.16*a*. Although the magnitude of the Py term compared to the Pe or $Hl_u/2$ moment is different, the two cases are closely related to each other. However, the infinitely stiff floors with zero joint rotations at A and B are not common.

If the joints at A and B rotate, Δ increases to Δ_2. This follows because the $Hl_u/2$ moment is unchanged and the curvature from A down to mid-height is changed only by the Py term which increases with Δ. Thus the

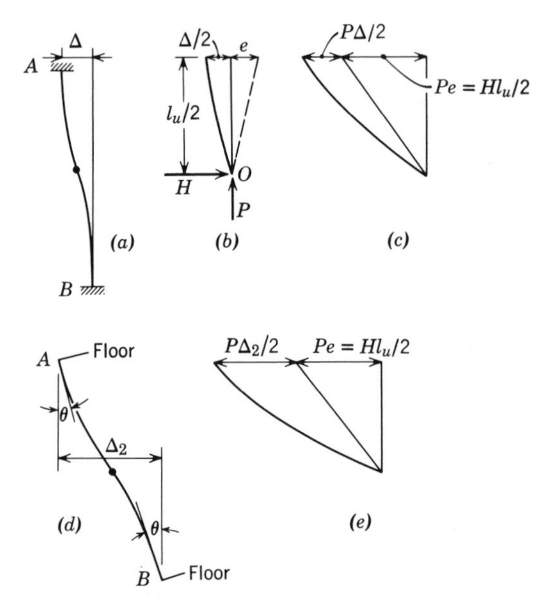

Fig. 14.18. Columns carrying horizontal shears. (*a*) Curvature. (*b*) Statics of a half height. (*c*) Moment. (*d*) With joints which rotate. (*e*) Moment for (*d*).

deflection of the cantilever in Fig. 14.18*b* changes from that of a member fixed at the end to one where deflection is augmented by joint rotation as in Fig. 14.18*d*. The moment diagram of Fig. 14.18*e* then involves a greater end moment. If the joint rotation is large, the end moment is multiplied several times, that is, $P\Delta_2$ is several times as much as the static force moment $Hl_u/2$. Hence the long column action can be very important in this case. The degree of importance is determined by the angle θ at each end. While this angle depends on the total moment that the column carries, for any given value of this moment the angle is controlled almost linearly by the stiffness of the beams; a beam half as stiff will double the angle.

14.14 FRAME LOADINGS FOR MAXIMUM COLUMN MOMENTS

The designer must distinguish sharply between columns where sidesway is prevented (diagonal bracing, shear walls, etc.) and those free to sway.

Three different frame loadings are involved in the three different curvatures of Sec. 14.12 and 14.13. Two of these cases on interior columns occur in a braced frame as shown in Fig. 14.19. For critical stresses in a relatively uniform frame at the point marked x, the loading of Fig. 14.19*a* gives both a maximum axial load and the maximum moment consistent therewith. The column behavior is essentially that of Fig. 14.17. There is a minimum of long-column effect. This case will be referred to as the reversed moment case, whether the reversal is small, as in the loading diagram, or a complete reverse, as it might be on an exterior column. The case shown in the loading diagram might also be called a column restrained at the far end (by the moment resisting the upper joint rotation).

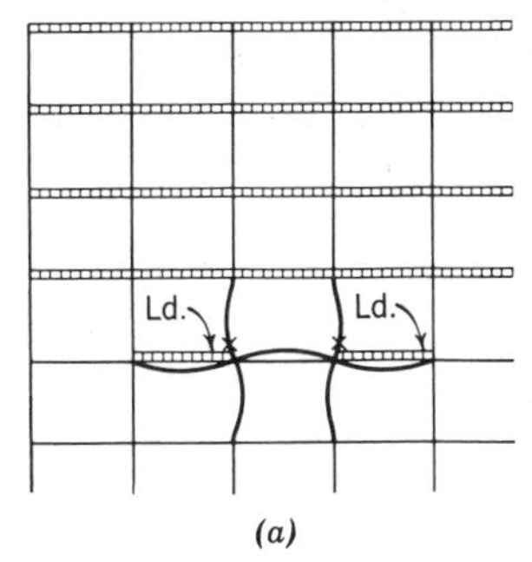

(*a*)

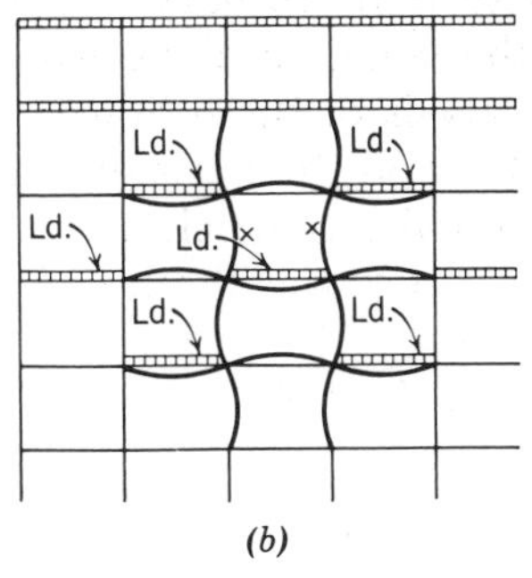

(*b*)

Fig. 14.19. Column moments without sidesway. (*a*) Maximum load and moment loading but minimum long column effect. (*b*) Single curvature loading.

A checkerboard loading adjacent to a column (Fig. 14.19*b*) produces single curvature in the columns, as in Fig. 14.16. Fortunately, this curvature requires that some potential load be omitted from the floor above and analysis also shows smaller end moments in this case than in the one previously discussed. Nevertheless, the long column effect is large at x and could make this govern on very long columns or some unequal span situations.

In the absence of shear walls, sidesway may permit a whole story height in Fig. 14.19*a* or *b* to deflect laterally and shift the moment conditions of Fig. 14.16 and 14.17 to one very nearly like Fig. 14.18. Symmetry of loading or frame delays this, but elastic analysis shows sidesway to be the expected collapse mode for the entire story. Length effects which might be small without sidesway would then become large when sidesway develops.

The sidesway case under shear load is worst when the shear is maximum. The distribution of these shears to the individual columns requires analysis of the entire story rather than individual columns and the Code specifies use of the entire story in computing the long column effects (Sec. 14.18e). Although the sidesway moment gives the most severe long-column effect, in a low structure the basic moments may still be small enough to make the vertical loading control the design.

14.15 CORRECTIONS NEEDED FOR FRAME MOMENTS ON ALL COLUMNS

Discrepancies between nominal and actual column moments cannot all be ascribed to column-length effects. A recent computer study[6,7,8] of braced frames, using fully cracked beams (with $\rho = \rho'$ over their full length) and columns programmed to their P-M-ϕ (curvature) response indicated that ordinary elastic analyses based on gross moments of inertia from the concrete alone were not totally adequate for design purposes. The analysis of these braced frames loaded for maximum moments as in Fig. 14.19, on the basis of gross I of concrete alone, led to what is termed the nominal eccentricity.

For a ratio of column-to-beam stiffness of $\psi = 1$ (based on I_g values) and nominal e/h of 0.10, the load-moment responses of two frames,[6] identical except for the beam steel, are plotted in Fig. 14.20. The lower curve represents $\rho = 0.005$ of Grade 40 beam steel and the upper curve $\rho = 0.023$ of the same steel. The resulting differencies are:

Ultimate loads:	309k	and	380k versus nominal 349k
Final e/h:	0.147	and	0.068 versus nominal 0.10
Initial e/h:	0.167	and	0.120 versus nominal 0.10

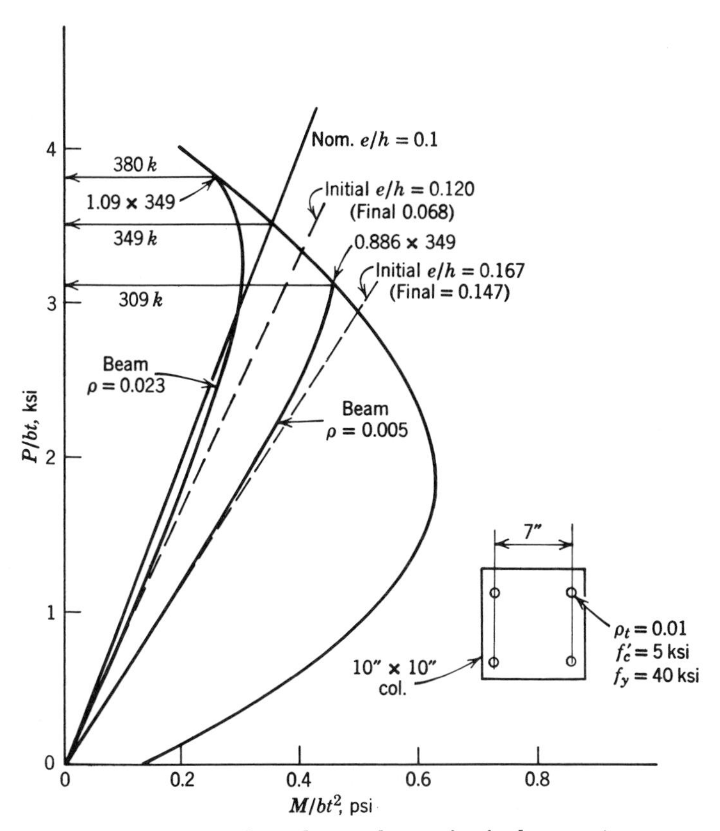

Fig. 14.20. Effect of beam steel on short columns in single curvature.

With the heavily reinforced beams the column became relatively less stiff as its compression steel yielded on one face (about 300k) and its resisting moment actually *decreased* somewhat as it went on to increased loading.

In the 121 frames of this type studied, the ratio of failure e/h to the nominal, which would also be the ratio of ultimate column moment to the nominal, varied from 0.53 to 2.12, with more cases increasing than decreasing. Column design moments thus need some correction,[7] although the matter is not yet codified.

The best "elastic-type" of analysis found for this loading considered beams cracked with a transformed steel area based on $4n$ while the columns were considered uncracked with $2.5n$ for the steel.

Alternatively, if the gross concrete I is used for both beam and column, for $\psi = K_{col}/K_{bm} = 1$, the resulting e/h or moment on the column can be multiplied by the empirical factor:

$$1.38 + 5.5\,\rho_t - 30\rho$$

where $\rho_t = A_{st}/bh$ for the column

$\rho = A_s/bd$ for the beam $= \rho'$

This factor plots as a plane surface in Fig. 14.21 where less steep planes are shown for larger ψ values. The horizontal axes are ρ_t for the column (toward the rear) and ρ for the beam (toward the right) and the vertical ordinates are the correction multiplier. For $\psi = 0.5$ the factor is much larger and erratic, suggesting that the combination is not a good one or the data are inadequate. The larger ψ becomes, the less the influence of the beam steel and the smaller the empirical correction.

The figure and numbering on it imply an accuracy that is not attainable. Multipliers shown are average values, not an upper bound, not always on the safe side. The actual increments added or subtracted (by the multiplier) will scatter in a typical ± 15 to $\pm 20\%$ range

Columns loaded by reversing moment diagrams (Fig. 14.17 type) are effectively stiffer and the multiplier is smaller, approximated by using ψ equal to two times the actual ψ.

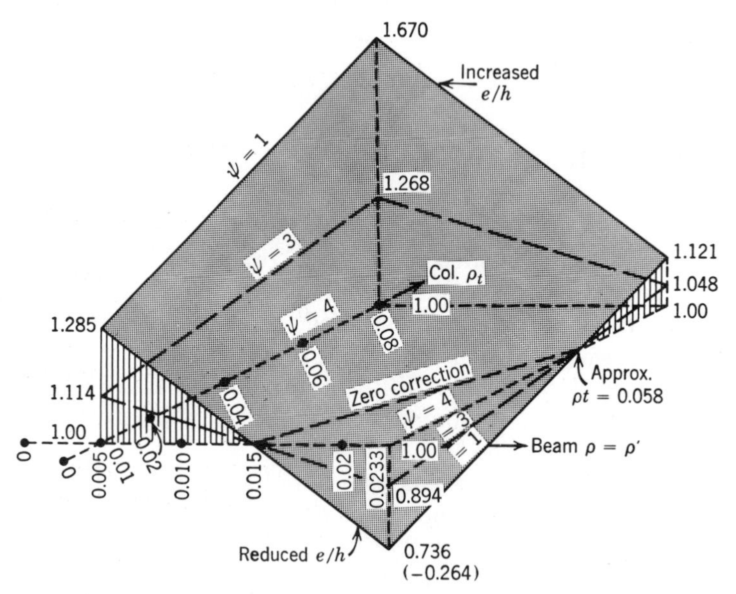

Fig. 14.21. Graphical presentation of empirical relationships for e/h multiplier, single curvature cases.

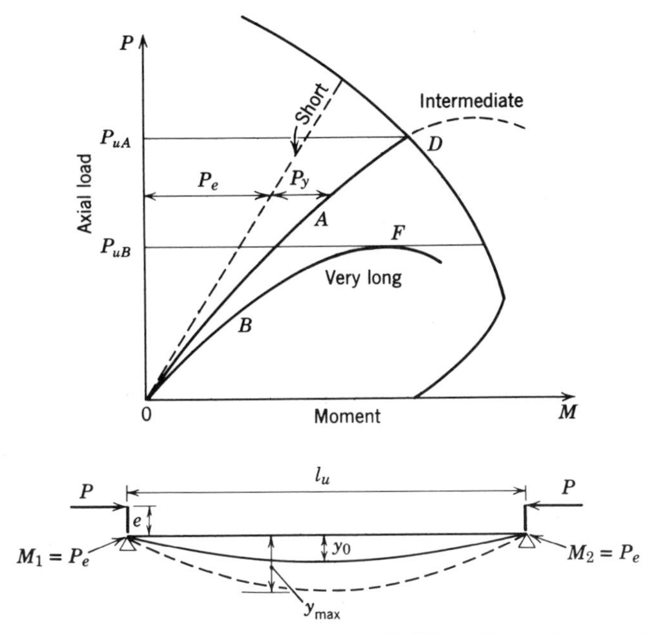

Fig. 14.22. (*a*) Behavior patterns for columns. (*b*) When the end moments are caused by an eccentric axial load the deflection resulting from M_1 and M_2 is increased from y_0 to y_{max}.

14.16 INFLUENCE OF LENGTH ON COLUMN BEHAVIOR

The behavior response of even the very short column to an increasing load, as already shown in Fig. 14.20 is not linear, possibly raising a question about the most appropriate design e/h to use. The discussion in that section selected the final value, which is usually a sound policy when one cannot extrapolate other values linearly. The Code committee has not yet undertaken the codifying of the design e/h, but in the area of length effects it has set up a specific procedure* which defines the increasing value of critical moment with increasing axial load (at a given nominal eccentricity).

Consider a column, say in single curvature as in Fig. 14.16. As one doubles the load, y_0 more than doubles, developing possibly a critical moment as indicated by curve *A* in Fig. 14.22*a*.

The amount of curvature in the curve for a particular case depends on the length of the member. The moment for a very short column might plot

*But see also the alternate *R* Method in the Commentary (Sec. 14.23).

essentially straight (y_0 negligible), following the dotted line to failure at the interaction diagram. The medium-length column of curve A would fail at point D. These failures would be in compression, and the author calls this failure at the interaction diagram a materials failure.

A longer column might give the curve B which turns horizontal before reaching the interaction diagram. This column fails from instability or buckling in the classical sense, the inability to resist more load, although no material has reached the failure point. Mathematically, some like to treat point D also as a special case of instability*, but the author believes the separate names for the two quite different phenomena help to clarify one's understanding of columns.

The method of the following sections permits one to compute the increased or "magnified" moment[9] and to plot this type of curve if he wishes. For design or analysis only the portion around D or F would be important; and at F the method only indicates a horizontal line, representing increasing moment with no increase in axial load.

14.17 MOMENT MAGNIFIER METHOD FOR LENGTH EFFECT

In the mechanics of elastic members a member that is bent in a specific shape, either from initial crookedness or some external loading, will have its deflection increased by the addition of an axial load P acting along the reference chord, from an initial y_0 to

$$y_{max} = y_0/(1 - P/P_c)$$

where P_c is the critical buckling load for that member (Euler load). In the hinged column of Fig. 14.22*b*, with $M_1 = M_2 = Pe$,

$$\begin{aligned}
M_{max} = M_1 + Py_{max} &= M_1 + Py_0/(1 - P/P_c) \\
&= [M_1(1 - P/P_c) + Py_0\,(M_1/Pe)]/(1 - P/P_c) \\
&= M_1(1 - P/P_c + y_0/e)/(1 - P/P_c) \\
\text{Since } y_0 &= M_1 l_u^2/8EI = Pel_u^2/8EI, \text{ and } P_c = \pi^2 EI/l_u^2, \\
M_{max} &= M_1(1 - P/P_c + Pl_u^2/8EI)/(1 - P/P_c) \\
&= M_1(1 - P/P_c + \pi^2 P/8P_c)/(1 - P/P_c) \\
&= M_1\,[1 + (P/P_c)\,(\pi^2/8 - 1)]/(1 - P/P_c) \\
&= M_1(1 + 0.23P/P_c)/(1 - P/P_c) \doteq M_1/(1 - P/P_c).
\end{aligned}$$

For the single curvature case the error in omitting $0.23P/P_c$ varies from 2.3% when $P/P_c = 0.1$ to 11.5% when $P/P_c = 0.5$; for other initial curvatures it can be either more or less.

*With the curve turning sharply downward at point D.

If $M_1 \neq M_2$, the maximum moment is not at mid-height initially, and the largest increase is near mid-height, resulting in a smaller total at the new maximum point. The Code cares for this by including the factor C_m which is to take care of different shapes of moment diagrams and shifts in points of maximum moment. Based on the larger end moment M_2,

$$M_{\max} = M_2[C_m/(1 - P/P_c)] = M_2\delta$$

where δ is the quantity in brackets and is called the moment magnifier or amplifier. In braced frames when $M_1 = M_2$, $C_m = 1.0$; and when $M_1 \neq M_2$,

$$C_m = 0.6 + 0.4M_1/M_2 \geqslant 0.4$$

Here M_1 is positive for a single curvature case, negative for a double curvature case, and the moment ratio itself is always in the range between +1 and −1.

With this procedure the long column must then be designed* for P_u and $M_2\,\delta = M_c$, where δ is always greater than unity but may be used as unity for "short columns," a term for which exact boundaries will be set in Sec. 14.19.

One important consideration in adopting this moment magnifier method was its present use in structural steel design. The designer must be alert, however, for detailed differences in values recommended for C_m and kl_u, and the addition here of a new term β_d (Sec. 14.18*c*) for creep.

14.18 ADAPTING THE MOMENT MAGNIFIER TO CONCRETE COLUMNS IN FRAMES

(a) General

The moment magnifier is based on an analysis of an elastic curve that increases amplitude but does not change in shape as axial load is applied at the column ends or at the joints of the frame. Although for concrete the deflected shape will change, this normally does not involve serious error.

In a general frame five important problems arise, the last three not being limited to reinforced concrete.

1. The effective EI of reinforced concrete is dependent on the magnitude and type of loading as well as the materials, and varies along the column length.
2. The creep of concrete occurs when load is sustained and creep modifies the effective EI needed in item 1.

*But note the limited alternate in Sec. 14.23.

3. The effective length of column for the calculation of P_c may be either more or less than l_u.
4. The column sway is limited to the story deflection.
5. The maximum moment does not always occur at mid-height.

The answers provided for these problems by any approximate method cannot give an exact result; there are too many basic variables. A somewhat forced fit of an elastic analysis to an inelastic material is necessary. For example, the amount of beam steel or the degree of beam cracking alone can create a scatter of at least 10% in column strength.

(b) Effective *EI*

Stiffness against curvature is measured by EI. In a reinforced concrete column, stiffness decreases (1) as concrete in compression approaches the flatter part of the stress-strain curve, (2) as creep develops under sustained load, (3) as compression steel yields, or (4) as the concrete cracks (for larger eccentricities). The varying degree of stress distribution over the column length means no single value of EI can be a true one for use under all types of loading.

The Code strongly backs a computer type of solution that can sum the behavior over many short elements of length and relate each element to its individual properties and loading. Aside from such a solution it provides a semiempirical value of EI by two separate equations.

Especially where ρ_t of the column is small (0.01, or up to 0.02 for small columns), and especially also for columns that nearly qualify as short columns (where the moment magnifier will be small anyway), the Code suggests

$$EI_1^* = (E_c I_g/2.5)/(1 + \beta_d)$$

where β_d is an allowance for creep discussed in (c). This equation would greatly underestimate EI where the value of ρ_t is large, leading to over-design of the column. For cases having a large ρ_t, unless the column length is still so short as to make the calculation unimportant, economy dictates the use of the other Code equation for EI:

$$EI_2^* = (E_c I_g/5 + E_s I_s)/(1 + \beta_d)$$

With either equation the scatter is broad, as shown in Fig. 14.23, especially so for EI_1. The scatter is essentially all on the safe side, since an EI value too small leads to a magnifier too large.

*The EI_1 and EI_2 subscripts are not in the Code but are convenient for reference.

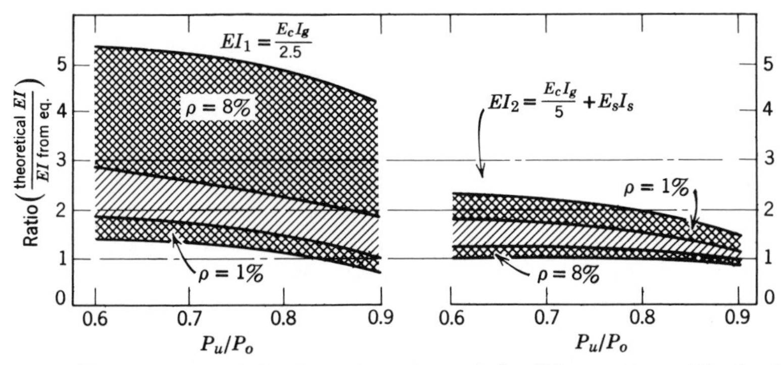

Fig. 14.23. The accuracy of the formula values of the EI equations. The horizontal line at unity would be a perfect fit. (Modified from Ref. 9, *ACI Jour.*)

Since the larger EI, either EI_1 or EI_2, gives the smaller δ multiplier, and since both are safe values, it is sometimes desirable to know the particular ρ_t that makes EI_2 the larger. The break-point will be established[12] by setting the ratio EI_2/EI_1 at unity, considering steel on two faces of a rectangular column:

$$1 = \left(\frac{E_c I_g/5 + E_s I_s}{1 + \beta_d}\right) \div \left(\frac{E_c I_g/2.5}{1 + \beta_d}\right) = \frac{bh^3/60 + (E_s/E_c)\rho_t\, bh(\gamma h)^2/4}{bh^3/30}$$

where γ (gamma) $= (h - 2d')/h$ = relative distance between the steel on the two faces. Multiplying top and bottom by $30/bh^3$:

$$0.5 + 7.5(E_s/E_c)\rho_t\gamma^2 = 1$$

Let $E_s/E_c = n$, as in transformed area calculations. The resulting equation

$$7.5n\rho_t\gamma^2 = 0.5$$

based on $EI_1 = EI_2$ indicates that EI_2 is the larger when ρ_t exceeds $1/(15n\gamma^2)$, that is, when Table 14.1 values of ρ_t are exceeded.

TABLE 14.1. Minimum Values of ρ_t Making EI_2 Govern over EI_1.

f_c' n	3 ksi 9	4 ksi 8	5 ksi 7
ρ_t	$1/(135\gamma^2)$	$1/(120\gamma^2)$	$1/(105\gamma^2)$
$\gamma = 0.6$	0.0205	0.0231	0.0264
0.7	0.0151	0.0170	0.0194
0.8	0.0115	0.0130	0.0149
0.9	0.0091	0.0103	0.0117

For design it is often more convenient to use the ratio of EI_2/EI_1 times an EI_1 value than to start fresh on an EI_2 calculation, that is, to use: $EI_2 = EI_1\,(0.5 + 7.5n\,\rho_t\gamma^2)$ or a similar equation for other type columns.

(c) The Creep Problem

At $l_u/h = 10$ creep effects tend to offset each other, the tendency toward increase of column curvature from creep being offset by the decreasing distribution factor to the column because of its reducing stiffness. Around l_u/h of 16 or 20 it becomes more important, except for columns with reversed moments in braced frames, that exhibit long-column behavior only in extreme cases.

The Code provides for creep by reducing the approximate EI by the $(1 + \beta_d)$ factor in the denominator, where β_d is the ratio of the maximum design dead load *moment* on the column to the maximum design total load moment on the column. This complex area of frame response deserves further study. (It would appear that the ratio of service load moments might be more appropriate, since creep is a function of sustained loading.) The author feels that this β_d correction is a little ahead of the state of the art and may be unimportant in the assessment of an EI value that probably involves a typical error of at least 30%. Fortunately, the effect of β_d (and even of EI) is far removed from the end result of a design. Experience may show that it is only on very unusual long columns that β_d is really of significance. However, β_d is an essential part of the fitting process in braced frames. Even with $\beta_d = 0.4$ the method tends to overestimate* the computer strength (which omits creep) by 8 to 10% when $P_u/f_c'\,bh$ is greater than 0.5 (Fig. 14.30 *b*).

(d) Effective Column Length, kl_u

In simple mechanics the concept of effective column length is well established; a fixed ended column has an effective length of half its overall height between fixed ends. In a frame it is a problem not so simple and, in the author's opinion, one not yet completely solved. If it were only a matter of the frame reaction to a statically determined moment applied directly as a load on the column, the effective length would be a relatively simple matter of frame or joint stiffness.

*But note the comment in the next to last paragraph of Sec. 14.22 indicating probable improvement if relative column stiffness ψ were based on better relative I values.

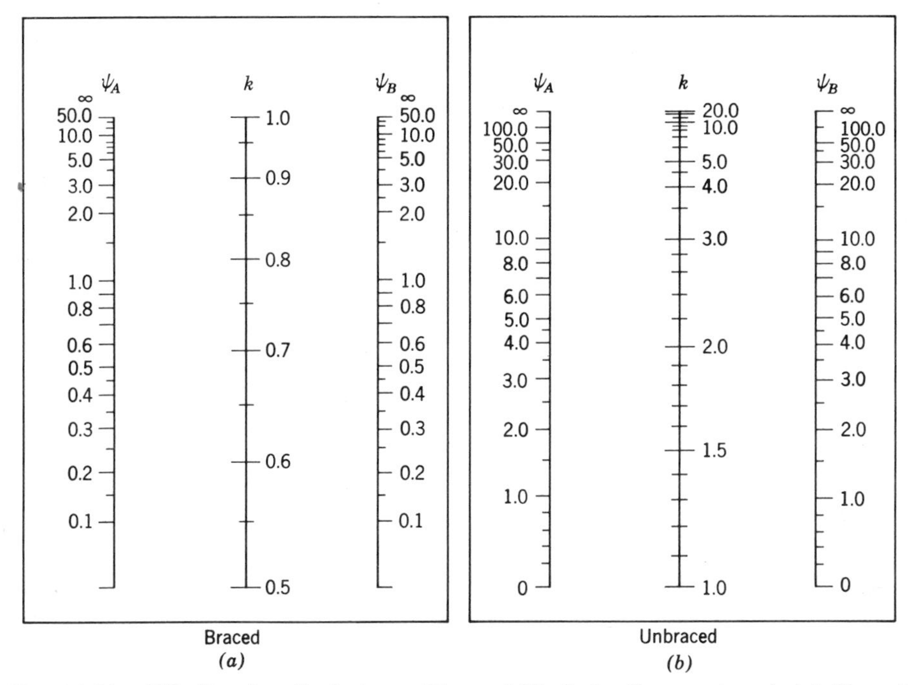

Fig. 14.24. Effective length factors. (From ACI Code Commentary.) (*a*) Braced frames. (*b*) Unbraced frames. ψ = ratio of $\Sigma EI/l_c$ of compression members to $\Sigma EI/l$ of flexural members in a plane at one end of a compression member; k = effective length factor.

Jackson and Moreland[10] solved the kl_u problem in terms of relative member stiffness and published the nomographs of Fig. 14.24. One enters with the values of the relative column stiffness ψ at each end of the column; a straight line between the two ψ values reads k on the center scale. A separate nomograph is required for the braced and unbraced frames. The greater k values for unbraced frames follow from the knowledge that final failure will tend to be a sidesway mode for the entire story. Table 14.2 tabulates some of the unbraced frame values.

One of the basic problems here is that in reinforced concrete, the true ψ changes as beams crack, as column deflection builds up, as column concrete in compression enters the flatter part of the f_c stress-strain curve, and as compression steel yields. These complications combine with the similar uncertainties in the value of EI just discussed.

The effect of *cracked* beams on the effective length is important in long columns, possibly in the sidesway case more important than its effect on

TABLE 14.2. Tabulation of k Factors for *Unbraced* Frames (by Dr. W. Furlong).

ψ	0.0	0.2	0.4	0.6	0.8	1.0	1.2	1.4	1.6	1.8	2.0
k	1.00	1.07	1.16	1.23	1.29	1.34	1.39	1.44	1.48	1.52	1.56

ψ	2.0	2.5	3.0	3.5	4.0	5.0	6.0	8.0	10.0	15.0	20.0
k	1.56	1.68	1.80	1.91	2.01	2.20	2.38	2.70	2.99	3.60	4.12

$$k = \frac{20 - \psi}{20}\sqrt{1 + \psi} \text{ for } \psi < 2$$

$$k = 0.9\sqrt{1 + \psi} \text{ for } \psi > 2$$

end moments generally, already discussed in Sec. 14.15. The code (10.11.3) specifies:

> For compression members not braced against sidesway, the effective length factor k shall be determined with due consideration of cracking and reinforcement on relative stiffffness, and shall be greater than 1.0.

The best thinking for this sidesway case at present* is that relative stiffness should be based on the I for the cracked beam (a simple transformed area) and for the column on EI_2 as in (b) above, with $\beta_d = 0$, and all divided by E_c, that is,

$$I_{\text{col}} = EI_2/E_c = (E_cI_g/5 + E_sI_s)/E_c = I_g/5 + nI_s$$

With ψ calculated on this basis, the value of k for kl_u in the sidesway case may be substantially higher. Since P_c varies with $1/(kl_u)^2$, the difference can be important.

For the braced frame the Code is less specific: ". . . k shall be taken as 1.0, unless an analysis shows that a lower value may be used." This seems to be encouraging the use of $k = 1.0$. Analysis might be simply the use of the Jackson-Moreland charts of Fig. 14.24 or might go so far as a ψ calculated similarly to that for the sway cases. The author feels that $k = 1.0$ may be the best answer, but the Sec. 14.21c example uses the chart k value.

(e) Column Sway Limited to Story Deflection

Except in isolated cantilever columns, when a single column sways, an entire floor or level must move relative to another. This story sidesway is

*Notice that use of beam I_g reduces safety (Sec. 14.22 and Fig. 14.20c). Appendix D, Fig. D.8, shows some ratios I_{cr}/I_g.

also a typical mode of failure in an unbraced frame under vertical load alone. A floor system is typically stiff enough to make all columns deflect alike, unless a torsional loading adds a rotation to the structure. Thus each column is not free to deflect independently as required to carry a fixed shear. Instead, the shear carried is a function of its stiffness relative to all the columns in the story. This discussion will omit the more complex effect of wind on unsymmetrical structures which can result in a torsional rotation; even this case marshals the unequal sway of the columns into an ordered, interrelated group movement that depends on the group stiffness.

Thus there are few cases where a column free to sway can be designed by itself; a single column supporting a hyperbolic paraboloid may qualify. Possibly also a bent or a series of single bents carrying a roof with overhanging ends and not much eccentricity of loading on the columns would qualify in that parts of the system might all tend to react alike. Typically there will be different column sizes, different beam restraints or, if nothing more, exterior columns braced by one beam combined with interior columns braced by two beams at each floor level.

In a story height the stiffest column tends to pick up horizontal load, possibly more than the designer tends to assign to it, while the flexible column is braced by the limited deflection of the stiffer column. Any design method for lateral loading involves an initial distribution of the horizontal shear to the individual columns, that is, some type of frame analysis. The 1971 Code introduces for the first time the further interaction betweeen the secondary moments on the various columns. Code 10.11.5.1 requires that a common moment magnifier δ be computed for the story and used for all columns in that story. This δ is to be based on $\Sigma P_u / \Sigma P_c$, with the Σ representing a summation over all the columns in that story, including any short columns. In determining k for use in this P_c calculation, the relative column stiffness ψ should be based on the cracked section of the beam and the reinforcement ratios of beam and column (10.11.3), as discussed in the previous subsection.

The actual data required for the P_c calculations are available only *after* the beams and columns have been designed, which means the design method must start more approximately. In a sense this is true of nearly all design situations, but here the approximations may initially have to be rougher than usual until design experience builds up in this area.

To start the column design in this sidesway case one could (1) make a rough estimate of all columns and beams sizes involved, (2) arbitrarily modify their relative stiffnesses at joints by some typical factor to represent beam cracking* and column reinforcement percentages expected to be used,

*Figure D.8, Appendix D, might help here.

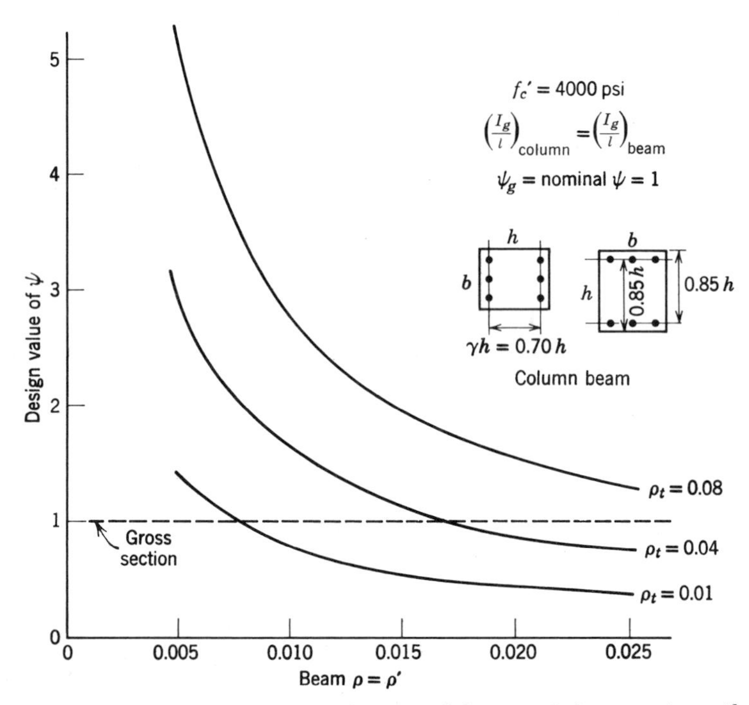

Fig. 14.25. Effect of percentage of column and beam reinforcement on the design ψ for $\gamma = 0.70$ and $f_c' = 4000$ psi. (Modified from chart by Dr. J. E. Breen.)

and on this basis (3) make an initial trial for $\Sigma P_u / \Sigma P_c$. A reference chart such as Fig. 14.25 would be helpful, although limited here to a particular f_c', γ, and column steel arrangement. The ψ design value of this chart is based on equal gross I/l of column and beam. Similar charts can be made for other gross ψ values, and this chart itself can be used as a multiplier of any nominal ψ in roughly estimating the second step above, since here it is in effect a ratio term.

Alternatively,* the designer might compare this story with some other he has computed and simply estimate a trial δ for initial use. Tentative column designs may then be made with the trial δ until it is possible to get a better feel for the real $\Sigma P_u / \Sigma P_c$ for the story; then the better δ value will determine whether some revisions are essential.

*The author also suggests in the middle of Sec. 14.22 that the R method may be a wise choice for preliminary sizing of unbraced columns. Although rather crude, the following has also been discussed as a starting point:

For preliminary design, using $0.5I_g$ for flexural members and I_g for compression members will usually result in reasonable estimates of member sizes.

This story system is more complex than the individual column designs in the past, but also more realistic, and possibly in the long run more economical. It is also probable that the story δ will act as a form of stabilizer in design calculation. A change in a single-column size will modify the story δ less than it would the individual member δ, and changes in all the column sizes should not usually be in the same direction.

The story procedure thus far described does not guarantee an individual slender column against overload. Such a column design might be too weak to handle the maximum moment and vertical load in a braced frame. The Code (10.11.5.1) requires this individual load check to be made (as if in a braced frame); of course, the larger of the two requirements controls that column. This check should only be necessary on the most slender column as a rule.

(f) Maximum Moment Away from Mid-height

The use of C_m has already been noted in Sec. 14.17 along with the governing equation, which is to be used only where sidesway is prevented and where no lateral load acts on the column. This equation is a straight-line approximation to a curve originally developed by Massonnet from steel column tests.

In sidesway under a shear loading, the worst moment stays at the end of the column and the maximum deflection is also there, making $C_m = 1$ appropriate.

In the reversed curvature case without sidesway (Fig. 14.17), which looks in the moment diagram slightly similar to the sidesway case, the deflection adds no *end* moment. Hence behavior is that of a short column until deflection can build up a larger moment away from the joint. The next section shows that most such practical columns would actually be short columns.

14.19 LIMITS FOR SHORT COLUMN CLASSIFICATION

A short column is simpler to design because its response normally follows the dotted line of Fig. 14.22*a*, for its particular eccentricity, directly to the interaction diagram. In the interest of simplifying design the Code has set short column limits that ignore any loss of strength up to 5 %. The limits set on this basis are detailed in the following paragraphs and displayed graphically in Appendix D, Fig. D.5.

In frames not braced, a short column is one with $kl_u/r \lesssim 22$. The radius of gyration r is to be taken as $0.3h$ for a rectangular column, with h taken in the direction considered, or $0.25h$ for a circular column. These limits indicate effective length-to-thickness ratios kl_u/h of 6.6 for rectangular columns and 5.5 for circular columns.

For braced frames the limit as a short column is

$$kl_u/r \lesssim 34 - 12M_1/M_2$$

For the single curvature case, $M_1 = M_2$ results in the same kl_u/r limit of 22 as above, but this limit increases as M_1 decreases to zero (a hinged end), with a ratio limit of 34. As M_1 becomes negative, it becomes about 39 for a far end restrained ($M_1 \doteq -0.4M_2$), or 46 if the member is fully reversed.

For the above ratios, Fig. 14.24 can be entered with the two column-beam stiffness ratios, ψ_1 and ψ_2. A straight edge across the chart between these two values reads k on the center scale. Notice that for braced frames, k will always be 1 or less, for unbraced frames 1 or more. Since r can be expressed in terms of h, one can convert the limiting kl_u/r into limiting l_u/h, and this has been done for some usual values of k in Table 4.2, using the relations

$$l_u/h = (\text{limiting } kl_u/r)(0.3/k) \text{ for rectangular columns}$$
$$l_u/h = (\text{limiting } kl_u/r)(0.25/k) \text{ for circular columns}$$
$$l_u = (\text{tabulated coefficient})h$$

TABLE 14.3. Typical Maximum Short Column Lengths in Terms of Thickness, l_u/h

		Rectangular Column			Circular Column		
Values of $\psi_1 = \psi_2 = \psi =$		0.5	1.0	2.0	0.5	1.0	2.0
Braced Frame k =		0.68	0.77	0.86	0.68	0.77	0.86
Single curvature	$M_1 = M_2$	9.7	8.6	7.7	8.1	7.2	6.4
	$M_1 = 0$	15.0	13.3	11.9	12.5	11.1	9.9
Reversed curvature	$M_1 = -0.4M_2$	17.1	15.1	13.6	14.2	12.6	11.3
	$M_1 = -M_2$	20.3	17.9	16.1	17.0	14.9	13.4
Unbraced Frame k =		1.17	1.31	1.59	1.17	1.31	1.59
All M values		5.6	5.0	4.1	4.7	4.2	3.4

The table shows for unbraced frames that only very stocky columns are classified as short. For braced frames where the usual worst moment condition is for far end restrained ($M_1 = -0.4M_2$), all ordinary columns are short columns. For single curvature columns in braced frames, columns are typically short unless the beams are light (larger ψ); also note that single curvature can accompany maximum axial load only when dealing with the unequal spans, and then usually with lower moments than for the case of far end restrained.

14.20 USE OF COLUMN INTERACTION DIAGRAMS

The ACI has a column design handbook[11] with many column interaction diagrams, each based on the concept in Fig. 14.6*b*. However, one major difference occurs in all charts. Instead of ordinates and abscissae in terms of $\overline{P}$ and $\overline{M}$, these charts have their scales adjusted to read $0.7\overline{P}$ and $0.7\overline{M}$ (or $0.75\overline{P}$ and $0.75\overline{M}$ for spiral columns) which are the same as P_u and M_u, since this introduces the value of ϕ. However, as discussed in Sec. 14.7, below $P_u = 0.10 f_c' A_g$, ϕ increases and the Code charts need modification.*

A single set of four charts for the usual range of steel placement ratios, described by four values or the ratio $\gamma = (h - 2d')/h$, for $f_c' = 4$ ksi and $f_y = 60$ ksi are given in Fig. 14.26*a* to *d*. Since each plot contains ratios for axial load, moment, eccentricity, and steel, any two of these, in the analysis of a given column, establish the other two. For example, P and M in a given short column fix e/t and the necessary $\rho_t m$† $= \rho_t(f_y/0.85 f_c')$.

14.21 DESIGNS WITH USE OF COLUMN INTERACTION DIAGRAMS

(a) Revised Format for Design Equations

A design problem starts with known, or easily established, data for f_c', f_y, P_u, e (or M_u), β_d (the ratio of dead load moment to total design moment), l_u, and k. The major unknowns are h (or b and h) and ρ_t, although there may be a desired ρ_t which can be used as a target.

For long-column design it is helpful to restate the EI and P_c equations in terms of the unknown column thickness h as Furlong has done[12]; this eliminates some awkward number repetitions. For square tied columns with steel on two faces:

$$EI_1 = \frac{E_c I_g/2.5}{1 + \beta_d} = \frac{57{,}000\sqrt{f_c'}\,h^4/(2.5 \times 12)}{1 + \beta_d}$$

$$= \frac{1900\sqrt{f_c'}\,h^4}{1 + \beta_d} = (\text{constant})h^4$$

*In this area the method of Sec. 14.9 (analysis) and Sec. 14.25c,d is particularly good.

†$m = f_y/0.85 f_c'$ is 1963 notation; m should now become a lower case Greek letter.

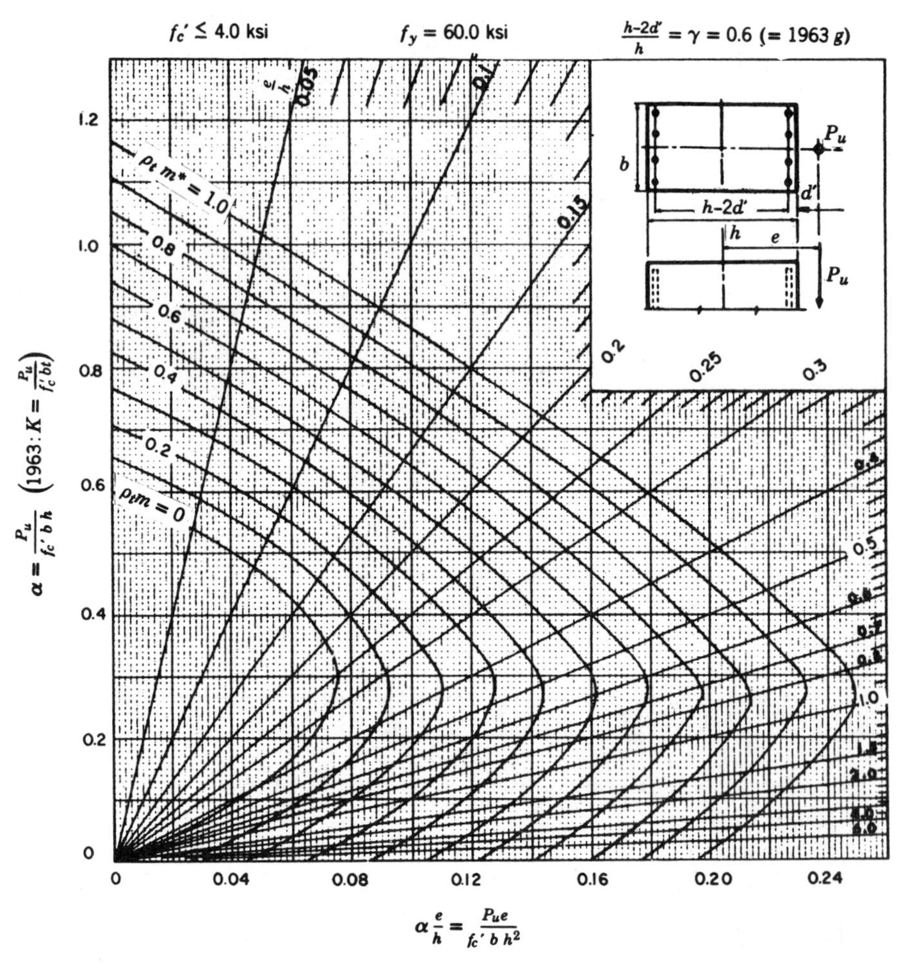

Fig. 14.26a. Column Interaction Chart. $\gamma = 0.6$. $m = f_y/0.85 f_c'$ is 1963 notation. m now should become a lower-case Greek letter.

$$\phi P_c = \phi \pi^2 EI/(kl_u)^2 = 0.7 \times 9.86\, EI_1/(kl_u)^2$$
$$= (\text{another constant})\, h^4$$

If ρ_t exceeds the values of Table 14.1 (Sec. 14.18b), the other equation for EI controls:

$$EI_2 = (E_c I_g/5 + E_s I_s)/(1 + \beta_d)$$

The ratio of this equation to the EI_1 value above was established in Sec. 14.18*b* as:

$$EI_2/EI_1 = 0.5 + 7.5 n \rho_t \gamma^2$$

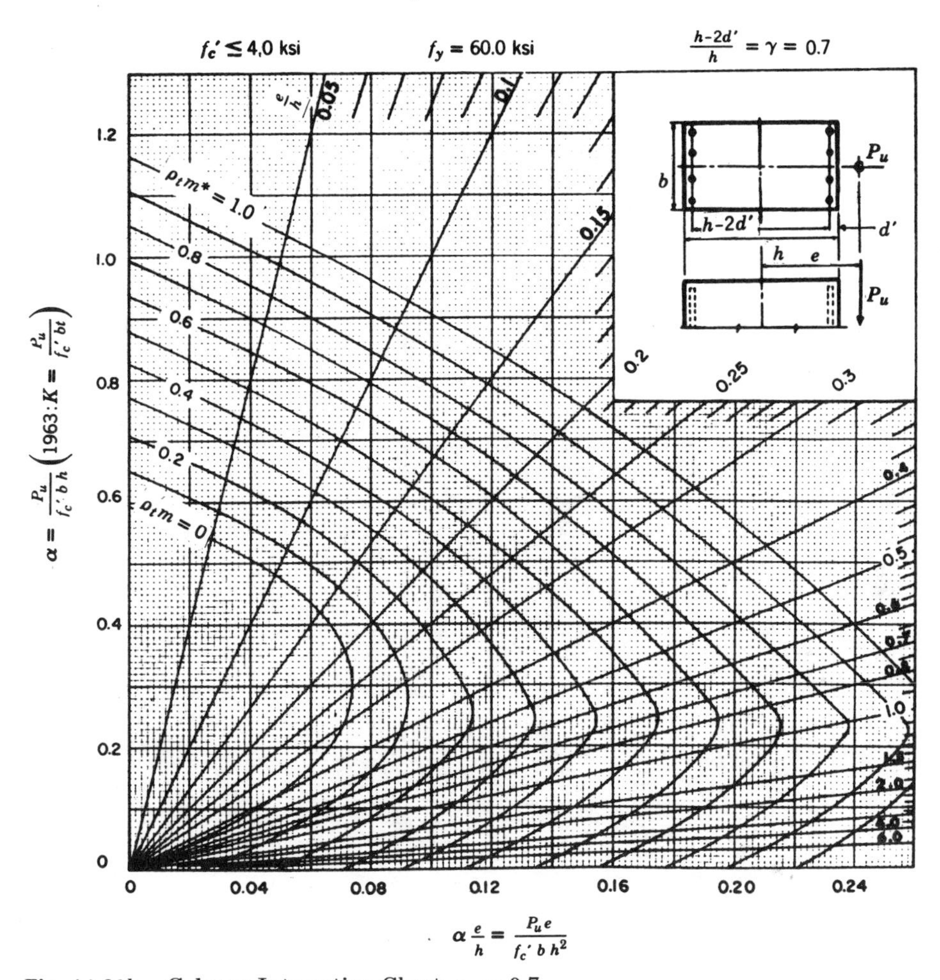

Fig. 14.26*b*. Column Interaction Chart. $\gamma = 0.7$.

Since ρ_t is not known in advance, it is frequently simplest to start with EI_1 and, when ρ_t has been roughly fixed, to find EI_2 and the new P_c with the above ratio applied to the old EI_1 values.

The curves of Fig. 14.27 are sometimes helpful instead of solving the equation for δ directly, particularly when $C_m \neq 1.0$.

(b) Short-Column Design

Design a square tied column with $l_u = 9$ ft, $f_c' = 4000$ psi, Grade 60 steel, minimum eccentricity with no dead load moment ($\beta_d = 0$), a total

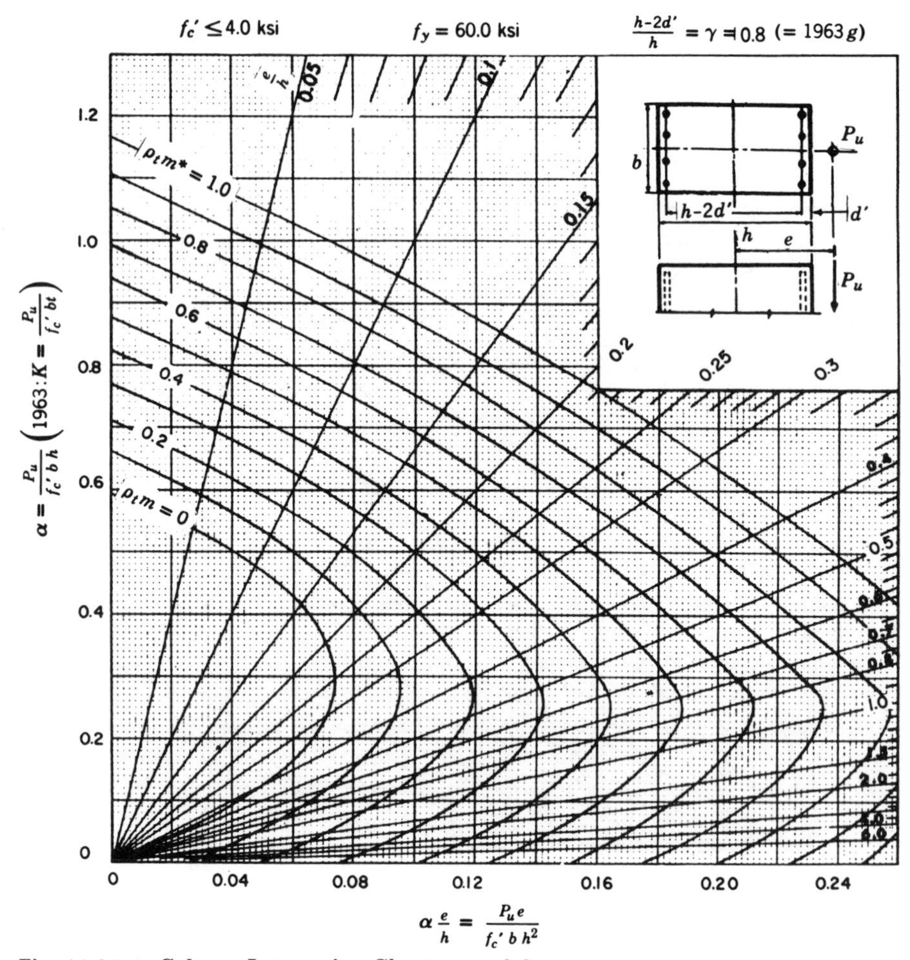

Fig. 14.26c. Column Interaction Chart. $\gamma = 0.8$.

dead load (including its own weight) of 200 k and a live load of 250k. Assume ψ (the ratio of column-beam stiffness) $= 2$ in a braced frame.* Try for ρ_t of about 0.02.

Solution

k (Fig. 14.24) $= 0.85$, $\quad kl_u = 0.85 \times 9 \times 12 = 91.8$ in.

*Note that a given ψ in an example is always an oversimplification. It assumes a specific relative column stiffness which the design may not provide.

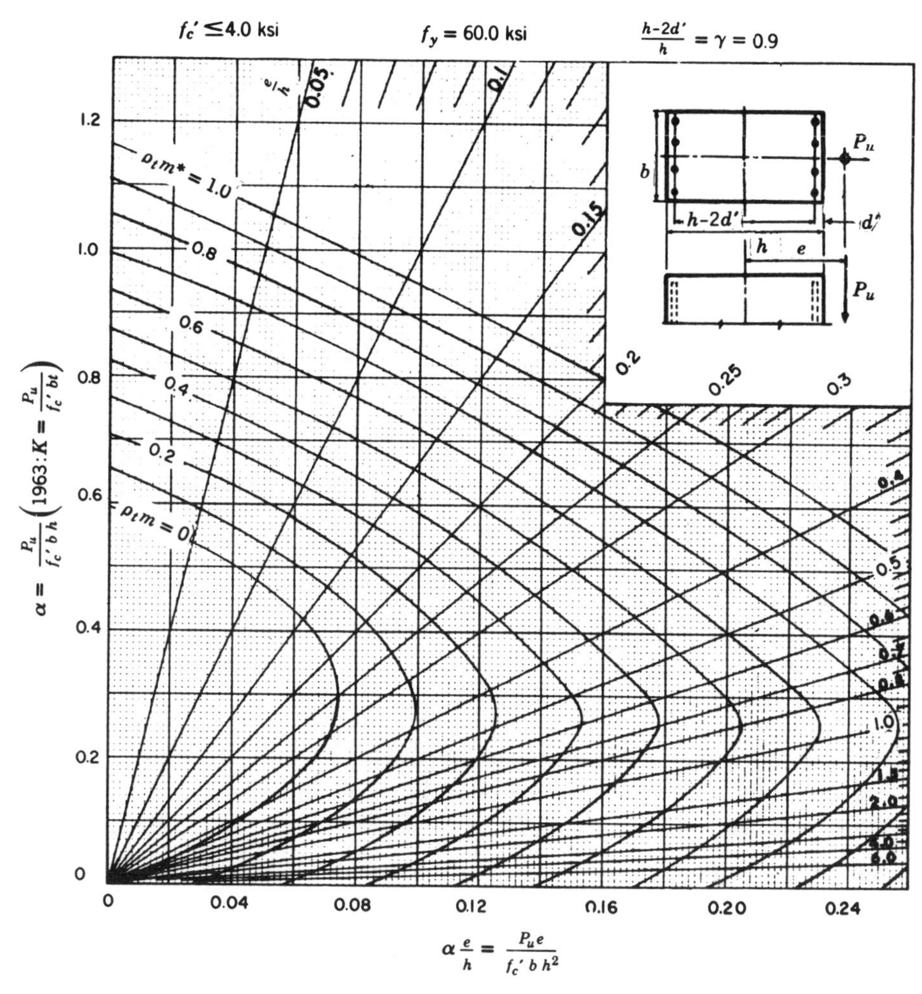

Fig. 14.26*d*. Column Interaction Chart. $\gamma = 0.9$.

The column must be assumed in single curvature, since no data on curvature are given; it is classified as a short column up to $kl_u/r = 22$. For the kl_u above and $r = 0.3h$, this means a short column if h is at least 13.7 in.

$P_u = 1.4 \times 200 + 1.7 \times 250 = 705\text{k}$

$m = f_y/0.85f_c' = 60/(0.85 \times 4) = 17.7$

Desired $\rho_t m = 0.02 \times 17.7 = 0.354$

Minimum $e/h = 0.10$ controls.

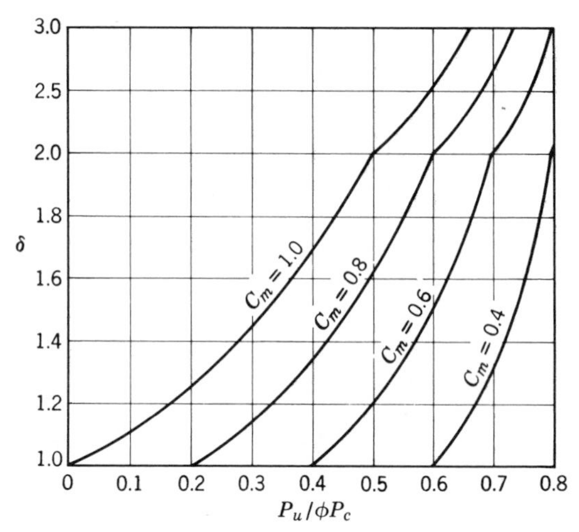

Fig. 14.27. Chart for determining the moment magnifier δ. (Adapted from Furlong, Ref. 12, ACI.)

Assume the steel layers spaced $\gamma h = 0.7h$ and enter chart in Fig. 14.26*b* at the intersection of this e/h and $\rho_t m$ to read on the left

$$\alpha = 0.64 = P_u/f_c'h^2 = 705/(4h^2), \qquad h = \sqrt{275} = 16.6 \text{ in}$$

One can choose a 16-in. square column with $\rho_t > 0.02$ or a larger column with $\rho_t < 0.02$. Column sizes are usually in full inches, often limited (to avoid so many different sizes) to even numbers.

USE 16in. $\times$ 16 in.

$h - 2d' = 16 - 2(1.5 \text{ clear} + 0.375 \text{ tie} + 0.5 \text{ half bar})$
$= 11.25$ in., or possibly less, $= 0.702\,h,\ \gamma = 0.70$

No interpolation for exact γ is needed here. Reenter chart with

Actual $\alpha = 705/(4 \times 16^2) = 0.687$

At $e/h = 0.10$ read required $\rho_t m = 0.48$, $\rho_t = 0.48/m = 0.0272$.

Reqd. $A_s = 0.0272 \times 16^2 = 6.96\text{in.}^2$

For one layer on opposite faces, this must mean an even number of bars. It is simplest, but not essential, if there is only one size of bar.

4- #11 = 6.24in.^{2} (too small).
6- #10 = 7.62
8- #9 = 8.00
USE 6- #10 with #3 ties at 16 in.

A single tie around all bars is satisfactory, since no bar is then more than 6 in. from a restrained corner bar. The spacing is governed by column h, since 48 tie diameters or 16 bar diameters is greater.

Actual $\gamma = (16 - 5)/16 = 0.687$

This γ is slightly less favorable than the 0.70 used, but the excess steel provided is more than ample to take care of this.

A complete design will also involve the moments about the perpendicular axis.

(c) Long Column in Braced Frame

Assume the column above with $P_u = 705\text{k}$ is in single curvature with $e/h = 0.10$, but now with $l_u = 20$ ft, $\psi = 1$ in a braced frame.* Design this column, again with ρ_t about 0.02.

Solution

k for ψ of $1 = 0.77$, $kl_u = 0.77 \times 20 \times 12 = 185$ in. This is a long column, and one must estimate the column size or e/h as a start. Since e/h is given, one cut-and-try step is eliminated. However, the moment will be magnified in a long column. Assume δ is 1.3, making $\delta e/h = 0.13$. $\rho_t m = 0.02 \times 17.7 = 0.354$. Since the column will also be larger than before with about the same cover, try Fig. 14.26c with $\gamma = (h - 2d')/h = 0.8$ and enter with $e/h = 0.13$ and $\rho_t m = 0.354$ to read

$$\alpha = 0.605 = P_u/f_c'h^2 = 705/4h^2, \qquad h = \sqrt{292} = 17.1 \text{ in.}$$

Try 18 in. $\times$ 18 in., since this was just a guess at δ. For this small ρ_t, try first $EI_1 = (E_cI_g/2.5)/(1 + \beta_d) = E_cI_g/2.5$, if it is assumed that the moment is wholly from live load, making $\beta_d = 0$. From Code 8.3.1,

$$E_c = 57{,}000\sqrt{f_c'} = 57{,}000\sqrt{4000} = 3.61 \times 10^6 \text{ psi} = 3.61 \times 10^3 \text{ ksi}$$

$$EI_1 = 3.61 \times 10^3 \times 18^4/(12 \times 2.5) = 12.6 \times 10^6 \text{ k-in.}^2$$

$$\phi P_c = \phi\pi^2 EI_1/(kl_u)^2 = 0.70\,\pi^2 \times 12.6 \times 10^6/185^2 = 2540\text{k}\dagger$$

$$C_m = 0.6 + 0.4M_1/M_2 = 1.0 \text{ (Code 10.11.5)}$$

$$\delta = C_m/(1 - P_u/\phi P_c) = 1/(1 - 705/2540) = 1/(1 - 0.277) = 1.38$$

$$M_c = M_2\delta \text{ or design } e/h = \delta e/h = 1.38 \times 0.10 = 0.138$$

$$\gamma = (h - 2d')/h = (18 - 2 \times 2.5)/18 = 0.72, \quad \rho_t m = 0.354$$

*See footnote in example of (b).

†Notice that if k were used as 1.0 (as suggested in Sec. 14.18d), l_u would be 240 in. and δ would increase sharply, probably calling for a column of 18 in. or 19 in. compared to 17 in. used.

From Fig. 14.26*b* with these e/h and $\rho_c m$ values, read

$$\alpha = P_u/f_c'h^2 = 0.585 = 705/4h^2, \qquad h = \sqrt{302} = 17.4 \text{ in.}$$

Although a reduction from the 18-in.-square column will increase δ a bit and will increase the required ρ_t,

USE 17 in. $\times$ 17 in.

$$EI_1 = 3.61 \times 10^3 \times 17^4/(12 \times 2.5) = 10.0 \times 10^6 \text{ k-in.}^2$$

Since ρ_t will be over 0.02, guess a value of 0.025 and calculate EI_2, using the ratio developed in (a) above. $\gamma = 12/17 = 0.708$.

$$EI_2 = EI_1\,(0.5 + 7.5n\,\rho_t\gamma^2)$$
$$= 10.0 \times 10^6\,(0.5 + 7.5 \times 8 \times 0.025 \times 0.708^2) = 12.5 \times 10^6 \text{ k-in.}^2$$

This larger value is a better value, since the formulas tend to underestimate EI. It happens that EI_2 is essentially the EI_1 used in the previous trial and, hence, δ will also be unchanged. With e/h given as a ratio, no recycling is necessary with the change in column size. Typically, EI will be modified and e/h will change with h (e being fixed by M_u/P_u), and a recycle will be necessary.

Actual $\alpha = 705/(4 \times 17^2) = 0.61$

Again in Fig. 14.26*b* for this α and the design e/h of 0.138, read $\rho_t m = 0.42$.

Reqd. $A_s = (0.42/17.7)(17^2) = 6.85$ in.^{2}

USE 2 – #9 + 4 – #10 (2.00 + 5.08 = 7.08 in.^{2})

Actual $\rho_t = 7.08/17^2 = 0.0245$ vs. 0.025 assumed for EI

Actual $h - 2d' = 17 - 2(1.5 + 0.375 + 0.63) = 12.0$ in. as assumed

At this e/h the charts are not sensitive to small changes in γ, but it should be noted that as e/h moves into the tension failure region it will typically be desirable to interpolate between two charts to reflect the exact steel location (that is, γ).

USE #3 ties at 17 in.

One tie should go around the whole bar group, and a single leg tie should go across the middle to make, at least, alternate bars tied.

A complete design would also involve the moment about the perpendicular axis.

(d) Long Column in Unbraced Frame—Typical Case

To indicate the more serious length problem in an unbraced frame, redesign the column in *c* considering the frame unbraced, $P_u = 705$ k, $e/h = 0.10$, $l_u = 20$ ft., $f_c' = 4$ ksi, $f_y = 60$ ksi, $\beta_d = 0$, and desired ρ_t about 0.02. Assume the story δ is either known or assumed at $\delta = 1.5$.

Solution

In actual unbraced frames one must start with a known or assumed δ and later verify this value for the entire story. With many columns in a story, the selected individual column size will have only a minor effect on δ, but the accumulative effect could be important. In this example the verification, or lack of verification, and its potential influence on the design cannot be resolved in terms of the single column. This design is thus oversimplified. (The design in *e* is at the other extreme and appears over complex.)

From Fig. 14.24, $k = 1.31$, $kl_u = 1.31 \times 20 \times 12 = 314$in.

$\delta e/h = 1.5 \times 0.10 = 0.15$, $m = f_y/(0.85 f_c') = 60/(0.85 \times 4) = 17.7$

Desired $\rho_t m = 0.02 \times 17.7 = 0.354$

Assume $\gamma = 0.80$ and use Fig. 14.26*c* to read α at intersection of $e/h = 0.15$ and $\rho_t m = 0.354$.

Required $\alpha = 0.580 = P_u/(f_c' h^2) = 705/4h^2$, $h = \sqrt{303} = 17.5$ in.

The column may be made 17 in. square with $\rho_t > 0.02$, with the only uncertainty (pending the final fixing of δ) being whether the γ of 0.8 is proper.

TRY 17 in. by 17 in. column

$\gamma h = 17 - 2(1.5 + 0.375 + 0.63) = 12$ in., $\gamma = 0.708$

Return now to Fig. 14.26*b* with $\gamma = 0.70$, to read:

Reqd. $\alpha = 0.57 = 705/4h^2$, $h = \sqrt{309} = 17.6$ in.

The choice is now between a 17-in. and an 18-in. column.

USE 17 in. by 17 in. column

At this $\alpha = 705/(4 \times 17^2) = 0.61$, Fig. 14.26*b* shows for $e/h = 0.15$:

Reqd. $\rho_t m = 0.46$, $\rho_t = 0.46/17.7 = 0.026$
Reqd. $A_{st} = 0.026 \times 17^2 = 7.50$ in.^{2}
$4 - \#10 + 4 - \#7 = 5.08 + 2.40 = 7.48$ in.^{2}
$4 - \#11 + 2 - \#10 = 6.24 + 2.54 = 7.78$ in.^{2}
$8 - \#9 = 8.00$ in.^{2}

USE 4 − #11 + 2 − #10 in single layers at opposite faces, because it permits single ties to be used.

USE single #4 ties (minimum for #11) at 17 in.

This design is subject to verification of δ and must also be checked about the other axis.

(e) Long Column in Unbraced Frame—Extreme Case

Assume now that the column initially described in *d* is one of only four columns in a single structure, so situated that all four columns (and their connecting beams) appear to be identical cases which would lead to identical δ values whether separate or in a group. Based on a known stiffness of the beams (cracked section) and an estimated column size of 18 in. square, with ρ_t about 0.02, the ψ is estimated on the basis of EI_1 to be 1.0. Load eccentricity is 2 in., for example simplicity assumed unchanged as column size varies in design. Redesign for these conditions.

Solution

Initial assumed column $EI_1 = EI_g/2.5 = E \times 18^4/(12 \times 2.5) = 3500\ E_2$ (lb-in.^{2})

$E = 57{,}000\sqrt{f_c'} = 57{,}000\sqrt{4000}/1000 = 3.61 \times 10^3$ ksi

For $\psi = 1$, Fig. 14.24 shows $k = 1.31$, $kl_u = 1.31 \times 20 \times 12 = 314$ in.

$\phi P_c = 0.7 \times \pi^2 EI/(kl_u)^2 = 0.7 \times 9.86 \times 3500 \times 3.61 \times 10^3/314^2$
$= 885\text{k}$

$\delta = 1/(1 - P_u/\phi P_c) = 1/(1 - 705/885) = 1/(1 - 0.795) = 4.87$

This δ is too large to be practical. Somewhere around 2, maybe 3, is a top practical limit.

Try 20 in. by 20 in. column
$EI_1 = (20/18)^4\ 3500\text{E} = 5320\text{E}$

Revised $\psi = 1 \times 5320/3500 = 1.53$, $k = 1.46$, $kl_u = 350$ in.

$\phi P_c = 0.7 \times 9.86 \times 5320 \times 3.61 \times 10^3/350^2 = 1090\text{k}$
$\delta = 1/(1 - 705/1090) = 1/(1 - 0.648) = 2.84$
$\delta e/h = 2.84 \times 2/20 = 0.284$
$\gamma = (20 - 2.5 \times 2)/20 = 0.75$
Actual $\alpha = P_u/f_c'h^2 = 705/(4 \times 20^2) = 0.442$

From Fig. 14.26*b*, $\gamma = 0.7$, reqd. $\rho_t m = 0.42$
Fig. 14.26*c*, $\gamma = 0.8$, $\rho_t m = 0.37$
$\gamma = 0.75$, $\rho_t m = 0.395$, $\rho_t = 0.0223$

This is near the desired ρ_t. Check EI_2 which, with this ρ_t, should be larger than EI_1.

$$EI_2 = (3.61 \times 10^3)\, 20^4/(12 \times 5) + 29 \times 10^3 \times 0.0223 \times 20^2\,(15/2)^2$$
$$= 9100 \times 10^3 + 14{,}500 \times 10^3 = 23{,}600 \times 10^3 \text{ k-in.}^2$$
$$I_2 = 23{,}600 \times 10^3/(3.61 \times 10^3) = 6540 \text{ in.}^4$$

Revised $\psi = 1 \times 6540/3500 = 1.87$, $k = 1.55$, $kl_u = 372$ in.
$$\phi P_c = 0.7 \times 9.86 \times 6540 \times 3.61 \times 10^3/372^2 = 1177 \text{ k}$$
$$\delta = 1/(1 - 705/1177) = 1/(1 - 0.600) = 2.50$$
$$e/h = 2.50 \times 2/20 = 0.25$$

Recalculate the required ρ_t, α remaining at 0.442.

$\gamma = 0.7$, $\rho_t m = 0.325$
0.8 0.28
0.75, $\rho_t m = 0.302$, $\rho_t = 0.0171$

This compares to 0.0223 in the last cycle, indicating the calculation is very sensitive when δ is this large. If one recycled with this lowered ρ_t of 0.0171, the I_s portion of EI_2 would drop to $(0.0171/0.0223)\ 14{,}500 \times 10^3 = 11{,}000 \times 10^3$, or the total EI_2 to $20{,}200 \times 10^3$ k-in.^{2}, I_2 to 5600, δ to 1.60, k to 1.48, $kl_u = 355$ in., giving:

$$\phi P_c = 1177 \times (5600/6540)(372/355)^2 = 1100\text{k}$$

This calculation would be about where it was with the earlier EI_1.

Since the use of $\rho_t = 0.0171$ results in overcorrecting, try an average $\rho_t = 0.5(0.0223 + 0.0171) = 0.0197$, say 0.02. Compared to the first EI_2 cycle, the I_s portion of EI_2 portion of EI_2 drops to $(0.020/0.0223)14{,}500 \times 10^3 = 13{,}000 \times 10^3$, EI_2 becomes 1500 less (or $22{,}100 \times 10^3$), I_2 becomes 6120 in.^{4}, $\psi = 1 \times 6120/3500 = 1.75$, $k = 1.52$, $kl_u = 365$ in.

$$\phi P_c = 1177(6120/6540)(372/365)^2 = 1150\text{k}$$
$$\delta = 1/(1 - 705/1150) = 1/(1 - 0.614) = 2.60,\ \delta e/h = 0.26$$

$\gamma = 0.7$, $\rho_t m = 0.36$
0.8 0.32
0.75 0.34, $\rho_t = 0.0192$

This shows $\rho_t = 0.02$ is close enough. USE 20 in. by 20 in. column.

Reqd. $A_{st} = 0.02 \times 20^2 = 8.00$ in.^{2}
Use 8 – #9 bars in single layers near opposite faces, #3 ties at 18 in., including an overall tie plus one crosstie to tie at least "alternate bars."

If, as stated initially, the structure is the same about both axes, A_{st} spread on all four faces would fit better, but for this arrangement a different set of column design charts is needed (available in Ref. 11). The effective γ would be reduced by this less effective steel arrangement (less effective for the single axis), and a little more total steel would be necessary.

Emphasis needs to be added as to the role of the beam in the unbraced frame. *All* the increased column moment must also be resisted by the beams. These increases are secondary beam moments to be added to the beam in addition to the first-order beam moments given by frame analysis. With such a large δ as in this example, there must be large resisting beam moments. The designer might consider whether a column larger than 20 in. square with a ρ_t lower than 0.02 and a δ lower than 2.60 might not be a stiffer and better design.

Although, in the particular case above, the 20-in. column is barely inside the limit for the R-method, the reader should compare the alternate design by the R-method in Sec. 14.24b. Within the boundaries suggested in the Commentary (and in Sec. 14.23), the author prefers the simpler approach of that method. Very slender columns *must* be designed by the moment magnifier method or a computer analysis; the R-method is not adequate beyond its stated limits.

14.22 ALTERNATE R METHOD VERSUS THE CODE MOMENT MAGNIFIER METHOD

The 1963 ACI Code used the R-method for all long columns. For a column long enough to be out of the short-column classification, a multiplier R was found that was less than unity. The permissible P_u and M_u were each reduced by this multiplier R, as sketched in Fig. 14.28, with no change in the nominal eccentricity. The use of R was designed to define a lower-bound measure of the column strength. Computer analyses and tests at the University of Texas at Austin verified that, although generally safe, the R-method was in some areas overconservative, was somewhat erratic for stability failures, and could be unsafe in some cases of instability.

Although not included in the 1971 Code, the R-method *with more limited boundaries and two new (modified) equations* is continued in the Commentary as a possible alternate. The details are presented in Sec. 14.23.

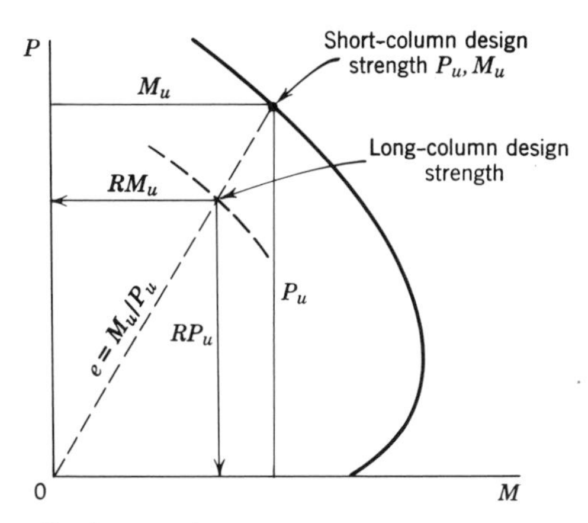

Fig. 14.28. The R-method concept.

For columns with sidesway under shear loading it is limited to columns of about the length used in the design of Sec. 14.21d and e; but within this limit it is as accurate as the magnifier method, as can be seen by comparing the R-method in Fig. 14.29*a* and the Code magnifier method in Fig. 14.29*b*. The special test value and all of the points noticeably low in *a* are eliminated by the Commentary restrictions, namely, limiting effective length l_{lu} to l_{lu}/r of 40 and requiring beam negative moment steel of at least $\rho = 0.01$.

For the R method the ψ value is simply that given by the ratio of *gross I* values; this value will here be designated as ψ_g for identification as such. Accurate values of I_g and ψ_g can enter into the design process much earlier than can the δ for the magnifier method. Apparently it is only on the longer columns that the physical difference is significant.

For the Code magnifier method the relative column stiffness ψ in an *unbraced* brame must be based on the cracked beam section and a column I consistent with the EI_2 equation (10.11.3). Such values lead to the excellent results plotted in Fig. 14.29*b*. The need for *this* value of ψ for the magnifier method is shown by the poorer results of Fig. 14.29*c* where ψ_g values were used in the calculations.

In the sidesway case the Code calls for consideration of the entire story in fixing the magnifier to be used uniformly for all the columns in that story. This requirement seems broad enough to require that columns which are officially "short columns" must go into the determination of the story δ and must have their design moments likewise increased by the factor. Although initiating this process into a design appears somewhat awkward, it probably provides some overall economy in that more flexible

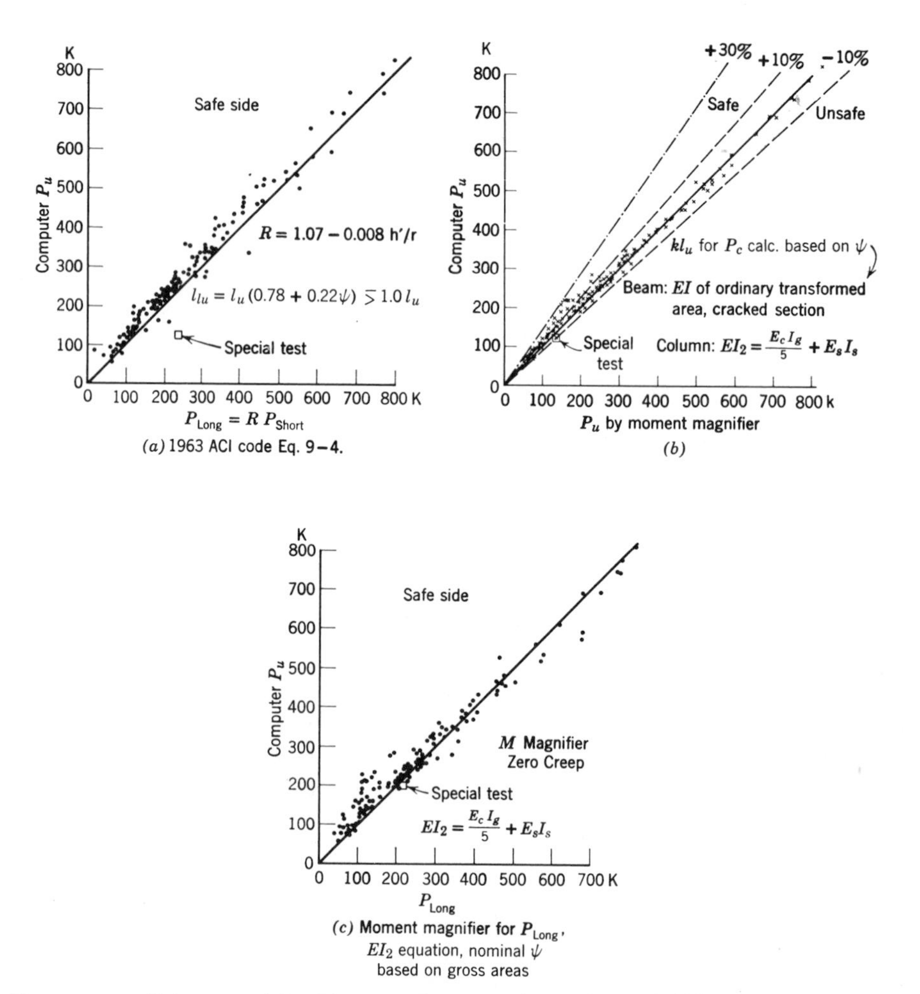

Fig. 14.29. Columns with *sidesway*. Comparative safety of different methods. All data are for 10 × 10 in. columns without allowance for creep.

columns will be designed for lower δ values than they individually would justify, and the increased δ for the heavier columns will mean only a small increase. Since current tests at the University of Texas seem to indicate that, after the most heavily stressed column reaches its maximum capacity, that column can maintain most of this capacity over a considerable added drift, it appears that either the Code approach or the single-column approach may prove adequate. The R method might be a wise choice (where columns are in its range) to use in sizing columns before a story δ can be established for the magnifier method.

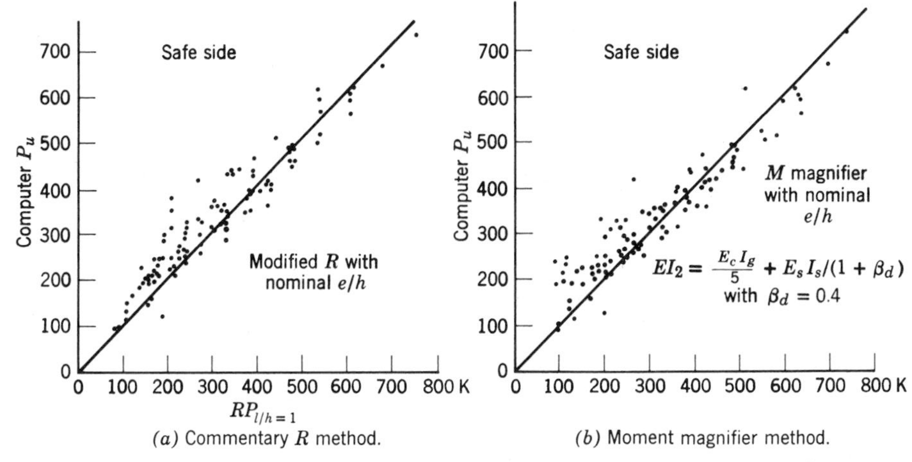

(*a*) Commentary *R* method. (*b*) Moment magnifier method.

Fig. 14.30. Comparative safety of two methods for single curvature columns in *braced* frames, nominal 10 × 10 in. columns. Both use ψ_g.

For braced columns a similar comparison of the two methods, using in both cases ψ_g ,is given in Fig. 14.30, where the magnifier method in *b* tends to predict loads a little high for the shorter columns, the upper half of the load range. The *R*-method in *a* shows slightly better for this case, in the author's opinion. If the ψ_g used in establishing *k* for Fig. 14.30*b* were made ψ based on the moments of inertia (cracked beam and EI_2/E_c) used in developing the data for Fig. 14.29*b*, it is probable that this excess would be reduced or eliminated. However, there is a Code difference here. For the sidesway case such a technique is required; for the braced frame this technique (by implication) is *not required*, but may be used. If $k = 1.0$ is used with the magnifier method in these braced frames, no study results are available, but this should offset the overestimate indicated.

Since the *R*-method as now limited still includes most practical columns not loaded at large eccentricities and is simpler and faster than the magnifier method, the designer's choice must depend on whether he wishes to keep familar with *two* methods: the *R*-method for the usual case, and the magnifier method for the very long column where moment is a major loading. The moment magnifier deals with more realistic behavior and properly emphasizes that strength is lower *because moment is higher*; it gives moment values easily available for possible revision of the beam strength in the sidesway case. Within the Commentary limits for *R*-design, the moment magnifier is *not safer* and, on the average for the University of Texas investigation, *not more economical* if one excludes from this cost comparison the shorter columns in braced frames where the magnifier slightly overestimates the design strength, represented by the upper points in Fig. 14.30*b*. This economy statement should also be discounted for very large columns (larger than 25 in.) and very high ρ_t values that have not been widely sampled.

14.23 THE ALTERNATE R-METHOD IN DETAIL

The R-method uses a reduction fastor R, linear with respect to the l_u/r ratio, which is applied to the short-column strength, P_u and M_u, to define the permissible long-column capacity. This operation is similar to the use of ϕ applied to the ideal column to give the permissible short-column load. Thus one can sketch three interaction diagrams in Fig. 14.31*a*. The middle one is the one used in the ACI column interaction charts, except below $P_u = 0.10 f_c'$ bt where ϕ has here been changed to agree with the ϕ factor changes in Sec. 14.7 and Fig. 14.9.

For a design loading P_l and M_l on a long column one can move radially from this point on the inner interaction diagram to $P_u = P_l/R$ and $M_u = M_l/R$ and, if he wishes, on to $\overline{P} = P_u/\phi$ and $\overline{M} = M_u/\phi$. With the ACI interaction charts, only the first shift is required to give P_u and M_u. (The moment magnifier method is a different concept and is not presented in terms of such an inner interaction diagram.)

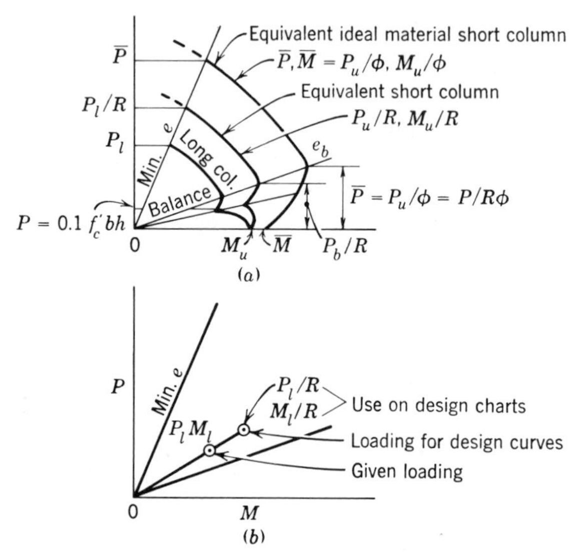

Fig. 14.31. Interaction diagram showing relation of long to short and ideal columns. (*a*) Note $\phi = 0.70$ or 0.75 only above $P_u = 0.10 f_c'bh$; below that see Fig. 14.9. (*b*) Shift from long column loading to design curve.

The required value of R is an empirically derived bounding straight line for each of three cases, two of which are subdivided.

1. In a braced frame for a column loaded by moment at one end but restrained at the far end such that a point of contraflexure occurs:

$$R = 1.32 - 0.006\, l_u/r \not> 1.0 \qquad \text{for} \qquad 54 < l_u/r < 100$$

Up to l_u/r of 54, or l_u/h of 16 for a rectangular column, this is a short column with $R = 1$

2. In a braced frame for a column bent in single curvature:

   a. With the nominal $e \not> 0.1h$

$$R = 1.23 - 0.008 l_u/r \not> 1.00 \qquad \text{for} \qquad 29 < l_u/r < 100$$

Up to l_u/r of 29, or l_u/h of 8.6 for a rectangular column, this is a short column with $R = 1$

   b. With the nominal $e > 0.10h$

$$R = 1.07 - 0.008 l_u/r \not> 1.0 \qquad \text{for} \qquad l_u/r < 100$$

Few practical columns are short columns in this case.

3. If an unbraced frame having $l_{lu}/r \not> 40$ *and* having beams with $\rho \not< 0.01$:

   a. For loads of short duration

$$R = 1.07 - 0.008\, l_{lu}/r \not> 1.0$$
$$\text{where } l_{lu} = l_u\,(0.78 + 0.22\psi_g{}^*) \not< 1.0\, l_u$$

   b. For loads of longer duration

$$R = 0.97 - 0.008 l_{lu}/r \not> 1.0$$

In the equation for l_{lu}, the ratio ψ of column-to-beam stiffness is the average value at the two ends based on gross I of members. When $l_{lu}/r > 40$, the moment magnifier should be used.

The above R equations are diagramed in Fig. D.6 in Appendix D.

Of these equations, the one in $2a$ is considerably liberalized compared to the 1963 Code, but it is for a limited e/h. Also the one in $3a$ is more restrictive in that it formerly applied only to sustained loads. The one is $3b$ is new.

In any case where $P_u \not> 0.10 f_c'bh$, the formula value of R, say R_f, should be increased to unity at P_u of zero in the fashion used for ϕ in Fig. 14.9. The result is: Eff. $R = 1 - 10(P_u/f_c'bh)\,(1 - R_f)$.

In Fig. 14.18 the column moment is shown much increased by the sidesway. This increased column moment must be resisted by the bracing

*The Code Commentary uses r' for ψ_g.

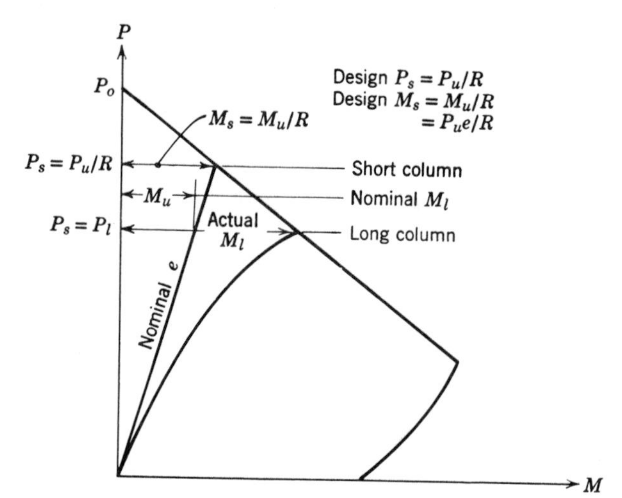

Fig. 14.32. Approximation of actual M_l for beam design. The Code Commentary notation has been altered here to keep P_u for the factored long column load (to match Code 9.2.1.2).

beams, a fact which designers have not always fully realized. The increase shown is largely from lateral load. However, much the same increase occurs at ultimate whether a lateral load is applied outside the structure or whether instability, the common mode of failure, develops internally in the frame. The moment can be estimated from Fig. 14.32 by using similar triangles that lead to

$$M_{bm} = \text{Col. } M_l = \text{Nom. } P_ue\,(1 - P_u/P_0)/(R - P_u/P_0)$$

The actual interaction diagram is curved upward above the straight line and the straight-line simplification is on the safe side for this calculation. $\bar{P}_0$ is defined in Sec. 14.5 and $P_0 = \phi\bar{P}_0$.

14.24 DESIGNS USING THE R-METHOD

(a) Long Column in Braced Frame

Design a square-tied column with $l_u = 20$ ft in a braced frame, $f_c' = 4$ ksi, Grade 60 steel, minimum eccentricity, a total dead load (including its own weight) of 200 k and live load of 250k. Try for ρ_t of about 0.02.

Solution

Assume $l_u/r < 100$, which is not seriously restrictive. To use the liberalized R (which is new in the Commentary), e must not exceed $0.10h$. Here $e = 0.10h$ and

$$R = 1.23 - 0.008 l_u/r = 1.23 - 0.008 \times 240/r = 1.23 - 1.92/r$$

Either column size h or R must be assumed to start. Assume $h = 15$ in.

$$R = 1.23 - 1.92/(0.3 \times 15) = 1.23 - 0.427 = 0.803$$
$$P_u = 1.4 \times 200 + 1.7 \times 250 = 705\text{k}$$
$$\text{Design } P_s = P_u/R = 705/0.803 = 876\text{k}$$

Assume $\gamma h = h - 2d' = 0.7h$ and use Fig. 14.26*b* with:

$$\rho_t m = 0.02 \times 60/(0.85 \times 4) = 0.02 \times 17.7 = 0.354 \qquad \text{and} \qquad e/h = 0.10$$
$$\text{Read } \alpha = 0.64 = P_s/f_c'h^2 = 876/4h^2, \qquad h\sqrt{342} = 18.5 \text{ in.}$$

With increased h, R increases a little and P_s goes down.

Try 18 in. × 18 in. column

$$R = 1.23 - 1.92/(0.3 \times 18) = 1.23 - 0.355 = 0.875$$
$$P_s = P_u/R = 705/0.875 = 806\text{k}$$
$$\gamma h = 18 - 2(1.5 \text{ cover} - 0.375 \text{ tie} - 0.6 \text{ half bar}) = 13 \text{ in.} = 0.724h$$

$$\alpha \text{ (unchanged)} = 0.64 = 806/4h^2, \qquad h = \sqrt{314} = 17.7 \text{ in.}$$

USE 18 in. × 18 in.

$$\text{Actual } \alpha = 806/(18^2 \times 4) = 0.620$$

γ is still close enough to 0.7 to avoid interpolation. Enter Fig. 14.26*b* with $\alpha = 0.620$ and $e/h = 0.10$, to read

$$\rho_t m = 0.30, \qquad \rho_t = 0.30/17.7 = 0.0170$$
$$\text{Reqd. } A_s = 0.0170 \times 18^2 = 5.50 \text{ in.}^2$$

USE 4 – #9 + 2 – #8 (5.58 in. ^{2})
with #3 ties at 16 in. ($16d_b$ governs, Code 7.12.3)

Since spacing of bars is greater than 6 in., a tie across the middle of column, as well as around all bars, is necessary.

Compare this design with that of Sec. 14.21c. The author considers this solution easier and more direct than the moment magnifier method. On the average it gives equal accuracy. The moment magnifier data in Fig. 30*b* include $\beta_d = 0.4$, and it is such data that have the equal average accuracy.

For e/h greater than 0.10 the other R equation (as in 1963 Code) must be used. It gives a smaller R; its economy has not been verified.

(b) Long Column in Unbraced Frame

Redesign the unbraced column of Sec. 14.21d, $P_u = 705\text{k}$, $e/h = 0.10$, $l_u = 20$ ft, $\psi_g = 1$, $f_c' = 4$ ksi, $f_y = 60$ ksi, $\beta_d = 0$, by the alternate R method.

Solution

Two requirements must be met to use this method with sway: (1) the beam negative moment steel must be at least $\rho = 0.01$; (2) the effective length $l_{lu} = l_u\,(0.78 + 0.22\psi) \gtrless l_u$ must not exceed $40r$. The first requirement should be a universal one for any bracing beam in the author's opinion, probably with substantial ρ' in addition. The second is restrictive, but columns not meeting this requirement are borderline cases that deserve special attention.

The following procedure is on the basis that the designer probably has freedom of choice in the section used. For the transient loading

$l_{lu} = l_u = 20 \times 12 = 240$ in. since $\psi_g = 1$, $m = f_y/0.85f_c' = 17.7$
$R = 1.07 - 0.008 \times 240/r = 1.07 - 1.92/r$

Since the column seems long for an unbraced one, assume $R = 0.6$

Design $P_s = P/R = 705/0.6 = 1170\text{k}$

Assume $\gamma h = 0.7h$, making Fig. 14.26*b* the chart to use initially. Enter with $e/h = 0.10$ and $\rho_t m = 0.02 \times 17.7 = 0.354$, to read

$\alpha = 0.64 = P_s/4h^2 = 1170/4h^2, \qquad h = \sqrt{457} = 21.4$ in.

Try 21 in. $\times$ 21 in. column, $R = 1.07 - 1.92/(0.3 \times 21) = 0.766$

Design $P_s = 705/0.766 = 920\text{k} < 1170$. Try smaller column.

Try 20 in. $\times$ 20 in. column, $R = 1.07 - 1.92/(0.3 \times 20) = 0.75$

Design $P_s = 705/0.75 = 940\text{k}$
$\alpha = 0.64 = 940/4h^2, \qquad h = \sqrt{367} = 19.1$ in.
USE 20 in. $\times$ 20 in. column

$l_u/r = 240/(0.3 \times 20) = 40$, just at the limit for R method. O.K. Except for this ratio limit, a 19-in. square column might be chosen, with $\rho > 0.02$.

Actual $\alpha = 940/(4 \times 20^2) = 0.587$
$\gamma h = 20 - 2(1.5 + 0.375 + 0.6) = 15 \text{ in.} = 0.75h$

From Fig. 14.26*b*, $\gamma = 0.70$, $\alpha = 0.587$, $e/h = 0.10$, $\rho_t m = 0.24$

From Fig. 14.26*c*, $\gamma = 0.80$, $\alpha = 0.587$, $e/h = 0.10$, $\rho_t m = 0.23$

For $\gamma = 0.75$, use $\rho_t m = 0.235$, $\rho_t = 0.235/17.7 = 0.0133$
Reqd. $A_s = 0.0133 \times 20^2 = 5.32 \text{ in.}^2$
USE 4 – #9 + 2 – #8 = 4.00 + 1.58 = 5.58 in.^{2}
with #3 ties, including a center tie.

Although the author's initial objective was to obtain a design by the R method comparable to that by the magnifier method of Sec. 14.21*e*, the omission of beam data (especially ρ, ρ', and cover) makes *exact* comparison of the two methods impossible. On the basis of Fig. 14.25, an assumed beam $\rho = \rho' = 0.01$, and the final column design with ρ_t near 0.013, it would appear that ψ of 0.8 or 0.9 might be equivalent to the ψ_g of 1 used here. Similarly for the final magnifier design of Sec. 14.21e with $\psi = 1.75$ and $\rho_t = 0.02$, the equivalent ψ_g might be around $1.75/1.1 = 1.6$. Since ψ_g is not influenced by ρ_t, the *equivalent* ψ value is senstive to the column steel used.

In spite of these differences the column by the two methods was chosen the same size and the steels were of the same order of magnitude. If the R-method design had been made with ψ_g of 1.6, its increased A_{st} would have been closer to the magnifier design. However, looking at the possible wide range of design values of ψ (relative to ψ_g) in Fig. 14.25, the author can make no helpful *general* comparison. Looking at Fig. 14.29*a* and *b*, he does not see any major differences which should result between the two methods, within the restricted usage limits of the R method. It should be noted that the "special test" point in this figure has $\rho = \rho' = 0.0074$, which is too small. Most practical cases are within the R method range.

14.25 DESIGNS WITHOUT THE USE OF THE INTERACTION DIAGRAM

(a) General Scope

The approximate procedures used for analysis is Secs. 14.9 and 14.10 are adaptable to design. For compression failures the equations of Sec. 14.9 are not as rapid or accurate as the interaction diagrams. For tension failures the internal force methods of Sec. 14.10 are *more* accurate than the usual interaction diagrams and give the designer a clearer feel for the problem. Expecially for these cases having axial load less than $0.10f_c'bt$ (with larger ϕ factors and larger R factors where the R-method is used), the direct approach is not distinctly longer than the interaction-diagram approach. The interaction diagrams are not so easily read for large eccentricities, and the required ρ_t is sensitive to γ values, often calling for interpolation between two charts.

Examples in each area follow: that for compression failure being for a short column, and the two for tension failures being for long columns, one each for the R-method and the moment magnifier.

(b) Short Column Failing in Compression

Design a short square tied column with a service load of 100 k dead load plus 105 k live load and an ultimate live load column moment of 30 k-ft. Assume $f_c' = 3000$ psi, $f_y = 60$ ksi. Try for ρ_t about 0.015.

Solution

$$P_u = 100 \times 1.4 + 105 \times 1.7 = 140 + 178 = 318\text{k}$$

(It is well to note here the other load factors of Code 9.3 which must be used with alternate loading involving wind or earthquake.) $\overline{P} = 318/0.70 = 454\text{k}$.

$$e = M_u/P_u = 30 \times 12/318 = 1.13 \text{ in.} < 0.1h \text{ (estimated)}$$

Because of its strong convergence, the Whitney formula (Eq. 14.2) will be used with an iteration type of approach. Convergence is rapid if an arbitrary column size is used in the denominator (only) and bh in the numerators is left as the unknown.

Try $h = 24$ in. (a deliberate poor choice to show convergence)

$d = 24 - 2.5 = 21.5$ in., $d - d' = 21.5 - 2.5 = 19.0$ in.

$b = h$, $e = 0.1h = 2.4$ in.

$$\overline{P} = \frac{A_s'f_y}{e/(d - d') + 0.5} + \frac{bhf_c'}{(3he/d^2) + 1.18}$$

$$454 = \frac{(0.5 \times 0.015h^2)60}{2.4/19.0 + 0.5} + \frac{h^2 \times 3}{\dfrac{3 \times 24 \times 2.4}{21.5^2} + 1.18} = \frac{0.450h^2}{0.626} + \frac{3h^2}{1.553}$$

$$= 0.718h^2 + 1.93h^2 = 2.65h^2$$

$$h = \sqrt{171} = 13.1 \text{ in.}$$

Since this is much smaller than assumed, the next cycle will probably give an intermediate value. Try $h = 14$ in., $e = 0.1h = 1.4$ in.

$$457 = \frac{0.450h^2}{1.4/9 + 0.5} + \frac{3h^2}{\dfrac{3 \times 14 \times 1.4}{11.5^2} + 1.18} = 0.687h^2 + 1.84h^2 = 2.53h^2$$

$$h = \sqrt{180} = 13.44 \text{ in.}$$

USE 14-in $\times$ 14-in. column, $0.1h = 1.4$ in. > 1.13 in., as assumed.

The same equation can now be used with A_s' as the only unknown.

$$457 = \frac{A_s' \times 60}{1.4/9 + 0.5} + 1.84 \times 14^2 = 91.5A_s' + 360$$

$$A_s' = 97/91.5 = 1.06 \text{ in.}^2$$

$$A_s = 2A_s' = 2.12 \text{ in.}^2, \ \rho_t = 2.12/14^2 = 0.0108$$

This ρ_t is lower than the projected 0.015 because of increasing the theoretical 13.44 in to 14 in. It would also be possible to use a 13-in. square column with ρ_t more than 0.015.

USE 4- #7 = 2.40 in. with #3 ties at 14 in. spcg.

$d = 14 - 1.5 \text{ cover} - 0.375 \text{ tie} - 0.44 (= 0.5d_b) = 11.69 \text{ in.} > 11.5$ assumed

Could revise, but say O.K.

The procedure above is direct enough that it is recommended for obtaining a column size even if the designer wishes to use one of the other formulas for final choice of steel. The design will now be checked using the sketches in Fig. 14.33, the idea of Fig. 14.14 and Eq. 14.3 (Sec. 14.9), which requires the tentative design.

$$c_b = \frac{87{,}000}{87{,}000 + 60{,}000} \times 11.69 = 6.94 \text{ in.}, \ a_b = 0.85c_b = 5.90 \text{ in.}$$

$$\epsilon_s' = 0.003 \times \frac{6.94 - 2.31}{6.94} = 0.00201 < 0.00207$$

$$f_s' = 0.00201 \times 29 \times 10^6 = 58.3 \text{ ksi}$$

$$\overline{C}_s = 2 \times 0.60(58.3 - 0.85 \times 3) = 66.9\text{k}$$

$$\overline{C}_c = 0.85 \times 3 \times 14 \times 5.90 = 211 \text{ k}$$

$$\overline{T} = 2 \times 0.60 \times 60 = 72.0$$

$$\overline{P}_b = 211 + 66.9 - 72.0 = 206\text{k}$$

$$\overline{M}_b = \overline{P}_b e_b = 206e_b = 211(7 - 2.95) + 66.9(7 - 2.31) + 72.0(7 - 2.31)$$

$$= 855 + 313 + 337 = 1505 \text{ k-in.}$$

$$e_b = 7.30 \text{ in.}$$

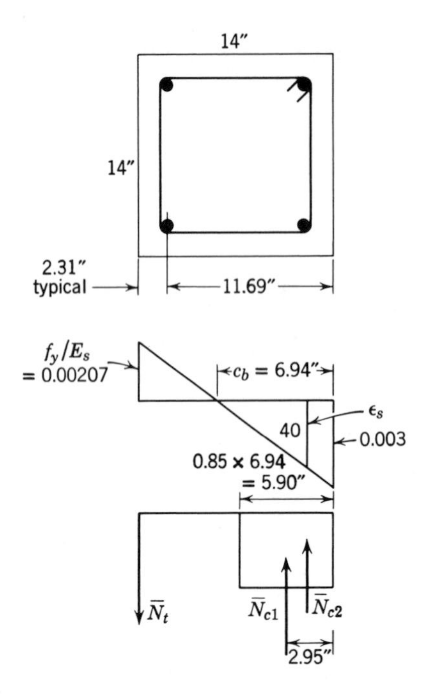

Fig. 14.33. Check on P_b for design of Sec. 14.25b.

$$\bar{P}_0 = 0.85 \times 3(14^2 - 4 \times 0.60) + 4 \times 0.60 \times 60 = 492 + 144 = 636 \text{ k}$$

$$\bar{P} = \frac{P_0}{1 + [(\bar{P}_0/\bar{P}_b) - 1]e/e_b} = \frac{636}{1 + (636/206 - 1)1.4/7.30} = \frac{636}{1 + 0.400} = 454 \text{ k}$$

$$P_u = \phi\bar{P} = 0.70 \times 454 = 318 \text{ k vs } 318 \text{ k}$$

One should not expect to agree so closely, possibly within a few percent.

If this method fails seriously to check the Whitney formula design the only theoretical remedy is to change the steel area slightly and try again, unfortunately rather a slow process. For this reason the author recommends using the Whitney equation as equally valid (without modification or further check) if design aids[11] are not available. The design charts,[11] for this column with 4- #7 bars, give $P_u = 354$ k, some 11% over the required 318 k.

(c) Long Column Design—Tension Failure—R Method

Design a square tied column 14 ft high with a service load of 20k dead load plus 22k live load and an accentricity of 24 in. Use $f_c' = 4$ ksi, $f_y = 60$ ksi, $\psi = 1$. The column is not restrained against sidesway and the R method for length is to be used. (The beams have the requisite 1% of longitudinal steel.) Try for ρ_t of about 0.02.

Solution

A 24-in. eccentricity and light axial loads indicate a probable tension failure where the internal stress distribution at ultimate is easy to handle by the method of Sec. 14.10. When the compression steel yields, the two couples of Fig. 14.13 balance each other. When the stress block depth a is so small that the compression steel does not yield, the designer can substitute $\Sigma M = 0$ about the compression steel; $\overline{N}_{c1}$ then will be so close to the moment center that an approximate value of $\overline{N}_{c1}$ and an approximate value of its arm will have only a small influence of the calculations. This is equivalent to noting that, when one cannot deal with simple couples, the problem almost resolves into a beam type of calculation for the steel, with a minor correction term included to account for the axial load. Note that a specification for a low ρ_t is equivalent to requiring a large arm for the tensile steel in the moment equation, that is, a relatively deep column.

$P_u = 1.4 \times 20 + 1.7 \times 22 = 28 + 38 = 66$ k

Since $\psi = 1$, the effective long-column length

$l_{lu} = l_u\,(0.78 + 0.22\,\psi) = l_u = 14$ ft $= 168$ in.

The R method may be used only if $l_{lu}/r \overline{<} 40$, which means for this length $r = 0.3h \overline{>} l_{lu}/40 = 168/40 = 4.2$ in. The minimum column for this method is 14 in. (A column size should not be selected on this basis, but the designer should be alert to the need for a change to the moment magnifier method if a smaller column seems to fit the loading.)
Try the 14-in. column.

$R = 1.07 - 0.008\,l_u/r = 1.07 - 0.008 \times 168/(0.3 \times 14) = 0.75$

Check whether ϕ and R can be increased because $P_u < 0.10 f_c'h^2$:

$P_u/f_c'h^2 = 66/(4 \times 14^2) = 0.084 < 0.10$
$\phi = 0.90 - 2 \times 0.084 = 0.90 - 0.168 = 0.732$ (Fig. 14.9)
Effective $R = 1 - 0.084 \times 10(1 - 0.75) = 0.79$ (Sec. 14.23)
$\overline{P} = P_u/R\phi = 66/(0.79 \times 0.732) = 114$k

Assume $f_s' = f_y$. The two couples in Fig. 14.34, which neglects displaced concrete, show that $\overline{N}_{c1}$ must equal $\overline{P}$, leading to

$a = 114/(0.85 \times 4 \times 14) = 2.40$ in.

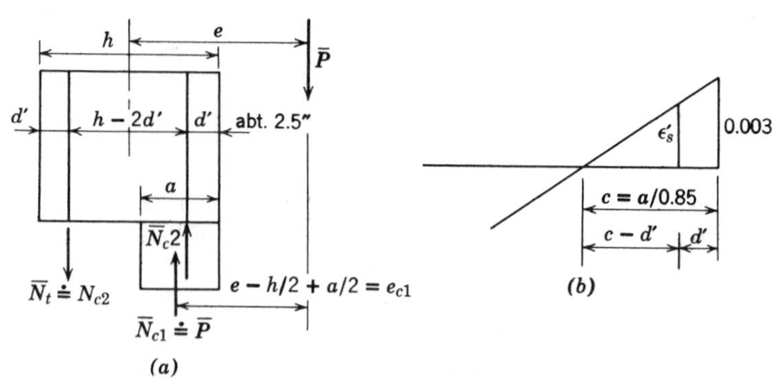

Fig. 14.34. Large eccentricity of load. (*a*) Equilibrium forces. (*b*) Compressive strains.

This places the compression steel very close to the neutral axis (low f_s'); but continue with the couples and the false assumption until a trial ρ_t is indicated. The method is not overly sensitive to shifting between $\overline{N}_{c1}$ and $\overline{N}_{c2}$ at this stage.

$$\overline{N}_t = 0.5A_sf_y, \qquad 0.5A_s60(14 - 5) = 114(24 - 7 + 0.5 \times 2.40)$$
$$270A_s = 2075, \qquad A_s = 7.65 \text{ in.}^2, \qquad \rho_t = 0.0392 > 0.02$$

Try 16-in. column and recycle, first without even revising the depth of stress block.

$R = 1.07 - 0.008 \times 168/(0.3 \times 16) = 0.79$

$P_u/f_c'h^2 = 66/(4 \times 16^2) = 0.0643$

$\phi = 0.90 - 2 \times 0.0643 = 0.771$

Effective $R = 1 - 0.0643 \times 10(1 - 0.79) = 1 - 0.135 = 0.865$

$\overline{P} = 66/(0.865 \times 0.771) = 99\text{k}$

$0.5A_s60(16 - 5) = 99(24 - 8 + 0.5 \times 2.40) = 1700$

$A_s = 1700/330 = 5.15 \text{ in.}^2, \qquad \rho_t = 0.0202$

USE 16 in. × 16 in. column

For A_s' at f_y this gives $a = 99/(0.85 \times 4 \times 16) = 1.93$ in. With this small value of a, the $\overline{N}_{c2}$ steel could be in the tension zone, but, if $\overline{N}_{c2}$ is missing, $\overline{N}_{c1}$ must increase to maintain the total, and this in turn increases a. Obtaining the exact $\overline{N}_{c1}$, a, and f_s' is almost a marginal need, since $\overline{N}_{c1}$ will act not far from A_s' (which will be used as the moment center) and will account for only a small part of the total moment. The designer might prefer to use a little more steel (say, 4- #9 + 2- #8 = 5.58 in.^{2}) and not refine the A_s requirement further.

More to demonstrate the stability of the solution than actually to refine it, investigate $\overline{N}_{c1}$, a, and f_s' further. As an extreme, if c falls at $\overline{N}_{c2}$, $f_s' = 0$, leaving $\Sigma F_y = 0$ or $\overline{N}_{c1} - \overline{P} - 0.5\, A_s f_y = 0$.

$$\overline{N}_{c1} = 99 + 0.5A_s f_y = 99 + 0.5 \times 5.15 \times 60 = 99 + 155 = 254\text{k}$$
$$a = 254/(0.85 \times 4 \times 16) = 4.66 \text{ in.}$$

The depth of the stress block must lie between 1.93 in. and 4.66 in. Try $a = 3.3$ in. (about the average), $c = 3.3/0.85 = 3.88$ in.

$$\epsilon_s' = 0.003(3.88 - 2.50)/3.88 = 0.00107$$
$$f_s' = 29 \times 10^3 \times 0.00107 = 31\text{ksi}$$
$$\overline{N}_{c1} = \overline{P} + \overline{N}_t - \overline{N}_{c2} + \text{correction for displaced concrete}$$

$$\overline{N}_{c1} = 99 + 0.5 \times 5.15(60 - 31) + 0.5 \times 5.15 \times 0.85 \times 4 \text{ for displaced concrete} = 99 + 74.6 + 8.8 = 182\text{k}$$

Take moments about $\overline{N}_{c2}$:

$$\Sigma M = 0 = 0.5A_s \times 60(16 - 5) - 99(24 - 0.5 \times 16 + 2.5) + 182(2.5 - 0.5 \times 3.3)$$

$$330A_s - 1830 + 153 = 0$$

$$A_s = 1677/330 = 5.06 \text{ in.}^2, \qquad a = 182/(0.85 \times 4 \times 16) = 3.34 \text{ in. vs } 3.30 \text{ assumed}$$

Although this agreement was only a fortunate guess, notice that the total influence of the $\overline{N}_{c1}$ term on A_s (in this ΣM equation) was less than 10%.

This looks like the steel should be 4- #9 + 2- #7 (5.20 in.^{2}) with #3 ties at 12 in. (16 d_b controls) without a crosstie. Note, however, that the required A_s is very sensitive to the distance between the steel layers.

$$h - 2d' = 16 - 2(1.5 + 0.375 + 0.5 \times 1.13) = 11.12 \text{ in.} > 11.00 \text{ used O.K.}$$

USE 4- #9 + 2- #7 (5.20 in.^{2})
with #3 ties at 16 in. (without crosstie)

(d) Long Column Design—Tension Failure—Moment Magnifier Method

Design the square tied column for the conditions of (c) except use the moment magnifier for the length effect. Note the special basis upon which ψ must be computed (Sec. 14.18e).

Solution

The comments which opened the solution in (c) are all appropriate here, but will not be repeated.

$P_u = 1.4 \times 20 + 1.7 \times 22 = 28 + 38 = 66\text{k}$

Since $\psi = 1$, Fig. 14.24 for the unbraced frames shows $k = 1.31$

$kl_u = 1.31 \times 14 \times 12 = 220$ in.

Assume $\delta = 1.1$ as a start, noting that this column load is very small and the moment large. This gives $\delta e = 24 \times 1.1 = 26.4$ in. Try a 14-in. column. Check whether ϕ may exceed 0.70, that is, whether $P_u/f_c'h^2 < 0.10$.

$$P_u/f_c'h^2 = 66/(4 \times 14^2) = 0.084 < 0.10$$
$$\phi = 0.90 - 0.084 \times 2 = 0.732$$
$$\overline{P} = P_u/\phi = 66/0.732 = 90.2\text{k}$$

Assume $f_s' = f_y$. The two couples of Fig. 14.34, which neglect displaced concrete, show that $\overline{N}_{c1}$ must equal $\overline{P}$. For $\overline{N}_{c1} = \overline{P}$,

$a = 90.2/(0.85 \times 4 \times 14) = 1.89$ in.

Although this indicates that f_s' is slightly negative and $\overline{N}_{c1}$ must be increased to make up the deficit, continue with the false assumption that $f_s' = f_y$ until a trial ρ_t is indicated. The method is not overly sensitive at this stage to shifting resistance between $\overline{N}_{c1}$ and $\overline{N}_{c2}$. With $\overline{N}_t = 0.5A_sf_y$,

$$0.5A_s60(14 - 5) = 90.2(26.4 - 0.5 \times 14 + 0.5 \times 1.89)$$
$$A_s = 1830/270 = 6.75 \text{ in.}^2, \qquad \rho_t = 6.75/14^2 = 0.0345$$

With ρ_t so large a larger column is indicated, although there are many assumed values in the calculations thus far. The most critical would be δ, and it cannot be signifiaently smaller.
Try a 16 in. column and check δ.

$$P_u/f_c'h^2 = 66/(4 \times 16^2) = 0.0645$$
$$\phi = 0.90 - 2 \times 0.0645 = 0.771$$
$$\overline{P} = 66/0.771 = 85.6k$$
$$E_c = 57\sqrt{4000} = 3.6 \times 10^3 \text{ ksi}$$

For $\beta_d = 0$, $EI_1 = E_cI_g/2.5 = 3.6 \times 10^3 \times 16^4/(12 \times 2.5) = 7.83 \times 10^6$ k-in.^{2}

$$kl_u = 220 \text{ in.}$$
$$P_c = \pi^2EI/(kl_u)^2 = 9.86 \times 7.83 \times 10^6/220^2 = 1590\text{k}$$
$$\delta = 1/(1 - 85.6/1590) = 1/0.941 = 1.06$$
$$\delta e = 1.06 \times 24 = 25.5 \text{ in.}$$
$$0.5A_s(16 - 5) = 85.6(25.5 - 8 + 0.5 \times 1.89)$$
$$A_s = 1580/330 = 4.78 \text{ in.}^2, \qquad \rho_t = 4.78/16^2 = 0.0187$$

Noting that the compression steel is so close to the neutral axis as to be of doubtful value, assume $\overline{N}_{c1}$ has to pick up all of $\overline{N}_{c2} = 0.5 \times 4.78 \times 60 = 143\text{k}$

$\overline{N}_{c1} = 85.6 + 143 = 229\text{k}$
$a = 229/(0.85 \times 4 \times 16) = 4.20$ in.

With this a, the steel becomes partially effective, which would decrease $\overline{N}_{c1}$ and a. Assume $a = 3.0$ in., $c = 3.0/0.85 = 3.53$ in.

$\epsilon_s = 0.003(3.53 - 2.5)/3.53 = 0.00088$
$f_s' = 29 \times 10^3 \times 0.00088 = 25.7\text{ksi}$
$\overline{N}_{c1} = 85.6 + 0.5 \times 4.78(60 - 25.7) + 0.5 \times 4.78 \times 0.85 \times 4$*
$\quad = 85.6 + 81.8 + 8.1 = 176\text{k}$
$a = 176/(0.85 \times 4 \times 16) = 3.23$ in.

Since this operation is not sensitive to $\overline{N}_{c1}$ and a, use $\overline{N}_{c1} = 170\text{k}$ and $a = (170/176)3.23 = 3.12$ in.

Write $\Sigma M = 0$ about compression steel as center:

$$0.5A_s \times 60(16 - 5) - 85.6(25.5 - 8 + 2.5) + 170(2.5 - 0.5 \times 3.12) = 0$$
$$330A_s - 1712 + 160 = 0$$
$$A_s = 1552/330 = 4.70 \text{ in.}^2, \qquad \rho_t = 4.70/16^2 = 0.0183$$

Another cycle with this A_s would reduce $\overline{N}_{c1}$ and a, a pair of changes that have opposite effects on the moment. Say O.K. as calculated.

USE 16 in. × 16 in. column
6- #8 (4.78 in.^{2})
#3 ties at 16 in.

Actual $h - 2d' = 16 - 2(1.5 \times 0.375 + 0.5 \times 1.00) = 16 - 4.75$
$\qquad = 11.25 \text{ in} > 11.0$ used

With this size of bars, the designer could reduce A_s about 2% below the 4.70 in.^{2}, but no apparent arrangement of bars seems to fit better.

14.26 BIAXIAL MOMENTS ON COLUMNS

Many columns are subject simultaneously to moments about both major axes, especially corner columns. The mathematics of such cases is quite involved, although for any given neutral axis the analysis can follow very

*To establish a, N_{c1} is increased (in lieu of reducing its area) to offset the concrete displaced by $0.5A_s$.

simple ideas like those of Sec. 14.6 for P_b and e_b. For example, in Fig. 14.35*a*, let the neutral axis be arbitrarily chosen as shown. One may construct, perpendicular to this axis, strain triangles precisely as done in Fig. 14.15*c* for the circular column. The maximum concrete strain of 0.003 is probably too small and the equivalent rectangular stress block may not be exactly $a = \beta_1 c$, but a thorough investigation[13] has indicated that in combination the two are satisfactory. Then $\overline{N}_{c1}$ is simply $0.85f_c'$ times the shaded area and acts at its centroid (or the area can be broken up into triangles and rectangles if preferred). The strains lead easily to $\overline{N}_t$, $\overline{N}_{c2}$, $\overline{N}_{c3}$ and $\overline{N}_{c4}$. $\Sigma F_y = 0$ leads to the value of $\overline{P}$, $\Sigma M_x = 0$ leads to x_P (equal to e_x along the x-axis), and $\Sigma M_y = 0$ leads to y_P (equal to e_y along the y-axis). This completely establishes the load which would create this neutral axis. The method is awkward chiefly because of the dimensions, which might be easily obtained graphically from a scale layout. The real problem is that one needs to start with given loads and eccentricities and not with a location of the neutral axis; and this neutral axis is not usually perpendicular to the resultant eccentricity.

For a circular column, the interaction diagram for moments about any axis is the same and no real problem exists. It is easy to visualize a three-dimensional interaction diagram for this case. It would be a surface of revolution obtained by rotating the interaction diagram about the $\overline{P}$-axis, a little like Fig. 14.35*b* except with the same diagram on each axis. With a square column having equal steel on each face, the interaction diagram on the x-axis is the same as on the y-axis; but when these are rotated they give values on the unsafe side for intermediate angles. Similarly, if a rectangular column is considered, it is easily to visualize a varying radius of rotation, as in Fig. 14.35*b*, creating an ellipse on any horizontal plane which connects the x- and y-axis diagrams. This would also indicate values on the unsafe side. In order words, both the circle for the square column and the ellipse for the rectangular column must be considered as upper bounds on proper values, the real three dimensional surfaces being a little flattened on the diagonals. This difference is quite small for the upper part of the diagrams and is maximum near the $\overline{P}_b$ level.

The author's present suggestion is that these surfaces be used as a starting point and then the moment capacity at any given $\overline{P}$ level be reduced, by a maximum (for ordinary amounts of reinforcing) of about 15% * when the resultant e is on the 45° line between the x- and y-axes. A study of Furlong's[14] and Pannell's[15] work will permit some refinements on this approximation.

*Pannell shows up to 32% for high-steel ratios.

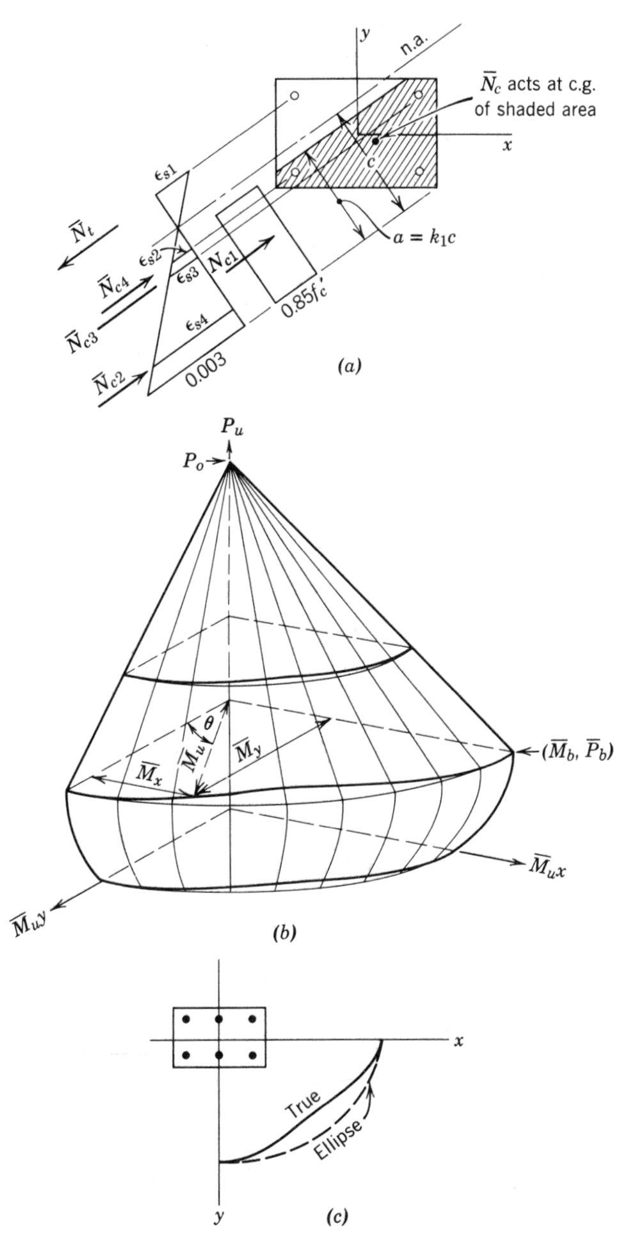

Fig. 14.35. Biaxial bending on column. (*a*) Forces acting on cross section. (*b*) Typical interaction surface (from ACI Ref. 14). (*c*) Contour at level $\overline{P}$ near $\overline{P}_b$.

For *square* columns loaded on the diagonal, interaction diagrams are now available in the ACI handbook[11]. More detailed diagrams have been published for most loading angles.[16]

For long columns the magnifier is to be separately calculated and applied to the moment about each axis independently. The relative values will then determine the plane of the principal moment. The minimum e/h applies about each axis, but not simultaneously.

14.27 DESIGN OF SPIRAL COLUMNS

With interaction diagrams the design of a spiral column differs from that of the rectangular column only in very minor ways. Three changes are necessary. (1) The ϕ factor is 0.75, unless $P_u/f_c'A_g < 0.10$. (2) Either an even or an odd number of bars may be used, with a minimum of six. (3) The spiral involves a strength design, covered in the next section.

Biaxial bending is not a complication when the spiral column is circular, since resistance is equal in all directions. However, spirals are often used in square columns and occasionally dual spirals (overlapping) are used in narrow rectangular columns. Interaction diagrams for the latter are not readily available.

COLUMN DETAILS AND GENERAL CONSIDERATIONS

14.28 SPIRAL DESIGN

The spiral must replace the strength of the concrete shell which can be expected to spall off when the column yields.

The spiral reinforcement is specified by ρ_s in Code 10.9.2:

$$\rho_s = 0.45(A_g/A_c - 1)f_c'/f_y \tag{2.3}$$

where ρ_s = ratio of volume of spiral reinforcement to volume of concrete core (out to out of spiral)

A_c = area of core of column

f_y = yield strength of spiral steel but no greater than 60,000 psi

This equation reflects the strengthening effect of the radial compressive forces from the spiral on the core of the column as it shortens enough to make the unsupported shell spall off. This relation can be arranged as follows:

$$\rho_s A_c f_y = 0.45(A_g - A_c)f_c'$$

$$2\rho_s A_c f_y = 0.90f_c'(A_g - A_c)$$

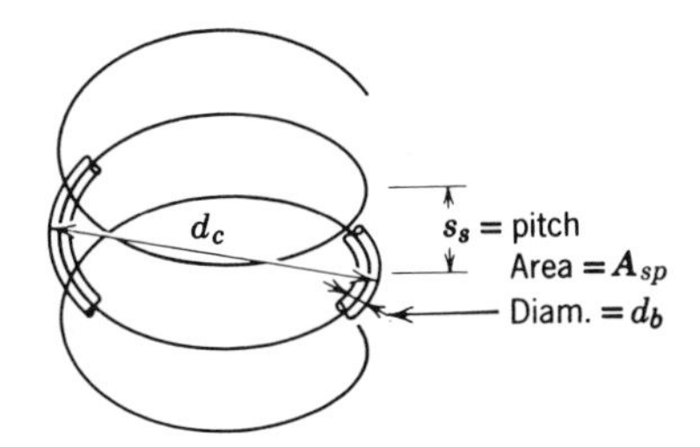

Fig. 14.36. Spiral notation.

Since $A_g - A_c$ is the shell area, which spalls off, the right-hand side represents the assumed ultimate strength of the shell with an increase in the coefficient from 0.85 to 0.90. The quantity $\rho_s A_c$ represents the spiral steel reduced to equivalent longitudinal steel. This steel can develop a stress f_y and in spiral form is roughly twice as effective as longitudinal steel. Thus the spiral is designed to replace the strength of the shell with a very slight extra allowance for safety. For a 21 in. diameter column:

$$\rho_s = 0.45\left(\frac{A_g}{A_c} - 1\right)\frac{f_c'}{f_y} = 0.45\left(\frac{21^2}{18^2} - 1\right)\frac{3500}{60{,}000} = 0.00944$$

By definition, ρ_s = (vol. of spiral in one round) ÷ (vol. of core in height s_s), where s_s is the pitch as shown in Fig. 14.36. The pitch s_s is small enough for the volume of one round or turn of the spiral to be taken as $A_{sp}\,\pi(d_c - d_b)$. The volume of the core in height s_s is $s_s\pi\, d_c^2/4$.

$$\begin{aligned}\rho_s &= A_{sp}\pi(d_c - d_b) \div (s_s\pi d_c^2/4)\\ &= 4A_{sp}(d_c - d_b) \div (s_s d_c^2)\end{aligned}$$

(Many designers ignore the difference between $d_c - d_b$ and d_c, and use $\rho_s = 4A_{sp} \div s_s d_c$.) It is recommended that the size of spiral wire be assumed and s_s then be calculated, since the wire will usually be $\frac{1}{4}$ in.,* $\frac{3}{8}$ in., $\frac{1}{2}$ in., or occasionally $\frac{5}{8}$ in., a rather narrow range of values. The spacing s_s can be specified as closely as desired, usually to quarter inches.

Try $\frac{1}{2}$-in. round spiral rod, $A_{sp} = 0.20$ in.^{2}

$$\rho_s = 0.00944 = 4 \times 0.20(18 - 0.5) \div (s_s \times 18^2)$$

$$s_s = 4.58 \text{ in. (4.08 in. clear)}$$

Specification maximum: 3 in. clear spacing

Specification minimums: 1 in. clear
1.33 max aggregate size, clear

It would be uneconomical to use the $\frac{1}{2}$-in. round at 3.00 in. on center.

* $\frac{3}{8}$ in. is the minimum for cast-in-place construction (Code 7.12.2).

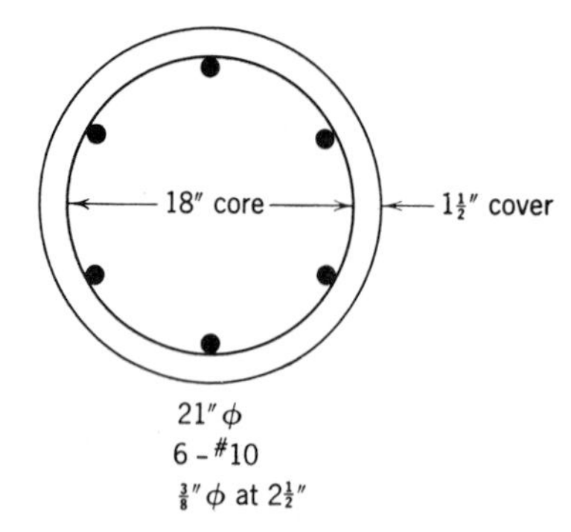

Fig. 14.37. Spiral column design of Sec. 14.28.

Try ⅜-in. round, $A_{sp} = 0.11 \text{ in.}^2$

$$\rho_s = 0.00944 = 4 \times 0.11(18 - 0.38) \div (s_s \times 18^2)$$
$$s_s = 2.54 \text{ in.}$$

USE ⅜-in. round at 2½-in.

The design would be as shown in Fig. 14.37, with the 6- #10 assumed in order to show a total column description in a form frequently used.

14.29 COLUMN TIES

Several of the earlier design examples have included the choice of ties. The details of Code 7.12.3 are rather simple, in effect requiring at least alternate bars to be braced and no bar more than 6 in. from a braced bar. The maximum tie spacing is the smallest of:

1. $16d_b$ for the column bars
2. $48d_b$ for the tie itself
3. The minimum column thickness.

A few typical tie layouts are shown in Fig. 14.38. The crosstie with a round hook on one end and a 90° hook on the other has been developed to facilitate placement when the vertical bars are already in place.

14.30 COLUMN SPLICES

The general subject of splice design is adequately covered in Chapter 5, but some further general comments are appropriate for columns.

Location of column splices in the past has typically been just above the floor level, in effect starting new column bars at the lowest possible

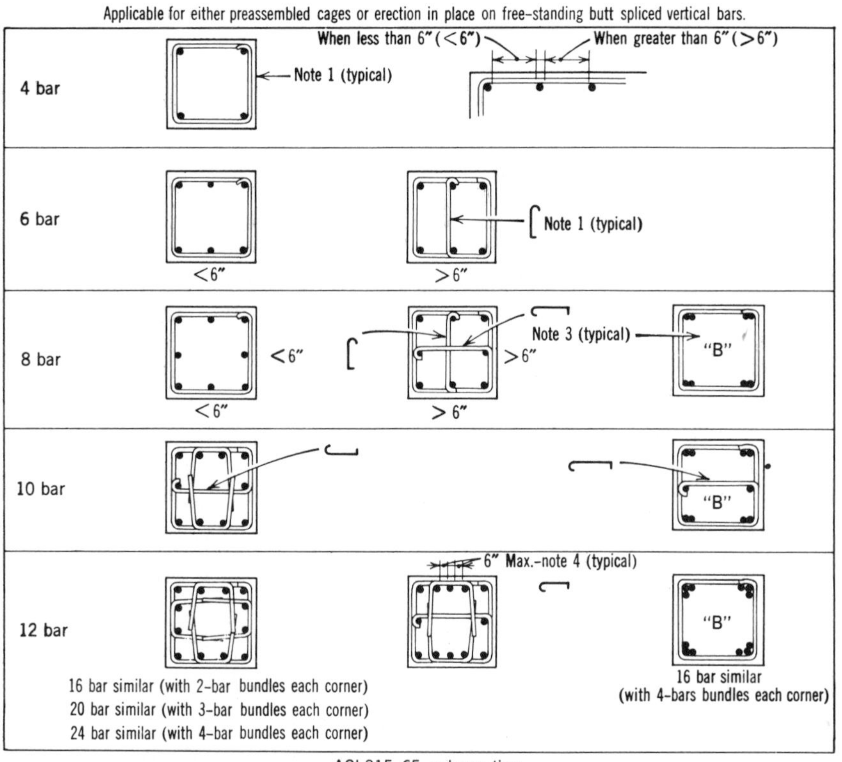

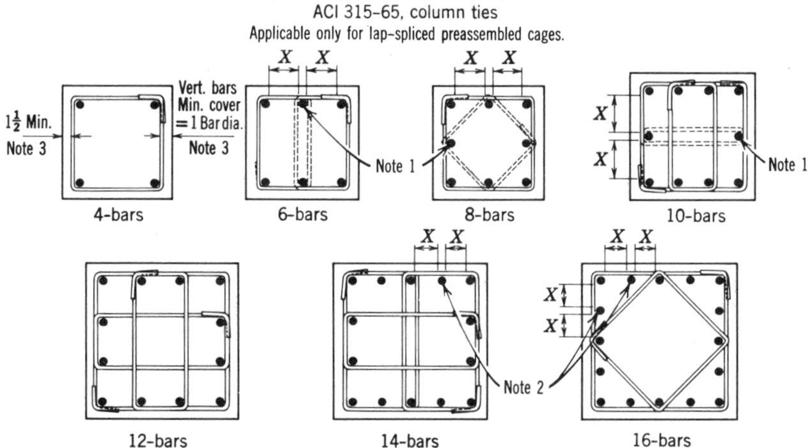

Fig. 14.38. Two column tie arrangements from ACI Committee 315.[18,19] Both groups are shown only in part.

point. Certain complications can result. If the dowel projections and the new column bars are placed side by side such that they are equidistant from the axis of the column, $2d_b$ must be deduced from the center-to-center spacing to find the clear bar spacing. When 5% or more of longitudinal steel is used, this spacing can become critical and limit the permissible column steel below what might be desired.

Several factors now tend on occasion to make other splice points look attractive. Code 7.10.5 requires at least 25% of the vertical steel on each face to be developed to f_y in tension (or more bars at a lower stress to give the same total tension capacity). If butt splices are not used, the compression lap may provide the required tension. Another way to achieve this is to run bars in two-story heights and stagger the splices. Although such bars are awkward to support on the job, this relieves the congestion in a single splicing section. Since tension splices tend to be long, elimination of some of these is also attractive. The awkwardness in construction lies not only in bracing the loose bars but also, after columns ties are in place, in interweaving the beam and girder steel through the column steel.

Appendix A of the Code, covering seismic design, limits the minimum splice to $30d_b \ngtr 16$ in. and the Commentary points out that splices at mid-height are preferable because there the moment is lowest. The maximum design moments for wind, earthquake, and even vertical loads (in typical multistory buildings) typically occur at the *end* (or ends) of the column; the splice at mid-height escapes some of this moment.

For compression splices, bars may be butted and held in place by pipe or commercial mechanical splices.

For tension splices, bars can be welded together, usually with butt welds, but it should be noted that a number of welding passes are required for large high strength bars. A Cadweld splice, using molten metal to hold the bar to a sleeve (deformed on its inside surface), in effect splices the bar to the sleeve and the sleeve to the next bar. The sleeve in this case is notably larger than the bar and uses up some of the clear bar spacing and may encroach on the specified clear cover unless this is specially considered in design.

14.31 BRIEF COMMENTS ON ECONOMY

Economy in column design favors the use of higher strength concrete on well-controlled jobs, since extra concrete strength costs only slightly more than low strength. Concrete will generally be cheaper than intermediate grade steel. Steel of Grades 50, 60, or 75 is cheaper than Grade 40 because of its higher f_y value. The higher strengths will often be as cheap as concrete in load-carrying capacity. Tied columns will generally be cheaper than spiral columns, particularly if square columns are needed rather than round. However, the cost of the column can rarely be considered alone. Spiral columns and heavy steel save floor space which has an annual rental value. Form costs are also a major item and beam forms can be reused from floor to floor most simply if column sizes are kept constant. Hence it is quite common practice to keep the column size constant over several stories and take care of increasing load with increasing steel, stronger concrete and steel, or the use of spiral steel.

14.32 REDUCTION IN COLUMN LIVE LOADS

Building codes specify uniform live loads large enough to represent ordinary local concentrations of loading. Over larger areas the probability that this same load intensity will exist everywhere is reduced. Codes usually permit some reduction in the assumed live load on floor areas in excess of 100 to 150 sq ft. Many codes permit such reduction only when the live load is 100 psf or less; heavier loads imply storage or machinery loads which are easily concentrated over large areas.

The live load on columns may also be reduced under the same reasoning. The reductions permitted vary considerably in the different city codes and it will be necessary for the designer to consult the particular code which governs his design. The magnitude of the reduction depends both on the occupancy, or size of the unit live load, and on the number of floors carried by the column; it may vary from none to as much as 60% of the total live load.

14.33 OTHER COLUMN SPECIFICATIONS

Although the ACI Building Code is the chief model of codes in the United States, there are other specifications in quite different terms. For example the 1969 AASHO Standard Specifications for Highway bridges[17] gives axial load capacities at service loads as the following for short columns with $l_u/h \lessapprox 10$:*

$$\text{Spiral column: } P = 0.225 f_c' A_g + A_s f_s$$

$$\text{Tied column: } P = 0.8(0.225 f_c' A_g + A_s f_s)$$

With moment giving $e/h \lessapprox 0.5$, approximately an elastic analysis of the uncracked transformed section is permitted. Above e/h of 0.5, a cracked transformed area section is used.

For work on highways structures the AASHO specification should be consulted and followed.

*The ACI notation is used here (but not in the actual specification).

SELECTED REFERENCES

1. F. E. Richart and G. C. Staehle, "Column Tests at University of Illinois," *Jour. ACI*, 2, Feb., Mar. 1931; *Proc.*, 27, pp. 731, 761; *Jour.*, *3*, Nov. 1931, Jan. 1932; *Proc.*, *28*, pp. 167, 279.
2. W. A. Slater and Lyse, "Column Tests at Lehigh University," *Jour. ACI*, 2, Feb., Mar. 1931; *Proc.*, 27, pp. 677, 791; *Jour.*, 3, Nov. 1931, Jan. 1932; *Proc.*, 28, pp. 159, 317.
3. Eivind Hognestad, "A Study of Combined Bending and Axial Load in Reinforced Concrete Members," Univ. of Ill. Eng. Exp. Sta. *Bull. No. 399*, 1951.
4. Frank E. Richart and Rex L. Brown, "An Investigation of Reinforced Concrete Columns," Univ. of Ill. Eng. Exp. Sta. *Bull No. 267*, 1934.
5. Frank E. Richart, Jasper O. Draffin, Tilford A. Olson, and Richard H. Heitman, "The Effect of Eccentric Loading, Protective Shells, Slenderness Ratios, and Other Variables in Reinforced Concrete Columns," Univ. of Ill. Eng. Exp. Sta. *Bull. No. 368*, 1947.
6. S. N. Pagay, P. M. Ferguson, and J. E. Breen, "Importance of Beam Properties on Concrete Column Behavior," *Jour. ACI*, 67, No. 10, Oct. 1970, p. 808.
7. H. Okamura, S. N. Pagay, J. E. Breen, and P. M. Ferguson, "Elastic Frame Analysis—Corrections Necessary for Design of Short Concrete Columns in Braced Frames," *Jour. ACI*, 67, No. 11, Nov. 1970, p. 894.
8. Phil M. Ferguson, Hajime Okamura, and S. N. Pagay, "Computer Study of Long Columns in Frames," *Jour. ACI*, 67, No. 12, Dec. 1970, p. 955.
9. J. G. MacGregor, J. E. Breen, and E. O. Pfrang, "Design of Slender Columns," *Jour. ACI*, 67, No. 1, Jan. 1970, p. 6.
10. B. C. Johnston (ed.), *The Column Research Council Guide to Design Criteria for Metal Compression Members*, 2nd ed., John Wiley and Sons, Inc., New York, 1966.
11. *Ultimate Strength Design Handbook, Vol. 2, Columns*, American Concrete Institute, SP-17*A*, Detroit, 1970.
12. Richard W. Furlong, "Column Slenderness and Charts for Design," *Jour. ACI*, 68, No. 1, Jan. 1971, p. 9.
13. Alan H. Mattock and Lidislay B. Kriz, "Ultimate Strength of Nonrectangular Structural Concrete Members," *Jour. ACI*, 32, No. 7, Jan. 1961; *Proc.* 57, p. 737.
14. Richard W. Furlong, "Ultimate Strength of Square Columns Under Biaxially Eccentric Loads," *Jour. ACI*, 32, No. 9, Mar. 1961; *Proc.* 57, p. 1129.

15. F. N. Pannell, "Failure Surfaces for Members in Compression and Biaxial Bending," *Jour. ACI*, No. 1, Jan. 1963, p. 129.
16. L. O. Bass, J. S. Ford, and R. L. Pinc, *Design-Analysis Graphs for USD Tied Columns with Biaxial Bending*, Oklahoma State University, School of Architecture, Stillwater, Okla., 1971.
17. *Standard Specifications for Highway Bridges*, AASHO, Washington, 10th ed., 1969.
18. ACI Committee 315, "New Developments in Detailing Practice," 64, No. 5, May 1967, p. 234.
19. *Manual of Standard Practice for Detailing Reinforced Concrete Structures*, ACI Standard 315-65, American Concrete Institute, Detroit, 1965.

PROBLEMS

Prob. 14.1. If $f_c' = 3000$ psi and bars are Grade 60 steel with the cover 2.5 in. to center of bars, find the permissible ultimate load in a braced frame by the moment magnifier method:

(*a*) For the column of Fig. 14.39*a* under a load on the *x*-axis 6 in. to the right of the center line of the column, $l_u = 10$ ft, single curvature, $C_m = 1$, $\psi = 1$, $\beta_d = 0$, $M_1 = M_2$.

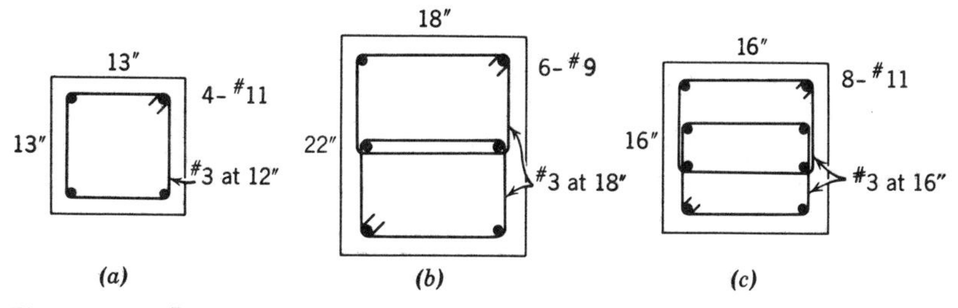

Fig. 14.39. Column cross sections for problems.

(*b*) For the column of Fig. 14.39*b* under a load on the *x*-axis 10 in. to the right of the center line of the column, $l_u = 10$ ft, single curvature, $C_m = 1$, $\psi = 2$, $\beta_d = 0.3$, $M_1 = M_2$.

(*c*) For the column of Fig. 14.39*c* under a load on the *x*-axis 8 in. off the center of the column, $l_u = 10$ ft, single curvature, $C_m = 1$, $\psi = 1.5$, $\beta_d = 0.4$, $M_1 = M_2$.

Prob. 14.2. Conditions same as in Prob. 14.1 except columns in reversed curvature with $M_1 = -0.4M_2$.

(*a*) As in Prob. 14.1*a*.
(*b*) As in Prob. 14.1*b*.
(*c*) As in Prob. 14.1*c*.

Prob. 14.3. Conditions same as in Prob. 14.1 except columns in an unbraced frame.

(*a*) As in Prob. 14.1*a*.
(*b*) As in Prob. 14.1*b*.
(*c*) As in Prob. 14.1*c*.

Prob. 14.4. Under the Code Commentary many, but not all, columns may be designed or checked by the R method. In each part below classify the columns as eligible for this method, not eligible, or with eligibility dependent on specific data not given. Also indicate any column permitting the use of $R = 1$.

(*a*) Columns of Prob. 14.1.
(*b*) Columns of Prob. 14.2.
(*c*) Columns of Prob. 14.3.

Prob. 14.5. By the R method find the ultimate column load permitted under the single curvature conditions of:

(*a*) Prob. 14.1a.
(*b*) Prob. 14.1b.
(*c*) Prob. 14.1c.

Prob. 14.6. By the R method find the ultimate column load permitted under the reversed curvature conditions of:

(*a*) Prob. 14.2*a*.
(*b*) Prob. 14.2*b*.
(*c*) Prob. 14.2*c*.

Prob. 14.7. By the R method find the ultimate column load permitted under the unbraced frame conditions of:

(*a*) Prob. 14.3*a*.
(*b*) Prob. 14.3*b*.
(*c*) Prob. 14.3*c*.

Prob. 14.8. If $f_c' = 4000$ psi and Grade 60 steel is used with the cover 2.5 in. to the center of steel, find the ultimate load permitted in a braced frame by the Code (moment magnifier) on the column of:

(*a*) Prob. 14.1*a* except eccentricity increased to 20 in.
(*b*) Prob. 14.1*b* except eccentricity increased to 20 in.
(*c*) Prob. 14.1*c* except eccentricity increased to 12 in.
(*d*) Prob. 14.1*c* except eccentricity increased to 18 in.

Prob. 14.9.

(*a*) Design a square tied column by the moment magnifier method for service loads of 120k dead plus 80k live and a single curvature moment of 20 k-ft dead and 60 k-ft live, using $f_c' = 4000$ psi, about 2% of Grade 60 steel with 2.5 in. cover to center of steel, $\psi = 1.2$, $l_u = 11$ ft, braced frame. Take $M_1 = M_2$.

(*b*) Same as (*a*) except reversed curvature moments with $M_1 = -0.5M_2$.

(*c*) Same as (*a*) except for an unbraced frame.

Prob. 14.10. Design a square tied column for a service load of 300k, service moment of 300 k-ft, each half dead and half live load, $f_c' = 4000$ psi, about 3% of Grade 60 steel, $l_u = 11$ ft in an unbraced frame, using $\delta = 1.3$. Cover assumed 2.5 in. to center of steel.

Prob. 14.11. Design a square tied column in an unbraced frame by the R method, assuming beam $\rho = 0.01$ and cover to center of steel is 2.5 in. Notice that l_{lu}/r must not be more than 40 for this method. Service axial loads of 150k dead and 150k live load, and service load moments zero for dead and live load and 75 k-ft from wind. (Note that factored axial load at minimum e/h must be compared with wind moment plus axial load under Code load factor Eq. 9.2 as given in Sec. 1.12d.) $l_u = 10$ft, $\psi_g = 1.2$, $f_c' = 4000$ psi, Grade 60 steel.

Prob. 14.12.

(*a*) Evaluate and locate the ultimate load which gives a horizontal neutral axis 6 in. below the top edge of the column of Fig. 14.39*c*. Assume cover 2.5 in. to center of steel, $f_c' = 4000$ psi, Grade 60 steel, and consider deformations to establish the steel stresses.

(*b*) Repeat for a horizontal axis 6 in. below the top edge of Fig. 14.39*b*.

Prob. 14.13. Design a column spiral of Grade 40 steel that will be adequate for an 18-in. square column with longitudinal steel in a circular pattern that permits the spiral to have an outside diameter of 15 in. $f_c' = 4000$ psi.

NOTE: *If column interaction curves similar to Fig. 14.26 are available for spiral columns, design problems 14.9, 14.10, and 14.11 are also suitable for assignment as spiral column designs.*

15
Footings

15.1 TYPES OF FOOTINGS

Although underground conditions call for many variations in foundation design, the majoirty of building footings can be classified as one of the following types.

Bearing walls may be supported on a continuous strip of concrete called a wall footing, as shown in Fig. 15.10.

Isolated column footings under individual columns may be square (as shown in Fig. 15.1*b*), rectangular, or round in plan. Alternatively, the column foundation may be a drilled pier, that is, a shaft drilled into the ground, sometimes flared at the bottom for greater bearing area, and finally filled with concrete.

When a single footing supports two or more column loads, as in Figs. 15.13 and 15.14, it is called a combined footing.

A cantilever footing, as in Fig. 15.16 also supports two or more columns. It is characterized by the fact that it is really two footings joined by a beam instead of by a bearing portion of the footing.

Under poor soil conditions it is sometimes desirable to support the entire structure on a single mat or slab. Such a foundation is often called a floating or raft foundation.

Any of these footings may be directly supported on the soil or they may rest on piling.

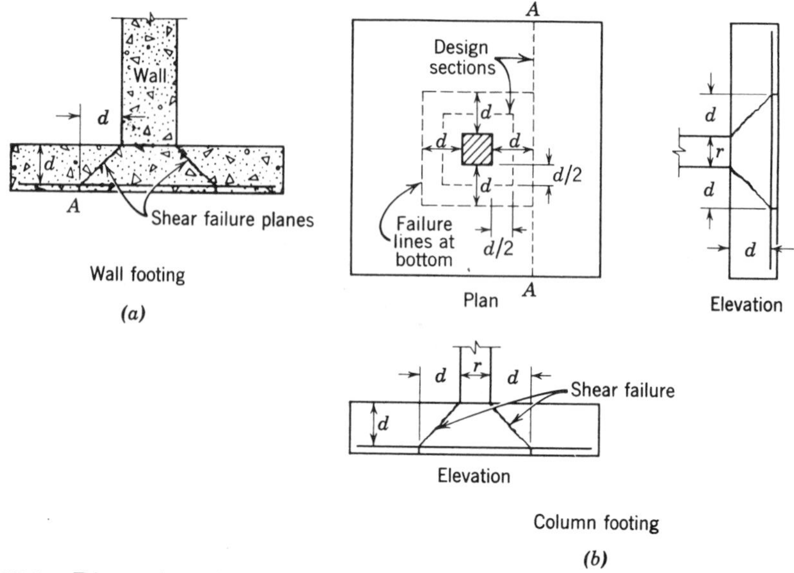

Fig. 15.1. Diagonal tension failure in footings.

15.2 TESTS ON FOOTINGS

Two noteworthy series of footing studies have influenced American practice, Talbot's tests[1] in 1907 and Richart's tests[2] in 1946. Moe's tests at the Portland Cement Association Development Laboratory have added more detail on the effect of openings near the column.

Shear failures never occur on vertical planes along the wall or around the column and the older design concept of punching shear around a column has been abandoned. For wall footings which fail in shear, a diagonal tension crack develops on an approximate 45° plane parallel to the wall, as shown at A in Fig. 15.1*a*. The shear causing this crack is that produced by the upward load to the left of A, that is, the load beyond a plane a distance d from the face of the wall. In an isolated square column footing a similar failure occurs, the column pushing ahead of it a truncated pyramid with an approximate 45° slope on all faces. This pyramid has a base width equal to the column width plus approximately twice the effective depth of the footing (Fig. 15.1*b*). The probable reason for this type of failure lies in the heavy vertical compressive stresses produced between the diagonal cracks by the load's spreading out directly beneath the column or wall, this vertical compression being accentuated by the upward soil reaction. In this zone the diagonal tension normally caused by shear stresses is somewhat counteracted or reduced by the vertical compressive stress. (Compare Fig. 4.3a and the related discussion in Sec. 4.4c). The 45° failure planes mark the approximate boundaries between which substantial vertical compression is developed.

In a column footing the intial diagonal cracks described above occur much before ultimate load and are not generally visible because they are in the interior of the concrete. The 1962 report[3] of the Joint ACI-ASCE Committee on Shear and Diagonal Tension recognized that *after* these shear failure planes first developed the shear was largely carried by the flexural compression area above the cracks in what can best be described as the old idea of punching shear of this limited compression depth around the perimeter of the column. Failure might then be described as a two-step process, initial diagonal cracking extending a distance d from the wall or column and finally a shearing failure at the face of wall or column. Usually, a considerable portion of the total load resistance develops after the diagonal crack is formed. Rather than introduce two checks for the two kinds of behavior the Committee recommended that the shear be calculated on a pseudo-critical plane between the two, that is, at a distance $d/2$ from the column. Then as a safety check, which probably controls only on long, narrow footings, the shear strength as a one-way slab should be checked on sections all across the footing at a distance d from the face of column.

With both walls and columns of concrete, the critical section for moment and bar development is found at the face of the wall or column. In the wall footing the one-way cantilever moment due to forces beyond the critical section is uniformly distributed along the wall (Fig. 15.10); any 1-ft wide strip can be used in this part of the design. In the column footing Richart found the distribution of this cantilever moment (in each direction) to be very nonuniform at working loads, being largest for strips passing under the column and least for strips near an edge. However, near ultimate loads the yielding of the steel in the central strips causes more moment to shift to the edge strips and moment failure does not occur until essentially all the steel has reached its yield point. Bond stress was found to be less critical than expected and maximum on the sections used for maximum moment. In this connection, the 1971 Code requires only that the bar length beyond the face of column be a full development length (Chap. 5).

15.3 SQUARE COLUMN FOOTINGS—ANALYSIS

(a) Critical Stresses

Column footing must be checked or designed for six strength conditions:

1. Bearing (compression) from column on top of footing.
2. Dowels into the footing.
3. Strength of soil beneath footing, soil pressure q_s.
4. Shear strength.
5. Reinforcement provided.
6. Development length of bars.

(b) Bearing under Column

The bearing from the column on the footing is permitted to be larger than the value of the same concrete in the column, that is, lower strength concrete is permissible in the footing without lowering the column capacity or necessarily increasing the dowels provided. The usual permitted bearing stress (Code 10.14) of $\phi \times 0.85f_c'$ may be multiplied by $\sqrt{A_2/A_1} \not> 2$, where A_1 is the bearing area and A_2 is the lower area sketched in Fig. 15.2. Area A_2 is concentric to and geometrically similar to A_1 and is established by going down and out from the loaded area at a slope of 1 vertically and 2 horizontally* until one intersects the boundary closest to the column. This boundary fixes one side of A_2 and it is simple to construct the complete area on the plan view similar to A_1. From similar areas

$$A_2/A_1 = (x_2/x_1)^2 \qquad \sqrt{A_2/A_1} = x_2/x_1$$

The total bearing value of the area A_1 of a steel bearing plate or a concrete column becomes:

$$\overline{N}_c = 0.85f_c'A_1\sqrt{A_2/A_1} = 0.85f_c'A_1(x_2/x_1)$$

where f_c' is the footing concrete strength and A_1 the bearing area. For the concrete column, this part of the load is also limited to the capacity of the concrete in the column, that is, $\overline{N}_c = 0.85f_c'A_1$ where f_c' is the column concrete strength.

*This slope is not intended to represent the direction of the pressure, but rather to insure some confining concrete around the actual pressure distribution.

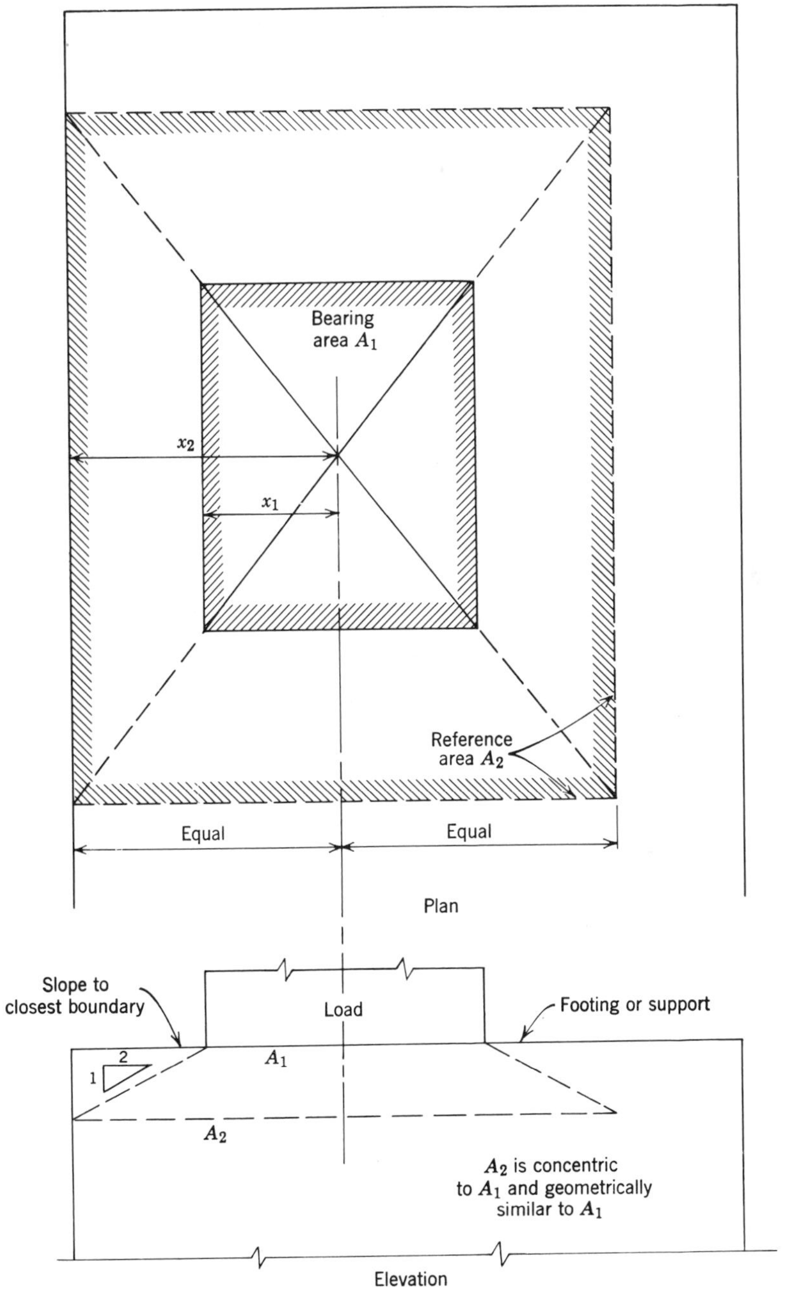

Fig. 15.2. Bearing pressure limit is based on $\sqrt{A_2/A_1} = x_2/x_1$.

(c) Dowels into the Footing

Since the 1971 Code permits the dowels into the footing to be designed for strength instead of requiring them to match the column steel, some savings in compression dowels are occasionally possible. However, any column bar which can be subject to tension must either be anchored into the footing or matched with dowels into the footing.

A small saving can occur if a column requires almost no steel to carry its load, but is reinforced with $\rho = 0.01$ to satisfy the Code minimum. The dowel minimum is now lowered to $\rho = 0.005$ of the column area with a minimum of four bars. Although not stated, the four bars should obviously be on four sides and probably at the column corners.

If the column concrete strength exceeds the footing concrete strength, the increased bearing strength could still mean that no *extra* dowels were required to transfer the difference in concrete strengths into the footing.

In designing compression dowels, one must consider both the footing below and the column above:

1. For the footing, dowels must carry the total load less the bearing value value of the footing concrete. More exactly stated, the dowels should fit the demands of the column interaction chart for the design eccentricity and A_1 when f_c' is used as the footing value times the ratio $\sqrt{A_2/A_1} \not> 2$.
2. For the column, the load, including moment from eccentricity, must be carried by the concrete in the column assisted by *only* those bars which are extended or doweled into the footing.

The second rule is necessary because any bars neither extended nor doweled into the footing have zero load capacity at the point where they are stopped and very little capacity nearby; strength is always limited to the lower available development length on the two sides of any design section.

It is quite important to note that the second condition says nothing done in the footing itself lowers the demand for effective column reinforcement, that is *developed* column steel.

(d) Soil Pressure

The discussion of actual soil pressure under a footing is beyond the scope of this book. This pressure commonly will be higher near the center of a column footing, but occasionally higher at the edges.

The concentration of pressure directly under the load when a footing is on rock can be reduced only by a very stiff footing as discussed in Sec. 15.10.

Although not necessary in simple cases, elastic analysis would be appropriate for foundations on rock and would be helpful in special cases of loading or footing shape. The same approach is usable in other stiff and well-consolidated foundation materials if there is enough uniformity to permit a meaningful determination of the foundation modulus.

Elastic analyses, however, are not a general answer to the pressure distribution problem. If measurable foundation settlement is expected, the soil is only momentarily elastic and the initial response to the load will be only a transient response. Time (and settlement) will smooth out this initial response, lowering peak pressures and building up low pressures. In such cases, the author considers elastic analyses of little value. More uniform (oversimplified) pressure distributions appear more appropriate as a design approach where significant settlement is expected.

The soil pressure is usually considered as uniform for a centered loading and trapezoidal or triangular for eccentric loadings. The assumed uniform pressure is usually on the safe side in the calculation of internal moments and shears in isolated footings, but a different situation in the middle of a combined footing is discussed in the last paragraph of Sec. 15.12.

The allowable soil pressure is also a matter of soil mechanics and cannot be discussed here. However, this again brings up the problem mentioned in connection with retaining wall design (Sec. 7.5*f*). Allowable soil pressure is usually given in terms of the service load permissible while footing design is by strength (at ultimate). The designer has at least two (equivalent) ways to keep the relationships proper in the case of isolated cloumn footings.

The Code (15.2.4) specifies, for the purpose of establishing the footing size, that the (unfactored) external forces and moments be used to match the allowable soil pressure; but also that soil pressures based on design (factored) loads be used to calculate design moment and shear.

The Commentary suggests an alternate giving the same answers, namely, increasing the allowable soil pressure in the ratio of the ultimate U values to the service loads involved, that is, multiplying the allowable soil pressure by

$$\frac{1.4D + 1.7L}{D + L}, \qquad \frac{1.4D + 1.7L + 1.7W}{D + L + W}, \qquad \text{etc.}$$

This method will be used here because the soil value thus computed is available for direct use in computing the strength demands for shears and moments in the footings.

Since the ratio of dead to live load varies from column to column, there is no unique relationship between soil pressure at service loading and soil pressure under ultimate loading. It appears to the author that design would be better served by the general use of an allowable or dependable *ultimate* soil pressure. It is noted, however, in settlement situations that a permissible pressure at *usual* loads might also be significant and possibly governing, just as deflection can govern over strength in a flexural member.

The footing base area must be adequate to care for the column load, footing weight, and any overburden weight, all within the permissible soil pressure, often assumed uniformly distributed* under the footing. Sometimes this over-all pressure is called the gross soil pressure to distinguish it from net soil pressure, which is a convenient *design* concept. Since the weight of footing and overburden is usually nearly uniform over the footing area, the design moment or shear on any section will be the result of the gross soil pressure upward less the design value of the footing and other overburden weight downward. It is convenient to think in terms of the resultant load or "net soil pressure" which is the difference between these upward and downward unit pressures. The net pressure is usually found most simply by dividing the design column load (alone) by the footing area.

(e) Shear Strength

Shear quite frequently controls the footing thickness.

The unit shear as a measure of the diagonal tension is first calculated for the shear caused by loads outside the inner dashed square in Fig. 15.1*b* and is resisted by a width equal to the perimeter of that square, $b_o = 4(r + d)$. The use of d_v as the average depth to the two layers of steel appears justified. There is two-way bending here with permissible unit shear of $4\sqrt{f_c'}$.

The diagonal tension on section AA at a distance d must also be checked, using permissible one-way slab shear of $2\sqrt{f_c'}$.

(f) Reinforcement Provided

A proper reinforcement design always gives an underreinforced member, automatically avoiding compressive failure.

The moment is critical on the section at the face of column and calls for two-way bottom bars. The moment in each direction is a matter of statics and can be varied only by a change in the distribution of soil pressure. The separate steel in each direction should be adequate for the moment in that

*But variation caused by moment loading must also be considered.

direction; excess steel in the y-direction cannot make up for a steel shortage in the x-direction. Hence it appears desirable that the smaller depth to the upper steel layer in the bottom of the footing be used for calculation of the moment resistance in a square footing. Economy will be achieved by using a z value corresponding to the actual percentage of steel used, this value usually being larger than $0.9d$.

The minimum reinforcement requirement of $200/f_y$, unless A_s is at least 4/3 that required by analysis, should be applied to footings. Although one could argue legally that a footing is a slab, and thereby exempt from this requirement, the combination of high shear and low ρ is not good.

(g) Development Length of Bars

In a square footing, the development of bars is less restrictive than under previous Codes. The available development length is from the face of column to the edge of footing (less the end cover). No bar requiring a longer development length can be used, unless it is hooked at the ends or is understressed at the maximum moment section.

15.4 ANALYSIS FOR MOMENT LOADING

Footings must often carry moment from a column or wall. Combined with direct load this gives the equivalent of an eccentric load, as in Fig. 15.3. If the moment were constant it would be desirable to put the center of the footing under this eccentric load. For the retaining wall (Sec. 7.5g) the adjustment of the base to make the resultant load fall between the center and third point was discussed. Usually the varying nature of the moment makes it impossible to avoid all eccentricity on the footing and a trapezoidal soil pressure results. This modifies the magnitude of the design moment and shears, but not the general design procedure. There might be

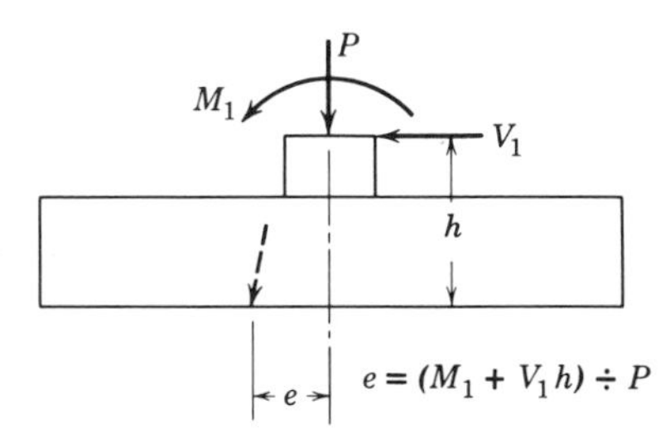

Fig. 15.3. Equivalent eccentricity of load.

doubts as to how to handle the two-way slab shear. For important eccentricities the method indicated in Sec. 4.6c is available. For minor eccentricities the author would simply apply the shear developed as in Fig. 15.5*a* from the heavier loaded half of the footing onto *one* half of the perimeter around the column at the distance $d/2$.

Although office practice in footing design has often neglected normal eccentricities of loading, considering them a minor matter, eccentricities are important and should more commonly be included. Columns are designed for eccentric loading and it is inconsistent to assume none of this moment acts on the footing. Accordingly, some of the examples of this chapter are first given in oversimplified form in order to concentrate attention on the general principles, and then a more correct solution follows which properly includes eccentricity considerations. The student might compare the results of the two solutions to build up an idea of the magnitude of eccentricity effects.

15.5 SQUARE FOOTING—ANALYSIS EXAMPLE

Check the footing of Fig. 15.4, assuming $f_c' = 3000$ psi for both column and footing, Grade 40 steel, and an allowable soil pressure of 3000 psf. The column dead load is 190 k and the live load 300 *k*. Assume no overburden of soil.

Solution

(a) Since the column is of the same strength concrete as the footing, the bearing on top of the footing will be adequate. Assuming the column fully stressed, dowels should be provided to match the column steel, extended vertically into the footing a compression development length l_d.

No moment is specified on the footing. Assume concentric loading.

Column design load $= 1.4 \times 190 + 1.7 \times 300 = 777\ k$

Average design soil pressure $= q_{net} = 777{,}000/(14 \times 14) = 3970$ psf

Allowable soil pressure $= 3000$ psf

Weight of footing $= 2.0 \times 150 = 300$

Allowable $q_{net} = 2700$ psf (at service load)

$$\begin{aligned} \text{Equivalent allowable at ultimate} &= 2700(\text{design load})/(\text{dead} + \text{live load}) \\ &= 2700 \times 777/(190 + 300) \\ &= 4270 \text{ psf} > 3970 \text{ psf actual} \qquad \text{O.K.} \end{aligned}$$

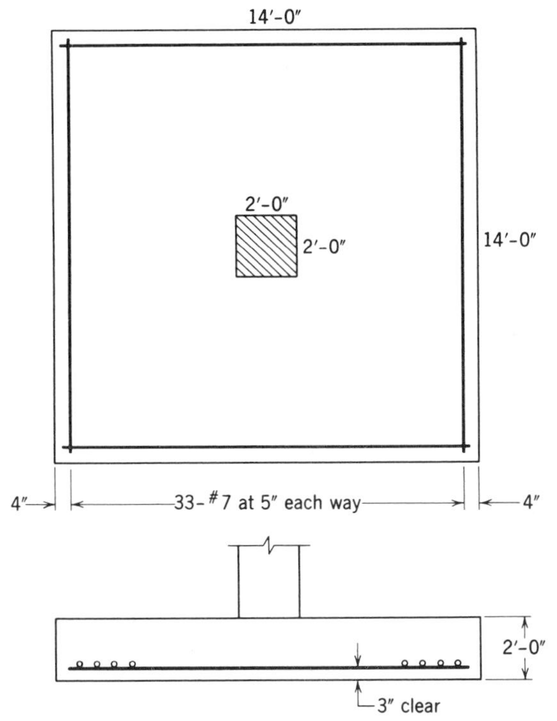

Fig. 15.4. Square column footing.

In a square footing the use of average d is recommended only for shear calculations around the column, with the depth to the upper steel layer for moment calculations and for shear all across the footing.

$$\text{Average } d_v = 24 - 3 - 0.88(\text{bar diam.}) = 20.12\text{in.}$$

Shear will first be checked for two-way action all round the column at a distance $d/2$ outside the column as indicated in Fig. 15.5*a*, with the net soil pressure used as the applied load. The net soil pressure omits that part of the reaction created by the footing weight because everywhere in moment and shear evaluation the downward weight of the footing balances this portion of the reaction.

$$V_u = 3970(14.0^2 - 3.68^2) = 3970(196.0 - 13.5) = 725{,}000 \text{ lb}$$

$$\overline{V} = V_u/\phi = 725{,}000/0.85 = 850{,}000 \text{ lb}$$

$$v = \overline{V}/b_o d = 850{,}000/(4 \times 44.1 \times 20.12) = 239 \text{ psi}$$

$$\text{Allowable } v = 4\sqrt{f_c'} = 219 \text{ psi} < 239 \qquad \text{N.G. } (v \text{ not O.K.})$$

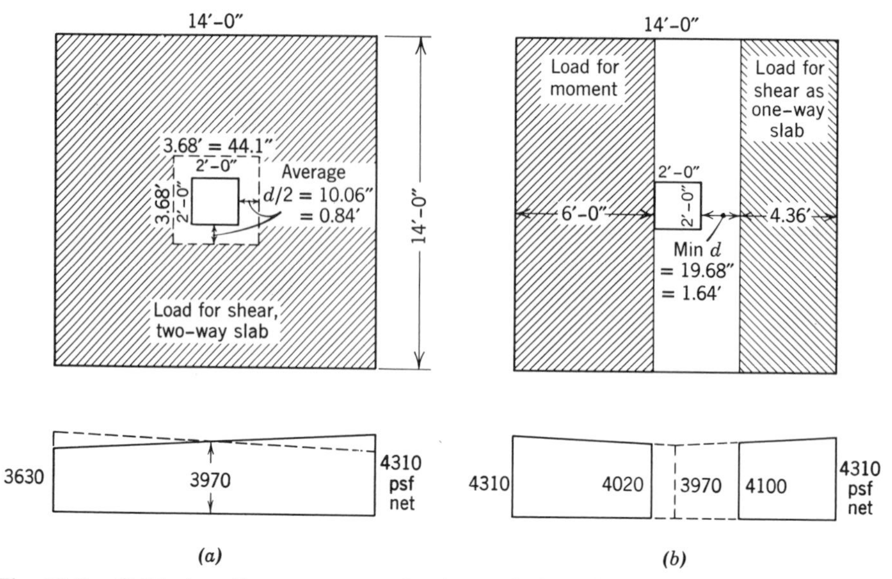

Fig. 15.5. Critical sections on square footing and, for solution (*b*) only, net pressures under eccentric loading. (*a*) Shear as two-way slab. (*b*) Moment and shear as one way slab.

Although it will normally govern only on long narrow footings, shear will also be checked on the full width as a one-way slab (Fig. 15.5b).

Min. $d_v = 24 - 3 - 1.5(0.88) = 19.68$ in.

$V_u = 3970 \times 4.36 \times 14 = 242{,}000$ lb, $\overline{V} = V_u/0.85 = 285{,}000$ lb

$$v = \frac{\overline{V}}{bd} = \frac{285{,}000}{(14 \times 12) \times 19.68} = 86 \text{ psi}$$

Allowable v on one-way slab $= 2\sqrt{3000} = 110$ psi > 86 O.K.

Moment must be calculated at the face of the column (Fig. 15.5b).

$M_u = 14 \times 6 \times 3970 \times 3 = 996{,}000$ ft-lb, $\overline{M} = M_u/1000\phi = 1110$ ft-k

$d_M = 24 - 3 - 1.5 \times 0.88 = 19.68$ in.

$$\rho = \frac{33 \times 0.60}{14 \times 12 \times 19.68} = 0.00600 \begin{array}{l} < \text{allowable Table 2.1 O.K.} \\ > \text{min. of } 200/f_y = 200/40{,}000 \\ = 0.005 \text{ O.K.} \end{array}$$

$a = A_s f_y/(0.85 f_c' b) = 33 \times 0.60 \times 40/(0.85 \times 3 \times 14 \times 12) = 1.85$ in.

$z = d - a/2 = 19.68 - 0.92 = 18.76$ in.

Reqd. $A_s = 1110 \times 12/(40 \times 18.76) = 17.7$ in.^{2} vs. $33 \times 0.60 = 19.8$ in.^{2} O.K.

The critical section for bar development is at the maximum moment section, where there is available for l_d 72 in. less end cover, say, 70 in.

$$\text{Reqd. } l_d = 0.04\,A_b f_y/\sqrt{f_c'} = 0.04 \times 0.60 \times 40{,}000/\sqrt{3000} = 17.5 \text{ in.}$$

A much larger bar could be used, but this spacing is quite reasonable. *Conclusions:* The footing must be deepened for shear. The A_s is about 10% more than needed for the present depth. A larger bar with wider spacing could be used. (These conclusions should be compared with those of the following solution.)

(b) *Analysis with moment loading.* In the absence of other information, assume the column was designed for a minimum eccentricity of $0.10h = 2.4$ in., since about 80 to 90% of column designs seem to be governed by minimum e/h. In terms of Fig. 15.3, assume also that the shear on the column leaves the base moment little changed from that on the column, with e on the footing of 2.4 in. = 0.20 ft insofar as the column loading is concerned.

Dowel requirements are unchanged from (a).

Based on the average design net soil pressure of 3970 psf in (a), the maximum edge pressures become

$$\begin{aligned} q_{net} &= 3970(1 \pm 6e/b)^* = 3970(1 \pm 6 \times 0.20/14) \\ &= 3970(1 \pm 0.086) = 4310 \text{ psf and } 3630 \text{ psf} \end{aligned}$$

This is in addition to the footing weight of 300 psf. At service load the maximum q_{net} reduces to $4310 \times 490/777 = 2720$ psf and gross $q = 2720 + 300 = 3020$ psf. versus 3000 psf allowable. Excess pressure is only 0.7% (compared to 7% below allowable in *a*). Say O.K.

For the one-way slab shear (Fig. 15.5b) the average net pressure is 4310 + 4100)/2 = 4200 psf. $\overline{V} = 4200 \times 14 \times 4.36/0.85 = 301{,}000$ lb.

$$v = \overline{V}/bd = 301{,}000/(14 \times 12 \times 19.68) = 91 \text{ psi} < 110 \text{ allow.} \qquad \text{O.K.}$$

For the two-way slab shear, since the eccentricity of loading is small, compute $\overline{V}$ on half the loaded area (heavy side) using the trapezoidal loading of Fig. 15.5a.

$$\begin{aligned} \overline{V} &= [0.5(4310 + 3970)\,14 \times 7 - (\text{estimated})\ 4050 \times 3.68 \times 1.84]/0.85 \\ &= (406{,}000 - 27{,}400)/0.85 = 447{,}000 \text{ lb} \\ v &= \overline{V}/bd = 447{,}000/(2 \times 3.68 \times 12 \times 20.12) = 251 \text{ psi} > 219 \text{ allow.} \end{aligned}$$

N.G.

*e/h would be better except that h has so many other meanings nearby.

The moment is calculated using two triangles for the trapezoid.

$\overline{M} = (0.5 \times 4310 \times 6 \times 4 \times 14 + 0.5 \times 4020 \times 6 \times 2 \times 14)/0.90$
$= (725{,}000 + 338{,}000)/0.90 = 1{,}186{,}000$ ft-lb $= 1186$ ft-k

Reqd. $A_s = 1186 \times 12/(40 \times 18.76) = 18.9$ in.$^2 < 19.8$ in.^{2} O.K.

Bar development length is unchanged from (a).

Conclusions: Soil pressure is increased about 8%, shear by 6%, and moment by 7%. Except for the original low soil pressure, a larger footing would be needed for soil pressure; a deeper footing is required for shear.

15.6 SQUARE COLUMN FOOTING—DESIGN

Design a square column footing using $f_c' = 3000$ psi, Grade 60 steel, and allowable soil pressure of 3500 psf, to carry an 18 in. square spiral column made of 5000 psi concrete and carrying a dead load of 200 k and live load of 350 k.

Solution

The load assigned to the concrete in the spiral column is based on a nominal stress of $0.85 f_c'\phi = 0.85 \times 5000 \times 0.75 = 3180$ psi. Since the footing A_2/A_1 is large, the allowable bearing is $2 \times 0.85 f_c'\phi = 2 \times 0.85 \times 3000 \times 0.70 = 3570$ psi > 3180 psi; no extra dowels are needed. Make dowels match column bars, at least in area. Required embedment length (Code 12.6) $= l_d = 0.02 f_y d_b/\sqrt{f_c'} = 0.02 \times 60{,}000\, d_b/54.8 = 21.9\, d_b$. The footing depth should accommodate this vertical bar length.

Col. $P_u = 1.4 \times 200 + 1.7 \times 350 = 280 + 595 = 875$k

Assume a 2 ft thick footing, weight $= 300$ psf, leaving allowable net pressure $= 3500 - 300 = 3200$ psf.* In terms of ultimate loads:

Allowable ult. $q_{net} = 3200 \times 875/(200 + 350) = 5080$ psf*
Reqd. A = col. load$/q_{net} = 875{,}000/5080 \doteq 172$ ft^2
Try 13 ft 2 in. square footing ($A = 173$ ft^2)
Actual $q_{net} = 875{,}000/173 = 5040$ psf

*It might appear at first glance that a 0.9 load factor should be used for the footing weight in establishing q_{net}, but the opposite conclusion is reached here. Whatever net soil pressure the column load produces is increased by the local footing weight. The weight of the footing and its reaction are two forces in equilibrium locally as well as overall; they therefore produce no moments or shears. If the designer estimates the footing weight incorrectly, this results in too much, or too little, *gross* reaction pressure computed; but, for the area of footing used, the moments and shears are unchanged by his bad estimate. It is thus appropriate to use the ratio of factored *column* load to service *column* load to establish the ratio of factored q_{net} to service load q_{net} for use in moment and shear calculations.

Since it is not typical that one can fit exactly the required A, it is desirable to use the *actual* q_{net} in moment and shear calculations. In this example the close fitting of the area makes it appear useless.

If h is assumed 24 in. and 3 in. clear cover is used as required when casting directly against the ground, the average $d = 24 - 3 - d_b =$ say, 20 in. The critical section for shear for the two-way slab action is then $d/2 = 10$ in. from the face of the column, as shown in Fig. 15.6a.

$$V_u = 5040(13.17^2 - 3.17^2) = 5040(173 - 10) = 822{,}000 \text{ lb}$$
$$\overline{V} = V/\phi = 822{,}000/0.85 = 967{,}000 \text{ lb}$$

Allowable $v = 4\sqrt{f_c'} = 4\sqrt{3000} = 219$ psi, $b_o = 4(18 + 20) = 152$ in.

$$d_v = \overline{V}/b_o v = 967{,}000/(152 \times 219) = 29.0 \text{ in.} > 20 \text{ in.} \qquad \text{N.G.}$$

The initial estimate of thickness was poor and must be revised upward. Since b_o increases on each side as much as d increases, the calculation for d_v is sensitive to changes in the assumed d. The footing size will also be increased by the increased weight of footing, but a better value of d will be found before attacking the problem of footing size. Assume h is increased about half as much as the deficiency in d to, say $h = 29$ in., making $d = 29 - 3 - d_b = 25$ in. From Fig. 15.6b,

$$b_o = 4(18 + 25) = 172 \text{ in.}$$
$$V_u = 5040(13.17^2 - 3.58^2) = 5040(173 - 13) = 805{,}000 \text{ lb}$$
$$\overline{V} = 805{,}000/0.85 = 946{,}000 \text{ lb}$$
$$d_v = 946{,}000/(172 \times 219) = 25.1 \text{ in. vs. assumed 25 in.}$$

This indicates d of around 25.5 in. is close. The increased footing weight will lower q_{net} and require a larger footing, but this will only slightly increase the required d_v. Continue with $h = 29.5$ in., average d approximately 25.5 in.

Footing weight $= 2.46 \times 150 = 369$ psf
$q_{net} = 3500 - 369 = 3130$ psf for service loads
Equiv. q_{net} for strength design $= 3130 \times 875/550 = 5000$ psf

Reqd. $A = 875{,}000/5000 = 175$ ft^2, say 13 ft 3 in. square (175 ft^2)

$$V_u = 5000(13.25^2 - 3.62^2) = 5000(175 - 13) = 810{,}000 \text{ lb}$$
$$\overline{V} = 810{,}000/0.85 = 955{,}000 \text{ lb}$$
$$d_v = 955{,}000/(174 \times 219) = 25.1 \text{ in. Aver. } d = 25.5 \text{ in.} \qquad \text{O.K.}$$

USE $h = 29.5$ in., $d = 25.5$ in. (assuming $d_b = 1$ in.)
USE 13 ft 3 in. square, as in Fig. 15.6c.

The check on one-way shear will be left to the end since it should not be significant on a square footing.

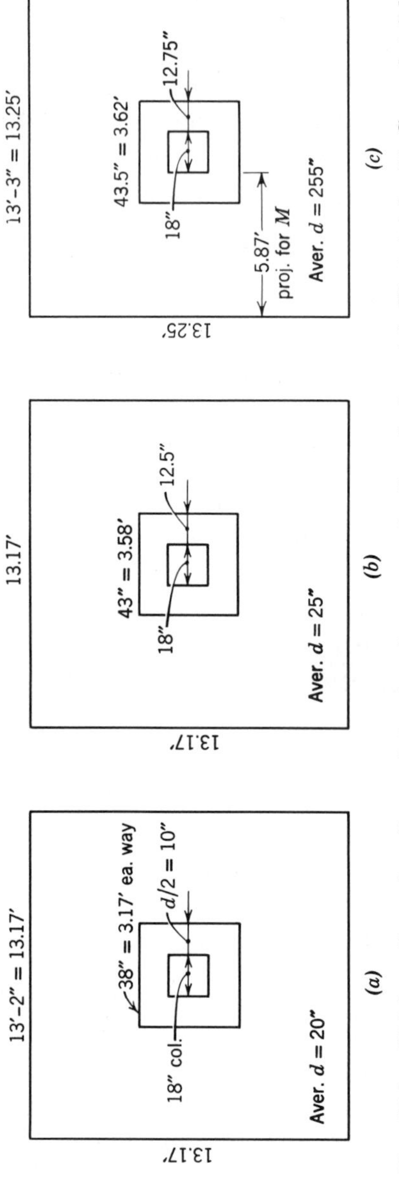

Fig. 15.6. Trial sections for diagonal tension in square footing design. (*a*) First trial. (*b*) Second trial. (*c*) Final design.

For moment the *lesser* depth controls, not the average.

$d_m = 29.5 - 3.0 - 1.5d_b = 26.5 - 1.5d_b = 25.0$ in. for $d_b = 1$ in.
$M_u = 5000 \times 13.25 \times 5.87 \times 2.93 = 1{,}140{,}000$ ft-lb
$\overline{M} = 1{,}140{,}000/0.90 = 1{,}266{,}000$ ft-lb $= 1266$ ft-k
Assume $z = 0.9d = 0.9 \times 25.0 = 22.5$ in.
Reqd. $A_s = \overline{M}/f_y z = 1266 \times 12/(60 \times 22.5) = 11.28$ in.^{2}
$a = 11.28 \times 60/(0.85 \times 3 \times 13.25 \times 12) = 1.67$ in.
$z = 25.0 - 0.83 = 24.2$ in.
Reqd. $A_s = 11.28 \times 22.5/24.2 = 10.50$ in.^{2}

$\rho = 10.5/(13.25 \times 12 \times 25) = 0.00265 <$ allowable Table 2.1 O.K
Min. $\rho = 200/f_y = 200/60{,}000 = 0.00333 > 0.0265$

This minimum controls especially since a footing commonly is a high shear member. Min. $A_s = 0.00333 \times 13.25 \times 12 \times 25 = 13.25$ in.^{2} The edge distance may be a half space, but should not be less than about 3 in. Assume first all bars uniformly spaced. The following are possibilities:

17- #8	13.4 in.^{2}	9.4 in. spcg.	d_m as assumed
14- #9	14.0	11.3	reduced to 24.8 in.
11- #10	13.97	14.5	reduced to 24.6
22- #7	13.2	7.3	increased to 25.2

The #7 bars look good. (The #8 bars are essentially as good because of fewer bars to handle.)

USE 22- #7 at 7.5 ± in. *each way*.

Note that in a footing or other slab a listing of this kind means 22 bars govern and the spacing is not exact, but is assumed uniform. For #7, $l_d = 0.04A_b f_y/\sqrt{f_c'} = 0.04 \times 0.60 \times 60{,}000/\sqrt{3000} = 26.4$ in.

Available $l_d = 68$ in. minus cover over end of bar. O.K.

Finally, shear on the one-way slab will be checked at $d = 25$ in. from face of column.

$v_u = 5000 \times 13.25(5.87 - 2.08) = 251{,}000$ lb
$\overline{V} = 251{,}000/0.85 = 295{,}000$ lb
$v = 295{,}000/(13.25 \times 12 \times 25) = 74$ psi $\ll 2\sqrt{f_c'} = 110$ psi

15.7 SQUARE COLUMN FOOTING—DESIGN WITH MOMENT

Redesign the footing of Sec. 15.5 to include at the base a moment equivalent to a 3 in. eccentricity of the column load.

Solution

The dowels would be unchanged, assuming the column to be reinforced adequately for the moment.

The footing is sized by iteration in a sequence which could be varied. As before:

Assume ftg. wt. 300 psf, leaving $q_{net} = 3500 - 300 = 3200$ psf
In terms of ultimate $q_{net} = 3200 \times 875/550 = 5080$ psf
Required area (rough without moment) $> 875{,}000/5080 = 172\ \text{ft}^2 > 13.2^2$

Temporarily, assume the footing size must be increased for moment to, say, 14 ft square. Then $e/b = 3/(14 \times 12) = 0.0179$, which increases the maximum soil pressure from the axial load by the factor $(1 + 6e/b) = 1 + 6 \times 0.0179 = 1.11$. This is equivalent, insofar as edge pressure is concerned, to an increase in axial load to $875{,}000 \times 1.11 = 975{,}000$ lb.

Reqd. $A = 975{,}000/5080 = 192\ \text{ft}^2 = 13.87^2$
Try 13 ft 11 in. square (193 ft²)

For $h = 24$ in., average $d = 24 - 3 - d_b =$ say, 20 in. for shear. The slab shear now is very nonuniform around the column. Assume half the critical section (Fig. 15.7*a*) carries the shear from the heavier half of the pressure.

$V_u = 0.5(5080 + 4520)13.92^2/2 - 4600(\text{approx.}) \times 3.17^2/2$
$\quad = 465{,}000 - 23{,}000 = 442{,}000$ lb
$\overline{V} = 442{,}000/0.85 = 520{,}000$ lb, $\quad b_o = 2(18 + 20) = 76$ in.
Allowable $v = 4\sqrt{f_c'} = 4\sqrt{3000} = 219$ psi
$d_v = \overline{V}/b_o v = 520{,}000/(76 \times 219) = 31.3$ in.

Since the design is sensitive to changing d, try about half the indicated change. Average $d = 26$ in., $h = 26 + 3 + d_b =$ say, 30 in., footing weight $= 375$ psf, which means the footing size will increase a little.

New $p_{net} = 3500 - 375 = 3125$ psf
For ultimate loads, $p_{net} = 3125 \times 875/550 = 4970$ psf
$A = 975{,}000/4970 = 195\ \text{ft}^2$

USE
~~Try~~ 14 ft square (196 *ft*²), as in Fig. 15.7*b*.

The eccentricity factor happens still to be good.

$b_o = 2(18 + 26) = 88$ in.
$V_u = 0.5(4460 + 4970)\ 14^2/2 - 4600(\text{approx.})\ 3.67^2/2$

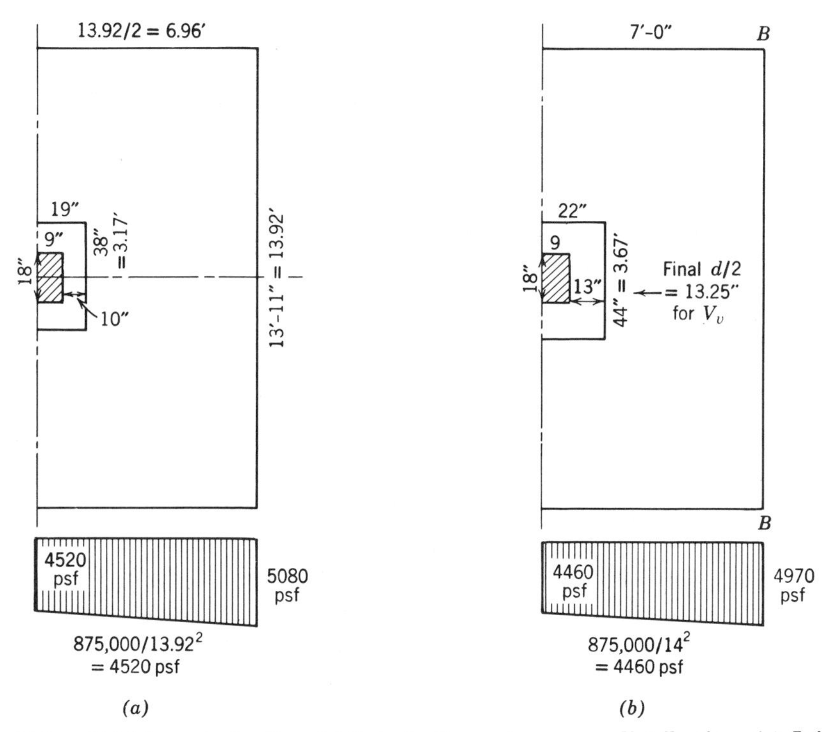

Fig. 15.7. Trial layouts for Sec. 15.7, including *net* pressure distribution. (*a*) Initial. (*b*) Second trial.

$$= 463{,}000 - 30{,}800 = 432{,}000 \text{ lb}$$

$$\overline{V} = 432{,}000/0.85 = 507{,}000 \text{ lb}$$

$$d_v = 507{,}000/(88 \times 219) = 26.4 \text{ in.}$$

USE aver. $d = 26.5$ in., $h = 30.5$ in., ftg. wt. $= 381$ psf

A quick mental check shows the added 6 psf would not change the footing size, but would reduce the q_{net} at ultimate roughly 10 psf, that is, negligibly insofar as other calculations are concerned. The chosen footing size is satisfactory.

If the moment is one which can be in any direction, the upper layer of reinforcement will again be critical. If otherwise, the more severe moment could be assigned the more favorable steel location in the bottom, but this would demand assured field control and inspection. Accept the first assumption, moment in any direction.

Min. $d = 26.0$ in. and q_{net} in Fig. 15.7b 10 psf lower, as found above.

$$M_u = (0.5 \times 4960 \times 6.25 \times 4.17 + 0.5 \times 4450 \times 6.25 \times 2.08)14$$
$$= (64{,}500 + 29{,}000)14 = 1{,}308{,}000 \text{ ft-lb} = 1308 \text{ ft-k}$$
$$\overline{M} = 1308/0.9 = 1460 \text{ ft-k} \qquad \text{Assume } z = 0.95d = 24.7 \text{ in.}$$
$$A_s = 1460 \times 12/(60 \times 24.7) = 11.80 \text{ in.}^2$$
$$\text{Min. } A_s = 200bd/f_y = 200 \times 14 \times 12 \times 26/60{,}000 = 14.55 \text{ in.}^2$$

The minimum controls since shears have required a large depth.

19- #8	15.0 in.^{2}	d_m as assumed
15- #9	15.0	reduced
12- #10	15.3	reduced
25- #7	15.0	increased

USE 19- #8 at 9± in. spcg. *each way*

The use of the same reinforcement in the other direction is obvious, since the eccentricity might be rotated. On the edge strip BB in Fig. 15.7b the uniform q_{net} of 4970 psf also acts almost as a uniform load without the 90° shift in the eccentricity. Although this appears more severe than on the average strip just computed above, tests under centered load show that the center strip moments become large well-ahead of the edge strip moments, which produce yielding only at the last, at just about failure. In this example, the edge pressure concentration is not severe and it appears that this normal behavior will offset it. If the pressure gradient were severely different, the designer might place an extra bar along each of the edges that could have the heavier loading.

$$l_d = 0.04 \times 0.79 \times 60{,}000/\sqrt{3000} = 34.5 \text{ in.} \ll \text{available} \qquad \text{O.K.}$$

$$\text{One way shear } V_u = 0.5(4960 + 4600\pm)14(7 - 0.75 - 2.17) = 272{,}000 \text{ lb}$$

$$v = (272{,}000/0.85)/(14 \times 12 \times 26) = 73 \text{ psi} < 2\sqrt{f_c'} = 109 \qquad \text{O.K.}$$

15.8 RECTANGULAR COLUMN FOOTINGS—ANALYSIS

Analyze the rectangular footing of Fig. 15.8 assuming $f_c' = 3000$ psi, Grade 60 steel, and an allowable soil pressure at service load of 3000 psf. The column service load is 200 k dead and 320 k live load without any moment and there is no overburden.

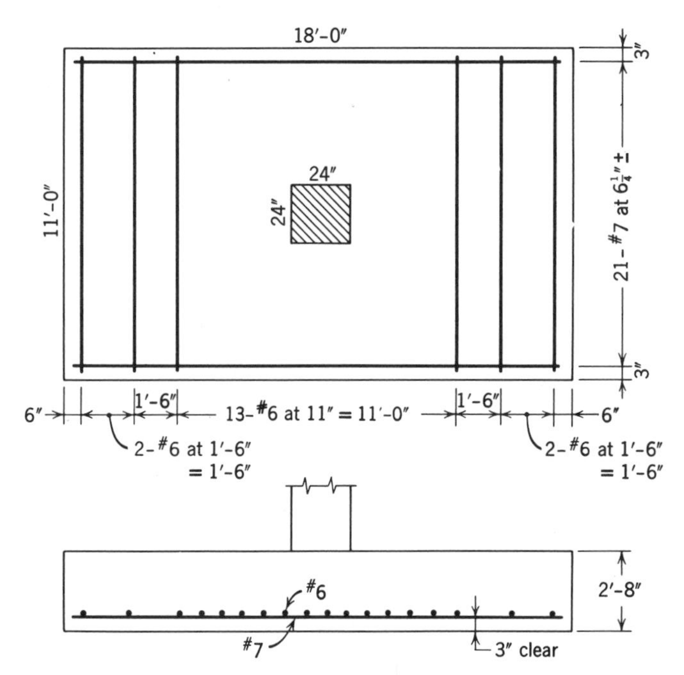

Fig. 15.8. Rectangular column footing.

Solution

Design $P_u = 1.4 \times 200 + 1.7 \times 320 = 824$ k

Net soil pressure at ultimate $= \dfrac{824{,}000}{11.0 \times 18.0} = 4160$ psf

p_{net} at service load $= 4160 \times 520/824 = 2620$ psf

Footing weight $= 150 \times 32/12 = 400$

Service load $p_{gross} = 3020$ psf > 3000

Say O.K. since excess is less than 1%.

No information on dowels is given, but these are obviously necessary. For the long projection, $d = 32 - 3 - 0.4 = 28.6$ in. For the short projection, $d = 32 - 3 - 0.88 - 0.38 = 27.7$ in. Diagonal tension around the column will be considered critical on the average depth of 28.1 in., say 28.0 in., at a distance $d/2$ beyond the column (Fig. 15.9) considering two-way slab action.

$b_o = 4(24 + 28.0) = 4 \times 52.0 = 208$ in.

$V_u = 4160(18 \times 11 - 4.33^2) = 4160(198 - 18.7) = 746{,}000$ lb

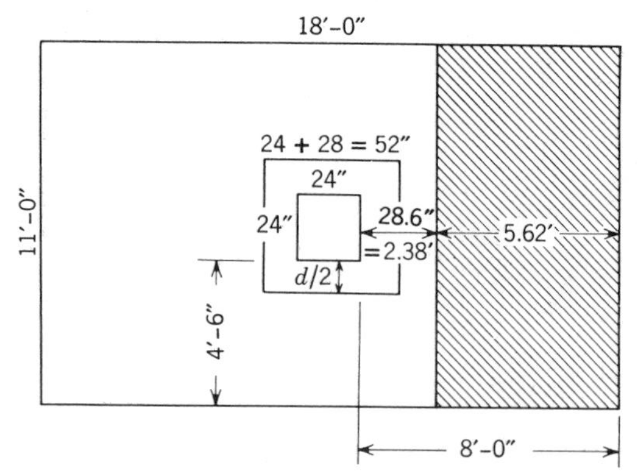

Fig. 15.9. Diagonal tension analysis for rectangular footing.

$\overline{V} = V_v/\phi = 746{,}000/0.85 = 880{,}000$ lb
Permissible $v = 4\sqrt{f_c'} = 4\sqrt{3000} = 219$ psi
Reqd. $d_v = \overline{V}/vb = 880{,}000/(219 \times 208) = 19.3$ in. $\ll 28.0$ O.K.

Check one-way slab shear on long projection at distance $d = 28.6$ in. from column, as shown in Fig. 15.9.

$V_u = 4160 \times 11(8.0 - 2.38) = 257{,}000$ lb
$\overline{V} = 257{,}000/0.85 = 303{,}000$ lb
Permissible $v = 2\sqrt{f_c'} = 109.6$ psi
Reqd. $d_v = 303{,}000/(11 \times 12 \times 109.6) = 21.0$ in. $\ll 28.5$ O.K.

The footing could be several inches thinner insofar as shear is concerned. In the long direction, $M_L = 8.0 \times 11 \times 4160 \times 4.0 = 1{,}460{,}000$ ft-lb

$\overline{M}_L = M_L/\phi = 1{,}460{,}000/0.90 = 1{,}622{,}000$ ft-lb

Assume $z = 0.92 \times 28.6 = 26.3$ in.

$$\text{Reqd. } A_s = \frac{1{,}622{,}000 \times 12}{60{,}000 \times 26.3} = 12.36 \text{ in.}^2 \text{ vs. } 21 \times 0.60 = 12.60 \text{ in.}^2$$

This is O.K. without a check on z.

Minimum A_s for flexure* $= (200/f_y)bd$
$= (200/60{,}000)\ 11 \times 12 \times 28.6 = 12.60 \text{ in.}^2 = 12.60 \text{ in.}^2$ as above. O.K.

*Minimum A_s will always be much under $0.75\ A_{sb}$ (balanced).

In the short direction, $\overline{M}_s = 4.5 \times 18 \times 4160 \times 2.25/0.9 = 842{,}000$ ft-lb

Assume $z = 0.94 \times 27.7 = 26.1$ in.

$$\text{Reqd. } A_s = \frac{842{,}000 \times 12}{60{,}000 \times 26.1} = 6.46 \text{ in.}^2 < 17 \times 0.44 = 7.48 \text{ in.}^2$$

$$\text{Min. } A_s = (200/60{,}000)bd = (200/60{,}000)18 \times 12 \times 27.7 = 20.0 \text{ in.}^2$$

Even for temperature and shrinkage, min. $A_s = 0.0018bh = 0.0018 \times 18 \times 12 \times 32 = 12.40$ in.^{2} Although exposure to temperature and shrinkage is usually not severe in a footing, this is still about the absolute minimum reinforcement anywhere. The 7.48 in.^{2} is grossly inadequate.

Code 15.4.4 requires much of the transverse reinforcement to be concentrated within the central width equal to the short side of the footing, the specified proportion being $2/(1 + \beta)$ where β is the ratio of the sides of the footing, $18/11 = 1.64$ in this case:

$2/(1 + 1.64) = 0.76$ vs. 13 bars/17 bars $= 0.77$ in the 11 ft width. This proportion is satisfactory, but the total is N.G.

For bar development the short projection appears the more critical, but probably adequate.

$$l_d = 0.04A_b f_y/\sqrt{f_c'} = 0.04 \times 0.44 \times 60{,}000/\sqrt{3000} = 19.3 \text{ in.}$$

The available is 54 in. less about 2 in. end cover.

15.9 WALL FOOTINGS

(a) Basic Concepts

The uniform nature of a wall loading leads to the use of a typical 1-ft width as a design strip, as in Fig. 15.10. For heavy walls, the design is quite similar to that of column footings, but simpler because the bending is in only one direction. For light walls, the design is frequently based on rather arbitrary minimum sizes. Here plain concrete is often the appropriate material.

When the wall is of reinforced concrete, the critical moment section is taken at the face of the wall (Code 15.4.2). With masonry walls a different critical section is specified, as illustrated in the following examples.

(b) Examples of Design

(1) Design a wall footing for a 13-in. brick wall carrying a service load per running foot of 6 k dead load plus 15 k live load using $f_c' = 3000$ psi,

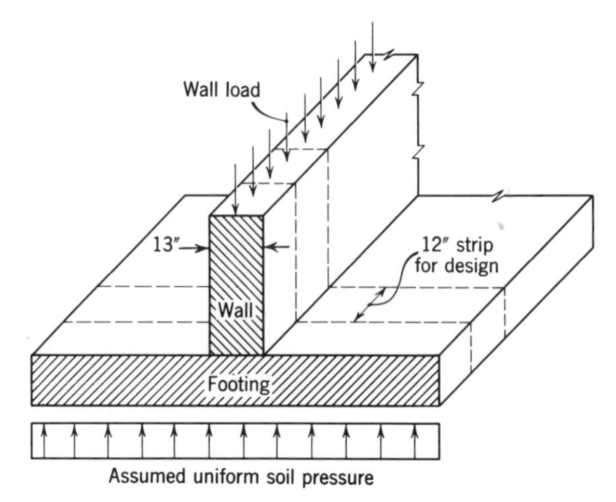

Fig. 15.10. Part of wall footing, with heavy loading.

Grade 60 steel, and an allowable soil pressure at service load of 4400 psf, negligible overburden and moment.

Solution

Assume the footing weight is 150 psf. Service $q_{\text{net}} = 4400 - 150 = 4250$ psf
Ultimate applied load $= 1.4 \times 6 + 1.7 \times 15 = 33.9$ k per running foot of wall

At ultimate, $q_{\text{net}} = 4250 \times 33.9/21 = 6850$ psf allowable
Footing width $= 33{,}900/6850 = 4.95$ ft, say, 5 ft 0 in. USE
$q_{\text{net}} = 33{,}900/5.0 = 6780$ psf

If d is assumed 8 in., the critical section for diagonal tension (Fig. 15.11a) lies 1 ft 3.5 in. or 1.29 ft from the edge.

$$V_u = 1.29 \times 6780 = 8750 \text{ lb/ft strip}$$
$$\overline{V} = 8750/0.85 = 10{,}300 \text{ lb}$$
$$\text{Reqd. } d_v = \overline{V}/bv = 10{,}300/(12 \times 2\sqrt{3000}) = 7.84 \text{ in.}$$

One might expect normally to need a second trial, but required d lies between 7.84 and 8 in., which is close enough. Required $h = 8 + 3$ cover $+ d_b/2 = 11 + d_b/2$, say 11.5 in.

USE $h = 11.5$ in., $d = 8.12$ in. if #6 bars are used

The footing weight changes so little it will be neglected. For a brick wall the critical section for moment lies at the quarter-point of the wall thickness, that is, in this case 1.08/4 = 0.27 ft inside the face of the wall. The use of this equivalent cantilever length is explained by Fig. 15.11*b*. There the shear diagram indicates that the maximum moment is under the middle of the wall, but if the cantilever projection were extended back to the quarter point of the wall thickness, as shown dotted, the area under the shear diagram would be nearly the same.

$$M_u = (1.96 + 0.27)6780 \times 2.23/2 = 16{,}800 \text{ ft-lb/ft strip}$$
$$\overline{M} = M_u/\phi = 16{,}800/0.90 = 18{,}700 \text{ ft-lb}$$

Assume $z = 0.94d = 0.94 \times 8.12 = 7.64$ in.

$$A_s = \overline{M}/(f_y z) = 18{,}700 \times 12/(60{,}000 \times 7.64)$$
$$= 0.490 \text{ in.}^2/\text{ft} = 0.0407 \text{ in.}^2/\text{in.}$$
$$a = 0.490 \times 60{,}000/(0.85 \times 3000 \times 12) = 0.96 \text{ in.}$$
$$z = 8.12 - 0.96/2 = 7.64 \text{ in.} \therefore \text{ estimate was good}$$

Min. $A_s = (200/60{,}000)\ 12 \times 8.12 = 0.325$ in.²/ft < 0.490 O.K.

#6 at 11.0 in. = 0.480 in.²/ft < 0.490

#5 at 7.5 in. = 0.495 in.²/ft (d becomes 8.18 in.)

For #6, $l_d = 0.04 \times 0.44 \times 60{,}000/\sqrt{3000} = 19.3$ in.

The projection of 23.5 in. beyond the wall is ample for l_d plus end cover for #6 bars or smaller. It should be noted that the full l_d is needed only from the point of maximum moment, but since the moment increases more slowly under the wall, the really critical l_d might be for the smaller f_s at face of wall. Here it was obviously on the safe side to compare the full l_d with the available projection in the footing.

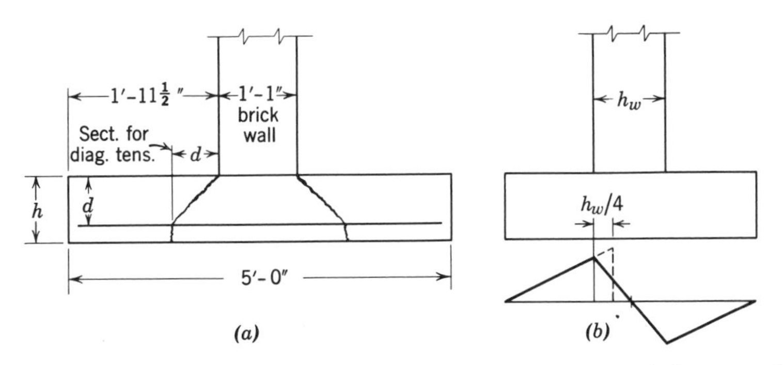

Fig. 15.11. Reinforced concrete wall footing, design. (*a*) Critical shear section. (*b*) Shear diagram and equivalent cantilever length.

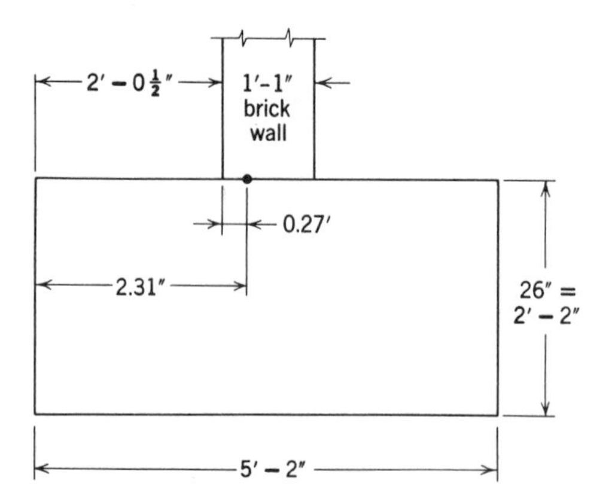

Fig. 15.12. Plain concrete wall footing, design.

USE #5 at 7.5 in. (with 4- #5 longitudinal bars as spacers)

$$\rho_{gr} = 1.24/(11.5 \times 60)$$
$$= 0.00179. \text{ Slightly less than reqd. } 0.0018. \qquad \text{Say O.K.}$$

(2) Redesign the wall footing of (1) as a plain concrete footing in accordance with Code 15.7.2.*

Solution

Ultimate applied load = 33.9 k per running foot, as before

Assume the footing weight is 300 psf

Service load $q_{\text{net}} = 4400 - 300 = 4100$ psf

At ultimate $q_{\text{net}} = 4100 \times 33.9/21 = 6620$ psf allowable

Footing width $= 33{,}900/6620 = 5.12$ ft, say, 5 ft 2 in.

$$q_{\text{net}} = 33{,}900/5.17 = 6560 \text{ psf}$$

Moment is taken at the quarter-point of the wall, Fig. 15.12.

$$M_u = 2.31 \times 6560 \times 1.16 = 17{,}600 \text{ ft-lb}$$

*It is usually not feasible to *combine* plain concrete concepts with reinforced concrete concepts. For example, a designer may see in a tower leg footing the possibility of uplift. If he is tempted to compute flexure from uplift on a plain concrete footing (having cared for positive load flexure with reinforcement), he should note that he has probably assumed tension cracks in the lower 80% of his footing and the *gross* area is *not* available for the reversed bending.

The allowable tension in such footings is given in Code 15.7.2 as $5\phi\sqrt{f_c'}$, with $\phi = 0.65$, or $3.2\sqrt{3000} = 175$ psi, to be used with M_u (not $\overline{M}$)

$17{,}600 \times 12 = f_t bh^2/6 = 175 \times 12h^2/6$
$h = \sqrt{605} = 24.7$ in.

The bottom inch of concrete placed against the ground should usually not be considered for strength. Hence a 26-in. overall depth is indicated, which requires a weight revision. Assume 26 in. thickness and a footing weight of 325 psf.

$q_{net} = 4400 - 325 = 4075$ psf
At ultimate, $q_{net} = 4075 \times 33.9/21 = 6600$ psf
Footing width $= 33{,}900/6600 = 5.14$ ft, say 5 ft 2 in. as above USE
USE $h = 26$ in. overall, 25 in. effective

Diagonal tension is rarely significant in plain concrete footings. If d is considered 25 in., the critical section for diagonal tension will fall 25 in. from the face of the wall or outside the footing. Unit shear in this footing is obviously lower than for the reinforced footing of Fig. 15.11. This more than offsets the fact that shear for the plain concrete must be calculated from $1.5V/bh$, as a homogeneous section. The author doubts that *average* shear (Code 15.7.2) should really be used, even at the specified stress of $2\sqrt{f_c'}$. Shear at the *face* of wall cannot be critical.

(3) Design a wall footing for a 13-in. brick wall carrying an ultimate load of 5 k per running foot using $f_c' = 3000$ psi and an allowable soil pressure at *ultimate* load of 5000 psf.

Solution

By inspection, soil pressure would call for a footing width less than the thickness of the wall. A minimum projection of 1 to 4 in. is generally used on walls.

USE footing width = 1 ft 3 in. $h = 8$ in.

Code 15.9.1 calls for the minimum thickness of 8 in. for plain concrete, which will be used. The moment will be calculated (at the quarter-point of wall thickness), although one senses that it will be negligible.

Actual $q_{net} = 5000/1.25 = 4000$ psf
$M_u = 0.36 \times 4000 \times 0.18 = 252$ ft-lb

This is of no significance, even on 7 in. of plain concrete, giving a calculated stress of around 30 psi.

15.10 FOOTINGS ON ROCK

Very often only a nominal footing is required on rock since good rock is usually at least as strong as concrete and, moreover, is usually loaded on only a small part of its surface. Where a significant footing is required, as under a heavy steel column, deformation and relative stiffness become more significant than calculated unit stresses. A reinforced concrete footing cannot function normally unless the steel elongates and the cantilever elements deflect. Since the underlying rock is stiffer against compressive deformation than the reinforced concrete footing is against bending deflection, such a footing cannot effectively distribute the load laterally to any significant extent. A reinforced concrete footing on rock is thus of little value. The designer must depend more on the transfer of stress through shear with its smaller deformations. This calls for a deep footing and suggests that plain concrete may be just as effective as reinforced. It should be noted that, on the usual *rough* rock surface, friction (and interlock) forces are ample to prevent any significant stretching of the bottom of the footing. A footing on rock requires a large depth to give it shear stiffness and little concern needs to be given to tension on its bottom face.

15.11 BALANCING FOOTING PRESSURES

Many years ago it was commonly accepted that footings would undergo essentially uniform settlements if they developed the same soil loadings per unit area under the usual loads on the structure. Such footings were said to have balanced footing areas.

Modern soil mechanics has greatly restricted the field of usefulness of this concept. It has shown in most practical cases that uniform soil pressure does not lead to uniform settlement. Proportioning for balanced footing pressures usually results in an increased size for outside wall footings. In a heavy structure on a thick stratum of material which consolidates slowly under additional load, such an increase in exterior footing size will actually *increase* the final differential settlement. In many areas, such as in the Southwest, foundation movement on ordinary residence and commercial structures is as likely to be upward due to expansive soils as downward due to applied loads. The engineer who uses the balanced footing areas should be certain that his special foundation situation justifies this approach. Frequently, it increases rather than reduces differential settlements.

For a balanced footing design, the term "usual load" will be applied to the average load or the load most commonly on the structure. This might vary from dead load alone for a church or a school, up to nearly full live load plus dead load for certain types of warehouses.

If all footings were proportioned for the usual loads, at full allowable soil pressure, the footing that cared for the greatest percentage of live load (compared to dead load) would most overstress the soil when full live load was acting. This gives the clue for balanced footing area design. That footing having the largest ratio of live load to dead load is designed first by the usual procedures, using maximum load and the full allowable soil pressure. The usual soil pressure under this footing is next established for the usual load, most frequently something between dead load alone and dead load plus half live load. The bearing areas of all other footings are then chosen to give this same usual soil pressure under the usual footing loads. (This will automatically give them less than the maximum permissible soil pressure under the maximum live load.) The individual footings are then designed for thickness and steel using these balanced areas and the full design live load.

Example

Balance the areas for footings given in Table 15.1 for dead load plus one-fourth live load if the allowable net soil pressure is 3 ksf at service load.

TABLE 15.1.

Footing	Dead Load	Live Load
A	100 *k*	200 *k*
B	75	200
C	50	150
D	100	150

Solution

Footing *C* has the largest ratio of live load to dead load, that is, 150/50 = 3.00. Choose the area of this footing for full dead load plus live load and the full allowable soil pressure.

Area $C = (50 + 150)/3 = 66.7\ \text{ft}^2$
Usual $q = (50 + 0.25 \times 150)/66.7 = 1.31$ ksf

Choose the other footing areas for this usual soil pressure and the specified usual load; see Table 15.2. The maximum pressure column is based on full dead load and live load and is shown only to demonstrate that under this procedure one need not consider maximum soil pressures after the usual pressure has been properly established and used as shown.

TABLE 15.2.

Footing	Dead Load	Live Load	Usual Load	Usual q	Area	Max q
A	100 k	200 k	150 k	1.31 ksf	114.5 ft^2	2.62 ksf
B	75	200	125	1.31	95.4	2.89
C	50	150	87.5	1.31	66.7	3.00
D	100	150	137.5	1.31	105.0	2.37

15.12 COMBINED FOOTINGS

When two or more columns are carried on a single footing, it is called a combined footing. Two circumstances, especially, call for a combined footing: (1) two columns so closely spaced that separate footings would overlap or be of uneconomic proportions, Fig. 15.13*a*; (2) an exterior column footing which cannot be made symmetrical because of property line limitations or other restrictions, Fig. 15.13*b*. In each case the centroid of the combined footing must coincide with the resultant of the two column loads, noted as *O*.

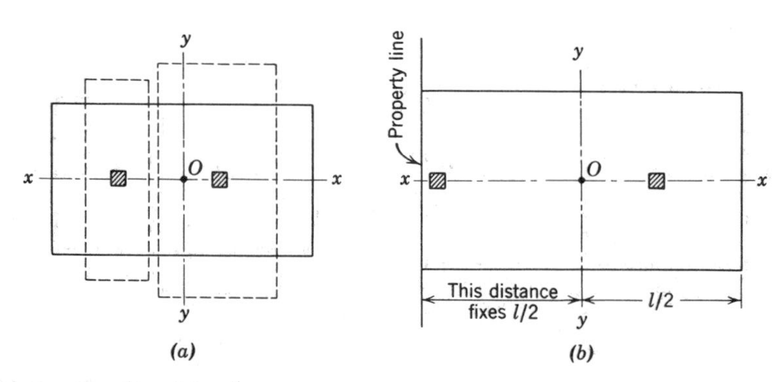

Fig. 15.13. Combined footings.

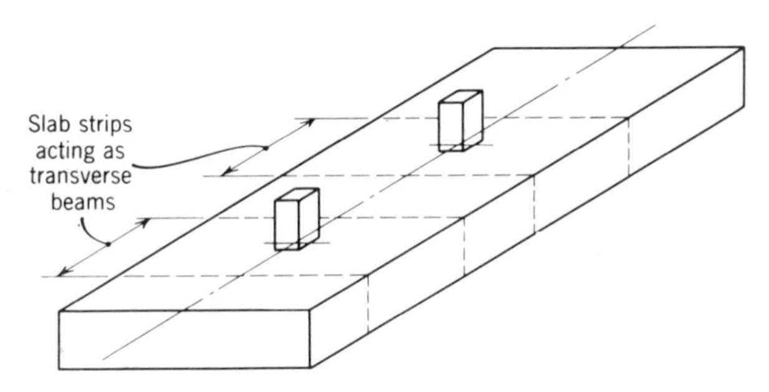

Fig. 15.14. Transverse "beams" in combined footing.

The point O would shift slightly as one considers the design (factored) loads,* the service loads, or the usual loads of Sec. 15.11. The designer may use as the centroid of his footing the one he feels most important in his design situation.

In Fig. 15.13*b*, the length of a rectangular footing is fixed by the fact that if rectangular it must be symmetrical about axis y-y through this resultant load point O. The width in turn is fixed by the required soil-bearing area to carry the combined column loads.

In Fig. 15.13*a*, the designer has some choice in the length and width he chooses, but the centroid of the area chosen must fall at O.

Design of combined footings, like the design of many other reinforced concrete structures, has not been standardized. One approach, sketched in Fig. 15.14, assumes structural behavior as consisting of transverse slab strips under the column which act to distribute load laterally and longitudinal slab strips which act like slabs supported on these "crossbeams." This type of analysis seems appropriate to long narrow footings. Another approach is to think of the projecting ends of the footing as similar to half of an isolated column footing and the slab between columns acting as a longitudinal beam. This method seems appropriate to footings having projections of such proportions as might be used for isolated columns, that is, projections somewhat of the same order of magnitude in the two directions. This type of design calls for a wider distribution of transverse steel than the transverse "beam" idea.

*Varying ratios between service dead and live loads on the two columns also invalidate the mathematical interchangeability of the two permissible soil pressures (at service and at ultimate conditions) discussed in Sec. 15.3d. The designer may take his choice and be consistent with it. The author considers the usual allowable soil pressure not so precisely established as to make the choice a critical one.

Shears should be checked as for other footings, which means a double check, around the column at a distance $d/2$ outside the column for two-way slab action, and at a distance d from the column all the way across the footing as a one-way slab.

Reasonably typical moment and shear diagrams for the longitudinal strips of a combined footing are shown in Fig. 15.15*a*. Design is concerned only with the portions outside the column widths since moments are considered critical at the face of column. The moment diagram can be visualized as a simple beam moment between columns (negative due to upward loads) superimposed on the positive moment diagram produced by the upward loads on the two end cantilevers. These two diagrams going into the total moment diagram of Fig. 15.15*a* are shown separately in Fig. 15.15*b*. These diagrams present no special design problem except one relating to the negative moment between columns. If the columns are closely spaced, no top steel may be required, because the simple span moment may be small compared to the cantilever moment.

Attention should be called to the fact that it is *not* on the safe side to assume a uniform soil pressure in the design of this center negative steel. If the pressure is nonuniform, it is probably larger near the center than it is toward the ends. A uniform load assumption gives too much positive cantilever moment and too little negative moment from the center span. These errors add to indicate a negative moment smaller than the actual value. The designer should be liberal in providing this negative moment steel, even introducing some steel when the uniform pressure assumption indicates a small remnant of positive moment.

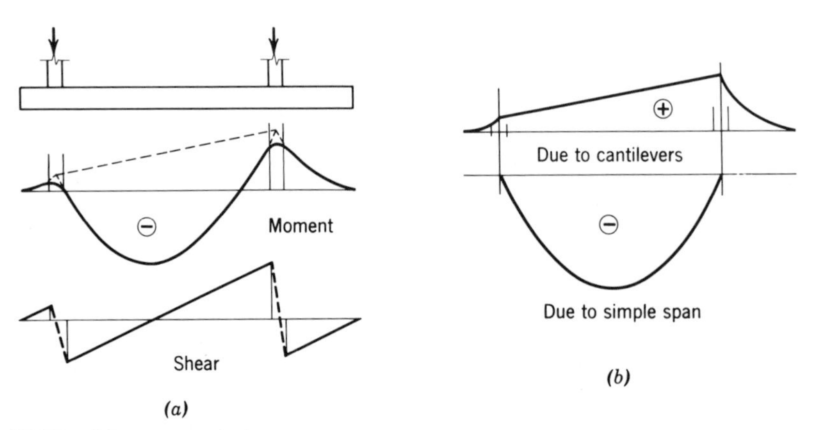

Fig. 15.15. Moment and shear diagrams for combined footing.

15.13 CANTILEVER FOOTINGS

When an exterior column footing, because of property line restrictions, calls for some kind of combined footing but when the nearest interior footing is some distance away, the ordinary combined footing becomes long and narrow and subject to excessive bending moment. The cantilever footing is then more economical.

The cantilever footing is really two separate column footings joined by a beam which preferably should not contribute to the bearing area. In Fig. 15.16*a*, footing A under the exterior column is eccentric by a distance e from the column. By itself it would produce a very unsatisfactory soil pressure distribution. The stiff beam C is attached to footing A to balance the overturning moment P_1e and is carried to the concentric footing B for its necessary balancing reaction. This beam must be stiff in order to function well. It functions most simply if it is relieved of soil pressure from below, except possibly enough to carry the beam's weight. This condition is approached in several ways, most commonly by loosening or spading the soil under the beam, occasionally by forming the beam free from the soil, and so forth. Each of these measures leaves some question as to its effectiveness, which causes some engineers to shun this type of footing.

Under the simplest assumption, that the beam weight is carried by the soil, the analysis of the cantilever footing is quite simple. The beam is then effectively weightless. The reaction from the exterior column alone is R_1 on footing A (Fig. 15.16*b*) and ΔR downward on the beam at the interior column, which means ΔR is an uplift on this footing at B (Fig. 15.16*c*). Hence

$$\Delta R = P_1e(l - e)$$

$$R_1 = P_1 + \Delta R$$

$$R_2 = P_2 - \Delta R$$

Hence footing A must be slightly larger than needed for P_1 alone and footing B can be slightly reduced in area below that needed for P_2 alone. The beam C must carry a constant shear ΔR and a moment as shown in Fig. 15.16*b*.

In sizing footings A and B, either the service loads with the usual allowable soil pressure or the design loads (fractored) with the increased soil pressure (Sec. 15.3d) may be used, R being computed on a comparable basis. At ultimate, the increased soil pressure will not be equal on A and B unless their ratios of dead to live load happen to be equal.

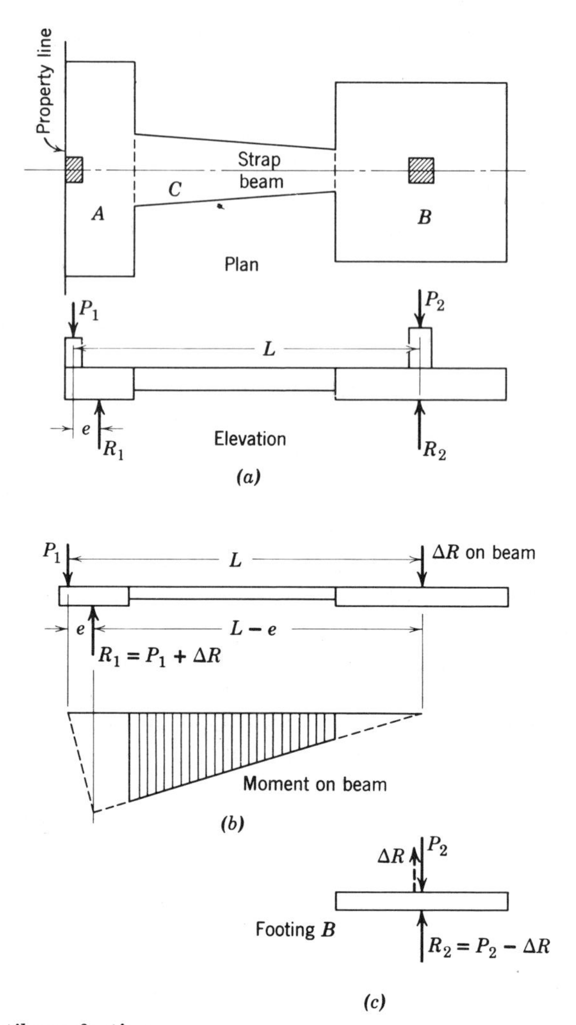

Fig. 15.16. Cantilever footing.

15.14 RAFT FOUNDATIONS

It is sometimes desirable to utilize the soil pressure under the entire building area by covering the site with a thick slab or mat foundation. Such a foundation is often called a raft foundation, particularly when enough soil is excavated to offset the increased load due to the structure. Such a foundation may consist of a uniform slab several feet thick, or a slab stiffened by beams (either above or below the slab), or an inverted flat slab floor (Chapter. 10).

No special design problems are involved in the concrete for such foundations other than those of a heavy floor design to resist upward pressures, with the columns furnishing the downward reactions. The distribution of the design upward soil pressures is the major problem here, this being dependent on the variability of the soil reaction properties and the resulting settlement response.

15.15 SLABS-ON-GROUND

A slab supported directly on a stable soil develops very little moment or shear in carrying a uniform load over its surface. When loads are unequally distributed or concentrated, the distributed reaction produced on the soil does create some moment and shear. (This is the principle used in pavement design.) On very stable soils, light loads such as residence loads can be carried on a simple 4-in. slab with a wire fabric reinforcement, providing no partition load is in excess of 500 plf. Chimmey and other heavy loads, such as columns, must be carried to separate footings and the slab should not rest on or be tied to these separate elements or footings, which often will not move exactly the same as the slab.

Many soil types are subject to shrinkage in drying and expansion in wetting. Slabs alone are not adequate on such soils and experience shows that one portion of the slab may be pushed upward or settle relative to another. If this is not to crack walls and partitions, the slab must be stiffened by monolithically cast beams in both directions. In effect, a residence may then be supported in the center and the stiffened slab will have to cantilever out in all directions to prevent excessive floor warpage; or the structure may be supported on two corner areas and the slab will have to span as a simple span between these. Thus the stiffened slab must ride like a boat over differential ground movements. Then the floor must be designed for the above loadings and a stiffening grid similar to that shown in Fig. 15.17 will be necessary. If such a system appears excessively costly, the final alternative is for spot footings to be carried deeper to a better soil and then to design a fully framed floor not in contact with the soil. Alternatively, soil stabilization measures are now being used on some bad soils.

15.16 FOOTINGS ON PILES

Many footings must be built on piles. The piles act as a series of relatively concentrated reactions for the footing and in design are usually considered as concentrated loads. Since it is usually difficult to drive

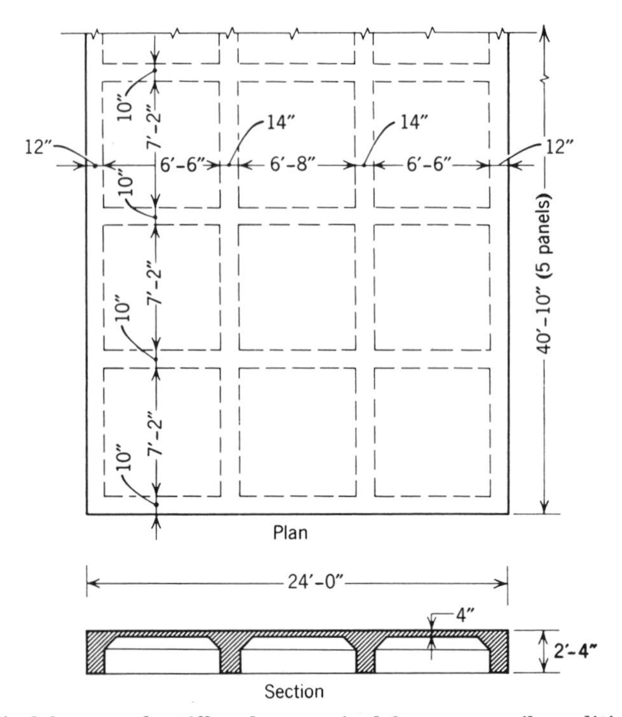

Fig. 15.17. A slab-on-grade stiffened as required for a poor soil condition.

piles exactly where they are desired, the designer must expect that piles will be from a few inches to a foot from the spot he assigns to them. Accordingly, the Code (15.5.5) provides specific rules for counting piles which are near the critical section and sets somewhat stricter minimum thicknesses in Code 15.9.2.

15.17 STEPPED AND SLOPED FOOTINGS

Specifications now permit thinner footings than were formerly used. There is thus less reason now for stepped or sloped footings and fewer are used. Such footings must definitely be cast as a monolith to avoid horizontal shearing weakness. Stresses must usually be checked at more than one section. For instance, steel stress at a step or development length beyond a step may be more critical than from the face of the column. Both these and shear must be checked from each step, the shear at a distance d or $d/2$ as in other cases.

Stepped footings are more appropriate for massive footings. A step can sometimes be used to advantage with a combined footing.

15.18 UPLIFT ON FOOTINGS

If uplift can exist on a footing, several special points must be considered. First, uplift probably results from lateral loading on the structure and Code Eq. 9.3 probably governs with its lower factor on dead load because it opposes the resultant uplift.

$$U = 0.9D + 1.3W$$

Second, the axial tension (plus the horizontal shear) may require tension splices and tension anchorage into the footing. Third, special design considerations are necessary. The footing must either be heavy enough in itself to balance the uplift or it must engage a mass of earth adequate for this balance. This reverses the type of bending on the footing and normally calls for some measure of top reinforcement. The footing will normally have some flexural cracks from service load positive moment. The assumption of an uncracked plain concrete section for resisting the negative moment from uplift would not be realistic.

15.19 DESIGN AIDS FOR SINGLE COLUMN FOOTING

A graph developed by Dr. R. W. Furlong is useful in shortening the cut-and-try process for footing thickness required for shear around the column. If all dimensions are in feet, the critical shear around the column is

$$[A_f - (r + d)^2]q_{\text{net}} = \phi v b_o d = \phi v_{\text{all}} 4(r + d)d \times 144$$

where r = width of square column
A_f = total footing area
q_{net} = P_u/A_f in psf
v_{all} = the allowable in psi
ϕ = 0.85 for shear

If this equation is divided by the column area r^2 and by q_{net}

$$\frac{A_f}{r^2} = \frac{\phi v_{\text{all}}}{q_{\text{net}}} \frac{576}{r^2} (r + d)d + \frac{r^2}{r^2} + \frac{2rd}{r^2} + \frac{d^2}{r^2}$$

$$= \frac{\phi v_{\text{all}}}{q_{\text{net}}} 576 \left(\frac{d}{r} + \frac{d^2}{r^2}\right) + 1 + 2\frac{d}{r} + \frac{d^2}{r^2}$$

For any given d/r this is a linear relation between A_f/r^2 and $\phi v_{\text{all}}/q_{\text{net}}$. As written, all ratios are dimensionless except that v_{all} is in psi and q_{net} is in psf, both as customarily used. This equation is the basis for Fig. 15.18.

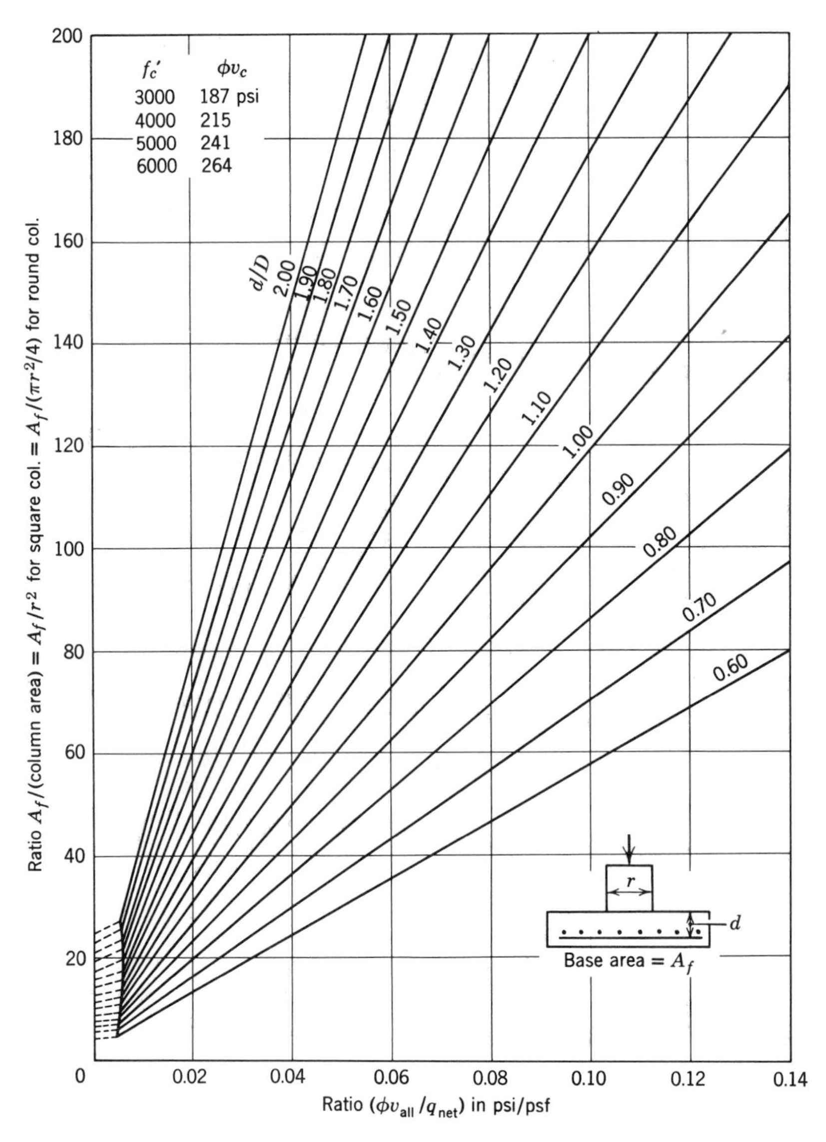

Fig. 15.18. Chart for establishing footing d for shear around column.

The design of Sec. 15.6 will be used to show how this technique applies. There the first assumed q_{net} was $3500 - 300 = 3200$ psf or 5080 psf for ultimate loads. The column r was 18 in. or 1.5 ft, required A_f was 172 ft², and allowable v was 219 psi.

$$A_f/r^2 = 172/1.5^2 = 76.6$$

$$\phi v/q_{net} = 0.85 \times 219/5080 = 0.0367 \text{ ft}^2/\text{in.}^2$$

These values in Fig. 15.18 intersect on d/r of 1.38 leading to $d = 1.38 \times 18 = 24.8$ in., $h = 24.8 + d_b + 3 \doteq 29$ in., $w = 2.42 \times 150 = 363$ psf, $q_{net} = 3500 - 363 = 3137$ psi $\times 875/550 = 5000$ psf at ult., $A_f = 875/5.00 = 175$ ft², $A_f/r^2 = 175/1.5^2 = 77.8$, and $\phi v/q_{net} = 0.85 \times 219/5000 = 0.0372$. These values in Fig. 15.18 show $d/r = 1.38$, leading to $d = 1.38 \times 18 = 24.8$ in. and to no change in h. This gives a footing 29 in. deep compared to 29.5 in. found analytically. This is about as closely as the printed chart can be read for d/r.

The lower dotted curves in Fig. 15.18 indicate a region where the one-way slab action governs the shear requirement.

Alternatively, Dr. Furlong's Table 15.3 may be used for the above design, as follows.

As before, assumed $q_{net} = 3500 - 300 = 3200$ psf or 5080 psf for ultimate loads. Reqd. $A_f = 875{,}000/5080 = 172$ ft²

$$\sqrt{f_c'}/q_{net} = \sqrt{3000}/5080 = 0.0108$$

Now enter the table in the left column at 0.0108 and (by interpolating) move to the right to $A_f/A_c = 76.4$:

At $d/r = 1.4$, $A_f/A_c = 73$
$\quad 1.5, \quad 81 \qquad$ At $A_f/A_c = 76.4$, $d/r = 1.44$

For column $r = 18$ in., reqd. $d = 1.44 \times 18 = 25.8$ in.

Reqd. $h = 25.8 + 3.0 + 1.5$ (for $1.5d_b$) $= 30.3$ in., say 30.5 in.

This increases the footing weight to 381 psf and makes $q_{net} = 3120$ psf or $3120 \times 875/(200 + 350) = 4960$ psf.

Reqd. $A_f = 875{,}000/4960 = 176$ ft², say 13 ft 3 in. square

$A_f/A_c = 176/1.5^2 = 78.3$

Again entering the table with the same 0.0108:

At $A_f/A_c = 78.3$, $d/r = 1.47$

Reqd. $d = 1.47 \times 18 = 26.5$ in. compared to 25.5 in. in Sec. 15.6.

TABLE 15.3. Footing Depth for Shear

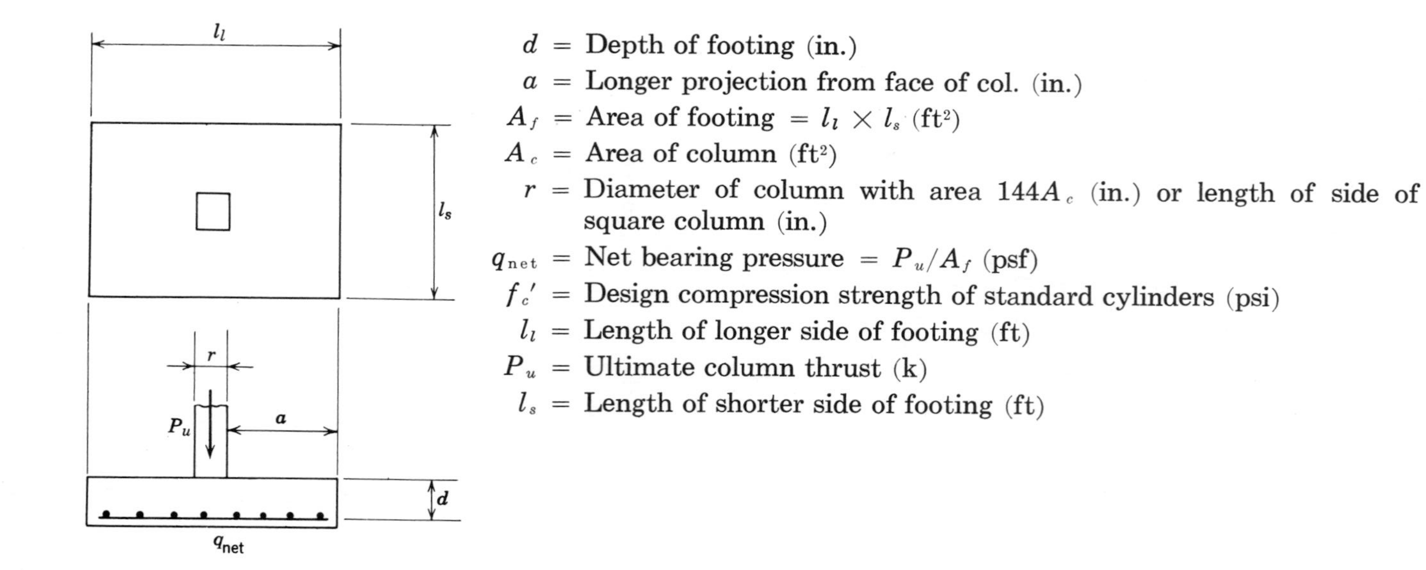

d = Depth of footing (in.)

a = Longer projection from face of col. (in.)

A_f = Area of footing = $l_l \times l_s$ (ft^2)

A_c = Area of column (ft^2)

r = Diameter of column with area $144A_c$ (in.) or length of side of square column (in.)

q_{net} = Net bearing pressure = P_u/A_f (psf)

f_c' = Design compression strength of standard cylinders (psi)

l_l = Length of longer side of footing (ft)

P_u = Ultimate column thrust (k)

l_s = Length of shorter side of footing (ft)

Values of A_f/A_c

$\frac{\sqrt{f_c'}}{q_{net}}$	Punching Shear d/r															Beam Shear a/d
	0.4	0.5	0.6	0.7	0.8	0.9	1.0	1.1	1.2	1.3	1.4	1.5	1.6	1.7	1.8	
0.000	1	2	2	2	2	2	2	2	2	2	2	3	3	3	3	1.00
0.002	4	4	5	6	7	9	10	11	13	14	16	18	19	21	23	1.49
0.004	6	7	9	11	13	15	18	20	23	26	29	31	35	39	42	1.98
0.006	8	10	13	16	18	22	26	29	33	37	42	46	52	57	62	2.47
0.008	10	13	17	20	24	28	33	38	44	49	55	61	68	75	83	2.96
0.010	12	16	20	25	30	35	41	47	54	61	68	75	84	93	102	3.45
0.012	15	19	24	30	35	42	49	56	64	73	81	90	100	111	121	3.93
0.014	17	22	28	34	41	49	57	65	75	84	94	105	117	129	141	4.42
0.016	19	25	32	39	47	56	66	75	85	96	108	119	133	147	161	4.91
0.018	21	28	35	44	52	62	72	84	95	108	121	134	149	165	180	5.40
0.020	23	31	39	48	58	69	80	93	106	120	134	148	165	183	200	5.89
0.022	26	34	43	53	64	76	88	102	116	131	147	163	182	200	220	6.38
0.024	28	38	47	57	69	82	96	111	126	143	160	178	198	218	240	6.87
0.026	30	41	50	62	75	89	104	120	137	155	174	192	214	236	259	7.36
0.028	32	44	54	67	80	96	112	129	147	166	187	207	231	254	279	7.85
0.030	34	47	58	71	86	102	120	138	157	178	200	221	247	272	299	8.34

One may also check the one-way slab shear depth from the same table. Entering with 0.0108 at the left as before proceed to the far right hand column and read $a/d = 3.45 + (8/20)0.48 = 3.64$

The given projection $a = 13.25/2 - 0.75 = 5.87$ ft

Reqd. $d = a/3.64 = 5.87/3.64 = 1.61$ ft $\ll$ 26.5 in. above

SELECTED REFERENCES

1. A. N. Talbot, "Reinforced Concrete Wall and Column Footings," Univ. of Ill. Eng. Exp. Sta. *Bull. No. 67*, 1913.
2. F. E. Richart, "Reinforced Concrete Wall and Column Footings," *Jour. ACI*, 20, Oct. and Nov. 1948; *Proc.*, 48, pp. 97, 237.
3. ACI-ASCE Committee 326, "Shear and Diagonal Tension," *Jour. ACI.*, *Proc.* 59, Jan. 1962, p. 30; Feb. 1962, p. 277; Mar. 1962, p. 1962.
4. ACI Committe 436, "Suggested Design Procedure for Combined Footings and Mats," *Jour. ACI*, *Proc.* 63, Oct. 1966, p. 1041.

PROBLEMS

Prob. 15.1. Redesign the rectangular footing of Sec. 15.8 and Fig. 15.8 as a square footing.

Prob. 15.2. Redesign the square column footing of Sec. 15.6 as a rectangular footing having a width of 11 ft 6 in.

Prob. 15.3. Design a square column footing for a 16-in. by 16-in. tied column (of 4000 psi concrete) to carry a service column load of 150 k dead load and 140 k live load using permissible *ultimate* soil pressure of 8000 psf, $f_c' = 3000$ psi, Grade 60 steel.

Prob. 15.4. Redesign the footing of Prob. 15.3 as a plain concrete footing of uniform thickness.

Prob. 15.5. Design a reinforced concrete wall footing to carry service loads of 6 klf dead and 4 klf live load from a concrete wall 12 in. thick. Use an allowable soil pressure of 2800 psf, $f_c' = 3000$ psi, Grade 60 steel.

Prob. 15.6. Redesign the footing of Prob. 15.5 using plain concrete.

Prob. 15.7. Resdesign the wall footing of Prob. 15.5 if the allowable soil pressure is only 2000 psf.

Prob. 15.8.

(a) Lay out the plan view of a combined footing for the columns of Fig. 15.19 where the service loads are:

A: 150 k dead plus 100 k live
B: 200 k dead plus 250 k live.

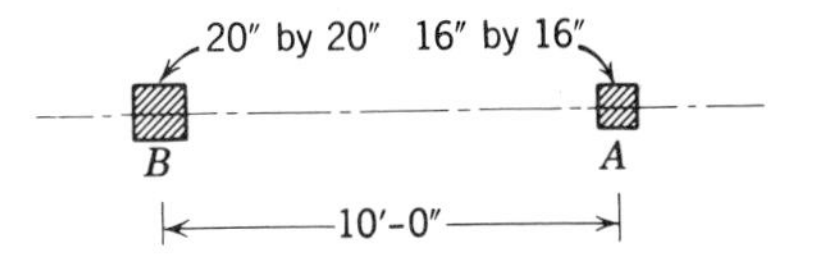

Fig. 15.19. Column layout for Prob. 15.8.

Based on an allowable *net* soil pressure of 3200 psf and a centroid based on full design loads, try to achieve equal projections around the heavier column.

(b) Lay out again on the assumption that there is a property line 4 in. beyond the outside face of column A.

(c) Lay out again for conditions of (a) on the assumption that lateral projections beyond column B are to be about half the longitudinal projection.

(d) Repeat (a) with the centroid based on full service loads.

Prob. 15.9. If the columns of Fig. 15.20 carry the service loads of Prob. 15.8(a), lay out the plan view of a cantilever footing (without dimensions on the beam) based on an allowable net soil pressure of 3200 psf. Draw the M and V diagrams for the strap beam which connects the two footing elements:

(a) If the footing under column A extends 4 ft from the property line.
(b) If the dimension in (a) is changed to 5 ft.
(c) If the dimension in (a) is changed to 6 ft.

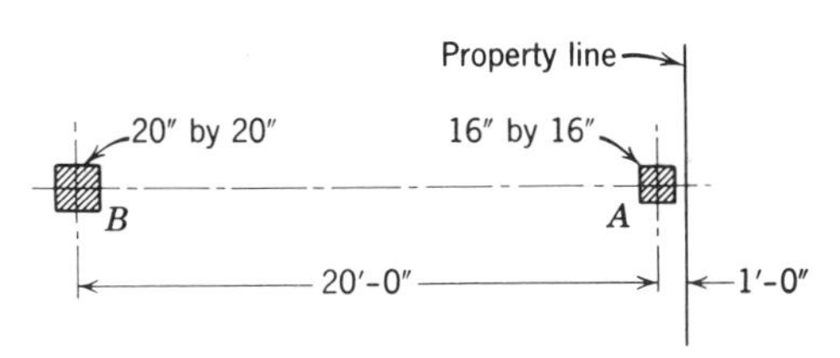

Fig. 15.20. Column layout for Prob. 15.9.

16
Prestressed Concrete Analysis

16.1 SCOPE OF PRESENTATION

This discussion of prestressed concrete covers some of the fundamental concepts and presents a general idea of how such concrete responds to loads. No attempt is made to develop design procedures. For the ease of presentation, simple rectangular sections are used in the examples rather than the more practical I or T shapes.

16.2 THE NATURE OF PRESTRESS

Reinforced concrete becomes prestressed concrete when permanently loaded in such a way as to build up initial stresses opposite to those which later will be developed by service loads.

A circular tank, designed to resist circumferential or ring tensile stresses, as in Fig. 16.1*a*, is prestressed by being placed in initial circumferential compression. This compression is produced by wrapping the tank with wire under high tensile stress as indicated in Fig. 16.1*b*.

A simple span beam is prestressed (Fig. 16.2) by introducing (1) a negative moment to offset the expected positive moment and at the same time (2) a longitudinal compression to offset the tensile stresses from bending moment. Both effects are obtained by embedding highly stressed tension members eccentrically below the middle of the beam, giving both

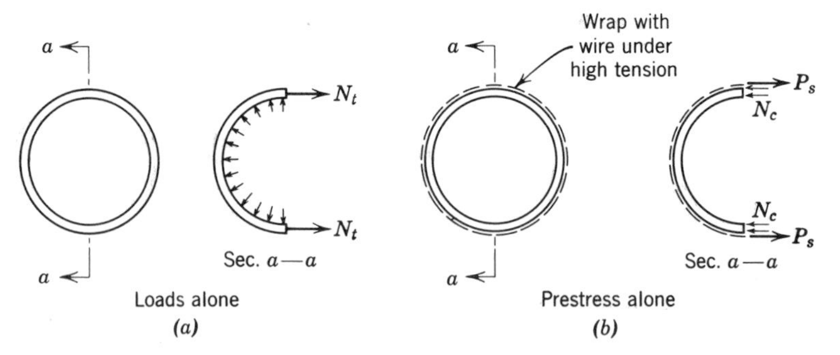

Fig. 16.1. Prestress largely eliminates ring tension in tanks.

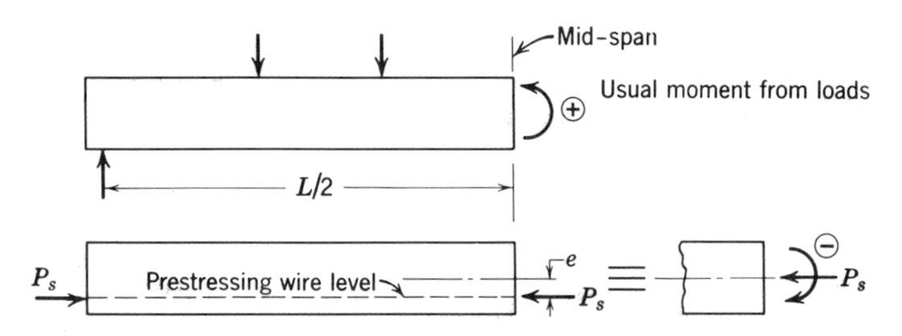

Fig. 16.2. Beam prestress produces axial compression and a moment opposite in sign to dead load moment.

a P/A effect from the load P_s and a negative bending moment effect measured by P_se. The tension prestressing element, called a tendon, may be wire, twisted wire strands, or rods.

Continuous beams are prestressed in similar fashion but for best results require an effective eccentricity above mid-depth in negative moment zones. Their analysis is more involved because the prestress may modify the external reactions, shears, and moments.

For piles, plank, columns, and other members subject to moments both positive and negative, or subject to loading on either face, longitudinal prestressing wires without any resultant eccentricity are used solely for their P/A effect, as in Fig. 16.3. Since eccentric prestressing wires in a simple span build up negative moment to offset the ordinary positive moment, such prestressing would actually weaken the member if it were inverted or put under negative moment.

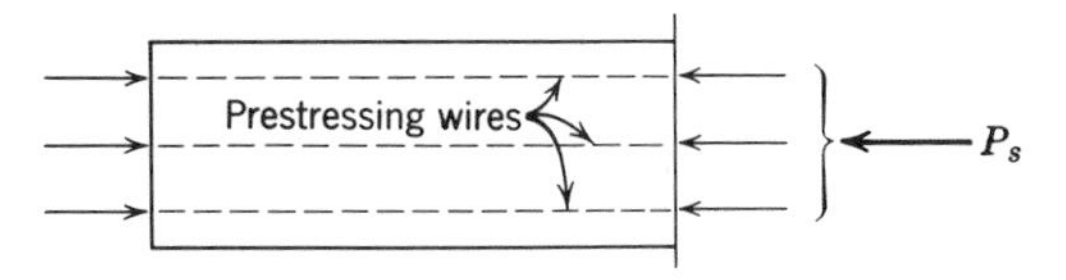

Fig. 16.3. Beam with uniform prestress.

16.3 OBJECTIVES OF PRESTRESSING

Proper prestressing ordinarily prevents the formation of tension cracks under working loads. This alone is adequate reason for prestressing tanks to avoid leakage, or for prestressing structural members subject to severe corrosion conditions.

With this lack of cracking, the entire cross section remains effective for stress; a smaller required section normally results. With lighter sections, precasting of longer members becomes possible. Precasting has been a most important economic factor in promoting prestressed concrete in the United States.

Since prestress puts camber into a member under dead loading, deflections at working loads are much reduced.

Because tension cracking of prestressed concrete at working loads is avoided, very high strength steel (and smaller steel areas) can be used to provide the necessary tension. In come countries prestressed concrete is thus a device to make small supplies of steel serve more efficiently. In this country prestress is a device to permit very high strength steel to be used economically and to increase permissible span lengths while keeping the concrete uncracked and hence more effective.

Prestress reduces diagonal tension stresses at working loads. This has led to the general use of modified I and T shapes with less web area than conventional beams, as in Fig. 16.4.

Most of the advantages of prestressed concrete are construction or service load advantages. Under ultimate load conditions the strength of the usual (underreinforced) prestressed beam is about the same as without prestress; its deflection increases sharply after cracking, and stirrups are then necessary. The Code requires at least minimum stirrups in all prestressed beams.

16.4 NOTATION AND SIGNS

The key to prestressed concrete analysis lies in a good bookkeeping system. All the usual calculations are individually simple, but they consist

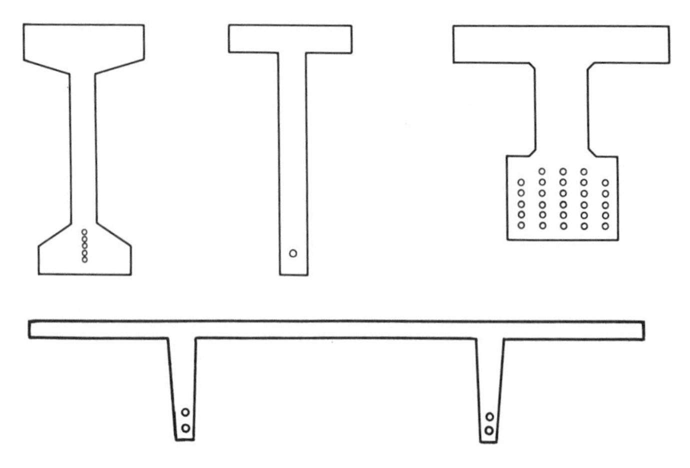

Fig. 16.4. A few of the cross sections used for prestressed members.

of a number of small pieces which must be assembled into various combinations. Good bookkeeping requires a consistent set of signs for stresses and an adequate notation. For the beginner numerous sketches, such as those in Figs. 16.7, 16.8, and 16.9, are helpful.

In this chapter a tensile stress will carry a plus sign and a compressive stress a negative sign. Although this is contrary to the signs used by many writers for prestressed concrete, it is in accord with general structural practice and is recommended also in Professor Lin's book.[1] No sign convention is needed for ordinary reinforced concrete since the kind of stress in such calculations is never in doubt; in prestressed concrete the concrete stress often changes from tension to compression (or the reverse) as the loading changes.

Much of the checking of a prestressed member is at service loads (unfactored loads) and thus requires the calculation of stresses; but ultimate strength must also be checked.

16.5 STRESSES BY SUPERPOSITION OF STRESS BLOCKS

In the following sections, detailed calculations show "exact" methods for computation of both bonded and unbonded tendons. A simpler, slightly approximate approach in terms of gross area of concrete and the final prestress (effective prestress) is often useful to the designer; it will also probably be easier to visualize here for the first numerical example.

Consider the simple span rectangular beam of Fig. 16.5*a* with effective prestress P_{se} of 120 k from a straight tendon 8 in. below middepth. Establish the service load stresses existing prior to the application of live load.

Solution

The effect of the prestress is an axial load from the 120 k prestress plus a negative moment from the eccentricity $= P_{se}e = -120 \times 8 = -960$

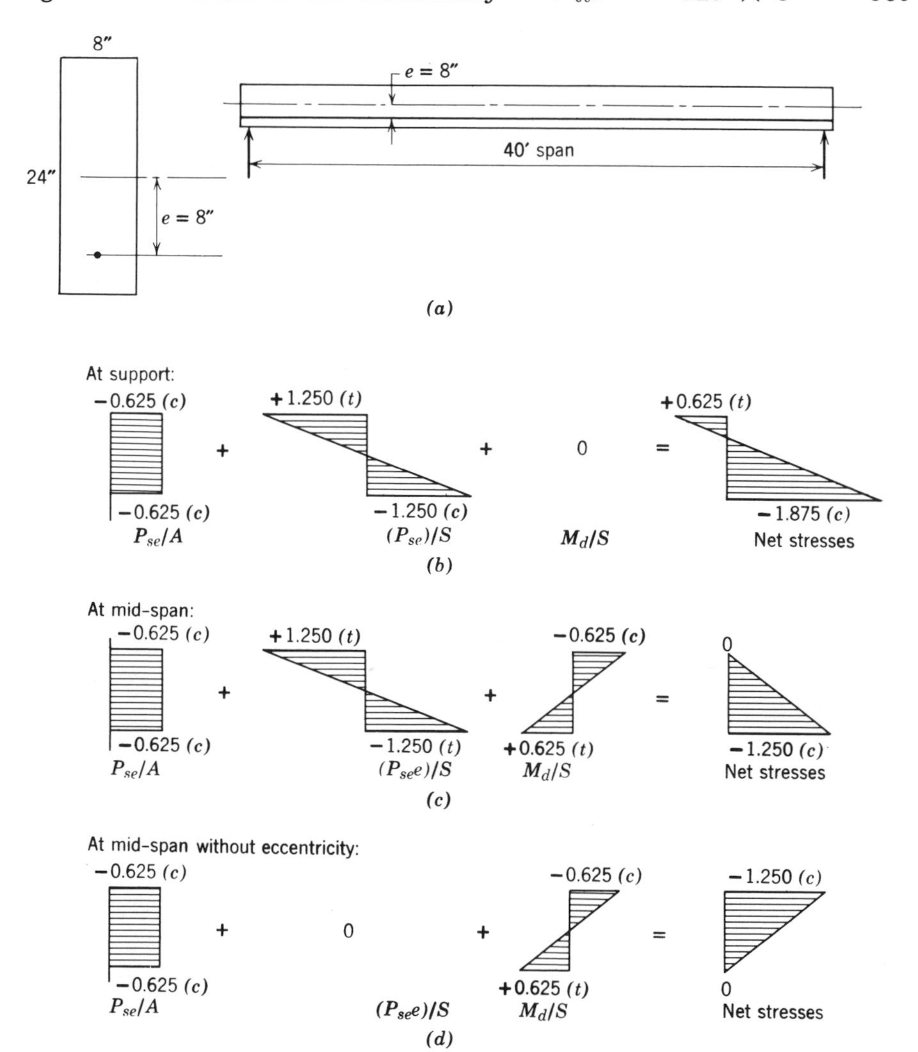

Fig. 16.5. Influence of tendon eccentricity on service load stresses. (*a*) Beam of Sec. 16.5. (*b*) Stresses at support. (*c*) Stresses at mid-span. (*d*) Stresses at mid-span without tendon eccentricity.

k-in. The axial load produces the uniform stress block of Fig. 16.5*b* with $f_c = -120/(8 \times 24) = -0.625$ ksi compression. The moment gives the antisymmetric triangles of that same figure.

Section modulus $S = bh^2/6 = 8 \times 24^2/6 = 768$ in.[3]*

$f_c = -960/768 = -1.25$ ksi compression on bottom

or $+ 1.25$ ksi tension on top

At the end of the beam the dead load moment is zero, leading to the combined stress block shown at the right of Fig. 16.5*b*. The 625 psi tension on the top is more than would be permissible because it exceeds the probable cracking stress. If the tendon were draped upward at the end, reducing the end eccentricity and thus $P_{se}e$, the stress could be made acceptable.

At midspan the dead load moment of the beam weight of $8 \times 24 \times 150/144 = 200$ plf is $M_d = (\frac{1}{8})200 \times 40^2/1000 = +40$ k-ft, leading to stresses:

$f_c = +40 \times 12/768 = +0.625$ ksi tension at bottom

or $-$ 0.625 ksi compression at top.

The superimposed stress blocks then show in Fig. 16.5*c*. It should be noted that this beam now has bottom compression all the way across the span and thus deflects upward under dead load and prestress. Also, this substantial compression stress at midspan indicates no tension would develop there until at least 400 plf of live load (twice the dead load) is added.

Had the prestress been added on the axis instead of at 8 in. eccentricity, the stresses in Fig. 16.5*b* at the support would have been only the rectangular stress block of -0.625 psi compression. (The cable could also be draped such as to obtain the same effect.) At the mid-span the stress blocks would then drop to those of Fig. 16.5*d*, with a zero stress at the bottom. Hence no live load could be added if one designed to the very strict requirement occasionally used of *no* tension under service load. Lowering the tendon at mid-span to the 8 in. eccentricity as above has increased the load capacity by 400 plf in this case, and at the same time increased the ultimate moment capacity because of the greater effective depth of the steel.

The solution is not as complete as in those examples which follow, but it is often used as being reasonably close. M_u must still be checked.

*With symmetry S is more convenient than the use of I/c.

16.6 PRE-TENSION METHOD OF PRESTRESSING

Since prestressing calculations follow closely the physical process of prestressing, both the pre-tensioning and post-tensioning procedures must be clearly understood. Sketches of the beam and stresses at each stage will be helpful to the beginner in making his summations. The individual calculations are quite simple.

In the pre-tension process the first step is to stretch high strength wires between abutments or end piers of a (long) prestressing bed, as indicated schematically in Fig. 16.6*a*. This total initial wire tension is here designated

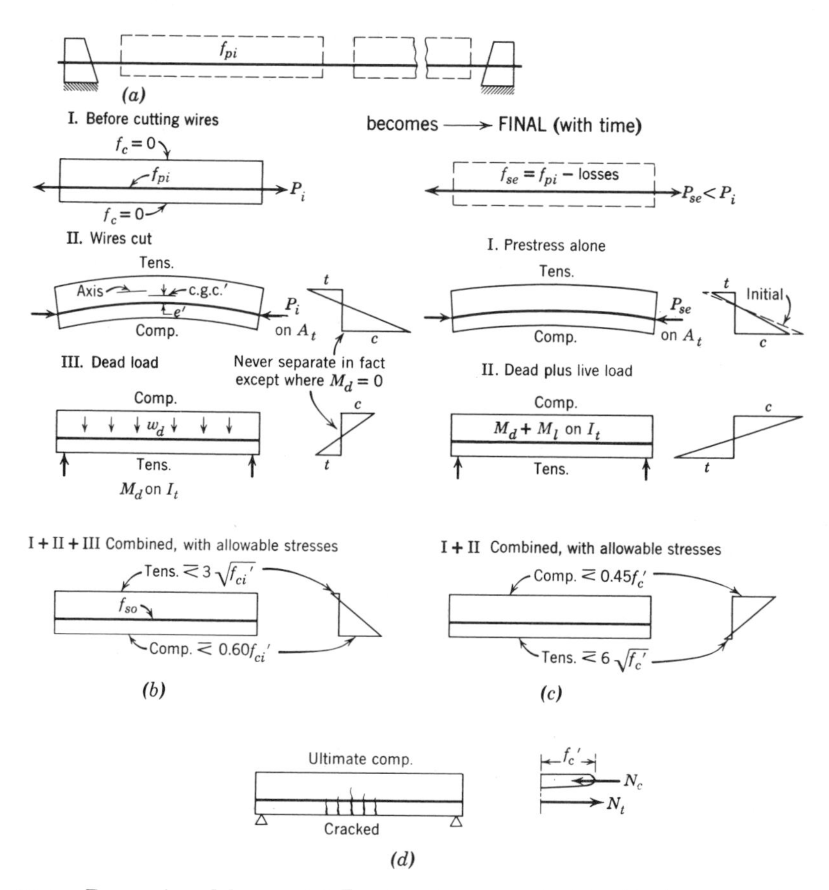

Fig. 16.6. Pretensioned beam. (*a*) Prestressing wire stretched between casting yard abutments. (*b*) Initial stresses at mid-span, separated for calculation purposes; also allowable concrete stresses. (*c*) Stresses at mid-span under full service load, separated for calculation purposes; also allowable concrete stresses. (*d*) Ultimate strength conditions.

P_i, the unit stress f_{pi}. The member forms are placed around the wire, with a number of such units along the total length, and in these forms the members are cast and cured (often by accelerated curing). The wires are then cut between members (Fig. 16.6*b*).

The resulting release of wire tension P_i is equivalent to P_i applied as an external compressive force on the entire transformed area (Sec. 6.4) of the member A_t. The wires are usually eccentric, below the centroid of the transformed area of the member, and thus the external force P_i introduces a negative moment $P_i e$, which causes the beam to curve upward and pick up its dead load moment M_d. Because of P_i alone, rather large calculated tensile stresses develop on the top beam fibers, but this need not be dangerous because these never exist without the counteracting compression from dead load moment.* This combined top tensile stress is often a critical design stress with a recommended limit of $3\sqrt{f_{ci}'}$ unless top steel is used, f_{ci}' being the cylinder strength at this early age. To turn such a beam on its side, or invert it, or lift it by a sling at the middle will be unsafe and may cause failure. The bottom stress in compression may also be a controlling stress, limited to $0.60f_{ci}'$. Shrinkage is usually omitted from this calculation. Although some shrinkage will have occurred by the time the wires are cut, and these calculations apply, the shrinkage reduces both these critical stresses.

With time, shrinkage of concrete and the creep of concrete and steel under stress reduce the length of the concrete holding the wires and thereby reduce P_i to a lower effective prestress P_{se}. Formerly the reduction was often taken as 35,000 psi but now it is commonly more carefully appraised. The loss tends to increase with small ratios of volume to surface area (thin members) and with exposure to lower humidities, but it varies considerably for different members in different regions and under different exposures. It increases if high initial stresses are used on the concrete. As a result of the lowered prestress the top tension and bottom compression are reduced below the initial values. It is upon these lowered stresses that live load stresses are superimposed, resulting in the bottom sketch of Fig. 16.6*c*. The final result is compression critical at the top and tension at the bottom, with the allowable stresses for buildings noted in the combined load sketch in Fig. 16.6*c*.

At ultimate load, the beam will be cracked as indicated in Fig. 16.6*d* and conditions will not differ greatly from those of an ordinary reinforced concrete beam.

*With straight wires and straight members, stresses would be critical near the ends where M_d is small. Present practice often reduces the eccentricity near the ends to keep the center stresses the critical ones (Sec. 16.13).

16.7 POST-TENSION METHOD OF PRESTRESSING

In the post-tension method the plain concrete member is first cast with a tube or slot for future introduction of steel as indicated at I in Fig. 16.7*a*. A variation in procedure is to embed the steel in the member but separate it from the concrete by an enclosing metal or plastic tube or mastic coating. After the concrete has cured, the steel tendons are stressed by jacking against the concrete and are then locked under stress by appropriate end anchorage or clamps. Since the concrete simultaneously shortens under compression, the jacking must account for both steel stretch and concrete shortening; and this is complicated by friction losses and by the fact that beam curvature under this stressing causes the dead load moment also to become active (Fig. 16.7*a*).

The figure shows the resultant steel tension as P_o acting on the plain concrete section A_c (compared to P_i on A_t for pre-tension). In the pre-tensioned case P_o could also have been calculated from the known value of P_i;* in the postensioned case there exists nothing comparable to P_i. The forces w_p between cable and concrete are discussed in Sec. 16.14. The allowable stresses are also marked at the bottom of Fig. 16.7*a*.

The calculations for full service load are diagrammed in Fig. 16.7*b*. These calculations should start with a loss of prestress slightly less than was used for the pre-tension case; some shrinkage will have occurred prior to prestressing and all the elastic deformation of the concrete occurs during jacking to the f_{so} stress. This loss, at one time taken as 35,000 psi, is now calculated more carefully, considering the volume-to-area ratio and exposure actually expected.

Unless tendons are grouted into place, the calculations for the effect of live load must also be based on the plain (net) concrete section. The live load increases the steel stress but less than n times the change in adjacent concrete stress and less than the earlier drop from f_{so} to f_{se}. The change in steel stress is usually not calculated.

If the tendons are grouted effectively, as is often specified, the live load stresses can be based on the entire transformed section area. The change in steel stress, although larger than when ungrouted, is still not important.

Ultimate strength of a grouted beam (Fig. 16.7*c*) is about the same as for a pretensioned beam. Without grouting the strength would be somewhat lower, with failure tending to occur in the conrete above a single wide crack, as in Fig. 16.7*d*. However, see Sec. 16.8 in this connection and note that Code 18.9 requires some bonded reinforcement to be added in the potential tension zone where prestressing steel is unbonded.

*In the pre-tension case P_o on the net concrete area would give the same result as P_i on the transformed area, but P_i is the known starting point.

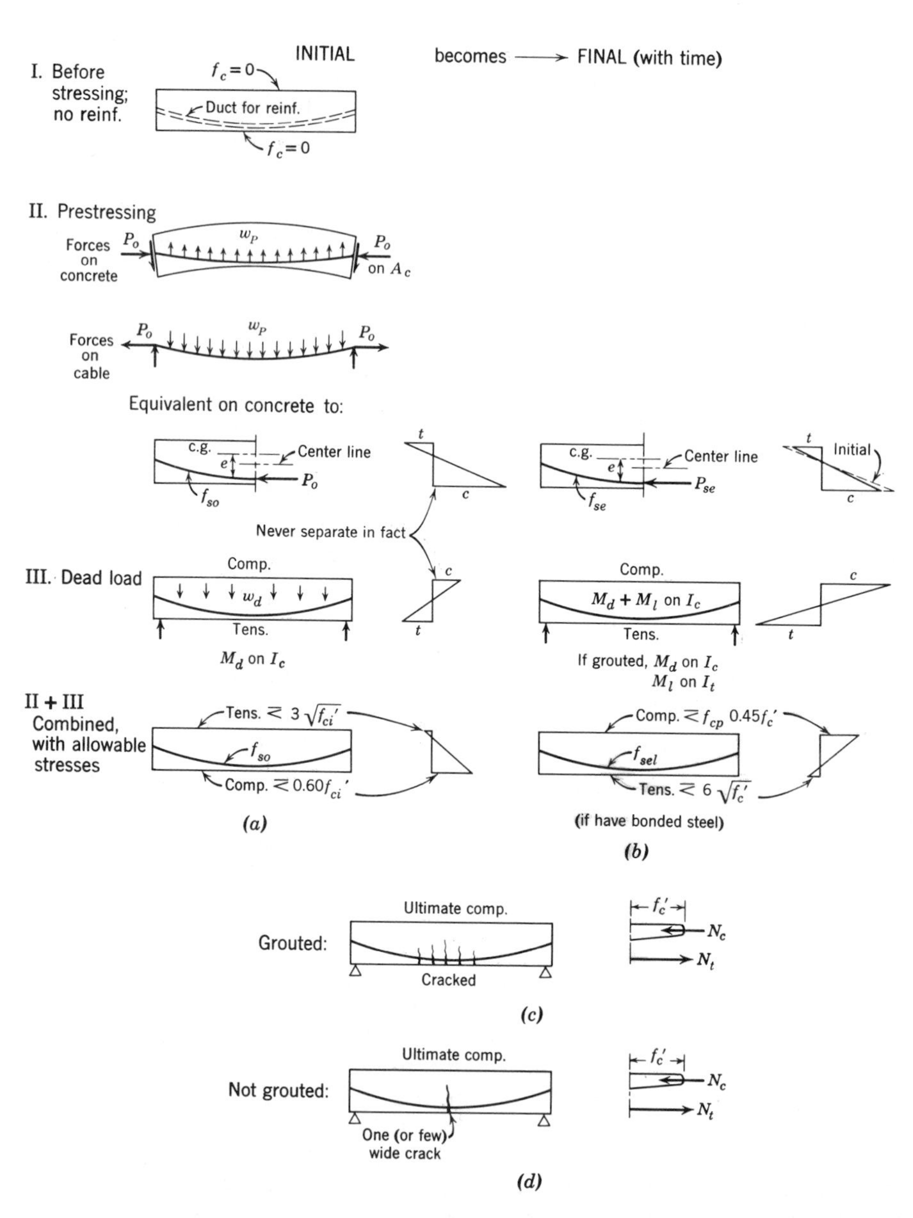

Fig. 16.7. Post-tensioned beam. (*a*) Initial stresses separated for calculation purposes; also allowable concrete stresses. (*b*) Stresses under full service load separated for calculation purposes; also allowable concrete stresses. (*c*) Ultimate strength when grouted. (*d*) Ultimate strength, when ungrouted, is lower.

16.8 OTHER PROCEDURES IN PRESTRESSED CONCRETE

The original sharp distinction between pre-tensioned and post-tensioned members is now less sharp than might be implied in the above discussion. Prestressing has become of age such that many variations on the original patterns are now common practice.

Precast systems may include either pre-tensioning or post-tensioning on a partial basis adequate for handling stresses with later additional post-tensioning used after members are in place to bring them to design capacity. This includes the possibility of adding such post-tensioning as is needed to assure continuity of adjacent members, the addition of inverted *U* strands over supports, and the like.

Both grouted and ungrouted systems are in use with strong differences of opinion over the merits of ungrouted tendons. The latter often consist of wires coated with a lubricating, protective mastic covering and so banded together or wrapped as to prevent intrusion of concrete. The wide crack formation at failure, indicated in Fig. 16.7*d*, can be corrected by casting in some ordinary reinforcing bars of small diameter for crack control. Likewise the ultimate strength of either a pre-tensioned or post-tensioned member can be raised where necessary by casting in ordinary reinforcement as supplemental steel. Supplemental steel may run only a limited distance, wherever reserve moment strength or crack control is needed.

The possibility of aligning the post-tensioned tendons in any desired shape gives this construction an advantage in taking care of shear and varying moments, especially in cast-in-place construction. For precast simple spans, however, it is now feasible with pre-tensioning to depress the strands in a precise fashion to accomplish much the same advantages.

16.9 CRACKING LOAD AND ULTIMATE STRENGTH

European specifications have placed considerable emphasis on the cracking load, requiring that no cracks be formed for a given (usually small) overload. American practice has placed greater emphasis on ultimate strength.

Each of these methods is an attempt to provide an adequate factor of safety, the first indirect, the second direct. The need for both working stress and ultimate strength calculations lies in the radical change in beam behavior when cracks form. Prior to cracking the gross area of the beam is effective. As a crack develops, all the tension from the concrete must be picked up by the steel. If the percentage of steel is small, there may be very little added capacity between cracking and failure. Cracking may be assumed to take place when the calculated tensile stress reaches $7.5\sqrt{f_c'}$.

16.10 EXAMPLE OF PRESTRESS WITHOUT ECCENTRICITY

The hollow member of Fig. 16.8*a* is reinforced with the equivalent of four wires of 0.10 in.[2] each, pre-tensioned to $f_{si} = 150$ ksi. If $f_c' = f_{ci}' = 5000$ psi, $n = 7$, determine the stresses when the wires are cut between members; also determine the moment that can be carried at a maximum tension of $6\sqrt{f_c'}$ and a maximum f_c of $0.45f_c'$. If 35 ksi of the prestress is lost (in addition to the elastic load deformation) determine this limiting moment.

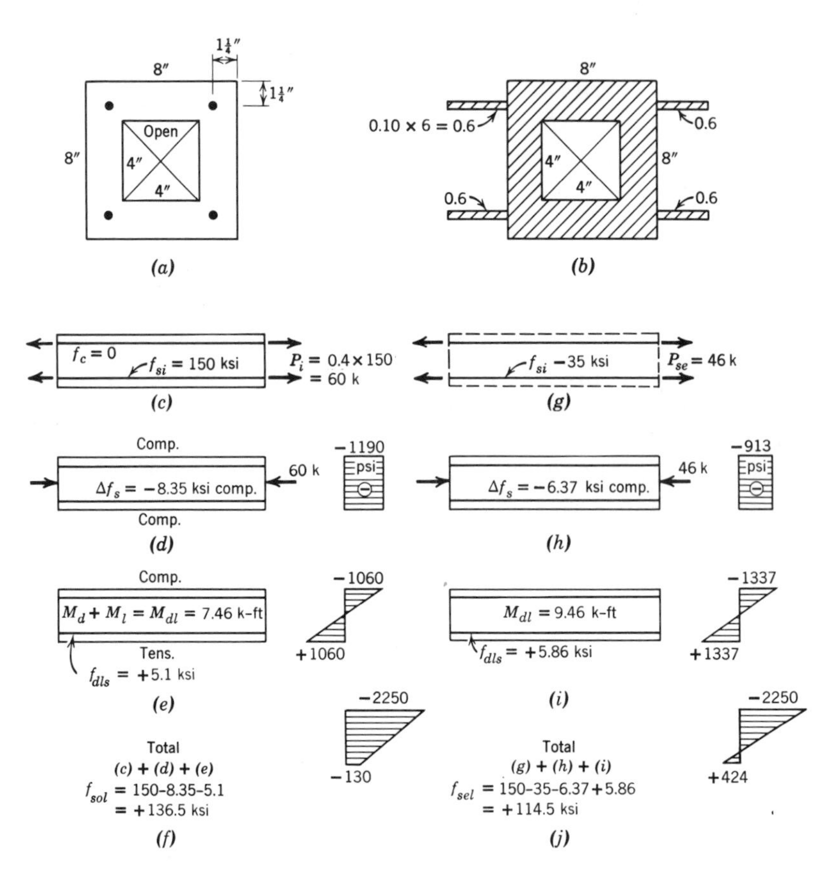

Fig. 16.8. Prestress without eccentricity, pretensioned. (*a*) Beam cross section. (*b*) Effective transformed area. (*c*) Wires under initial tension. (*d*) Effect of cutting wires. (*e*) Permissible moment for initial loading. (*f*) Total stresses for initial loading. (*g*) Equivalent initial tension after losses. (*h*) Equivalent reduction due to compression of concrete. (*i*) Permissible moment for later loading. (*j*) Total stresses for later loading.

Solution

Transformed area (Fig. 16.8*b*) $= A_t = 64 - 16 + 4 \times 0.6 = 50.4$ in.^{2}
$I_t = 8 \times 8^3/12 - 4 \times 4^3/12 + 4 \times 0.6 \times 2.75^2 = 342 - 21 + 18 = 339$ in.^{4}

The starting stresses of Fig. 16.8*c* are $f_c = 0$ and $f_{si} = 150$ ksi. When the wires are cut, the 60-k force acts on the transformed area (Fig. 16.8*d*) to add these stresses:

$f_c = -P_i/A_t = -60 \times 1000/50.4 = -1190$ psi compression
$\Delta f_s = nf_c = -7 \times 1190 = -8350$ psi $= -8.35$ ksi compression

The total stresses before loading are thus:

$f_c = 0 - 1190 = -1190$ psi compression
$f_{so} = 150 - 8.35 = +141.6$ ksi tension

Before any prestress losses take place, $M_d + M_l$ can be enough to bring f_c at the top to $0.45f_c' = 2250$ psi compression or to bring f_c at the bottom to $6\sqrt{f_c'} = 424$ psi tension.

This leaves for dead and live load moment, $M_d + M_l = M_{dl}$:

On top, $f_c = -2250 + 1190 = -1060$ compression
On bottom, $f_c = +424 + 1190 = +1614$ tension

The smaller value limits M_{dl}, as shown in Fig. 16.8*e*.

$M_{dl} = 1060 \times I_t/y_t = 1060 \times 339/4 = 89{,}500$ in.-lb $= 7.46$ k-ft

Under this loading:

Total concrete stress on top $= -1190 - 1060 = -2250$ psi compression
Total concrete stress on bottom $= -1190 + 1060 = -130$ psi compression
Total steel stress $= +150 - 8.35 + n \times 1.060 \times 2.75/4$
$= +141.6 + 7 \times 1.060 \times 2.75/4 = +136.5$ ksi tension

After 35 ksi of prestress is lost, P_i is, in effect, reduced to $(150 - 35)0.4 =$ 46.0 k, as in Fig. 16.8*g*,*h*, giving uniform stresses:

$f_c = -1190 \times 46.0/60 = -913$ psi compression
$f_{se} = +115 - 7 \times 0.91 = +108.7$ psi tension

M_{dl} may safely change the top stress by $-2250 + 913 = -1337$ psi and the bottom stress by $+424 + 913 = +1337$ psi. It is a coincidence that the two stress increments are equal; the smaller controls.

$M_{dl} = 1337 \times 339/4 = 113{,}500$ in.-lb $= 9.46$ k-ft

The lower M_{dl} value of 7.46 k-ft controls (Fig. 16.8*e*,*f*).

The final stresses, if the higher M_{dl} could be used, would have been:

$$\text{Concrete on top} = -913 - 1337 = -2250 \text{ psi} = 0.45f_c'$$
$$\text{Concrete on bottom} = -913 + 1337 = +424 \text{ psi} = 6\sqrt{f_c'}$$
$$\text{Steel stress} = +108.7 + 7 \times 1337 \times 2.5/4.0$$
$$= +108.7 + 5.86 = 114.6 \text{ ksi tension}$$

Ordinarily there is no need to calculate the steel stresses. The values given indicate steel stresses cannot become higher than initial values.

The capacity cannot be accepted as above until the value of M_u is also established as shown in Sec. 16.13.

16.11 EXAMPLE OF PRE-TENSIONED BEAM WITH ECCENTRICITY

The pre-tensioned beam of Fig. 16.9 will be investigated for $f_c' = 5000$ psi, $n = 7$, $f_{si} = 135$ ksi, and a creep and shrinkage loss of 35 ksi. $M_d = 65$ k-ft, $M_l = 80$ k-ft. The concrete strength when wires are released is assumed to be $f_{ci}' = 4000$ psi, but for simplicity the same n will be used. Assume that the beam axis (or wire) is curved so that stresses under dead load and prestress do not require a check other than at mid-span. Allowable concrete stresses are shown in Fig. 16.6*b*,*c*.

Solution

$$A_t = 10 \times 20 + 9.00 = 209 \text{ in.}^2$$
$$\bar{y} \text{ from mid-depth} = -9.0 \times 7/209 = -0.30 \text{ in. (below)}$$
$$I_t = 10 \times 20^3/12 + 200 \times 0.30^2 + 9.00 \times 6.70^2$$
$$= 6667 + 18 + 402 = 7087 \text{ in.}^4$$

Figure 16.9*c* shows forces before cutting wires and Fig. 16.9*d* and *e* shows the stress changes occurring after cutting wires. Actually, these two sets of changes occur together and cannot be separated, except in the calculations.

From cutting the wires:

$$f_c \text{ at top} = -P_i/A_t + P_i e y_t/I_t = -202/209 + (202 \times 6.70)$$
$$\times 10.30/7087 = -0.976 + 1.970 = +0.994 \text{ ksi tension}$$
$$f_c \text{ at bottom} = -202/209 + (202 \times 6.70)(-9.70)/7087$$
$$= -0.976 - 1.855 = -2.831 \text{ ksi compression}$$
$$\Delta f_s = nf_c \text{ at same level} = -7(0.976 + 1.855 \times 6.70/9.70)$$
$$= -7(0.976 + 1.285) = -15.8 \text{ ksi compression}$$

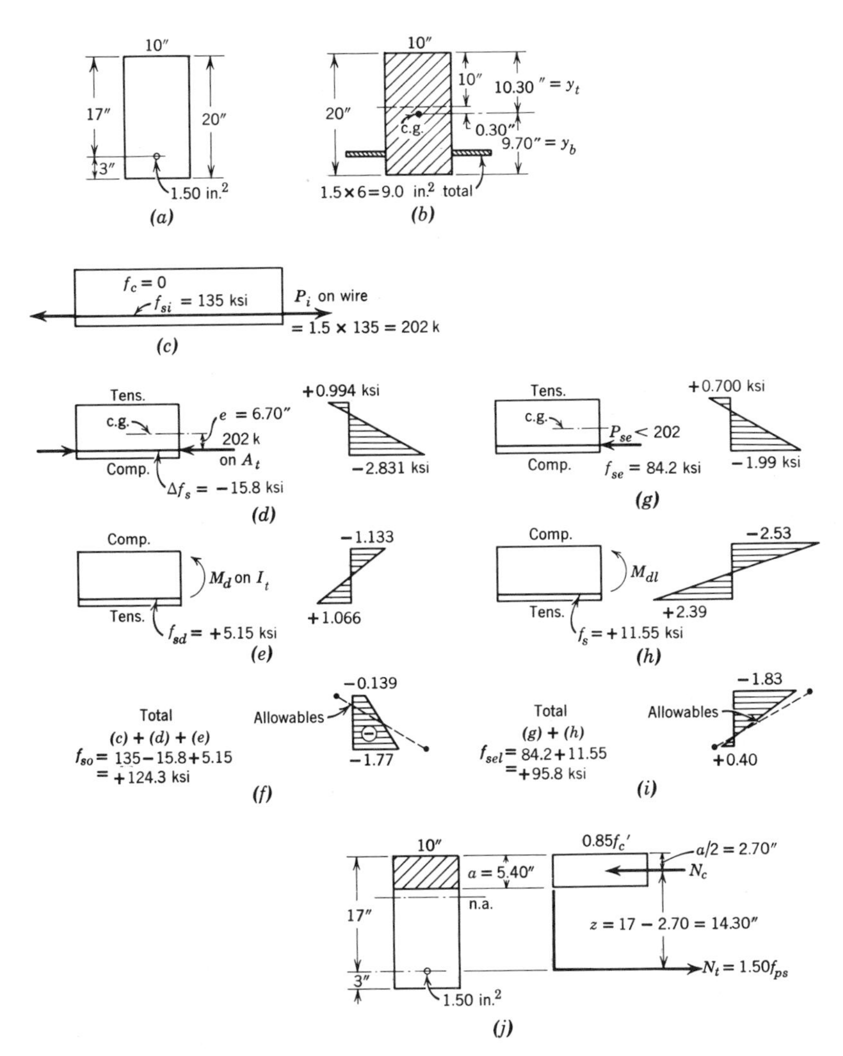

Fig. 16.9. Pretensioned beam. (*a*) Cross section. (*b*) Transformed area, A_t. (*c*) Wires under initial tension. (*d*) Effect of cutting wires, eccentric load on A_t. (*e*) Dead load stresses. (*f*) Total initial stresses. (*g*) Effective pretension stresses are 84.2/119.2 times those in (*c*) and (*d*). (*h*) Dead load plus live load stresses. (*i*) Total stresses under full service load. (*j*) Ultimate strength calculation.

From dead load moment:

f_c at top $= M_d y_t / I_t = -(65 \times 12)10.30/7087 = -1.133$ ksi compression

f_c at bottom $= +(65 \times 12)9.70/7087 = +1.066$ ksi tension

$f_s = 7(+1.066 \times 6.70/9.70) = +5.15$ ksi tension

Total stresses without live load:

f_c at top $= 0 + 0.994 - 1.133 = -0.139$ ksi compression $< 3\sqrt{f_{ci}}$ tension

f_c at bottom $= 0 - 2.831 + 1.066 = -1.765$ ksi compression $< 0.6f_{ci}'$

$f_s = +135 - 15.8 + 5.15 = +124.3$ ksi tension

A prestress some 35% larger would be possible since both stresses in Fig. 16.9*f* are low. (One objection to the rectangular beam shape is the large prestress required for full effectiveness.)

When f_s drops 35 ksi (due to losses) from $135 - 15.8 = 119.2$ ksi to 84.2 ksi tension, as in Fig. 16.9*g*, the prestress effect, without M_d, drops to:

f_c at top $= +0.994 \times 84.2/119.2 = +0.700$ ksi tension

f_c at bottom $= -2.831 \times 84.2/119.2 = -1.990$ ksi compression

Total moment $= M_d + M_l = 65 + 80 = 145$ k-ft, giving stresses:

f_c at top $= -145 \times 12 \times 10.30/7087 = -2.530$ ksi compression

f_c at bottom $= +145 \times 12 \times 9.70/7087 = +2.390$ ksi tension

Total f_c at top $= +0.700 - 2.530 = -1.830$ ksi compression $< 0.45f_c'$

Total f_c at bottom $= -1.990 + 2.390 = +0.400$ ksi tension $< 6\sqrt{f_c'} =$ 424 psi.

Total $f_s = +84.2 + 7(2.390 \times 6.70/9.70) = 84.2 + 11.55 = 95.8$ ksi tension. The total moment could be increased about 1% or the prestress could be reduced nearly 1% and still keep the bottom tension within the allowable.

The check cannot be considered complete without an investigation of the ultimate moment capacity (Sec. 16.13).

16.12. EXAMPLE OF POST-TENSIONED BEAM WITH ECCENTRICITY

The post-tensioned beam of Fig. 16.10*a* will be investigated for $f_c' = f_{ci}' = 5000$ psi, $n = 7$, $f_{so} = 135$ ksi, 25-ksi loss in prestress with time, $M_d = 60$ k-ft, $M_l = 70$ k-ft. It is assumed that the cables are positioned along the span so as to make the mid-span section the critical one. The cables will first be considered as ungrouted; then the effect of grouting will be investigated. Allowable concrete stresses are shown in Fig. 16.7*a*,*b*.

Solution

$A_c = 10 \times 20 - 4 \times 2.5 = 190$ in.^{2} (Fig. 16.10*b*)

$\bar{y}$ from mid-depth $= (-4 \times 2.5) \times (-7.25)/190 = +0.38$ in. (above)

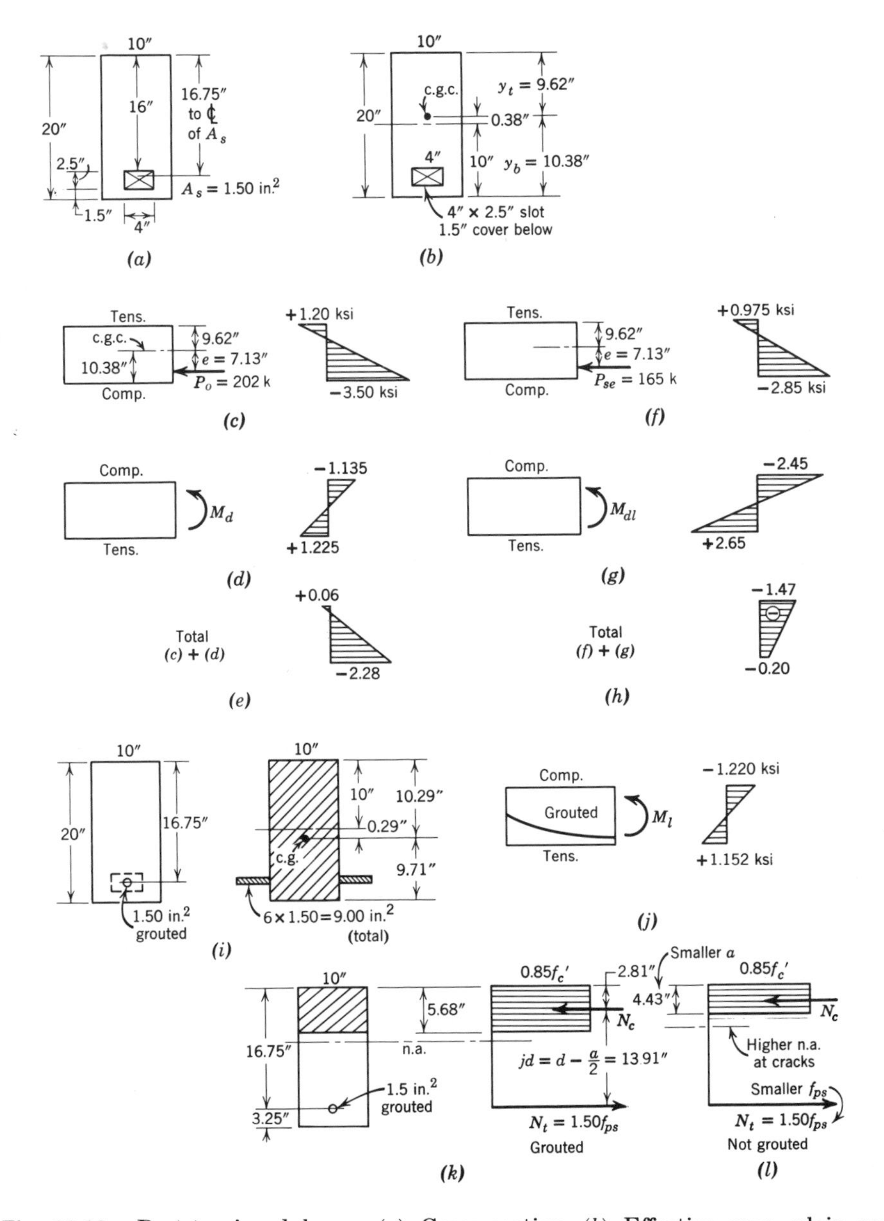

Fig. 16.10. Post-tensioned beam. (*a*) Cross section. (*b*) Effective area, plain concrete only, A_c. (*c*) Initial prestressing. (*d*) Dead load stresses. (*e*) Total initial stresses. (*f*) Prestress effects after losses. (*g*) Dead plus live load service stresses (ungrouted). (*h*) Total service load stresses (ungrouted). (*i*) Effective section for live load when effectively grouted. (*j*) Service live load stresses, effectively grouted. (*k*) Ultimate strength calculation, grouted. (*l*) Effect of lack of grouting on ultimate strength conditions.

$I_c = 10 \times 20^3/12 + 10 \times 20 \times 0.38^2 - 4 \times 2.5^3/12 - 4 \times 2.5 \times 7.63^2$
$= 6110 \text{ in.}^4$

$e = 16.75 - 10.0 + 0.38 = 7.13$ in.

Figure 16.10*c* and *d* shows prestress effects artifically separated from M_d stresses.

For prestress alone:

$P_o = 1.5 \times 135 = 202$ k

At top, $f_c = -P_o/A_c + P_o e y_t/I_c = -202/190 + 202 \times 7.13 \times 9.62/6110 = -1.060 + 2.26 = +1.20$ ksi tension

At bottom, $f_c = -202/190 - 202 \times 7.13 \times 10.38/6110 = -1.060 - 2.44 = -3.50$ ksi compression

For M_d alone:

At top, $f_c = -60 \times 12 \times 9.62/6110 = -1.135$ ksi compression

At bottom, $f_c = +60 \times 12 \times 10.38/6110 = +1.225$ ksi tension

Combined prestress and dead load (Fig. 16.10*e*):

At top, $f_c = +1.20 - 1.135 = +0.06 \text{ ksi} < 3\sqrt{f_{ci}'} = 0.212$ ksi O.K.

At bottom, $f_c = -3.50 + 1.225 = -2.28 \text{ ksi} < 0.60 f_{ci} = 3.00$ ksi O.K.

With time, f_{so} drops to $f_{se} = 135 - 25 = 110$ ksi, and the effect of prestress drops in like ratio, as indicated in Fig. 16.10*f*.

At top, $f_c = +1.20 \times 110/135 = +0.975$ ksi tension

At bottom, $f_c = -3.50 \times 110/135 = -2.85$ ksi compression

For dead and live load, $M_{dl} = 60 + 70 = 130$ k-ft; these stresses are:

At top, $f_c = -130 \times 12 \times 9.62/6110 = -2.45$ ksi compression

At bottom, $f_c = +130 \times 12 \times 10.38/6110 = +2.65$ ksi tension

Total at top, $f_c = +0.975 - 2.45 = -1.47$ ksi compression $< 0.45 f_c' = 2.25$ ksi

Total at bottom, $f_c = -2.85 + 2.65 = -0.20$ ksi compression vs. allowable tension $= 6\sqrt{f_c'}$. O.K.

For an ungrouted beam the changes in f_s due to live load are small (significantly less than n times the concrete live load stress at that level). These changes in f_s are not usually calculated.

Neglecting the complications the grouted section may involve in the matter of loss of prestress, the only change in working stress calculations resulting from grouting is that the live load stresses will be based on the transformed area.

If the beam were effectively grouted, and again neglecting the variation in n, the transformed area would become that of Fig. 16.10i.

$$A_t = 20 \times 10 + 6 \times 1.50 = 209 \text{ in.}^2$$
$$\bar{y} \text{ from mid-depth} = -9.00 \times 6.75/209 = -0.29 \text{ in.}$$
$$I_t = 10 \times 20^3/12 + 10 \times 20 \times 0.29^2 + 9.00 \times 6.46^2$$
$$= 6667 + 17 + 376 = 7060 \text{ in.}^4$$

At top, $f_l = -70 \times 12 \times 10.29/7060 = -1.220$ ksi comp.

At bottom, $f_l = +70 \times 12 \times 9.71/7060 = +1.152$ ksi tension

Total on steel, $f_s = 110 + 7 \times 1.152(6.75 - 0.29)/9.71 = 110 + 5.35 = +115.4$ ksi

Total at top, $f_c = +0.975 - 1.135 - 1.220 = -1.380$ ksi compression $< 0.45f_c'$ O.K.

Total at bottom, $f_c = -2.85 + 1.225 + 1.152 = -0.47$ ksi compression $<6\sqrt{f_c'}$ tension

These totals put together stresses from Fig. 16.10f, 16.10d, and 16.10j.

The differences as a result of grouting are small at working stress, about 0.09 ksi less compression on the top and 0.27 ksi more compression on the bottom. The ultimate moment is more noticeably improved by grouting, as noted in Sec. 16.13.

16.13 ULTIMATE STRENGTH OF PRESTRESSED BEAMS

Since the high strength steels lack a sharp and distinct yield point (Fig. 16.11) and since prestress may modify the neutral axis location at failure, it is not possible to use ultimate strength relations developed for ordinary beams without some concern or caution. Theoretical considerations seem to call for a calculation process correlating strains on steel and concrete at failure.

Nevertheless, for beams with bonded wire and a precentage of steel low enough to ensure a tension failure, the strength method of analysis (Sec. 2.9) gives approximate values which seem to agree fairly well with tests. Since there is no sharp yield point, the Code, following a report by the ACI-ASCE Joint Committee on prestressed concrete, uses steel stress f_{ps} at design (factored) beam load in this calculation rather than f_y. Fundamentally, the determination of f_{ps} depends on stress-strain diagrams and

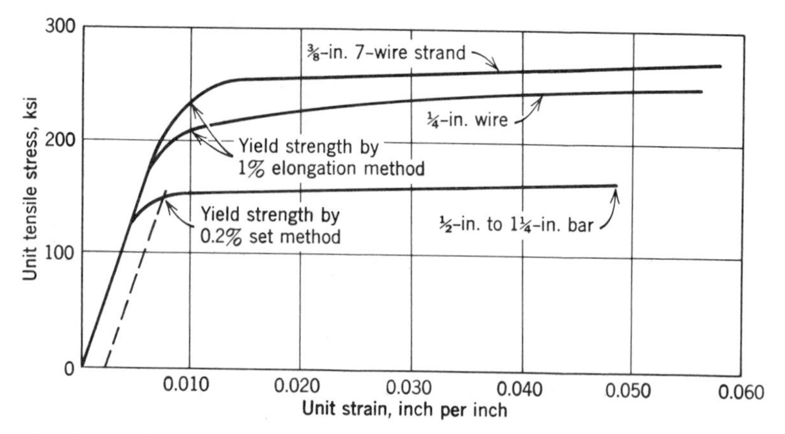

Fig. 16.11. Typical stress-strain curve for prestressing steels. (From Lin.[1])

strain relationships, which means that f_{ps} is not a quantity easily established without detailed data. This situation is eased by two approximate formulas for f_{ps} for use in typical cases:

For bonded members, $\quad f_{ps} = f_{pu}(1 - 0.5\rho_p f_{pu}/f_c')$

For unbonded members, $\quad f_{ps} = f_{se} + 10{,}000 + f_c'/(100\rho_p)$

but for unbonded members not more than the specified yield strength f_{py} or f_{se} +60,000 psi. The ultimate moment for rectangular beams, or for flanged beams with the neutral axis within the flange, is expressed by the relation

$$M_u = A_{ps}f_{ps}d(1 - 0.59\rho_p f_{ps}/f_c')$$

which is the relation given in Sec. 2.8 for ordinary reinforced concrete except that f_{ps}, A_{ps}, and ρ_p take the place of f_y, A_s, and ρ.

To avoid approaching the overreinforced beam condition, steel for rectangular sections must not exceed that giving $\rho_p f_{ps}/f_c' = 0.30$. For larger steel ratios the value of M_u is limited more by compression conditions; the Commentary suggests $M_u = \phi(0.25f_c'bd^2)$ is reasonable.

As an example of approximate ultimate strength analysis, the pretensioned beam of Sec. 16.11 and Fig. 16.9 will be investigated assuming $f_{pu} = 195$* ksi. The failure conditions of Fig. 16.9*j* give:

$$\rho_p = 1.5/(10 \times 17) = 0.00883$$

$$f_{ps} = f_{pu}(1 - 0.5\ \rho_p f_{pu}/f_c')$$

$$= 195(1 - 0.5 \times 0.00883 \times 195/5) = 162 \text{ ksi}$$

*This ultimate and the initial wire stress of Sec. 16.11 are roughly 70% of those commonly available in 1972(Fig. 16.11).

$\rho_p f_{ps}/f_c' = 0.00883 \times 162/5 = 0.28 < 0.30$ Usual M_u eq. O.K.
$a = A_{ps}f_{ps}/(0.85\,f_c'b) = 1.5 \times 162/(0.85 \times 5 \times 10) = 5.70$ in.
$\overline{M} = A_{ps}f_{ps}(d - a/2) = 1.5 \times 162(17 - 5.70/2) = 3440$ k-in.
$M_u = \phi\overline{M} = 0.9 \times 3440 = 3100$ k-in. $= 258$ k-ft
Reqd. $M_u = 1.4\,M_d + 1.7\,M_l$
$= 1.4 \times 65 + 1.7 \times 80 = 227$ k-ft < 258 O.K.

This comparison shows about 14% surplus strength, compared to 1% at service load.

For the post-tensioned beam of Sec. 16.12 and Fig. 16.10*a* with $f_{pu} =$ 195 ksi and the steel well grouted, the failure conditions of Fig. 16.10*k* give:

$\rho_p = 1.5/(10 \times 16.75) = 0.00895$
$f_{ps} = f_{pu}(1 - 0.5\rho_p f_{pu}/f_c')$
$= 195\,(1 - 0.5 \times 0.00895 \times 195/5) = 161$ ksi
$a = 1.5 \times 161/(0.85 \times 5 \times 10) = 5.68$ in.
$\overline{M} = 1.50 \times 161(16.75 - 0.5 \times 5.68) = 3370$ k-in. $= 283$ k-ft
$M_u = \phi\overline{M} = 0.9 \times 281 = 253$ k-ft
Reqd. $M_u = 1.4 \times 60 + 1.7 \times 70 = 203$ k-ft < 253 k-ft

The section provides about 25% more capacity than required compared to 37% under the service load analysis. The ultimate strength analysis is the more restrictive in this example.

If the beam of Sec. 16.11 were left ungrouted, a somewhat careless strength analysis might give the same M_u value as for the bonded case just calculated. Nevertheless, ungrouted beams without crack control steel are weaker because the unbonded steel leads to failure after only one crack opens, or at most a few. These few cracks open widely, crowd the neutral axis upward, and thereby cause a secondary compression failure at lower moment, as indicated in Fig. 16.10*l*. Some tests by Baker indicate cracks three to ten times as wide as for the bonded case where many small cracks occur. The formula for f_{ps} for this case gives a lower f_{ps}.

For the beam just investigated, if grouting is omitted (Fig. 16.10*l*), the calculations become:

$\rho_p = 1.50/(10 \times 16.75) = 0.0090$
$f_{ps} = f_{se} + 10{,}000 + f_c'/(100\rho_p)$, in psi
$f_{ps} = (135 - 25) + 10.0 + 5.0/(100 \times 0.0090) = 125.5$ ksi
$\rho_p f_{ps}/f_c' = 0.0090 \times 125.5/5 = 0.224 < 0.30.$ Usual M_u eq. O.K.
$a = 125.5 \times 1.5/(0.85 \times 5 \times 10) = 4.43$ in.
$\overline{M} = 125.5 \times 1.5(16.75 - 4.43/2) = 2730$ k-in.

$M_u = \phi\overline{M} = 0.9 \times 2730 = 2460$ k-in. $= 205$ k-ft
Reqd. $M_u = 1.4 \times 60 + 1.7 \times 70 = 203$ k-ft < 205

The beam is just adequate at ultimate when ungrouted, which can be compared with the 25% excess strength when grouted. Note that the Code permits the unbonded tendons only when some bonded reinforcement is cast into the tension area; the yield strength of these bars would automatically be available for increasing the available M_u and would control the cracking.

If calculations for ungrouted sections are based on correlating strains, as suggested in the second paragraph of this section, it is not difficult to introduce a factor to represent relative crack width. The effective steel strain (beyond initial prestress) can simply be taken as about one-fifth of the strain indicated by the strain triangles. This reduced steel strain, when added to the initial prestress strain, gives the total to be used with the stress-strain curve.

16.14 CABLE EFFECT CONSIDERED AS A NEGATIVE LOADING

Since simple beam cable eccentricity develops a negative moment which offsets much or all the dead load moment, this eccentricity must be reduced where dead load moment is small; otherwise a negative moment failure can occur as prestressing is applied. With post-tensioned steel the tendons or cables can be formed to any desired shape. With pre-tensioned wires or strands practice is often using hold-down devices to give the effect of bent-up steel. Other pre-tensioned beams have been made with a curved soffit, the curved beam axis thus producing a varying eccentricity.

A complete analysis can be made by considering all critical sections with their respective eccentricities. The relieving moment is (almost) proportional to the eccentricity, since the slope angles are small and the horizontal component of the prestress is essentially a constant. Hence the relieving moment diagram has the same shape as the curve of plotted e values.

A more convenient way for some to visualize the effect of curvature is in terms of a negative or upward load on the beam. A curved cable under heavy tension requires vertical load to deflect it from a nearly horizontal line. A concentrated load at the center will deflect the cable as in Fig. 16.12*a*, loads at both third points as in Fig. 16.12*b*, and a uniform load as in Fig. 16.12*c*, a parabolic curve. When embedded in a concrete beam in a slot having one of these shapes, the cable after prestressing exerts upward forces on the concrete opposite to those required to deform the free tension cable into the same shape.

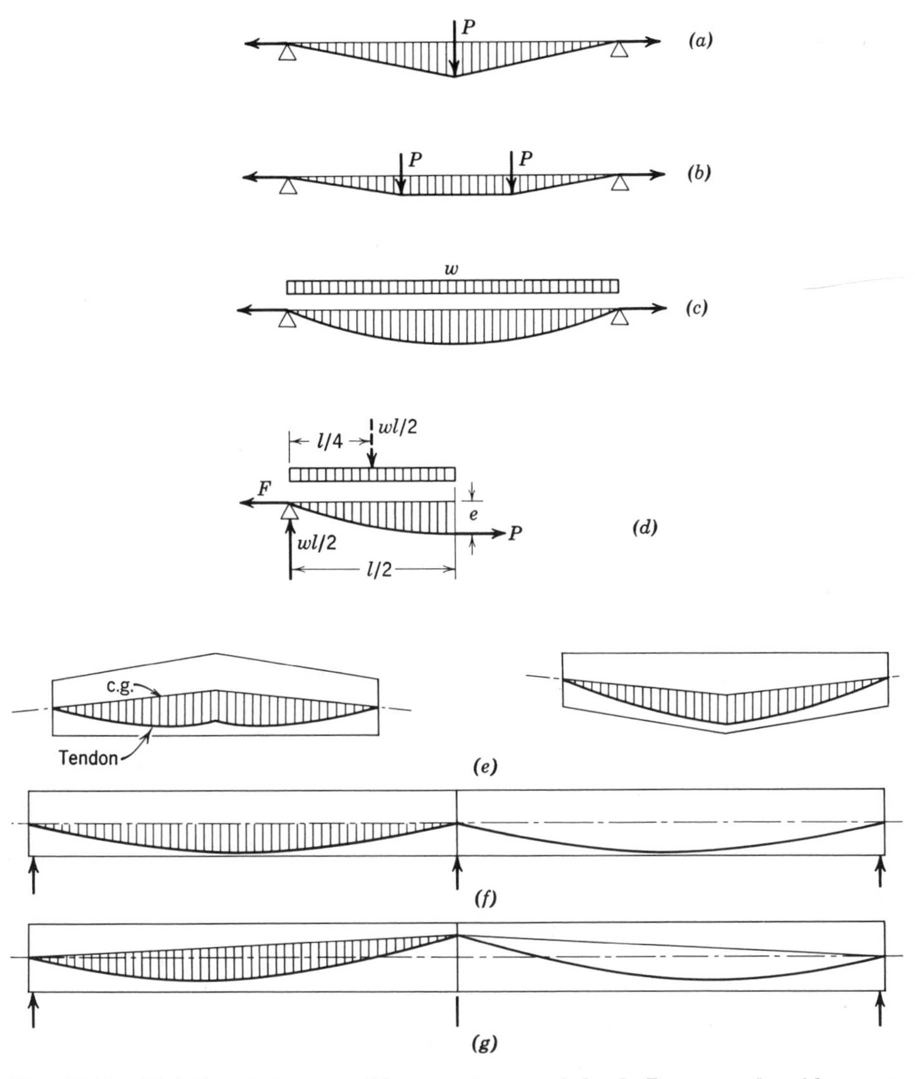

Fig. 16.12. Relation between cable curvature and load. Prestressed cable exerts similar upward forces on (plain) concrete. (*c*) through (*g*) give uniform upward loads.

If the parabolic cable shape is used, the free body of Fig. 16.12*d* indicates from ΣM that $Pe = (w_pl/2)(l/4) = w_pl^2/8$, or $w_p = 8Pe/l^2$. Hence the cable can be visualized as contributing an axial load and an upward uniform load w_p, both acting on the plain concrete section. The cable can be shaped in such a way as to offset nearly completely any dead load distribution which exists.

If the member is inclined instead of horizontal, the eccentricity is from the gravity axis of the member, not from a horizontal reference line. Likewise, if the member is of variable depth, the eccentricity is from the centroidal axis. The eccentricity may then be defined as the vertical distance between tendon and centroidal axis and Fig. 16.12*e* indicates the tendon shape which would correspond to the uniform upward load of Fig. 16.12*c* when the member is of varying depth. The shaded ordinates in (*e*) and (*c*) must be identical.

16.15 LOAD BALANCING

Lin[1] has expanded the negative loading concept to a design technique which balances with prestress any desired portion of the load, such as most or all of the dead load, or possibly under other conditions, dead load plus some portion of the live load.

The balanced load could ideally* provide no moment and no curvature or deflection. Hence, for continuous beams or slabs the unbalanced load alone may be used with ordinary elastic theory to compute the bending moments acting on the (nearly) uniform concrete section. This in the case of ungrouted tendons would be approximate in neglecting concrete displaced for ducts or tendons. Thus a continuous member might call for tendons, as in Fig. 16.12*f* or, what can be shown to be equivalent, the so-called "linearly transformed" tendon profile of Fig. 16.12*g*. The bend over the support obviously cannot be as sharp as shown and separate cables for each span, possibly with inverted *U* strands over the reaction, or other modification may be needed. Such is beyond the scope of this text.

16.16 SHEAR AND DIAGONAL TENSION

A cable with varying eccentricity acts somewhat as a suspension cable, partially relieving the concrete not only of bending stress but also of shear stress. The shear thus carried by the cable can be calculated either as the vertical component of the cable pull or as the shear created by the equivalent upward loads discussed in Sec. 16.14. The remainder of the external load creates a shear which must be resisted by the concrete.

The axial compression from prestress reduces the diagonal tension stresses so long as the beam is uncracked from moment stress. Cracking

*Obviously changes in prestress from time effects of creep and relaxation modify the effective negative loading and upset the initial assumptions slightly.

reduces this beneficial effect and inelastic steel deformations at loads near the ultimate largely eliminate it.

The 1971 Code considers two types of situations which can lead to shear failure. One is the combined stress in the web which can lead to a diagonal crack starting in the web before there is any flexural cracking of the nearby flange. The other is the shear condition which exists after the tension flange has cracked and thereby produces diagonal tension rather closely related to that typical in an ordinary reinforced concrete member. The equations presented for the shear that the concrete can take are checked against extensive tests at the University of Illinois and the Portland Cement Association. It is further provided that shear reinforcement shall always be used in a specified minimum amount, excepting only instances where the design without stirrups has been verified by tests which developed the full moment and full shear capacities. The minimum stirrups shall provide $A_v = 50b_w s/f_y$, just as in reinforced concrete. Alternatively, however, for prestressed members having an effective prestress of at least 40% of the tensile strength of the flexural reinforcement, the 1963 provision is retained and will often permit less stirrup area:

$$A_v = \frac{A_{ps}}{80} \cdot \frac{f_{pu}}{f_y} \cdot \frac{s}{d} \sqrt{\frac{d}{b_w}}$$

where f_y relates to the stirrups.

The maximum stirrup spacing is limited to $0.75h$ but not more than 24 in.

When prestressed strands are stopped (or left unbonded and unstressed) for a part of the member length, a problem in shear strength results very similar to that of bars cut off within a moment region of reinforced concrete. Until more is known about this phenomenon, excess stirrups should be used near these "cutoff" points.

16.17 DEVELOPMENT OF REINFORCEMENT

On pre-tensioned beams the self-anchorage of wires and strands at the ends of the beams leads to high bond stresses. This end anchorage requirement has received much attention and Lin's treatment[1] of the subject is quite complete.

For three- or seven-wire prestressing strand the Code (12.11.1) specifies a development length

$$l_d = (f_{ps} - 2f_{se}/3)d_b$$

where d_b is the nominal diameter in inches and the calculated stress f_{ps} and f_{se} (effective prestress after losses) are in ksi. The specified length must be available between the end of the strand and the closest point where full strength is required, which should govern only in cantilever and short span members.

Where bonding does not extend to the end of the member, the development length provided must be doubled.

In post-tensioned beams there is no problem of development because end anchorages are used. Even when flexural cracking occurs, the stresses adjacent to cracks when tendons are grouted are less important than in ordinary reinforced concrete where they are not separately computed.

16.18 DESIGN CONSIDERATIONS

Design considerations will be treated very briefly. Allowable working load stresses have already been mentioned in analysis examples.

Since the optimum total prestress approximates half the allowable f_c multiplied by the total concrete area, economy is provided by reducing the concrete area near the centroid, where it is less effective in resisting moment, and by increasing the area near the extreme fibers. Hence I, T, and ⊓ shapes are the general types which prove economical. Tables of section modulus values for such shapes are helpful in picking a size.

It is sometimes stated that the prestress carries the service dead load and that the size of the beam depends only on the live load. The negative moment $P_s e$ at transfer can be made to offset the dead load moment. Of course, as P_s decreases in time to P_{se}, part of this offsetting moment is lost. Furthermore, at ultimate strength the effect of prestress is largely lost and the beam must carry all the moment, essentially as an ordinary beam.

Referring to Fig. 16.6*b*, it will be noted that the top stress under prestress and dead load may be a tension as large as $3\sqrt{f_{ci}}$. Under the full load (Fig. 16.6*c*) this stress may be the full allowable compression $0.45f_c'$. Thus, except for the effect of prestress losses, the full range of $3\sqrt{f_c'} + 0.45f_c'$ is available to take care of M_l. Likewise, on the bottom the stress range for M_l can be from $0.60f_{ci}$ to $6\sqrt{f_{ci}}$, except for the effect of prestress losses. Magnel[4] has thus worked out design requirements for a section modulus established from M_l alone divided by the major part of this stress range. This approach is most suitable when M_l is large compared to M_d. It indicates the saving sometimes possible in beam size compared to an ordinary beam which must be sized for $M_d + M_l$.

Lin has developed[1] design procedures for elastic designs based on visualizing the internal resisting couple. These are very straightforward in application. He has also devised a "load balancing" method for slabs and beams, already mentioned in Sec. 16.15.

16.19 CONTINUOUS BEAMS

Continuous beams cannot be considered here except to mention one special condition which often controls. Since an eccentrically placed prestressing tendon itself develops bending moment and curvature in a beam, the prestressing operation may well require reactions (up or down) to hold the member in contact with its supports. These reactions change the prestressing moment from the simple $P_s e$ value to something more involved. Cables can be so arranged as to cause zero external reactions, but this appears to be an unnecessary restriction. Lin[1] gives a complete discussion of this problem following basic methods originally suggested by Guyon[3] and has simplified it considerably in connection with his "load balancing" method.

SELECTED REFERENCES

1. T. Y. Lin, *Design of Prestressed Concrete Structures*, 2nd ed., John Wiley and Sons, New York, 1963.
2. James R. Libby, *Modern Prestressed Concrete*, Van Nostrand, Princeton, N. J., 1971.
3. Y. Guyon, *Prestressed Concrete*, John Wiley and Sons, New York, 1953.
4. Gustave Magnel, *Prestressed Concrete*, Concrete Publications, Ltd., London, 2nd ed., 1950.
5. N. Khachaturian and G. Gurfinkel, *Prestressed Concrete*, McGraw-Hill Co., New York, 1969.

PROBLEMS

Prob. 16.1. The rectangular beam of Fig. 16.13 has pre-tensioned steel at two levels as shown with $f_{pi} = 170$ ksi. Consider $f_{ci}' = f_c' = 5000$ psi, n = 7, and a simple span of 40 ft.

(*a*) Calculate concrete stresses at top and bottom and both steel stresses when prestressed wires are first cut between units.

(*b*) Recalculate stresses after shrinkage and creep have decreased f_s by 30 ksi below the value found in (*a*).

(*c*) Calculate all stresses when a live load of 600 plf is added to the conditions in (*b*).

Prob. 16.2. The post-tensioned beam of Fig. 16.14 has a 40-ft simple span. It has two holes as shown for wires with a total area of 2.10 in.^{2}, which after tensioning will be 1.65 in. below the top of beam. $f_c' = 5000$ psi, $f_{so} =$ 175 ksi. Considering the holes ungrouted and without using the approximation of area as the gross area of the concrete, calculate:

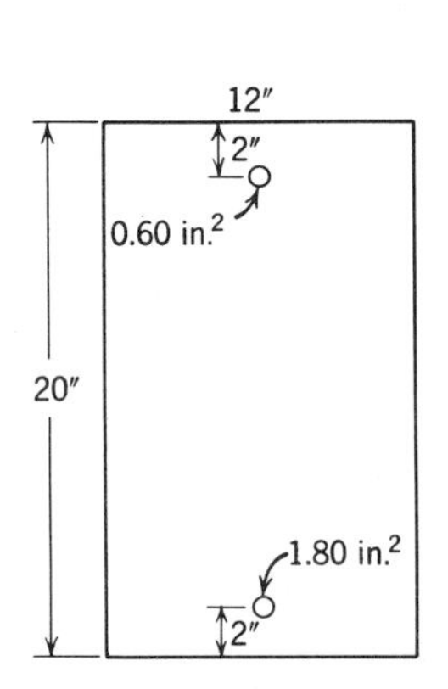

Fig. 16.13. Beam for Prob. 16.1.

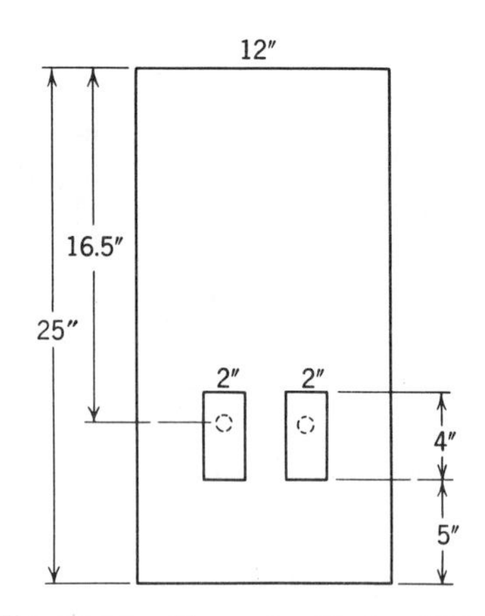

Fig. 16.14. Beam for Probs. 16.2 and 16.3.

(*a*) Stresses under dead weight alone after post-tensioning.
(*b*) Stresses after shrinkage and creep reduce the steel stress to $f_{se} = 155$ ksi and a live load of 625 plf acts on the beam.

Prob. 16.3.

(*a*) Recalculate condition (*b*) of Prob. 16.2 if the beam hole is well grouted.
(*b*) Calculate the ultimate moment capacity of this beam and its over-all factor of safety assuming it is well grouted and has $f_{pu} = 250$ ksi.

Prob. 16.4.

(*a*) If the pre-tensioned beam of Fig. 16.9 is changed to $f_c' = 5000$ psi, $f_{si} = 150$ ksi, $n = 7$, $A_s = 1.10$ in.^{2} with M_d still 65 k-ft, calculate all stresses after the wires are cut.
(*b*) If the wires are straight and concrete stresses are limited to a (bottom) compression of 3000 psi and a top tension of $3\sqrt{f_c'}$, what is the maximum wire eccentricity that can be used? [The student should note that M_d at the support would be zero. The change in c.g. from that used in part *a* may be ignored.]
(*c*) If the wires are draped such that the condition at the end of the span does not control, and if f_{so} decreases to f_{se} of 120 ksi, what live load moment is permissible with the allowable f_c of $0.45f_c'$ and allowable tension of $6\sqrt{f_c'}$?

17
Composite Beams

17.1 THE NATURE OF COMPOSITE BEAMS

Composite beam is the name given the combination of a steel beam with a concrete slab, or the combination of a precast concrete beam with a cast-in place concrete slab, when the two are so connected together that they act as a single unit in resisting flexure. Since every reinforced concrete slab or beam is really a composite member of steel and concrete, the basic theory for flexure requires almost no new concepts. Likewise the bonding together of the two units is theoretically a simple problem of horizontal shear. A roughened concrete surface has a considerable bonding strength if the slab is held tightly in place by stirrups extending into the slab. Shear lugs can be welded to steel beams to perform the same function.

Composite construction is economically important because the steel beams or precast concrete beams* can furnish supports for the slab forms and the dead weight of the concrete while, at the same time, the beams can be made lighter because the slab forms a stiff attached flange which helps in resisting live loads. The primary bending moment in a slab calls for slab steel transverse to the beam and hence neither this slab steel nor slab bending seriously complicates the analysis of the composite beam.

*The majority of pretensioned prestressed concrete beams are cast with stirrups extending out the top to bond into a cast-in-place slab to form a composite T-beam.

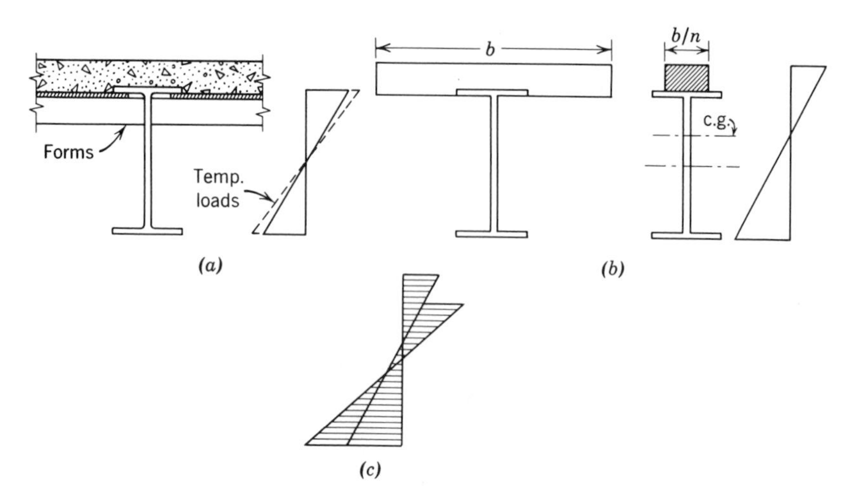

Fig. 17.1 Stresses in an unshored composite beam. (*a*) Dead load alone. (*b*) Live load alone. (*c*) Total stresses.

The engineering problems hinge around three considerations: flexural strength, horizontal shear strength, and deflection. The latter may call for shoring under the beams during construction since this makes the entire section available for resisting dead load as well as live load. The ultimate strength is not significantly influenced by shoring, shrinkage, or creep.

17.2 FLEXURAL STRENGTH WITH SLAB CAST ON STEEL BEAM

Flexural strength could be considered at three different stages of loading, that is, under dead load prior to the time when the slab is effective for strength, under live load with the slab acting effectively, and at ultimate under overload conditions.

Consider first a symmetrical steel beam with the cast-in-place slab. If there is no shoring the beam acts to carry its own weight, the slab forms, the slab concrete, and all the superimposed construction load. As the concrete sets, the steel beam carries Mc/I stresses for the steady portion of this unfactored load, a very simple calculation, as indicated in Fig. 17.1*a*. After the concrete gains strength, the section becomes the transformed area of Fig. 17.1*b* for any further load, it being simplest to transform the concrete into equivalent steel by the relation A_c/n or b/n. This results in the addition of stresses as shown in Fig. 17.1*c*.

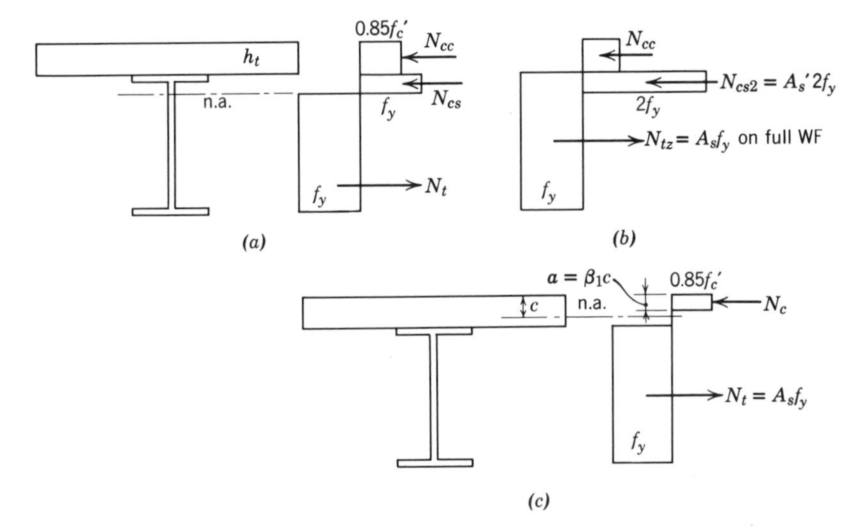

Fig. 17.2. Ultimate stress conditions. (*a*) Neutral axis below slab. (*b*) Equivalent to (*a*) using N_t on full A_s. (*c*) Neutral axis within slab.

For long-time loads some multiple of n may be used to represent the effects of creep and shrinkage, up to $3n$.

Since neither creep nor shrinkage influences ultimate strength significantly, the ultimate design condition will first be discussed.* The neutral axis in deep beams will usually fall below the slab leading to the stress distribution shown in Fig. 17.2*a*. Viest, Fountain, and Singleton[2] have pointed out that this is equivalent to Fig. 17.2*b*, which is considerably simpler to use.

The same authors point out, especially in buildings, that the neutral axis may fall in the slab, as in Fig. 17.2*c*, in which case the tension on the concrete below the neutral axis is ignored and the concrete compression above the neutral axis is handled as in any T-beam.

The later discussion on deflection will point out that beams are frequently shored enough to avoid deflection and stresses of consequence from the dead weight of the slab until after the concerete slab is strong enough to act as a flange. In this case the dead load stresses of Fig. 17.1*a* are small, usually only from the weight of the beam itself, and the composite beam stresses of Fig. 17.1*b* becomes the major part. This shoring does not influence the ultimate moment capacity significantly.

*Although the AISC specification is written in terms of service loads, the specification is an adaptation from behavior at ultimate.

17.3 EXAMPLE OF FLEXURAL CALCULATIONS

For a simple span of 26 ft check the flexural capacity of a $f_y = 36$ ksi steel* W18 × 50 on 7-ft center with a 4-in .slab to carry a service load of its own weight and a live load of 440 psf. Assume the beam is shored at the quarter points so as not to pick up the weight of the concrete until it reaches its intended f_c' of 3000 psi. The beam properties include $A = 14.7$ in.^{2}, $I = 802$ in.^{4}, $b_f = 7.50$ in., and flange thickness = 0.57 in. (The beam in such a case is usually designed by a service load procedure covered in the AISC Specifications.)

Solution

A composite beam based on a steel shape is usually designed under the AISC Specification,[3] which is really an ultimate design restated in terms of service load stresses. The effective flange width used here will be taken as that recommended by AISC, that is, the smallest of:

1. Spacing of beams = 84 in.
2. Flange width of steel section plus 8 times slab thickness as the overhang on each side $= b_f + 16h_f = 7.50 + 64 = 71.5$ in.
3. Beam apan/4 = 6.5 ft = 78 in.

The 71.5 in. width controls.

This text section relates to flexural strength only. The necessary shear connectors are covered in Sec. 17.6 and deflections are discussed in Sec. 17.5.

Here the given section will first be checked by applying ACI Code strength concepts to the analysis of Fig. 17.2. It will then be rechecked by the AISC procedures.

Whether the beam is or is not initially shored, the ultimate condition includes all loads on the full span with the steel shape considered as yielding throughout its depth, or nearly so. This total tension fixes the depth of the compression block, as in a homogeneous section.

$$\overline{N}_c = \overline{N}_t = 14.7 \times 36 = 529 \text{ k}$$
$$a = \overline{N}_c/(0.85 f_c' b) = 529/(0.85 \times 3 \times 71.5) = 2.90 \text{ in.}$$
$$c = a/0.85 = 2.90/0.85 = 3.42 \text{ in.}$$

The neutral axis is in the concrete, but the sketch in Fig. 17.3*a* indicates it is too low to put the upper steel flange at the full f_y of 36 ksi. Since $\overline{N}_t$ is then less, try $c = 3.20$ in. and check whether the tension justifies this value.

*The AISC uses $F_y = 36$ ksi.

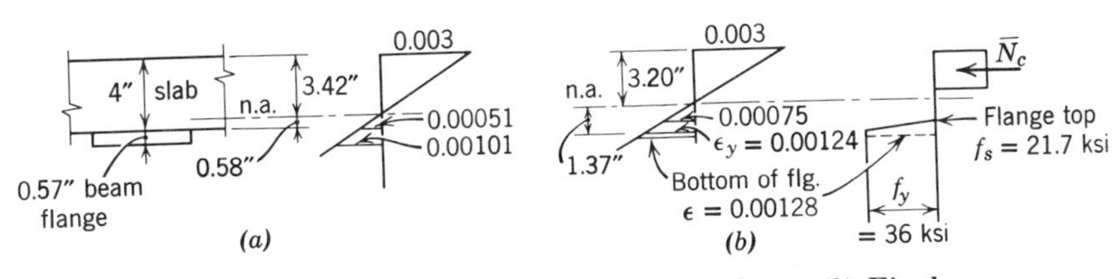

Fig. 17.3. Ultimate analysis of Sec. 17.3. (*a*) First trial n.a. (*b*) Final n.a.

Tension is reduced (Fig. 17.3*b*) by roughly $0.5(36 - 21.7)(7.50 \times 0.57) = 30$ k

$$\overline{N}_t = 529 - 30 = 499 \text{ k}, \quad a = (499/529)2.90 = 2.74 \text{ in.}$$
$$c = (499/529)3.42 = 3.23 \text{ in. O.K.}$$
$$\overline{M} = 529(9.0 + 4 - 1.37) - 33(4.0 - 1.37 + 0.57/3)$$
$$529 \times 11.63 - 30 \times 2.82 = 6160 - 85 = 6070 \text{ k-in.} = 505 \text{ k-ft}$$

The small influence of the reduction term confirms that its exact evaluation is not critical.

$$\begin{aligned}
&\text{Wt. beam} = && 50 \text{ plf} \\
&\text{Wt. slab} = 7 \times 50 = && \underline{350} \\
& && 400 \times 1.4 = 560 \text{ plf} \\
&\text{L.L.} = 7 \times 400 = && 2800 \times 1.7 = \underline{4760} \\
& && \text{Total} = 5320 \text{ plf}
\end{aligned}$$

$$M_u = (1/8)\, 5.32 \times 26^2 = 450 \text{ k-ft}$$
$$\overline{M} = M_u/\phi = 450/0.9 = 500 \text{ k-ft} < 505 \text{ k-ft} \quad \text{O.K.}$$

The beam is safe at ultimate, by 1%.

The beam will now be checked for strength under the service load procedure of the AISC Specification,[3] with allowable $f_s = 24$ ksi and allowable $f_c = 0.45f_c' = 1350$ psi. Although using the service load format, this calculation derives its validity directly from ultimate strength calculations. Shrinkage and creep strains do not influence ultimate strength significantly and hence are not introduced in these strength calculations for either shored or unshored beams. (These strains do influence service load deflections; see Sec. 17.5.)

Since this is a shored* beam, the total moment applies directly to the composite section. For 3000 psi concrete, $n = 9$, which permits the flange

*If the beam had been unshored initially, the stresses would have to be combined from two calculations: first, those for the total dead load on the steel shape alone and, second, the live load on the total transformed area.

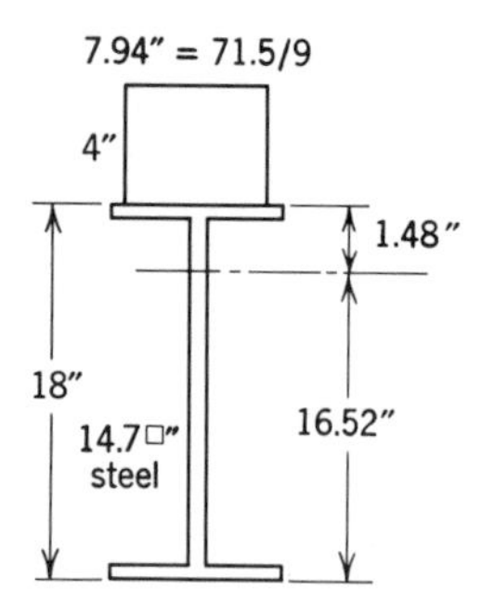

Fig. 17.4. Transformed area for analysis in Sec. 17.3.

to be considered as transformed into steel of width 71.5/9 = 7.94 in. as in Fig. 17.4. Moments about the bottom of the flange establish the neutral axis:

$$(7.94 \times 4)2 - 14.7 \times 9 = +63.5 - 132.3$$
$$= y(31.8 + 14.7)$$
$$\bar{y} = -68.8/46.5 = -1.48 \text{ in.}$$
$$I \text{ of slab} = (1/12)7.94 \times 4^3 = 42$$
$$+ 31.8(2 + 1.48)^2 = 385$$
$$I \text{ of } W = \quad = 802$$
$$+ 14.7(9 - 1.48)^2 = 833$$
$$I = 2062 \text{ in.}^4$$

$$M = (1/8)(400 + 2800)26^2/1000 = 270 \ k\text{-ft}$$
$$f_s = Mc_s/I = (270 \times 12)(18 - 1.48)/2062 = 25.9 \text{ ksi} > 24 \text{ ksi}$$

The section is overstressed by about 8% in tension. Compression is usually satisfactory:

$$f_c = (Mc_s/I)/n = (270 \times 12)(4 + 1.48)/(2062 \times 9) = 0.958 \text{ ksi} < 1350 \text{ psi} \quad \text{O.K.}$$

Attention is called to the AISC limitation on unshored beams expressed by limiting the section modulus S_{tr} (referred to the bottom flange stress) of the transformed composite section to

$$S_{tr} \lessapprox (1.35 + 0.35\, M_L/M_D)S_s$$

where M_L and M_D are live and dead load moments and S_s is the section modulus of the steel section alone. In this example:

Actual $S_{tr} = 2062/(18 - 1.48) = 125 \text{ in.}^3$

Limiting $S_{tr} = (1.35 + 0.35\, M_L/M_D)S_s = (1.35 + 0.35 \times 2800/400)(809/9) = 338 \text{ in.}^3$

Since 125 ≪ 338, this loading does not require shored beams. This provision is used by AISC to guard against excessive tensile stress at service loads, stresses not fully reflected in the nominal stress calculation above.

The AISC has many suggestions and rules[4] to cover composite construction. It is the purpose here only to illustrate the calculations for a simple case.

17.4 FLEXURAL STRENGTH WITH SLAB CAST ON PRECAST CONCRETE BEAM

When the basic element is a precast beam, either with or without prestressing, the details of the calculations change, but not the general ideas. For example, without prestressing, the precast beam by itself must carry its own weight and this is a larger item than with a steel beam. If it is partially shored such that the slab weight is carried by shores until the slab reaches its design strength (say, $0.75f_c'$), the analysis proceeds as above, with the transformed area expressed as equivalent concrete instead of steel. In this case the concrete must be considered cracked wherever tension strains exist, but this is not a serious problem to handle.

The chief difference between using reinforced concrete and steel as the basic beam lies in the horizontal shear provisions discussed in Sec. 17.6.

17.5 DEFLECTIONS

For the uniform load and steel section of Sec. 17.3, deflections would be calculated at service loads by $(5/384)wl^4/\text{EI}$, using the transformed area (in steel) for the stress calculation. In this case, however, the long-term dead load deflection must be separately computed with n assigned at least twice the nominal value. The live load deflection would be based on the I with the usual value of n. The AISC suggests as a starting point in design a depth-span ratio of 1/22 for $f_y = 36$ ksi and 1/16 for $f_y = 50$ ksi, but deflection must still be calculated.

For composite construction involving precast reinforced concrete beams, the Code (9.5.5.1) provides that a composite member with shored construction may be considered as a cast-in-place member with the minimum thickness requirements of Code Table 9.5a* controlling unless deflections are computed. In unshored members the precast element existing before the slab is cast is the one to meet this thickness requirement; the composite

*See Table 2.2 in Sec. 2.8.

member with slab may then be considered satisfactory* without deflection check. The precast portion may need investigation for long-time deflection occurring prior to the time the slab becomes effective.

The requirements of Code Table 9.5a may be waived if deflections are calculated and these are less than the values in Code Table 9.5b. Such calculations must involve differential shrinkage between precast and cast-in-place parts of the member. Otherwise the requirements and methods discussed in Chapter 6 are appropriate.

Composite beams with prestressed, precast sections are quite common, but are not covered here. The Code requires that deflections *always* be computed for these members. The Code Commentary suggests the calulation method of References 5, 6, and 7 as a suitable one.

17.6 HORIZONTAL SHEAR

(a) General

With steel beams all the vertical shear should be carried by the web of the beam. With reinforced concrete precast elements the vertical shear may be considered carried by the precast elements or by the entire composite member as in monolithic construction.

The special problem of composite beams is the horizontal shear between flange and beam below, whether it is a steel beam or a precast concrete beam. Steel beams may be used with either full or partial shear transfer; the reader is referred to the Steel Manual[3] for the partial shear transfer case. Full shear transfer is required by the ACI Code for reinforced concrete members.

Theoretically the horizontal shear varies along the beam length as the external shear varies and could be calculated on a $v = VQ/Ib$ basis. However, both in steel and concrete composite beam work it is recognized that other distributions of the horizontal shear resistance will also serve adequately, except in the case of large concentrated loads in a positive moment region near a point of inflection or near the end of a simple span.†

*This Table 9.5*a* does not apply when members are attached to or support partitions likely to be damaged by large deflections. For a short summary of Code Table 9.5*b* requirements governing in this case, see footnote at close of Sec. 6.5.

†The partial shear transfer with the steel beam is an extreme case of such reassignment of shears.

(b) Steel-Concrete Composite Beams

Although spirals or short transverse lengths of small steel channels welded to the flange make adequate shear connectors, the stud connector is most commonly used. The stud is round, automatically welded to the flange, and is topped with a hook or flange making a head which can engage the concrete slab and prevent any vertical separation which may tend to occur between slab and beam. The stud height must permit at least 1 in. of concrete cover over the head.

The AISC Manual[4] tabulates the shear load permitted on each stud or channel connector, for example, 11.5 k for a 3/4 in. × 3 in. stud in 3000 psi concrete made from ordinary aggregate. This is a service load equivalent, since AISC uses a service load computation (derived from ultimate strength values). The total (service load) horizontal shear between points of maximum and zero moment is taken as 0.5 of the ultimate strength of the concrete flange in compression, or 0.5 of the tensile yield strength of the entire steel section, whichever is smaller, that is, the smaller of

$$0.85f_c'bh_f/2 \qquad \text{or} \qquad A_sf_y/2.$$

The shear is considered fully transferred if the connectors used have a total shear capacity of this amount. The connectors may be spaced uniformly from maximum moment point to zero moment point.*

(c) Reinforced Concrete Composite Beams

The designer may assume full shear transfer if four requirements given in Code 17.5.2 are met:

1. Contact surfaces are clean and intentionally roughened (full amplitude approximately 1/4 in.).
2. Minimum ties are provided extending into the slab at a specing not more than four times the least dimension of the element or 24 in. ($A_v = 50b_ws/f_y$)
3. Web below the slab is designed to resist the entire vertical shear.
4. The ties in item 2 are fully anchored in both elements.

In all other cases the horizontal design shear v_{dh} must be computed and satisfied.

*With some further check needed if a large moment closer to the zero moment point can occur.

The horizontal shear may be computed at any cross section from the relation

$$v_{dh} = V_u/\phi(b_v d)$$

where ϕ = 0.85
b_v = width being investigated
d = effective depth (compression face to centroid of tension reinforcement) for the entire composite member.

Alternatively, the compression (or tension) in any segment may be computed, divided by 0.85, and the resulting force considered as the total horizontal shear to be transferred. The two alternates permit either a v_{dh} varying as the external shear or considered as uniformly distributed. In the latter case special consideration appears appropriate where large concentrated loads fall close to the end of a simple span or close to the point of inflection in a positive moment region of a continuous member.

For clean surfaces the resisting v_h may be evaluated on the basis of the provision of minimum ties, or of an intentionally roughened surface, or both. The allowable v_h is 80 psi for *either* minimum ties or intentionally roughened surface *alone*, or 350 psi for *both together*. These shears act over the contact area between slab and precast element where flexural strains might otherwise cause horizontal slip to occur. If the horizontal shear exceeds 350 psi, the concepts of shear-friction discussed in Sec. 4.17 should be used.

17.7 CONTINUOUS BEAMS—AASHO SPECIFICATION

When beams are continuous over a support, the steel in the slab parallel to the beam may be considered a portion of the transformed area, the slab concrete being neglected because it is in tension. In such a case, the requirements for connectors are computed separately for positive and negative moment regions. The AASHO Specification is an ultimate strength one insofar as connectors are concerned. Hence the connectors in the negative moment region are required to resist horizontal shear equal to the area of the longitudinal slab steel (within the effective width of the slab) times its yield stress. The AASHO Specification limits the *total* effective slab width to $12h_f$, more restrictive than the AISC value of $16h_f$ added to the steel flange width.

Since there is some objection to welding connectors to the tension flange in the region of maximum tension, the AASHO Specification provides that the pitch may be modified to avoid connectors at locations of high tension in the flanges. It allows closer spacings to be used away from the maximum negative moment point (closer to the point of inflection) to maintain the total number required while avoiding the highest flange stresses. Alternatively, the total negative moment may be assigned to the steel beam alone, but in such a case added connectors are required at the dead load point of inflection (within a length equal to one-third of the effective flange width, on either side of or centered on the *P.I.*), based on a fatigue requirement discussed in the next paragraph.

The basic design for connectors in the AASHO Specification is based on fatigue and the ultimate strength design above is used as a check rather than as the chief consideration. The equation used (in AASHO notation) is:

$$S_r = V_r Q/I$$

where S_r = range in horizontal shear stress pli at the connection

V_r = range in shear due to live load plus impact, that is, the difference between maximum and minimum shear envelopes, excluding dead load.

Q, I = statical area moment above the junction and moment of inertia of the total transformed area, both about the neutral axis

The reader is referred to that specification for the allowable shear on a connector, which is written in terms of length of channel or square of the diameter of stud multiplied in each case by factors which vary substantially with the design number of cycles.

It is noted that AASHO requires 2 in. of concrete over the connectors and requires that they penetrate at least 2 in. into the slab.

SELECTED REFERENCES

1. ACI-ASCE Committee 333, "Tentative Recommendations for Design of Composite Beams and Girders for Buildings," *Jour. ACI*, 32, No. 6, Dec. 1960; *Proc. 57*, p. 609.
2. Ivan M. Viest, R. S. Fountain, and R. C. Singleton, *Composite Construction in Steel and Concrete*, McGraw-Hill Book Co., New York, 1958.
3. "Specifications for the Design of Structural Steel for Buildings," American Institute of Steel Construction, New York, 1969.
4. *Manual of Steel Construction*, American Institute of Steel Construction, New York, 7th Edition, 1970.
5. Subcommittee 5, ACI Committee 435, "Deflections of Prestressed Concrete Members," *Jour. ACI*, 60, *No. 12*, Dec. 1963, p. 1697.

6. Subcommittee 2, ACI Committee 209, "Prediction of Creep, Shrinkage, and Temperature Effects in Concrete Structures," *Designing for the Effects of Creep, Shrinkage, and Temperature in Concrete Structures*, SP-27, American Concrete Institute, Detroit, 1971.
7. D. E. Branson, B. L. Meyers, and K. M. Kripanarayanan, "Time-Dependent Deformation of Noncomposite and Composite Prestressed Concrete Structures," Symposium on Concrete Deformation, *Highway Research Record 324*, Highway Research Board, Washington, D.C., 1970, p. 15.
8. Roger G. Slutter and John W. Fisher, "Fatigue Strength of Shear Connectors," *Highway Research Record 147*, Highway Research Board, Washington, D.C., 1966, p. 65.

PROBLEMS

Prob. 17.1. Calculate the ultimate moment capacity of a structural steel W 24 × 76 beam spaced at 8 ft on centers and adequately anchored to a 5-in. slab, assuming structural grade steel and $f_c' = 4000$ psi.

Prob. 17.2. In Prob. 17.1, assume the beam is not shored. Calculate the AISC service load stresses for the dead load including 20 psf for forms and a construction live load equal to 30 psf.

Prob. 17.3. Calculate the AISC design horizontal shear to be developed between beam and slab in Prob. 17.1 under full live load.

18
Detailing for Seismic Resistance

18.1 SEISMIC AREAS

A growing awareness exists among code authorities that seismic areas are not restricted to California and the West. Instead, some hazard exists over major portions of the United States, as indicated on the map of Fig. 18.1 taken from the Uniform Building Code of 1970. For the purpose here, the zone designations are more simply and less sharply characterized than in the Code.

Zone 0—no damage
Zone 1—minor damage
Zone 2—moderate damage
Zone 3—major damage

The Zone 0 area is quite limited, which means there must be a more general design acceptance of some earthquake hazard over most of the United States.

18.2 SEISMIC DESIGN—GENERAL OBJECTIVES

Under the present state-of-the-art, earthquake hazard has to be treated as a matter of probability, based on past history of shocks, geologic structures, and evidence of previous faulting. The map of Fig. 18.1 does not consider frequency of occurrence.

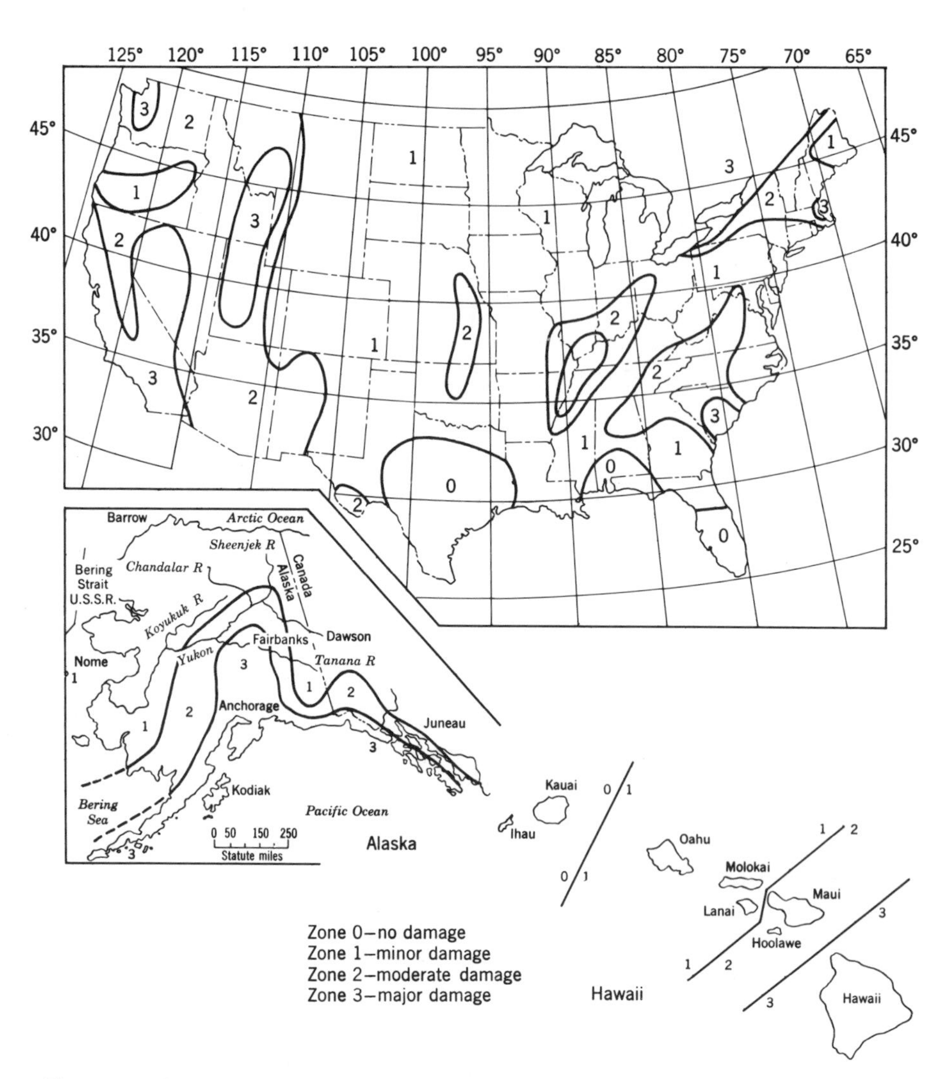

Fig. 18.1. Seismic zone map of United States, from 1970 Uniform Building Code.

Blume, Newmark , and Corning[1] consider several different design criteria appropriate in seismic design. First, earthquakes of an intensity which might be expected to occur several times during the life of the structure should be so covered by the design that no damage will result, or at most that minor repairs will suffice if an unusual *number* of such earthquakes should occur.

Second, under the most severe *probable* earthquake, the structure should not collapse with possible loss of life, and severe damage should not occur. Design against this earthquake involves incorporating a reserve of strength in the structure beyond the elastic range.

Third and finally, even in an unusual earthquake, somewhat greater than that probable during the structure's life, the building must not collapse, even though under these conditions it may require major repair.

The uncertainty in any probability evaluation is such that the second and third conditions must be accepted as potentially possible. Also, it is not economically feasible to design to avoid all such damage. These conditions imply reversal of axial forces, reversal of shears, and reversal of moments. The ductility of a structure then becomes the governing consideration, ductility even in the presence of reversal of stresses.

It is not within the scope of this text to discuss the natural frequency of the structure nor the methods used in defining design forces. But the basic requirements which beams, columns, joints, and walls must include for adequate service in a ductile moment-resisting space frame will be summarized as Code Appendix A presents them. Appendix A applies where both (1) major damage has a high probability of occurring and (2) the total lateral seismic forces have been *reduced* because of using the "ductile moment-resisting space frame" with or without special shear walls. Seismic resistance is a complex area and Appendix A represents minimum requirements, not a guarantee of performance. (See Sec. 18.8.)

18.3 LIMITATIONS ON MATERIALS

Minimum concrete strength is specified as f_c' of 3000 psi. No maximum is set because ductility is primarily achieved by the use of members under-reinforced, a method which eliminates compression failure until ductility limits have been exceeded. The Uniform Building Code requires that connections and panel joints be designed to allow movement between stories of not less than twice the story drift from computed wind or seismic forces. Blume et al[1] recommend a ductility factor of at least 4 (measured as the ratio of maximum deflection to deflection at first yield) and discuss ductility ratios from 4 to 6 as reasonable in the context* of the then (1959) Structural Engineers' Association of California design code.

The maximum specified yield strength of reinforcement is limited to 60 ksi. If the design is based on Grade 40 bars, a higher grade such as Grade 60 may not be substituted.

*Any variation in code rules shifts the computed point at which first yield occurs; any earthquake code thus presents its own relatively arbitrary reference point.

18.4 REQUIREMENTS FOR FLEXURAL MEMBERS

(a) Primary Reinforcement Requirements

The maximum steel ratio ρ is limited to $0.5\rho_b$ in order to insure a definite ductility, just as the nominal yield stress itself is limited.

The various minimum requirements for A_s are summarized in Fig. 18.2. These minimums are:

1. Minimum top and bottom steel for the full length of beam, at least two bars top and bottom; on each face at least $\rho = 200/f_y$, which is 0.5% for Grade 40 and 0.33% for grade 60.
2. For top steel, a minimum of one-fourth of the larger negative moment A_s must be continuous all across the span.
3. At each support minimum bottom steel must equal half of the negative moment steel.
4. *All* bars at the face of column, top and bottom, shall extend *through the support into the adjacent beam*, if possible. Where, because of a change in member size this requirement is impossible for some bars, any bars stopped must be anchored into the column for their full f_y.
5. At least one-third of the negative moment tension bars must be carried to the extreme range of the *P.I.* and in no case less than $0.25l_n$ from the support.

(b) Bar Development and Splicing

Where a beam frames into a column with no beam continuing on the other side, all bars shall be extended to the far face of the confined region (Sec. 18.6) and anchored to develop f_y from the near face of column. "Every

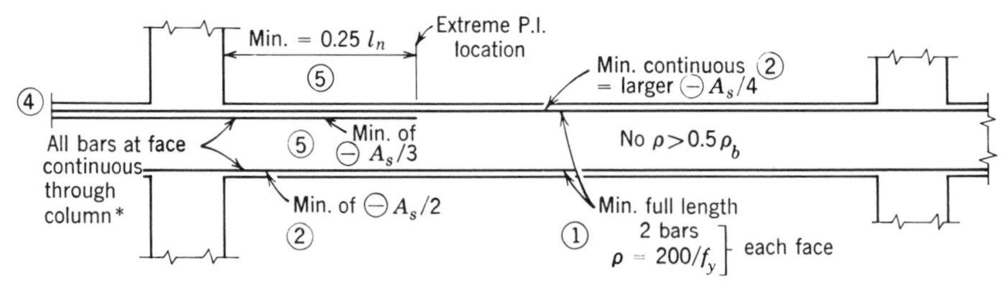

Fig. 18.2. Minimum requirements for arrangement of reinforcement in beams. (*If impossible, see text.)

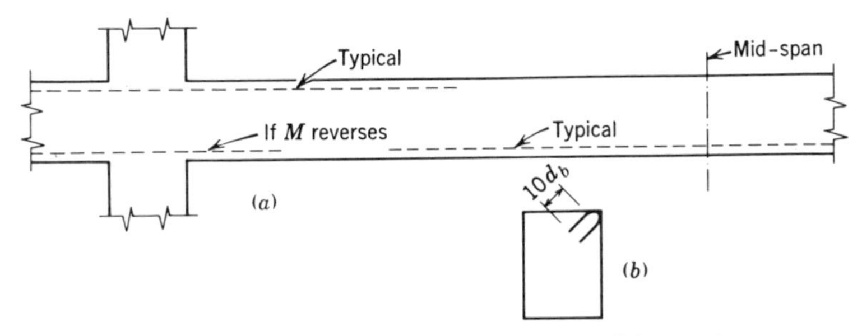

Fig. 18.3. (*a*) Tensile zones shown by dotted lines. (*b*) Stirrup-tie.

bar shall terminate with a standard 90 degree hook or with a hook and additional bar extension where needed to provide the required development length."* In the confined region the use of two-thirds the required development length l_d of Sec. 5.10 (Code 12.5) is authorized; elsewhere the usual development length is required except that the minimum is set at 16 in. instead of the usual 12 in.

No lap splices are permitted in a tensile zone or a zone of reversing stress (Fig. 18.3), unless stirrup-ties are provided at the lap as required around compression bars. This requirement is for stirrup-ties spaced at not more than $16d_b$ or 12 in. At all beam ends, such stirrup-ties must extend at least $2d$ from the column face.

At all tension splices, at at least two stirrup-ties shall be provided and splice lengths must be at least $24d_b$ or 12 in.

No welded splice must fall within an effective depth d of a plastic hinge.

(c) Shear Type Reinforcement for Flexure

Ordinary web reinforcement must be increased by designing for the sum of gravity load shears and the shears necessary to balance hinging moment at both ends of the beam, that is, an extra $V_h = 2M_h/l$ of either sign, since hinging moments M_h tend to be additive on the beam as a free body and may increase either positive or negative shear. The Code suggests that M_h may be taken the same as the usual M_u which would come from analysis of the cross section.

Web reinforcement of at least #3 at a spacing of $d/2$ is required over the full length of the member.

*The author is somewhat skeptical as to the effectiveness of extensions beyond the 90° hook, except possibly at the extreme state where the concrete at the start of the hook is seriously damaged.

Within the region near each end of the member for a length of $4d$, web reinforcement shall be at least

$$A_v d/s = 0.15A_s \text{ (or } 0.15A_s' \text{ if larger)}$$

and the spacing there shall not exceed $d/4$. The first stirrup shall be placed within 3 in. of the column face and the first two shall be stirrup-ties. Stirrup-ties (Fig. 18.3*b*) are closed stirrups which end in 135 degree bends (hooks) hooked around longitudinal bars and extended an additional $10d_b$ past the bend.

Wherever inelastic deformation may cause the development of ultimate M_u *away* from the end of a member (as at a concentrated load near mid-span) this point is to be treated as a column face insofar as applying the requirements of the paragraph just above. In addition, the spacing of these stirrup-ties shall not exceed $16d_b$ or 12 in. for at least $2d$ from this point.

18.5 REQUIREMENTS FOR COLUMNS

(a) General

Joints between beams and columns are covered in Sec. 18.6.

Columns must carry from 1 to 6% of vertical reinforcement and no excess concrete may be disregarded in satisfying this minimum. Unless the strength of the column cores is sufficient to resist the design load, the sum of the moment strengths of the column at the joint (above and below) must exceed the sum of the beam moment strengths. This distribution of strength is intended to force hinging into the beams. A particular beam-column joint may be exempted if the others are able to handle the full shear without help from the exempted column.

Column detailing is divided into two sets of rules. If the maximum design load during an earthquake P_e is not more than $0.4P_b$, the member should be detailed as a flexural member.

(b) Confinement Reinforcement

When $P_e > 0.4P_b$, confinement of the core is required, by spirals or hoop reinforcement, for a distance above and below the joint equal to the overall h of the column (the larger h if a rectangular column), but not less than one-sixth of the clear height or 18 in. The requirements outlined in Sec. 18.6 in effect continue a similar reinforcement through the joint.

The Commentary mentions two objectives of the confinement reinforcement:

1. To care for hinging if it should occur in the column.
2. To compensate for spalling which occurs at overload.

In addition, the confinement steel provides some or all of the shear reinforcement over these lengths.

If a spiral is used, the ratio ρ_s is that normally used for a spiral column, where the spiral replaces the strength which was originally carried by the column shell. A second lower limit of not less than $0.12f_c'/f_y$ is also set. If rectangular hoops, either separate or continuous, are used, the hoop is required to be

$$A_{sh} = l_h \rho_s s_h/2$$

where l_h is the maximum unsupported length of rectangular hoop measured between perpendicular legs or crossties, ρ_s is the volumetric steel ratio required for a spiral when the area of the rectangular core area A_{ch} is substituted in the formula instead of A_c of the circular core. This relation is a semirational one presented in detail in the Commentary; some related tests in the Portland Cement Association laboratory have given good results.[2] The formula assumes the hoops about half as effective as a circular spiral. The hoop pitch or spacing center to center must not exceed 4 in.

The concept of supplementary crossties to reduce the length of the straight hoop side is new, an attempt to increase the confinement of the element. A long straight side of a hoop cannot be very effective in developing restraint on the hoop core. The supplementary crosstie must be of the same diameter as the basic hook and must have a standard semicircular hook at each end. These hooks must engage the hoop at each end and also must be fastened to a longitudinal bar at each end. The previous formula accepts a single supplementary tie as reducing l_h to half the main hoop value, thus permitting A_{sh} half as large. Since there would then be three cross legs, each half as large, a 25% reduction in required steel area results. Two supplementary crossties double the number of legs but each is only one-third as heavy, a saving of 33%. Two details of column cross sections detailed under this concept are shown in Fig. 18.4.

Supplementary ties may be used in addition to crossties for shear resistance, these hooking around primary vertical bars rather than hoops.

Minimum cover over supplemental crossties may be reduced to 0.5 in.

Where a wall or stiff partition is carried by columns, the Code requires that the entire column height have confinement reinforcement, because these often tend to pick up heavy loading from the wall as a cantilever.

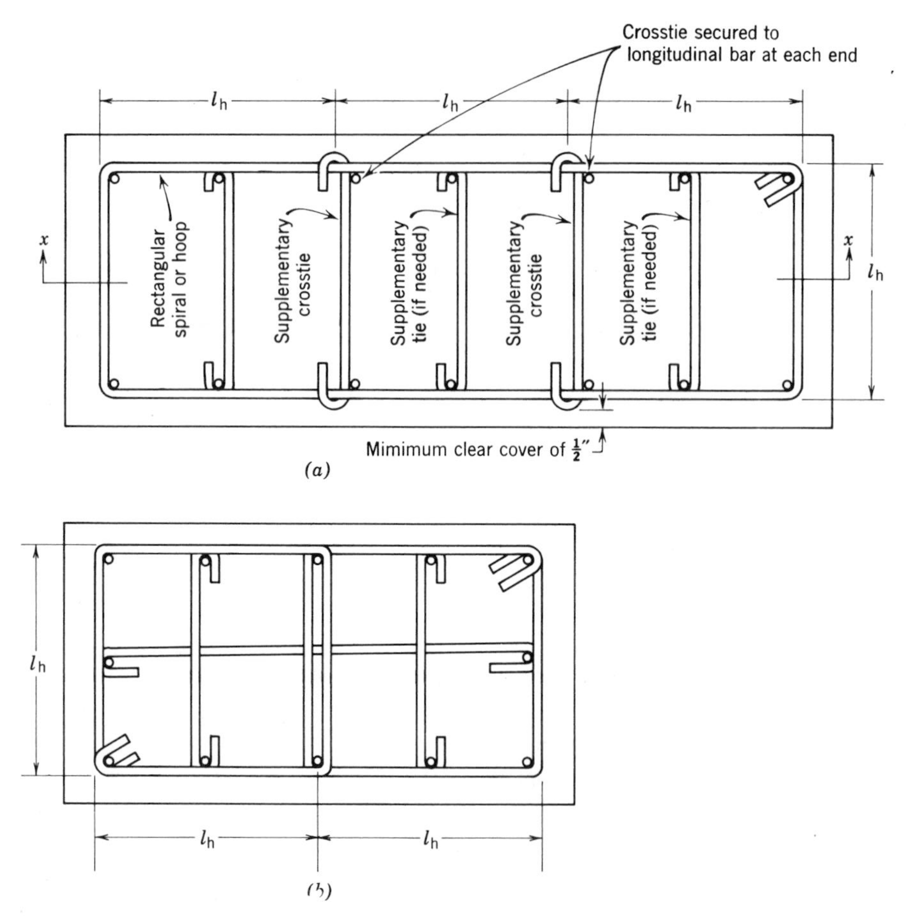

Fig. 18.4. Two columns with supplementary ties. (*a*) With supplementary crossties also to reduce l_h. (*b*) With overlapping hoops to reduce l_h (From Code Commentary.)

If it can be shown that the column loads are not significant, the confinement can be omitted, according to the Commentary.

(c) Shear Reinforcement

Lateral forces and joint displacement both contribute to shear forces on the columns. The shear the frame can deliver when hinging occurs should be considered. Figure 18.5 is from the Commentary, suggesting the possible combinations of hinges around the column joints. The hinging of the column

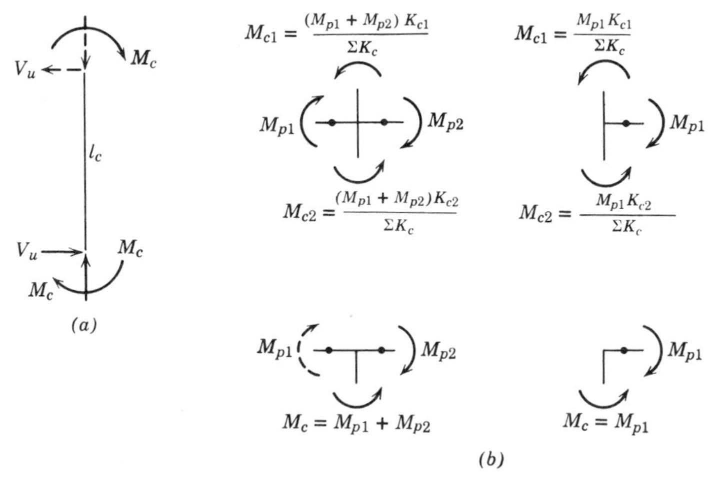

Fig. 18.5. Column under seismic loading. (*a*) Maximum total column shear under seismic loading. (*b*) Column moments in joints where plastic hinges develop in beams (From Code Commentary.)

itself produces the worst shear, if it is assumed possible, because of the relative strengths specified and already discussed in (a). It would always be safe to use shear reinforcement required by hinging moments at each end of the column.

The allowable shear on the concrete is that which includes the effect of axial tension or compression as may be appropriate (Sec. 4.8d). Shear reinforcement shall not be spaced farther apart than $d/2$.

(d) Column Bar Splices

Splices must conform to the usual standard for compression bars, but a stricter minimum of $30d_b$ or 16 in. is set for the seismic situation. Splices near midheight of the column seem desirable.

The Appendix also requires that mechanical or welded splices be staggered, not more than 25% of the bars spliced at any one level, and at least 12 in. of stagger between the levels used.

18.6 BEAM-COLUMN CONNECTIONS

The column spiral or the hoop confinement must extend through the joint between beams and column, with consideration given to both axial load and shear. Usually the shear will be the critical one.

The joint receives a shear from the column above and below. Within the joint depth there will be moments from the adjoining beams, as shown on the joint in Fig. 18.6. If one assumes a hinging moment in each beam, there is on each side a tension pull of $A_s f_y$, to the right at the top (on the right of the figure) and to the left at the bottom. The total compression, although differently constituted in each case, matches the tension in amount, since axial force at most is small in comparison. These forces permit the shear diagram to be constructed and horizontal shear reinforcement to be computed. The necessary hoops or spiral demands may change as different sections are considered.

Where four beams frame in at a joint, the transverse reinforcement may be reduced 50% provided the beams have widths at least half that of the column and have heights such that the most shallow one is at least 75% the height of the deepest one.

The Code calls attention to the torsion, shears, and moments that occur when beams frame in to the column but off of its axis; these require consideration.

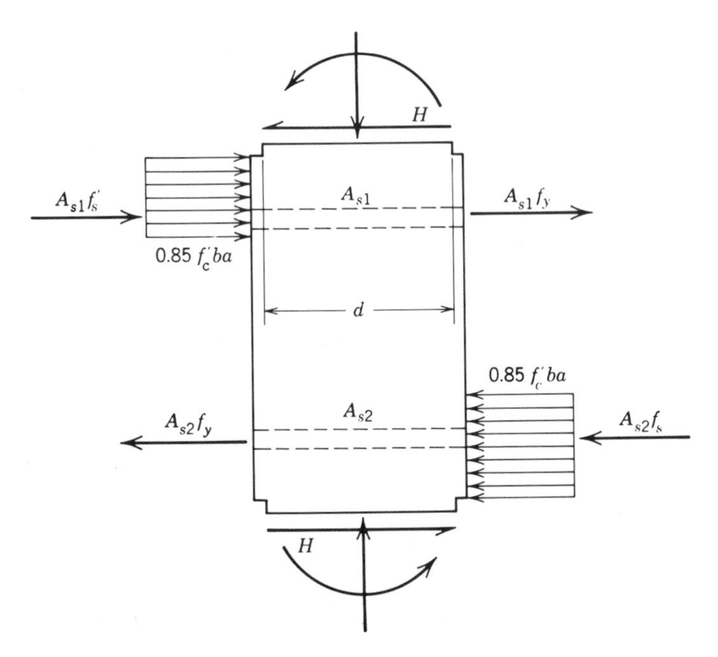

Fig. 18.6. Horizontal shearing forces acting on a connection under seismic loading (From Code Commentary.)

18.7 SPECIAL SHEAR WALLS

(a) General

Horizontal and vertical reinforcement in the wall shall provide at least $\rho = 0.0025$ on the gross area in each direction. Reduced horizontal force factors permitted for a ductile moment resisting space frame are *not* to be applied to the shear walls.

As in the case of columns, where $P_e \lesseqgtr 0.4P_b$ the wall is designed primarily for flexure, but must be good for shear, axial load, and moment. Based on the wall (gross area) as an elastic homogeneous member, wherever the extreme fiber stress exceeds $0.15f_r$ (f_r being the modulus of rupture), the minimum area of vertical reinforcement must be

$$A_s = (200/f_y)hd$$

Here the reinforcement is concentrated near the ends of the wall, and d is the effective depth to the centroid of that steel and h is the member thickness.

(b) Walls with $P_e > 0.4P_b$

The Appendix suggests a design "based on accepted engineering principles" or the specific provisions of A.8.5.1 and A.8.5.2 in the Code. These sections call for vertical boundary elements designed to carry all vertical stresses resulting from the design wall load, tributary dead and live loads, and horizontal forces. The author envisions the wall thickened at the ends, or an I shape, with the end sections detailed as columns with confining reinforcement *for the full height* just as specified over a part of the height for *end* confinement on the columns.

18.8 GENERAL COMMENT

The general purpose of Appendix A is to add toughness or ductility to the frame, to provide for reversal of stresses, to guard against sudden shear failure, and to leave a confined core on columns even when some concrete has spalled off. The end result should be a greatly improved frame. But there appear to be traps for the careless designer. This system is much more complex than normal present-day construction; the designer must visualize his erection problem clearly. The author feels this paragraph from the Commentary is quite appropriate as a closing comment:

Great care is needed to avoid impractical placement problems in seismic design. This is particularly true for the reinforcing steel in the connections of special ductile frames. If high percentages of reinforcement are used in beams and columns, it may be physically impossible to place the beam and column reinforcement and the required ties in the connections; or if the reinforcement is placeable, it may be impossible to place the concrete in the connection or to get a vibrator into it. It may prove economical to construct a full size model of the reinforcement in a typical connection to investigate its constructibility, unless the designer has had considerable experience in this type of work.

SELECTED REFERENCES

1. J. A. Blume, N. M. Newmark, and L. H. Corning, *Design of Multistory Reinforced Concrete Buildings for Earthquake Motions*, Portland Cement Association, Skokie, 1961.
2. N. W. Hanson and H. W. Conner, "Seismic Resistance of Reinforced Concrete Beam-Column Joints," ASCE, *Proc.*, 93, ST5, Oct. 1967, p. 533.
3. ACI Committee 315, "Seismic Details for Special Ductile Frames," *Jour. ACI*. 67, May 1970, p. 374.

APPENDIX

A

Deflections: Area-Moment Method

A.1 DEFLECTIONS RELATIVE TO A TANGENT; THE AREA-MOMENT METHOD

The area moment method, developed by Professor Charles E. Greene of the University of Michigan in 1873, determines deflections *relative* to a specific *tangent* to the elastic curve, not absolute deflections. These relative moments are here called tangential deviations to emphasize that absolute deflections can be directly obtained only when a fixed reference tangent is available. However, the absolute deflection can be found for any beam by a two-step process, since the movement of certain reference tangents can be established simply. The area-moment method is not always the shortest one; but it keeps the physical picture to the fore, is readily usable for variable section members, and is quite useful for proofs. In its usual form it applies only to members within their elastic range.

The two principles can be stated briefly as follows:

> The total angular movement θ, or the change in slope of the elastic curve due to load, between any two points is equal to the area of the M/EI diagram between these two points.

A positive θ indicates a predominant positive moment with the right end rotating counterclockwise with respect to the left end or the left end rotating clockwise with respect to the right end.

> The tangential deviation t_{BC} of any point B on the elastic curve of a member, measured from a reference tangent to the elastic curve at C, is equal to the moment about B of the area of the M/EI diagram between C and B.

The calculated deviation is perpendicular to the arm used in calculating the moment of the area. Thus the deviation in the y-direction is given by using moment arms in the x-direction. If the arm is automatically taken as positive and the usual sign convention for M is used, a positive t will indicate that the predominant moment is positive and thus the elastic curve lies above the reference tangent; a negative t will indicate an elastic curve below the reference tangent.

A.2 CHOICE OF REFERENCE TANGENT

Any beam reference point fixed against both deflection and rotation provides the simplest possible reference tangent. This includes the fixed end of a cantilever beam or the fixed end of a continuous beam. The next simplest case is that of a symmetrical beam where the tangent at the point of symmetry remains horizontal, but may deflect vertically. For other cases, a tangent through a supported point (reaction point which does not move up or down) is recommended.

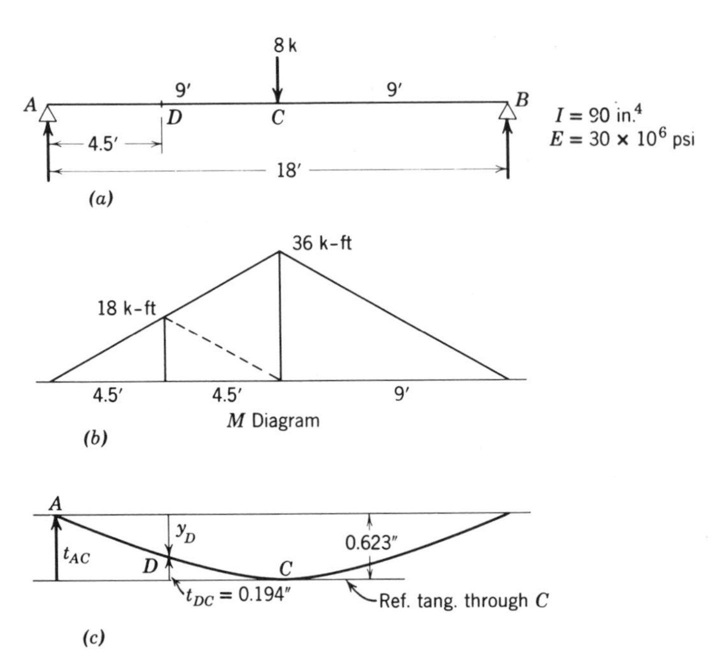

Fig. A.1. Deflection problem with symmetry.

A.3 DEFLECTION OF A SYMMETRICAL BEAM AND LOADING

The quarter point deflection of the beam of Fig. A.1 illustrates the use of a reference tangent which deflects but does not rotate. From symmetry, the elastic curve is horizontal at mid-span.

Use a reference tangent through C and locate it by calculating t_{AC} at A where the real vertical movement is known to be zero.

$$t_{AC} = \frac{1}{EI}\text{(mo. abt. } A \text{ of area of } M \text{ diag. } A \text{ to } C)$$

$$= \frac{1}{EI}\left(+\frac{36}{2} \times 9 \times 6.0\right) = \frac{+972 \times 1000 \times 1728}{30 \times 10^6 \times 90} = +0.623 \text{ in.} \uparrow$$

Since A lies 0.623 in. above the reference tangent through C, the reference tangent must have deflected 0.623 in. downward.

$$t_{DC} = \frac{1}{EI}\text{(mo. abt. } D \text{ of area of } M \text{ diag. } D \text{ to } C)$$

$$= \frac{1}{EI}\left(+\tfrac{36}{2} \times 4.5 \times 3.0 + \tfrac{18}{2} \times 4.5 \times 1.5\right)$$

$$= \frac{+(243 + 60.8) \times 1000 \times 1728}{30 \times 10^6 \times 90} = +0.194 \text{ in.} \uparrow$$

The net deflection, as shown in Fig. A.1*a*, is

$$y_D = \text{defl. of ref. tang.} + t_{DC}$$

$$= -0.623 + 0.194 = -0.429 \text{ in.} \downarrow$$

A.4 GENERAL CASE OF DEFLECTION

Calculate the deflection of mid-span C and of the free end F of the beam of Fig. A.2*a* for $E = 29 \times 10^6$ psi and $I = 100$ in.^{4}

Solution

The moment diagram is shown in Fig. A.2*b*. The two separate diagrams making up this total are shown in Fig. A.2*c* and these will be used in the calculations. The student should note the sequence of calculations to be given now and follow it in other problems. Use a reference tangent through

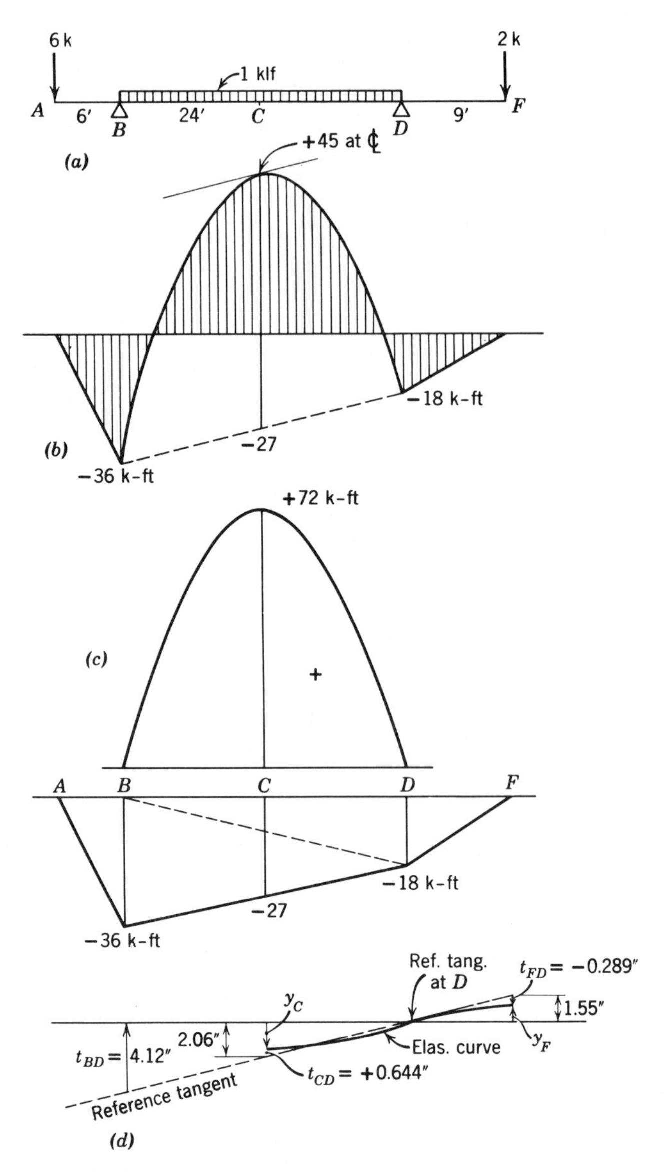

Fig. A.2. General deflection problem.

D, which does not deflect vertically. Locate this tangent by calculating the apparent deflection of point B, which actually does not move.

$$t_{BD} = \frac{1}{EI} \text{ (mo. abt. } B \text{ of area } B \text{ to } D)$$

$$= \frac{1}{EI} \left(-\tfrac{36}{2} \times 24 \times 8 - \tfrac{18}{2} \times 24 \times 16 + \tfrac{2}{3} \times 72 \times 24 \times 12 \right)$$

$$= \frac{(-3460 - 3460 + 13{,}850) \times 1000 \times 1728}{29 \times 10^6 \times 100} = +4.12 \text{ in.} \uparrow$$

Since B lies above the reference tangent, the tangent must be shown below B, as in Fig. A.2d. At point C, by proportion, the tangent lies $4.12/2 = 2.06$ in. below the original position of C.

$$t_{CD} = \frac{1}{EI} \text{ (mo. abt. } C \text{ of area } M \text{ diag. } C \text{ to } D)$$

$$= \frac{1}{EI} \left(-\tfrac{27}{2} \times 12 \times 4 - \tfrac{18}{2} \times 12 \times 8 + \tfrac{2}{3} \times 72 \times 12 \times \tfrac{3}{8} \times 12 \right)$$

$$= \frac{(-648 - 856 + 2590)\, 1000 \times 1728}{29 \times 10^6 \times 100} = +0.644 \text{ in.} \uparrow$$

Based on Fig. A.2d,

$$y_C = -2.06 + t_{CD} = -2.06 + 0.644 = -1.42 \text{ in.} \downarrow$$

The same reference tangent can be used to calculate y_F. In fact, its convenience for this purpose was the reason for originally using it in preference to one through B. At F the reference tangent, by similar triangles, lies $+4.12 \times 9/24 = +1.55$ in. above the original position of F.

$$t_{FD} = \frac{1}{EI} \text{ (mo. abt. } F \text{ of area of } M \text{ diag. } D \text{ to } F)$$

$$= \frac{(-\tfrac{18}{2} \times 9 \times 6)\, 1000 \times 1728}{29 \times 10^6 \times 100} = -0.289 \text{ in.} \downarrow$$

Again referring to Fig. A.2d, one finds

$$y_F = +1.55 + t_{DF} = +1.55 - 0.289 = +1.26 \text{ in.} \uparrow$$

A.5 MOMENT AND SHEAR DIAGRAMS FOR CONTINUOUS BEAMS

The combination and subdivision of moment diagrams as discussed in this section will often be helpful in connection with area-moment calculations.

For any member as a whole or for any shorter length, the complete moment diagram can be constructed if one knows the loads and the end moments. In Fig. A.3*a* the uniformly loaded section is in equilibrium with the known end moments M_1 and M_2 and some necessary end shears. It is convenient to divide this system into two sets of forces (Fig. A.3*b* and *c*), each in equilibrium. The first set of forces represents the external loading without end moments, in effect, a simply supported span leading to the simple beam M and V diagrams as shown. The second set of forces consists of the end moments and the end shears necessary for their equilibrium. The end shears represent a constant shear from end to end, here designated as the continuity shear. The continuity shear can be calculated from the equilibrium of moments or from the slope of the moment diagram. The total moment is the sum of the moment ordinates for the two cases, as in Fig. A.3*d*. The total shear is similarly obtained and enables the point of zero shear and point of maximum moment to be established.

It follows that if a chord is cut across any moment diagram, as in Fig. A.4, the portion above the chord has vertical ordinates which correspond exactly to those of a simple span moment diagram for a beam of length a. For a uniform load this small diagram is a parabola. It should be noted that the zero base line of the original moment diagram can also be be considered as such a chord. It follows that the portion of the moment diagram above the base line is a symmetrical parabola with maximum ordinate $wL_0^2/8$, where L_0 is the distance between points of inflection. If the maximum positive moment is known, the points of inflection can be located easily as symmetrical distances each side of the point of maximum moment.

Example

Plot the moment and shear diagram for a 20-ft span carrying a uniform load of 3 klf and having $M_{AB} = -100$ k-ft, $M_{BA} = -50$ k-ft, as shown in Fig. A.5*a*.

Solution

The simple span has reactions of 30 k each, a maximum moment of $wL^2/8 = 3 \times 20^2/8 = +150$ k-ft, and end shears of $+30$ k and -30 k as shown in Fig. A.5*a*.

The end moments as loads require a positive V_c for equilibrium equal to $(100 - 50)/20 = +2.5$ k. This is also the slope of this part of the moment diagram (Fig. A. 5*b*).

The two sets of M and V diagrams can be combined for the total.

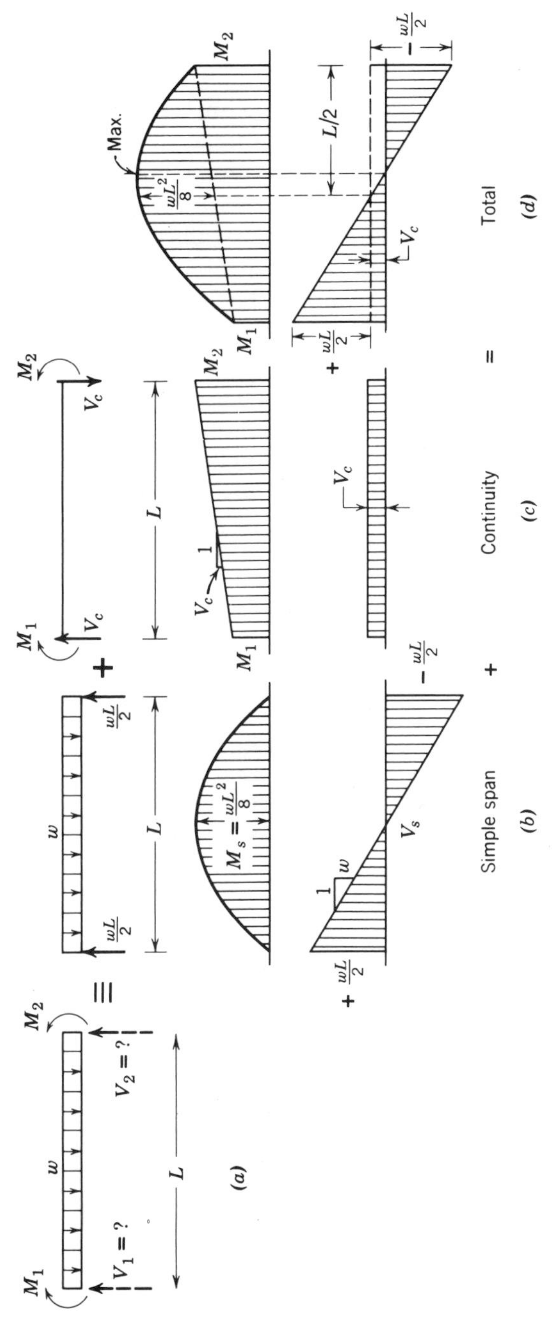

Fig. A.3. Moment and shear diagrams as the sum of two parts.

It is just as simple to build this total diagram without sketching the separate diagrams in such detail. First, plot the end moments and complete this part of the moment diagram with a dashed line as shown in Fig. A.5*c*. The slope of this line is V_c, in this case, $+50/20 = +2.5$ k. This continuity shear can also be plotted as a dashed line for the start of the shear diagram. The simple beam moment diagram can be plotted as vertical ordinates from the sloping (dashed) continuity moment line as a base line. This gives an ordinate of $-75 + 150 = +75$ k-ft at mid span. The slope of the moment diagram at mid-span is parallel to the dashed continuity moment line and the maximum moment obviously lies to the right of the mid-span. The shear diagram can also be completed by adding the simple beam shears to the dashed value of V_c.

The point of maximum moment can be located at the point of zero shear, a distance e from mid-span. The small shear triangle at mid-span shows that $e = V_c/w = +2.5/3.0 = +0.83$ ft and the increase in moment over that at mid-span is given by the area of this triangle, that is, $\Delta M = 2.5 \times 0.83/2 = +1.04$ k-ft. Therefore the maximum moment is $+75$ (at mid-span) $+ 1.04 = 76.04$ k-ft. As an alternate, ΔM is the simple beam moment on a span of $2e$, that is, $\Delta M = w(2e)^2/8 = 3 \times 1.66^2/8 = +1.04$ k-ft.

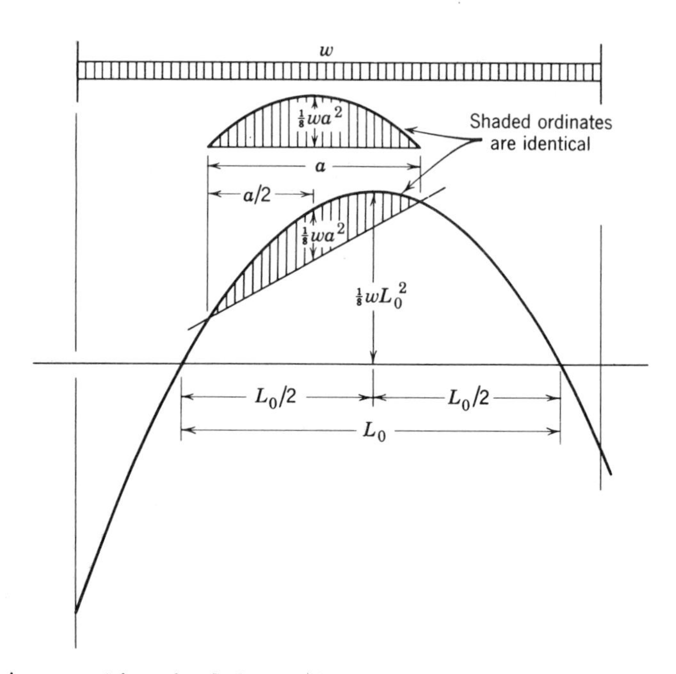

Fig. A.4. A segment is a simple beam diagram.

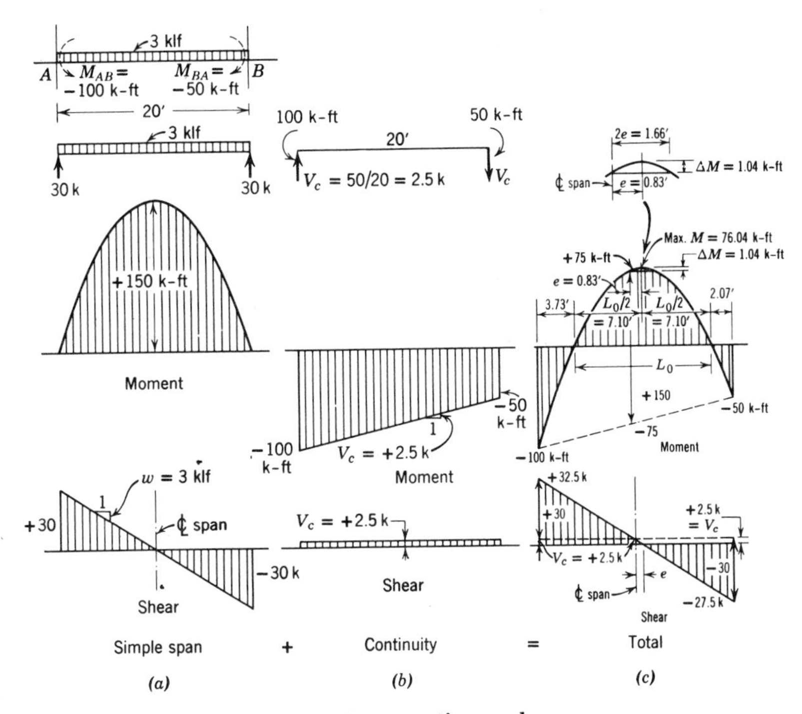

Fig. A.5. Moment and shear diagrams for a continuous beam.

The points of inflection are located by equating maximum positive moment to $wL_0^2/8$.

$$76.04 = \tfrac{1}{8} \times 3 \times L_0^2$$

$$L_0 = \sqrt{\frac{8 \times 76.04}{3}} = \sqrt{202.5} = 14.20 \text{ ft}$$

The points of inflection lie at $L_0/2 = 7.10$ ft to the right and left of the point of maximum moment.

APPENDIX

B

Frame Analysis: Moment Distribution

B.1 THE GENERAL PROCESS OF MOMENT DISTRIBUTION

Moment distribution is a method of analysis for frames in the elastic range which was originated by Professor Hardy Cross and is often referred to as the Hardy Cross method. It involves a series of simple steps taken in sequence such that the moments shown to be acting on the several joints approach closer and closer to the true values resulting from the given loads. Since the calculated moments converge (usually rapidly) on the true moments, any degree of accuracy desired is obtainable.

Consider the frame in Fig. B.1*a* with lateral movement of all joints prevented, that is, with no sidesway. It is assumed that each joint holds the members at fixed angles to each other and that any rotation of a joint requires all attached members to rotate through the same angle. This means the joints are rigid compared to the stiffness of the members.

The moment distribution process follows these steps:

1. Lock or clamp all joints in their unloaded position. Temporarily, this fixes every member at each end.

2. Calculate and record the fixed-end moments due to loads, that is, the moments at each end of each member. With the ends fixed, this calculation is relatively simple. Unloaded members have zero fixed-end moment.

3. Unlock or unclamp one set of alternate joints. Where unbalanced fixed-end moments exist, joints will rotate until they build up resisting moments in adjacent members sufficient to bring them into equilibrium. It should be noted that this release of alternate joints forms a checkerboard

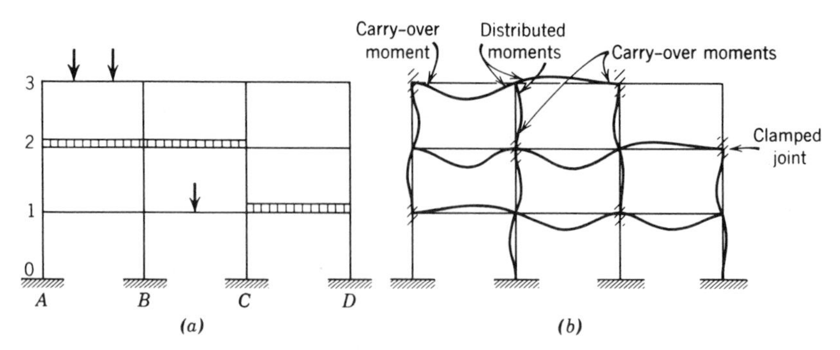

Fig. B.1. Joint rotations with alternate joints clamped.

pattern with each rotating joint surrounded by members having their far ends still fixed at locked joints, as shown in Fig. *B*.1*b*.

4. Write the distributed moments around each rotating joint, that is, the changes in end moments necessary to bring the joint into equilibrium. Each member framing into the joint shares in the distributed moment.

5. Write the carry-over moments, that is, the changes in moments that occur at the far (fixed) ends of each member because of the above joint rotations. These are directly related to the distributed moments at the rotating ends.

6. Now relock or reclamp this first set of alternate joints in their rotated or equilibrium positions.

7. Unlock or release the other set of alternate joints. These will rotate to equilibrium under the combined influence of fixed-end moments and such carry-over moments as have already been developed there.

8. Write distributed moments for all members at these rotating joints and the carry-over moments at the far ends of such members.

9. Again lock or clamp this second set of alternate joints in these equilibrium positions.

10. Release the first set of joints, which will now rotate again under the influence of the moments carried over to them in step 8. Write distributed moments at the rotating joints and carry-over moments out at the fixed ends. Then clamp this set of joints in its equilibrium position.

11. Release the second set of joints and continue the process until the unbalanced moments are neglibigle (for the accuracy desired).

12. At each end of each member add the original fixed-end moment and all subsequent distributed moments and carry-over moments, with due regard to signs. The total is the final moment for this end of the member.

It will be noted that one requires only a few simple quantities to use this process:

1. Fixed-end moments for various loadings.
2. Distribution factors, to establish the proper allotment of distributed moment to each adjacent member when a joint rotates.
3. Carry-over factors, to establish the moment at one fixed end of a member when the joint at the other end is rotated.
4. The stiffness of each member, as a step in establishing the distribution factors in item 2.

It is also helpful to have:

5. A convenient set of signs for moments.
6. A convenient form for tabulating the various moments acting at each end of each member.

B.2 SIGN CONVENTION FOR MOMENT DISTRITUBION—JOINT SIGNS

Ordinary bending moment signs are useful in describing a beam because *plus* always means compression in the top, and *minus* always means tension in the top. Such a sign convention is essential to drawing a moment diagram and designing a beam. It is the only convention usable with the area-moment method of calculating deflections.

In frame analysis, however, the determination of the moments is simpler if one deals entirely with the moments exerted by the various members *on the joints*; and for these, joint signs are recommended. (The final joint moments must eventually be interpreted in terms of the usual bending moment signs to make them usable for M diagrams and design purposes.) For joint moments in frame analysis the following convention is recommended:

> When the moment a member exerts *on a joint tends* to rotate the *joint* in a clockwise direction, this joint moment is designated as positive; when it tends to rotate the joint in a counterclockwise direction the joint moment is negative, as indicated in Fig. B.2.

Under the usual vertical loads, a fixed-ended beam has a negative *bending* moment at each end, producing tension in the top of the beam, as indicated in Fig. B.3. The moment at A tends to rotate the joint clockwise, making M_{AB} positive by joint signs. Likewise, the moment at B tends to rotate the joint counterclockwise, making M_{BA} negative by joint signs.

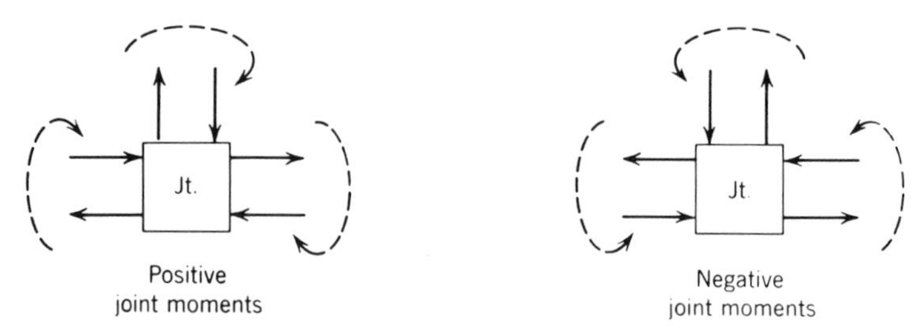

Fig. B.2. Signs for joint moments.

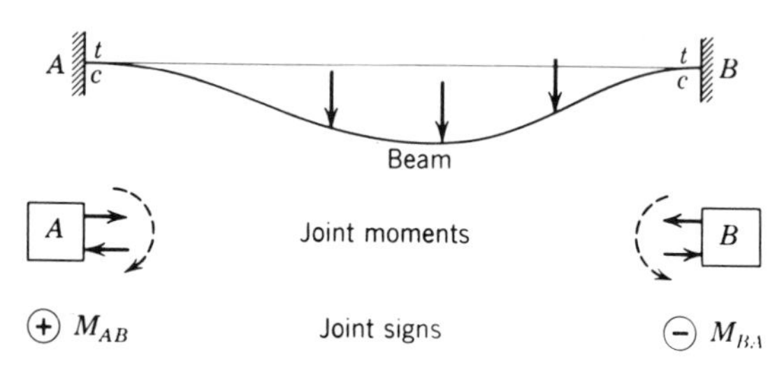

Fig. B.3. Joint signs for fixed-end moments.

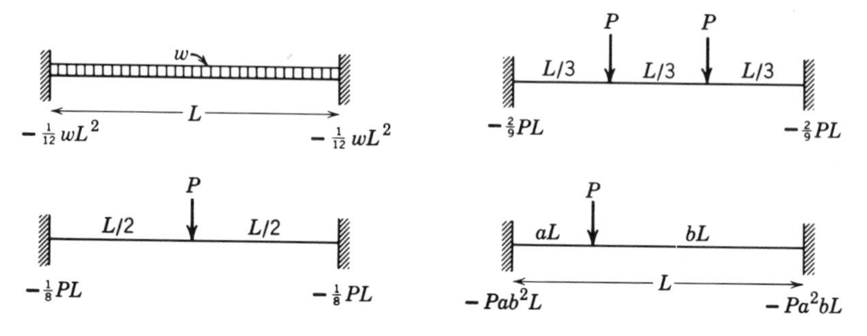

Fig. B.4. Fixed-end moment values, signs according to beam bending moments.

B.3 Fixed-End Moments

Fixed-end moments can be calculated from deflection methods such as the area moment relations or from the three moment equation. Many handbooks tabulate them for a variety of loads. A few cases, for uniform I and E, are given in Fig. B.4. The following symbols are all in use: $M_{AB}{}^F$, $M_{f,AB}$, F.E.M., F.

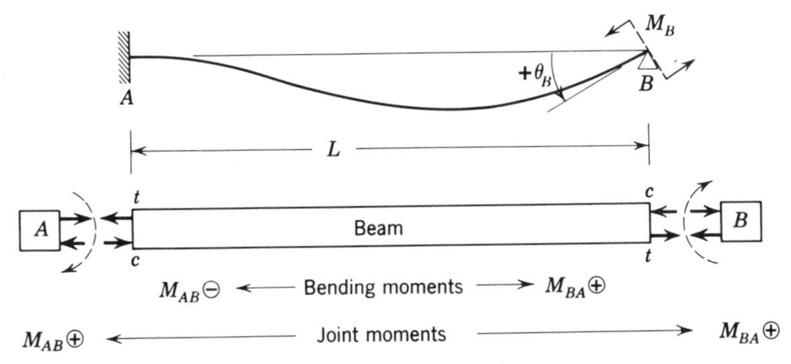

Fig. B.5. Joint moment signs for a member rotated at one end.

B.4 FACTORS USED IN MOMENT DISTRIBUTION

The beam AB of Fig. B.5 is fixed at A and supported, but free to rotate, at B. Assume that a moment M_B acts at B, rotating the end of the beam and setting up bending moments M_{AB} and M_{BA} *in the beam* at A and B. The simple use of area moments to calculate the deflection at B leads to the fact that $M_{AB} = -M_{BA}/2$ in terms of beam bending moment signs. With the joint signs, Fig. B.5 indicates that this becomes $M_{AB} = +M_{BA}/2$. The moment M_{AB} is called the *carry-over movement* (C.O.M.) and the ratio $+\frac{1}{2}$ is called the *carry-over factor* (C.O.F.).

The angle θ_B can also be evaluated by area-moment relationships and this leads to the relationship that $M_{BA}/\theta_B = 4EI/L = 4EK$, where $K = I/L$. The ratio of M_{BA} to θ_B defines the *stiffness of a beam* (against rotation at one end) when fixed at the far end, that is, $4EK$ is the stiffness. For many uses only *relative stiffness* is significant and K alone may be used.

If θ_B is considered positive when counterclockwise (as in trigonometry), a positive θ causes a positive joint moment to be built up, as shown in Fig. B.5.

If a joint B is rotated by some applied moment, each attached member rotates through the same end angle, building up a resisting moment $4EK\theta_B$. Equilibrium of moments leads to the equation $\Sigma M_B + M_0 = 0$, to the rotation angle $\theta_B = -M_0/\Sigma 4EK$, and to the moment distributed to the members as $-M_0K_1/\Sigma K$ to member 1, $-M_0K_2/\Sigma K$, etc., where ΣK is the total stiffness and K_1, K_2, etc., are the individual relative member stiffnesses. The ratios $K_1/\Sigma K$, $K_2/\Sigma K$, . . . are called *distribution factors* for member 1, member 2 . . . It is also true that the moments are distributed in proportion to their stiffnesses, that is.

$$\frac{M_{B1}}{M_{B2}} = \frac{K_1}{K_2}, \ldots$$

B.5 TABULATION FORM

Moment distribution calculations can best be superimposed on a skeleton picture of the frame, grouping the moments around the joints where they apply. For this purpose the skeleton frame need not be to scale.

The Portland Cement Association has used a tabulation form at each joint which has great merit. Its use as an arbitrary standard form should be unvarying. Figure B.6 shows this form superimposed on a typical joint.

B.6 EXAMPLES OF MOMENT DISTRIBUTION—BASIC PROCESS

(a) Alternate Joint Procedure

Find the joint moments for the continuous beam of Fig. B.7*a*.

Solution

Fixed-end moments (signs are obtained from the joint sketches in Fig. B.7*a*):

$$M_{AB}^F = +Pab^2L = +40 \times \tfrac{8}{20} \times (\tfrac{12}{20})^2 \times 20 = +115.2 \text{ k-ft}$$
$$M_{BA}^F = -Pa^2bL = -40 \times (\tfrac{8}{20})^2 \times \tfrac{12}{20} \times 20 = -76.8 \text{ k-ft}$$
$$M_{BC}^F = +\tfrac{1}{12}wL^2 = +\tfrac{1}{12} \times 2 \times (25)^2 = +104.2 \text{ k-ft}$$
$$M_{CB}^F = -\tfrac{1}{12}wL^2 = -104.2 \text{ k-ft}$$
$$M_{CD}^F = +\tfrac{1}{2}wL^2 = +\tfrac{1}{2} \times 2 \times 6^2 = +36 \text{ k-ft}$$

Distribution factors:

Joint A: Since the fixed end at A is infinitely stiff, no rotation occurs and no D.F. is needed for the beam. Technically, beam D.F. $= (I/L)/(\infty + I/L) = 0$.

Joint B:

	K	Rel. K	D.F.
BA	$I_{AB}/20$	1.00	$1.00/2.60 = 0.384$
BC	$I_{BC}/25 = 2I_{AB}/25$	1.60	$1.60/2.60 = 0.616$
		$\Sigma K = 2.60$	ΣD.F. $= 1.000$

Joint C: The stiffness of a cantilever overhang is zero, since this length does not *resist* any rotation about C.

	K	D.F.
CD	0	0
CB	$I_{BC}/25$	1.00

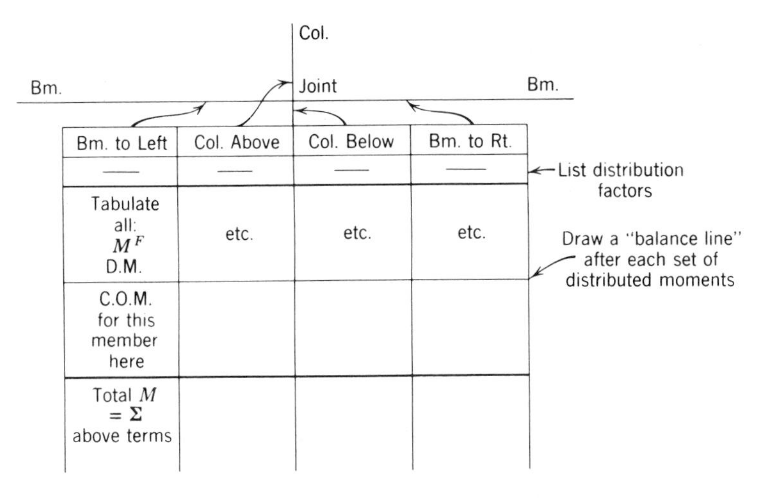

Fig. B.6. Standard tabulation form for each joint.

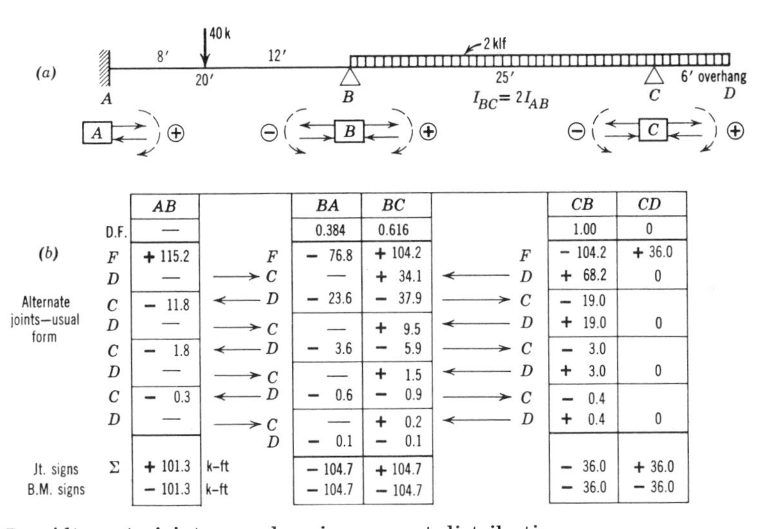

	AB		BA	BC		CB	CD
D.F.	—		0.384	0.616		1.00	0
F	+ 115.2	F	− 76.8	+ 104.2	F	− 104.2	+ 36.0
D	—	→ C	—	+ 34.1	← D	+ 68.2	0
C	− 11.8	← D	− 23.6	− 37.9	→ C	− 19.0	
D	—	→ C	—	+ 9.5	← D	+ 19.0	0
C	− 1.8	← D	− 3.6	− 5.9	→ C	− 3.0	
D	—	→ C	—	+ 1.5	← D	+ 3.0	0
C	− 0.3	← D	− 0.6	− 0.9	→ C	− 0.4	
D	—	→ C	—	+ 0.2	← D	+ 0.4	0
		D	− 0.1	− 0.1			
Jt. signs Σ	+ 101.3	k-ft	− 104.7	+ 104.7		− 36.0	+ 36.0
B.M. signs	− 101.3	k-ft	− 104.7	− 104.7		− 36.0	− 36.0

Fig. B.7. Alternate joint procedure in moment distribution.

The moment distribution is carried out in Fig. B.7*b*. Since there are no vertical members, the form of Fig. B.6 becomes somewhat simplified. The choice of which set of alternate joints to unlock and balance first is arbitrary. Usually the convergence is more rapid if the set with the largest unbalances is used first.

The initial unbalance at C is $-104.2 + 36.0 = -68.2$; therefore $+68.2$ is distributed, in this case entirely to CB because of the D.F. of 1.00. Half of this distributed moment is carried over to BC with the sign unchanged. The fixed end at joint A gives no rotation nor distribution.

Joints A and C are next locked and B is released. The unbalanced moment of $-76.8 + 104.2 + 34.1 = +61.5$ is distributed $-0.384 \times 61.5 = -23.6$ to BA and $-0.616 \times 61.5 = -37.9$ to BC. Half of these distributed moments are carried over to the far ends of members.

The process is continued until the carry-over moments are negligible or small enough to neglect for the accuracy desired. The columns are totaled to obtain final moments (joint signs). These must then be changed into beam or bending moment signs.

(b) Simultaneous Joint Release

The same results are obtained by a slightly different sequence of calculations which many prefer. In this sequence the distributions are first made at all joints (with the distribution factors unchanged) as though all joints were simultaneously released. The carry-over moments which occur simultaneously are not written until all these distributed moments have been calculated and recorded. Then all carry-over moments are written, another complete distribution is made, another set of carry-over moments, and so forth. This sequence eliminates the need for keeping track of alternate joints, but is often slightly slower in converging. Figure B.8 shows this solution form for the same problem.

(c) Example with Vertical Members

Find the joint moments in the double box frame for Fig. B.9 subject to the loads shown. Consider I as constant and any sidesway prevented.

Solution

For AB, DE, and EF, $M^F = \frac{1}{12}wL^2 = \frac{1}{12} \times 2.5 \times (20)^2 = \pm 83.3$ k-ft.
for BC, $M^F = \frac{1}{8}PL = \frac{1}{8} \times 50 \times 20 = \pm 125$ k-ft.

Corner joints:

	K	Rel. K	D.F.
Beam	$I/20$	1.00	0.375
Wall	$I/12$	1.67	0.625
		$\Sigma = 2.67$	1.000

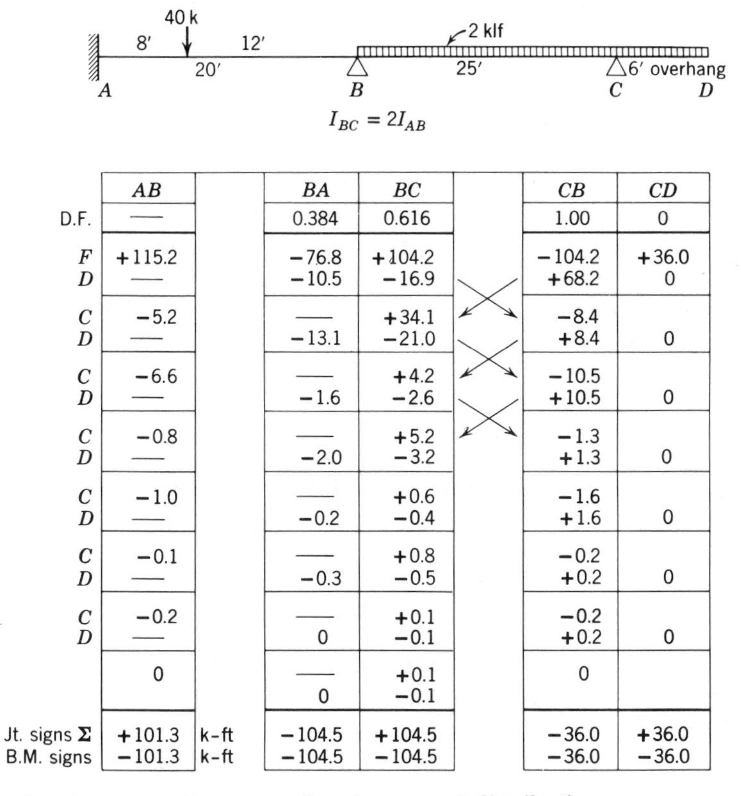

	AB		BA	BC	CB	CD
D.F.	—		0.384	0.616	1.00	0
F	+115.2		−76.8	+104.2	−104.2	+36.0
D	—		−10.5	−16.9	+68.2	0
C	−5.2		—	+34.1	−8.4	
D	—		−13.1	−21.0	+8.4	0
C	−6.6		—	+4.2	−10.5	
D	—		−1.6	−2.6	+10.5	0
C	−0.8		—	+5.2	−1.3	
D	—		−2.0	−3.2	+1.3	0
C	−1.0		—	+0.6	−1.6	
D	—		−0.2	−0.4	+1.6	0
C	−0.1		—	+0.8	−0.2	
D	—		−0.3	−0.5	+0.2	0
C	−0.2		—	+0.1	−0.2	
D	—		0	−0.1	+0.2	0
	0		—	+0.1	0	
			0	−0.1		
Jt. signs Σ	+101.3	k-ft	−104.5	+104.5	−36.0	+36.0
B.M. signs	−101.3	k-ft	−104.5	−104.5	−36.0	−36.0

Fig. B.8. Simultaneous release procedure in moment distribution.

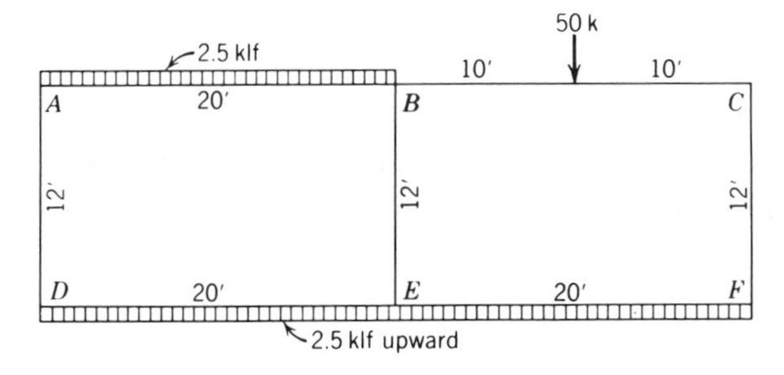

Fig. B.9. Moment distribution problem with vertical members.

Center joints:

Beam	$I/20$	1.00	0.272
Beam	$I/20$	1.00	0.272
Wall	$I/12$	1.67	0.456
		$\Sigma = 3.67$	1.000

Figure B.10 shows the solution carried out until the unbalance appears to be negligible. It should be noted that, with this form, carry-over moments on vertical members are always from lower left to upper right, or upper right to lower left in columns of data adjacent to the vertical member. No carry-over moves from second to fourth quadrant or from fourth to second. The blank columns can be omitted from the tabulation completely if desired, as illustrated in Fig. B.18.

B.7 PIN ENDS AND ENDS WITH MOMENTS STATICALLY DETERMINED

(a) Simplified Procedure

The convergence in the example of Figs. B.7 and B.8 is slow because the distribution factor is 1.00 at joint C. A slight modification of the general procedure may be used to shorten this process. This modification consists of leaving joint C unlocked throughout the process and modifying the fixed-end moments, stiffness factors, and carry-over moments as necessary to agree with this condition.

(b) Modified Fixed-End Moments

Modified end moments for member BC are needed for B fixed and C not fixed, and this automatically involves member CD as well. These moments can be established by a simple (separate) moment distribution as shown in Fig. B.11. The first line of fixed-end moments is the same as in Fig. B.7. The release of C establishes the moments needed for starting the analysis with joint C unlocked. The author sometimes uses the notation $M_{BC}{}^{FH}$ for this modified moment.

For beams without overhangs, handbooks show these modified moments for many loadings. However, the above process is so simple that a handbook is not necessary.

Joints A, B, C:

	AD	AB	BA	BE	BC	CB	CF
	0.625	0.375	0.272	0.456	0.272	0.375	0.625
F	0	+83.3	−83.3	0	+125.0	−125.0	0
D	−52.0	−31.3	−11.4	−19.0	−11.3	+46.7	+78.3
C	+26.0	−5.7	−15.6	0	+23.4	−5.6	−26.0
D	−12.7	−7.6	−2.1	−3.6	−2.1	+11.8	+19.8
C	+8.1	−1.0	−3.8	+2.2	+5.9	−1.0	−12.3
D	−4.4	−2.7	−1.1	−2.0	−1.2	+5.0	+8.3
C	+1.6	−0.6	−1.4	+1.0	+2.5	−0.6	−3.5
D	−0.6	−0.4	−0.6	−0.9	−0.6	+1.5	+2.6
C	+0.5	−0.3	−0.2	+0.4	+0.8	−0.3	−1.5
D	−0.1	−0.1	−0.3	−0.4	−0.3	+0.7	+1.1
C	0	−0.2	0	+0.2	+0.4	−0.2	−0.5
D	+0.1	+0.1	−0.2	−0.2	−0.2	+0.3	+0.4
Jt. Σ	−33.5	+33.5	−120.0	−22.3	+142.3	−66.7	+66.7
B.M.	−33.5	−33.5	−120.0	−22.3	−142.3	−66.7	+66.7

Joints D, E, F:

	DA	DE	ED	EB	EF	FE	FC
	0.625	0.375	0.272	0.456	0.272	0.375	0.625
F	0	−83.3	+83.3	0	−83.3	+83.3	0
D	+52.0	+31.3	0	0	0	−31.3	−52.0
C	−26.0	0	+15.6	−9.5	−15.6	0	+39.2
D	+16.2	+9.8	+2.6	+4.3	+2.6	−14.7	−24.5
C	−6.4	+1.3	+4.9	−1.8	−7.4	+1.3	+9.9
D	+3.2	+1.9	+1.2	+2.0	+1.1	−4.2	−7.0
C	−2.2	+0.6	+1.0	−1.0	−2.1	+0.6	+4.2
D	+1.0	+0.6	+0.6	+0.9	+0.6	−1.8	−3.0
C	−0.3	+0.3	+0.3	−0.4	−0.9	+0.3	+1.3
D	0	0	+0.3	+0.4	+0.3	−0.6	−1.0
C	0	+0.2	0	−0.2	−0.3	+0.2	+0.6
D	−0.1	−0.1	+0.1	+0.3	+0.1	−0.3	−0.5
Σ	+37.4	−37.4	+109.9	−5.0	−104.9	+32.8	−32.8
B.M.	−37.4	+37.4	+109.9	+5.0	+104.9	+32.8	+32.8

For bending moment signs on vertical members: View from right side.

Fig. B.10. Solution of frame of Fig. B.9 using simultaneous joint releases.

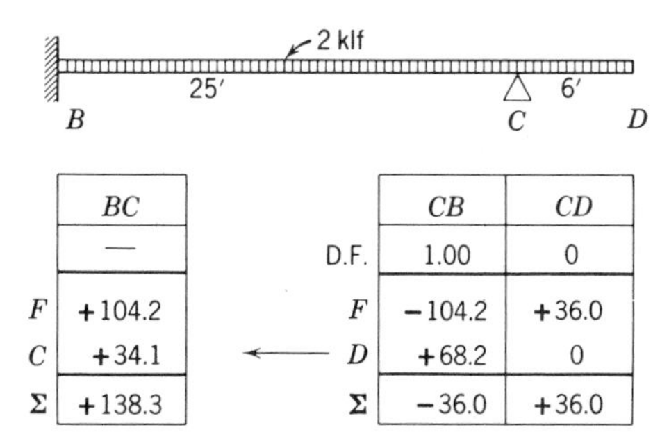

	BC
	—
F	+104.2
C	+34.1
Σ	+138.3

	CB	CD
D.F.	1.00	0
F	−104.2	+36.0
D	+68.2	0
Σ	−36.0	+36.0

Fig. B.11. Modified fixed-end moment with far end statically determinate.

(c) Modified Stiffness

The overhang CD does not affect the moment required to rotate end B of beam BC, either with joint C fixed or not fixed. The modified stiffness can be established by another simple moment distribution. Section B.4 established the stiffness with far end fixed as $4EK$. In Fig. B.12, C is first considered fixed as B is rotated to give $M_{BC}{}^F = +4EK\theta_B$ and a carry-over moment $M_{CB}{}^F = +2EK\theta_B$. When joint C is released, the final moment at B becomes $+3EK\theta_B$ and the modified stiffness is $M_{BC}/\theta_B = +3EK$. This is three-fourths of the stiffness with end C fixed. Hence, when relative stiffness is used, the modified value may be taken as $\frac{3}{4}(I/L)$. Obviously, this relationship applies to all pin end members as well as to those with overhangs. It is not limited to this particular example.

(d) Carry-Over Moments

When the far joint is never locked, its moment remains statically determined. The joint thus receives *no* carry-over moment as a result of rotations at the other end of the beam.

(e) Final Solution of Example of Sec. B.6a

In Fig. B.13, the fixed-end moments for member AB are taken from the original solution and those for BCD from Sec. B.13. New distribution factors must be calculated at B:

	K	Ref. K	D.F.
BA	$I_{AB}/20$	1.00	0.455
BC	$\frac{3}{4}\frac{I_{BC}}{25} = \frac{3}{4}\frac{2I_{AB}}{25}$	1.20	0.545
	$\Sigma K =$	2.20	1.000

This solution is usually short because only one joint must rotate to equilibrium.

(f) Example with Columns

Find the joint moments for the frame of Fig. B.14.

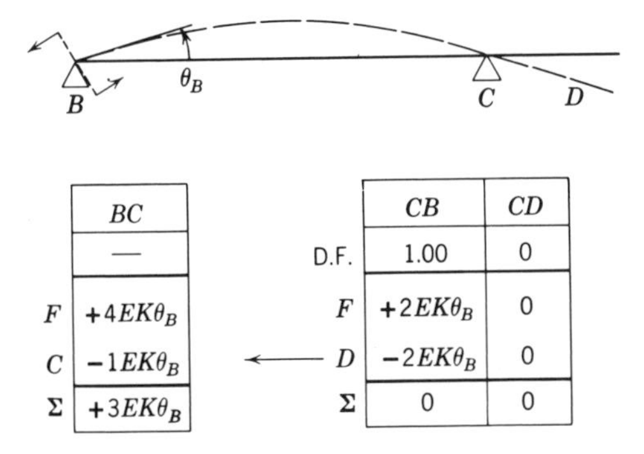

	BC
	—
F	$+4EK\theta_B$
C	$-1EK\theta_B$
Σ	$+3EK\theta_B$

	CB	CD
D.F.	1.00	0
F	$+2EK\theta_B$	0
D	$-2EK\theta_B$	0
Σ	0	0

Fig. B.12. Stiffness with far end hinged or on knife edge.

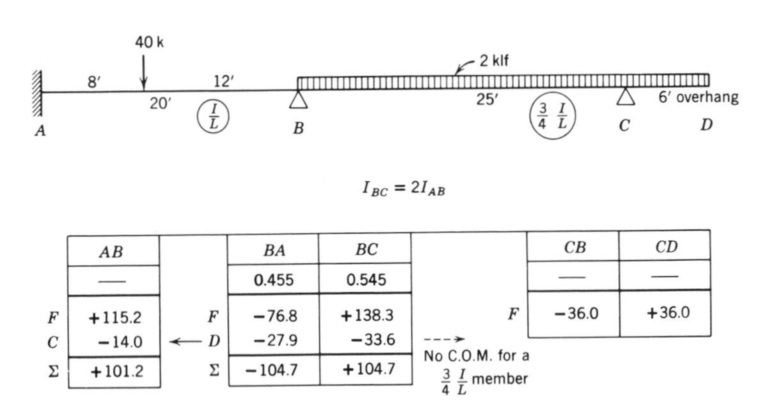

	AB
	—
F	+115.2
C	−14.0
Σ	+101.2

	BA	BC
	0.455	0.545
F	−76.8	+138.3
D	−27.9	−33.6
Σ	−104.7	+104.7

	CB	CD
	—	—
F	−36.0	+36.0

Fig. B.13. Example showing use of modified stiffness and modified end moments. The solution is unusually short, as noted in text.

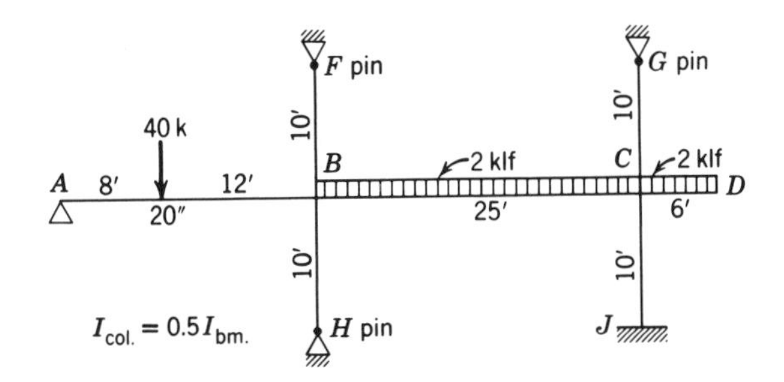

Fig. B.14. Frame to illustrate modified stiffness.

Solution

All members except BC, CD, CJ are pinned at their outer ends. Hence the modified stiffness procedure with joints A, F, G, and H remaining unlocked will be much shorter.

Fixed-end moments (and modified values):

$$M_{AB}{}^F = +Pab^2L = +40 \times \tfrac{8}{20} \times (\tfrac{12}{20})^2 \times 20 = +115.2 \text{ k-ft}$$
$$M_{BA}{}^F = -Pa^2bL = -40 \times (\tfrac{8}{20})^2 \times \tfrac{12}{20} \times 20 = -76.8 \text{ k-ft}$$

These are modified in Fig. B.15a to $M_{AB} = 0$ and $M_{BA}{}^{FH} = -134.4$ k-ft

$$M_{BC}{}^F = +\tfrac{1}{12}wL^2 = +\tfrac{1}{12} \times 2 \times (25)^2 = +104.2 \text{ k-ft}$$
$$M_{CB}{}^F = -\tfrac{1}{12}wL^2 = -104.2 \text{ k-ft}$$
$$M_{CD}{}^F = +\tfrac{1}{2}wL^2 = +\tfrac{1}{2} \times 2 \times 6^2 = +36 \text{ k-ft}$$

Distribution factors:

Joint A: none.

Joint B:

	K	Rel. K	D.F.
BA	$0.75I/20$	0.75	0.246
BF	$0.75 \times 0.5I/10$	0.75	0.246
BC	$I/25$	0.80	0.262
BH	$0.75 \times 0.5I/10$	0.75	0.246
	$\Sigma K =$	3.05	1.000

Joint C:

	K	Rel. K	D.F.
CB	$I/25$	0.80	0.313
CG	$0.74 \times 0.5I/10$	0.75	0.294
CD	0	0	0
CJ	$0.5I/10$	1.00	0.393
	$\Sigma K =$	2.55	1.000

The final solution is shown in Fig. B.15b. Advantage has been taken of the fact that joint J does not rotate to make one total carry-over of the distributed moments at CJ rather than several individual carry-over operations.

B.8 MEMBERS OF VARIABLE CROSS SECTION

For members of variable cross section, the fixed-end moments are different, the stiffness is greater than $4EI_0/L$ (where I_0 is the minimum I),

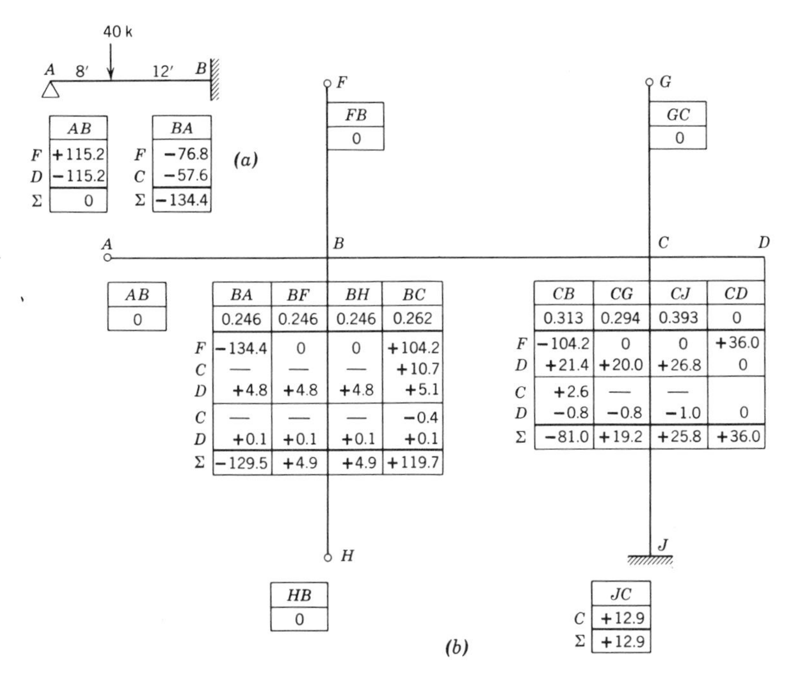

Fig. B.15. Solution of frame of Fig. B.14.

and the carry-over factor is different from 0.5. Unless the member is symmetrical about mid-span, the stiffness and carry-over factors are different from each end of the member. However, the procedures of moment distribution are unchanged for frames containing such members, as in the analysis of Secs. 10.17 and 10.18 and especially Fig. 10.24.

The constants for fixed-end moment, stiffness, and carry-over factor for many variable section members are tabulated in the Portland Cement Association pamphlet "Handbook of Frame Constants."[9]

B.9 FIXED-END MOMENTS FOR DEFLECTION OF JOINTS

Consider the beam AB of Fig. B.16, with constant I, deflected upward a distance Δ at B, but without any joint rotation. The moment diagram is a straight line, since the shear is constant. Since there is no angle change from A to B, the net area of the M/EI diagram is zero and this requires that $M_{AB} = -M_{AB}$ (bending moment signs). Thus the M diagram must be that of Fig. B.16. By the area-moment principle,

$$\Delta = \frac{I}{EI} \text{ (moment abt. } B \text{ of } M \text{ diag. from } A \text{ to } B)$$

$$= \frac{I}{IE}\left(\frac{M_{AB}}{2}\frac{L}{2}\frac{5}{6}L + \frac{M_{BA}}{2}\frac{L}{2}\frac{L}{6}\right)$$

$$= \frac{1}{EI}\left(\frac{5}{24}M_{AB}L^2 - \frac{1}{24}M_{AB}L^2\right) = \frac{1}{EI}\frac{M_{AB}L^2}{6}$$

$$\left.\begin{aligned} M_{AB} &= +6EI\Delta/L^2 = +6EK\Delta/L \\ M_{BA} &= -M_{AB} = -6EK\Delta/L \end{aligned}\right\} \text{Bending moment signs}$$

$$M_{AB} = M_{BA} = -6EK\Delta/L \quad \text{Joint signs}$$

Here Δ is positive when it "rotates" the member in counterclockwise fashion as shown. It should be noted that this relation requires absolute or real values of E and K, not the relative values which are often used elsewhere.

B.10 FRAMES WITH SIDESWAY

For frames carrying lateral load and for many frames under vertical load, sidesway cannot be ignored. Generally, this sidesway effect must be calculated in a separate operation. The general procedure will be developed on the basis of a numerical example.

Consider the frame of Fig. B.17 with all members of equal I. The relative stiffness of members is indicated in a circle alongside each.

The effect of loads between joints, with sidesway prevented, will be considered first, as Case I. This requires an external holding force at B or C, called a restraining force. By the method presented here, the magnitude of this restraining force is of no interest. Figure B.18*a* shows the distribution for Case I.

Without the restraining force at C, the frame would sway, probably to the right (but it is not necessary to know the direction). When the correct amount of sway is added, the frame will be in equilibrium without any external restraints. The correct sway moments might be calculated separately and added as indicated diagrammatically in Fig. B.19*a*. As an alternate, and a more practical one, the frame might be given an arbitrary sway Δ, and the equilibrium statement could then be that X parts of these arbitrary sway moments would put the frame in equilibrium, as diagrammed in Fig. B.19*b*. In either case the correct sway moments must be established from an equilibrium equation of statics, in the case, $\Sigma F_x = 0$.

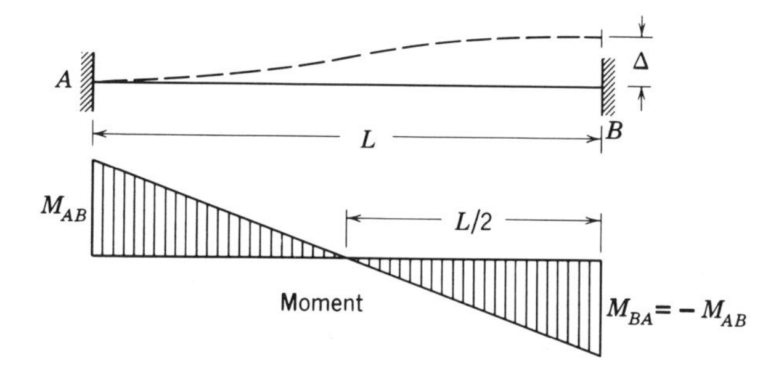

Fig. B.16. Moments due to joint movement perpendicular to member axis.

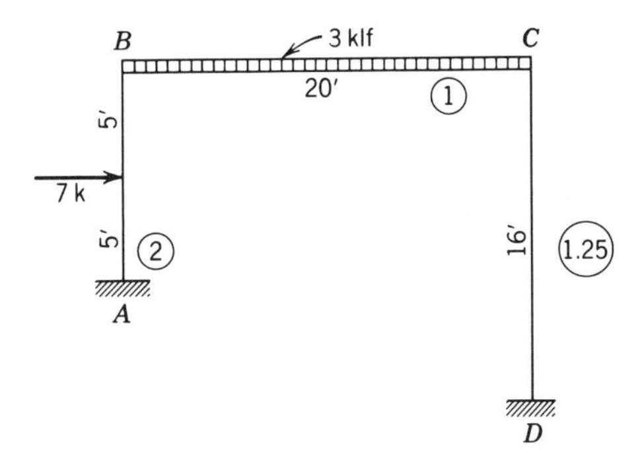

Fig. B.17. Frame where sidesway is significant.

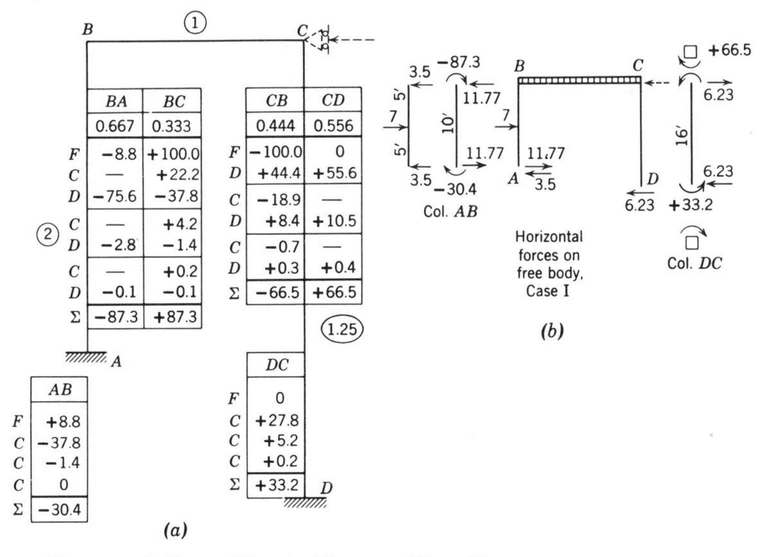

Fig. B.18. Frame solution without sidesway (Case I).

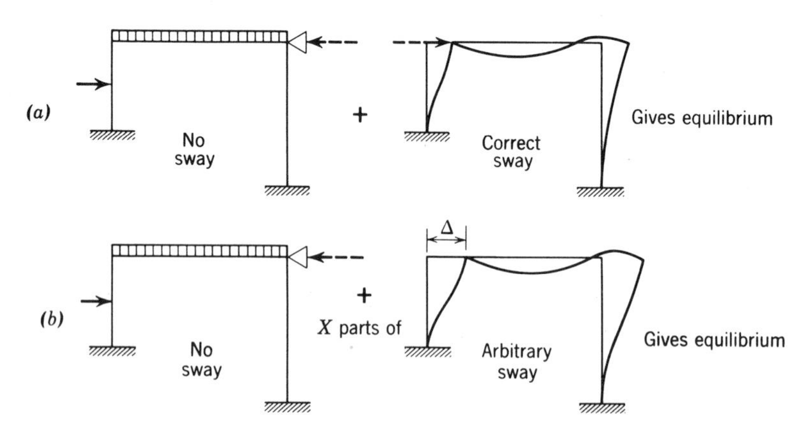

Fig. B.19. Corrections for sidesway.

In Fig. B.20*a* the moment distribution (Case II) for an arbitrary sidesway, enough to make $M_{AB}{}^F = M_{BA}{}^F = -1000$ k-ft, has been completed. A sway to the left, contrary to the apparent real movement, has been used to indicate that one is not required to know the direction of sway.

For E, I, and Δ in foot units:

$$M_{AB}{}^F = -6EK\,\Delta/L = -\,6E\frac{I}{10} \times \frac{\Delta}{10} = -1000 \text{ k-ft}$$

$$M_{DC}{}^F = -6E\frac{I}{16} \times \frac{\Delta}{16}$$

$$\frac{M_{DC}{}^F}{-1000} = \frac{-6E(I/16) \times (\Delta/16)}{-6E(I/10) \times (\Delta/10)} = \frac{1/256}{1/100} = \frac{100}{256}$$

$$M_{DC}{}^F = -391 \text{ k-ft}$$

The equilibrium equation $\Sigma F_x = 0$ will now be applied to the entire frame as a free body. Figure B.18*b* shows the horizontal forces for Case I, omitting vertical and moment reactions which are not required for this equation. Note that the end shears on CD form a couple to balance the end moments such that $V_{DC} = V_{CD} = (M_{CD} + M_{DC})/L$, using the *joint* signs for moments. Likewise, the shear on AB is the simple beam shear plus

$$V_{AB1} = (M_{AB} + M_{BA})/L = (-30.4 - 87.3)/10 = -\,11.77\text{k}$$

Similarly, in Fig. B.20*b* for Case II:

$$V_{AB} = (-367 - 682)/10 = -104.9 \text{ k}$$

$$V_{DC} = (-262 - 327)/16 = -36.7 \text{ k}$$

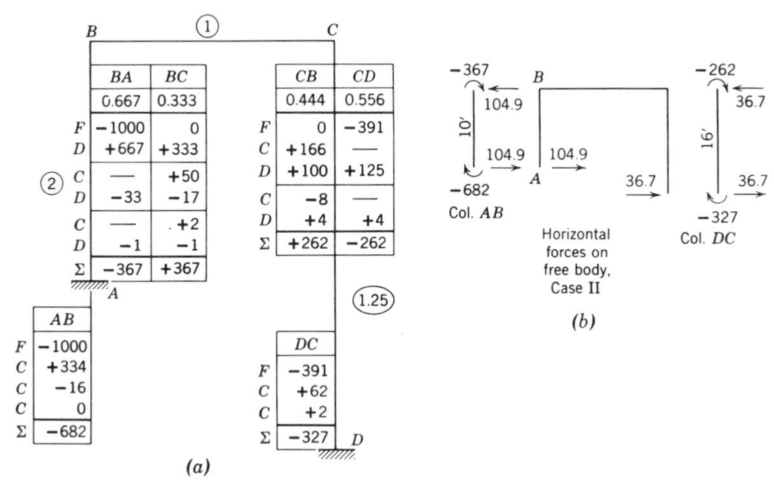

Fig. B.20. Arbitrary sidesway (Case II).

Figure B.21*a* shows the forces finally used in the equilibrium equation:

$$(\text{Case I}) + X\ (\text{Case II}) = 0$$

Based on horizontal forces, this becomes

$$(+7 - 3.5 + 11.77 - 6.23) + X(+104.9 + 36.7) = 0$$

$$X = -9.04/141.6 = -0.0637$$

The final shears, moments, reactions, and so forth, are each given as (Case I) + X(Case II) values, that is, as (Case I) − 0.0637 (Case II) values. Moments calculated on this basis are tabulated in Fig. B.21*b*.

In multiple-story structures sidesway involves simultaneous equations or an entirely different procedure. It is fortunate that sidesway is not very significant in multibay frames under vertical loads. It must, of course, be considered wherever wind force analysis is required.

SELECTED REFERENCES

1. Hardy Cross and Newlin D. Morgan, *Continuous Frames of Reinforced Concrete*, John Wiley and Sons, New York, 1932.
2. L. E. Grinter, *Theory of Modern Steel Structures*, Vol. 2, The Macmillan Co., New York, rev. ed., 1949.

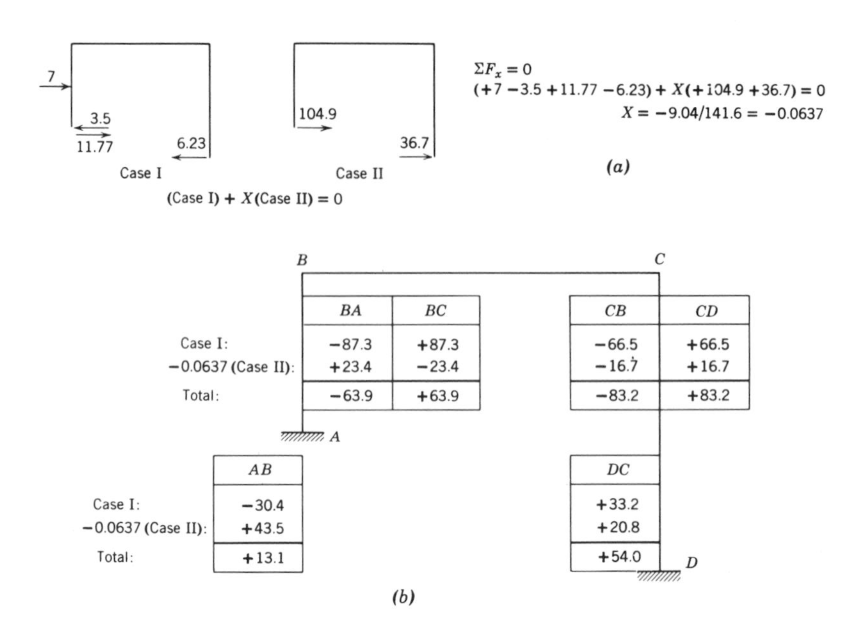

Fig. B.21. Final moment determination for frame of Figs. B.17, B.18, and B.20.

3. John I. Parcel and Robert B. B. Moorman, *Analysis of Statically Indeterminate Structures*, John Wiley and Sons, New York, 1955.
4. Chu-Kia Wang, *Statically Indeterminate Structures*, McGraw-Hill Book Company, New York, 1953.
5. Phil M. Ferguson and Ardis H. White, "The Statics Ratio for Analysis of Frames that Deflect," Univ. of Tex. Bur. of Eng. Res. *Bull. No. 45*, 1950.
6. "Moment Distribution Applied to Continuous Concrete Structures," Portland Cement Association, Chicago.
7. "Concrete Building Frames Analyzed by Moment Distribution," Portland Cement Association, Chicago.
8. "Continuity in Concrete Building Frames," Portland Cement Association, Chicago, 3rd ed.
9. "Handbook of Frame Constants," Portland Cement Association, Chicago.
10. "Frame Analysis Applied to Flat Slab Bridges," Portland Cement Association, Chicago.

APPENDIX

C

Summary Treatment of Service Load Analysis and Design for Flexure

C.1 GENERAL FIELD OF USE

Elastic analysis and the transformed area concept for the purpose of calculating moments of inertia relating to deflections and shrinkage stresses was introduced in Sec. 6.4. In building design and analysis the old working stress design method is gradually being phased out. For slabs and beams the 1971 Code classifies it as an alternate design method with a unity load factor. For columns it is so restricted as to be essentially the strength design of Chapter 14.

In the highway bridge field so much emphasis is necessary on service conditions that the 1969 AASHO specifications[1] (with 1970 and 1971 Interim Specifications) still relate to what formerly was called working stress design.

The presentation here will be restricted to simple shapes, with and without compression steel, using AASHO allowable stresses.

There is very little basic difference between calculating beam stresses resisting a given moment, for comparison with allowable stress values, and calculating an allowable moment based on these allowable stresses.

The *design* of beams for moment, on the other hand, is a process considerably different from analysis. Design is treated in the second part of this Appendix.

ANALYSIS

C.2 COMPRESSION REINFORCEMENT

The elastic analysis of Sec. 6.4 accepted compression reinforcement as a portion of the elastic member with a steel stress $f_s' = nf_c$ and a transformed area nA_s' or an effective addition of $(n - 1)A_s'$ to the transformed area and $(n - 1)f_c$* to the stress, after accounting for displaced concrete.

For strength, the creep of concrete when under load (or the flattening of the stress-strain curve for concrete under even short time high stresses), shortens the compressive concrete and steel more than an elastic analysis would suggest. Analysis or design by the working stress method recognizes this by using $2nA_s'$ as the transformed area and $2nf_c$ as the stress. It would be logical to use $(2n - 1)A_s'$ and $(2n - 1)f_c$ to recognize the displacement of useful concrete, but this will sometimes be simplified to $2n$ instead of $2n - 1$ in recognition that the factor 2 applied to n is not really accurate enough to justify a 5 to 8% correction.

C.3 AASHO ALLOWABLE STRESSES

In flexure AASHO allows a maximum fiber stress in compression $f_c = 0.4f_c'$ with zero in tension (i.e., cracked over tension zone) except in plain concrete (as in some footings). On reinforcing bars it allows in tension 20 ksi on Grade 40 and 24 ksi on Grade 60 steel, and in compression $2n$ times the concrete stress at that level, but not greater than the allowable in tension. The AASHO value of n is 10 for f_c' of 3000 to 3900 psi, 8 for f_c' of 4000 to 4900 psi, 6 for f_c' of 5000 psi or more; however, for deflections n is always 8.

C.4 RECTANGULAR BEAM ANALYSIS

Since Sec. 6.4 used only the moment of inertia approach, this analysis will now be checked by looking directly at the internal stress system. This stress system is not an improvement for analysis but it is important as a better approach for design and for easy visualization. The data already used are $M = 100$ k-ft, $f_c' = 4000$ psi, Grade 40 steel, $c = 9.95$ (Sec. 6.4d)

*Here f_c is at the level of the compression steel, a value less than the maximum stress on the extreme fiber.

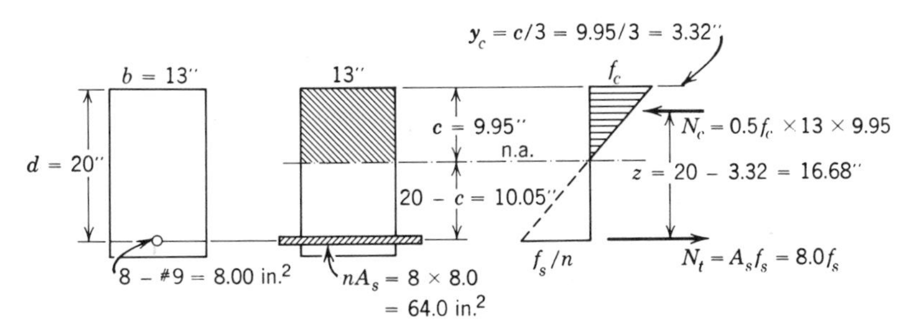

Fig. C.1. Analysis of rectangular beam.

and the other dimensions of Fig. C.1 which is a copy of Fig. 6.4 with stress triangles added. The resultant compression N_c and tension N_t form a couple with arm $z^* = d - c/3 = 16.68$ in., as noted on the figure.

$$N_t = N_c = M/z = 100{,}000 \times 12/16.68 = 71{,}900 \text{ lb}$$
$$N_t = A_s f_s$$
$$f_s = N_t/A_s = 71{,}900/8.0 = 8990 \text{ psi}$$

or
$$f_s = \frac{M}{A_s z} = \frac{100{,}000 \times 12}{8.0 \times 16.78} = 8990 \text{ psi}$$

$$N_c = \frac{f_c}{2} bc$$

$$71{,}900 = 0.5 f_c \times 13 \times 9.95$$
$$f_c = 1111 \text{ psi}$$

Check:
$$\frac{f_c}{f_s/n} = \frac{c}{d - c} \qquad \frac{f_c}{8990/8} = \frac{9.95}{20 - 9.95}$$
$$f_c = 1113 \text{ psi vs. } 1111 \qquad \text{O.K.}$$

Under the AASHO allowable stresses of $f_c = 0.4 f_c' = 0.4 \times 4000 = 1600$ psi and $f_s = 20$ ksi, the allowable moment on this beam may now be found by proportion from the stresses already established. Based on compression alone, f_c may be increased from 1111 psi to 1600 psi:

$$\text{Allowable } M_c = (1600/1111)100 = 144 \text{ k-ft}$$

Based only on tensile stress:

$$\text{Allowable } M_t = (20{,}000/8990)100 = 223 \text{ k-ft}$$

*In the older notation, c was kd, z was jd, and y_c was sometimes designated as z.

The lower "allowable" is the real limit, in this case a limit on compression.

$$\text{Allowable } M = 144 \text{ k-ft}$$

If the calculated stresses had not already been available, the limiting stress could have been established as soon as the neutral axis was found. Consider *either* triangle in Fig. C.2 (considering both would be a waste of time). From the triangles in Fig. C.2*a*:

$$\text{If } f_c = 1600 \text{ psi}, f_s/n = (10.05/9.95)1600 = 1615 \text{ psi}$$
$$n = 8, f_s = 8 \times 1615 = 12{,}930 \text{ psi} < 20{,}000 \qquad \text{O.K.}$$

Or, from the triangles of Fig. C.2*b*:

$$\text{If } f_s/n = 2500 \text{ psi}, f_c = (9.95/10.05)2500 > 1600 \text{ N.G.}$$

Either trial establishes that the moment is limited by f_c to:

$$\text{Allowable } M = N_c z = (0.5 f_c bc)z = (0.5 \times 1600 \times 13 \times 9.95)16.68$$
$$= 1{,}720{,}000 \text{ lb-in.} = 143 \text{ k-ft}$$

This example has covered the operating procedures for analysis. The details but not the principles change in the case of compression steel, T-beams, or irregular sections.

C.5 T-BEAM ANALYSIS

(a) Compared to a Rectangular Beam

The AASHO specification for flange width is the same as in the ACI Code (Sec. 2.11) except for limiting the flange overhang to $6h_f$ instead of $8h_f$, where h_f is the flange thickness.

The flange area is usually more than adequate to care for compressive stresses; hence allowable moments are usually limited by the steel stress. Furthermore, the large compression area in the flange often pulls the centroid of the transformed area up into the flange itself. When this

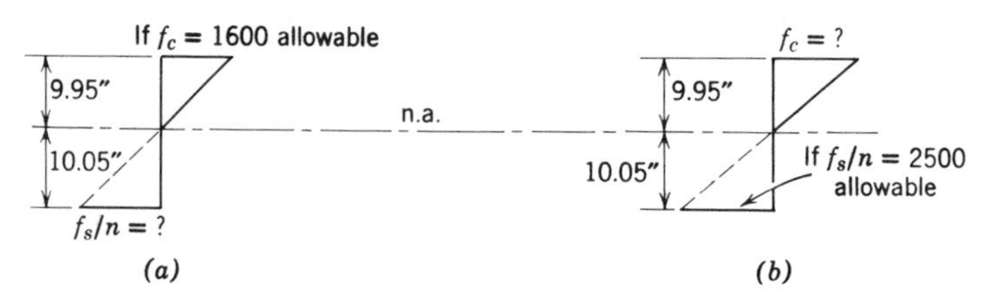

Fig. C.2. Determination of limiting or governing stress.

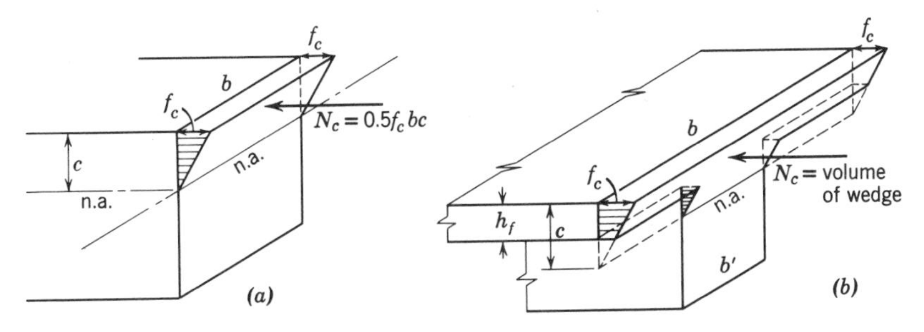

Fig. C.3. Wedge of stress for (*a*) rectangular beam and (*b*) T-beam.

occurs, the analysis for moment is exactly that for a wide rectangular beam of width b, since the missing area below the flange would have been in tension and considered as cracked.

The neutral axis does more often fall below the flange with working stress analysis than in strength design, where it rarely does so. In this case the evaluation of N_c in terms of f_c and the location of N_c are no longer possible by simple inspection. The student will probably find it helpful to think in terms of the wedge of stress. For a rectangular beam, this wedge of stress is a simple triangular wedge (Fig. C.3*a*). The volume of the wedge gives the magnitude of N_c and the centroid of the wedge marks the location of the resultant N_c. When this concept is applied to the T-beam, as in Fig. C.3*b*, neither the volume nor the centroid can be determined except by a summation process, as in the following examples. The first example is calculated by the so-called exact analysis; the second example uses the approximate method which neglects the web area between neutral axis and flange. The approximate method fits much easier into a solution by formulas, charts, or curves, but has little advantage for a basic analysis such as is used here.

(b) Exact Analysis

Find f_c and f_s for the T-beam shown in Fig. C.4*a* under a moment of 200 k-ft, assuming $f_c' = 3000$ psi and Grade 60 steel, AASHO specification.

Solution

The AASHO specification for f_c' between 3000 psi and 3900 psi uses $n = 10$ for strength and $n = 8$ for deflection calculations, making here $nA_s = 10 \times 6.24 = 62.4$ in.^{2}. Flange overhang of 10 in. is less than $6h_f = 24$ in. O.K.

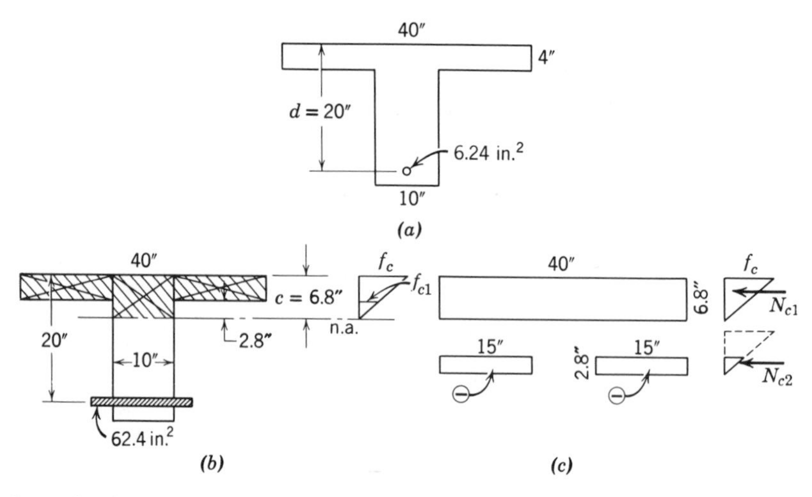

Fig. C.4. Analysis of T-beam, exact method. (The 4 in. slab is too thin for highway use, but here it helps to emphasize the influence of *n.a.* location.)

First make a quick check on where the neutral axis lies with respect to the bottom of the flange, using area moments about the bottom of flange.

$$40 \times 4 \times 2 \stackrel{?}{=} 62.4(20 - 4)$$

$$320 < 998, \; n.a. \text{ is well below flange.}$$

Take area moments about the neutral axis using the area of Fig. C.4*b*.

$$(10c)c/2 + (40 - 10)\,4(c - 2) = 62.4(20 - c)$$

$$c^2 + 36.5c + 18.2^2 = 298 + 18.2^2 = 622$$

$$c = 25.0 - 18.2 = 6.8 \text{ in.}$$

At the bottom of flange the stress triangle gives

$$f_{c1} = (2.8/6.8)f_c = 0.41f_c$$

The total compression can be calculated by subdividing the areas, and if necessary the stresses, into pieces that can be visualized simply (Fig. C.4*b*). In this case, the simplest pieces are N_{c1} on a rectangle 40 in. wide and extending all the way to the neutral axis and a negative N_{c2} deducting the surplus below the flanges. Sketches, as in Fig. C.4*c*, and a tabular form are desirable. To locate the resultant N_c, moments are taken about some convenient axis, by custom about the top of beam.

	N_c	Arm abt. top	Moment m abt. top
$N_{c1} = (40 \times 6.8)f_c/2$	$= 136.3f_c$	$6.8/3 = 2.27$ in.	$308f_c$
$N_{c2} = -(30 \times 2.8)0.41f_c/2$	$= -17.3f_c$	$4 + 2.8/3 = 4.93$ in.	$-86f_c$
	$N_c = 119.0f_c$		$m = 222f_c$

$$y_c = m/N_c = 222f_c/119.0f_c = 1.86 \text{ in.}$$

$$z = 20 - y_c = 20 - 1.86 = 18.14 \text{ in.}$$

$$f_s = \frac{N_t}{A_s} = \frac{M}{A_s z} = \frac{200{,}000 \times 12}{6.24 \times 18.14} = 21{,}200 \text{ psi vs. } 24{,}000 \text{ psi allowable}$$

O.K.

$$N_c = M/z = 200{,}000 \times 12/18.14 = 132{,}000 \text{ lb} = 119.0f_c \text{ (from above)}$$

$$f_c = 1110 \text{ psi vs. } 0.40 \times 3000 = 1200 \text{ psi allowable} \quad \text{O.K.}$$

(c) Approximate Analysis

Same beam as in (*b*) above, Fig. C.5*a*.

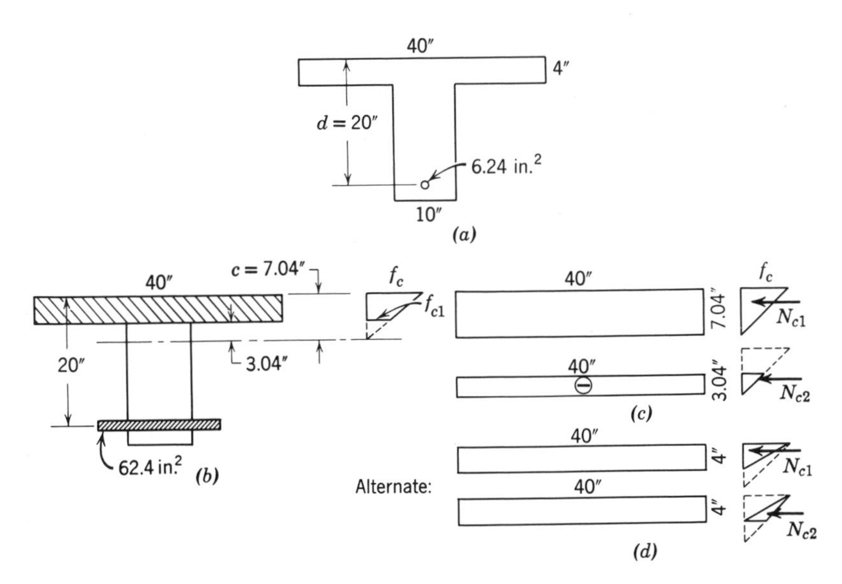

Fig. C.5. Analysis of T-beam, approximate method.

Solution

As before, the neutral axis falls below the flange. The approximate method neglects the small compressive area (and small unit stress) below the flange, Fig. C.5*b*. Then area moments about the neutral axis give:

$$(40 \times 4)(c - 2) = 62.4(20 - c)$$
$$222.4c = 1658$$
$$c = 7.04 \text{ in.}$$

At the bottom of the flange, $f_{c1} = f_c \times 3.04/7.04 = 0.432f_c$.

The areas and stresses shown in Fig. C.5*c* lead to:

		N_c	Arm abt. top	Moment m abt. top
$Nc_1 =$	$(40 \times 7.04)f_c/2$	$= 141.0f_c$	$7.04/3 = 2.35$ in.	$331f_c$
$N_{c2} =$	$-(40 \times 3.04)0.432f_c/2$	$= -26.3f_c$	$4 + 3.04/3 = 5.02$	$-132f_c$
	$N_c =$	$114.7f_c$		$m = 199f_c$

$$y_c = 199f_c/114.7f_c = 1.74 \text{ in.}$$
$$z = 20 - 1.74 = 18.26 \text{ in.}$$

The student should see clearly that there are many different patterns for N_{c1} and N_{c2} which lead to identical values of N_c and m. For example, the pattern of Fig. C.5*d* where the unit stress picture has been subdivided rather than the area. It could have been subdivided into a rectangle and triangle of stress or the trapezoid could have been used undivided, provided its centroid was calculated as the lcoation of N_c. The subdivisions to be used are a matter of convenience, but it should be noted that *in the exact solution* in (*b*) no other subdivision leads as directly to the desired result.

$$f_s = \frac{M}{A_s z} = \frac{200{,}000 \times 12}{6.24 \times 18.26} = 21{,}300 \text{ psi vs. } 24{,}000 \text{ psi} \qquad \text{O.K.}$$
$$N_c = M/z = 200{,}000 \times 12/18.26 = 131{,}000 \text{ lb} = 114.7f_c$$
$$f_c = 1150 \text{ psi vs. } 1200 \text{ psi} \qquad \text{O.K.}$$

(d) Allowable Moment

Just as for the rectangular beam in Sec. C.4, the allowable moment may be found by proportion from stresses already calculated; or, the governing stress can be determined from the stress triangles and the allowable moment then calculated.

Although this particular beam with its narrow flange will be limited by compression, this is a rare case for a real T-beam. Hence an approximate allowable moment can usually be estimated as $A_s f_s$, with the stress at the allowable, times an estimated arm z which will usually be in the neighborhood of $0.9d$.

C.6 ANALYSIS WITH COMPRESSION STEEL

(a) General

The use of $2n$ with A_s' as discussed in Sec. C.2 is direct and reasonable for design and fairly satisfactory (but occasionally awkward) for allowable moments. In the calculation of stresses from a given moment, the concepts become less significant and the stresses become only nominal ones.

(b) Analysis

Assume 2.0 in.^{2} of compression steel is added at 2 in. below the top of the rectangular beam of Sec. C.4a and Fig. C.1a, with $f_c' = 4000$ psi, $n = 8$, Grade 40 steel. Find stresses for $M = 150$ k-ft. Also find the allowable moment.

Solution

The transformed area is shown in Fig. C.6b, using $(2n - 1)A_s'$ for the compression steel.

Area moments about neutral axis:

$$(13c)c/2 + 30(c - 2) = 64(20 - c)$$
$$c^2 + 14.46c + 7.23^2 = 206 + 7.23^2 = 258$$
$$c = 16.06 - 7.23 = 8.83 \text{ in.}$$

Based on Fig. C.6c:

$$f_{cs}' = f_c \times 6.83/8.83 = 0.776 f_c$$

	N_c	Arm abt. top	Moment m abt. top
$N_{c1} = (13 \times 8.83) f_c/2$	$= 57.3 f_c$	$8.83/3 = 2.94$ in.	$168.0 f_c$
$N_{c2} = 30.0 \times 0.776 f_c$	$= 23.2 f_c$	$d' = 2.00$	$46.4 f_c$
	$N_c = 80.5 f_c$		$m = 214.4 f_c$

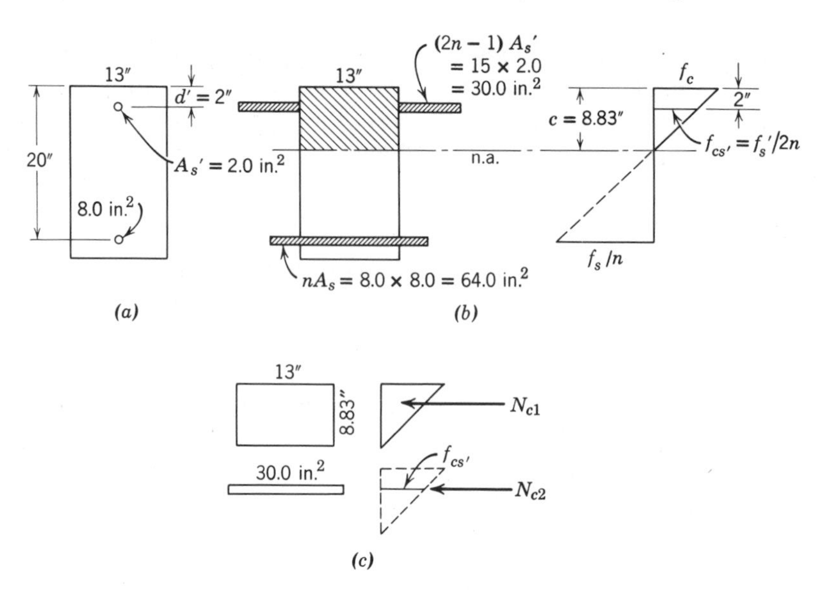

Fig. C.6. Analysis of double-reinforced beam.

$$y_c = 214.4f_c/80.5f_c = 2.65 \text{ in.}$$

$$z = 20 - 2.65 = 17.35 \text{ in.}$$

$$f_s = \frac{M}{A_s z} = \frac{150{,}000 \times 12}{8.0 \times 17.35} = 13{,}000 \text{ psi vs. } 20{,}000 \text{ psi allowable} \qquad \text{O.K.}$$

$$N_c = M/z = 150{,}000 \times 12/17.35 = 103{,}500 \text{ lb} = 80.5f_c \text{ (from above)}$$

$$f_c = 1290 \text{ psi vs. } 0.40 \times 4000 = 1600 \text{ psi} \qquad \text{O.K.}$$

Check:

$$f_c = \frac{13{,}000}{8} \times \frac{8.83}{11.17} = 1294 \text{ psi} \qquad \text{Say O.K.}$$

$$f_s' = 2nf_{cs}' = 2 \times 8 \times 0.776 \times 1290$$

$$= 16{,}000 \text{ psi vs. } 20{,}000 \text{ psi allowable} \qquad \text{O.K.}$$

For allowable moment: If $f_c = 1600$ psi, similar stress triangles give:

$$f_s = nf_c \frac{20 - 8.83}{8.83} = 8 \times 1600 \times 11.17/8.83$$

$$= 16{,}180 \text{ psi vs. } 20{,}000 \text{ psi O.K. Compression controls.}$$

$$f_s' = 2nf_{cs}' = 2 \times 8 \times 0.776 \times 1600$$

$$= 19{,}900 \text{ psi vs. } 20{,}000 \text{ psi O.K.}$$

With $M = A_s f_s z$ and $z = 17.35$ in. as above:

Allowable $M = 8 \times 16{,}180 \times 17.35/12{,}000 = 187$ k-ft

If f_s' were excessive, it would not require that the allowable f_c be lowered, as would be necessary if f_s were excessive. What it would mean is that the compression steel could not be counted on as contributing more than 20,000 psi. This would upset the initial assumption that f_s' is $2nf_{cs}'$ and the effective transformed area $(2n - 1)A_s'$. The effective compression area would be something else, a little difficult to assess closely because a change in the area would shift the neutral axis. Hence an "exact" theoretical solution would require a new beginning.*

C.7 IRREGULAR-SHAPED BEAMS—ANALYSIS

Beams of irregular shape are often used for special purposes, such as curved-top or triangular-top shapes for railings, curbs over exterior bridge girders, beams with continuous side brackets, or beams with recesses for supporting a future slab or masonry. Where the shape is entirely bounded by vertical and horizontal surfaces, the location of the resultant N_c as for T-beams and double-reinforced beams, is a practical procedure. Where sloping faces or curved surfaces occur, the use of the moment of inertia and $Mc = fI$ is generally simpler.

It should be noted that many irregular-shaped beams are not symmetrical and are not loaded along their principal axes. Fortunately, most of these are restrained against lateral deflection by a monolithic slab. In such cases, lack of symmetry can be ignored and the bending axis will be essentially horizontal.

DESIGN

C.8 DESIGN VERSUS ANALYSIS

In analysis, whether for actual stresses or for allowable moments, the engineer deals with given beams, known both as to dimensions and steel. He has no control over the location of the neutral axis, which lies at the centroid of the transformed area.

In design, loads and allowable stresses are known and some or all the dimensions remain to be fixed. In this case the designer has some control over the location of the neutral axis. He can shift it where he wants it,

*A theoretical solution is scarcely worthwhile, but it is not difficult. One must substitute the fundamental relationship $N_c = N_t$ for the derived idea that the neutral axis is at the centroid.

to the extent that his change in dimensions can shift the centroid of the transformed area.

The student should understand clearly this fundamental difference between design and analysis problems.

C.9 RECTANGULAR BEAMS—MOMENT DESIGN

(a) The Balanced Beam in Design

A beam which reaches its allowable f_c and its allowable f_s under the same working moment is called a balanced beam.* Such a beam will often constitute the most economical construction; it usually represents the smallest desirable beam.

Although a balanced beam is frequently the design objective, the balanced depth will only occasionally be a practical dimension. Usually, over-all beam depths are specified in full or half inches. If the balanced depth (in decimals of an inch) is adjusted to a more practical dimension, the beam will no longer be exactly balanced for the given moment. Hence practical design is concerned with the balanced beam only as an approach to practical beam dimensions. The design of the reinforcing steel should be based on the actual dimensions used.

Slightly deeper beams are in the direction of economy. Since less steel is required with the larger z, such a beam is termed an underreinforced beam. On the other hand, slightly shallower beams are also practical, but these require a considerable increase in steel area. Such overreinforced beams are somewhat more expensive and become impractical when the depth is reduced more than 5 or 10%. For larger reductions in depth compression steel is indicated.

(b) Balanced Beam Design

The design procedure for a beam exactly balanced *at working load* will be illustrated by the design of a rectangular beam for $M = 88.5$ k-ft, $f_c' = 3000$ psi, Grade 60 steel.

*The student should note that a beam thus balanced for allowable working stresses will not be balanced at failure. The balance at failure calls for a much larger A_s.

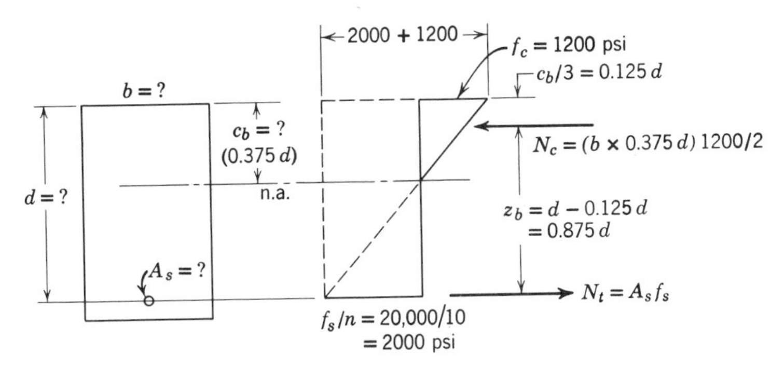

Fig. C.7. Design of a balanced rectangular beam.

Solution

Since the steel is still to be selected, the neutral axis can be made to fall at any convenient location. That location corresponding to balanced stresses* is desired. In this example, the AASHO specification establishes: allowable $f_c = 0.40 \times 3000 = 1200$ psi, $n = 10$, allowable $f_s = 20{,}000$ psi.

The beam, with dimensions still as symbols, with subscript b emphasizing balanced, can be sketched as in Fig. C.7 along with the unit stress triangles which must accompany a straight-line distribution of stress. When simultaneous values of allowable f_c and f_s are assigned to the maximum stress values, the similar triangles locate the proportionate depth c_b to the desired neutral axis. It is convenient to consider the large dashed triangle of height d and horizontal dimension 2000 + 1200 as one of the similar triangles.

$$\frac{c_b}{d} = \frac{1200}{2000 + 1200} = 0.375, \ c_b = 0.375d$$

$$z_b = d - c_b/3 = d - 0.375d/3 = 0.875d$$

The resultant compression N_c can now be sketched in its proper location and evaluated in terms of b amd c_b.

$$N_c = (b \times 0.375d)1200/2$$

$$M = N_c z_b = (b \times 0.375d \times 1200/2)0.875d = 197bd^2 = k_b bd^2$$

where k_b represents the constant $197 = M/bd^2$ for these balanced stresses.

*Notice that the balanced stress condition at working load is unrelated to the strength concept of balanced beam at failure.

This constant, $k_b = 197$, is a function of the unit stresses and not of the loading. Hence every balanced rectangular beam designed for these materials (or these unit stresses) will be concerned with this constant. One calculates it once on a given job and then uses if for many designs. It will usually be found tabulated in all handbooks.*

The size of the beam can be found by equating the maximum bending moment to this resisting moment, noting that units must be consistent, usually inches and pounds.

$M = 88{,}500 \times 12 = k_b bd^2 = 197\ bd^2$

Reqd. $bd^2 = 5380$ in.^{3}

This requirement can be met by a wide range of sizes, for example:

If $b = 7.5$ in., $d = \sqrt{718} = 26.8$ in. $b = 11$ in., $d = \sqrt{490} = 22.1$ in.
$b = 8,\quad d = \sqrt{674} = 26.0\qquad b = 11.5,\quad d = \sqrt{468} = 21.6$
$b = 9,\quad d = \sqrt{600} = 24.5\qquad b = 12,\quad d = \sqrt{448} = 21.2$
$b = 9.5,\quad d = \sqrt{567} = 23.8\qquad b = 13,\quad d = \sqrt{415} = 20.4$
$b = 10,\quad d = \sqrt{538} = 23.2\qquad b = 14,\quad d = \sqrt{385} = 19.6$

USE $b = 9.5$ in., $d = 23.8$ in.

The depth of 23.8 in. is probably not entirely practical; it is not a dimension one would turn over to a carpenter or steel worker. But either more or less depth than 23.8 in., with this width, would *not* give a balanced beam. This theoretical balanced depth gives:

$$\text{Balanced } A_s = \frac{N_t}{f_s} = \frac{M}{f_s z} = \frac{88{,}500 \times 12}{20{,}000(0.875 \times 23.8)} = 2.55 \text{ in.}^2$$

The design should be changed to a practical depth before choosing bars.

(c) Beams Deeper Than a Balanced Section—Underreinforced

When dimensions must be modified to get practical sizes, the usual procedure is to add an allowance for depth from the center to the bottom of steel and for clear cover over the steel and then to use the next larger practical over-all dimension. Beams are also made deeper because such a size may be required for shear or to match a level established by heavier loaded beams.

Redesign the beam of (*b*) for $b = 9.5$ in., $d = 24.5$ in., in this case an arbitrary choice to show the method.

*Under older notations it would probably show as *R* or *K*.

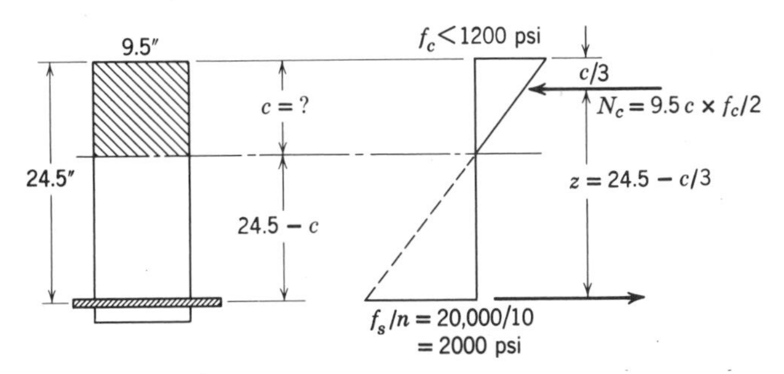

Fig. C.8. Stress triangles for beam deeper than a balanced beam.

Solution

The steel area is the only element left to design. Since the beam is larger than the balanced section, f_c tends to be less than allowable and f_s would also be less than allowable if the 2.55 in.^{2} steel area calculated above for the balanced beam were used. For economy, reduce A_s until f_s is the full allowable. The unit stress diagram in Fig. C.8 is thus shown with f_s = 20,000 psi and with f_c unknown, but known (assumed) to be less than allowable. From similar triangles, the unknown f_c can be expressed in terms of the known f_s and unknown c. Then N_c and M can be calculated in terms of c.

$$f_c = 2000 \frac{c}{24.5 - c}$$

$$N_c = 9.5c \times \frac{2000}{2} \times \frac{c}{24.5 - c}$$

$$M = 88{,}500 \times 12 = N_c z = 9.5c \times \frac{2000}{2} \times \frac{c}{24.5 - c}\left(24.5 - \frac{c}{3}\right)$$

$$1{,}062{,}000 = 9500 \frac{c^2}{24.5 - c}\left(24.5 - \frac{c}{3}\right)$$

$$\frac{c^2}{24.5 - c}\left(24.5 - \frac{c}{3}\right) = 112.0$$

This is a cubic equation which can be solved by trial from the present form about as easily as if it were simplified. The value of c will be slightly less than the balanced beam value of $0.375 \times 24.5 = 9.20$ in.

$$\text{Try } c = 9.0 \text{ in.,} \quad \frac{9.0^2}{15.5}(24.5 - 3.00) = 112.5 > 112.0$$

$$c = 8.95 \text{ in.,} \quad \frac{8.95^2}{15.4}(24.5 - 2.98) = 110.8 < 112.0$$

USE $c = 8.99$ in.

$$z = 24.5 - 8.99/3 = 21.5 \text{ in.}$$

$$\text{Reqd. } A_s = \frac{M}{f_s z} = \frac{88{,}500 \times 12}{20{,}000 \times 21.5} = 2.46 \text{in.}^2 \leftarrow$$

Check by area moments about the neutral axis:

$$9.5 \times 8.99 \times 8.99/2 = 10A_s(24.5 - 8.99)$$

$$A_s = 2.48 \text{ in.}^2 \qquad \text{Say O.K.}$$

Verify f_c by similar triangles:

$$f_c = \frac{20{,}000}{10} \times \frac{8.99}{15.51} = 1160 \text{ psi} < 1200 \qquad \text{O.K.}$$

Since the cubic equation is awkward, an empirical approximation is usually substituted in the calculation for A_s, that is, the use of the balanced z for the given increased d.

$$\text{Reqd. } A_s = \frac{88{,}500 \times 12}{20{,}000(0.875 \times 24.5)} = 2.48 \text{ in.}^2$$

Since less steel is used than would be required for a balanced beam, the neutral axis is higher and the real z value is always greater than the balanced z would be for the given increased d. Thus this approximation for A_s is always on the safe side. For small increases in d it is not wasteful and is recommended. When d is 20 to 50% greater than the balanced d, some steel can be saved by using a more exact z.

(d) Beams Shallower Than a Balanced Section—Overreinforced

It is rarely feasible to reduce d below the balanced depth by more than possibly 10%, unless compression steel is used. A reduction in depth greatly increases the required steel and makes this type of construction uneconomical except for very small deficiencies in d.

Redesign the beam of Sec. C.9b for $b = 9.5$ in., $d = 23$ in.

Solution

In this case, the reduced depth tends to cause increased f_c and f_s. A reasonable addition to A_s would bring f_s within the allowable, but would leave f_c too large. The only way to remedy the overstress in compression

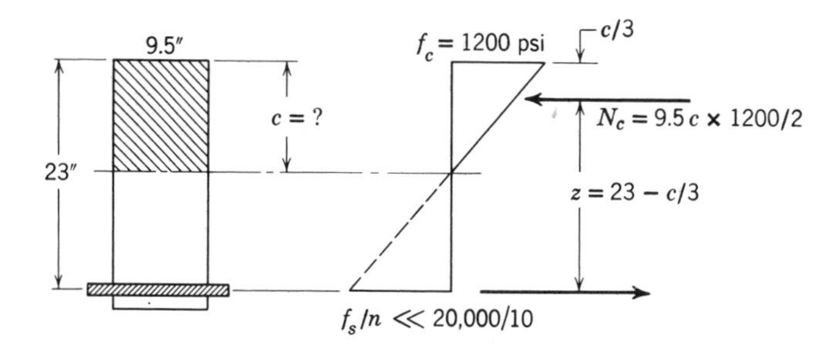

Fig. C.9. Stress triangles for beam shallower than a balanced beam.

(other than with compression steel) is to increase the compression area by lowering the neutral axis. Extra A_s will lower the neutral axis in this fashion, but a large amount is required; this operation tends to be inefficient since the steel has to work at a lowered unit stress. The unit stress diagram in Fig. C.9 is thus shown with f_c at the allowable and f_s below the allowable.

$$N_c = 9.5c \times 1200/2$$

$$M = 88{,}500 \times 12 = N_c z = 9.5c \times \frac{1200}{2}\left(23 - \frac{c}{3}\right)$$

$$c^2 - 68.9c + 34.45^2 = -560 + 34.45^2 = 615$$

$$c = \pm\sqrt{615} + 34.45 = \pm 24.9 + 34.45 = 9.55 \text{ in.}$$

$$z = 23 - 9.55/3 = 19.82 \text{ in.}$$

By similar triangles,

$$f_s/n = 1200 \times \frac{23 - 9.55}{9.55} = 1690 \text{ psi}$$

Limiting $f_s = 10 \times 1690 = 16{,}900$ psi (to protect against overstress in f_c)

$$\text{Reqd. } A_s = \frac{M}{f_s z} = \frac{88{,}500 \times 12}{16{,}900 \times 19.82} = 3.17 \text{ in.}^2$$

Check by area moments about the neutral axis:

$$9.5 \times 9.55 \times 9.55/2 = 10A_s(23 - 9.55)$$

Reqd. $A_s = 3.18$ in.^{2} Say O.K.

Note that a 3.4% reduction in d has required a 24% increase in A_s compared to the balanced condition in Sec. C.9*b*. There is no accepted approximate solution for this case.

(e) Reinforcing Steel for Beam of Given Size

Design procedures for rectangular beams can be summarized as indicated in Table C.1. In this table c_b and z_b represent the balanced condition for the beam on which they appear, for instance, for the stresses of Sec. C.9b, always $c_b = 0.375d$ and $z_b = 0.875d$.

The first step after the beam size is known or has been established is to identify the beam as underreinforced, balanced, or overreinforced. As noted under "Identification," the actual moment may be compared to the balanced moment $M = k_b bd^2$. An alternate equivalent calculation is to compare the actual k = actual M/bd^2 with the balanced value of k_b. The comparison can also be made between the actual d and the balanced $d = \sqrt{\text{actual } M/k_b b}$.*

If the actual d is less than 90 to 95% of the balanced d, it is more economical to design as a double-reinforced beam, as in the following section.

C.10 BEAMS WITH COMPRESSION STEEL—MOMENT DESIGN

(a) General

Although concrete is usually cheaper for carrying compression than is steel, economy is often served by using compression steel over a short critical length. Compression steel increases the toughness of a beam, reducing the possibility of a sudden type of failure. It also reduces the creep and the creep-and-shrinkage deflection of a beam.

Design procedures are simple and are best shown by examples. In all cases the total design moment is subdivided into two parts, M_1 which is the allowable moment on the balanced beam without compression steel ($M_1 = k_b bd^2$), and M_2 which is the surplus moment to be resisted with compression steel and extra tension steel (Fig. C.10). The use of A_s' with $2n$ leads to a semielastic analysis where $f_s' \leq$ allowable f_s; if the strains should then indicate $f_s' >$ allowable f_s, complete validity requires that a nonelastic analysis be substituted.

(b) Semielastic Case

Redesign the steel for the beam of Sec. C.9d, using cover concrete $d' =$ 2.5 in.

*In analysis problems (not design) the simplest identification is by the steel ratio ρ, below ρ_b being underreinforced, over ρ_b overreinforced.

TABLE C.1 Rectangular Beams

Identification:		
Actual $M < (\text{bal. } k)bd^2 = k_b bd^2$	Actual $M = k_b bd^2 = \text{bal. } M$	Actual $M > (\text{bal. } k)bd^2 = k_b bd^2$
Actual $k = M/bd^2 < k_b$	Actual $k = M/bd^2 = k_b$	Actual $k = M/bd^2 > k_b$
Actual $d > \sqrt{M/k_b b}$	$d = \sqrt{M/k_b b} = \text{bal. } d$	Actual $d < \sqrt{M/k_b b}$
Designation:		
Underreinforced	Balanced	Overreinforced (If $d \ll \text{bal. } d$, use A_s' with balanced stresses.)
Design:		
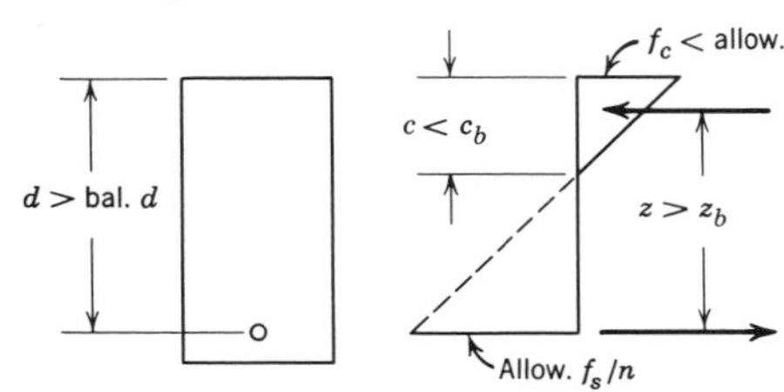	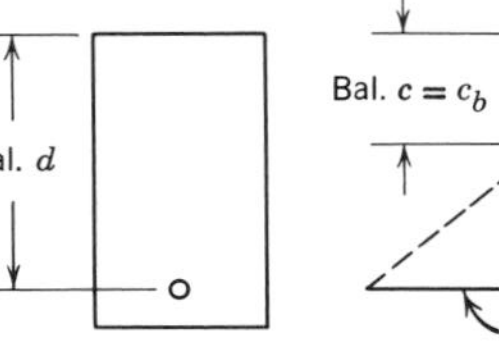	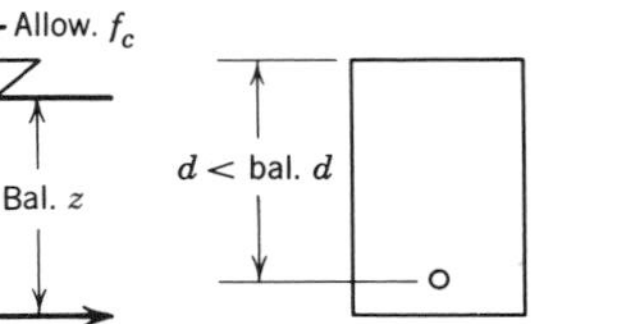
$f_c = \dfrac{f_s}{n} \dfrac{c}{d - c}$	c from similar triangles	c from $M = N_c z$(quadratic eq. in c)
c from $M = N_c z$(cubic eq. in c)	$M = k_b bd^2$	Reduced $f_s = nf_c \dfrac{d - c}{c} \ll \text{allow. } f_s$
$A_s = \dfrac{M}{(\text{allow.})f_s \times z(\text{from above})}$	$A_s = \dfrac{M}{f_s z_b}$	$A_s = \dfrac{M}{(\text{reduced})f_s \times z(\text{requires exact } z)}$
Approx. $A_s = \dfrac{M}{(\text{allow.})f_s \times z_b \text{ for this depth}}$	*Alternate* for any of the three types: Find A_s by area moments about n.a. after c is established: $nA_s(d - c) = bc^2/2$	

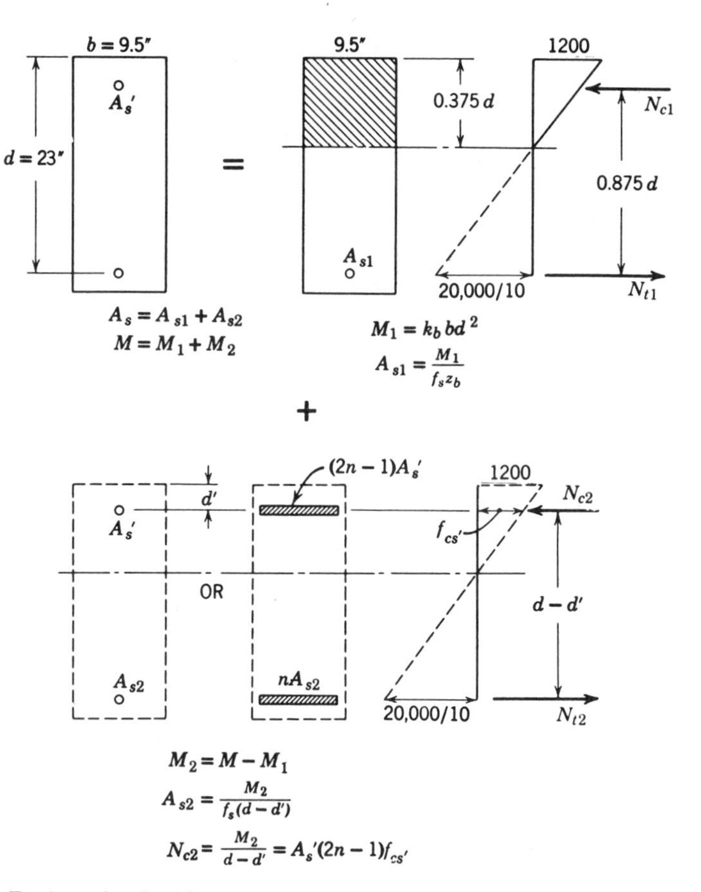

Fig. C.10. Design of a double-reinforced beam.

Solution

Compression steel is to be valued at twice the stress an elastic analysis would indicate, provided this does not exceed the allowable tension of 20,000 psi in this case. The requirement for ties around the compression steel must not be overlooked.

The procedure sketched in Fig. C.10 will be followed. As a balanced beam the moment capacity is $M_1 = k_b bd^2$. In Sec. C.9b, for $f_c = 1200$ psi, $f_s = 20{,}000$ psi, $n = 10$, k_b has been established as 197, $c_b = 0.375d$, $z_b = 0.875d$.

$$M_1 = k_b bd^2 = 197 \times 9.5 \times 23^2/12{,}000 = 82.5 \text{ k-ft}$$

$$A_{s1} = \frac{M_1}{f_s z_b} = \frac{82{,}500 \times 12}{20{,}000 \times 0.875 \times 23} = 2.46 \text{ in.}^2$$

$$M_2 = M - M_1 = 88.5 - 82.5 = 6.0 \text{ k-ft}$$

$$A_{s2} = \frac{M_2}{f_s(d - d')} = \frac{6000 \times 12}{20{,}000(23 - 2.5)} = 0.18 \text{ in.}^2$$

$$\text{Total } A_s = A_{s1} + A_{s2} = 2.46 + 0.18 = 2.64 \text{ in.}^2$$

$$N_{c2} = M_2/(d - d') = 6000 \times 12/20.5 = 3520 \text{ lb} = A_s'(2n - 1)f_{cs}$$

$$c = 0.375d = 0.375 \times 23 = 8.64 \text{ in.}$$

$$f_{cs}' = \frac{c - 2.5}{c} f_c = \frac{6.14}{8.64} \times 1200 = 854 \text{ psi}$$

$$f_s' = 2nf_{cs}' = 2 \times 10 \times 854 = 17{,}100 \text{ psi} < 20{,}000 \text{ psi} \qquad \text{O.K.}$$

$$\text{Effective } f_s'' = (2n - 1)f_{cs}' = (20 - 1)854 = 16{,}200 \text{ psi}$$

$$A_s' = N_{c2}/16{,}200 = 3520/16{,}200 = 0.22 \text{ in.}^2$$

This theory is identical with the semielastic analysis of Sec. C.6b. Check A_s' by area moments about the neutral axis with the effective transformed areas nA_{s2} and $(2n - 1)A_s'$:

$$10 \times 0.18(23 - 8.64) = (2 \times 10 - 1)A_s'(8.64 - 2.5)$$

$$A_s' = 0.22 \text{ in.}^2$$

It might be noted that the total steel used in tension and compression is 2.86 in.^{2} compared to 3.17 in.^{2} when tension steel alone was used (Sec. C.9d). A double-reinforced beam is usually cheaper than the use of tension steel alone when the beam must be smaller than a balanced beam.

(c) Nonelastic Value of f_s'

Redesign the steel for the beam of (*b*) above, using $d' = 1.25$ in. (This is a rather impractical assumption taken to illustrate a method.)

Solution

The values of M_1, M_2, A_{s1}, and c can be taken unchanged from (*b*) above:

$$A_{s2} = \frac{M_2}{f_s(d - d')} = \frac{6000 \times 12}{20{,}000(23 - 1.25)} = 0.17 \text{ in.}^2$$

$$\text{Total } A_s = A_{s1} + A_{s2} = 2.46 + 0.17 = 2.63 \text{ in.}^2$$

$$N_c = 6000 \times 12/(23 - 1.25) = 3320 \text{ lb} = A_s'(2n - 1)f_{cs}'$$

$$f_{cs}' = \frac{8.64 - 1.25}{8.64} \times 1200 = 1028 \text{ psi}$$

$$f_s' = 2nf_{cs}' = 2 \times 10 \times 1028 = 20{,}560 \text{ psi} > 20{,}000$$

$$\text{USE } f_s' = 20{,}000 \text{ psi}$$
$$\text{Effective } f_s'' = 20{,}000 - f_{cs}' = 20{,}000 - 1028 = 18{,}970 \text{ psi}$$
$$A_s' = N_{c2}/18{,}970 = 3320/18{,}970 = 0.18 \text{ in.}^2$$

In this case the 20,000-psi limit on f_s' has no relation to the elastic properties of the materials. Since total N_t = total N_c, there is no reason to shift the original neutral axis, but *it no longer falls at the centroid of the usual transformed areas.* Hence the check on A_s' based on area moments about the neutral axis as used in Sec. C.10b is invalid here.

C.11 T-BEAMS—MOMENT DESIGN

The choice of b_w (web width) and d is excluded since these are usually governed by factors other than flexure, although b_w is often modified to give more width for bars.

The flange ordinarily provides enough compression area to keep f_c low, and the check on this can be postponed until one of the last steps. The design for A_s is thus assumed to be for a beam much deeper than a balanced section. For a T-beam the algebraic approach used in Sec. C.9b for a similar rectangular beam becomes too involved for practical use. Instead, a cut-and try procecedure is recommended, based on the fact that the value of z can be estimated quite closely.

Find the required area of Grade 60 steel (f_y = 60,000 ksi) for the T-beam of Fig. C.11 if f_c' = 4000 psi and M = 200 k-ft, AASHO stresses.

Solution

Both the shape of the compression area and the high neutral axis resulting from a low f_c tend to locate the resultant N_c high in the beam. On this basis, an average value of z could be taken as $0.9d$. If the small amount of com-

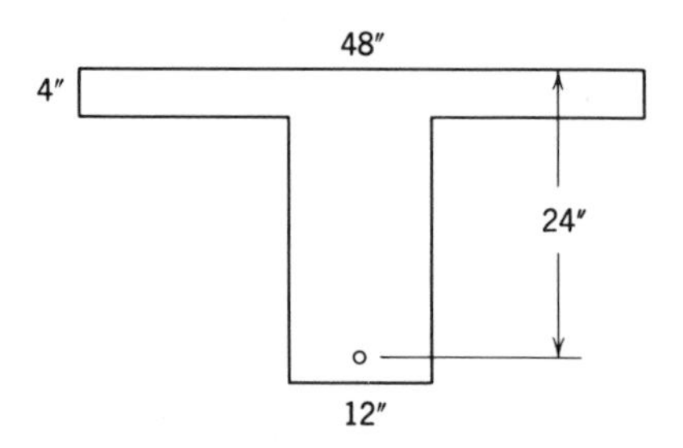

Fig. C.11. T-beam for calculation of A_s.

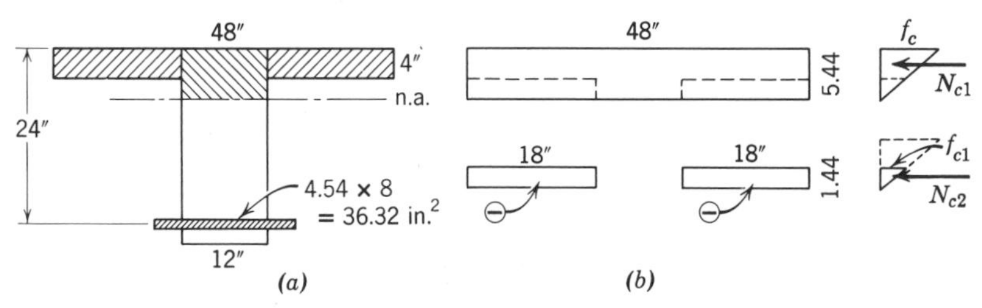

Fig. C.12. Areas used in T-beam calculations. (*a*) For location of neutral axis. (*b*) For location of N_c.

pression in the web below the flange is neglected, the trapezoidal stress distribution locates the resultant N_c higher than the middle of the flange. This means that z is greater than $d - h_f/2$, where h_f is the flange thickness. For the initial approximation, it is recommended that z be taken as the larger of $0.9d$ or $d - h_f/2$, this giving a value probably within 1 to 3% of the exact one.

$0.9d = 0.9 \times 24 = 21.6$ in.

$d - h_f/2 = 24 - 2 = 22.0$ in. USE for trial z

Allowable $f_s = 24{,}000$ psi for steel with $f_y = 60{,}000$ psi

$$\text{Approx. } A_s = \frac{M}{{}_s z} = \frac{200{,}000 \times 12}{24{,}000 \times 22.0} = 4.54 \text{ in.}^2$$

This approximate A_s will be used to establish c and a better (almost exact) value of z. The "exact" method for T-beams, Sec. C.5b, will be used (an arbitrary choice), as indicated in Fig. C.12*a*.

Area moments about the neutral axis:

$(12c)c/2 + 36 \times 4(c - 2) = 4.54 \times 8(24 - c)$

$6c^2 + 144c - 288 = 871 - 36.3c$

$c^2 + 30.0c + 15.0^2 = 193.2 + 15.0^2 = 418$

$c = 20.44 - 15.0 = 5.44$ in.

f_{c1} at bottom of flange $= (1.44/5.44)f_c = 0.265f_c$

Forces shown in Fig. C.12*b*:

		Arm abt. top	m
$N_{c1} = (48 \times 5.44)f_c/2$	$= 130.0f_c$	$5.44/3 = 1.81$ in.	$235f_c$
$N_{c2} = -(36 \times 1.44)0.265f_c/2$	$= -6.9f_c$	$4 + 1.44/3 = 4.48$ in.	$-30.8f_c$
N_c	$= 123.1f_c$	m	$= 204f_c$

$$y_c = \frac{204f_c}{123.1f_c} = 1.65 \text{ in.}$$

$$z = 24 - 1.65 = 22.35 \text{ in.}$$

$$A_s = \frac{200{,}000 \times 12}{24{,}000 \times 22.35} = 4.47 \text{ in.}^2$$

The calculation of y_c above used $N_c = 123.1f_c$:

$$f_c = \frac{N_c}{123.1} = \frac{M}{123.1z} = \frac{200{,}000 \times 12}{123.1 \times 22.35} = 868 \text{ psi} \ll 0.40 \times 4000 \qquad \text{O.K.}$$

An estimate of f_c based on approximate theory requires less preliminary calculation and is often adequate. A trapezoidal stress on the flange gives:

$$f_c < 2 \times \text{average } f_c, \quad \text{since average } f_c > f_c/2$$

$$f_c < 2M/(bh_f z)$$

$$f_c < 2\,\frac{200{,}000 \times 12}{48 \times 4 \times 22.35} = 1065 \text{ psi}$$

$$f_c < 1065 \text{ psi} < 0.40 \times 4000 \qquad \text{O.K.}$$

The weakness of this check lies in the fact that twice the average f_c might exceed the allowable and still be satisfactoy. Such a condition calls for the more exact calculation procedure.

C.12 USE OF DESIGN AIDS

Only a few simple charts or curves are presented here:

Fig. C.13. Values of c/d and z/d for rectangular beams.
Fig. C.14. Coefficients of resistance for rectangular beams.
Fig. C.15. Values of c/d and z/d for T-beams.

(a) Rectangular beams

The allowable moment on the beam of Sec. C.4 and Fig. C.1 will be checked using the curves of Fig. C.14. Allowable $f_c = 1600$ psi, $n = 8$, allowable $f_s = 20{,}000$ psi.

$$\rho = A_s/bd = 8/(13 \times 20) = 0.0307$$

(The neutral axis distance c and internal arm z could now be read directly from Fig. C.13, but these are not needed to find the allowable moment.) Enter the upper chart ($n = 8$) of Fig. C.14 and note that it fails to extend

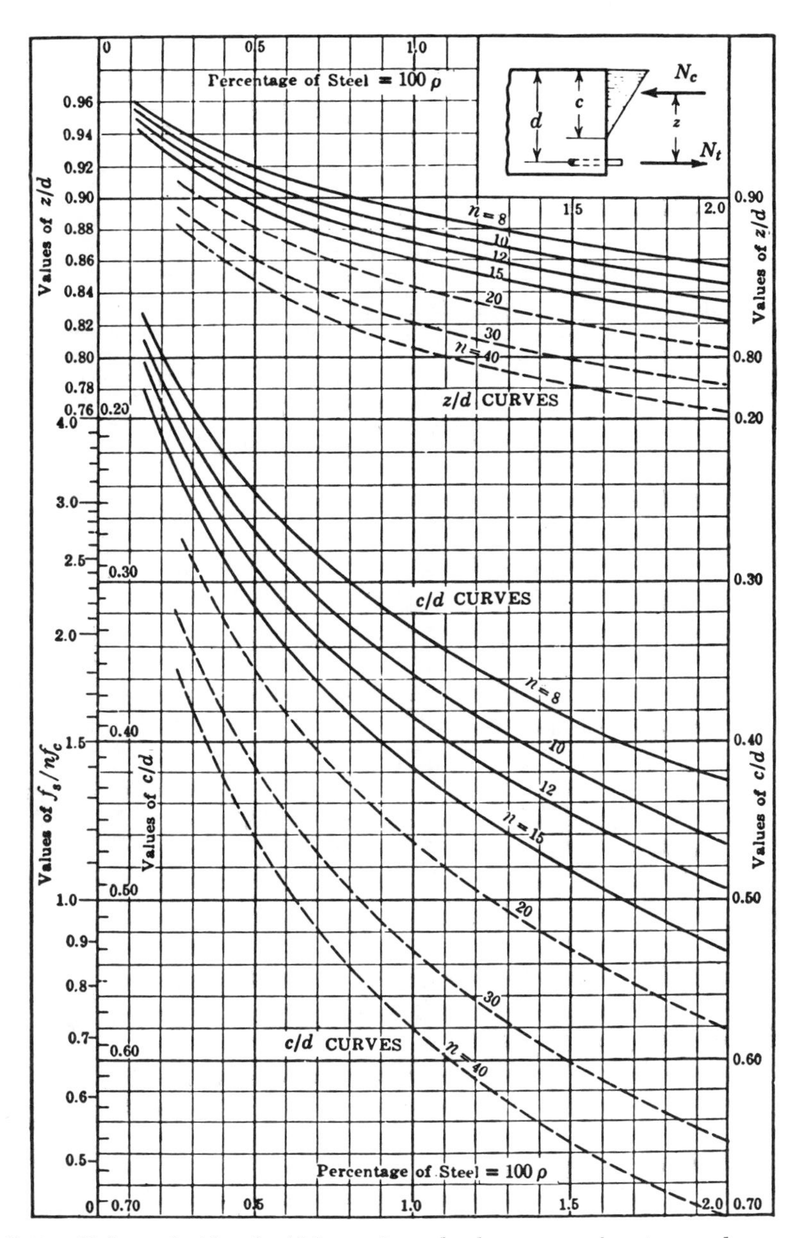

Fig. C.13. Values of c/d and z/d for rectangular beams, service stress values.

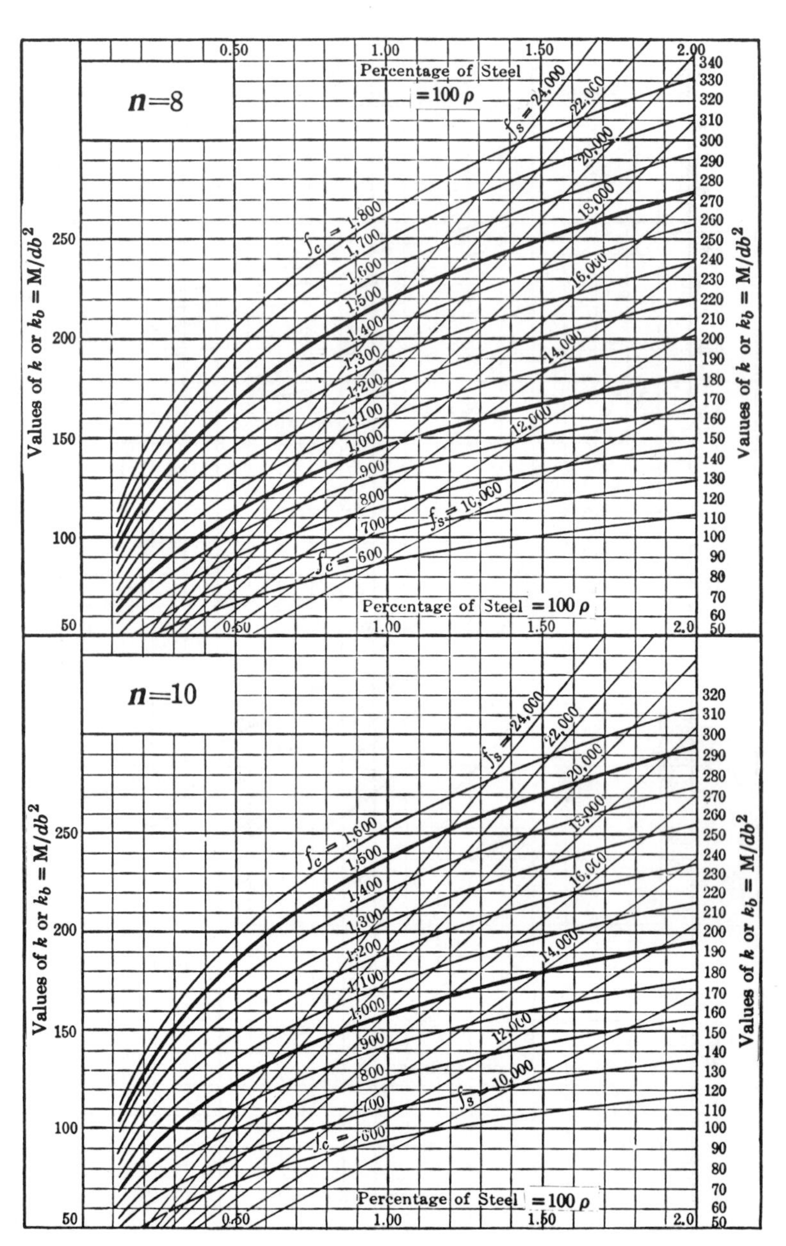

Fig. C.14. Coefficients of resistance of rectangular beams, $k = M/bd^2$, for service stress design.

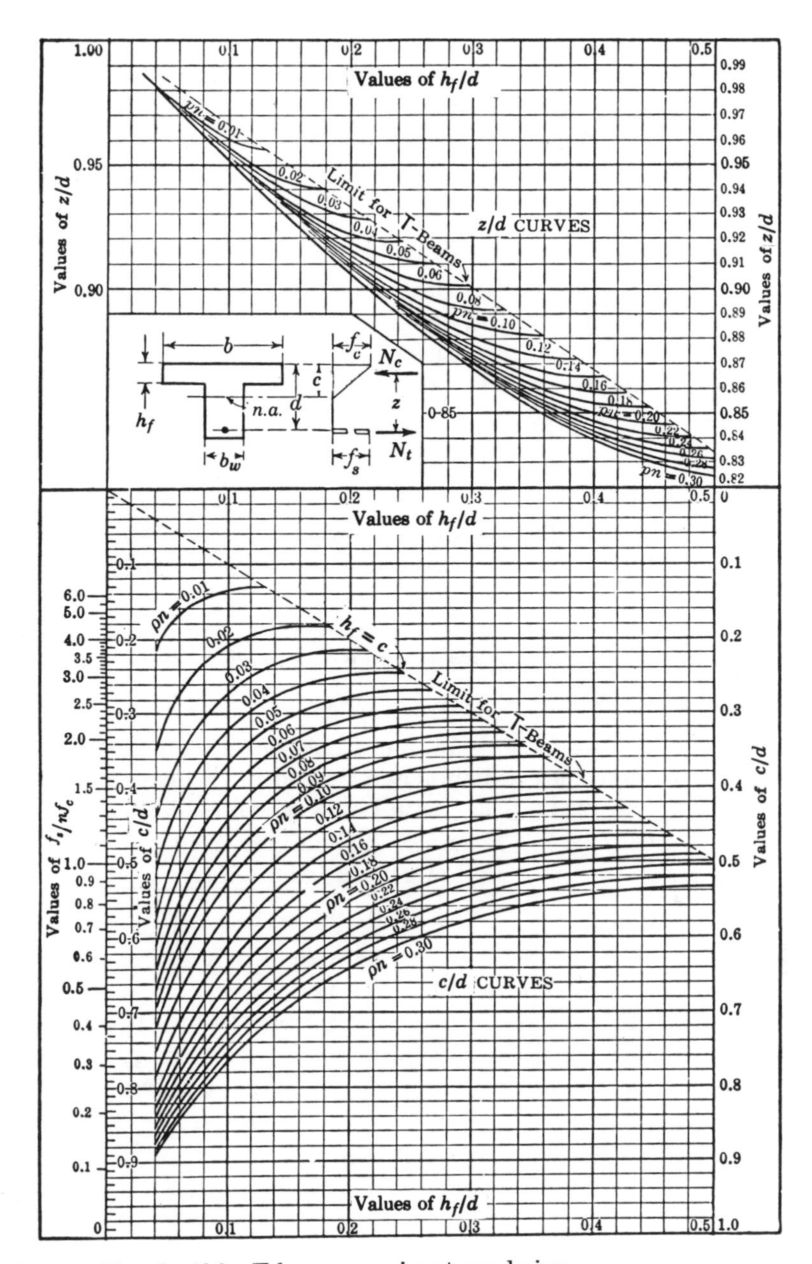

Fig. C.15. c/d and z/d for T-beams, service stress design.

to the 100ρ value of 3.07, which is an unusually high value. However, continue approximately. Going up vertically at $100\rho = 2.0$, one first crosses $f_c = 1600$ psi, with $f_s = 20{,}000$ psi far higher and the two curves diverging. If one roughly sketches an extension of the $f_c = 1600$ psi curve (by guess) to $100\rho = 3.07$, he would find a k ordinate reading possibly 320 (a very rough approach to the true value).

$$\text{Allowable } M = kbd^2 = 320 \times 13 \times 20^2/12{,}000 = 132 \text{ k-ft}$$

This is obviously a crude guess, compared to the better value of 144 k-ft, but for the reasonable plotted values of ρ sketching errors would be eliminated and the process would be entirely adequate.

For a design example, consider the rectangular beam of Sec. C.9b:

$$M = 88.5 \text{ k-ft}, f_c = 1200 \text{ psi}, f_s = 20{,}000 \text{ psi}, n = 10.$$

Start in Fig. C.14 on the lower chart for $n = 10$ and find the intersection of curves of $f_c = 1200$ psi and $f_s = 20{,}000$ psi at a level of $k = 197$, which is the balanced k_b; also the horizontal coordinate shows $100\rho_b = 1.13$.

Reqd. $bd^2 = M/k = 88.5 \times 12{,}000/197 = 5380$ in.^{3} as in the analytical solution with $d_b = 23.8$ in. for $b = 9.5$ in.

Bal. $A_s = \rho_b bd_b = 0.0113 \times 9.5 \times 23.8 = 2.57$ in.^{2} (vs. 2.55 in.^{2})
For $b = 9.5$ in., $d = 24.5$ in. (deeper than balanced section):

$$\text{Actual } k = M/bd^2 = 88.5 \times 12{,}000/(9.5 \times 24.5^2) = 187$$

At $k = 187$ in Fig. C.14 for $n = 10$, move horizontally until both f_c and f_s are within their allowables, which is determined in this case by the intersection with the $f_s = 20{,}000$ psi curve, below which read $100\rho = 1.06$.

Reqd. $A_s = 0.0106 \times 9.5 \times 24.5 = 2.47$ in.^{2} (vs. 2.48 in.^{2})
For $b = 9.5$ in., $d = 23$ in. (shallower than balanced section):

$$\text{Actual } k = 88.5 \times 12{,}000/(9.5 \times 23^2) = 212$$

At $k = 212$ in Fig. C.14 ($n = 10$) move horizontally across $f_s = 20{,}000$ psi to intersect $f_c = 1600$ psi at $100\rho = 1.42$.

Reqd. $A_s = 0.0142 \times 9.5 \times 23 = 3.10$ in.^{2} (vs. 3.18 in.^{2}) Note that the f_c curve is flattening when the member is shallower than a balanced section, making an exact reading of 100ρ very sensitive and less accurate.

(b) T-Beams

For the analysis example of Sec. C.5b, $M = 200$ k-ft, $n = 10$, $A_s = 6.24$ in.^{2}, $b = 40$ in., $b_w = 10$ in. $h_f = 4$ in., $d = 20$ in., allowable $f_c = 1200$ psi, and allowable $f_s = 24{,}000$ psi. Stresses are to be calculated.

$$n\rho = 10 \times 6.24/(40 \times 20) = 0.078$$
$$h_f/d = 4/20 = 0.20$$

The intersection of the $n\rho = 0.078$ curve (interpolation) with $h_f/d = 0.20$ in the lower curves is within the limit for T-beams, which indicates a neutral axis below the flange. At the left one reads $f_s/nf_c = 1.88$. In the upper curves the intersection is at $z/d = 0.914$.

$$f_s = M/(A_s z) = 200 \times 12{,}000/(6.24 \times 0.914 \times 20) = 21{,}100 \text{ psi} \quad \text{(vs. 21,300 psi)}$$

With $f_s/nf_c = 1.88$, $21{,}100/(10 \times 1.88) = 1120$ psi (vs. 1110 psi)

The design example of Sec. C.11 for allowable $f_c = 1600$ psi, $n = 8$, and $M = 200$ k-ft on the beam of Fig. C.11 would start with an estimate of $A_s = 4.54$ in.^{2}, possibly just as before.

$$\text{Approx. } \rho n = [4.54/(48 \times 24)]\,8 = 0.0316$$
$$h_f/d = 4/24 = 0.167$$

With these two values, Fig. C.15 gives $z = 0.933$d from the upper curves and $f_s/nf_c \doteq 3.3$ from the lower curves.

$$A_s = 200 \times 12{,}000/(24{,}000 \times 0.933 \times 24) = 4.47 \text{ in.}^2 \text{ (as earlier)}$$
$$f_s/nf_c = 24{,}000/8f_c = 3.3$$
$$f_c = 3000/3.3 = 910 \text{ psi} < 0.4 \times 4000 \quad \text{O.K.} \quad \text{(vs. 868 psi)}$$

SELECTED REFERENCE

1. *Standard Specifications for Highway Bridges*, AASHO, Washington, 10 ed., 1969.

PROBLEMS

Note: Sec. C.3 states the AASHO allowable stresses. When stresses are calculated for a given moment, compare these with the allowables.

Designs in this series of problems are to be partial designs, for flexure alone; bond, shear, and bar spacing are not a part of these problems.

When the actual moment exceeds the balanced beam moment, the student should decide whether excess tension steel or compression steel is preferable.

Prob. C.1.

(*a*) For the rectangular beam of Fig. C.16, $f_c' = 4000$ psi, Grade 40 steel, calculate f_c and f_s from the internal couple when $M = 130$ k-ft.

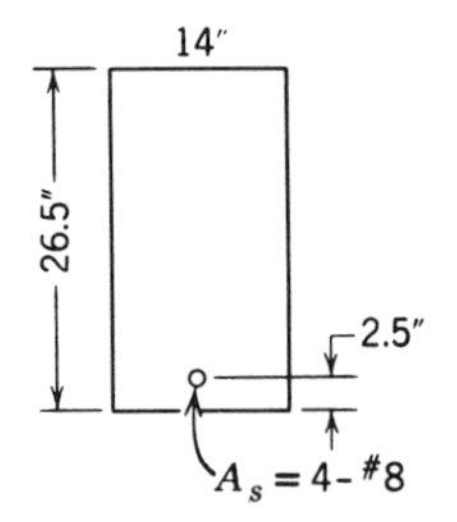

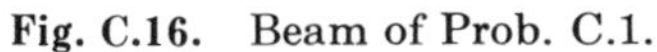

Fig. C.16. Beam of Prob. C.1.

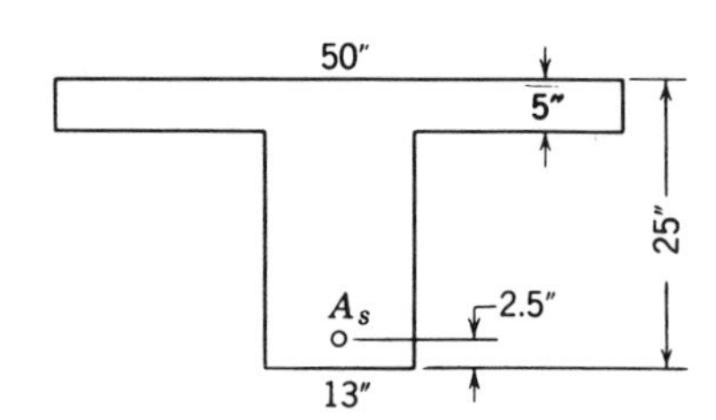

Fig. C.17. T-beam for Prob. C.4.

(*b*) Calculate f_c and f_s from $Mc = fI$. (See Sec. 6.4d)

(*c*) What is allowable (working) moment on this beam under AASHO Specification.

Prob. C.2. Same as Prob. C.1 except A_s is 4- #10 bars of Grade 60 steel.

Prob. C.3. Same as Prob. C.1 except A_s is 4- #10 bars and A_s' of 2- #7 bars has been added, centered 2.5 in. below the top, and steel is Grade 60 steel.

Prob. C.4.

(*a*) For the T-beam of Fig. C.17, $f_c' = 4000$ psi, A_s of 3- #10 bars, intermediate grade steel, calculate f_c and f_s from the internal couple when $M = 100$ k-ft. Use the approximate method (Sec. C.5c).

(*b*) Calculate f_c and f_s from $Mc = fI$.

(*c*) What is the allowable moment?

Prob. C.5. Repeat Prob. C.4 except change A_s to 4- #10 bars and use the exact method of analysis.

Prob. C.6. Find the allowable moment on a 1-ft strip of the slab of Fig. 2.9 if the bars are made #7 at 7 in. on centers and $d = 6$ in., $f_c' = 4000$ psi, Grade 60 steel.

Prob. C.7. Find the allowable moment about the horizontal axis of the regular hexagonal pile of Fig. C.18 in the position shown, if $f_c' = 4000$ psi and steel is Grade 60. (*Suggestion*: Use $Mc = fI$).

Prob. C.8. If the hexagonal pile of Fig. C.18 and Prob. C.7 is rotated 90°, find the allowable moment.

Prob. C.9. Neglecting the lack of symmetry (because the flange is part of a continuing slab), find f_c and f_s in the beam of Fig. C.19 if $f_c' = 3000$ psi and $A_s = 2$- #9 bars of Grade 60 steel. $M = 70$ k-ft.

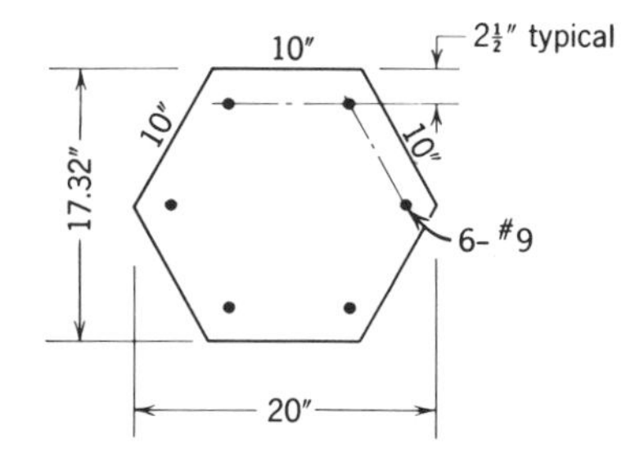

Fig. C.18. Hexagonal pile for Probs. C.7 and C.8.

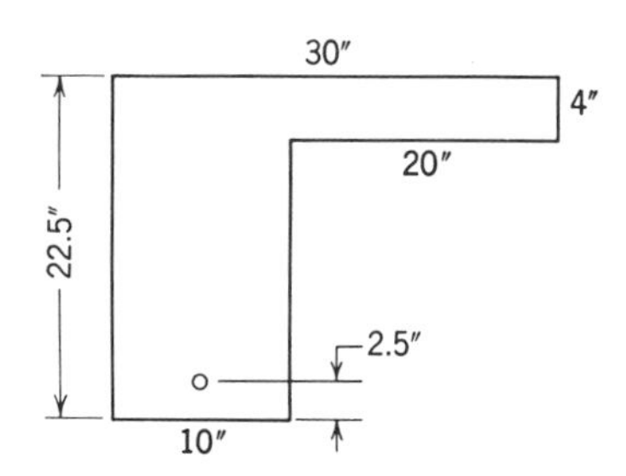

Fig. C.19. Spandrel T-beam for Prob. C.9.

Prob. C.10.

(*a*) Design a rectangular beam exactly balanced for a total $M = 195$ k-ft using $f_c' = 4000$ psi, Grade 60 steel. After the required bd^2 has been established, use $b = 15$ in.

(*b*) Redesign steel if d is made 26 in. and b is maintained at 15 in.

(*c*) Redesign steel if d is made 23 in. and b is maintained at 15 in.

Prob. C.11. A rectangular beam 12 in. wide by 20 in. deep to center of steel must carry a total moment of 100 k-ft with $f_c' = 3000$ psi and Grade 40 steel. Find the required steel.

Prob. C.12. A slab 7.5 in. thick with 1.5-in. cover to center of steel must carry a total moment of 6.0 k-ft per foot width. Find the required A_s per foot if $f_c' = 3000$ psi and steel is Grade 60.

Prob. C.13. A rectangular beam 14 in. wide by 25 in. deep over-all with 3-in. cover to center of steel must carry a total moment of 220 k-ft. Find the necessary steel if $f_c' = 4000$ psi and steel is Grade 60.

Prob. C.14. From basic principles establish the value of the balanced steel ratio ρ_b for a rectangular beam with $f_c' = 4000$ psi and $f_s = 20{,}000$ psi.

APPENDIX D
Design Tables and Curves

TABLE D.1 Flexural Resistance Coefficients for Rectangular Sections without A_s' (Modified from ACI *Ultimate Strength Design Handbook*)

	f_c' = 3000 psi			4000 psi			5000 psi			Any $f_c' \leqslant 5000$*		
	f_y	40 ksi	60	—	40	60	—	40	60			
$\rho f_y/f_c' = \omega$	k_m	ρ	ρ	k_m	ρ	ρ	k_m	ρ	ρ	c/d*	a/d	z/d
0.020	53	0.0015	0.0010	71	0.0020	0.0013	89	0.0025	0.0017	0.028	0.024	0.988
.030	80	.0023	.0015	106	.0030	.0020	133	.0038	.0025†	.042	.035	.982
.040	105	.0030	.0020	141	.0040	.0027†	176	.0050	.0033	.056	.047	.976
.050	131	.0038	.0025	175	.0050	.0033	218	.0063	.0042	.069	.059	.971
.060	156	.0045	.0030	208†	.0060	.0040	260	.0075	.0050	.083	.071	.965
.070	181	.0053	.0035	242	.0070	.0047	302	.0088	.0058	.097	.083	.959
.080	206	.0060	.0040	274	.0080	.0053	343	.0100	.0067	.111	.094	.953
.090	230	.0068	.0045	307	.0090	.0060	384	.0113	.0075	.125	.106	.947
.100	254	.0075	.0050	339	.0100	.0067	423	.0125	.0083	.139	.118	.941
.110	278	.0083	.0055	370	.0110	.0073	463	.0138	.0092	.153	.130	.935
.120	301	.0090	.0060	401	.0120	.0080	502	.0150	.0100	.167	.142	.929
.130	324	.0098	.0065	432	.0130	.0087	540	.0163	.0108	.180	.153	.923
.140	347	.0105	.0070	462	.0140	.0093	578	.0175	.0117	.194	.165	.917
.150	369	.0113	.0075	492	.0150	.0100	615	.0188	.0125	.208	.177	.912
.160	391	.0120	.0080	522	.0160	.0107	652	.0200	.0133	.222	.189	.906

.170	413	.0128	.0085	551	.0170	.0113	688	.0213	.0142	.236	.201	.900
.180	434	.0135	.0090	579	.0180	.0120	724	.0225	.0150	.250	.212	.894
.190	455	.0143	.0095	607	.0190	.0127	759	.0238	.0158	.264	.224	.888
.200	476	.0150	.0100	635	.0200	.0133	794	.0250	.0167	.278	.236	.882
.210	497	.0158	.0105	662	.0210	.0140	828	.0263	.0175	.292	.248	.876
.220	517	.0165	.0110	689	.0220	.0147	862	.0275	.0183	.305	.260	.870
.230	537	.0173	.0115	716	.0230	.0153	895	.0288	.0192	.319	.271	.864
.240	556	.0180	.0120	742	.0240	.0160	927	.0300	.0200	.333	.283	.858
.250	575	.0188	.0125	767	.0250	.0167	959	.0313	.0208	.347	.295	.853
.260	594	.0195	.0130	792	.0260	.0173	991	.0325	.0217	.361	.307	.847
.270	613	.0203	.0135	817	.0270	.0180	1022	.0338	.0225	.375	.319	.841
.280	631	.0210	.0140	841	.0280	.0187	1052	.0350	.0233	.389	.330	.835
.290	649	.0218	.0145	865	.0290	.0193	1082	.0363	.0242	.403	.342	.829
.300	667	.0225	.0150	889	.0300	.0200	1111	.0375	.0250	.416	.354	.823
.310	684	.0233	.0155	912	.0310	.0207	1140	.0388		.430	.366	.817
.320	701	.0240	.0160	935	.0320	.0213	1168	.0400		.444	.378	.811
.330	718	.0248		957	.0330		1196	.0413	$>0.75\rho_b$	.458	.389	.805
.340	734	.0255		978	.0340		1223	.0425		.472	.401	.799
.350	750	.0263	$>0.75\rho_b$	1000	.0350	$>0.75\rho_b$	1250			.486	.413	.794
.360	766	.0270		1021	.0360		1276	$>0.75\rho_b$		.500	.425	.788
.370	781	.0278		1041	.0370		1302			.514	.437	.782

*For $f_c' = 5000$ psi, the given c/d must be multiplied by 0.85/0.80.
†Adove these heavy lines $\rho < 200/f_y$, for f_y in psi.

TABLE D.2. Ultimate Moment Coefficients for Rectangular Sections

$$\text{Coef.} = \frac{M_u}{f_c'bd^2} = \omega\,(1 - 0.059\,\omega),\; k = f_c'(\text{coef.})$$

$$\omega = \frac{\rho f_y}{f_c'} \text{ or } \rho = \frac{\omega f_c'}{f_y}$$

ω	.000	.001	.002	.003	.004	.005	.006	.007	.008	.009
0.0	0	0.0010	0.0020	0.0030	0.0040	0.0050	0.0060	0.0070	0.0080	0.0090
.01	0.0099	.0109	.0119	.0129	.0139	.0149	.0159	.0168	.0178	.0188
.02	.0197	.0207	.0217	.0226	.0236	.0246	.0256	.0266	.0275	.0285
.03	.0295	.0304	.0314	.0324	.0333	.0343	.0352	.0362	.0372	.0381
.04	.0391	.0400	.0410	.0420	.0429	.0438	.0448	.0457	.0467	.0476
.05	.0485	.0495	.0504	.0513	.0523	.0532	.0541	.0551	.0560	.0659
.06	.0579	.0588	.0597	.0607	.0616	.0625	.0634	.0643	.0653	.0662
.07	.0671	.0680	.0689	.0699	.0708	.0771	.0726	.0735	.0744	.0753
.08	.0762	.0771	.0780	.0789	.0798	.0807	.0816	.0825	.0834	.0843
.09	.0852	.0861	.0870	.0879	.0888	.0897	.0906	.0915	.0923	.0932
.10	.0941	.0950	.0959	.0967	.0976	.0985	.0994	.1002	.1011	.1020
.11	.1029	.1037	.1046	.1055	.1063	.1072	.1081	.1089	.1098	.1106
.12	.1115	.1124	.1133	.1141	.1149	.1158	.1166	.1175	.1183	.1192
.13	.1200	.1209	.1217	.1226	.1234	.1243	.1251	.1259	.1268	.1276
.14	.1284	.1293	.1301	.1309	.1318	.1326	.1334	.1342	.1351	.1359
.15	.1367	.1375	.1384	.1392	.1400	.1408	.1416	.1425	.1433	.1441
.16	.1449	.1457	.1465	.1473	.1481	.1489	.1497	.1506	.1514	.1522
.17	.1529	.1537	.1545	.1553	.1561	.1569	.1577	.1585	.1593	.1601
.18	.1609	.1617	.1624	.1632	.1640	.1648	.1656	.1664	.1671	.1679
.19	.1687	.1695	.1703	.1710	.1718	.1726	.1733	.1741	.1749	.1756
.20	.1764	.1772	.1779	.1787	.1794	.1802	.1810	.1817	.1825	.1832
.21	.1840	.1847	.1855	.1862	.1870	.1877	.1885	.1892	.1900	.1907
.22	.1914	.1922	.1929	.1937	.1944	.1951	.1959	.1966	.1973	.1981
.23	.1988	.1995	.2002	.2010	.2017	.2024	.2031	.2039	.2046	.2053
.24	.2060	.2067	.2075	.2082	.2089	.2096	.2103	.2110	.2117	.2124
.25	.2131	.2138	.2145	.2152	.2159	.2166	.2173	.2180	.2187	.2194
.26	.2201	.2208	.2215	.2222	.2229	.2236	.2243	.2249	.2256	.2263
.27	.2270	.2277	.2284	.2290	.2297	.2304	.2311	.2317	.2324	.2331
.28	.2337	.2344	.2351	.2357	.2364	.2371	.2377	.2384	.2391	.2397
.29	.2404	.2410	.2417	.2432	.2430	.2437	.2443	.2450	.2456	.2463
.30	.2469	.2475	.2482	.2488	.2495	.2501	.2508	.2514	.2520	.2527
.31	.2533	.2539	.2546	.2552	.2558	.2565	.2571	.2577	.2583	.2590
.32	.2596	.2602	.2608	.2614	.1621	.2627	.2633	.2639	.2645	.2651
.33	.2657	.2664	.2670	.2676	.2682	.2688	.2694	.2700	.2706	.2712
.34	.2718	.2724	.2730	.2736	.2742	.2748	.2754	.2760	.2766	.2771
.35	.2777	.2783	.2789	.2795	.2801	.2807	.2812	.2818	.2824	.2830
.36	.2835	.2841	.2847	.2853	.2858	.2864	.2870	.2875	.2881	.2887
.37	.2892	.2898	.2904	.2909	.2915	.2920	.2926	.2931	.2937	.2943
.38	.2948	.2954	.2959	.2965	.2970	.2975	.2981	.2986	.2992	.2997
.39	.3003	.3008	.3013	.3019	.3024	.3029	.3035	.3040	.3045	.3051
.40	.3056									

TABLE D.3. Tension Bar Development Lengths, Inches

Tension bar l_d (in.)-Ordinary concrete-f_c' = 3000 psi*

Bar	f_y = 60 ksi, $s < 6$ in.		f_y = 60 ksi, $s \geqslant 6$ in.		f_y = 40 ksi, $s < 6$ in.		f_y = 40 ksi, $s \geqslant 6$ in.		Hooks, f_y = 60 ksi, f_h in ksi	
	Other	Top	Other	Top	Other	Top	Other	Top	Other	Top
# 3	12.0	12.6	12.0	12.0	12.0	12.0	12.0	12.0	29.5	29.5
# 4	12.0	16.8	12.0	13.4	12.0	12.0	12.0	12.0	29.5	29.5
# 5	15.0	21.0	12.0	16.8	12.0	14.0	12.0	12.0	29.5	29.5
# 6	19.3	27.0	15.4	21.5	12.9	18.0	12.0	14.3	29.5	24.7
# 7	26.3	36.9	21.0	29.5	17.6	24.6	14.0	19.7	29.5	19.7
# 8	34.6	48.4	27.7	38.8	23.1	32.2	18.5	25.9	29.5	19.7
# 9	43.8	61.4	35.0	49.0	29.3	41.0	23.3	32.7	29.5	19.7
#10	55.6	77.9	44.5	62.4	37.1	51.8	29.7	41.6	26.3	19.7
#11	68.3	95.5	54.6	76.5	45.6	63.7	36.4	51.0	23.0	19.7
#14	93.0	130.0	75.0	104.0	62.0	87.0	50.0	70.0	18.2	18.2
#18	121.0	169.0	97.0	135.0	81.0	113.0	65.0	90.0	12.2	12.2

*For other f_c' multiply by $\sqrt{3000/f_c'}$ with min. l_d still 12 in.

Tension bar l_d (in.)-Lightweight concrete-f_c' = 3000 psi* – f_{ct} *not* specified

Bar	f_y = 60 ksi, $s < 6$ in.		f_y = 60 ksi, $s > 6$ in.		f_y = 40 ksi, $s < 6$ in.		f_y = 40 ksi, $s > 6$ in.		For sand-lightweight multiply table values by factor 1.18/1.33 but min. l_d still 12 in.
	Other	Top	Other	Top	Other	Top	Other	Top	
# 3	12.0	16.8	12.0	13.5	12.0	12.0	12.0	12.0	
# 4	16.0	22.4	12.8	17.9	12.0	14.9	12.0	12.0	
# 5	20.0	28.0	16.0	22.4	13.3	18.7	12.0	14.9	
# 6	25.7	36.0	20.5	28.8	17.2	24.0	13.7	19.2	
# 7	35.0	49.0	28.0	39.2	23.3	32.7	18.7	26.1	
# 8	46.0	64.4	37.0	51.5	30.7	43.0	24.7	34.3	
# 9	58.3	81.5	46.7	65.3	38.9	54.4	31.2	43.5	
#10	74.0	103.5	59.2	83.0	49.3	69.0	39.5	55.4	
#11	90.8	127.0	72.8	102.0	60.5	84.8	48.5	68.0	
#14	124.0	173.0	99.0	138.0	82.0	115.0	66.0	92.0	
#18	161.0	225.0	129.0	180.0	108.0	150.0	86.0	120.0	

*For other f_c' multiply by $\sqrt{3000/f_c'}$ with min. l_d still 12 in.

TABLE D.4. Compression Bar Development Lengths, Inches

	f_y = 60 ksi		f_y = 40 ksi		
Bar	l_d, f_c' = 3000	Min. l_d	l_d, f_c' = 3000	Min. l_d	
# 3	8.2	8.0	8.0	8.0	For $f_c' \neq 3000$ psi multiply l_d values by $\sqrt{3000/f_c'}$, but l_d must not be below minimum tabulated.
# 4	11.0	9.0	8.0	8.0	
# 5	13.7	11.3	9.2	8.0	
# 6	16.4	13.5	10.9	8.0	
# 7	19.2	15.8	12.8	8.0	
# 8	21.9	18.0	14.6	10.5	
# 9	24.8	20.3	16.5	13.6	
#10	27.8	22.9	18.5	15.3	
#11	31.0	25.4	20.6	16.9	
#14	37.0	30.5	24.6	20.4	
#18	49.5	40.7	33.0	27.1	

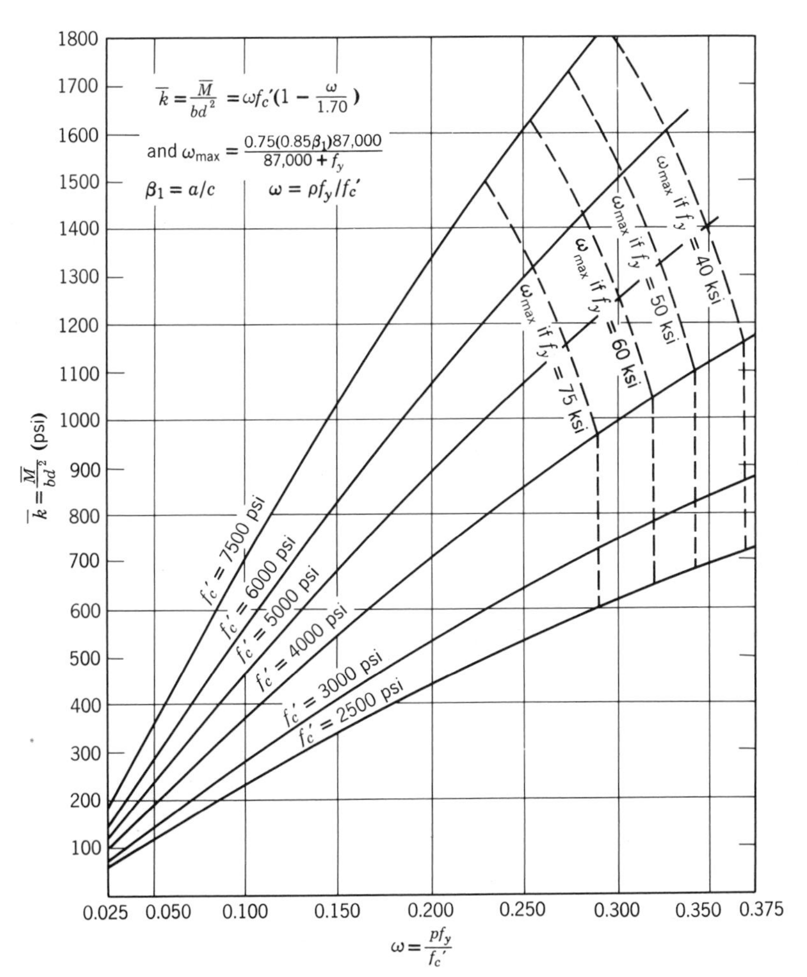

Fig. D.1. Values of $\bar{k}$ for various reinforcement ratios. (Chart by Dr. R. W. Furlong.)

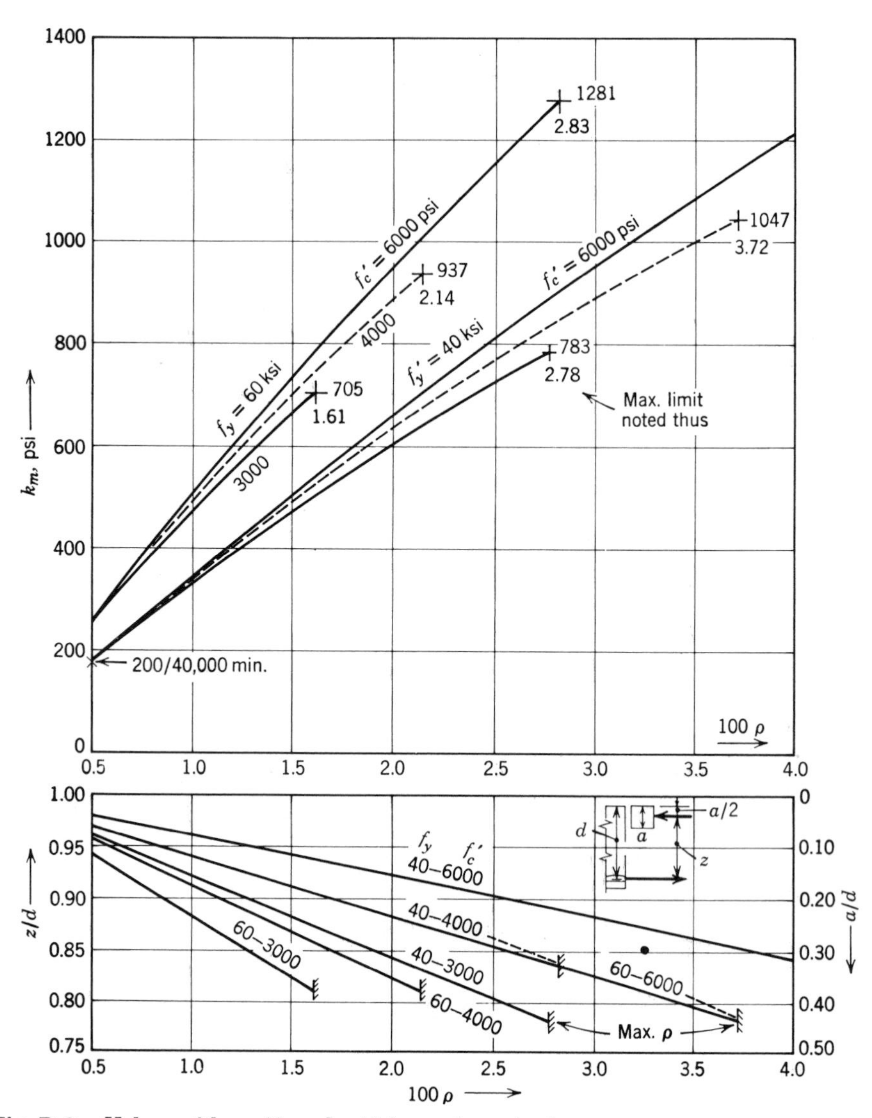

Fig. D.2. Values of k_m, z/d, and a/d for rectangular beams.

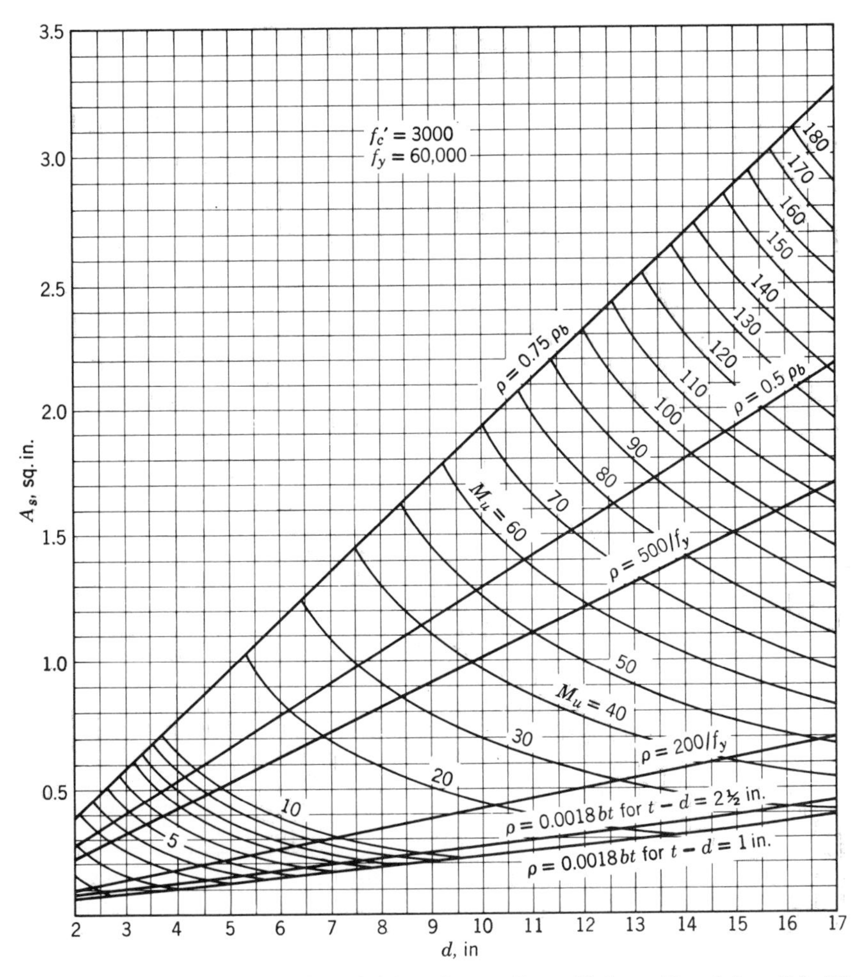

Fig. D.3. Resisting moments M_u, ft-kips, for sections 12 in. wide, slabs. (Modified from ACI *Ultimate Strength Design Handbook*.)

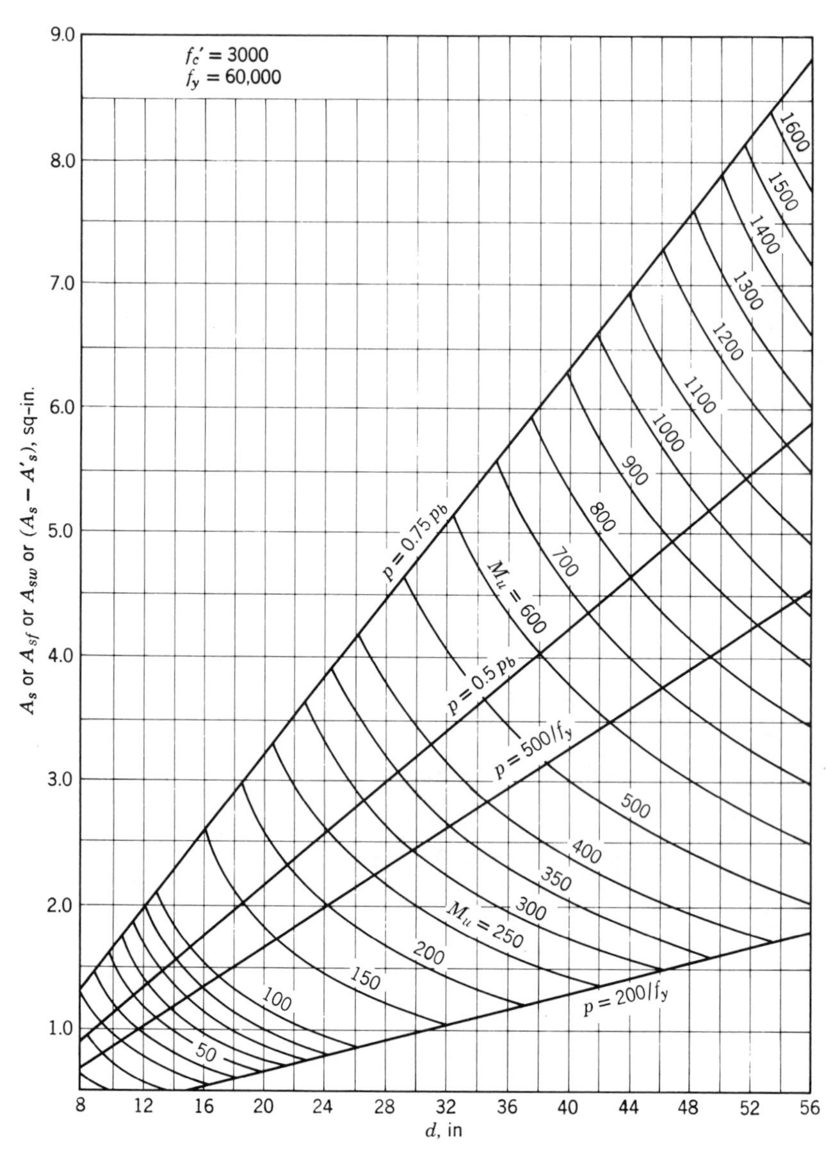

Fig. D.4. Resisting moments M_u, ft-kips, for sections 10 in. wide, beams. (Modified from ACI *Ultimate Strength Design Handbook.*)

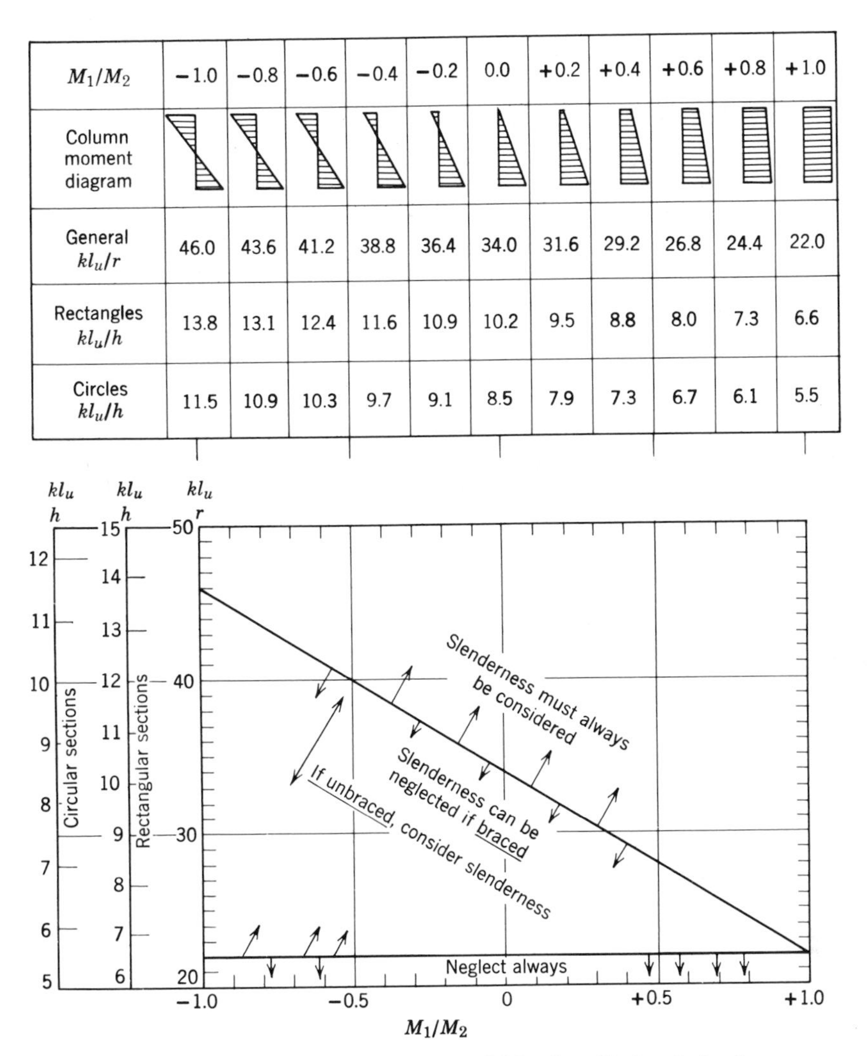

M_1/M_2	−1.0	−0.8	−0.6	−0.4	−0.2	0.0	+0.2	+0.4	+0.6	+0.8	+1.0
Column moment diagram											
General kl_u/r	46.0	43.6	41.2	38.8	36.4	34.0	31.6	29.2	26.8	24.4	22.0
Rectangles kl_u/h	13.8	13.1	12.4	11.6	10.9	10.2	9.5	8.8	8.0	7.3	6.6
Circles kl_u/h	11.5	10.9	10.3	9.7	9.1	8.5	7.9	7.3	6.7	6.1	5.5

Fig. D.5. Column slenderness ratios below which the effects of slenderness may be neglected. (Basic chart by Dr. R. W. Furlong.)

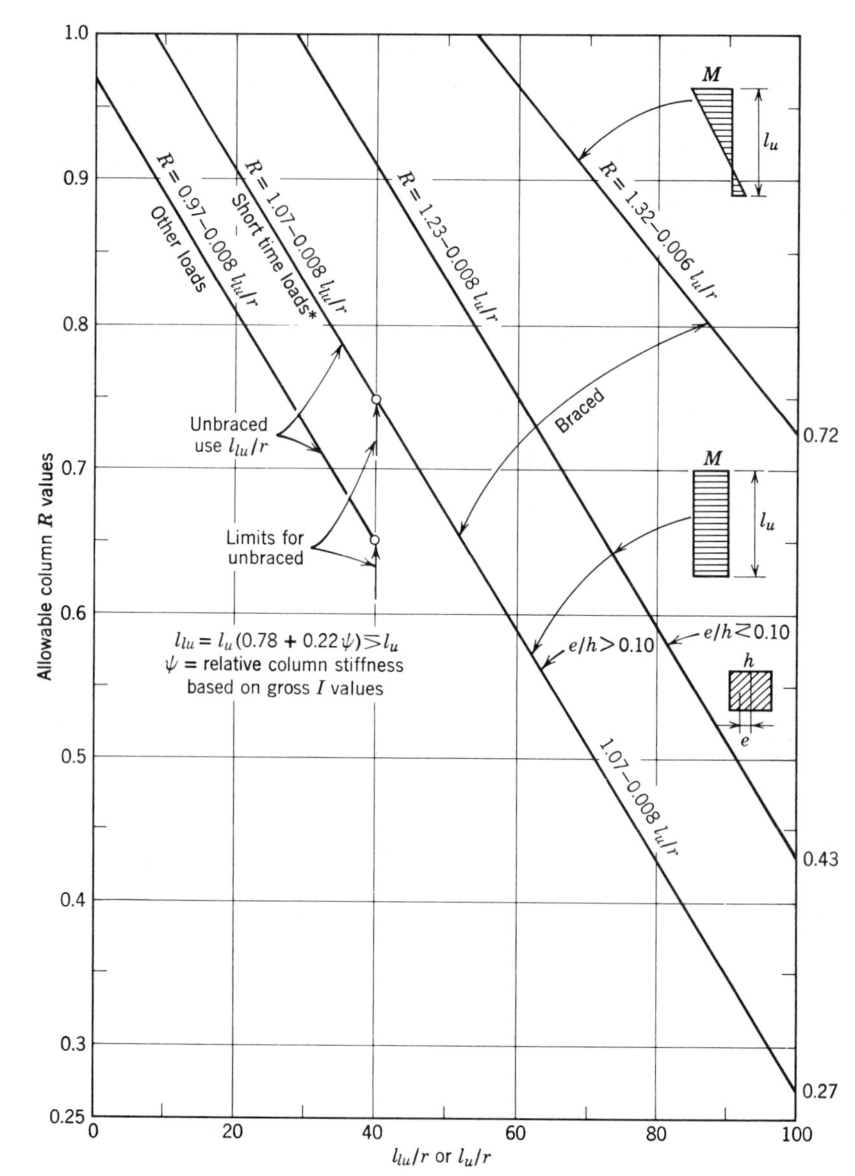

Fig. D.6. Column R values for alternate column design method. Design $M = M_u/R$ *and* $P = P_u/R$; e unchanged at $M_u/P_u = M/P$. (*Min. beam $\rho = 0.01$ for this R eq.)

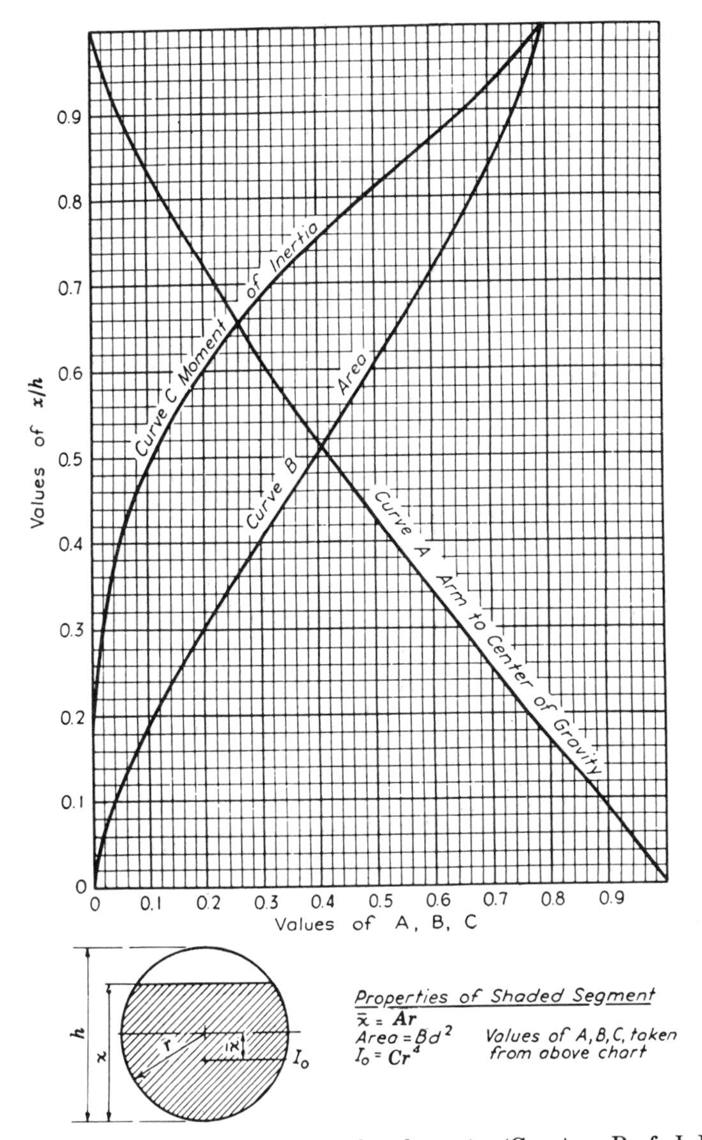

Fig. D.7. Constants for properties of circular elements. (Courtesy Prof. J. R. Shank, Ohio State University.)

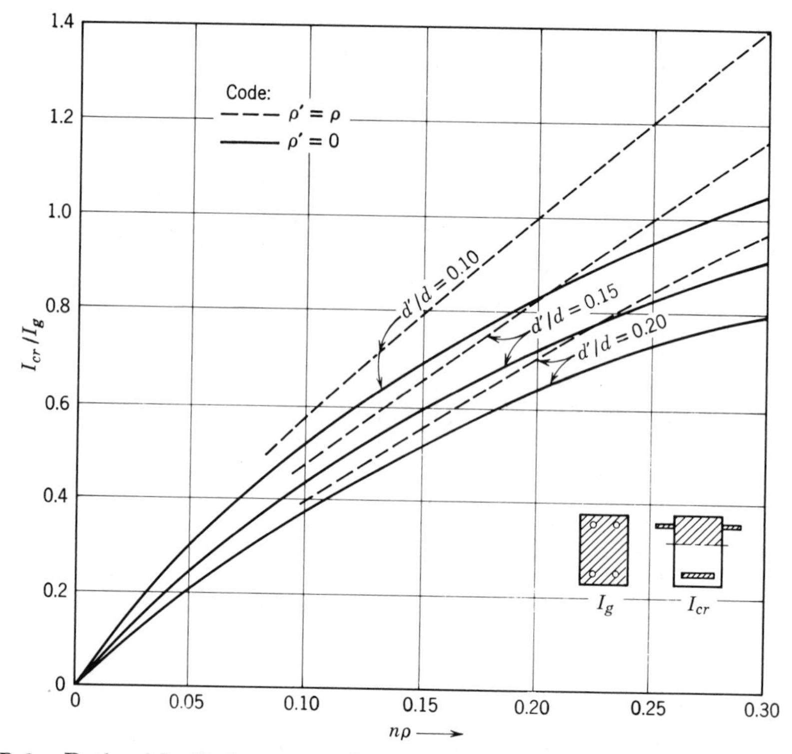

Fig. D.8. Ratio of I_{cr}/I_g for rectangular sections having $d' = h\text{-}d$

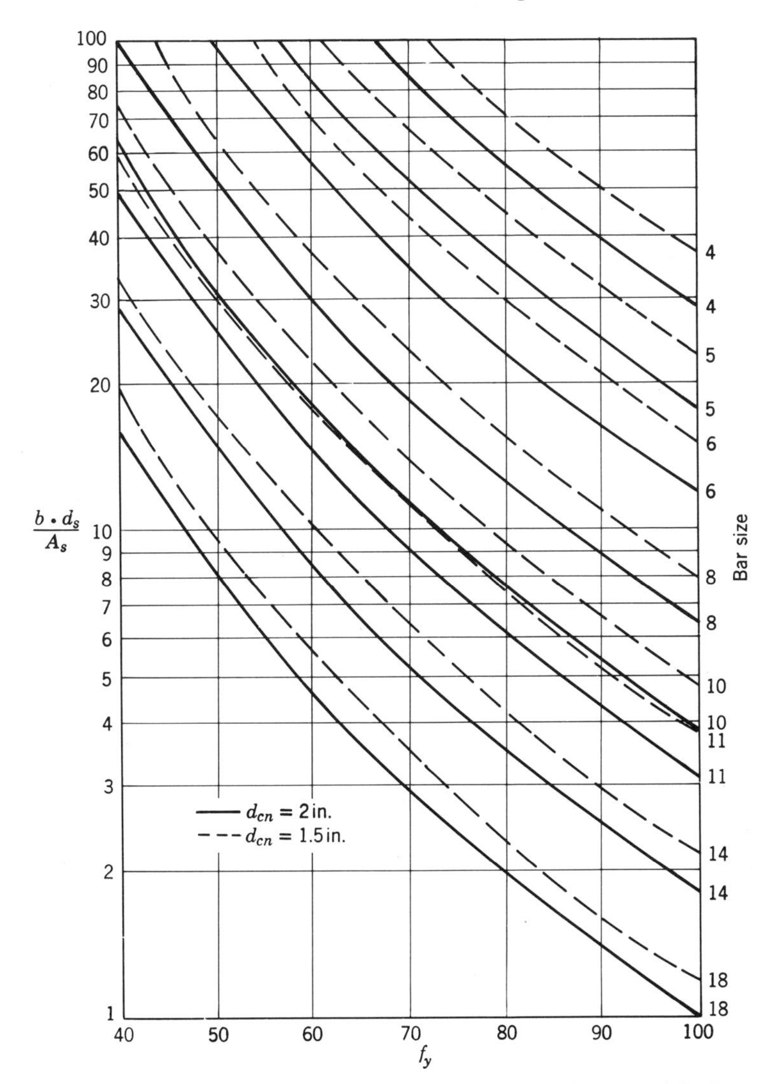

Fig. D.9. Variation of maximum bd_s/A_s versus f_y for crack control in beams with interior exposure. See Fig. 6.15 for symbols used. Assumes $f_s = 0.6f_y$. (From ACI Ref. 11, Chap. 6.)

Index